C0-ATR-922

Retrofit Opportunities for Energy Management and Cogeneration

RETROFIT OPPORTUNITIES
FOR
ENERGY MANAGEMENT
AND
COGENERATION

RETROFIT OPPORTUNITIES FOR ENERGY MANAGEMENT AND COGENERATION

Library of Congress Catalog Card No. 88-82219

Published by
THE FAIRMONT PRESS, INC.
700 Indian Trail
Lilburn, GA 30247

Printed in the United States of America

10 9 8 7 6 5 4 3 2 1

ISBN 0-88173-057-2 FP

ISBN 0-13-779042-2 PH

While every effort is made to provide dependable information, the publisher, authors, and editors cannot be held responsible for any errors or omissions.

Distributed by Prentice Hall
A division of Simon & Schuster
Englewood Cliffs, NJ 07632

Prentice-Hall International (UK) Limited, London
Prentice-Hall of Australia Pty. Limited, Sydney
Prentice-Hall Canada Inc., Toronto
Prentice-Hall Hispanoamericana, S.A., Mexico
Prentice-Hall of India Private Limited, New Delhi
Prentice-Hall of Japan, Inc., Tokyo
Simon & Schuster Asia Pte. Ltd., Singapore
Editora Prentice-Hall do Brasil, Ltda., Rio de Janeiro

CONTENTS

PREFACE — CONFERENCE SPONSOR'S STATEMENT

The World Energy Engineering Congress is the most important industry-wide energy conference in the nation. This comprehensive work, based on the WEEC, reflects the accomplishments of hundreds of energy engineers and managers who have contributed as speakers, chairmen, and reviewers. We congratulate our sponsors—the U.S. Department of Energy, (Office of Institutional Programs), Gas Research Institute, Mission Energy Company, Alliance to Save Energy, and Oglethorpe Power Corp.—for supporting this program and playing a major role in fostering technology transfer.

We also congratulate the Association of Energy Engineers and its 6,000 members for providing the leadership for the industry to develop.

For eleven years, the World Energy Engineering Congress has provided the essential forum for industry. The sharing of information is important to the continued growth of the energy engineering profession. AEE is proud to play a major role in sponsoring this vital conference.

Albert Thumann, P.E., CEM
Executive Director
Association of Energy Engineers

CONTRIBUTORS

Abramson, Alan B., *Syska and Hennessy*
Ahuja, Anil, *Los Angeles Unified School District*
Backlund, Lennart, *University of Linköping (Sweden)*
Baldi, Robert W., *General Dynamics Space Systems Division*
Bantz, Tom, *Stanford University*
Bartley, Robert P., CEM, *Energy Office of Delaware*
Beck, Richard V., *Rockwell International Corporation*
Behrendt, John, *John Deere Harvester Works*
Bergoust, Donald G., P.E., *Bergoust Engineers & Co., Inc.*
Burger, Robert, *Burger Associates, Inc.*
Buttorff, Leslie A., *Stone & Webster Management Consultants, Inc.*
Carver, Gary F., P.E., *Orlando Utilities Commission*
Cathcart, Thomas M., *Tennessee Valley Authority*
Chappell, R. Harold, *IllumElex Corporation*
Cilia, John P., *International Business Machines Corporation*
Collett, Chris W., *Theodor D. Sterling and Associates Ltd. (Canada)*
Connor, S. F., *Cleaver-Brooks*
Crew, Michael A., Ph.D., *Rutgers University*
Crocker, Keith J., Ph.D., *Pennsylvania State University*
Culpepper, R. Lee, *Tennessee Valley Authority*
Curtin, Deborah, *Bailey Controls Company*
Davis, Jack E., P.E., *Arizona Public Service Company*
De Simone, Lawrence E., *Energy Management Associate, Inc.*
Debban, G. D., *Bailey Controls Company*
Dubin, Fred S., P.E., *Dubin-Bloome Associates*
Dwyer, Michael J., Jr., *University of Arkansas for Medical Sciences*
Fetters, John L., CEM, *AT&T Network Systems*
Feldman, Roger D., P.C., *Nixon, Hargrave, Devans & Doyle*
Fierce, Warren, *Deere & Company*
Filler, Victor, Ph.D., *Stanford University*
Fine, Robert, *Associations of Municipalities of Ontario (Canada)*
Fleisher, John G., *Orgontz Corporation*
Friedle, Martin A., *Engineering Measurements Company*
Ghiya, Naresh, *New York State Energy Office*
Giles, John H., *QualitAire, Inc.*
Goodwin, Lee M., *Wickwire, Gavin & Gibbs, P.C.*
Gould, M. Scott, CEM, *Stanford University*
Grove, Manfred, *Steuler International Corporation*
Hahn, Warren G., P.E., *Naval Facilities Engineering Command*
Hansen, Shirley J., Ph.D., *Hansen Associates, Inc.*
Harmon, Kermit S., P.E., CEM, *Texas Energy Engineers, Inc.*
Heinz, Donna, *Randolph Air Force Base*
Heis, Mel, P.E., CEM, *John W. Galbreath & Co.*
Heller, Robert J., *Johnson Controls, Inc.*
Heselton, Kenneth E., P.E., CEM, *Power & Combustion, Incorporated*
Hills, Alan L., *Cogeneration Capital Associates, Inc.*
Hinson, Fletcher, *Bailey Control Company*
Hong, Shuibo, *Oklahoma State University*
Hoshide, Robert K., CEM, *Rockwell International*
Jendrucko, Richard J., Ph.D., *University of Tennessee*
Jhaveri, Arun G., *U.S. Army Corp of Engineers*
Johnson, R. A., *General Dynamics Space Systems Division*
Kai, Gui, *World Coal Technology (China)*
Kapur, Arjun, Ph.D., *North Carolina A & T State University*
Karlsson, B., *Linköping Institute of Technology (Sweden)*
Kaya, A., Ph.D., P.E., *The University of Akron*
Kenney, Thomas M., *Orgontz Corporation*
Kimmy, E. R., *General Dynamics Space Systems Division*
King, William R., *Office of Public Affairs, Oklahoma City*
Kirn, Alan J., CEM, *Sachs Energy Management Systems, Inc.*
Knable, Steven M., *Gas Research Institute*
Kogan, Peter L., P.E., CEM, *YEI Engineers, Inc.*
Lewald, A., *Linköping Institute of Technology (Sweden)*
Lincoln, Allen A., *Lincoln Associates*
Lincoln, Rann C., *Consumers Power Company*
Ling, Ed, P.E., *Randolph Air Force Base*
Loyd, D., *Linköping Institute of Technology, Sweden*
Mallik, Arup K., Ph.D., P.E., *North Carolina A & T State University*
Mathews, Jessica Tuchman, Ph.D., *Word Resources Institute*
Matousek, E. J., *Energy Systems Engineers, Inc.*
Maxwell, Charles L., *Mesquite Independent School District*
McBride, Peter A., *Arizona State University*
McLean, Thomas J., Ph.D., P.E., *The University of Texas*
Mehta, D. Paul, Ph.D., *Bradley University*
Meredith, Jack, P.E., *British Columbia Building Corporation (Canada)*
Miller, Phylissa S., *University of Tennessee*
Moorhouse, John C., Ph.D., *Wake Forest University*
Morris, R. J., CEM, *Young & Pratt Service, Inc.*
Morrow, William S., P.E., *Tennessee Valley Authority*
Mulloney, Joseph A., Jr., P.E., *EA-Mueller, Inc.*
Neal, Charles E., Jr., P.E., *Sheppard Air Force Base*
Nelson, Martin E, CEM, *U. S. Postal Service*
Nelson, William G. Jr., *ENERMAX Corporation*
Panich, Michael T., P.E., *EA-Mueller, Inc.*
Parmer, J. R., *General Dynamics Space Systems Division*
Partridge, A. James, P.E., *James Partridge Associates, Inc.*
Pate, Marvin E. III, P.E., *Engineering Sciences, Inc.*
Persons, Robert W., P.E., *Tecogen, Inc.*
Persson, Jorgen, *Linköping Institute of Technology (Sweden)*
Peterson, Kent W., *Sosoka & Associates*
Poulos, Jim, *Jones, Nall & Davis, Inc.*
Rayburn, Carolyn, *Consumers Power Company*
Rivenaes, Ulf, Ph.D., *Halfslund Engineering, Norway*
Robertson, Gray, *ACVA Atlantic Inc.*
Rock, William F., Jr., P.E., *Rockwell International*
Rodriguez, J. Alfredo, *J. A. Rodriguez Associates*
Rose, Patricia H., *U. S. Department of Energy*
Rosner, Carl. H., *Intermagnetics General Corporation*
Ross, James R., P.E., *University of Tennessee*
Sears, Robert L., P.E., *Arizona State University*
Shapiro, Robert F., Esq., *Chadbourne & Parke*
Sinclair, Ken, *Sinclair Energy Services Ltd. (Canada)*
Singh, Harmohindar, Ph.D., P.E., *North Carolina A & T State University*
Skjaeggestad, Pal, *Hafslund Engineering (Norway)*
Smith, Dennis A., *Atlanta Gas Light Company*
Smithart, Eugene L., P.E., *The Trane Company*
Solmar, A., *Linköping Institute of Technology (Sweden)*
Sosoka, John R., P.E., CEM, *Sosoka & Associates*
Stebbins, W. L., *Hoechst Celanese Textile Fibers*
Sterling, Elia M., *Theodor D. Sterling and Associates Ltd. (Canada)*
Sterling, Theodor D., Ph.D., *School of Computing Science (Canada)*
Sterrett, Richard H., P.E., *Science Applications International Corp.*
Steudtner, Edward, *Southern California Edison*
Stewart, Alan F., P.E., *U. S. Air Force, Randolph Air Force Base*
Stewart, Robert P., *New York State Energy Office*
Sturm, Werner, *Steuler Industriewerke, GmbH (W. Germany)*
Swift, Andrew, H. P., *University of Texas*
Teji, Darshan S., P.E., CEM, *CIty of Phoenix*
Todd, Tom R., P.E., *Engineering Sciences, Inc.*
Tucker, R. Arnold, P.E., CEM, *GTE Products Corp.*
Tuckner, James R., P.E., *YEI Engineers Inc.*
Turner, Wayne C., P.E., Ph.D., CEM, *Oklahoma State University*
Ventresca, Joseph A., *City of Columbus*
von Paumgartten, Paul, *Johnson Controls, Inc.*
Ward, Kristin J., *Tennessee Valley Authority*
Watt, John R., Ph.D., *Lincoln Associates*
Weaver, Terry R., P.E., *Johnson Controls, Inc.*
Weisman, Dennis L., *Southwestern Bell Telephone Company*
Whaley, Michael B., *Rockwell International Corporation*
Windingland, Larry, *U. S. Army Construction Engineering Laboratory*
Wong-Kcomt, Jorge B., *Oklahoma State University*
Wood, Byard D., Ph.D., P.E., *Arizona State University*
Yunming, Zhang, P.E., *Guangxi University, Nanning, China*
Zietlow, David C., P.E., *Bradley University*

ACKNOWLEDGEMENTS

Appreciation is expressed to all those who have contributed their expertise to this volume, to the conference chairmen for their contribution to the 11th World Energy Engineering Congress, and to the officers of the Association of Energy Engineers for their help in bringing about this important conference.

The outstanding technical program of the 11th WEEC can be attributed to the efforts of the 1988 Advisory Board, a distinguished group of energy managers, engineers, consultants, producers and manufacturers:

Richard Aspenson
3M Company

David L. Burrows
TVA

Karlin J. Canfield
Naval Civil Engineering Lab

Keith Davidson
The Gas Research Institute

Steve Dickson
Deere & Company

Wilbur J. Funk, CEM
Witco Corporation

Jon R. Haviland, P.E., CEM
Robinson's

Konstantin K. Lobodovsky
Pacific Gas & Electric

Dilip Limaye
Synergic Resources Corp.

William H. Mashburn, P.E.
Virginia Polytechnic Institute

Malcolm Maze
Abbott Laboratories

Harvey Morris
Cogeneration Partners of America

Martin A. Mozzo, Jr., P.E.
American Standard Inc.

Guy Nelson
Western Area Power Administration

Patricia H. Rose
U.S. Dept. of Energy

Frank Santangelo, P.E., CEM
Johnson & Johnson

Walter P. Smith, Jr.
BASF Fibers

Albert Thumann, P.E., CEM
Association of Energy Engineers

Wayne C. Turner, Ph.D., P.E., CEM
Oklahoma State University

The following organizations have given outstanding support in promoting principles and practices to help achieve energy savings, greater productivity, and lower operating costs.

CORPORATE MEMBERS
Association of Energy Engineers

Abbott Laboratories
Alliance to Save Energy
American Gas Association
American Medical International
Bell South Services
Binswanger Management
Brakes India Limited
C C & F Asset Management
Central Maine Power Company
CIBA Vision Care
Coggins Systems Ltd.
C. P. National/Trident Energy Group
Crane & Company
Edens and Avant, Inc.
Energenics, Inc.
Energy Automation Systems, Inc.
Engineers Digest
Esterline Angus Instrument Corporation
Frito-Lay, Inc.
Florida International University
Florida Power & Light Company
Georgia Power Company
HVAC Control Systems, Inc.
H. F. Lenz Company
Hewlett Packard
Home Savings of America
Honeywell, Inc.
Kirtland AFB, Civil Engr. Sq.
Kowloon-Canton Railway Corporation
Kuempel Services Inc.
Los Angeles Dept. of Water & Power
Mohawk College of Applied Arts & Technology
National Electrical Contractors Assn.
National Energy Management Institute
National Independent Energy Producers
National Wood Energy Association
Naval Aviation Depot
New Jersey Transit Bus Operations, Inc.
Niagara Mohawk Power Corporation
Northern Illinois Gas Company
Ogden Allied Service Corporation
Orlando Utilities Commission
Palomar Community College
Procter and Gamble
Softaid, Inc.
Solar Engineering & Contracting
So-Luminaire Corporation
Southern Alberta Institute of Technology
Southwest Gas Corporation
Stone & Webster Engineering Corp.
Tennessee Valley Authority
3 M Company
University of Michigan Hospitals
University of Missouri
US Army Logistics Evaluation Agency
Utah Association of Municipal Power Systems
Utilities Systems/Rates Auditing, Inc.
VEI, Inc.
Verle A. Williams & Associates, Inc.
Virginia Power
Viron Corporation
Western Area Power Administration

CORPORATE MEMBERS
The Cogeneration Institute of AEE

Anker Energy Corporation
Bailey Controls
Ballard Engineering
CMS Generation
Cogeneration Partners of America
Diesel & Gas Turbine Publications
Electronic Monitoring & Controls
Energy Initiatives, Inc.
Energy Management Specialists Inc.
Energy Networks, Inc.
Gas Research Institute
G & W Electric Company
General Physics Corporation
HDR Techserv, Inc.
Jacobs Engineering Group, Inc.
Johnson Controls, Inc.
Mission Energy Company
National Energy Systems, Inc.
NSS, Inc.
Nuclear Energy Consultants, Inc.
Reading Anthracite Company
Rochester Gas & Electric Corp.
San Diego Gas & Electric
Sierra Energy Systems, Inc.
Sverdrup Corporation
Tecogen Inc.
Thermo Engineering, Inc.
Turner Power Group, Inc.
US Army Construction Engineering Research Laboratory
Virgin Islands Water & Power Authority

INTRODUCTION

Today there is more opportunity to reduce utility costs by applying new efficiency improvement products, negotiating the best electric and gas rates, installing cogeneration facilities, and through modernization projects.

For example, according to a recent study by Lawrence Berkeley Laboratory, 50 percent of the electrical energy consumed by lighting, or 12 percent of the total national electrical energy sales, could be saved by replacing existing lighting with new energy efficient systems.

This comprehensive reference represents the latest methodologies used to improve efficiency and lower operating costs for facilities large and small. Contributed by noted authorities, the information contained is highly practical and will serve as a valuable reference source to improve an organization's overall efficiency.

SECTION 1
LIGHTING SYSTEM RETROFITS

Chapter 1

THE IMPACT OF ENERGY-RELATED LIGHTING RETROFITS ON YOUR BUSINESS

R. H. Chappell

The impact of energy related lighting retrofits on American business has changed immensely, both in dollars and cents and in the nature of the strategies being used. It is time for the industry that implements these retrofits--the lighting management industry--to examine the nature of those changes and what impact they now have on our retail, commercial, and industrial customers.

To do that we need to go back to the early 70s and see where we have been and how we got where we are today. We will see how we moved from "black box" types of energy related products into simple lighting retrofits in order to reduce the demand for energy use and the bottom line impact of energy cost today.

Beginning in approximately 1972 with the oil embargo, an immediate need for reducing energy consumption was thrust upon us overnight. To solve this problem as it applied to retail, commercial, and industrial businesses, manufacturers quickly responded with the so-called "black box," a device or microprocessor that effects electrical demand. Most "black box" purchases were made based on promises rather than on proven performance, because there was little experience upon which to draw conclusions. The building owner was often times left with a little understood and often misapplied apparatus. Throughout the 70s, specialty manufacturers began to appear all over the country. Technology continued despite the fact that the building owner's confusion was growing. Conventional lines of distribution were often not qualified to represent these specialty manufacturers and their products, only adding to the dilemma facing the building owner. As all of these "special products" were being developed, the lamp and ballast manufacturers also responded with their own energy saving products. Confusion continued to expand and there seemed to be no simple solutions to the problem of reducing energy costs in the area of lighting.

Then an industry began to emerge providing focus to this confusing issue -- The Lighting Management Industry. This new discipline applies five important considerations in evaluating the retrofitting of a lighting system:

1) The evaluation of the building

2) An objective recommendation

3) Broad menu of products

4) Provision of professional installation

5) On-going maintenance

A close look at the most common retrofit ideas being used today really provides insight into the importance of these strategies and how they are applied in any building. The most used products today include the following:

1) Reflectors

2) Power reducers or dimmers

3) Reduced wattage lamps

4) Energy efficient ballast

5) Occupancy sensors

All of these products are simple retrofit strategies that building owners can understand. But on the flip side of the strategies, the misapplication of these products is evident. Outrageous and unbelievable claims are being made about savings and the corresponding light levels. There are <u>no</u> magic solutions. Yes, lighting does account for 30-50% of a building's electrical energy consumption and should probably be the first item evaluated when trying to reduce consumption, but there must be a practical approach to what the building owner is buying when retrofitting his lighting system.

The retrofitted lighting system will usually not contain the excess that is prevalent in most buildings today. The key strategy to reducing lighting costs is simply to remove the maintenance cushion afforded by overlighting. Most lighting systems are over-designed approximately 30% to allow for what most designers and engineers know up front: The lack of professional maintenance will result in light losses of 30-50% over the first three to five years of the system's life. This need not be true. Simply put, you pay for lack of maintenance in increased energy costs. Let's evaluate the significance of that statement by better understanding the cost of light.

According to a General Electric study, "The Cost of Light," 88% of the cost of light is energy (See Figure 1). For example, if you pay $1.00 for an F40CW and $2.00 for labor to install that lamp, you will spend $44.00

FIGURE 1

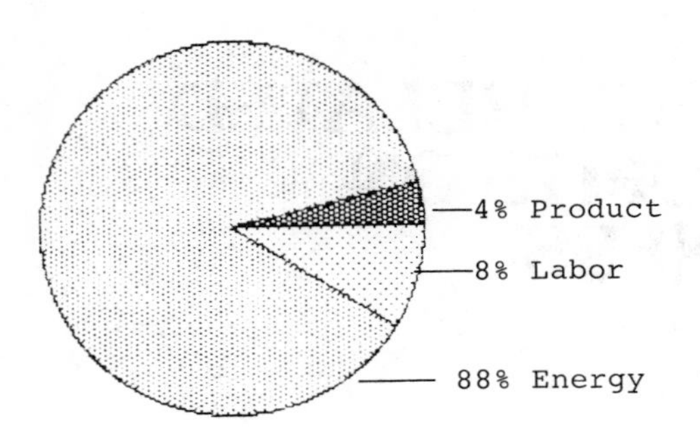

"THE COST OF LIGHT"

in energy to burn out that lamp. Our society is so commodity-oriented that we remain focused on how we can reduce the cost of that $1.00 lamp or the $2.00 we are spending to get the lamp installed instead of giving proper attention to our real cost of light, energy.

So why not just remove lamps? Remember that lighting is a very emotional and personal issue. Often the delamped fixture gives the appearance of a poorly maintained property or it will give the employee an impression of insufficient light even though the light meter might tell a professional that sufficient light is available for the task. The five strategies previously discussed solve these problems with a simple, understandable product offering. Remember, there is no magic device that can create light. Fixtures can be made more efficient with better light distribution, and more efficient components can be utilized by retrofitting existing fixtures, but it is pretty safe to say that you will lose light levels by reducing energy consumption. The most important factor to keep in the forefront, however, is that with proper maintenance, combined with the best of retrofit strategies, there does not have to be a loss of lighting levels! But it is only the combination of proper strategies and adequate on-going maintenance that can provide this winning combination.

While energy related lighting retrofits can provide upwards of 50% energy savings and paybacks that typically range between 9 months and 2 years, there are eventual losses of lighting levels of 30-40%, on the average. But this cloud does have a silver lining! Maintenance properly executed can regain the loss incurred by the retrofit. Of course there are the cases we in the lighting business all dream about where the light loss is unimportant because the facility has 150FC. However, these cases rarely exist. And there are other instances where the CRT usage is high and replaces the pencil pushers of the 60s and 70s, and in these cases there is a need to provide improved lighting quality at reduced lighting levels. We all read about the truthfully successful jobs, but are usually brought back to reality when faced with reducing costs of our own lighting systems.

The most important considerations when evaluating a lighting retrofit project, then, can be summed up very simply by asking yourself the following questions:

* Has a fair evaluation of my building been done?

* Are the choices and recommendations being offered objective?

* Are the product selections geared to only one manufacturer or have several been evaluated?

* Is professional and timely installation being provided that will not disrupt the work force in the property?

* Has an on-going maintenance program been considered and are we committed to continue the proper program long-term?

The lighting industry of the late 80s remains fragmented. We as a whole have not figured out the best way of presenting our products and services to the building owner/engineer. The lighting management industry can provide solutions for retrofitting and maintaining existing lighting systems.

Thus far we have discussed what the lighting retrofit strategies have been, and what strategies are currently available. We have also discussed the critical need for maintenance of the retrofitted system. We have established that there are no magic solutions and that these products do not create light. You, hopefully, have been provided an objective view of lighting retrofit and the challenges of reducing operating costs in the area of lighting. But let's take a closer look at the "bottom line" impact of lighting retrofit, and you will realize why, if you have not already done so, you should consider lighting retrofit.

FACT #1 - Lighting accounts for 30-50% of electrical consumption in the average building.

FACT #2 - 80% of existing buildings in the U.S. are over 10 years old, providing the greatest opportunity for energy savings.

FACT #3 - Most buildings have lighting systems that are over-designed to provide for the 30% light loss that will occur during the first 3-5 years of life.

FACT #4 - Utility rates are expected to rise at a rate 10% faster than inflation.

FACT #5 - Many states are beginning to legislate the allowable watts per square foot (usually under 2 watts per sq. ft.).

FACT #6 - Lighting retrofits usually provide a payback on investment of under two years.

FACT #7 - A properly applied lighting retrofit strategy will enhance the building's environment rather than detract from it.

FACT #8 - The energy savings from lighting retrofit are easily measurable and verifiable without sophisticated monitoring devices.

FACT #9 - The new lighting retrofit products are simple in their design and intended application, but dynamic in the results provided, and their impact on the bottom line is immediate and on-going.

In summary, lighting retrofit when properly applied and managed is indeed management's new cost saving tool. In today's competitive environment, it is one of the truly remarkable tools for reducing costs immediately as well as long-term. To enjoy these cost savings to the full, however, means a commitment to professional maintenance.

Chapter 2
LIGHTING IN SCHOOLS

A. Ahuja

BACKGROUND

Electrical bill for the Los Angeles Unified School District runs at about $24 million annually. This represents 70% of the total energy and water bill. Lighting and air conditioning costs are evenly split. In the lighting area alone, there is a huge potential to save money from energy efficient lighting and at the same time improve the quality of light in most cases.

Even though efficient lighting plays a major role in improving building operating efficiency, not much has been done in schools as commercial building standards do not directly apply to school construction, a cohesive policy on building standards is absent and efficient systems require high initial investment.

Los Angeles Unified School District has over 12,000 buildings which require adequate lighting. A recent survey revealed that 18,700 classrooms, 7,500 kitchens and offices, 16,500 auditoriums and cafeterias and 300 gymnasiums require replacement of lights with energy efficient lights. It will cost $84.5 million to replace lights in 18,700 classrooms alone.

Goal

Goal of school lighting is the creation of an effective total visual environment for learning, dependent on a three dimensional pattern of brightness and colors. It takes into account emotional and aesthetic values. The District emphasis is not only in cost-effective efficient lighting but it also addresses the need for effective lighting. Being efficient shows how well lighting equipment converts the electrical power input to light output which alludes to the relationship between output and input. Being effective requires the use of the right kind of equipment for a specific task which would promote a conducive work environment.

Criteria for Lighting Changes

As replacement of all the lights will require huge resources, a criteria for replacement of lighting has been developed as under:

1. Classrooms and libraries
2. Offices
3. Gyms
4. Auditoriums and cafeterias
5. Other, such as storerooms

Amongst the classrooms, following priority ranking has has been developed:

- Classrooms with incandescent lighting
- Classrooms with hazardous/obsolete fluorescent fixtures
- Sites with high student density

Lighting Types

There are four types of lighting in wide use at Los Angeles Unified School District:

1. Incandescent lighting has the lowest efficacy (9-21 lumens/watt) and shortest lifetime (750-2000 hours). It is widely used due to low initial cost and superior color rendition.
2. Fluorescent lamps have good efficacy (65-85 lumens/watt) and a long lifetime (10,000-20,000 hours). Their color rendition is acceptable to most people. They are widely used in classroom and office settings.
3. Metal halide lamps have good efficacy (65-80 lumens/watt) and lifetime (7,500-20,000 hours). Good color rendition makes this lamp an excellent candidate for gym and stadium lighting.
4. High pressure sodium lamps have highest efficacy (80-105 lumens/watt) and a very long lifetime (12,000-24,000 hours). Poor color rendition limits their utility to parking lots and outdoor security lighting.

Lighting Design Criteria

New Construction: The school district has established a policy to provide adequate and quality lighting in schools. This policy contributes towards aesthetics, pleasant surroundings and reduction in operating costs. It promotes an effective lighting design which must balance many different and sometimes conflicting criteria. It ensures that the design engineer does not ignore the criteria and provides efficient and effective lighting. Serious attempts are made to include the concerns of both energy management and lighting professionals.

Motion sensors have been included in all new construction design. These controls allow the maximum wattage per square foot to increase and thus provide greater flexibility in design within the interior lighting power allowance under the Title 24 system performance method.

Progress is being made to coordinate the design of HVAC, electrical and lighting experts to ensure that whole building energy allowance is met.

We are considering the use of luminaires in lighting design. By having the photometric data of several lighting equipment manufacturers stored in computer memory, analysis of the proposed design for quality lighting becomes a simple matter. A perspective graphic package enhances this capability.

Retrofit: The school district requires that lighting level of 70 foot candles from fluorescent lighting be maintained except the following major areas:

Locations	Foot Candles	Light Source
Drafting, sewing, conference, and sight saving rooms and shops	100	Fluorescent
Corridors	20	Fluorescent
Auditoriums	30-70	Incandescent
Gymnasiums	30-70	Metal Halide
Athletic Field (Stadium)	45	Metal Halide
Parking Lots	2+	High Pressure Sodium
Building Exterior (Security)	35-100 watts	High Pressure Sodium

Few Pointers on Fluorescent Lighting

- Natural lighting: Use as much natural lighting without glare to provide high visual comfort probability (VCP). Glare is minimized by adjusting the sun angle.

- Refractors: Light at a particular angle of refraction provides exceptional brightness control. The prismatic lens create uniformity of light cutting down the glare.

- Prismatic shielding: Provides optimum efficiency.

- Diffusers: Help to spread light and provide uniform and soft illumination by eliminating shadows.

- Surface mounted/flush mounted fixtures: Both types are capable of providing superior photometric performance. However, surface mounted offers ease of installation and maintenance and flush mounted gives out less glare and a more refined appearance and may require higher number of fixtures.

Terms to Learn

- Maintenance factor (MF) = Luminaire Dirt Depreciation (LDD) factor x Lamp Lumen Depreciation (LLD) factor.

- Coefficient of Utilization (CU): Directly proportional to the amount of reflectance.

- Foot Candles = $\frac{\text{Number of Lamps x Lumens per lamp x MF x CU}}{\text{Area}}$

CASE STUDIES ON RETROFIT

Case Study 1

Stadium lighting at Birmingham High School

Criteria for Selection: High energy usage

High potential for energy savings

Public relation value (Mayor's Stadium)

Old System

160 incandescent lamps of 1500 watts each = 240kw

Light levels measured on field = 12 Footcandles

110 foot high poles made maintenance difficult

Resulting in high number of burned out lamps

Lamp life expectancy = 750 hours

New System

60 super metal halide lamps of 1000 watts each = 60kw

New light levels = 35 Footcandles

This light level calculated at mean lumen output of lamp, rather than initial lumen output.

Lamp life expectancy = 12,000 hours

Savings Per Year

240kw - 60kw = 180kw x 1664 hours = 299,520kwh

Energy Cost @ 7 cents per kwh	= $20,966 year
Replacement Cost	= $ 5,984 (labor on boom truck)
Total Savings	= $26,950
Cost of the Project	= $33,000
Payback	= 1.2 years

Additional benefits: Safety, replacement of two PCB transformers with one, yielding savings of $50,000 and improved lighting level.

Source of Funds

- Utility budget

- Maintenance

- Student body funds

- Operations

- School budget (IMA)

Case Study 2

Gymnasium lighting

Old System

30 incandescents at 1000 watts each	= 30kw

New System

30 high pressure sodiums at 400 watts each	= 12kw
Savings/Year	= 18kw x 3600 hrs. = 64,800kwh
	= 64,800kwh x 7 cents/kwh = $4,536
Cost	= $7,000
Payback	= 1.5 years

Poor color rendition from high pressure sodiums has forced the District to consider metal halides for future projects.

Special Applications

Visually Handicapped Schools: Many low vision children have special and unusual sensitivity to illumination levels. Most commonly strong lighting is preferred but in some instances dim illumination is required for best performance. Thus there is a need to ensure flexibility to cater for the wide variety of lighting needs.

Many commonly used fluorescent tubes have a disproportionate amount of their luminous energy in the blue region of the spectrum and this needs to be avoided. Many low vision children will have some haziness of the normally clear media of their eyes; because blue light is scattered more than longer wavelengths, it is disadvantageous to use blue rich lighting when there is haziness of the ocular media. Incandescent or red-rich fluorescent is more desirable in such cases.

Daylighting: This is the best source of light if there are no glare problems. Blinds or other light shielding can remove glare whenever necessary. Further control of light can be achieved with automated dimmer controls. Application of these controls to fluorescents is in its infant stage but could bring about promising results to balance the light in a classroom and at the same time reduce electrical lighting costs.

Chapter 3

LIGHTING RETROFITS THAT MAKE SENSE AND SAVE DOLLARS

J. L. Fetters

INTRODUCTION

The AT&T Manufacturing and Development Center at Columbus, Ohio was built as a Western Electric crossbar manufacturing plant in 1958. The conversion of this 2-million square foot facility in recent years to a modern, world-class manufacturing and software development center has provided the opportunity to convert older, inefficient lighting systems to more efficient equipment. Lighting equipment has been developed in the past few years that makes the retrofit option attractive to both energy management and lighting professionals. Replacing old or existing equipment decreases energy costs and, at the same time, provides an improved workplace by improving lighting quality. The conversion of a few lighting systems will be described.

RETROFIT ECONOMICS

Several factors are used to qualify potential projects for retrofit. The age of the installation, the current operating costs of energy and maintenance, and how well the particular installation is meeting the current lighting requirements.

Lighting costs must account for the cost of the electrical energy to operate the system, the initial cost of the luminaires and lamps, the replacement costs of the lamps, and the maintenance costs. Current lighting costs at the Columbus AT&T location divide in this way: energy costs 86%, maintenance costs 11% and replacement lamp costs 3%. This division establishes the ranking of retrofit investments with a priority for energy costs first, the maintenance costs considered next and with least attention given to the cost of the lamps.

Personal computer spreadsheets are used to calculate present costs and proposed retrofit solution savings to determine if the lighting project is economical. An example is shown in figure 1. Three sections provide calculations for: 1) Energy savings, 2) Lamp savings, and 3) Labor savings. A summary at the bottom of the worksheet sums the three types of savings. The example shown is the actual cost reduction worksheet for the cooling tower retrofit described later in the paper.

The 9 cases which follow are examples of proven and successful approaches to lighting retrofits that have improved quality of life, enhanced productivity and reduced operating costs of energy and maintenance. All these cases have savings that have immediate, measurable, bottom line improvements to profitability.

INTERIOR CASES

INDUSTRIAL PROJECTS

Case 1 - High Bay Retrofit:

A 244,000 square foot high bay space, originally lighted by 819 twin 400 watt mercury vapor fixtures, provided an average of 23 footcandles at task surfaces. Most of the fixtures had been changed to self-ballasted lamps increasing the lighting load to 800 KW, the unit power density (UPD) to 3 w/sf, and the annual energy costs to $240,000!

The retrofit design required integrating five design criteria:

1) Increase light levels and provide good color rendition,
2) Provide a more pleasant, productive work environment,
3) Increase system efficiency and reduce operating costs,
4) Locate the retrofit system in existing pendant positions, and
5) Provide a flexible design to accommodate future changes.

A 400 watt, clear, 36,000 lumen metal halide lamp was selected to achieve the desired light output and color rendering properties. High pressure sodium sources were rejected by the owner because many of the high bay tasks are color sensitive. The commercial luminaire selected has a prismatic glass reflector with a spacing criteria of 1.7, allowing it to be mounted in existing pendant positions and still meet uniformity and light level requirements. This unique glass reflector provides 20% up light that converts the previously unused ceiling space into a luminous cavity. The resulting reflected light visually opens the formerly dark overhead space and provides a diffuse source to eliminate shadows and to redirect part of the direct component on the horizontal surfaces to an indirect component that provides the added vertical illumination needed for better visibility.

COST REDUCTION WORKSHEET FOR LIGHTING PROJECTS

COOLING TOWER PROJECT

ENERGY SAVINGS *************** BEFORE LIGHTING PROJECT ************* *************** AFTER LIGHTING PROJECT *********

Before: HRS/DAY= 18 DAYS/WK= 7 WKS/YR= 52 — After: HRS/DAY= 12 DAYS/WK= 7 WKS/YR= 52

	Before	* WATTS/FIXTURE *		TOTAL		After	* WATTS/FIXTURE		TOTAL	
LOCATION	#FIXTURES	LAMP	BALLAST	WATTS	HRS/YR	#FIXTURES	LAMP	BALLAST	WATTS	HRS/YR
2 CELL TOWER SOUTH	2	100	0	200	6552	2	35	15	100	4380
2 CELL TOWER POSTS	4	200	0	800	6552	4	100	20	480	4380
3 CELL TOWER STEPS	1	150	0	150	6552	1	70	20	90	4380
3 CELL TOWER POSTS	4	200	0	800	6552	4	100	20	480	4380
4 CELL TOWER SOUTH	2	300	0	600	6552	2	50	33	166	4380
4 CELL TOWER NORTH	2	300	0	600	6552	2	400	50	900	100
4 CELL TOWER POSTS	10	300	0	3000	6552	10	100	20	1200	4380
BOIL HSE, B42, ETC	15	150	0	2250	6552	15	35	15	750	6552
			TOTAL KWH BEFORE =		55037			TOTAL KWH AFTER =		16025
			TOTAL KW =	8.4				TOTAL KW =	4.2	

LAMP SAVINGS *************** BEFORE LIGHTING PROJECT ************* *************** AFTER LIGHTING PROJECT *********

LOCATION	#FIXTURES	LAMPS/ FIX	COST/ LAMP	LAMP LIFE*	# LAMPS PER YR	#FIXTURES	LAMPS/ FIX	COST/ LAMP	LAMP LIFE	# LAMP PER YR
2 CELL TOWER SOUTH	2	1	$4.10	2450	5	2	1	$24.73	16000	0.55
2 CELL TOWER POSTS	4	1	$3.89	2450	11	4	1	$21.85	28500	0.61
3 CELL TOWER STEPS	1	1	$3.16	2450	3	1	1	$20.52	28500	0.15
3 CELL TOWER POSTS	4	1	$3.89	2450	11	4	1	$21.85	28500	0.61
4 CELL TOWER SOUTH	2	1	$5.79	2450	5	2	1	$22.10	28500	0.31
4 CELL TOWER NORTH	2	1	$5.79	2450	5	2	1	$24.78	24000	0.01
4 CELL TOWER POSTS	10	1	$5.79	2450	27	10	1	$21.85	28500	1.54
BOIL HSE, B42, ETC	15	1	$3.16	3500	28	15	1	$24.73	16000	6.14
		TOTAL LAMP COSTS BEFORE =			$419	TOTAL LAMP COSTS AFTER =				$236

* 3500 HR LIFE DERATED BY 30 % DUE TO TOWER VIBRATION

LABOR SAVINGS *************** BEFORE LIGHTING PROJECT ************* *************** AFTER LIGHTING PROJECT *********

LOCATION	#FIXTURES	# LAMPS PER YR	HRS/ LAMP	LABOR$/ HOUR	TOTAL$ LABOR	#FIXTURES	# LAMPS PER YR	HRS/ LAMP	LABOR$/ HOUR	TOTAL$ LABOR
2 CELL TOWER SOUTH	2	5.35	1.0	$28.38	$152	2	0.55	0.6	$28.38	$9
2 CELL TOWER POSTS	4	10.70	1.0	$28.38	$304	4	0.61	0.6	$28.38	$10
3 CELL TOWER STEPS	1	2.67	1.0	$28.38	$76	1	0.15	0.6	$28.38	$3
3 CELL TOWER POSTS	4	10.70	1.0	$28.38	$304	4	0.61	0.6	$28.38	$10
4 CELL TOWER SOUTH	2	5.35	1.0	$28.38	$152	2	0.31	0.6	$28.38	$5
4 CELL TOWER NORTH	2	5.35	1.0	$28.38	$152	2	0.01	0.6	$28.38	$0
4 CELL TOWER POSTS	10	26.74	1.0	$28.38	$759	10	1.54	0.6	$28.38	$26
BOIL HSE, B42, ETC	15	28.08	0.6	$28.38	$478	15	6.14	0.6	$28.38	$105
			TOTAL LABOR BEFORE =		$2,376			TOTAL LABOR AFTER =		$169

** SUMMARY OF LIGHTING COST REDUCTION SAVINGS **

CASE 341428-071

COOLING TOWER PROJECT

	BEFORE	AFTER	SAVINGS
ENERGY COSTS	$2,752	$801	$1,951
LAMP COSTS	$419	$236	$183
LABOR COSTS	$2,376	$169	$2,207
TOTAL COSTS	$5,547	$1,206	$4,340

FIGURE 1

Power demand was reduced from 800 KW to 230 KW, a dramatic 70% reduction! The resulting UPD is 0.8 W/SF compared to the original 3 W/SF. First year energy savings totalled $171,400. Now with only 498 luminaires in place of the original 819, horizontal light levels have been increased by 50%, resulting in an average of 30 footcandles on the work surfaces. The fact that the occupants have expressed the appearance of a much higher light level, suggests that the reflected light from the ceiling cavity contributes to the quality workplace beyond a simple quantity of light measure.

The use of the glass reflector has not only provided high efficiency and superior light control, but it stays cleaner and is easier to maintain. This was verified at the first group relamping. Worker acceptance has been high, resulting in lower absenteeism and higher morale. The project was paid back by savings from lower operating costs in 8 months.

The high bay lighting retrofit was awarded the Edwin F. Guth Memorial International Lighting Design Award of Merit by the Illuminating Engineering Society of North America in July, 1985.

Case 2 - Plating Room Retrofit:

AT&T manufactures several different types of telephone apparatus, each with its own finishing specification. To ensure durability and long life, many of the metal parts are electro-plated. The plating room is the action center for many plating processes. It was originally lighted with eight foot, slimline fluorescent fixtures having porcelain-enamel reflectors. The corrosive environment of this area had taken its toll on the lighting equipment and its arrangement made it difficult to maintain. In addition, the wall and ceiling surfaces were dirty and reflected no light.

When AT&T decided to modernize this facility in 1987 it was done as a part of a plant modernization project to improve the quality of its plated parts and to provide a better place to work. A combined painting and lighting project was developed with the goal of improving product and workplace quality.

Criteria was established to: 1) Increase light levels and minimize glare from bright metal parts, 2) Use no more energy than the original lighting system, and 3) Improve access to the equipment for maintenance.

The lighting was installed in several stages, due to the limited availability and access to the room. This limitation actually helped by allowing minor equipment relocations to assure that the selected locations did not produce glare on the difficult visual tasks of handling and inspecting specular metal parts.

A mix of high-output (HO) fluorescent and metal halide sources was chosen for their efficiency and life characteristics. The fluorescent luminaire selected had a porcelain-enamel reflector with slots to allow 20% up light to take advantage of the newly painted ceiling. By using an HO fluorescent lamp instead of the slimline, 15% fewer luminaires were required to provide the design illuminance of 60 footcandles, maintained, in the north half of the room.

In the south half of the room, the three plating machines had the original slimline, fluorescent lighting embedded over the machines, as well as being located in rows parallel to the machines. These were removed and replaced with 175 watt, metal halide lamps in enclosed luminaires with glass prismatic refractors. The superior light control of this luminaire was the key to achieving glare-free lighting for viewing the metal parts. By placing the luminaires around the periphery of the machines, the machines, the tanks and their liquid contents, and the racked parts moving on conveyors can now be easily seen. Many of the visual tasks at the plating machines demand good vertical illumination and the prismatic refractor achieves this while controlling glare and surface brightness. Combined with proper placement, the result is a pleasing, quality workplace.

Two bench operations, previously employing open slimline fixtures before the retrofit were replaced with HO luminaires with acrylic, prismatic lenses. Not only was the glare controlled, but there was a noticeable difference in both productivity and quality.

A total of 13, 175 watt small optic luminaires and 3, 250 watt, larger optic units, and 95, HO fluorescent luminaires were used for a total lighting load of 30 kw. This is slightly less than the original system which measured 32 kw. For slightly less power, the light levels have nearly tripled the values measured before retrofit; from an average of 25 footcandles (FC) on the horizontal work surfaces to an average of 70 FC. More importantly, the lighting quality has had a significant effect on the product quality. By improving the maintenance access, lighting quality can now be maintained more easily.

Case 3 - Hazardous Material Storeroom Retrofit:

Originally built as an oil storage room, this 60 foot by 56 foot storeroom was lighted with 16, 500 watt incandescent glass enclosed, porcelain-enamel, canopy reflector fixtures. The space is now used to store volatile, flammable liquids that need special handling, such as paint, solvents, and lubricating oils. With short operating hours, the energy costs were not the main concern with this system. Instead, the concern was for inadequate light levels and poor visibility for reading labels and for the safe handling of the stored liquids. A second concern was for high maintenance labor costs. Because of the volatile nature of the materials handled, an enclosed and gasketed, UL rated class 1, division 2 luminaire is required, increasing relamping time and therefore labor costs for the short life incandescent lamps.

The objectives of this lighting retrofit were to: 1) Increase light levels and improve visibility, especially on vertical surfaces, 2) Reduce maintenance and energy costs, and 3) Improve safety lighting, especially for power outage emergencies. By timing this project to coincide with a scheduled repainting, the retrofit design was able to take advantage of the improved reflectances of the walls and ceiling.

Sixteen class 1, division 2, enclosed and gasketed luminaires with prismatic glass refractors and 175 watt metal halide lamps provide a light level of 50 footcandles, maintained. The refractor directs more light downward and outward to distribute more light on the vertical surfaces and this property allowed the original mounting centers to be reused. Three of the fixtures were equipped with a quartz restrike feature for lighting while the metal halide lamps warm up. In addition, an external battery unit powers a separate sealed light at the doorway which comes on in the event of a power failure to provide a safe emergency exit. A sealed, two lamp fluorescent luminaire was added at the storekeeper's desk.

Visibility has been improved for the safe handling of hazardous liquids and operating costs have been reduced to pay for this retrofit in less than four years.

COMMERCIAL PROJECTS

Case 1 - Main Cafeteria Retrofit:

The main employee cafeteria was built as a functional, institutional cafeteria, replete with silver-bowl indirect and recessed incandescent fixtures that were inefficient and expensive to maintain. The original lighting system had a UPD of 3.4 w/sf. A more modern, efficient environment that was both functional and more appealing to its employees would improve quality of life and cut operating expenses at the same time. A remodeling project was planned, with lighting to play a key role to compliment the new interior by giving each area its own kind of lighting.

The new parabolic, 3 lamp fluorescent luminaires with high efficacy, high color rendering, 3000K, tri-phosphor lamps achieve this effect in the dining room by enhancing the softer colors and textures. The dining area supplements the natural daylighting provided by the surrounding windows. Matching, smaller parabolics illuminate the lowered ceiling perimeter area.

The food service areas are highlighted with 175 watt, 3000K metal-halide recessed luminaires, set into lowered soffits, adding sparkle and attraction. Contrast is provided by lowered lighting levels in the circulation areas. At the food service counters, lighting attracts customers by highlighting and rendering the natural foods colors and enhancing the food presentation under the low brightness, prismatic luminaires. Backlighting in the areas behind the counters matches the circulation area fluorescent lighting. In the salad bar area, 3000K fluorescent lighting mounted over the food presents an attractive contrast with the surrounding lower general illumination.

Both project budget constraints and power budget targets were achieved. The UPD after retrofit was 1.6 w/sf and energy costs are further contained by using a commercial lighting control, scheduling 14 independent zones.

Case 2 - Small Cafeteria Retrofit:

A small food service area in the north part of the building provided a vending area and tables for use by employees at break times and meal times. Like most of the manufacturing area, it was originally lighted by 8 foot, 2 lamp F96 slimline fluorescent lamps in open reflector fixtures. The lighting load for this 3200 square foot area was 6.4 kw, for a UPD of 2 w/sf.

When AT&T decided to redecorate the area to provide a modern, attractive space and to expand food services, effective lighting was considered a key part of the new interior. The objectives of the design were to: 1) Tailor the lighting to the needs of three functional areas, 2) Reduce the UPD to 1.6 w/sf, and 3) Provide lighting to compliment the new look.

A separate vending machine area, located along the north wall was lighted to increase the attraction of the vending machines. A low ambient level of 35 FC of diffuse overhead light, provided by 2 X 2 fluorescent luminaires, equipped with energy-saving U-tube lamps, supplements the vending machine lighting. The contrast of the two lighting sources directs attention to the machines. This technique is both attractive and energy efficient at 1.2 w/sf, and is a technique used in all new vending areas at this facility.

The dining room has two sections, divided by a center counter area where food is served at meal times. Four, 175 watt metal halide luminaires equipped with prismatic lenses highlight the food with 85 FC at mealtimes. At all other times, the center area is lighted with two 2 X 2 U-tube fluorescent luminaires, with lenses matching the metal halide units, which are connected to the emergency lighting circuit for continuous operation. A timing switch controls the serving lights by turning them off automatically after they have been on a preset time.

The two dining sections are illuminated to an average of 55 FC, maintained, from high efficiency recessed downlights mounted in the new 12 foot ceiling. Compact metal halide, 150 watt, 11,250 lumen lamps with a color temperature of 4300K and a CRI of 85 provide efficiency and color rendition needed to enhance and compliment the new, cooler look. An electronic time clock turns off the dining room lighting when not in use.

The result of this retrofit is an attractive space where the energy budget of 1.6 w/sf was achieved without any sacrifice in light level or quality.

<u>Case 3 - Toilet Rooms Retrofit</u>:

AT&T wanted to improve the looks of its highly visible, public toilet rooms at this location, without spending a fortune operating and maintaining them. Compared to the cost of operating the rest of its 2-million square foot facility, the cost of operating these public spaces is small, but an important consideration here is the impression made on employees and customers alike. If people see a toilet room lighted to 100 FC, or leaking water, they may go away with the impression that we don't care how wasteful we are. This is a hard impression to correct. So, the user impressions of the facility were an important aspect of relighting the renovated spaces.

The original facilities were lighted with 2 X 4, 2 lamp fluorescent troffers, all located in a row, down the center of the room. Although illuminance measurements showed 50 FC, the central placement of the luminaires left the walls dark which resulted in a gloomy and confined feeling. The UPD was 2 w/sf and the lights burned constantly, resulting in higher than necessary operating costs.

The objectives for the lighting retrofit design were to: 1) Change the dark and confining space impressions to bright and spacious, 2) Reduce operating costs and restrict UPD to 1.2 w/sf, and 3) Project an image of efficiency and quality.

The lighting concept was drastically changed from a direct system to an indirect system, from uniform to non-uniform, and the design illuminance value was reduced to 30 FC. To provide the impressions of being bright and spacious, recessed fluorescent wall slot luminaires were installed along the long walls. Single, T8, 3000K, CRI 80, lamps, chosen for their efficiency and color rendering characteristics, placed end-to-end, continuously light the walls to provide an average illuminance of 30 FC. Since the lamps are recessed above the ceiling line, no lamps are visible, eliminating the objectionable troffer surface brightness and reducing glare. The wall lighting technique is especially effective above the wash basins where the reflected light from the wall mirrors supply an average illuminance of 60 FC on the counter top.

An ultrasonic motion sensor controls the lights off when there is no occupancy. Two 4 foot lamps, one at each wall are powered from the emergency lighting circuit to provide continuous lighting in the event of a power failure.

The objectives of the retrofit were achieved including the power budget of 1.2 w/sf, by changing the lighting concept and using modern lighting controls and equipment. The use of the walls as part of the design was the key to this effective lighting retrofit.

EXTERIOR CASES

Case 1 - Walkway Retrofit:

Lighting for the pedestrian walkways, front drive, and two small parking lots for visitors and staff was originally provided by 32, 500 watt incandescent glass globes mounted atop 20 foot poles. The system was powered by a 480 volt underground distribution wiring system and converted to 120 volts for the lamps by individual step-down transformers in each pole base. Short lamp life caused high maintenance costs, in addition to high energy costs. Before retrofit, total annual operating costs were $6,000. The symmetric, incandescent globes, mounted on wide pole spacings did not provide adequate lighting for safety and security. The retrofit design had to: 1) Provide adequate illumination for the walkway and the adjacent roadway, 2) Increase light levels to improve safety and security, and 3) Decrease operating costs. An asymmetric, prismatic glass refractor, housed in an attractive, traditional postop luminaire to compliment the existing architecture was chosen to meet the design criteria while providing a pleasing daytime appearance. Existing poles were used in their original positions, but the base transformers were removed to eliminate a source of system inefficiency. The new luminaire operates directly from the 480 volt line. A 70 watt high pressure sodium (HPS) source was selected for its high efficacy, long life and lumen output. The asymmetric distribution puts the light where it is needed and improves uniformity.

After retrofit, annual operating costs totaled $1,000, a savings of $5,000 per year. The project was paid from savings in less than three years.

Case 2 - Roadway Retrofit:

The roadway at the rear of the facility was lighted with 30, 150 watt mercury-vapor, cobra-head luminaires mounted on 30 foot poles on 200 foot centers. Deterioration caused higher than normal maintenance and outages became a problem. In addition, the mercury-vapor lamps seemed to burn forever, although not much light was provided on the roadway, causing concerns for safety and security.

Design criteria for the retrofit were to: 1) Increase light levels to improve safety and security, 2) Lower maintenance costs, and 3) Reduce roadway glare and provide a more modern daytime look. A high performance, prismatic glass refractor packaged in a modern roadway housing, replaced the cobra-head on existing mounting arms. A 250 watt high pressure sodium (HPS) source was chosen for higher lumen output (27,500) and lower lumen depreciation. Visibility was improved by using a refractor design that provides sufficient high-angle candlepower for uniform pavement brightness, while limiting roadway glare. The vertical burning HPS lamp distributes over 90% of its lumen output to the side so that the glass refractor can efficiently control the light away from the base of the pole without the use of a reflector.

This retrofit project was done chiefly for the benefit of improving safety and security, while replacing a deteriorating system, with some tangible benefits from reduced maintenance and energy savings.

Case 3 - Cooling Tower Retrofit:

Three cooling towers, located on the north side of the boiler house, were all originally equipped with incandescent lamps in standlights on the top of the towers. The function of the cooling tower lighting is to provide area light for evening and nighttime tower maintenance and routine operational tasks. The original systems had deteriorated and the lighting was no longer capable of providing adequate light for personnel to see well enough to do their assigned tasks at night. In addition, the operating costs of energy and maintenance were higher than necessary. Part of the high energy cost was attributed to the lack of any control to keep the tower lights off during the daytime.

The objectives of the retrofit were to: 1) Reduce operating costs, 2) Provide higher levels of illumination on all task surfaces, 3) Improve visibility of the towers from the boiler house, and 4) Improve safety and security. A total of 25 tower and 15 boiler house building fixtures of various wattages, ranging from 100 to 300 watts, were replaced with high pressure sodium (HPS) sources in low brightness prismatic glass luminaires. HPS sources were chosen for their long life and high efficacy, resulting in reduced labor costs for lamp changes and reduced lighting energy costs. The original symmetrical standlights on the top of the towers were replaced with pendant hung prismatic luminaires with a long and narrow distribution to more uniformly light the sides of the tower.

Lighting energy before the retrofit was 55,000 kwh per year. After retrofit, it was 16,000 kwh per year. Annual labor and lamp cost savings total $2,400 and energy cost savings are $2,000 per year, paying for the project in 2 years. The economics of this retrofit are shown in detail in Figure 1.

SUMMARY

Nine different lighting retrofit projects have been described that have measurably improved the quality of the workplace and reduced operating expenses for the AT&T Manufacturing and Development Center in Columbus, Ohio.

These lighting retrofit projects demonstrate that good lighting practice does not have to be sacrificed for energy efficiency. Corporations like AT&T are discovering that innovative lighting solutions that deliver both energy efficiency and high quality,result in the most effective use of both energy and human resources.

Chapter 4

A SPREADSHEET TEMPLATE FOR SIMPLE LIGHTING DESIGNS USING LOTUS® 1-2-3®

J. L. Fetters

LDW: The Lighting Design Worksheets

The filename on the disk storing the template is LDW and the file contains three sections called worksheets. The template was made user friendly with a menu driven command selection so it only requires that the user have a little working knowledge of Lotus 1-2-3 commands. Lotus is loaded in drive A. The disk with LDW is placed in drive B and the command "/FR" is typed to retrieve the LDW file or any other file that the user has stored on the data disk.

A title screen is presented to the user with instructions (Figure 1), along with a menu displayed at the top of the screen for the user to select a worksheet. Selecting the "LUMINAIRES" worksheet allows the user to determine the number of luminaires required to light a given space with a certain light source. Selecting the "FOOTCANDLES" worksheet provides the user with an illuminance level for a given space using a certain number of luminaires and a given light source. Selecting the "GRID-LAYOUT" assists the user with simple rectangular grid layouts. Other menu selections are provided to "PRINT" the worksheets, "SAVE" the files on the data disk and "QUIT" when finished.

After the user selects a worksheet, a new menu will be displayed at the top of the screen. The user is asked to "ERASE OLD WORKSHEET" or "MODIFY EXISTING WORKSHEET" (Figure 2). This provides an easy method to retrieve and modify an existing worksheet without reentering all the data again. The spreadsheet will automatically date and time stamp the current document and the user is prompted to enter a name. This name will appear on the worksheet near the bottom of the space identified as "WORKSHEET PREPARED BY:". The program then moves the cursor to each data cell required for the calculations.

A convention has been used to help identify those cells used for inputting data from those cells that operate on data cells to calculate a result. Note that all data input cells are marked by a ":" at the right end of the label, while those calculated cells are identified with a "=".

To help the user identify their lighting project, a two line space is provided at the cells to the right of the label "PROJECT IDENTIFICATION:". Next, the cursor moves to a cell labeled "LENGTH:" where the user is expected to enter the dimensions of the room length. After each data entry, the user strokes the enter key. Most PC's now mark this key with a 90 degree bent arrow. (Some call it a carriage return; a throwback to ancient typewriters, no doubt!) At each enter stroke, the program moves the cursor to the next data entry cell of the worksheet.

At this point it is assumed that the user understands how to enter the data required of the template where the cursor has guided them. But what if an incorrect entry is made? With these worksheets, it is easier to make all corrections after all data is entered.

After all data for the worksheet is entered, the user may "disconnect" from the automatic cursor advancement by choosing "QUIT" from the command menu. Then the user simply moves to the cell requiring correction using the arrow keys, enters the correct data, and resumes the design process.

The "LUMINAIRES" Worksheet (Figure 3)

The "LUMINAIRES" worksheet is used when the designer wants to determine how many luminaires are required to develop an average illuminance in a space. The basis of this worksheet is the IESNA zonal cavity method (also called the Lumen method) and requires that the user provide coefficient of utilization (CU) values obtained from the luminaire manufacturer's catalog. It is intended for simple, direct lighting applications.

After the user has entered values for the room dimensions, the worksheet calculates a value for the Room Cavity Ratio (RCR) and enters this number in the cell labeled "Room Cavity Ratio (RCR)=".

```
T1:                                                                      MENU
LUMINAIRES  FOOTCANDLES  GRID-LAYOUT  SAVE  PRINT  QUIT
Calculates # luminaires for a given FC
       T          U          V          W          X          Y          Z
1
2
3
4                                    LIGHTING DESIGN WORKSHEETS
5                                    SPREADSHEET TEMPLATES for
6                                           LOTUS 1-2-3
7
8
9                                              PLEASE
10                                        SELECT FROM MENU
11
12
13                                   TO RESTART THE SPREADSHEET,
14                                            USE ALT R
15
16
17                                          TO QUIT 1-2-3,
18                                   USE THE 1-2-3 /QUIT command
19
20                                             FIGURE 1
04-May-88  05:53 PM                              CMD                    CAPS
```

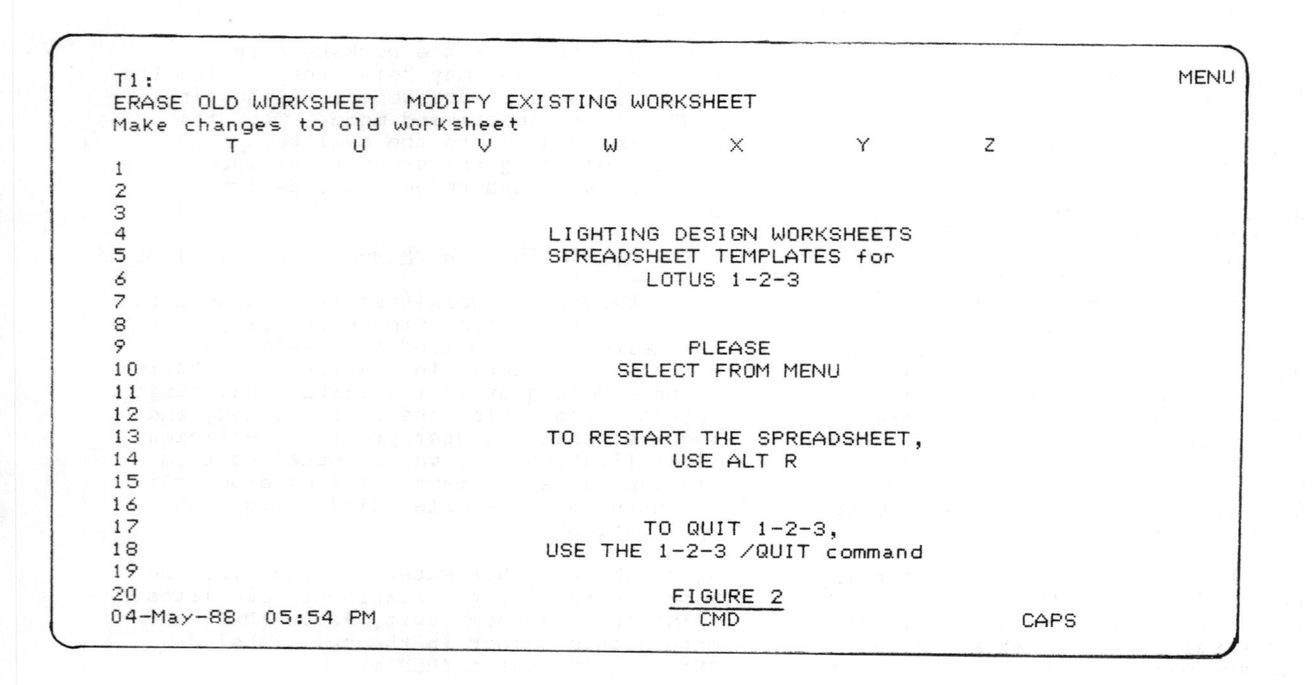

```
T1:                                                                      MENU
ERASE OLD WORKSHEET  MODIFY EXISTING WORKSHEET
Make changes to old worksheet
       T          U          V          W          X          Y          Z
1
2
3
4                                    LIGHTING DESIGN WORKSHEETS
5                                    SPREADSHEET TEMPLATES for
6                                           LOTUS 1-2-3
7
8
9                                              PLEASE
10                                        SELECT FROM MENU
11
12
13                                   TO RESTART THE SPREADSHEET,
14                                            USE ALT R
15
16
17                                          TO QUIT 1-2-3,
18                                   USE THE 1-2-3 /QUIT command
19
20                                             FIGURE 2
04-May-88  05:54 PM                              CMD                    CAPS
```

```
                              LUMINAIRES
                       LIGHTING DESIGN WORKSHEET
                     USING IES ZONAL CAVITY METHOD
  PROJECT IDENTIFICATION:PART 5                    SAVED AS:P5DWR2.WK1
                            DISHWASHER ROOM
 ROOM DIMENSIONS  LENGTH: 45.0 FEET                  REFLECTANCES
                   WIDTH: 19.3 FEET                   FLOOR:       0.20
         MOUNTING HEIGHT:  9.0 FEET                 CEILING:       0.70
HEIGHT (FIXTURE TO TASK):  6.5 FEET                    WALL:       0.50

 ROOM CAVITY RATIO (RCR)= 2.41     LIGHT LOSS FACTOR:              0.80
                                             UPD LIMIT:            1.7 W/SF
DESIGN ILLUMINANCE (AVG):    55 FC
                 INITIAL=    66 FC              LAMP TYPE:  FLUOR
                                                  CATALOG: F40LWSS
  LUMINAIRE MANUFACTURER:DAY-BRITE        INITIAL LUMENS:       2925
               CATALOG #:CG142-CO2A           LAMP WATTS:         34
                                        LAMPS/LUMINAIRE:           2
      CU (FROM CATALOG): 0.57              BALLAST WATTS:        6.0
                                          BALLAST FACTOR:       0.90
SPACING CRITERIA(L-WISE): 1.20         WATTS/LUMINAIRE=           74
SPACING CRITERIA(W-WISE): 1.20             LUMENS/WATT=           79 per
                                                            LUMINAIRE
MAXIMUM SPACING (L-WISE)=  7.8 FEET
MAXIMUM SPACING (W-WISE)=  7.8 FEET
                                          TOTAL LOAD =          1.5 KW
   THEOR # OF LUMINAIRES=   20                   UPD =          1.7 W/SF
  ACTUAL # OF LUMINAIRES:   20              WHICH IS       100.5% OF
   SQUARE FEET/LUMINAIRE=   43                            THE UPD LIMIT
                                                 DATE: 13-May-88
  WORKSHEET PREPARED BY:J.L.FETTERS              TIME:  05:19 PM
```

FIGURE 3

Reflectance values are entered into the worksheet but are not used for computation. These values are recorded only as a handy place to record them as a reminder of what values were used to look up a CU (coefficient of utilization) value in the luminaire catalog, using the computed RCR and the reflectance values. A light loss factor is estimated, based on considerations found in the IES Handbook. Using the light loss factor, the worksheet will calculate the "INITIAL=" illuminance in the cell directly below the cell where the user entered the "DESIGN ILLUMINANCE (AVG):". In the newest version of the template, a separate cell is dedicated to "BALLAST FACTOR:" to show the effect of this variable on the initial light level.

Lamp data, obtained from the lamp supplier's catalog, is entered. "LAMP WATTS:" and "BALLAST WATTS:" values are used in the power calculations; the "UPD=", the "WATTS/LUMINAIRE=", the "LUMENS/WATT=", and the "TOTAL LOAD=" calculated cells. This information is provided to help the designer determine if the lighting design will come in under the power budget set at the "UPD LIMIT:" cell. If the user does not wish to invoke the UPD calculation, they should enter an @NA in the "UPD LIMIT:" cell.

Next, the user, while consulting the luminaire catalog, enters data for "LUMINAIRE MANUFACTURER:", "CATALOG #:", the "COEFFICIENT OF UTILIZATION (CU):", and the "SPACING CRITERIA:". Spacing criteria inputs are used to calculate the maximum spacing values. ("MAXIMUM SPACING (_-WISE):" When the user enters the CU value, the worksheet calculates a theoretical number of luminaires and entered it in the cell labeled "THEOR # OF LUMINAIRES=". The cursor now stops at the cell labeled "ACTUAL # OF LUMINAIRES:" so that the user may enter the integer number of luminaires, rounding or making the number even, if required. Although it may not be possible for the designer to know the exact value of this integer on the first pass, a number close to the theoretical number of luminaires will allow the worksheet to calculate the "SQUARE FEET/LUMINAIRE=", the "TOTAL LOAD=", and the "UPD=". Examination of these calculated values will permit the designer to decide if the design at this point meets some of the design criteria.

The program prompts the user is choose another type of worksheet, choose to "PRINT", "SAVE", or "QUIT".

The "FOOTCANDLES" Worksheet (Figure 4)

The "FOOTCANDLES" worksheet is used when the designer wants to determine the illuminance of a space for a given number of luminaires, using a particular luminaire and light source combination. For example, it would be used to compute the actual value of illuminance resulting when the actual number of luminaires in the "LUMINAIRES" worksheet is different than the theoretical number calculated by the worksheet. This worksheet also uses the IESNA zonal cavity method and, like the "LUMINAIRES" worksheet, also requires the user to supply a CU value obtained from the luminaire manufacturer's information. Data values are entered in the same way as described for the "LUMINAIRES" worksheet. The major difference between the "FOOTCANDLES" worksheet and the "LUMINAIRES" worksheet is that the calculation for illuminance and number of luminaires is reversed, and in this worksheet, an average footcandle level is calculated from lamp and luminaire data. This make it convenient for the user to try different lamp and luminaire combinations to test the sensitivity of the selected equipment on light levels and power levels.

The "GRID-LAYOUT" Worksheet (Figure 5)

The IES Zonal Cavity method assumes a uniform layout to achieve a uniform average light level in the space. Many task/ambient lighting designs will require the ambient portion be uniform. Since many of the simpler, fluorescent ambient designs will be accomplished using troffers in a suspended grid ceiling, the Grid-Layout worksheet is used to provide a quick troffer layout.

```
                              FOOTCANDLES
                       LIGHTING DESIGN WORKSHEET
                     USING IES ZONAL CAVITY METHOD
   PROJECT IDENTIFICATION:PART 5                     SAVED AS:P5DWR2.WK1
                              DISHWASHER ROOM
  ROOM DIMENSIONS  LENGTH: 45.0 FEET                   REFLECTANCES
                    WIDTH: 19.3 FEET                FLOOR:        0.20
          MOUNTING HEIGHT:  9.0 FEET              CEILING:        0.70
 HEIGHT (FIXTURE TO TASK):  6.5 FEET                 WALL:        0.50
                                        LIGHT LOSS FACTOR:        0.80
  ROOM CAVITY RATIO (RCR)= 2.41
                                               UPD LIMIT:          1.7 W/SF
   ACTUAL # OF LUMINAIRES:    18
     SQUARE FEET/LUMINAIRE=    48              LAMP TYPE:   FLUOR
                                                 CATALOG:  F40LWSS
   LUMINAIRE MANUFACTURER:DAY-BRITE        INITIAL LUMENS:      2925
                 CATALOG #:CG142-C02A          LAMP WATTS:        34
                                          LAMPS/LUMINAIRE:         2
        CU (FROM CATALOG): 0.57             BALLAST WATTS:       6.0
                                           BALLAST FACTOR:      0.90
 SPACING CRITERIA(L-WISE): 1.20           WATTS/LUMINAIRE=        74
 SPACING CRITERIA(W-WISE): 1.20               LUMENS/WATT=        79 per
                                                                LUMINAIRE
 MAXIMUM SPACING (L-WISE)=   7.8 FEET
 MAXIMUM SPACING (W-WISE)=   7.8 FEET
                                             TOTAL LOAD =        1.3 KW
       ILLUMINANCE (AVG)=    50 FC                  UPD =        1.5 W/SF
                 INITIAL=    60 FC             WHICH IS        90.5% OF
                                                            THE UPD LIMIT
                                                 DATE: 13-May-88
  WORKSHEET PREPARED BY:J.L.FETTERS              TIME:  11:50 AM
```

FIGURE 4

AN EXAMPLE DESIGN SOLUTION

To illustrate how the LDW template is used, an actual design problem will be shown in detail. The space to be illuminated is a remodeled 45 ft by 19 ft dishwasher room, with a new suspended ceiling at 9 ft. The task at the dishwasher, loading and unloading areas is 30 inches off the floor. The reflectance value of the ceiling tile is .70, the ceramic floor tile is .30, but the effective cavity value due to the equipment in the room is reduced to .20, and the wall reflectance is estimated to be .50.

An adjacent area uses a luminaire that the client wishes to use again to keep the ceiling look the same. The luminaire to be used is a 2 lamp, 1 X 4 fluorescent troffer. The client prefers to use energy saving lamps, since this is their standard lamp in this facility. To keep their energy savings program on track, the client demands that the unit power density (UPD) not exceed 1.7 watts/sf. In the old dishwasher room the lighting was described as "dim" and was measured to be 40 footcandles.

In this example, the type of luminaire is known and the solution becomes one of determining how many luminaires are required and where to place them. The grid and room size determine the maximum number of luminaires that will practically fit into the room. The design solution was begun by using the "LUMINAIRES" worksheet first (Figure 3), choosing a target footcandle illuminance of 55 fc. After the project description "Part 5 Dishwasher Room" is entered, the room dimensions (LENGTH at 45 ft, WIDTH of 19 ft 4 in), mounting height of 9 ft, and the height of the fixture to the task of (9 ft minus 30 inches) are entered, and the template calculates a Room Cavity Ratio (RCR) of 2.41. The reflectance values (.2, .7, and .5), LLF of .8, UPD LIMIT of 1.7 W/SF, and AVG DESIGN ILLUMINANCE of 55 FC are entered.

Lamp data, found in the lamp manufacturer's catalog and the value of 2 LAMPS/LUMINAIRE are entered next. The Day-Brite catalog number is entered and the catalog data is looked up to find the CU value. Part of the Day-Brite catalog sheet is shown reprinted in Figure 7. The photometric data CU table shows that for a 2-lamp luminaire, with an RCR of 2.41 (approximately halfway between RCR = 2 and RCR = 3), and floor reflectance (pfc) of 20% (.20), ceiling reflectance (pcc) of 70% (.70), and wall reflectance (pw) of 50% (.50), the CU value can be interpolated between 54 and 60 to be 57 (.57). Spacing to mounting height (S/MH) value of 1.2 is entered in place of the spacing criteria. An energy saving ballast was used that has a loss of 6 watts and a ballast factor of .90. These values are found in the ballast manufacturer's catalog.

The spreadsheet template computes the "Theor # OF LUMINAIRES=" as 20, and 20 is then entered in the cell after "ACTUAL # OF LUMINAIRES:" to provide the template with an integral number of luminaires. When 20 is entered in at "ACTUAL # OF LUMINAIRES", the "UPD" calculates to 1.7 W/SF, "WHICH IS 100.5% OF THE UPD LIMIT". This shows that the maximum number of luminaires used that will not exceed the UPD limit is 20. In addition, the template calculates a "MAXIMUM SPACING" of 7 ft, 9 in, 43 "SQUARE FEET/LUMINAIRE", 74 "WATTS/LUMINAIRE", 79 "LUMENS/WATT per LUMINAIRE", and a "TOTAL LOAD" of 1.5 KW. These calculated values provide the designer with comparative figures for other design iterations.

Since the luminaires are to be installed in a suspended grid ceiling, the "GRID-LAYOUT" worksheet is selected next. The worksheet shown in Figure 5 titled "LUMINAIRE LAYOUT WORKSHEET for SPACING IN REGULAR ROWS" comes up on the screen. The room dimensions are copied to this worksheet automatically by the template and the user begins data entry at "LENGTH of LUMINAIRE:" Here the designer must choose how to orient the 1 X 4 luminaires for the tasks to be illuminated. The long dimension of the luminaires is to be parallel with the short dimension of the room, because of equipment layout and to provide good side illumination for the racking operations. This decision requires the entry of 1 foot for "LENGTH of LUMINAIRE:" and 4 foot for "WIDTH of LUMINAIRE:"

Figure 6 shows this orientation. Since 2 X 4 panels are to be used between the luminaires an integral number of 2 or 4 foot panels must be considered when choosing the "LENGTHWISE SPACE BETWEEN LUMINAIRES:" of 6 foot. The worksheet calculates a "CALCULATED LUMINAIRES per ROW=" of 6.4. The user then enters the integer value of 6, since partial luminaires are not allowed! When the value of 6 is entered at the "ACTUAL # of LUMINAIRES per ROW:", the worksheet calculates a value of 9.0 feet "at the ends" This is the total length remaining at both ends of the 45 foot length after 6, 1 foot luminaires have been placed 6 foot apart. The worksheet calculates a value of 3.3 "CALCULATED THEORETICAL # of ROWS=", and the user enters the integer value of 3 for "ACTUAL # of ROWS:" 18 "ACTUAL # of LUMINAIRES=" is automatically calculated and 48 " s.f./luminaire." The user must now choose a value for "WIDTHWISE SPACE BETWEEN LUMINAIRES:". If the value chosen causes the "leaving __ feet at the ends" value to go negative, the user chooses a smaller spacing, in 2 or 4 foot spacing, until the total length remaining at both ends of the 19 foot, 4 inch width is positive and of an acceptable value. Recall from the "LUMINAIRES" worksheet that a value for "Maximum Spacing" was calculated (Figure 3) and the spacings of the luminaires on this Luminaire Layout worksheet must be less than these values for an even distribution of light.

Figure 6 shows the final layout of 3 rows of 6 fixtures per row. This results in 18 luminaires which is less than the 20 required to meet the target illuminance of 55 FC. To determine what the actual average illuminance of the space would be using 18 luminaires, the designer chooses the "FOOTCANDLES" worksheet (Figure 4). Data entry is the same as described for the "LUMINAIRES" worksheet, except that for the "ACTUAL # OF LUMINAIRES:" 18 is entered, instead of the theoretical value of 20 calculated previously. The worksheet calculates values for "ILLUMINANCE (AVG)=" of 50 FC and for "INITIAL=" of 60 FC. These actual illuminance values are acceptable for this design, so the design process is terminated by printing and saving the three worksheets. If the actual illuminance values were not acceptable, the process would have been repeated until an acceptable layout with acceptable values of illuminance resulted.

SUMMARY

The LDW spreadsheet template has been valuable for simple lighting designs to enable the author to calculate several different solutions for a given lighting problem in a short design interval. Follow up measurements made after the lighting design has been implemented show good correlation. In some cases where the results were different than expected, the major cause is due to reflectance values which are usually chosen too high for the final finish selection, which was probably not made by the lighting professional.

Of further benefit, a large library of lighting solutions has now been saved, for preferred luminaire and lamp combinations that further cuts the design interval and frees the designer to spend more time with the quality aspects of the design.

REFERENCES

IES Lighting Handbook, 1984 Reference Volume, John E. Kaufman, PE, FIES, Editor

```
                 LUMINAIRE LAYOUT WORKSHEET
                for SPACING IN REGULAR ROWS

       PROJECT IDENTIFICATION:PART 5 DISHWASHER ROOM

                   ROOM LENGTH:45.0
                    ROOM WIDTH:19.3
                     ROOM AREA= 866   sq ft

          LENGTH of LUMINAIRE: 1.0   feet
           WIDTH of LUMINAIRE: 4.0   feet

  ROOM LENGTH/LUMINAIRE LENGTH=  45
      LENGTHWISE
      SPACE BETWEEN LUMINAIRES: 6.0
    CALCULATED LUMINAIRES per ROW= 6.4
   ACTUAL # of LUMINAIRES per ROW:   6  ,leaving 9.0 feet at the ends.

 CALCULATED THEORETICAL # of ROWS= 3.3
               ACTUAL # of ROWS:   3

       ACTUAL # of LUMINAIRES=  18  , making  48 s.f./luminaire.
      WIDTHWISE
      SPACE BETWEEN LUMINAIRES: 2.0  ,leaving 3.3 feet at the ends.

       WORKSHEET PREPARED BY:J.L.FETTERS  DATE:      13-MAY-88
                                          TIME:       11:43 AM
```

FIGURE 5

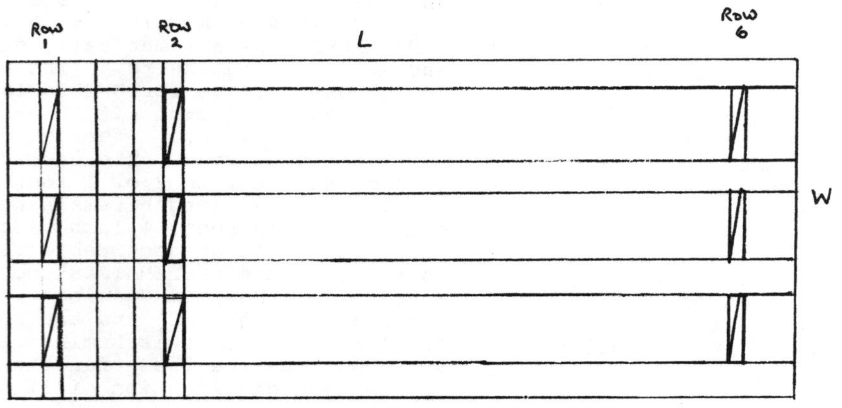

Row 1
Row 2
L
Row 6
W

FIGURE 6

PHOTOMETRIC DATA

Coefficient of Utilization-Zonal Cavity Method

2-Lamp

pfc	20							
pcc	80			70			50	
pw	70	50	30	70	50	30	50	30
RCR								
0	76	76	76	74	74	74	71	71
1	71	68	66	69	67	65	64	63
2	66	61	58	64	60	57	58	55
3	61	55	51	59	54	50	52	49
4	56	50	45	55	49	44	47	44
5	52	45	40	51	44	39	43	39
6	48	41	35	47	40	35	39	35
7	45	37	32	44	36	31	35	31
8	41	33	28	40	33	28	32	28
9	38	30	25	37	30	25	29	24
10	35	27	22	35	27	22	26	22

Test #7630-2 S/MH=1.2

3-Lamp

pfc	20							
pcc	80			70			50	
pw	70	50	30	70	50	30	50	30
RCR								
0	67	67	67	66	66	66	63	63
1	63	60	58	61	59	57	57	55
2	58	54	51	57	53	50	51	49
3	54	49	45	52	48	44	46	43
4	50	44	40	49	43	39	42	38
5	46	39	35	45	39	35	38	34
6	43	36	31	42	35	31	34	30
7	39	32	28	39	32	28	31	27
8	36	29	25	36	29	25	28	24
9	34	26	22	33	26	22	25	21
10	31	24	20	31	24	20	23	19

Test #7644 S/MH=1.2

Average Brightness (Footlamberts) with 3150 Lumen Lamps

2-Lamp			3-Lamp	
End	Cross	Angle	End	Cross
1316	1242	45	1756	1604
950	866	55	1276	1130
743	698	65	996	931
747	698	75	1009	943
702	635	85	922	815

Light Loss Factor Data

LLF = 0.77
LLF = Light Loss Factor
LDD = Luminaire Dirt Depreciation
IES Category V Clean Annually
LLD = Lamp Lumen Depreciation
BF = Ballast Factor (commercial ballast performance relative to reference ballast)

Light Loss Factor (LLF) = LDD×LLD×BF
LDD = Very Clean 0.93 Clean 0.88 Medium 0.82
LLD = 0.88 @ 40% Rated Lamp Life
BF = 0.94 (Std. Ballasts & Lamps)
Relamp @ 70% Lamp Life

FIGURE 7

Chapter 5

NEW LIGHTING OPTIONS FOR STATE BUILDINGS

J. B. Wong-Kcomt, W. C. Turner, S. Hong, W. R. King

ABSTRACT

A large energy management project is underway where Oklahoma State University and the Office of Public Affairs for the State of Oklahoma are working together to reduce energy costs for state buildings under the control of Office of Public Affairs. One of the biggest cost components is lighting and the vast majority of lighting is standard fluorescent tubes. Much technological development has occurred recently in fluorescent tubes so there are many retrofit options.

The Energy Division of the Office of Public Affairs wanted to find out which options can be cost effective and, at the same time, be accepted by the building occupants. This paper presents 1) the conclusions from an economic analysis of the most relevant fluorescent lighting alternatives for the State Buildings and 2) the results from an experimental test carried out in an occupied building. The economic evaluation has been performed by using a life cycle approach and the computation for the costs and savings of each option has been done on an incremental basis. Results on the measures of energy usage and readings of light levels on the tested options are included.

Significant energy cost reduction will be realized when the recommended lighting options are fully implemented in all buildings of the State Capitol Complex, Oklahoma City. These recommendations can be extended to all state buildings of the State of Oklahoma.

INTRODUCTION

The Office of Public Affairs (OPA) of the State of Oklahoma in conjunction with the School of Industrial Engineering of Oklahoma State University (OSU) is carrying out a large energy management project for the state office buildings. An important part of this project constitutes both the technical and economical evaluation of more efficient fluorescent lighting systems.

A variety of new lighting technologies have entered the market. In most existing buildings and facilities, when confronted with specifying a fluorescent bulb and/or accessories for energy efficiency, building and plant Managers are usually confused with such diversity.

Considering that many of the actual lighting systems were not designed for energy conservation, the new options for fluorescent lighting may lead the state buildings to opportunities for energy and dollar savings. A selected set of these opportunities is presented in this paper.

To evaluate the actual energy savings attainable from some of these options, and to gather the opinions of building occupants about the new options, the OSU research team conducted an experiment in one of the State buildings. Results of this experiment are presented as well.

Because of configurations of space or use, some options could be more advisable, even though they are less efficient. For instance, historical buildings and art exhibits usually require lamps with a minimum emission of ultraviolet rays to prevent art-work deterioration.

AVAILABLE TECHNOLOGIES

Among the products offered in the market as "fluorescent lighting energy savers," for a four-40-watt-lamp fixture (2 ft x 4 ft), we have selected four different categories:

1) Energy efficient fluorescent lamps, EEFL
2) Electronic fluorescent ballast, EFB
3) Power reducing devices, PRD
4) Improved fluorescent reflectors, IFR

Energy efficient fluorescent lamps consume less energy than standard (F40) lamps while giving nearly the same light levels. They cost more initially, but the incremental cost will be recovered through energy savings. Essentially, there are two kinds of EEFL, 1) those that do not alter the color rendition and visual definition, and 2) those that have a higher lumen output and an improved color rendition. Within the first kind of EEFL we consider energy efficient (EE), energy efficient "plus" (EE+) and Octron (T8) lamps. Some examples of new lamps with improved color rendering illumination and higher lumen rating are the Aurora IV from VL Service Lighting Corporation, and the Advantage X from North American Philips Lighting Corporation.

Ballast designed with solid state electronics have recently become commercially available. Their high frequency operation (near 25,000 Hz) allows fluorescent lamps to operate more efficiently and still provide a light level equal to that of traditional electromagnetic ballast. Some energy efficient lamps, as the Octron, require special magnetic ballast, but the manufacturer recommends the use of electronic ballast to maximize energy savings.

Power reducing devices basically "regulate" the flow of current going into the bulb once the lamp is lit. These devices can also reduce the lumen depreciation rate, allowing a longer bulb life. One example of these devices is the Edison 21 fluorescent monitor. The Edison 21 comes in two variations. One saves 30% and the other 50% of energy consumption [1]. A variation of PRDs is a "power reducing fluorescent bulb", marketed under the commercial name of "Thrift/Mate" by Sylvania. Two Thrift/Mate lamps are required in four-lamp fixtures, one on each ballast circuit. This lamp accomplishes the same purpose as a PRD. One type of Thrift/Mate reduces power and energy by 33% and another type by 50%. The manufacturer claims that lumens-per-watt efficiency remains the same after replacement. Hence, illumination levels are reduced proportionally to the power/energy reduction [2]. Therefore, this option may be used in rooms with excessive illumination. This is the case in many Oklahoma State buildings. In general, PRDs may be considered as an alternative to EEFL.

Fluorescent reflectors can be improved by using surfaces or films with better reflectivity. In addition, the position of the reflective surfaces with respect to the bulb may affect the performance of a given fixture. Improved lighting levels can be achieved with a good combination of high reflective surfaces and well designed reflector geometry. A vendor claims a "completely-lit" impression for a four-bulb fixture, when two lamps are removed, the two existing lamps are relocated and an "optimum bend reflector" is installed in the fixture [3]. In this paper we only consider reflectors for four-lamp fixtures.

ECONOMIC ANALYSIS

From the four categories mentioned above, eight alternatives have been considered as "challengers" to the "do-nothing" option, e.g. keep using a 4-tube fixture with 4 standard 40-watts lamps and standard ballast. The alternatives are listed as follows:

1. Use 4 energy efficient fluorescent lamps (EE) and keep standard ballast.
2. Use 4 EE lamps and electronic ballast.
3. Use 4 energy efficient "plus" lamps (EE+) and keep standard ballast.
4. Use 4 "Octron" (T8) lamps with electronic ballast.
5. Replace 4 standard lamps by 2 Aurora IV lamps (or similar) and keep standard ballast.
6. Install improved reflectors, remove 2 lamps, relocate the existing 2 lamps in the fixture and keep one standard ballast.
7. Install PRD (Edison 21) and keep the standard lamps and ballast.
8. Replace two standard lamps with two Thrift/Mate lamps, e.g., one replacement per ballast circuit.

There could be other alternatives but time constraints prohibited considering all possibilities. To evaluate the 8 alternatives listed above, we have considered a life cycle approach. In this way, we consider every relevant cost item throughout the service life of a given alternative. In addition, we account for the time value of the money by using a discount rate. Also, we use the rated service life as the planning horizon for the cash flow analysis. Since bulb life and/or ballast life are not the same for different lamps and are dependent on the intensity of usage (whether one, two or three shifts), we use the Annual Worth method to evaluate alternatives with different service lifes. This method is based on the assumption that lighting systems will be replaced by identical models possessing the same cost [4].

We have developed a generalized life-cycle cost profile for all the alternatives. This profile is depicted in Figure 1. We have assumed that change over will be carried out in an "as-lamps-burn-out" basis. Thus, the first cost is the increment "above" standard lamps and ballasts. It includes material and labor to install a given alternative. Similarly, any maintenance costs are included in an incremental basis. Cost figures have been obtained from trade literature and vendor quotations. Table 1 shows the total wattage per fixture and percent of energy savings for each alternative.

Table 2 depicts the lamp cost, the ballast cost and the total installed cost for each alternative. An alternative's incremental initial cost is the difference between its initial cost and alternative's 0 (do nothing) replacement cost. For alternatives that use existing standard lamps and ballasts, it has been assumed that they can be used for the remaining half of their service life.

Our economic analysis is rather conservative since it does not consider the air conditioning impact of lighting reduction. This could be justified whenever a building's conditioning savings (due to

YES
End of Year
1
2
3...
L-1
L
L+1...
B
N
ILRC
IBLC
IIC
AW = YES - [IIC + ILRC (P/A i-eff1,N/L) + IBRC (P/A i-eff2,N/B)] (A/P i,N)
Where:
AW = Annual Worth, $/year-fixture
YES = Yearly Energy (and demand) Savings, $/year-fixture
IIC = Incremental Initial Cost, $/fixture
ILRC = Incremental Lamp Replacement Cost, $/fixture
IBRC = Incremental Ballast Replacement Cost, $/fixture
L = Lamp service life, 5-6 years
B = Ballast service life, 10-23 years
N = Planning horizon, years
i = Minimum attractive rate of return
i-eff = effective interest rate of i compounded in N/L or N/B years (during lamps or ballast service life respectively).
Note 1: When an alternative proposes a reduction on the number of lamps/ballast per fixture, ILRC and IBRC may become avoided costs (savings). In this case, ILRC and/or IBRC are negative.
Note 2: Typically, the planning horizon is the same as the ballast life, N=B (in general B>L). But when an alternative includes devices with longer life than B, the devices' longer life is used as planning horizon. For instance, for Alternative 7 (PRD: Edison 21), N=21 years; and for Alternative 6 (reflectors), N=15.

FIGURE 1. CASH FLOW PROFILE AND ANNUAL WORTH EQUATION

lighting reduction) are approximately the same as the additional cost of heating (during winter) required to make up for lighting heat reduction. This is generally the case in "thermally light buildings". But, in "thermally heavy buildings" and/or buildings with excessive lighting, additional savings can be realized from cooling load reduction.

Also, this study assumes that electricity prices and replacement cost for lamps and ballasts will remain constant throughout the planning horizon of each alternative.

Example

For Option 2, energy efficient lamps and electronic ballasts, the Annual Worth is:

AW = YES + [-IIC - IRC1 (P/A i-eff,4) + IRC2 (P/F 10,11)] (A/P 10,23)

= 10.24 + [-19.52 -3.52(1.39) + 30(.3505)] (0.1126)

= $8.67/year

In this example, YES=$10.24 is the yearly energy savings; i-eff is the effective rate for a yearly rate i=10%, compounded during a period equal to the lamp service life, L=6 years. E.g. i-eff= (1+i)**L -1, [5]. IIC is the incremental initial cost over the standard option. IRC1 and IRC2 are the incremental replacement costs, for lamps and for ballasts respectively.

The savings due to energy and power reduction are computed in terms of a weighted average demand charge ($6.05/Kw-month) according to Oklahoma Gas & Electricity's PL-1 schedule (service level 5). Consumption charge is $0.03639/kWH. [6]

Most of the state office buildings operate during 8 to 10 hours per day, but are occupied 2 to 4 additional hours by custodial personnel. Hence, we consider 12 hours of operation per day. In addition we have assumed a 10% interest as a minimum rate of return for each alternative's cash flow.

Annual Worth (equivalent net dollar savings per year) is the measure of merit for each alternative. Note, this means a higher number is preferred over a lower number if everything else (other valuation factors as lumens/watt, color rendition, building occupants' opinion, etc.,) is equal. However, everything else is not equal, and a particular building, or even room application may require a specific alternative. But, in general OPA will select the two or three "best" alternatives to simplify purchasing and replacement. Annual Worth results are listed in Table 3.

TABLE 1

WATTAGE PER FIXTURE AND PERCENT OF ENERGY SAVINGS

ALTERNATIVE	Total Wattage* (watts)	Energy Savings (%)
0. Four Std. Lamps & 2 Ballasts (do nothing)	174	0.0
1. Four Energy Eff Lamps & 2 Std. Ballasts	155	10.9
2. Four Energy Eff Lamps & 2 Elect. Ballasts	119	31.6
3. Four Energy Eff+ Lamps & 2 Std. Ballasts	144	17.2
4. Four Octron Lamps & 1 Elec. Ballast**	106	39.0
5. Two Aurora IV Lamp & 1 Std. Ballast	92	47.1
6. Reflectors with two Std. Lamps & 1 Ballast	87	50.0
7. Edison 21 with 4 Std. Lamps & 2 Ballasts	122	30.0
8. Two Thrift/Mate Lamps, 2 Std. Lamps & Ballasts	117	33.0

*Source: "Fixture Comparison Data," (Sylvania EB-0-362)
**A four Octron lamps fixture needs one specifically designed electronic ballast instead of two.

TABLE 2

LAMP, BALLAST AND INSTALLED COST FOR FLUORESCENT LIGHTING ALTERNATIVES (2 X 4 FT. FLUORESCENT FIXTURE)

ALTERNATIVE	COST Lamp	COST Ballast ($/unit)	COST Other	Installed Cost ($/fixt.)
0. Four Std. Lamps & 2 Ballasts (do nothing)	2.27	15	--	39.00
1. Four Energy Eff Lamps & 2 Std. Ballasts	3.15	15	--	42.60
2. Four Energy Eff Lamps & 2 Elect. Ballasts	3.15	23	--	58.60
3. Four Energy Eff+ Lamps & 2 Std. Ballasts	3.48	15	--	43.92
4. Four Octron Lamps & 1 Elec. Ballast**	3.67	39	--	53.68
5. Two Aurora IV Lamp & 1 Std. Ballast	7.00	15	--	29.00
6. Reflectors with two Std. Lamps & 1 Ballast	2.27	15	40.00	59.54
7. Edison 21 with 4 Std. Lamps & 2 Ballasts	2.27	15	30.00	69.00
8. Two Thrift/Mate Lamps, 2 Std. Lamps & Ballasts	2.27	15	11.34	57.22

*These figures include labor cost for installation.
**A four Octron lamps fixture needs one specifically designed electronic ballast instead of two.

TABLE 3

ANNUAL WORTH FOR FLUORESCENT LIGHTING ALTERNATIVES (Equivalent Net Annual Savings)

Alternative	Annual Worth ($)
1. Energy Eff Lamps & Std. Ballast	2.66
2. Energy Eff Lamps & Elect. Ballast	8.67
3. Energy Eff+ Lamps & Std. Ballast	4.37
4. Octron & Elec. Ballast	11.31
5. Aurora IV Lamp & Std. ballast	16.34
6. Reflectors, Std. Lamp & ballast)	14.79
7. Edison 21, Std. Lamp & ballast (30% savings)	8.00
8. Thrift/Mate, Std. Lamp & ballast (33% savings)	6.16

LIGHTING EXPERIMENT IN THE DEPARTMENT OF TRANSPORTATION BUILDING, OKLAHOMA CITY

To gain insight on the actual performance of fluorescent lighting systems available to the State Buildings, we asked vendors to install several fluorescent lighting options in a large drafting room of the Department of Transportation at the State Capitol Complex, Oklahoma City. Due to time constraints and sample availability, we were able to test only a limited number of options. The selected room had a number of different circuits with fourteen four-40w-lamp fixtures each (2 x 4 ft).

Table 4 lists the results of the experiment on options available from local vendors at the time we started the study. This table also includes average Amps and kW readings. In addition, after measuring during a period of time the kWH consumed by each circuit, we computed the relative energy consumption of each one of the options with respect to the current situation (standard lamps and ballasts). Measured line voltage at the breaker panel was about 270 volts.

Finally, we measured the average illumination level under each circuit's area. For this purpose foot-candle readings were taken for each individual circuits (the other circuits in the same room were turned off). The readings were taken during night at "drawing board height" (about 3 ft above the floor and 9 ft below the fixture level). The foot candle figures in Table 4 are mean values and have certain bias due to light reflected from partitions and walls.

CONCLUSIONS

When using Annual Worth as the evaluation criteria, the Aurora IV, or equivalent (improved-color-rendition-with-higher-lumen lamps), appear to be the best. However, we need much more experimentation with this new kind of lamp. We will use them on an experimental basis in some of the smaller buildings with current excessive illumination. For other buildings, the selection will be improved reflectors. The replacement lamps will be energy efficient and/or Octron lamps (with electronic ballasts). The specific recommendation will depend on the actual illumination levels and the requirements of a given building.

TABLE 4

RESULTS OF FLUORESCENT LIGHTING EXPERIMENT IN THE DEPARTMENT OF TRANSPORTATION BUILDING

No of Circuit	Option	Current (Amps)	Power (kW)	Energy Index	Illum. (fc)*
19	Standard lamps & ballasts (40w)	9.0	2.40	1.00	103
14	Energy eff lamps elec. ballasts (36w)	4.8	1.31	0.57	65
23	Reflector, half std lamps & ballasts	4.7	1.20	0.52	92
17	Power reducer, 33%, std lamps & ballasts**	5.5	1.54	0.67	82
16	Improved color rend lamps & std ballasts**	4.2	1.23	0.53	91
22	Octron Lamps with elec. ballast***	5.5	1.54	0.70	152

*A footcandle (fc) is the illumination on a surface one square foot in area on which there is a uniformly distributed flux of one lumen [7].

**All options, except these two systems -with 2 lamps per fixture, had 4 lamps/fixture.

***This option requires one especially designed electronic ballast per fixture.

By the time these recommendations are fully implemented, significant savings will be realized by each one of the state buildings in the Capitol Complex. The savings can be from 30 to 50% of the actual lighting energy cost. Further cost reduction can be attained, since other energy cost reduction measures are being recommended on heating, ventilation and air conditioning systems. Nevertheless, savings have started to show up in all those buildings that are carrying out our energy cost reduction recommendations. As a part of an ongoing energy management program, Office of Public Affairs will make available the results of this and other studies, being carried out by the Oklahoma State University team, to other Oklahoma State agencies.

ACKNOWLEDGEMENTS

The Energy Division of the Office of Public Affairs and the research team from Oklahoma State University are grateful to the firms and individuals that supplied cost and technical information, and/or participated in the lighting experiment. We also express our appreciation to building management personnel, especially from the Department of Transportation Building, for their valuable cooperation.

REFERENCES

1. Edison 21 Fluorescent Monitor, National Energy Research Corporation, brochure NE-101, Newport Beach, California 92663

2. Thrift/Mate Fluorescent Lamps, Sylvania Industrial/Commercial Lighting, GTE, brochure FL-803, Danver, Mass. 01923

3. Silverlux Fluorescent Reflectors Turn On Energy Savings, Energy Control Products, 3M Center, St. Paul, MN 55144-1000.

4. Riggs, J.L. and T.M. West: Engineering Economics, 3rd ed., McGraw-Hill, New York, 1986.

5. Oklahoma Gas and Electric Company, Standard Rate Schedule, PL-1 (service level 5), effective on and after July 2, 1987. Oklahoma City, OK 73101-321.

6. White, J.A., M.H. Agee, and K.E. Case: Principles of Engineering Economic Analysis, 2d ed., Wiley, New York, 1984

7. Illuminating Engineering Society: IES Lighting Handbook, 5th ed, New York, 1972.

Chapter 6

LIGHTING RETROFIT OPPORTUNITIES MAKE GOOD BUSINESS SENSE

R. A. Tucker

ABSTRACT

Energy management should be applied using the same basic business principles as one would apply to administration, finance, purchasing or production. Corporations have undertaken lighting retrofits because installing energy saving lamps makes good lighting sense and good business sense. Energy efficient lamps provide a return on investment and make a meaningful contribution to energy conservation.

This paper discusses the yardsticks used in business to measure the value of a lighting retrofit. Both the energy and economic performances of installing more efficient lighting products will be highlighted using two case studies. The economic justification for lighting systems in these facilities varies significantly depending on the building design parameters, equipment performance and local utility rates and rebates. These case studies on completed facilities will clearly document the projected cost benefits and other lighting details.

RETURN ON INVESTMENT

The strategy is to raise the perception of efficient lighting from a commodity item to one that can help a business improve its bottom line by lowering operating expenses. By communicating in financial terms, the level of lighting awareness could move upward within the organization hierarchy. Energy efficient lighting can provide returns that are more attractive than traditional investments such as CDs, mutual funds, stocks and bonds and real estate.

Lighting in a facility should be looked at as an investment instead of just a cost. Technologically-advanced light sources offer significant returns on lighting investment. An end user's objective should be to replace existing lamps or to equip new facilities with energy-efficient lighting that will give the best return on investment.

Return On Investment (ROI) is an annual measure, as a percentage, of how that product is benefiting you. ROIs over 150% are not uncommon with today's advanced, energy-efficient lighting products. In other words, for every dollar invested in energy-saving lighting, you may be able to recoup two dollars and a half primarily through lower electricity usage.

The formula for calculating return on investment is:

$$\text{ROI} = \frac{\dfrac{\text{Energy Savings - Investment}}{\text{Equipment Life in Years}}}{\text{Investment}} \times 100$$

Return on Investment has long been a measure used by management in making business decisions. Lighting is no different an investment. If the lighting is handled wisely, it can give a high ROI with no risk.

FLUORESCENT LAMP TECHNOLOGY

Using the standard energy-saving T-12 (one and one-half inch diameter) fluorescent lamp with standard or energy-saving ballast systems has reduced light output when operated in the same luminaire. The T-8 (one-inch diameter) fluorescent lamp, however, is an energy-efficient, fluorescent lighting system that delivers the same light output from the luminaire as a standard F40T12 lamp/ballast combination, while reducing overall energy requirements and providing good color rendering at color temperatures of 3100, 3500 and 4100K.

The first T-8 lamp was a 4-foot, 32-watt unit. Because of optical and thermal improvements when operated in a luminaire as well as improved phosphor technology, the T-8 system is able to provide the same amount of light as a 40-watt, T-12 system. The T-8 lamp uses medium bi-pin bases and is physically compatible with luminaires designed for 4-foot, T-12 lamps. However, a new ballast is required for the T-8 lighting system since it operates at 265ma as opposed to 430ma for T-12 lamps. Lamp lumen depreciation of the T-8 system is also increased to 90% as compared to 88% for a T-12 system because of the reduced current and the use of rare earth phosphors.

Other versions of this lamp type include 2-foot, 3-foot and 5-foot straight lamp units carrying wattage ratings of 17, 25 and 40 watts, respectively.

Smallest of the T-8 lamp family, the 17-watt, 2-foot lamp has the same overall length as the commonly used 20-watt, T-12 preheat lamps, but offers some distinct advantages. It produces more light output, consumes less power and has a life rating of 20,000 hours, which is more than twice the 9000-hour life rating of the 2-foot preheat lamp.

U-shaped lamps provide added design flexibility for this family of T-8 lamps. New fixture and lighting system designs now are available for three new wattages: 16, 24 and 31 watts. These tight-bend lamps have a leg spacing of 1-5/8 inches. Equipped with standard medium bipin bases, they are used with common U-lamp sockets and operate on the 265ma ballasts available for comparable wattage straight T-8 lamps. These U-lamps have a rated life of 20,000 hours, excellent lumen depreciation characteristics, and are available in three color temperatures.

The 31-watt lamp has an overall length of 22-1/2 inches. Its compact size allows for the design of more efficient 2-foot by 2-foot luminaires using up to four lamps inside a modular fixture.

The 16-watt U-lamp is only 10-1/2 inches long and it will fit in a compact 1-foot by 1-foot square fixture. Two of these lamps with ballast in such a fixture consume only about 40 watts of power but deliver approximately the same amount of light as a 150-watt incandescent lamp while lasting 20 times as long.

Development of solid-state, high-frequency ballasts offer opportunities for operating T-8 fluorescent lamps at reduced power consumption. Presently, there is not an industry standard for electronic ballasts; however, their use is becoming widespread, especially as their parameters become more clearly defined. Using T-8 fluorescent lamps with a four-lamp, instant-start, high-frequency ballast is second to none in its ability to achieve the highest system efficiency today.

THE T-8 SYSTEM AS A RETROFIT

A growing number of lighting users are recognizing the benefits of renovating their "old" fluorescent lighting systems to take advantage of some of these new technologies, thereby improving the energy efficiency of their systems, and in many cases the quality of the lighting as well. One of the most significant developments has been the increased use of one-inch diameter, T-8 lamps.

Because the reduced diameter lamps require a different ballast, they are often thought of as mainly a new construction product while T-12 energy saver types are emphasized for retrofits. However, in many cases, a conversion to T-8 lamps with electronic ballasts would be the most effective for the user. This type of retrofit offers the following features and benefits:

1. Reduced wattage: A standard four-lamp recessed troffer with an acrylic lens and F40 lamps consumes 174 watts. That same fixture converted to a T-8 system with electronic ballasts consumes only 106 watts. This offers the user the benefit of significant energy cost reductions.

2. Similar light distribution: Because the fixture is still equipped with four lamps and the same lens, the optical characteristics will not be altered. This gives the user the benefit of maintaining the uniformity of his lighting system; the conversion will not introduce the possibility of dark spots developing between fixtures or at walls.

3. Maintained light output: Although the T-8 lamp has a lower lumen rating than an F40 type, the actual light output from a fixture will be nearly the same with these two systems. This is due to optical gains made possible by the smaller T-8 diameter, and the thermal gains afforded by the 265 ma characteristic of the system. The benefit to the user is that their lighting levels in the work areas will not be diminished, as they would with most other retrofit alternatives.

4. High CRI: The dual phosphor coating on the T-8 lamps, utilizing rare-earth triphosphors, provides a Color Rendering Index of 75, much higher than the traditional cool white lamp's rating of 62. As a result, the appearance of colored objects, walls, floors, etc. will be enhanced.

5. New ballast: A final feature of this retrofit is the installation of a new high-frequency electronic ballast. The user is thus installing a state-of-the-art lighting system, and will not normally be faced with replacing ballasts again for 10 or more years.

These five features of the reduced diameter lamp and electronic ballast retrofit make it a very attractive alternative for the user who is considering renovating a fluorescent lighting system. Another alternative sometimes considered today is the installation of a special reflector in the fixture, and the removal of two lamps from a four-lamp fixture. Although this approach also offers reduced wattage, it falls short of the system in terms of the other four features. Thus, the use of T-8 lamps in a fluorescent lighting system retrofit assures the user not only of increased energy efficiency, but also of improved lighting quality in the area, resulting in a more satisfactorily lighted environment.

LIGHTING MAINTENANCE

Another consideration in any lighting design is the maintenance and reliability of the system. The T-8 bent lamp, introduced in 1985, is designed to operate on the same ballasts developed for the straight lamps. The ballasts for the 31-watt T-8 U-shaped lamp have been available since 1981 when the 4-foot straight T-8 lamp was introduced. The ballasts have a proven track record.

If T-8 lamps are used in both 2x2 and 2x4 luminaires on the same project, the system may use the same ballast for both, thus simplifying replacement ballast inventory and eliminating the possibility of installing the wrong replacement ballast. The ballast designed to operate the 40-watt biaxial lamp is dedicated to that system, and is not listed to operate any other lamp type.

The T-8 U-shaped lamp with its medium bipin bases uses the same sockets as ordinary U-lamps. The sockets were introduced in the 60s when the U-lamps were announced, and are readily available both for new installations and as replacements. The T-8 U-shaped lamp is a natural extension of the family of straight T-8 lamps.

In terms of system efficiency the magnetically ballasted T-8 lamp will equal the efficiency of an electronically ballasted T-12 system. There are electronic ballasts available to operate the T-8 and T-8 U-shaped lamps as well. The combination of high frequency electronic ballasts and T-8 lamps provides the highest system efficiency for a general lighting system.

CASE STUDIES

Case Study 1 - In renovating its office and manufacturing facility in Waltham, MA, Hewlett Packard put a high priority on the type of lighting selected. Efficiency and return on investment received the greatest attention. The new lamps and ballasts in the existing fixtures are T-8 fluorescent lamps and electronic ballasts throughout the approximately 460,000 square foot facility.

Compared to the standard T-12, 4-foot fluorescent lamps, the T-8 fluorescent systems make possible significant energy savings of up to 46%. This combination provided the most efficient 4-foot system available anywhere at any cost. With the highest fluorescent lamp efficiency, the T-8 lamp and electronic ballast system qualified for the lighting rebate offered by Boston Edison. The rebate reduced the overall cost in labor and materials to wash, relamp and reballast 1,710 fixtures. The fixtures were lamped with 34-watt, energy saving lamps and they were driven by standard magnetic ballasts. Fixture wattage measured 155 watts and the lamps were operated 5,408 hours per year. Electric rate for this facility was $.072 per kilowatt hour. Replacing 6,840 34-watt lamps with 5,130 32-watt, T-8 lamps, an ROI for this facility was calculated to be 22.5%. (See Table I.)

TABLE I
ENERGY ANALYSIS DONE FOR HEWLETT PACKARD

ANNUAL SAVINGS		
LIGHTING		
TOTAL PRESENT LIGHTING LOAD FOR 6,840 - F40/SS	264.71	KW
TOTAL NEW LIGHTING LOAD USING 5,130 - F032 W/ELECTRONIC	143.64	KW
LOAD REDUCTION DUE TO LIGHTING	121.07	KW
ANNUAL LOAD REDUCTION FOR 5,408 HRS/YR	654,747	KWH
ANNUAL SAVINGS AT $.0720 PER KWH	$47,141.78	
RETURN ON INVESTMENT		
TOTAL ENERGY SAVINGS FOR 1,815,613 KWH SAVED OVER LAMP LIFE	$130,724.16	
LESS TOTAL INVESTMENT FOR 5,130 NEW LAMPS AT A $15.67 INVESTMENT PER LAMP	$80,387.10	
NET RETURN ON INVESTMENT	$50,337.06	
ANNUALIZED NET RETURN FOR 2.8 YEARS	$18,147.98	
RETURN ON INVESTMENT	22.50%	
PAY BACK PERIOD FOR PROPOSED SYSTEM	20 MONTHS	

Case Study 2 - After extensive evaluation, the T-8 fluorescent system was selected for an energy-saving relamping program at 17 McDonald's locations, owned and managed by McCopco. With the initial presentation made in Atlanta, the relamping will eventually extend to some 60 other McDonald's operations around the country.

Objectives were threefold: to meet present lighting standards, to provide good color rendering, and to give a good return on their investment. The existing lighting consisted of 2-foot and 4-foot fluorescent fixtures equipped with standard F20T12 and F40T12 fluorescent lamps and ballasts. Fixtures were cleaned and instant start electronic ballasts were used with 4-lamp, 4-foot fixtures and magnetic, rapid start T-8 ballasts were used with 2-lamp, 4-foot fixtures and the 2-foot fixtures. The existing lamps and ballasts were replaced with 17-watt and 32-watt, T-8 lamps. The T-8 conversion was a natural because it made good practical sense, i.e., it was easily accomplished. T-8 lamps have medium bi-pin bases as do the T-12 lamps and that meant no general replacement of sockets was necessary. Lamps fit into the T-12 lampholders. And the T-8 lamps had overall end-to-end lengths that matched the overall lengths of the standard 2', 3' and 4', T-12 lamps. Ballasts for the T-8 had identical mounting dimensions as the T-12 ballasts.

Making good lighting sense and being practical in terms of ease of installation is not the bottom line, however. To make good sense the conversion must make good financial sense and be a sound investment. An ROI analysis revealed a total facility return on investment of over 130%. (See Table II.) No utility company lighting rebates were considered for the first seventeen locations. This T-8 retrofit indeed made good financial sense.

TABLE II

ROI ANALYSIS FOR MCDONALD'S

EXISTING LAMP: F40CW	
PROPOSED LAMP: F032/41K ON STANDARD BALLAST	
WATTS PER LAMP - EXISTING	46.00 WATTS
WATTS PER LAMP - PROPOSED	34.00 WATTS
WATTS SAVED DUE TO USING PROPOSED LAMP	12.00 WATTS
ANNUAL LOAD REDUCTION FOR 6,552 HRS/YR	78.62 KWH
ANNUAL ENERGY RETURN (SAVINGS) AT $.0650 PER KWH	$ 5.11
TOTAL ANNUAL ENERGY RETURN WITH PROPOSED LAMP	$ 5.11
THE BOTTOM LINE FOR MCDONALD'S AT A 0.0% RATE OF ESCALATION:	
ENERGY COST TO OPERATE EXISTING LAMP OVER LIFE	$59.80
ENERGY COST TO OPERATE PROPOSED LAMP OVER LIFE	44.20
TOTAL ENERGY RETURN (SAVINGS) OVER 20,000 HOURS OF LAMP LIFE	$15.60
INVESTMENT FOR PROPOSED LAMP	2.46
NET RETURN OVER LAMP LIFE	$13.14
ANNUALIZED NET RETURN FOR 3.1 YEARS	$ 4.30
RETURN ON INVESTMENT	174.9%
EXISTING LAMP: F20T12	
PROPOSED LAMP: F017/41K ON STANDARD BALLAST	
WATTS PER LAMP - EXISTING	27.50 WATTS
WATTS PER LAMP - PROPOSED	21.50 WATTS
WATTS SAVED DUE TO USING PROPOSED LAMP	6.00 WATTS
ANNUAL LOAD REDUCTION FOR 6,552 HRS/YR	39.31 KWH
ANNUAL ENERGY RETURN (SAVINGS) AT $.0650 PER KWH	$ 2.56
TOTAL ANNUAL ENERGY RETURN WITH PROPOSED LAMP	$ 2.56
THE BOTTOM LINE FOR MCDONALD'S AT A 0.0% RATE OF ESCALATION:	
ENERGY COST TO OPERATE EXISTING LAMP OVER LIFE	$35.75
ENERGY COST TO OPERATE PROPOSED LAMP OVER LIFE	27.95
TOTAL ENERGY RETURN (SAVINGS) OVER 20,000 HOURS OF LAMP LIFE	$ 7.80
INVESTMENT FOR PROPOSED LAMP	4.73
NET RETURN OVER LAMP LIFE	$ 3.07
ANNUALIZED NET RETURN FOR 3.1 YEARS	$ 1.01
RETURN ON INVESTMENT	21.1%

FACILITY RECAP

Lamp	Energy Return		Number of Lamps		Total Energy Return	Years		Annualized Energy Return
F032	$15.60	x	128	=	$1,996.80	/3.1	=	$644.13
F017	7.80	x	20	=	156.00	/3.1	=	50.32
			Total		$2,152.80			$694.45

Lamp	Invest-ment		Number of Lamps		Total Invest-ment	Years		Annualized Investment	ROI
F032	$ 2.46	x	128	=	$ 314.88	/3.1	=	$101.57	174.9%
F017	4.73	x	20	=	94.60	/3.1	=	30.52	21.2%
			Total		$ 409.48			$132.09	

INITIAL COST CONSIDERATIONS FOR NEW DESIGN

In recent years, an alarming trend has been developing where new lighting systems are first priced and then designed by the specifier or designer. This situation is created when a developer or general contractor contacts an electrical designer or contractor and requests a quotation for installing a lighting system.

The information on which the contractor bases his estimate usually includes the projected size of the building, its function and some other general specifications.

Heavy emphasis is placed on low initial cost as a primary consideration. With these considerations in mind, the specifier, designer or contractor then quotes a price for lighting. The estimate is then included in a total building bid package and becomes locked in as a maximum value for the electrical systems costs.

The major problem with this approach is that the initial investment is not the only cost that a building owner must bear in relation to lighting systems. There are several other direct and indirect expenses associated. The installation, operation, and maintenance of lighting systems must be considered.

A major cost factor is monthly payments to the local utility for the energy required to operate this equipment. Poorly designed lighting systems can result in spending a lot more on utilities.

Maintaining lighting systems is also a major cost item. Inadequate design consideration for maintenance often leads to higher expenditures than necessary. This can reduce the overall profitability of the operation.

One not so apparent cost is that many improperly designed systems fail to provide an appropriate lighting environment throughout the facility. If the building is owner-occupied, the result is a loss of worker productivity. In leased facilities, tenants are becoming more critical of these concerns and are making lease renewal decisions, based on the lighting level provided. The problem is that many building owners fail to appreciate the interaction between properly lighted and maintained facilities and possible future difficulties in paying for their operation and maintenance.

Developers, architects and general contractors bear a primary responsibility for teaching facility owners and managers that initial investment is not the only cost. Mechanical and electrical contractors must assist in this effort by providing information to the developers, architects, and general contractors to use in presenting alternatives to the owner before the building is constructed.

One way out of this dilemma is to use a total owning and operating cost approach for designing a new building. This technique will project cost factors such as annual utility expenditures and the return on investment for improving such things as the proposed building's lighting and HVAC system.

This will allow optimization between such factors as the lighting system and the size of the mechanical systems required to provide a comfortable environment. This approach can definitely result in a properly designed and constructed building at a reasonable initial cost and often the lowest.

An underlying factor in creating the "price first, design later" mentality is a general lack of appreciation for the value of good lighting system design in today's marketplace. Lamps are treated as just another commodity in the overall building price. Often the lowest bid price is the sole determiner as to which firm gets the design contract. This approach is rarely in the best long-term interest of the enterprise that will own and operate the building.

All the participants in the process of designing and installing the lighting and the HVAC systems in new buildings have a vital interest in seeing to it that this equipment will perform well with a reasonable initial cost. When owners place too much emphasis on just the lowest initial cost, then specifiers, designers and contractors have an important responsibility to educate them on the long-term dangers of such a philosophy.

For example, for every 50 standard F40T12 fixtures in the original design of a new facility that you redesign to the T-8 lamp/electronic ballast lighting system, you can save one ton of air conditioning capacity or 12,000 BTUs. The typical contractor cost to install HVAC in a facility in California is $1800 per ton.

Additional fixture cost for a T-8 lamp/electronic ballast system over the F40T12/standard magnetic ballast is approximately $25. Therefore, a 1,000 fixture facility will save 20 tons of HVAC at $1800/ton.

Reduced Mechanical Initial Cost	$36,000
Increased Electrical Initial Cost	25,000
Total Bid Reduction	$11,000

A 5,000 fixture project will show the following savings:

Reduced Mechanical Initial Cost (Reduction in HVAC of 100 tons)	$180,000
Increased Electrical Initial Cost	125,000
Reduction in Total Bid	$ 55,000

Another example for estimates only using cost figures from the St. Louis area will analyze three lamp/ballast combinations. (See Table III.) The facility is 100,000 square feet (200 x 500) and designed to be a low rise office building with packaged rooftop VAV cooling system. All lighting fixtures are 3-lamp, 18 cell recessed 2x4 parabolic troffer return air type and will be operated 3,000 hours per year to maintain 75 footcandles. The electrical labor rate is $37/hour and the energy cost is $.08/KWH. System 1 is an F40T12 lamp with a standard magnetic ballast, System 2 is an energy saving T-12 lamp with energy saving magnetic ballast, and System 3 is the T-8 lamp with electronic ballast.

TABLE III

	SYSTEM 1	SYSTEM 2	SYSTEM 3
LUMINAIRE TYPE:	LITHONIA 2PM3G340	18 277 2X4 3 LAMP	PARABOLIC F032/41K
LAMP TYPE:	F40CW	F40CW/RS/SS	F032/41K
BALLAST TYPE:	STD MAGNETIC	E S MAGNETIC	ELECTRONIC
LIGHTING DATA:			
No. Luminaires Required	1,160	1,380	1,170
Watts Per Luminaire	140	109	88
Relative Light Output	100	91	98
KWH Consummed per Year	487,200	451,260	308,880
Annual Ltg. Energy Cost	$ 38,976.00	$ 36,100.80	$ 24,710.40
Wiring & Dist. Cost	$ 20,056.40	$ 18,576.87	$ 12,715.56
Installation Cost	$ 42,091.00	$ 50,549.00	$ 42,857.00
Luminaire & Lamp Cost	$113,529.00	$147,163.00	$140,518.00
First Cost Ltg. Premium		$ 40,612.47	$ 20,414.16
Annual Energy Savings		$ 2,875.20	$ 14,265.60
Lighting Watts/Sq.Ft.	1.62	1.50	1.03
HVAC DATA:			
Heat Load (BTU)	554,271	513,383	351,402
Tonnage	46.19	42.78	29.28
Tonnage Saved vs. Standard		3.41	16.91
Operating Cost Savings		$ 831.38	$ 4,125.00
HVAC First Cost Savings		$ 6,473.89	$ 32,120.88
Ltg. Heat Load Watts/Sq.Ft.	.51	.47	.32
TOTAL SYSTEM DATA:			
First Cost Savings		($34,138.58)	$ 11,706.72
Annual Energy Savings		$ 3,706.58	$ 18,390.60
Watts/Sq.Ft. Due to Ltg.	2.13	1.97	1.35

SUMMARY

Fluorescent lighting has long been the choice for the illumination of commercial, industrial, office and retail spaces. The reasons: it's efficient, the equipment is affordable and fluorescent lighting has, in general, very good maintenance characteristics. As energy costs began to climb, it became apparent that wattage reductions in the lamp, ballast or lamp/ballast combination, could provide the owner of the fluorescent system with measurable and often significant savings.

The energy challenge resulted in the creation of an entirely new fluorescent T-8 lamp for general lighting applications. It is an extremely energy efficient fluorescent lighting system that delivers the same light output from a 2x4 luminaire as a standard F40 while providing exceptional color rendering properties. Developing the T-8 fluorescent into today's most innovative, most efficient system for new construction also led naturally to many retrofit applications, i.e., the replacement of standard T-12 lamps and ballasts with T-8 lamps and ballasts in existing, in-place fluorescent lighting fixtures. In all ways, the T-8 system is a natural for this conversion, and it's a natural for business reasons. It makes good lighting sense, good financial sense, and practical sense.

Chapter 7

FLUORESCENT REFLECTORS: THE MAIN CONSIDERATIONS

P. von Paumgartten

The benefits of replacing lamps in fluorescent light fixtures with reflector panels have been well documented. Articles have touched on everything from energy savings and reduced lamp and ballast maintenance expenses to lower air conditioning costs, extended lens life and reduced glare. These benefits have helped make reflector retrofitting the fastest growing segment of energy management.

At a glance, it appears simple enough. Investment in the retrofit and reap the return. But, in choosing to retrofit a building's fixtures, building owners and managers must closely examine a variety of considerations that will affect the budgetary savings, as well as the overall satisfaction of the building occupants. This article will spell out those technical factors that building managers and owners should evaluate when considering a fluorescent fixture retrofit plan.

Among the key considerations are:

* Light Losses. Replacing lamps with reflectors decreases light levels, but they may be maintained within current lighting standards and not adversely affect worker productivity.

* Fixture Appearance. The appearance of the fixture is critical, because people respond to lighting emotionally, as well as physically.

* Designs. The styles vary from full reflectors to partial, with varying effects on the lighting system.

* Material. Each have their own reflective qualities, affecting reflector output, appearance, and durability.

* Costs. The various reflective materials and designs present a cost-benefit dilemma.

* Heating, Ventilating and Air Conditioning System Changes. Fewer lights mean less heat, reduced cooling load and increased heating load.

LIGHT LOSSES

Logically, there is light list when a lamp is removed and reflectors are installed. If two lamps from a two-by-four-foot, four-lamp troffer are replaced with optical reflectors, light losses may be 10 to 40 percent.

But, things are not always as they appear at first glance. Light losses vary from project to project depending on the fixture type and condition. Dirty lenses or old lamps cause fixtures to operate well below their designed efficiencies; they are dim to begin with. Buildings with poor lighting maintenance may be retrofit with reflectors with very little, if any, difference in light level. The illumiance that would be lost in an average reflector installation can be made up by cleaning the fixture and replacing the old lamps with new ones during the installation. (See Figure 1.) Installers can do this during a retrofit.

That's what makes the reflector concept work. Generally, a 10 to 25 percent light loss can be expected with a reflector retrofit.

In addition, the light loss may not be significant to the needs of the building occupants. Before the move to conserve energy, many building designer over specified for lighting. That may leave room to reduce lighting without cutting productivity.

As it is, new lighting standards suggest lower lighting levels are acceptable. Previously, most buildings were designed for light levels between 70 and 100 footcandles (FC). But, according to the Illuminating Engineering Society of North America lighting recommendations, office need only 30 to 50 FC for Video Display Terminal usage; 50 to 70 FC for reading, writing, and typing; and 70 to 100 FC for accounting and drafting. In general lighting areas, office need 10 to 20 FC for circulation areas, corridors and lobbies; 25 to 35 FC for conference rooms and non-task areas; and 30 to 40 FC for filing areas. (See Figure 2.) Task lighting can meet the needs of workers in many areas, while lower level ambient lighting is sufficient elsewhere.

Previously, building designers used 2.5 to 3 watts per square foot for lighting. Today, the standards have been reduced to 1.5 watts per square foot -- without necessarily reducing occupant satisfaction or employee performance.

In addition to loss of light, reflectors can change the light distribution from a fixture. Reflectors direct light downward, creating more light directly under the fixture and less light between fixtures.

This light redirection is similar to that of the low-brightness louvers, which eliminate the high-angle light. This brightness control can reduce glare in offices where computer video display terminals are in use. (See Figure 3.) Spacing requirements may change because of less light between fixtures.

If the two-by-four-foot, four-lamp fixtures are providing more light than necessary, redesigning the fixture for reflectors could be readily acceptable to office workers. It is also less expensive than installing a new lighting fixture completely. But, several other key conditions remain in selecting the appropriate type of lighting designs using reflectors.

FIXTURE APPEARANCE

Aesthetics, though difficult to quantify, are very important to a building environment. Many are obsessed with how much light is produced from a fluorescent reflector, but they may be overlooking an important factor: the overall acceptance of a retrofit by the building occupants. Appearance can make a big difference in working conditions and worker satisfaction.

That is why delamping alone is not recommended for energy savings in an office setting. Delamping leaves a dark spot in the fixture where the lamps have been removed. This is aesthetically unacceptable to building occupants.

It has been shown that a great looking retrofit will more often prove to be successful than a bright one. Ambient lighting from overhead fixtures, task lighting for specific jobs or even accent lighting for visual affect must be done in an appropriate, tasteful manner.

DESIGNS

The reflector design can have an impact on aesthetics. As a minimum, a good looking retrofit fluorescent fixture should do one of two things: Create an image that the fixture contains a full complement of lamps or diffuse the light so no lamps can be seen in the fixture whatsoever. Different reflector shapes and materials will yield these results.

Some designs involve moving the lamp holders and repositioning the remaining lamps. Some virtually line the inside of the fixture with reflective materials. Others involve simply replacing existing lamps with reflectors. The reflectors may replace the in-board or the out-board lamps in a four-lamp fixture. Some have bends that help create the image of a lamp where one has been removed. And some are partial reflectors, which leave the ballast uncovered for easy access. (See Figure 4.)

MATERIALS

There are several different materials that can be used. Their reflective qualities differ, with silver film on aluminum rating the highest with a 90 to 95 percent reflexivity. (See Figure 5.) from a building occupant's perspective, however, the difference between materials may be imperishable. Depending on the reflector design, the material used may have little effect on light output.

Perhaps the best way to get a true impression of how a reflector design will look is to perform a mock-up. Have it installed in the ceiling to see the actual fixture where it will actually be used and measure the performance. Do not compare the measurements to the light levels that existed previously; instead, compare them to current lighting standards. The key question is: Will the reflector yield an acceptable light level

Some types of reflectors will change in appearance over time. While aluminum reflector have been proven durable in use for over 30 years, the thin anodized coating on aluminum reflectors -- generally about two microns -- may crack when the reflector is bent, causing possible oxidation of the aluminum.

While aluminum films have a long history, silver films have little track record. But research groups have run accelerated aging tests on silver films show that films with back reflector materials age faster, losing more than 20 percent of their reflectivity over 500 hours. Front reflectors maintained their reflectivity through the course of the study.

In time, temperature swings in the fixtures can also cause silver film to crack or buckle. Some film manufacturers, however, have added "ultraviolet inhibitors" to counter the effects of temperature swings.

Over time, a greater attention will have to be paid to lighting maintenance following a retrofit. Considerations should be make to group relamping on a regular, timely basis.

COSTS

While the effects of reflector designs and materials may have considerable impact on appearance, the retrofit/s costs will have the greatest bottom-line impact. A high-cost retrofit could double the payback period.

The costs of the materials can vary greatly. Silver film reflectors are about $10 more expensive per fixture than aluminum. The average cost for a silver film reflector for a two-by-four-foot fluorescent fixture is $50 to $75 installed, compared to $40 to $65 for aluminum. Partial reflectors with aluminum film cost between $35 and $45. (See Figure 6.)

Installation accounts for 25 to 40 percent of the cost. But lowest installation price may not be the best option, since proper installation is critical to the success of the retrofit. the installation contract should include new lamps and a thorough fixture cleaning to minimize the light loss in the retrofit.

The installation should also be done when it will least disrupt the building occupants. If performed after hours, the lamps can be removed and reflectors installed without the building occupants realizing there has been a change.

HVAC IMPLICATIONS

Improper installation or design can also mean additional hidden costs. Many light fixtures serve as return air ducts, an integral part of the heating, ventilating and air conditioning system. Full reflectors may cover the vents, impeding proper operation. Partial reflectors can render the necessary levels of illuminance without interfering with the air duct systems.

Also, building managers may not want to cover the ballast compartments. That increases the compartment temperature and shorten the life of the ballast. It also makes it more difficult to reach the ballast for maintenance procedures.

There are several considerations in HVAC systems that can drastically affect a building operation expenses as well as occupant comfort.

Reduced lighting causes a corresponding reduction in the cooling load for the air conditioning equipment, especially in the interior zones, where outdoor conditions have little influence. Demand for winter space heating may increase incrementally with reduced building lighting. This decreases the savings from the light reduction program by the amount of energy that must be added to offset the loss of heat. However, a building's heating system generally provides spare heat much more efficiently than the heat given off by excessive lighting.

For example, in a terminal reheat system, a change in lighting could require as much additional energy to reheat the duct air as is saved by reducing the lighting. The reheat requirement, however, can be minimized by raising the cool supply air temperature so comfort conditions in the room with the maximum cooling load are satisfied without reheating the air going to other rooms.

In the variable air volume system, a reduced cooling load would reduce the amount of cool air that is distributed through the building. This reduction may present an opportunity to replace the supply fan motors with smaller motors, saving additional energy. An HVAC expert is necessary to evaluate the retrofit savings potential.

A reflector retrofit also present an opportunity to change the entire lighting system. Different lamps for better color rendition or greater efficiency could be installed. Parabolic louvres replacing the old lenses would reduce glare on office Video Display terminals. And high-efficiency ballasts would afford even greater energy savings.

SUMMARY

With lights burning between 30 and 50 percent of a building's electrical load, many building owners and managers have chosen to install reflectors in their fluorescent fixtures as a quick-fix way to cut energy expenses. But the decision to install reflectors should involve more than simply changing a fixture. Aesthetics, designs, materials, costs and the retrofit implication to the HVAC system all need to be considered carefully. It also gives building management an opportunity to redesign the entire lighting system.

Careful selection of a reflector system and installer can help building management meet its desires for energy savings while increasing the satisfaction and productivity of building occupant.

REFLECTOR CHARTS

Fixture Efficiency

Fixture Condition	Efficiency
Seasoned and cleaned white troffer with new lamps and reflectors. (Retrofit fixture)	65 - 85 percent
Seasoned and cleaned white troffer with new lamps. (New fixture)	50 - 65 percent
Seasoned and uncleaned white troffer with seasoned lamps. (Old fixture)	35 - 45 percent

FIGURE 1

Recommended Light Levels

Task Lighting Requirements	Footcandles
VDT Usage	30 - 50
Reading, writing, typing	50 - 70
Accounting, drafting	70 - 100
General Lighting Requirements	
Circulation areas, corridors, lobbies	10 - 20
Conference rooms, non-task areas, work stations	25 - 35
Filing areas	30 - 40

FIGURE 2

Material Reflectances

Material Types	Total Reflectivity
Silver film (on aluminum)	90 - 95 percent
Aluminum film (on aluminum)	80 - 85 percent
White paint (on aluminum)	75 - 90 percent
Anodized aluminum sheet	70 - 85 percent

FIGURE 5

Cost Comparisons

Reflector type	Typical Installed Cost	Payback*
Silver Film (Full)	\$50 - \$75	2.3 - 3.4 years
Anodized Aluminum (Full)	\$40 - \$65	1.8 - 3.0 years
Aluminum Film (Partial)	\$35 - \$45	1.6 - 2.0 years

* Based on 90 watts saved time 3,500 hours/per time seven cents/KwH or \$22.05/year.

FIGURE 6

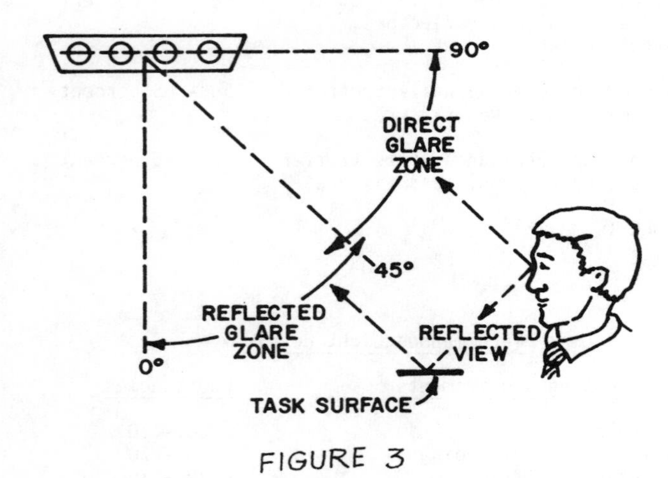

FIGURE 3

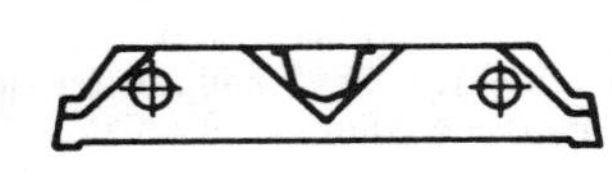

Simple bend reflector with 2 lamps outboard.

Complex bend reflector with 2 lamps inboard.

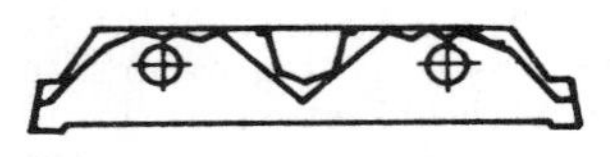

Complex bend reflector with 2 lamps relocated.

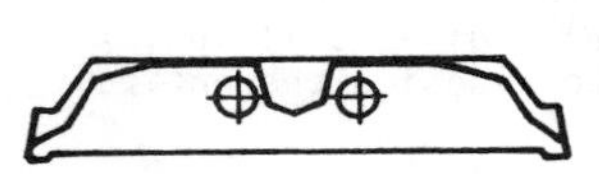

Partial reflectors with 2 lamps inboard.
Ballast cover is not covered.

FIGURE 4

Chapter 8

LIGHTING APPLICATION: THE FIRST SUCCESSFUL FEDERAL SHARED ENERGY SAVINGS PROJECT

M. E. Nelson

ABSTRACT

The U. S. Postal Service, along with every other branch of the Federal Government, is continually looking for ways to reduce cost without negatively impacting operation or production. Since April 1986 when President Reagan signed into law the Consolidated Omnibus Budget Reconciliation Act, there has been a tool available that could reduce cost by as much as 40% and improve the quality of the facilities at the same time. That tool is Shared Energy Savings, which by now is a familiar concept to most people. However, not a single contract was awarded within the Federal sector until January 1988, when the U. S. Postal Service San Diego Division, awarded it's contract for a very basic lighting retrofit. What took so long, and why was the U. S. Postal Service the first to have a contract award that succeeded?

OPPORTUNITIES IN WAITING

San Diego has one of the highest electric rates in the nation, causing utility costs to continually increase at an alarming rate which directly impacted the profit and loss statement each and every accounting period. The San Diego Division, General Manager/ Postmaster, Margaret Sellers was confident that utility costs could be controlled and reduced by at least 30% to 40% if basic energy savings retrofits could be implemented. The challenge: How to fund these projects in a time of extreme fiscal restraint.

There was no doubt that energy conservation paid for itself or that the General Mail Facility (GMF) was the best candidate in the San Diego Division. The San Diego GMF is a 458,911 square foot facility with a 24 hour per day mail processing operation and an annual electric utility cost of $859,681 for Fiscal Year 1987.

The lighting audit of the GMF revealed that out of a total of 3,284 fluorescent fixtures in the facility, 2292 fixtures should be retrofitted with optical reflectors and energy efficient ballasts. The remaining 992 fixtures which were located primarily in mechanical rooms, mezzanines, and storage tunnels, all of which were severely over lit, were selectively removed. This basic lighting retrofit calculated out to an annual savings of 1,264,227 Kilowatt or $138,968.17 dollars.

PRE AUDIT 1,927,229 KWH x $0.11 = $221,995.19

POST AUDIT 663,882 KWH x $0.11 = $73,027.02

ANNUAL SAVINGS: 1,263,347 KWH x $0.11 = $138,968.17

The very best option for the Postal Service would have been to fund this project using in-house funds; however, the same Omnibus Reconciliation Act that allowed federal agencies to participate in Shared Energy Savings contracts also restricted the available funds needed for facility improvement.

WHAT TOOK SO LONG?

The primary hurdles that had to be overcome were that no existing Shared Energy Savings contracts existed in the Federal sector to use as guides. The Department of Defense had tried several times to adapt performance contracts into Shared Energy Savings contracts, but the projects became so cumbersome that they could not be implemented. The popular train of thought was to establish an energy baseline using either metered data or a very complicated mathematic algorithm, neither of which proved to be very accurate or available. Many governmental facilities do not have utility meters and fewer still have loads that can be completely isolated for meters to be installed. Additional burdens were added by good intentioned contract specialists trying to infuse as much of the Federal Acquisition Regulation (FAR) as possible into the contracts for familiarity sake.

However, the greatest barrier to Shared Energy Savings within the Federal sector was the blanket approach to the entire cost reduction issue. Many solicitations were drafted requiring the contractor to solve every problem in the facility. The scopes of some of the early Federal contracts ranged the entire technological spectrum, from cogeneration systems mixed with thermal energy storage systems, lighting retrofits and variable air volume controls, all combined in the same project, all needing quantifiable results.

THE USPS, SAN DIEGO DIVISION SHARED ENERGY SAVINGS CONTRACT

A Headquarters sponsored National Task Force, consisting of representatives from contracting, engineering, legal, and labor relations was formed to develop the Shared Energy Savings contract for the San Diego GMF. This combined effort resulted in a contract that had several unique features which allowed it to become the first feasible project in the Federal sector.

The first two major changes to the previous Shared Energy Savings strategy was selecting a basic energy conservation technology that was already well proven, and the method that should be used to quantify the results. Several smaller offices within the Division had previously been retrofitted with reflectors and ballasts, all having excellent results. However, lighting loads are very difficult to isolate for metering purposes and run times are generally not constant for any specific zone. After a very brief discussion by the Task Force Engineering staff, it was decided that the very nature of a reflector and ballast retrofit needed no metering to determine the savings. For this project, a calculated watt/hour constant, times an average run time for each zone type, times the number of fixtures in each zone, would be used to determine the pre-retrofit energy consumption. The post retrofit energy consumption was then calculated to be 51% less than the original consumption. Specific offices were identified as test sites and metered before and after the retrofit to validate the calculated savings.

After resolving the two major issues of what technology to use, and how to quantify the savings, the problem of not having a contract that fit nicely into one of the already established contract groups, still remained. Generally speaking, government contracts are written by the pound! After many hours of intense discussion with task force contracting specialists and lawyers, a contract was drafted having less than 100 pages, including all mandatory contract enclosures, and a detailed lighting matrix of the facility, all weighing in at less than 2 pounds, including the binding. The actual project scope and specification section consisted of only thirteen pages.

THE FINISHED PRODUCT

This new generation contract was designed around a very basic energy savings program which did not require metering or complicated base lines to calculate the cost avoidance. The scope and specification were straight forward, they required the contractor to furnish all materials, labor, tools, and equipment necessary to install a reflector and ballast retrofit project at the San Diego GMF. Detailed specifications were written requiring minimum performance levels for the reflectors and ballasts, thus insuring quality material. The lighting matrix specified the number and location of fixtures that were to be retrofitted or removed and the minimum lighting level that would be accepted after the retrofit. The contract also specified that all fixture maintenance, including a yearly cleaning, and two group relampings every third year, would be the responsibility of the contractor. This requirement reinforced the responsibility of the contractor to insure retrofitted fixture performance.

THE BOTTOM LINE

Award criteria was based on the highest net present value offered to the U. S. Postal Service. The contract was awarded to a joint venture of American Illuminetics, Inc./Ascot Electric Inc./Co-Energy Group, on their offer of $386,872.58 dollars net present value over a seven year term. The contractors total construction cost was $164,714 dollars with a seven year maintenance cost of $66,384 dollars. The percentage split for the entire seven year term was approximately 60% to the U. S. Postal Service and 40% to the contractor. The construction portion of the contract was completed April 5, 1988, almost 20 days ahead of schedule.

The Shared Energy Savings Task Force is now carefully evaluating the entire project and will be reporting its findings and recommendations to the Assistant Postmaster General in the latter part of 1988. Preliminary indications are that Shared Energy Savings projects should be implemented throughout the Postal Service using the San Diego contract as a guide. Although utility bills do not provide an accurate means of evaluating the success of the project, they can be useful as a general indicator for consumption trends. The following graph shows the electrical consumption for the San Diego GMF during Fiscal Year 1987 and 1988.

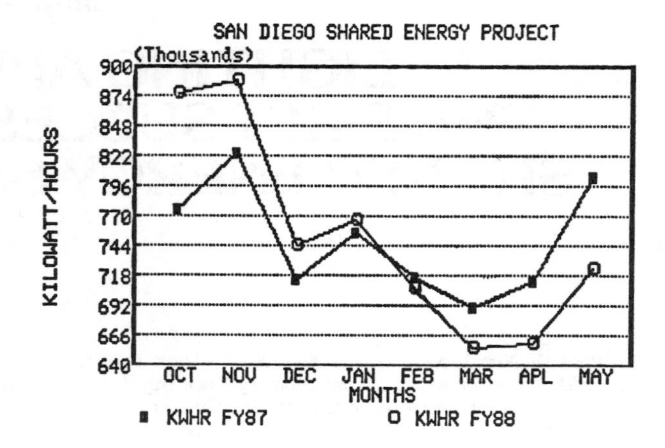

An additional benefit which resulted from the project was a better quality of light in all areas, especially the administrative offices where fixture maintenance had been allowed to deteriorate. Many employees commented that the lighting in their offices appeared much better after the retrofit, and in many areas lighting levels were improved.

IN CONCLUSION

The San Diego Shared Energy Savings Project was a success due to three primary reasons. The first being the continual interest and support of San Diego Division General Manager/Postmaster Margaret Sellers. Her commitment to this project and energy conservation in general provided the essential enthusiasm from upper management that all innovative projects require in order to overcome the well established barriers that are encountered whenever new territory is explored. The second reason the San Diego project succeeded was the cooperation and determination of the Shared Energy Savings Task Force. Their combined effort developed a specification that was both technically and contractually superior to anything previously tried within the Federal sector. And last, but not least, was the working relationship between the contractor and the San Diego Postal staff. This contract required an attitude of cooperation in order for it to succeed. The general nature of most contracts is for both parties to try to achieve some advantage over the other party. The character of this contract was such that both parties would win, or both would loose, depending on the success or failure of the entire project.

SECTION 2
ENERGY SYSTEM RETROFITS

Chapter 9

COOLING TOWERS, THE NEGLECTED ENERGY RESOURCE

R. Burger

SUMMARY: Loving care is paid to the compressors, condensers, and computer programs of refrigeration and air conditioning systems. When problems arise, operators and engineers run around in circles with expensive "fixes", but historically ignore the poor orphan of the system, the cooling tower perched on the roof or located somewhere in the backyard. When cooling water is too hot, high temperature cut-outs occur and more energy must be provided to the motors to maintain the refrigeration cycle.

COOLING TOWERS: -

1) ... are just as important a link in the chain as the other equipment,
2) ... are an important source of energy conservation,
3) ... can be big money makers, and
4) ... operators should be aware of the potential of maximizing cold water.

Most towers were designed over 20 years ago and were inefficiently engineered due to cheap power and the "low bidder gets the sale" syndrome. Operating energy costs were ignored and purchasing criteria was to award the contract to the lowest bidder. All too often the low bidder - even though some of the most respected firms were involved - cut thermal corners for the sale.

This paper investigates the internal elements of the typical types of cooling towers currently used, delineates their functions and shows how to upgrade them in the real world for energy savings and profitability of operation. Hard before and after statistics of costs and profits obtained through optimization of colder water by engineered thermal upgrading will be discussed. Salient points will be reenforced with on-the-job, hands-on, slides and illustrations.

1. HISTORICAL NEGLECT

From the very beginning, cooling towers were ignored. Figure 1 is an illustration from a basic refrigeration training manual /1/ that delineates the

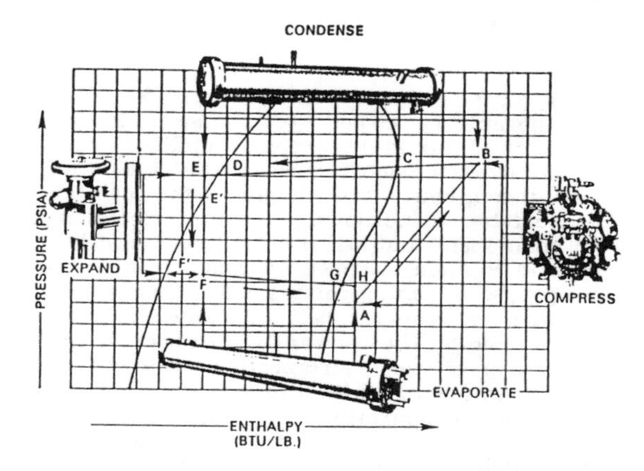

Figure 1 - Basic refrigeration cycle neglecting cooling towers importance.

refrigeration cycle consisting of compression, evaporation, expansion, and condensing. The discussion continues on and explains that heat is not generated or destroyed, it is just moved from one location to another. What is ignored though, is "How is that waste heat dissipated?" As we know, it is picked up by the circulating water and brought to the cooling tower where it is discharged into the atmosphere and the cooling water returns for another heat pickup.

While the cooling tower seems like a simple mechanism, it is just as important to the refrigeration cycle as any of the other units. If the cooling tower does not function properly, high head temperatures require that additional electricity be pumped into the system to make it operate and, at a critical point, the equipment will shut the system down if the cooling tower cannot produce sufficient cold water for the refrigeration machinery.

2. REFRIGERATION ECONOMICS

A cost-effective approach to the solution of conserving energy (and therefore, money) would be to reduce the power input to your system while maintaining maximum efficiency. The power is either purchased from a public utility in the form of electricity or steam, or is generated by the facility by purchasing fuel oil for a diesel engine to power the system.

The following basic principles explain how colder water from the cooling tower conserves energy to create a cost-effective rapid dollar return for cooling tower upgrading expenditures.

Whether it be heat rejection from compressors, electric motor, or chemical process equipment, the cost of "hotter" cooling water is expensive in requiring additional energy to run the equipment at efficient levels to reduce head pressures and temperatures. Excessive heat will create maintenance problems, deteriorate the equipment, and cause shut downs of the process.

A typical example of this is where a refrigerant is cooled (condensed) in the condenser and in turn cools "chilled" water to reduce the temperature of the circulating air throughout the facility to maintain comfortable conditions. Input (electricity or steam) and output (tons of refrigeration) depend upon speed (rpm) of the compressors and refrigerant temperature (condenser temperature). At any particular speed, both the power requirements and capacities of the refrigeration machine will vary significantly with the refrigerant pressure and temperature. These refrigerant conditions are determined by the cooling in the refrigerant condenser. The quantity and temperature of the condenser water (tower water) determines the available cooling. When operating at full condenser water flow a reduction in condenser water temperature will reduce refrigerant temperatures and pressures. This will permit producing similar refrigeration capacities at lower machine speeds (rpm) and lower power (steam and electricity requirements.)

Enthalpy charts for refrigerants indicate that for every degree of colder water to the equipment, a 3½% reduction in energy input can be attained. /2/

If, for example, the refrigeration system utilizes $400,000.00 of electricity per year typical for a 2,000 ton installation and a 4°F (2.2°C) reduction of cooling tower water is obtained, this will lower the utility bill by $56,000.00 ($400,000.00)(4)(.035). With utility costs soaring, this savings will increase each year.

The facts are readily available in most operating and maintenance departments to determine the cost of energy utilized, the percent reduction colder water can be obtained, and the cost savings involved. By comparing the cost savings as against the retrofit cost, a payback will usually occur within 6 months to a year and a half depending upon the conditions of the cooling tower.

Many authorities point out that lower cooling tower water can produce significant savings for refrigeration equipment. The cooling tower plays one of the key roles in the efficiency of your air conditioning machines. Efficiency, or energy consumption, is measured in kilowatt usage. If your tower does not create the proper heat transfer, your machine will work harder to compensate for the loss and inefficiency /3/. Figure 2 clearly indicates that colder condensing temperatures will improve the performance of a compressor significantly. This can result in a substantial energy and dollar savings.

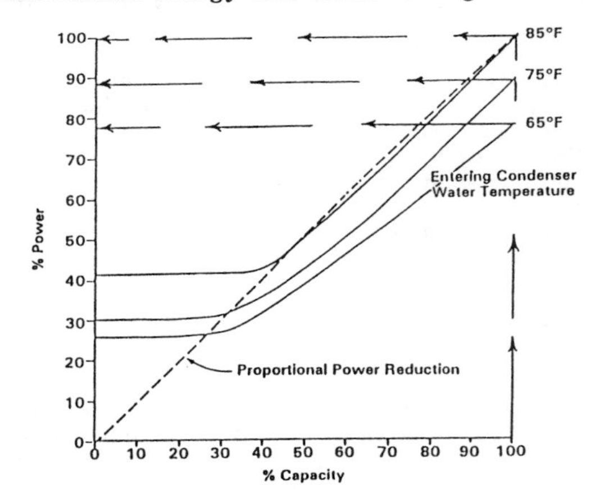

Figure 2 - Relationship of colder water and electric power in the above diagram, (energy) conservation for 2000 ton centrifugal system.

The role of the cooling tower is to remove waste heat in the refrigeration or chemical reaction. The degree of elevation of the discharge temperature above ambient conditions is the sum of the tower's approach of the cold water to the wet bulb temperature, the cooling range (which equals the temperature rise in the heat exchanger), and the terminal difference in the exchanger. A reduction in operating temperature, always desirable for economic reasons, may be obtained by increasing the capability of the cooling tower's performance /4/.

The relationship between percent power for a 2,000 ton centrifugal machine and condenser water temperature is shown on this graph developed by Mr. T. L. White of the Monsanto Chemical Company. At 85°F (29.4°C) entering condenser water temperature, 100% capacity of the unit is developed, utilizing 100% of the motor horsepower. When the condenser water temperature is 10° colder, or 75°F (23.9°C), 88% of the power is required with a resulting savings of energy and kilowatt input of 12%. In other words, the colder the water leaving the cooling tower, the less energy is required to operate the refrigeration system.

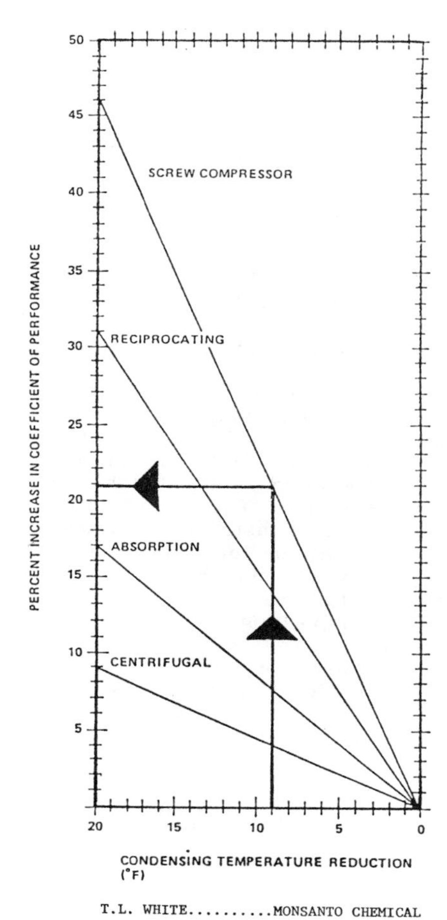

T.L. WHITE.........MONSANTO CHEMICAL

3. REAL WORLD PRACTICALITIES

All too often, due to the inaccessible location of the cooling tower, usually installed to conserve real estate on the top of a building, or in the backyard somewheres, maintenance personnel are many times hard pressed to adequately service the equipment.

Training and misplaced priorities tend to keep the operating engineers more closely attuned to the requirements of the compressors, condensers, and evaporators rather than the quality of cold water being discharged from the cooling tower.

Since the cooling tower is open to the atmosphere, it can quite readily deteriorate due to corrosion of the ferrous parts, bacteriological and chemical attack of the wood. Many maintenance people feel that if it is ignored, it will go away. But, of course, being familiar with Murphy's Laws - this does not happen.

4. COOLING TOWER MODERNIZATION

A great majority of cooling towers operating today, even though some are newly installed, have been engineered with techniques over 20 years old. Today's state of the art can be utilized to retrofit practically all towers and upgrade their capability in producing colder water or cooling greater volumes of circulating water. /4/

Listed below in ascending order of cost are some of the major components that can be upgraded. It is axiomatic that the greater the dollar input, the more rapid and profitable the return can be obtained.

a. Air Handling - More air volume results in better thermal transfer and colder water discharge. By pitching the fan blades up to a higher angle, which is determined by the plate amperage, additional air can be generated for the same horsepower. Velocity regain (VR) Venturi Stacks should be investigated for increasing air flow through the tower while reducing fan motor horsepower.

b. Drift Eliminators - Conventional two-pass wood slat herringbone or steel "zig-zag" configurations usually have a higher pressure drop then the new PVC cellular units, Figure 3. By eliminating the solid droplets of water to a higher degree and at a lower static pressure loss, the cellular drift eliminator provides additional cooling air through the tower for colder temperatures.

Figure 3 old-fashioned herringbone heavy wood slat drift eliminators.

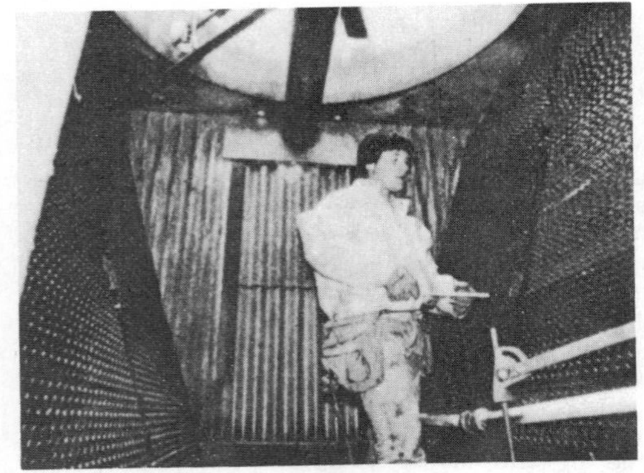

modern high efficiency PVC cellular replacements on typical crossflow tower.

c. Water Distribution Systems - By installing metered orifice target nozzles in older crossflow towers, a more uniform water pattern is obtained and the resulting uniformity will improve the tower's performance. Counterflow towers with spray systems can be greatly improved by installing the new square spray ABS practically non-clogging nozzles, Figure 4. An added advantage of this newer type nozzle is that maintenance and cleaning are greatly simplified due to upwards of 75% fewer nozzles required than the conventional small orifice conical pattern spray units.

d. The greatest improvement in performance modernization is obtained by changing out the old wood splash bars and installing self-extinguishing PVC cellular film fill, together with the new efficient nozzles. These are calculatable values /5/.

5. EXAMPLES OF UPGRADING

a. Blow Through Squirrel Cage Tower

The subject three cell metal tower had a water distribution system of 960 small orifice nozzles on 1½" pipe circulating 1800 GPM, Figure 5, and was operating at high head temperatures. The clogged and corroded water distribution system was removed from the tower and the rusted clogged steel plate corrugated wet decking fill was also disposed of. After sandblasting and coating with moisture cured urethane, the new spray system consisting of 36 nozzles on 3" diameter PVC pipes was installed together with new PVC cellular fill, Figure 5.

Figure 4 - Cellular fill and square pattern nozzles provide startling improvement in colder water and volume capacity.

Figure 5 - By changing 388 small orifice nozzles to 36 non-clogging square spray type and installing cellular fill, 5°F colder water was obtained.

BLOW THRU

WATER IN @ 100°F OUT @ 90° @ 78 F°WBT
960 NOZZLES ON 1 1/2" PIPE
OPERATING WITH HIGH HEADS

36 NOZZLES + CELLULAR FILL
PRODUCES DESIGN OF 95°– 85°– 78

5 APPR x 3 1/2% SAVES 17% ENERGY

17% x $250,000 = $43,750 /YEAR
10 YEARS = $437,500 EARNED

This chart illustrates the rapid payback for converting the tower illustrated in Figure 5. Enthalpy charts for Freon indicate that 3½% of the electrical energy to the compressors and condensers can be saved for every 1°F colder water. /2/ The subject cooling tower was tested in accordance with the Cooling Tower Institute Acceptance Test Code 105 and indicated a 5°F colder water was obtained after retrofit. Since the rebuilding for colder water cost $38,500.00, the return on the investment was realized in approximately 9 months with a projected ten year savings of more than $500,000.00 considering the escalation of electric costs.

b. Crossflow Air-Conditioner Tower

i. The conventional deteriorating wood splash bar slats together with the leaking fiber drift eliminators were taken out of the tower and disposed of, Figure 6.

ii. Cellular drift eliminators and PVC splash bars were installed to obtain higher levels of heat transfer, Figure 7.

iii. A new basin lining was also installed, consisting of exterior plywood, polyurethane caulking and 25 to 30 mil. moisture cured urethane coating sprayed to provide a monolithic rubber waterproof membrane, Figure 8.

iv. Before retrofit, operating records indicated that the 1,250 gallons per minute of circulating water was entering the tower at an average of 105°F, leaving at 90°F, during a 78°F Wet Bulb. After retrofit and certified testing, /6/, the same 1,250 gallons of circulating water was now entering the cooling tower at 95°F, be discharged at 85°F, during a 78°F Wet Bulb. Besides reducing the approach to the Wet Bulb by 5°F, the entire system temperature was reduced by 5°F. This saved the Company approximately $45,000.00 in electrical costs per year.

Figure 7 - Cellular drift eliminators and PVC splash bars were installed to obtain higher levels of heat transfer.

v. Inefficient wood fill and leaking drift eliminators were retrofitted with new state of the art self-extinguishing PVC high heat transfer cellular fill and drift eliminators, which reduced the discharge water temperatures by 5°F, representing an approximate 600,000 kilowatt hour and $45,000 energy savings at 7½¢ KwH utility charge.

Figure 6 - Inefficient wood fill and eliminators retrofitted with PVC high heat transfer fill and eliminators 5°F colder water, or 600,000 kilowatt hour and $45,000 energy savings at 7½¢ KwH.

Figure 8 - New basin lining consisting of exterior plywood, Urethane caulking and 25 to 30 mils of Urethane coating sprayed to provide a monolithic water proof membrane.

6. CONCLUSION

If a refrigeration/air-conditioning system is operating marginally due to high head temperatures, it behooves the owner and operator to investigate the possibility of upgrading the existing cooling tower rather than installing another O.E.M. unit which may or may not solve the problem of requiring colder water. It should be well understood that colder water can save energy and create an operating profit. Cooling towers are hidden bonanzas for energy conservation and dollar savings when properly engineered and maintained. In many cases, the limiting factor is the quality and quantity of cold water coming off the cooling tower /7/.

It would be prudent for the engineer with responsibility for the efficient operation of the refrigeration/air conditioning system to have a professional inspection of the cooling tower by a consultant who can analyze the energy savings potential of his installation.

In these days of high energy costs, the savings accrued from a well engineered and retrofitted cooling tower bringing it into the 1980's can make a significant impact on a company's profit and loss statement.

7. LITERATURE CITED

1. Trane Corporation, Air Conditioning Training Manual.
2. Allied Chemical Corp., "The Pressure Enthalpy Diagram, Its Construction, Use, and Value".
3. Charles Weiss, Technical Engineering Instructor, "Cooling Tower Inspection and Economies", National Engineer, (April, 1985).
4. ASHRAE Equipment Handbook, Chapter 21, Page 11, (1975)
5. Robert Burger, "Cooling Tower Technology", Chapter 8, "Thermal Engineering", (1979)
6. A Texas College, test report and location upon request.
7. Jim Willa, "Cooling Tower Operations", Cooling Tower Institute Proceedings, (1980).

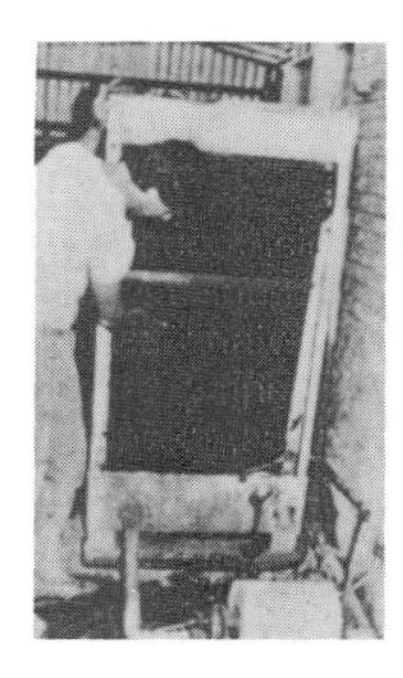

Figure 9 - Upgrading technology is the same for all types of towers, only the size is different.

Chapter 10

SELLING ENERGY EFFICIENCY TO MUNICIPALITIES

R. Fine

Ontario's municipal governments spend up to 15% of their annual budgets on energy consumption. In times of fiscal restraint, controlling energy budgets can result in a reduction of operating costs, not services. In 1981, in an attempt to promote greater energy efficiency in Ontario, the Ministry of Energy, in conjunction with the Association of Municipalities of Ontario (AMO), created the Municipal Oil Conversion and Energy Conservation Program (MOCECP) and the Municipal Energy Conservation Program (MECP). Both programs were effective through their early years, but the majority of participants were larger municipalities with sufficient staff to handle the investigation of energy efficiency. Changes in the marketing and focus of the programs resulted in a new direction in the "selling" of energy efficiency to municipalities.

Background

The Province of Ontario, located in the eastern part of Canada between the Great Lakes and Hudson Bay, contains vast regions and varied topography. Encompassing an area of approximately 365,000 square miles, Ontario contains over one third of Canada's population and borders on several U.S. States including New York, Michigan, and Minnesota. There are 839 municipalities in the province, with large cities such as Toronto (capital of Ontario; population of 3 million) and Ottawa (capital of Canada; population of 350,00) and 700 municipalities with populations under 10,000. The size and location of Ontario's municipalities creates problems for AMO, in the delivery of programs to its members. The energy program is no exception to this phenomenon. For example, the average high temperature in Toronto, the southern part of the province, during March is 42°F. In northwestern Ontario, near the Minnesota border, the municipality of Fort Frances registers and average high temperature of 20°F. This wide variance in temperature is but one factor affecting the successful implementation of an energy program. The MOCECP and MECP experience has demonstrated this factor.

MOCECP - The Beginning

In late 1981, in an attempt to make Ontario municipalities more energy-efficient, the Ontario Ministry of Energy created the Municipal Oil Conversion and Energy Conservation Program (MOCECP) to give advice, financial and technical assistance on energy conservation measures. Original assistance came in the form of a 30% grant to cover the retrofit cost of upgrading or converting an oil furnace to another source of heat. By mid-1982, the program was expanded to include any project that would significantly reduce energy consumption. Grants were paid if the project had a payback (defined as the projects projected savings - the project's cost) period of less than six years. The percentage of grant would drop if the payback was longer. (For a complete layout of the program operation see Figure 1 below.)

Figure 1 Funding Levels for Municipalities

	Project Cost	Percent Grant by Payback	Maximum	
Fuel Conversions and energy conservation retrofits		1 - 6 year	6 - 10 years	
	First $200,000	30%	15%	$100,000

	Percent Grant	Maximum
Technical Assistance	50%	$10,000

The program was promoted through mailings, speeches from the Minister of Energy and his staff at annual municipal associations and conferences, and was supported through the Municipal Energy Conservation Program (MECP) which was established through a Ministry of Energy annual operating grant to AMO. The program was designed to give information, technical assistance and training to municipalities through an activity co-ordinator and through NETWORK, a bi-monthly publication utilized to promote energy efficiency, MOCECP, and to highlight the experience and successes of other municipalities who had delved into the energy game.

Early Program Results

The first two years of MOCECP were less than overwhelming in terms of the number of projects and the number of municipalities who participated by applying for grant dollars. By the end of 1983, 375 projects involving 231 municipalities for $2,032,726 were implemented. The Ministry of Energy continued its support to municipalities by creating, in association with the Ministry of Municipal Affairs, a program designed to assist in the funding of hiring an energy auditor for interested municipalities. MEAP, the Municipal Energy Audit Program, provided municipalities with 50% of the cost associated with hiring an energy auditor over three years. As well, a number of technical training sessions were held throughout the year on topics of interest that often resulted from experiences within municipal facilities.

The MEAP initiative was an attempt to get local town and city councils involved by appointing an energy advisor and committee. The net result would be greater uptake in grant dollars and participation in energy retrofit projects in municipalities across Ontario. Sixty energy auditors were hired by municipalities under MEAP, with AMO assisting in the co-ordination of seminars and training sessions. By 1985, results were encouraging with a much greater uptake in energy grants and projects. By the end of 1985, 581 projects by 216 municipalities were undertaken, with a annual cost avoidance of $5, 156,142. This represented an increase of 61% over the same period from 1981-1983. Even so, there were still over 623 or 74% of all Ontario municipalities who did not get involved with MOCECP. Users in the under 50,000 population group, roughly 780 of Ontario's municipalities, were still not getting involved in energy efficiency projects. Why was this happening? Two main issues seem to arise after studying the MOCECP pattern.

1) Smaller municipalities did not have the expertise or staff to investigate the possibilities of applying for and applying MOCECP funding to possible energy projects. There appeared to be difficulty accessing both grant dollars and information. More times than not,they often talked of difficulty in understanding grant forms and determining which projects were acceptable and which were not, from a grant and efficiency perspective.

2) Even though the programs (MOCECP and MECP) were supposedly well publicized, a high percentage of smaller municipalities had no real knowledge of the program's existence.

THE NEW MOCECP AND MEAP

In an attempt to solve the lack of access to the programs by potential users, the Ministry of Energy decided to increase the amount of funding available to smaller municipalities for energy retrofits. If a given municipalities population was less than 50,000, up to 50% of the cost of the project would be made available. (For a complete grant payment outline, see Figure 2). As well, a grant of up to 75% was added for technical assistance. Assistance would now be available for funding energy audits and advice; evaluating maintenance and operating procedures affecting energy consumption, and also life cycle cost analysis for new building construction. Clearly, if the program was not being utilized effectively then perhaps no interest existed within Ontario municipalities to reduce energy consumption.

The second major improvement involved the creation of the Small Municipalities Advisory Service (SMAS), designed to provide direct technical assistance to municipalities with populations under 50,000. A technical advisor was hired at AMO to directly assist municipalities with hands-on training and to give advice on the types of projects available to improve

Figure 2 Funding Levels for Small Municipalities

	Project Cost	Percent Grant by Payback	Maximum	
		1 - 6 year	6 - 10 years	
Fuel Conversion and energy conservation retrofits	First $200,000	50%	25%	$100,000
	Above $200,000	30%	15%	$100,000

	Percent Grant	Maximum
Technical Assistance	75%	$10,000

energy efficiency and most importantly, to show how to access and submit grant forms. A new column, called Small Talk, was created in NETWORK to demonstrate that no municipality was too small to get involved in energy efficiency. Case studies from small municipalities were also included as a regular feature. For example, Brethor, with a population of only 123, received $1,204 from the grant program.

Results

In the year and a half after the change to MECP, MOCECP became a tremendous success. The program was originally intended to end in March of 1988, but this deadline was changed when grant requests outstripped available funding. This was clearly a testament to the changes in the program and marketing carried out through the Small Municipalities Advisory Service. In the year and a half following program changes, 246 municipalities applied for $2,688,374 in grants. This represented a 34% (based on funding) increase over the previous period, and an increase in new municipal participation of 84%. By the end of MOCECP, almost half of all Ontario municipalities had moved towards improving their facilities through the grant program.

LEARNING FROM THE MOCECP EXPERIENCE

With MOCECP over, the program staff at AMO began to examine trends and patterns that had developed over the life of the program in an attempt to determine why it did, and did not, work. The following represents the results of the investigation:

1) Don't Mail Your Message, Hand Deliver It.

One of the earliest problems demonstrated in the marketing of MOCECP related to the lack of direct contact with our client group. A series of brochures and letters were mailed to every municipality in the Province, but as is often the case, most material was buried or lost in the shuffle. It was obvious that direct contact was needed. As a result, an attempt was made for MECP staff to make energy wise presentations to as many municipally-oriented groups as possible. A preliminary list was drawn up to highlight key meetings of interest to politicians and staffs throughout Ontario. The following groups were contacted:

i) The Association of Municipal Clerks and Treasurers of Ontario (AMCTO) - Clerks and treasurers who were directly involved in the daily operation of all facilities in municipalities were spoken to in hour-long sessions, at 9 different zone meetings across the province.

ii) Ontario Municipal Recreation Association (OMRA), the Society for the Directors of Municipal Recreation of Ontario (SDMRO), and the Northern Ontario Municipal Recreation Association (NOMRA) - Recreation managers, who work in those facilities utilizing the greatest amounts of energy in every area of Ontario, were contacted and accessed through demonstrations and presentations throughout the year.

iii) The Association of Municipalities of Ontario (AMO) - The Association's annual conference attracts 1600 administrative and political representatives from across the Province. In essence there is no greater opportunity to spread "the word" than at this annual function. An "energy drop" in wine and cheese afternoon is held, often demonstrating projects undertaken by other municipalities, in which municipal officials are invited to drop by and discuss their ideas and problems.

Thirty seminars aimed at clerks, treasures, council members and recreation staffs were also held across the province in 1986-87. The impact of such sessions was tremendous, with over 300 municipalities participating. Upwards of 62% of all seminar attendees submitted projects for grant consideration within six months of hearing the word. The figure ran as high as 80% in northern Ontario. Also, the degree of networking that took place between attendees in a given area, and non-attendees, also increased participation in the grant program. At a regional council meeting a mayor or reeve would stand up and tell of the energy session held in his town, and the benefits of accessing MOCECP grants.

2) Know Your Client Group

Marketing a product or idea involves identifying your potential user group. The last two years have provided many examples of knowing, and not knowing a client group. For example, material was sent and presentations were made in eastern Ontario, in the Ottawa area of the province, yet little response was seen from the area in terms of program participation. What had occurred related to language. There are a great number of bilingual and predominantly French-speaking municipalities in this particular part of the province. To correct the inadequacy on our part, brochures in French were sent, and the net result was greater program adoption. A complete list of French-speaking municipalities was obtained from the Ministry of Municipal Affairs, and future correspondence was sent in both English and French.

Another example involving a lack of client knowledge revolved around a seminar held in Sault Ste. Marie in a northern part of the province, adjacent to the state of Michigan. Of the 30 municipal officials invited, only two indicated they would attend. Did this mean that the Sault area should be abandoned in terms of future attempts to promote energy efficiency? Of course not. It turned out that November is moose hunting season in the Soo; a time when municipal staff and politicians leave work to enjoy the great outdoors with their families.

In terms of content, at a seminar in Sudbury, the nickel and rock centre of Ontario, MECP staff espoused the benefits of solar pool blankets, not realizing there are very few outdoor pools because of the cost of digging through the heavy rock basins.

One effective way to gage your client group is to look at previous work that was done in a particular area to see if any trend exists. Speak to the host community to find out what interests them, and shape a presentation to reflect the subtle difference from one area to another. One way to ensure that your message is heard is through higher attendance. One way to ensure this is to invite the mayor and members of council. In doing so, this will ensure that the item appears on the councils agenda, and will be discussed at council. Inviting a staff member from the public works department does not necessarily guarantee participation but if council is aware, there is a greater likelihood of seeing someone from that particular municipality. It also helps to hold the session in a municipal facility rather than a motel or service club. People feel more at ease in facilities they are familiar with. By using a municipal office, a greater sense of camaraderie is created. The use of a local contact on the invitation will also ensure greater attendance and eventual program participation.

3) Target Your Audience

A subsection of knowing your client group relates to targeting of the material. Delivering seminars are fine, but of greater importance is who attends. There is little point in getting a road engineer to represent the municipality at a seminar because his/her interest will obviously vary from what one is trying to present. Target your message to the right audience. In Ontario, after analyzing those who got involved with energy efficiency projects, amazing conclusions were drawn. If you want success, invite a woman to the seminar. Of the communities who received grant dollars from the Ministry of Energy, there was four times greater likelihood of energy projects being implemented, if the clerk, treasurer or recreation director, often female-dominated fields, were contacted. While this may be more reflective of where and what positions in the municipality women have, it none the less demonstrates the importance of targeting activities. If for example, a new grant program were to be implemented, one would want to implement the program in those municipalities where woman are involved in daily operations.

4) Talk Dollars($), Not GigaJoules (GJ)

Remember who your audience is! Selling energy in municipalities involves keeping concepts as simplistic as possible, highlighting paybacks not technicalities. Describing how a low emissivity ceiling functions through the use of detailed light defraction, heat gain and loss calculations will not encourage a municipal clerk or recreation director to get further involved with energy conservation. But talk of return on investment, dollars to be saved, and cost to be avoided will help convince a municipality that energy efficiency is the route to go. Detail the ways to fund energy projects, such as lease to own, shared savings arrangements, or grant programs. Talk about dollars first, then branch out into other benefits such as increased comfort.

5) Back up Concepts With Services

It is one thing to promote energy efficiency activities but what real impact will this have if there is no back up support. For example, if a municipality is encouraged to have its facilities audited, it may not mean a great deal if a comprehensive list of consultants is not included. In the case of MOCECP, a grant for up to 75% of the cost of an audit was made available with the changes in the program in 1986. To assist in accessing the grants, MECP staff began surveying engineering and consultant firms around the province, hoping to provide any given municipality the opportunity to find a qualified consultant to provide audits. The 107 consultants who responded had to identify the types of energy projects they have been involved with, and names of municipalities they have worked for. A computer at AMO contained the data, and the data was made available upon request. The service has been well utilized, with over 300 requests being handled since the spring of 1986. This listing is currently being updated.

6) Program Success = Employee Benefit

In selling energy efficiency, attempt to highlight the importance of the projects to municipal employees - they will have to live with the changes in the future. Try to tie energy savings to employee benefits. For example, th creation of an employee awards program or activity days tied to the successful implementation of energy conservation programs and projects. Some municipalities have stated recreation staffs may receive larger pay increases if they could be justified by facility savings. Energy conservation allows such opportunities and gives employees a greater stake in implementing programs and projects. For example, a municipality in eastern Ontario implemented an energy efficiency program based on using manual measures to prove the eventual effectiveness in installing automatic controls. The program met with success in part due to the stakes that were mentioned for staff members. As the recreation director for the municipality stated in NETWORK, "successful implementation of the manual systems is dependent upon staff understanding the philosophy behind the procedures and having a stake in the outcome. Tying wage increases and lieu time to this program is essential!"

7) Get Political Support

Politics is a way of life in today's world. Without strong political support, any energy management plan is likely to fail. Politicians control the purse strings of a municipality, and if council will not support energy management with funding, there is little point in attempting to implement projects. Municipalities who have successful energy management plans are municipalities whose staff have a strong alliance with members of council. If you are attempting to initiate any energy program in a particular community, have the mayor speak to the group that has been organized. Often this will get the host municipality involved through potential embarrassment. If a host community has not followed through on any potential projects, sometimes an energy seminar can bring about rapid change.

CONCLUSION

This paper has been an attempt to outline how a good energy program became a great one through ongoing analysis of events, and the application of seven basic steps. The energy management edict is one that should be practiced by all levels of government, not just municipal. There is tremendous potential for more efficient use of scarce resources throughout the western world. By examining the user group, understanding their abilities and disabilities, strength and weaknesses, energy management can be adopted successfully. Using these seven steps should break down the barriers of inactivity in the energy field.

Chapter 11

EVAPORATION/HEAT RATE—A SIMPLE BOILER EFFICIENCY MEASUREMENT METHOD

K. E. Heselton

Evaporation Rate and Heat Rate are representative of average boiler efficiency, and are easily determined by you and your boiler operator. The simple math (subtraction and division) places this measurement of efficiency in the hands of someone who can do something to improve it. It is an effective tool for operation. It can, for most operations, be converted to a boiler efficiency. Even when your plant does not have a steam flow meter or Btu meter you can determine a Rate for a guide to efficient operation.

You all can relate to "miles per gallon" being used as a measure of the energy efficiency of an automobile. Evaporation Rate and Heat Rate are similar. Instead of mileage we are interested in boiler output in pounds of steam evaporated, Btu added to the working fluid, or a count of production that utilizes the energy added at the boiler. The method of calculation is identical to determining automobile mileage and the comparison of results between your boiler and the other person's boiler must consider fundamental differences.

Evaporation Rate is identified as "pounds per gallon," "pounds per thousand cubic foot," "pounds per pound," or "pounds per ton," depending upon the fuel you are burning in your boiler. When firing oil, you calculate Evaporation Rate as the number of pounds of steam generated per gallon of fuel oil burned. When firing gas, you calculate Evaporation Rate as the number of pounds of steam per thousand cubic feet of gas burned. When firing coal or other solid fuel it is steam generated per pound or ton of fuel burned. The value is normally subjected to a decimal place shift to produce a number between 100 and 1000 which facilitates comparisons.

Your hot water boiler may be equipped with a Btu meter permitting you to calculate your Heat Rate as "thousand Btu (or therms)" relative to one of the fuel units listed above.

Output Rate can be calculated using any number of bases. Which one you use is dependent upon what is measurable. Output Rate is a term the author uses to identify any system utilised for relative efficiency other than Evaporation Rate or Heat Rate. Some of the possible measurements are pounds of product per gallon of fuel oil, truck loads per ton, degree days per gallon of oil, gallons of wash water per cubic foot of gas, and revolutions per gallon of oil.

If you do not have meters for fuel and steam or Btu measurement, consider their worth. You would expect a variation in automobile mileage depending on the manner in which the driver operated the vehicle. When recording mileage a driver may alter his driving habits to produce maximum mileage. Problems with the vehicle's carburetor or engine are indicated with a decrease in mileage. For similar reasons, the value of having a measurement of boiler efficiency is that efficiency could be improved by as much as 20%. The Boiler Efficiency Institute indicates improvements of 6% to 10% are not uncommon with conscientious maintenance. If the cost of installing meters is less than 12% of your annual fuel bill you should not be without them; they should pay for themselves within two years. Metering methods can be inexpensive. A number of low cost methods are described later.

The primary reason to determine your Rate is to compare current operation to historical data. If the Rate is higher, you are doing a better job of utilizing energy. If the Rate decreases, something is going wrong. The actual value is not important when used in this manner. That is its primary benefit. You may choose to use Rate in its simple form and never calculate the representative boiler efficiency. If so, you will still benefit from the knowledge of the factors that are required to determine efficiency because several of those factors affect Rate.

You should determine your Rate as frequently as possible. In medium and large plants the Rate should be calculated each shift. It permits each operator to measure his own ability to operate efficiently and generates competition between the operators as each one attempts to achieve a higher Rate than the others. In small plants with restricted measurement capability, a daily Rate calculation should be made when measurement by shift is impractical. The calculated Rate should be made a matter of record.

For purposes of testing a procedure or method of operation, Rate can be calculated over a much shorter period of time. A boiler operator may use Rate to determine the effect of an adjustment to the boiler controls. He can tell a short time after making the adjustment if he improved the Rate.

This paper will be read by management personnel more than by the boiler plant operators. A caution is appropriate for those managers to ensure that the emphasis on heat rate improvements is not placed before the priorities of safety and reliability. Most boiler plant operators identify maintaining steam to the facility as their top priority. That should never be foremost in their minds but it normally is; they never hear from the boss unless they lose the plant! Now the boss may show an interest in Rate and that becomes the new priority in their minds. As a manager you must ensure that you are not misunderstood by the boiler operators and that you prescribe priorities in the following order:

1) Operator safety (his own).
2) Safety of other people.
3) Safety of the plant and its equipment.
4) Reliability in operation.
5) Efficient operation.

While the efficient operation of the plant is the last priority on the list, it is the one that the operators should normally be spending the most time on. If maintenance of the first four priorities consumes most or all of your operator's time you should be giving serious consideration to a major revamping of the plant.

CALCULATING RATE

When calculating automobile mileage you record the quantity of gasoline or diesel fuel you put in the tank and the odometer reading at the same time. The key here is time. In normal boiler plant operation fuel is burned to generate heat on a continuous basis. To ensure an accurate calculation of Rate the readings of heat production and fuel consumption must be made simultaneously or nearly so.

Rate determinations require reading the output meter or identifying the output count, and determining fuel consumed at the same time. If the two sources of information are not convenient to each other, establish a procedure of reading one and then proceed to read the other so the interval between the readings is as consistent as possible. The output / steam generated / Btu absorbed, and fuel burned is determined by subtracting the most recent reading from a prior reading. The Rate is calculated by dividing the output, steam flow or Btu difference by the fuel difference. Use meter reading differences directly, with correction factors as multiples of 10 to produce a value between 100 and 1000. That simplifies the calculation for a boiler operator and eliminates confusion. The calculated Rate should then be entered in the plant log book.

Some would argue that calculating by meter differences only, without correcting for meter multipliers, etc., produces a value that is not truly representative of Rate. That's true; however, incorporating constant values of meter correction, etc. does not necessarily produce a truly representative value either. By making the calculation of a Rate as simple as possible errors are reduced and there is no illusion of absolute accuracy. The primary use of the Rate is comparison - are we doing better or worse? We will cover conversion of Rate to boiler efficiency in a few moments.

RATE AS A FUNCTION OF LOAD

Just as an automobile's mileage will vary between city and highway driving; a boiler's performance varies with load. A generally accepted theory that boiler efficiency decreases with load is disproved by many Rate calculations. Utilizing Rate calculations tabulated relative to load permits simplified analysis of boiler performance over the load range.

By developing a curve similar to that of Figure 1 for your boilers, you can make more intelligent choices in operating your plant for optimum efficiency. When each Rate is calculated, divide the difference in output readings by the time between readings to obtain an average output. Plot the average output and Rate on a graph. One of the plant engineers or your professional consultant can prepare graph paper with the recorder chart or load indicator indexed to average output.

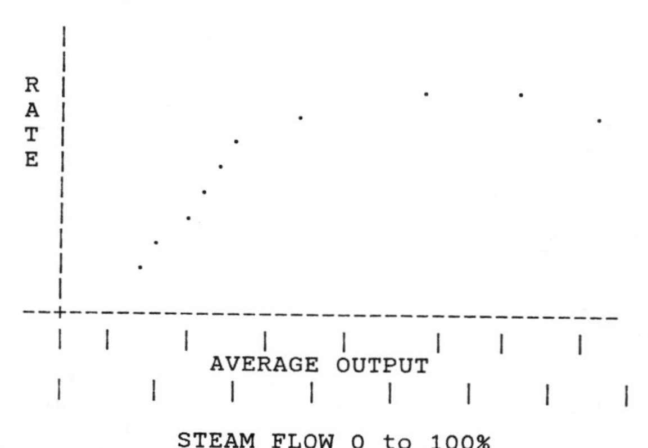

FIGURE 1. RATE TO LOAD PERFORMANCE CURVE

The development of a curve similar to that of Figure 1 will permit determination of the best way to distribute loads to multiple boilers. If the curve is not uniform, with only one peak and always a down slope from that point, it can indicate control malfunctions or non-linearity in the vicinity of loads where the other peaks occur. [Refer to the author's paper of the 1986 WEEC on solutions for non-linearity.]

INEXPENSIVE OUTPUT MEASUREMENTS

Most small steam heating boilers are operated without a means of measuring output. One argument is that the fuel consumption is too low to warrant the expense of a meter. Another, and very true, argument is that metering steam flow at low pressures requires sophisticated meters due to the dramatic variations in steam density with small pressure changes. A simpler and low cost measure of steam flow can be accommodated by using a feedwater flow meter. Another simple measurement that can be used is an operating hour timer (hardware cost of $35 to $40) connected across the starter coil or pump motor feed wiring of boiler feed pumps that are operated by boiler water level switches. Any measurement of feed water flow includes water used for continuous blowdown and bottom blowoff.

Reconsider metering on small hot water heating boilers that did not utilize enough fuel to justify metering a year or two ago. The new microprocessor based instruments are substantially less expensive than older devices.

Any relatable, measurable value representative of output could be applied to calculate Output Rate. Try one or more to see if they appear to be proportional to fuel consumption. Heating plants can use degree days. Laundries can use washloads or count or weight of goods processed. Manufacturing processes can use the same measure as production. These Rates are a measure of the efficiency of the boiler plant as a whole and require allowances for variation of make-up water quantity and temperature to be representative of boiler efficiency.

Measuring fuel used is seldom a problem. The supplier will always measure it because the fuel bill is based on what you use. You can read the gas company's meter; they will even teach you read it. Fuel oil tank levels can be measured with little difficulty to determine oil consumption over a shift or day. Coal and other solid fuels can be measured by the elevator bucket (using a counter), by hopper load, front loader bucket, or by weight.

The best meter is one that measures the flow with counts of more than 100 per hour.

CONVERTING RATE TO EFFICIENCY

The computational exercise required to determine the conversion factor that generates an efficiency from Rate by simple multiplication is not a one time activity. Conditions change to alter the factor as efforts to increase the Rate approach the optimum for your plant the improvements are made in smaller and smaller increments. When the operators are doing their level best, there will appear to be uncontrolled variations in the Rate (in the order of tenths of a percent). Those variations are normally due to the conditions that must be considered when converting Rate to efficiency. The conversion factor cannot be a constant, it must be adjusted as conditions warrant.

Under ideal situations, conversion of a Rate to a representative boiler efficiency is a simple process; however, an ideal situation seldom exists. If plant conditions are consistent with design, the plant operates with constant operating conditions, the instruments maintain calibration, and fuel suppliers provide the design fuel then conversion is as simple as multiplying the Rate by a known constant. Now let us evaluate the real world.
First, consider the output measurement. The meter was originally designed to measure a specific fluid at specific operating conditions and calibration. The effect of variation in steam pressure on a steam flow meter is considerable. All too many energy engineers are not cognizant of that deviation and erroneously attribute the higher flow indication of a steam meter to savings in energy attributable to reducing operating pressure. Obtain the original design conditions and calibration of the meter. If the normal meter operating conditions are different, then determine the correction factor that should be applied to the meter readings. Values to consider include operating pressure, operating temperature, pipe schedule and upstream and downstream piping. Note that the conditions of concern are at the meter, not at the boiler.

Heat in soot blowing steam, bottom blow off and continuous blowdown is considered heat output for steam boilers. The energy added to continuous blowdown and bottom blowoff is normally estimated by extrapolation from boiler water chemical analysis.

Second, consider calibration. A meter calibration should be checked on a quarterly basis as a minimum. Instrument calibration can suddenly shift due to an inadvertent bump or a temporary unusual operating condition. The technician who checks the meter calibration should provide "before and after" data when a meter is recalibrated to provide some indication of the error in the meter data prior to recalibration. A meter recalibration should be recorded in the boiler plant log. If operating and design conditions vary considerably, you may want to have the output meter recalibrated for actual conditions.

Your fuel meter readings should be consistent with the fuel supplier's readings, provided you are burning a liquid or gas fuel. When burning a solid fuel, calibrate the measuring device as carefully as possible.

Finally, maintain a record of the heating value of the fuel. There is a considerable difference in fuel oil supplies within a locality. A truckload of oil seldom has the same specifications as the prior five or six loads. If you are purchasing fuel oil from more than one supplier the two can be supplying oil with significantly different heating values. Local gas suppliers purchase natural gas from more than one pipeline, depending on current market prices, which will result in slight variations in the heating value of the gas. Coal can come from different mines and wood from different stands of timber. Monitoring the quality of fuel you are purchasing is as important as the price, and the purchasing agent should be able to provide you with regular analysis of the fuel quality.

Boiler efficiency determined using Rate is comparable to the ASME Power Test Code's abbreviated "input/output" efficiency. The author prefers to call it "output/input" efficiency because it is calculated as written; it is 100 times the measured heat output divided by measured heat input. The heat output is the heat added to the "working fluid" (ASME definition) and the heat input is the heat generated by burning the fuel. A complete ASME input/output efficiency is determined by including all energy delivered to the boiler, including electric power for driving combustion air and induced draft fans, heat added to the fuel (in the case of heavy fuel oils), and other sources of energy supplied to the boiler. The abbreviated method is the same as Rate because it measures only the fuel input and heat added to the working fluid.

The conversion factor to be multiplied by Rate to obtain boiler efficiency consists of the values required to accurately convert the meter readings to energy output divided by input then multiplied by 100 to convert the value to a percentage.

Output factors:

The value required to convert a steam flow meter reading to an output in Btu consists of:

1) The meter constant; a value that converts the 100% value of the meter to the pounds of steam that full scale represents. The new microprocessor based recorders display the actual values eliminating this factor.

2) A correction for steam pressure or temperature which are different than design; a value that is the square root of the ratio of the specific volume of steam at the design and average actual operating pressure.

3) The heat added to a pound of boiler feedwater to convert it to steam; the difference between enthalpy of steam at the boiler drum pressure for saturated boilers and the enthalpy of a pound of boiler feedwater at the boiler inlet (economizer inlet if the boiler has an economizer). For superheated steam boilers, the steam enthalpy is determined using the pressure and temperature of the steam at the superheater outlet.

4) The heat added to generate soot blowing steam. Soot blowers do not represent a substantial portion of the boiler output but the heat added can be estimated by determining the combined area of the soot blower nozzles, calculating their capacity using Napier's equation, and timing the operation of the soot blowers. The estimated quantity is added to the meter quantity.

5) The heat added to boiler feedwater that exits the boiler in the form of bottom blow-off and continuous blowdown. The quantity of boiler feedwater that leaves the boiler as bottom blow-off and continuous blowdown can be estimated as a percentage of the steam generated utilizing water chemistry analysis. The analysis method is prone to error if carryover is a problem. For boilers with high condensate returns that do not require continuous blowdown and utilize infrequent blowdown a fixed quantity can be estimated based on the boiler drum or shell dimensions. The heat added to the blow-off and blowdown water is the quantity of that water multiplied by the difference in enthalpy of saturated boiler water at drum pressure and the enthalpy of the feedwater.

To determine Btu output requires a flow meter plus temperature inputs and computation relays to determine the difference between the two temperature readings and multiply the result by the flow signal. The newer microprocessor based meters for this purpose are inexpensive relative to a few years ago, and more reliable. A Btu meter requires factors for all the flow rate adjustments similar to steam flow meters and require more care in calibration.

Using a feedwater flow meter for a steam boiler includes the blowdown and blow-off water as well as the steam. The factor used for steam flow meters is used except it is subtracted, not added. These meters normally read flow in gallons. A gallon of water contains less mass at 180 degrees than it does at 70 degrees. Avoid using standard weight conversions; look up the specific volume for your feedwater temperature in steam tables.

Utilizing pump operating hour time does not provide the accuracy required to make a precise determination. Maintaining good Rate records for a period when the boiler is subjected to a heat loss efficiency check will permit you to "back-in-to" a factor that will be reasonable for your plant. Lacking any other means, select the flow value at normal operating conditions from a pump curve.

Fuel input factors:

The value required to convert the metered fuel input to an input value in Btu is determined by:

1) The meter constant; a value that converts the 100% value of the meter to the gallons of fuel oil, cubic foot of gas or pound or ton of solid fuel that full scale represents. Typically a fuel oil meter reads directly in gallons. For oil and gas flow the new microprocessor based recorders normally display the actual values eliminating this factor.

2) A meter correction factor; a value that corrects liquid meters for leakage that is

associated with variations in viscosity. That factor is normally published in the meter manufacturer's instruction manual.

3) A tare factor; for solid fuel metering when the recorded weight includes the weight of the container.

4) An expansion factor; to correct the meter reading for variations in the volume of the fuel due to thermal expansion. The American Petroleum Institute has standard formulas for correcting fuel oil. Utilizing perfect gas laws will provide a thermal expansion correction factor for natural gas.

5) Compression factor; for gas to correct for a difference between the calibrated meter pressure and the actual average pressure at the meter. The perfect gas laws will permit determination of this correction factor with reasonable precision.

6) The heating value of the fuel; the number of Btu released per unit mass or volume (gallon, cubi
c foot, ton) of the fuel.
The efficiency conversion factor is determined by dividing the product of 100 and all the correction factors to convert the output meter reading to Btu by the product of all the correction factors to convert the input measurement to Btu. Multiplying the efficiency conversion factor and the Rate will produce a value of output/input efficiency for your boiler.

It is appropriate to go through the exercise to determine your Rate to efficiency conversion factor whenever you "go public" with your boiler efficiency.

REAL WORLD VALUES

Output Rates must be accepted as determined. Unless they are compared to other similar or identical operations there is no way of determining if they are true representative values. Evaporation Rate and Heat Rate, because they are convertible to efficiency, should fall within a range of normal values. Provided the units of output measurement are pounds of steam or thousand Btu, the values can be compared to those of the real world. One Rate value that is always suspicious is the one that is constant. Many a plant has a logged Evaporation Rate of 122 pounds per gallon every shift for years. When you see this, it is apparent that the Rate is not calculated; the operators simply enter it on the log as a constant.

The amount of heat required to convert a pound of water to a pound of steam is always close to 1000 Btu. Evaporation and Heat Rate should, therefore, be in the same range. The heating values of fuel vary so a real world value is dependent on the fuel. Knowing the typical heating value of a fuel permits you to readily evaluate a calculated Rate. To evaluate a Rate, assume an output unit of 1000 and estimate what it would be at 80% boiler efficiency for the given fuel.

Light fuel oil has a higher heating value close to 140,000 Btu/gallon. If the calculated Rate is over 140 it indicates a boiler efficiency of more than 100% which is highly improbable. At 80% efficiency, the Rate would be 112 (140 x 0.8). Rate values of 100 to 120 are representative of the real world.

Heavy fuel oils have higher heating values close to 150,000 Btu/gallon to produce an 80% efficient Rate of 120.

Natural gas has a heating value that is always close to 1000 Btu/cubic foot which permits a quick mental analysis of the Rate because Btu/MSCF (Thousand Standard Cubic Feet) is ten times efficiency. A Rate of 800 = 80% efficiency.

The higher heating value of coal ranges from 8,000 to 16,000 Btu/lb. A Rate for 80% efficiency would be either 6.4 to 12.8 per lb, 6,400 to 12,800 per thousand pounds or 12,800 to 25,600 per ton. The fuel units used must be noted to compare a calculation with a real world value.

Chapter 12

OUT OF THE MECHANICAL ROOM AND INTO THE WORKPLACE: HOW TO MOTIVATE EMPLOYEES TO CONSERVE ENERGY

M. S. Gould

ABSTRACT

Technical engineering techniques do not exhaust all the energy saving opportunities available to the energy manager. At Stanford University, a public awareness program has successfully promoted the wise use of energy for the last eight years. A quarterly newsletter, individual electricity consumption reports, and a catchy advertising campaign are prepared by the central Energy Management Group. Each building or department has a designated Building Energy Manager (BEM) who provides local assistance, keeping track of over 400 campus buildings. Many BEMs are enthusiastic participants in the program, tracking energy consumption patterns and promoting conservation to others in their building. In addition, the Energy Managers help schedule the centralized Energy Management and Control System (EMCS), a vital energy saving task when building schedules change quarterly.

HISTORY OF THE ENERGY MANAGEMENT GROUP

When energy prices rose dramatically in the 1970's, the Stanford community was called on to make sacrifices to avoid financial hardship. Staff, faculty and students responded well to these initial efforts. Local news media reinforced the importance of saving energy. At the time, energy conservation was an important priority. Most were willing to withstand a small amount of discomfort to benefit the campus community. Economic motivations were not yet necessary. The imminent depletion of nonrenewable energy resources such as oil was an incentive for most to conserve. Simple changes, such as vacation shutdowns and different scheduling, were implemented.

Keeping conservation awareness in the limelight is a relatively difficult task: other important issues compete for faculty, staff and student attention. The mission of the university is, after all, education and research, not energy conservation. Some would argue (fortunately very few) that any interruption from their work is a nuisance and not their responsibility. However, most people realize that utility costs must be paid, as opposed to discretionary items in the budget. Money paid for energy directly competes with funds that could be used for teaching, research or other educational needs. Because Stanford is large the energy cost figures are impressive--annual energy costs exceed $12,000,000. A large portion of this cost is difficult to reduce. For example, several of the lab buildings require 100 percent outside air due to the types of research being done. Other areas have mainframe computers that have large power and cooling loads.

In 1980 an energy group was formed with the specific charter of reducing campus energy consumption. A two-pronged approach was taken. The first part was technically based. Engineers familiar with building systems such as lighting and H.V.A.C. develop and implement modifications to improve the efficiency of building and utility energy systems. The second part of the group was responsible for public awareness and the implementation of the Building Energy Managers Program.

PUBLIC AWARENESS HURDLES

A good deal of attention has been given to the difficulty of public awareness or informational programs for energy conservation. Several factors contribute to this problem. Perhaps the most significant is the lack of direct accountability for energy costs. At Stanford, individual campus departments are not charged directly for energy. Ways must be found to motivate employees to save energy without using economic incentives. Direct feedback such as meter readings is the most effective tool available. Continuing expansion of facilities, weather variables and occupant changes complicate attempts to calculate the success of an awareness program. Although most people on campus are rational and understand that saving energy is saving money, most underestimate the effect of their individual participation.

BUILDING ENERGY MANAGERS PROGRAM

Because Stanford has over 400 buildings of varying age and usage, keeping track of building schedules, maintenance conditions and energy usage is difficult for a small energy management group. This problem is amplified since schedules change quarterly. Even continually updated records kept by the EMCS and maintenance groups cannot reflect completely how the building is performing. Still missing are the human factors or the attitudes occupants have about their building. To assist the EMG in monitoring local conditions and attitudes, an individual was chosen in each building to act as an "energy representative," later named Building Energy Manager. Those chosen come from a variety of backgrounds depending on what department they work for or what type of building they work in. For example, in the humanities building energy managers usually are secretaries or administrative assistants. Many of the laboratory buildings, however, have more technical personnel as the BEM, since building systems are complex.

All of the BEMs have other duties and can only allot a certain amount of time to this responsibility. It is important that the BEM know how the building is operated and what the specific needs of the occupants are. These needs might include humidity requirements, special scheduling requirements or emergency power provisions. To be successful, a building energy manager should have an interest in energy conservation, be skilled in persuading others, have the support of those responsible for running their department, and be able to allow for time to promote energy conservation.

RESPONSIBILITIES OF THE BUILDING ENERGY MANAGER

Appealing to a sense of community or "team spirit" is much more effective than mandating participation. A sense of individual commitment comes from several sources. These include the personal sense of responsibility and encouragement from outside sources such as the EMG or their own department. The list below is an outline of the recommended tasks of the Building Energy Managers.

- o Update EMG on schedule changes to make most efficient use of the EMCS.
- o Identify areas of energy waste. Refer technical problems to the EMG.
- o Update graphs describing their building's energy usage. Both the graphs and consumption reports are supplied by the EMG.
- o Post and distribute flyers and a quarterly newsletter provided by the EMG.
- o Act as a project liaison for work done by the EMG in their building. For example, BEM's will know when rooms are available for lighting changes or when fan systems can be shut down.
- o Act as a building energy educator by keeping co-workers up to date on campus energy problems.
- o Establish new patterns of use by personally turning off lights or equipment.

Even the ideal BEMs' interest and commitment can vary. It is the responsibility of the EMG to keep cooperation levels high.

To assist the BEM's the EMG issues an easy-to-read booklet describing basic energy saving techniques. Chapters include how the EMCS works, energy efficient lighting practices, and how energy costs vary from season to season and by time of day. Also included is a description of the type of support the BEM can expect from the EMG. A glossary is included to help readers get through some of the energy jargon that is often taken for granted.

One of the most common excuses for not conserving energy is lack of funds. Many departments have restricted budgets that make even small conservation fixes hard to carry out. To solve this problem the EMG set aside $450 per BEM to be spent on projects that they identify. The only requirement was that the project save energy and have a two year or better payback. In most instances lighting projects were chosen. In any case, money that can be spent from a different budget (in this case the EMG's budget) is a positive encouragement for doing projects. If BEMs identify larger, more complex projects, the project is referred to the technical part of the EMG.

The ability to react quickly to a variety of problems is another benefit of having a Building Energy Managers Program. For example, last winter was particularly dry in northern California, causing a drought during the summer. Mandatory reductions have been placed on the University requiring prompt action. Since the BEM program is in place, distribution of water conservation materials has been made much easier. A great deal of time has been saved using established building contacts.

THE *ENERGY EXCHANGE*

To keep the BEMs current on campus energy issues and activities, a packet of information is sent out quarterly by the Energy Management Group. Included in this packet is a four-page newsletter called the *Energy Exchange*. The purpose of this is threefold. First, this is an informal way to promote achievements of the Energy Program and help to make BEMs feel they are a part of these accomplishments. Other developments such as grants or rebates successfully obtained are described. Second, in each issue a BEM profile describes how different BEMs tackle energy waste in their building or how others handle the problems of common interest. Last, it is a way to alert BEMs to seasonal steam shutdowns or other changes which might affect building occupants.

Included with each quarterly mailing of the *Energy Exchange* are energy consumption reports. Most of the buildings on campus have their own meter. Most of the larger buildings also have steam and chilled water meters that are tracked by the EMCS. These meters are read monthly for accounting purposes, and the data is passed on for analysis and summary. A spreadsheet program (in our case, Lotus 1-2-3) makes data transfer to a format that the BEMs can read quickly and understand. Each BEM gets a readout for the building with a comparison with the same period one year earlier. An added benefit of this mailing is the opportunity to identify sudden increases that may be the result of HVAC or equipment schedule problems. The EMG encourages the BEMs to chart this data and leave it in a visible place for others to see.

The last component of the quarterly mailing is a packet of public awareness posters. Each quarter the EMG designs a poster to remind faculty, staff and students to save energy. Most campus members are receptive to a light, humorous approach. College campuses are a popular place for advertisers, so there is a challenge for the EMG to come up with unique slogans and ideas. Early ads focused on simple conservation ideas such as turning off lights. Several years ago some thought that turning on and off fluorescent lamps used more energy than leaving them on. One poster attempted to dispel this notion.

Don't Be Afraid to Conserve Energy

In order to provide a consistent approach, a theme is usually chosen for the public awareness posters. The current theme is *Don't be Afraid to Conserve Energy*. Monsters were selected from famous horror movies and used with energy-related slogans. For example, the Mummy is shown with the slogan "Sometimes you get all wrapped up in yourself and forget to save energy." Another poster has the Wolfman with the caption "Some things change every full moon, energy conservation is here to stay." Several other monsters, such as the Bride of Frankenstein and Dracula, have been used. In all cases the message is upbeat and positive.

To complement the monster posters campaign, a brochure was developed and given to all campus employees. Using the same monster theme, an explanation of how energy costs affect the university's operation, along with phone numbers of people who answer energy questions, is provided. To help develop good energy habits early, this brochure is also given to new employees. To make sure the brochure is more useful, it folds out into a large map of the campus. In Stanford's case, a map that is attractive can be hung in offices or on cubicle walls and help newcomers find their way amidst the unfamiliar buildings and roads. Others more familiar with the campus can use it as a reference or to give directions. For other companies or institutions a calendar with important dates might be as useful.

While the "monster" posters are visible to most full-time faculty and staff, students can be a difficult audience to reach. At Stanford, energy conservation is encouraged in the dormitories and other campus residences. Some students, however, live off campus and need to be reached in another way. Fortunately, there is a daily student paper where advertisements can be printed. To be effective, the ads must be unique. Repeating ads more than two or three times diminishes impact. Using the same "monster" poster sent to the BEMs as an ad at the beginning of each quarter is enough to reach our target student group. Reinforcement is an additional benefit of advertising for those have already seen the posters.

RECOMMENDATIONS

- Support from upper management is required to encourage BEM participation. Without solid support, participation will be uneven. A kickoff letter from a position of authority requesting support can do this effectively.
- Building energy managers must be selected who can devote a least some time to the program. Fortunately, even a busy BEM who is enthusiastic will make time for energy conservation.
- Find a way to acknowledge the active BEMs. For example, a short biography in the newsletter or a letter of thanks. Last year the EMG sent T-shirts to the BEMs as a thank you.
- An energy fair or other display should be presented annually to gain a wider audience.
- Most importantly, the public awareness effort must communicate that conservation is in the interest of each member of faculty, staff and student body. Explaining energy costs in terms that everyone can understand is vital.

BENEFIT ANALYSIS

Quantifying the energy savings of a public awareness program is a challenging task. Normally energy conservation paybacks are calculated by verifying engineering assumptions. As discussed, meter readings cannot be used to calculate savings unless weather conditions, occupant changes and technical improvements are factored. BEMs were interviewed to determine techniques used in conservation. Some of the techniques were delamping, controlling thermostats and updating scheduling on the EMCS. If the net energy saving coincides with changes in behavior suggested by the BEM, then this saving can be attributed to the BEM program. Savings varied from building to building, but overall program impact is estimated to be one percent of total campus usage, or $120,000 per year. And $120,000 is an energy term everyone can understand!

Chapter 13

EFFICIENT BOILER OPERATION

S. F. Connor

Cleaver-Brooks, as you may or may not know, has been in the business of building packaged steam and hot water generators, both Firetube and Watertube, for over fifty-five years and as such we have developed considerable experience in the area.

And I suppose it's like anything else -- the more you work with something and get used to the concept, you begin to take it for granted and presume that everyone is as familiar with the subject as you may be.

So at the risk of becoming too basic, I thought the best way to present this topic today would be to first define the differences, the advantages and probable application for the two basic types of boilers (Firetube/Watertube) -- then go into a brief discussion of efficiency, what it means and how its achieved and finally go into the various ways to improve efficiency, and in effect, take a major step toward the optimization of the entire system within your building or plant.

Firetube boilers, as a class, normally range in size from 15 - 800 H.P. and as such are without doubt the most often used steam or hotwater generator for larger commercial or industrial applications.

Just as the name implies, in the Firetube design, the radiant and convective energy passes through the tubes which are secured within a steel cylindrical shell. These tubes then are surrounded by water and the heat is exchanged through the combination of temperature difference and velocity.

The Watertube, on the other hand, normally ranges in size, anywhere from 10 - 3,000 H.P. and again, as the name implies, the water (just the opposite from the Firetube) is now inside the tubes and the radiant and convective heat passes around and through these tubes to generate the required steam or hot water.

So essentially through different means, the same result is attained and it's, therefore, the features and benefits of the two designs which dictate the ultimate choice.

For instance, in the case of the Watertube, if you have a load which includes constant swings back and forth -- this design -- because of its large furnace volume and minimal water content will respond more quickly to these swings, while maintaining stable water levels and delivering pure steam.

Another advantage of the Watertube over the Firetube is that of space.

Although the Watertube is physically higher, it offsets considerably less floor space than a comparably sized Firetube when one considers the room necessary for operator attendance and door swing.

The Watertube is also lighter by about 20 - 30% -- and finally the Watertube -- horsepower for horsepower and viewing it at the same load point as a Firetube -- will have significantly less draft loss which reduces the fan horsepower requirement needed for combustion.

Turning now to the advantages of the Firetube design, one finds a number of them as well.

Beginning with price and considering sizes of 250 HP and larger, you will find that the Firetube is probably less expensive size for size. One of the major reasons for this lies in the labor manufacturing hours. It's far easier to drill tube sheets instead of drums, to roll tubes from the outside rather than the inside and to use straight tubes rather than a larger number of bent tubes with special shapes.

So initial price is one of the first major benefits the Firetube design offers.

Secondly, I mentioned earlier that the Watertube had an advantage over the Firetube when it came to swing loads -- and that's true! The reason, as mentioned, is the abundance of radiant heating surface and minimal water content.

The Firetube though -- has the advantage when it comes to sudden steam demand with an associated drop in pressure.

In this case -- and because of the Firetubes' much larger mass of water which is at or near saturation temperature -- a sudden demand for steam with an associated drop in pressure would allow this large mass of water to flash, thus providing the steam requirement with acceptable purity. It's a distinct advantage depending on the system to which the boiler is applied.

Lastly, for the purposes of our discussions today, and again comparing size for size -- the Firetube unit which is properly designed will be anywhere from 4 - 5% more efficient.

Now when your looking at fuel prices, as high as they are today -- even in light of the recent relief where experiencing -- that's a major consideration and something that we will explore further, a bit later on.

But first let's spend a little time now talking about efficiency in general, which if not properly understood can be misleading, and can result in confusion and possible mistakes when sizing system loads.

To begin with, we should recognize that there are three basic types of efficiencies which are most often discussed when referring to boilers. They are:

- Combustion Efficiency
- Thermal Efficiency
- Fuel-to-Steam Efficiency

Combustion Efficiency relates to the effectiveness of the burner and its ability to burn the fuel completely. If a burner could consume all of the fuel without the need for any excess air, and if it could produce the ultimate CO_2 readings with no carbon monoxide, then it would have reached 100% combustion efficiency.

For example, No. 2 oil consists of approximately 85% carbon and 14% hydrogen. The highest theoretical CO_2 obtainable would be 15%. In actual practice, however, excess air must be used to completely mix with the fuel oil, so that it will burn without forming residue. A typical burner, therefore, will burn No. 2 oil completely and without any carbon monoxide when set to achieve about 13% CO_2. This would involve about 20% excess air and a combustion efficiency ofapproximately 95%.

As an oil comparison -- No. 6 oil is composed of approximately 87% carbon and 13% hydrogen. The ultimate CO_2 of this oil would be 16.5%. Again excess air is required in actual practice and a value of 20% excess air would give us a flue gas analysis of 13.8% CO_2 and a combustion efficiency of approximately 95%.

The constituents of natural gas vary slightly depending upon it's source, but generally speaking we're looking at 80.5% methane, 18.2% ethane and 1.3% hydrogen. The ultimate CO_2 of this fuel is 12.1%; however, normally a gas burner is set up to burn at 15% excess air to produce a 10 - 11% CO_2 reading, which gives a combustion efficiency of 95.5%.

Now be mindful that in all of these cases, we are talking only of the effectiveness of the burner to efficiently combust the fuel without any consideration being given to heat transfer -- which leads us to the second efficiency term -- Thermal Efficieny.

Thermal Efficiency is quite broad, but in a general sence relates to the effectiveness of heat transfer within a heat exchanger.

The main point to recognize here is that Thermal Efficiency does not take into account the various radiation and convection losses from the shell and piping of the boiler.

The term Thermal Efficiency strictly means that all the BTU's going in are extracted as usable energy less the stack losses. We know, however, that this isn't true as every boiler radiates some energy away and based on boiler design, in some cases, this can be considerable.

So what then is the most meaningful criteria under which Boiler Efficiency can be evaluated? -- It's without doubt Fuel-to-Steam Efficiency because it does account for radiation and convection losses.

To put things in perspective -- Fuel-to-Steam Efficiency is the true value that determines the actual amount of fuel that must be burned to meet a given load demand. It is simply the ratio of BTU output to BTU input expressed as a percent and since it can mean the savings of thousands of dollars per year -- it is a function that should be completely understood by those responsible for boiler selection and operation.

For example -- let's assume for the moment that we have an existing 600 HP boiler which operates at 15 PSIG and burns natural gas. Let's further assume that a stack sample finds a CO_2 reading of 8% and a gross stack temperature including ambient of 430°F. This all equates to a flue gas loss of 19.8%. Then if we add in the additional losses for radiation and convection which with many boilers is in the 2 - 2½% range -- we have a total loss of 22% or a 78% F.T.S.E.!

Now let's look at an efficient package and what a few percentage points in efficiency gain can mean in dollars and cents.

The CB 600, when firing natural gas will deliver a nominal 10% CO_2 value in the stack and at 15 PSIG will register a gross stack temperature of approximately 370°F. These figures would then equate to a total loss of approximately 18% or an overall F.T.S.E. of 82%.

A 600 HP Boiler burning natural gas at 78% efficiency consumes 257.5 therms/hr. Conversely, an 82% efficient boiler burns 245 therms. The difference, of course, is 12.5 therms/hour.

Now let's assume that the comparative boilers will run 4,000 hrs./yr. at 100% of rating, and the cost per therm is 55¢. That's 12.5 therms per hour times 4,000 hrs./yr. is 50,000 therms/yr. times 55¢ per therm or $27,500 saved with a 4% more efficient boiler!

Another interesting point that is made in comparisons is the fact that the percent of fuel costs increases at a faster rate than the percent of efficiency loss. For example, fuel costs increase 6.5% with a 5% drop in efficiency -- increase 9.5% with a 7.5% drop and 14.3% with a 10% drop.

So it goes without saying that a proper understanding of efficiency and what it means is extremely important when looking at ways to significantly reduce cost and/or accelerate return on capital.

Now before leaving the subject of Efficiency in general and before moving on to the subject of Efficient Boiler Design -- I would be remis if I didn't touch on three (3) key elements which affect efficiency and to which I've earlier referred to. They are:

- % O_2
- % CO_2
- % CO

The slide on the screen now shows the relationship of excess air (% O_2) to CO_2 when burning natural gas and No. 2 or 6 oil.

Referring to the natural gas curve, you will find that if we were able to burn the fuel with 0% excess air or 0% O_2, we could achieve the highest % CO_2, thus the highest efficiency.

Practically speaking, however (and as mentioned earlier burning fuel in a standard burner with 0 excess air is not advisable because it could lead to a very dangerous situation. Dangerous because we would be burning the fuel with no margin for error -- error which can be caused by variations in temperature, humidity, barometric pressure, BTU value of the fuel, etc.

The resultant fuel rich condition manifests itself with the formation of noxious CO which besides being lethal if high enough levels are generated -- can be explosive because CO is nothing more than unburned fuel.

Therefore, it's for this reason that standard burners are normally setup at 15% excess air to accommodate these potential variations.

As you can see, 15% excess air equates to about 10% CO_2 -- as you can also see, the higher the percent excess air, the lower the % CO_2, thus the lower the combustion efficiency. Conversely, the lower the percent excess air -- the higher the CO_2 and, therefore, the higher the efficiency.

This next slide is interesting because often times people automatically equate high CO_2 levels with good combustion -- but -- as you can see it's important to know what side of the curve you're on. In this example, we're burning gas and although it's possible to achieve a CO_2 reading of approximately 10% -- we could also be generating CO at or near 200 PPM.

So it's essential that we always keep in mind that although CO_2 normally relates to good combustion and high efficiency -- unless we are also aware of and check for CO, unnecessary fuel dollars could be going up the stack.

Having looked at efficiency generally, and covered what it means -- let's now look at what goes into the making of a highly Efficient Packaged Boiler.

For optimum design results and highest efficiencies, we begin with the total packaged concept which doesn't just mean that everything is provided -- but -- rather everything is compatably fit to provide maximum output.

The Efficient Boiler Designer builds both the boiler and the burner and designs the burner to precisely meet the combustion pattern and firing rate that would be most compatable with the pressure vessel, and result in the optimum efficiency.

For instance -- this is our 4-pass -- 5 sq. ft. -- dryback Firetube boiler which encompasses our own fan, burner and fuel/air metering system.

You'll note in this slide that the furnace is located low in the shell. This is for two (2) distinct reasons:

- No. 1 -- For safety in the event of a low water condition -- and --
- No. 2 -- For efficiency reasons because this is the area of highest heat release and we want to assure ourselves of having the coolest outside temperatures around the furnace to give us the highest temperature differential to maximize heat transfer. This low position in the shell will accommodate that.

Next you'll notice that the area for heat transfer gets successively reduced as we proceed through the passes.

By reducing the area in each successive pass, we are able to maintain high gas velocities (for example 50 FPS), thereby again increasing heat transfer to maximize efficiencies as high as 88%!

The burner portion of the package is designed around the principal of high pressure drop.

High pressure drop which offers the advantage of being able to operate the unit with back pressures in the magnitude of 0.5" - 1.0" of WC with no adverse affect on combustion efficiency.

Also the high pressure drop design provides a turbulent environment to maximize fuel/air mixing for high CO_2 and low or undetectable CO readings.

This highly efficient forced draft design begins with the integral fan and motor assembly which again are compatably mated to deliver the required air volume for maximum output.

The fan itself is likened to a radial blade centrifugal air pump.

The air is then pulled through the top of the unit through this acoustical lined duct and actually pressurizes the front door. It then enters the air control device which we call our rotary damper. This you'll notice is located at the outlet side of the fan for superior control and distribution.

The rotary damper itself is carefully machined -- a sleeve within a sleeve if you will -- which allows for properly metered air throughout the entire turndown range of the burner. This is very important as mentioned earlier because too much air can adversely affect combustion as can too little.

The next very essential part of an efficient metering system is to develop a means of controlling the air opening within the damper while also metering the proper amount of fuel for optimum combustion efficiency

At Cleaver-Brooks, we have developed our own single point positioning system using a single jackshaft with dedicated linkages going to the fuel valve and damper.

The heart of the system is this infinitely adjustable contoured cam containing twelve (12) allen head screws for precise tuning of fuel and air over the entire operating range.

So now we have fuel and air -- it's properly metered -- but what else is required to achieve maximum combustion efficiency?

The answer is proper mixing.

I mentioned before that a high pressure drop burner design was essential for peak performance -- but unless we have something to create the turbulent atmosphere to intimately mix fuel and air, we have defeated the purpose.

The answer here, as far as Cleaver-Brooks is concerned, is the diffuser.

This galvanized metal device contains many precisely sized open vanes throughout its circumference -- and -- because of its position slightly behind the point of fuel entry -- it creates the turbulent, spinning affect caused by air and fuel flowing across its face at fairly high pressure.

The diffuser then, is the mixing device which is so essential in achieving high combustion efficiency.

The last major consideration in Optimum Boiler Design, is the ease of the unit to be inspected and maintained.

It goes without saying that heat transfer surface fowling is a major deterrant to ongoing F.T.S.E.

It's therefore our philosophy that an easy to maintain boiler will be maintained, thus assuring the customer of ongoing efficiency year after year.

To briefly recap -- so far we've discussed the basic types of boilers and what makes them unique -- we've looked at efficiency and what it means -- and just now we reviewed how an efficient boiler is designed.

Next, let's look at some enhancements which can further improve efficiency -- and as long as we were just talking about the burner, we might just as well stay with it and look at least what we did to further enhance what many claim to be the most efficient combustion system on the market today.

Our design objectives started with improving turndown which in our standard design is limited to approximately 4:1. The intent was to accomplish 8:1 while maintaining flame stability -- and achieve an acceptable excess air level throughout the firing range.

The other objective was to reduce emissions especially as it related to NO_x on gas.

As you probably know -- the problem with turndown, once the burner decreases to 30% or below, is flame stability and the ability to maintain proper fuel/air ratios.

Many burners on the market today can achieve respectable CO_2 levels at 50% and above, but once the burner turns down -- the lack of a proper air/fuel metering device causes the ratios to skew thus reducing efficiency.

Additionally, the flame can become very unstable -- pulling back on the burner itself causing serious degradation.

Our particular problem was not one of controlled fuel/air metering as we had an efficient system for that. We did, however, have to concern ourselves with the problem of stability at firing rates of 25 - 30% and below. We also wanted to do better with NO_x emissions on gas as mentioned previously.

Ultimately our solution was the development of a burner, which in affect satisfied the design objectives while providing some side benefits.

Let's now look at the unit and in so doing, I'll point out the various features that account for higher combustion efficiency -- higher turndown and lower emissions.

As you can readily see, the burner is markedly different from our standard. The differences begin with this third air supply which is referred to as tertiary air.

This tube ducts directly into the front head and supplies air to and through these tertiary air cones which are also new.

You'll also notice the target plate which in combination with the precisely machined air cones, replace the standard diffuser.

Essentially then, these are the major differences from our standard burner configuration and its these changes or features which account for the benefits we were striving for.

Specifically, the ability to achieve high turndown while maintaining flame stability is accompished through the interplay between the air cones, the tertiary air tube and the rotary damper.

As the rotary damper closes, the close tolerances of the machined stainless steel air cones provide a stable flame front as the fuel is turned down. The inner and outer cone air flows are straight (parallel) flow, and provide fuel mix control at high pressures (anywhere from 4 - 8" WC more than standard) while also providing air to cool the oil nozzle.

The intermediate cone furnishes high spin vortex type air flow to provide control of the flame core and further establish flame stability. This controlled central core of flame exists throughout the firing range.

When the damper is fully closed, the primary source of combustion air is through the tertiary air tube. Quite frankly, once the damper is fully closed and we turndown beyond 4:1 to let's say 8:1, the excess air does increase because the tertiary air is not regulated. The flame stability is there, however, and we feel this is a very important point which I will address shortly.

Here is how the flame appears when it's turned down from 8:1.

- Beginning at about 50% firing rate and coming down -- remember over 90% of the air is parallel flow.
- Note the tight ball of flame at the core. This is the high spin core which is vortex flow.
- Note the stability -- no burnback or degradation.

The 8:1 turndown with flame stability has a distinct benefit and it's this point which I referred to earlier and said we would discuss.

The distinct benefit of high turndown regardless of excess air level, is the fact that it significantly reduces burner cycling, and thereby reduces air purge heat losses compared to a unit that cycles frequently.

As you know, many plants have insulated their steam lines, have replaced faulty traps, return more condensate and insulated their buildings -- all of which result in an oversized boiler cycling on and off.

With higher turndown, we can reduce cycling and save fuel due to unnecessary air purge heat losses. High turndown will also maintain a more constant steam pressure and provide faster response to changing load conditions.

I mentioned before that lower emissions are also achieved with this design and to put things in perspective -- we're talking about less than 60 PPM of NO_x (0.07#/MM BTU) when burning natural gas -- this compares to results on our standard burners of approximately 150 PPM.

The reason for this improvement, simply is the high pressure air and excellent mixing environment which allows us to efficiently burn the fuel.

It's also due to the inner and outer cones which deliver high pressure air in a parallel flow forcing the hot gases out of the high temperature zone where most of the NO_x is formed. In other words, the residence time is reduced.

And so -- through all of this, overall efficiency is gained because of:

- Better mixing -- better combustion.
- Cleaner burning -- giving us less emissions and deposition on the fireside surfaces which inhibits good heat exchange.
- Better air control -- the tertiary air tube giving us lower excess air capability.

Now -- when talking about excess air and being able to perform at relatively low levels throughout the turndown range -- any responsible burner manufacturer will admit that although their equipment may be capable of this performance -- there are a number of variables that intervene, which if gone unmonitored, could cause an unsafe or inefficient condition to exist.

I refererd to these before and they are:

- BTU value of the fuel.
- Barometric pressure.
- Ambient temperature.
- Humidity.
- Etc.

To counter these natural phenomena and assure ourselves of optimum efficiency and safety, one should consider a vigilant monitor or what we refer to today as an oxygen-trim system.

The purpose of an oxygen-trim system is to maintain an optimum fuel/air ratio in the combustion process.

By providing just enough air to completely burn the fuel, energy losses due to high excess air levels and incomplete combustion are reduced.

The typical boiler uses mechanical linkage to maintain a desired fuel/air ratio, but as mentioned -- changes in the fuel, air-or-normal linkage wear can cause the actual ratio to change at any time.

An O_2 Trim System then, measures the O_2 level in the flue gas which is directly related to the combustion process.

Using this value, and comparing it to a predetermined optimum value (setpoint), the system will vary either the fuel or air to make the necessary correction.

The system begins with a zirconium oxide cell which is mounted in the boiler stack. This cell then measures the amount of O_2 present in the flue sample which in turn is converted to an electrical signal that is proportional to the percent of O_2 in the gas.

With the Cleaver-Brooks DATA 1 Microprocessor based unit, the O_2 cell is located in this sensor head which is bolted to the stack proper. In addition to the O_2 cell, the head also contains the stack thermo couple for measuring temperature, the air outlet tube because this is an aspirating system, the heater for the O_2 cell -- and finally the inlet tube through which the stack sample is drawn.

Essentially this is how the system works:

The sample is taken through the inlet tube and passes across the zirconium oxide cell which is being held at a temperature of approximately 1,400°F by the integral heater. The O_2 IONS present are then converted into an electrical single which is fed to the central control panel. This output is then converted to a 3 - 15 PSI pneumatic signal that is used to operate an actuator which in turn adjusts the air flow to match the fuel.

A major benefit of the DATA 1 system is response time or it's ability to measure the % O_2 and adjust within one second.

The key to this benefit is the way we take our direct sample as opposed to convection or diffusion methods which are often employed.

The problem here is that if the control cannot respond in a very short period of time and if the boiler is in a modulating mode, the trim system will never catch up to match proper fuel/air ratios for maximum efficiency.

Other things one should look for in an O_2 trim system -- besides quick response would be appropriate annunciation and maximum safety.

The DATA 1 for instance, clearly indicates the essential information required to assure the user of maximum performance. This includes a digital readout of:

- Stack Temp
- % O_2
- O_2 Setpoint
- Firing Rate
- % F.T.S. Efficiency

As far as safety is concerned -- and when dealing with high combustion efficiencies with low excess air, one can't be too careful -- this system upon denoting that a fuel rich condition is occurring will automatically drive the damper to the full open position, shutdown and annunciate the condition. This, we feel, is essential for any O_2 trim system to provide.

So far, we've been talking about efficiency primarily from the standpoint of improving combustion efficiency -- Let's now spend a few minutes looking at ways to maximize or improve the heat transfer process.

The system shown in this sketch lacks a fundamental device: a water softener. In order to maintain the efficiency built into the boiler, the heating surfaces must be kept clean. Scale from hard water that forms on the water side of a boiler's heating surface is a thermal insulator, which prevents heat from flowing through the metal of the boiler, to the water and causes the flue gas to leave the boiler at a higher temperature. This increases stack losses and reduces boiler efficiency. As a matter of fact, an 1/8" accumulation of scale reduces efficiency by a minimum of 5%. That's why the water going into any steam boiler should contain no hardness. Unless you have an unusual water source, all make-up water to a steam system should be softened.

The effects of scale in boiler water are cumulative so they should be prevented, either by a water softener or by adding chemicals (sulphites) to the water. But chemicals are expensive operating costs and increase the rate of blowdown. Consequently a water softener is usually the best approach.

Now, if a steam system operates with a significant amount of make-up and/or if the water used has a significant amount of impurities -- like alkalinity, TDS, or silica -- then consideration should be given to further pre-treating of the make-up water by adding a dealkalizer or a reverse osmosis unit.

The price of the dealkalizer or reverse osmosis unit depends upon flow rate and water quality, so prices can vary widely. The payback -- often attractive -- accrues from three factors: sharply less use of water treatment chemicals, significantly less blow-down, and improved fuel-to-steam efficiency because of clean waterside surfaces.

Maintaining clean waterside surfaces is a constant concern with boiler operation. Essentially it has to do with the control of the total dissolved solids within the boiler water, and most commonly this is accomplished through a combination of proper pretreatment and/or controlling the T.D.S. level in the boiler water through surface blowdown. Surface blowdown is also an excellent source for recoverable heat.

Before consideration can be given to any recovery of otherwise wasted heat, the use and timing of the recovered heat must be considered. If the heat cannot be used at the time and rate at which it is available, it must be stored until the time it can most effective waste heat recovery is that which avoids storage by using the heat at the time it is available. And that's the case with a surface blowdown heat recovery system.

Returned condensate is quite pure whereas make-up water added to a steam system usually contains impurities. These impurities concentrate in the boiler. The concentration must then be reduced by using surface blowdown. Make-up and blowdown are, therefore, cause and effect.

A heat exchanger in the make-up line complies with the use requirement of effective heat recovery. However, the effectiveness of blowdown heat recovery could be further improved by improving the timing. One type of blowdown heat recovery system improves the timing by thermostatically sensing the flow of make-up water and causing blowdown only when make-up is flowing. Such devices will recover upward of 90% of the heat in the process.

The final element we will be looking at today to improve overall fuel-to-steam efficiency is the Economizer.

The major heat loss from any boiler is the stack loss. One way to increase boiler efficiency then would be to recover some portion of the heat lost out of the stack and add it to the feedwater. This recovery is possible if certain conditions pre-exist in a steam system.

First of all -- Economizer performance depends upon the temperature difference between the flue gas and the feedwater. Since the feedwater normally comes from a Deaerator, that temperature is relatively constant: about 220°F. The temperature of the flue gas depends upon the operating pressure of the boiler. In practice, the pay back of an economizer is not attractive if the boiler operates below 100 PSI if the boiler is smaller than 100 HP. These are ballpark figures based on today's prices of oil and natural gas.

Another factor enters into the consideration of whether to use an economizer: sulfur in the fuel being burned. If sulfur is present, the feedwater temperature must be increased to prevent cooling the flue gasses below the sulfur dew point and forestalling sulfur corrosion from attacking the outside of the tubes. But this also reduces the temperature differential across the economizer and its payback.

However, if conditions are right with regard to boiler size and respective temperatures -- economizers can easily improve efficiency from 5 - 10%, and in most cases pay for themselves in a relatively short period of time.

Through this presentation today, we have covered considerable ground as we looked at the major types of boilers and how they're applied -- we looked at efficiency -- what it means and how it's achieved -- and finally, we looked at ways to further improve efficiency through better combustion, control and heat recovery.

It is my sincere hope that what I have said will be of some help to you -- or -- at the very least will stimulate thought as to where you might begin to improve efficiency within your boiler room and, thereby conserve our most valuable finite resources while generating considerable dollar savings in the process.

# # # #

Chapter 14

AUTOMATING THE CONTROL OF FLUID HEAT TRANSFER SYSTEMS

T. M. Kenny, J. G. Fleischer

The ability to automate the control of your heat transfer systems with a self-contained thermal actuated control valve will result in:

* Energy cost savings
* Improved system performance
* System life extension
* Reduced maintenance and labor costs
* Piece of mind

The following outlines some of the applications and economics associated with system automation.

Winterization steam tracing lines have been employed for years to prevent a variety of products from freezing in lines or equipment whenever ambient temperatures drop to relatively low levels. The old, accepted method for operating these tracing lines was manpower - maintenance personnel were required to manually turn on all of the steam tracing valves at the beginning of the "cold season" (typically October) and leave them on throughout the winter period. When the danger of low temperature passed, maintenance personnel went out and manually shut off all of the steam tracing valves.

Recently, energy-conscious companies have begun to question the wisdom of leaving steam lines open through an entire season - particularly since steam is not really required on a continuous basis during the entire winter. Actually, steam is only needed a fraction of the time in most tracing lines, when the ambient temperature falls below the freezing point. In the St. Louis area, for example, a manually operated steam tracing system would be wasting steam 49 percent of the time during the six-month winter season. Steam would flow regardless of ambient temperature changes - resulting in tremendous energy waste.

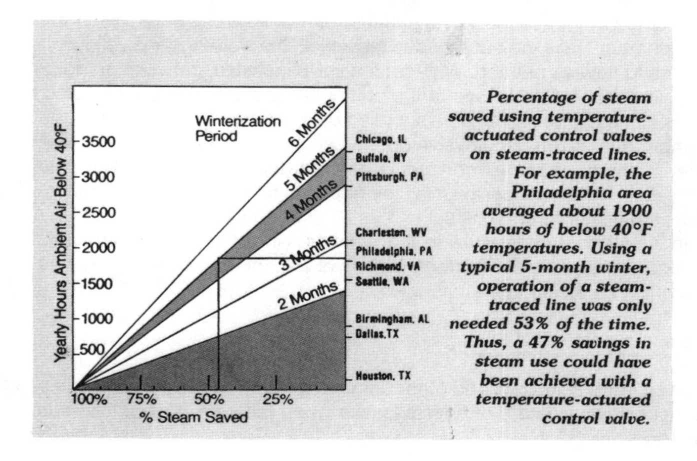

Percentage of steam saved using temperature-actuated control valves on steam-traced lines. For example, the Philadelphia area averaged about 1900 hours of below 40°F temperatures. Using a typical 5-month winter, operation of a steam-traced line was only needed 53% of the time. Thus, a 47% savings in steam use could have been achieved with a temperature-actuated control valve.

Another basic problem with a manually controlled steam tracing system is the labor involved. Not only is it time consuming for maintenance crews to physically locate all valves and turn them on or off based on temperatures conditions, but personnel may be too busy with regular scheduled maintenance or emergencies - or they simply forget.

A Reliable Method Using Ogontz TL AND CTV Valves

Automatic regulation of steam flow is the only way to maximize both operational efficiency and energy use in steam tracing systems. The automatic tracing system shown in Figure 1 has been adopted by numerous major steam energy consumers around the U.S. The system is gaining recognition as a reliable and cost-effective way to automatically control steam flow, which results in boiler base load reductions and the virtual elimination of steam waste. This automated steam tracing system is relatively inexpensive, easy to install, and has proven that it will operate reliably with minimal maintenance.

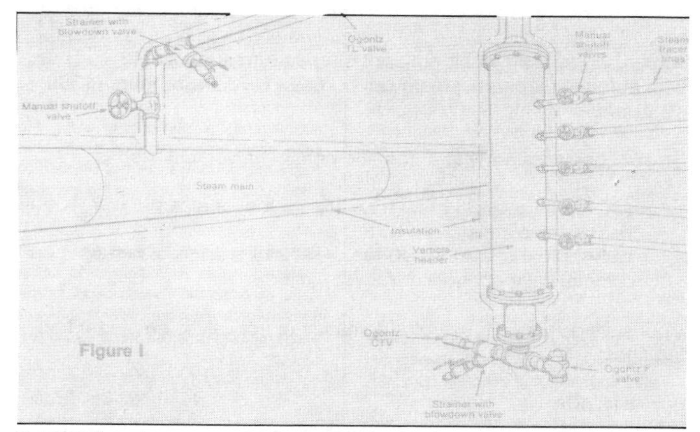

Detailed above is an automatically controlled vertical tracer line header. A thermostatic steam trap is installed at the low point of the header to discharge condensate.

The key to the operation of this automated system is the ambient temperature sensing control valve shown installed in the steam supply line. This valve is direct acting and thus eliminates the need for pilot valves and their potential problems. The valve is also totally self-contained, requiring no external power or operating signal. To prevent system shock, the valve port is designed to open gradually during operation. Reliability of the valve in severe service is due to its use of a self-contained, solid-liquid thermal actuator - this rugged thermostatic device eliminates the unreliability common with chemically filled diaphragms, delicate capillaries, thin walled bellows and other valve actuation methods. This thermal actuator senses ambient temperature changes and automatically opens and closes the valve, allowing steam to flow only when ambient temperature falls below 40°F (or some other preset temperature).

The thermostatic steam traps, also shown in Figure 1, complete the automatic steam tracing system by controlling the discharge of condensate based on the optimum discharge temperature. These traps create condensate legs which eliminate live steam losses. In addition, these traps also allow effective use of the sensible heat in the tracing system - providing additional energy savings and further reducing the boiler base load.

Steam Tracing Economics

A realistic analysis of costs and potential cost savings from automation can be made for any winterization steam tracing system, no matter how complex. All that is needed is U.S. Weather Bureau data, a table on the thermodynamic properties of steam, a few relevant operating factors and formulas, and some simple calculations.

A Case Study

A Baltimore, Maryland plant required a retro-fitted winterizing steam system for all pertinent product lines, service lines and equipment. The following calculations compare the cost of a manual versus an automatic steam tracing system.

* Calculation of Total Heating Load - heating load for winterizing season (October 1 to March 31) for five separate systems requiring tracing was calculated to be a total of 425 lbs/hr.

* Calculation of Seasonal Steam Consumption and Costs - with manual control of steam supply, "on" October 1 and "off" March 31, steam consumption is:

  425 lbs/hr x 4368 hrs. (6 mo. @ 24 hrs/day)
  = 1,856,400 lbs. of steam.
  And, @ $7/1000 lbs. steam cost is:
  1,856,400 lbs. x $7/1000 lbs. = $12,995 seasonal steam cost on manually controlled tracing system.

* Energy Savings With Automatic Control of Steam Tracers - ambient temperature sensing control valves will turn off steam when the ambient temperature is above 42°F. From U.S. Weather Bureau data for Baltimore, Maryland, there are approximately 1933 hours during the winterizing season when ambient temperature is above 42°F. Thus, with automatic control, steam will be used only 2435 hours (4368 - 1933 hrs.)

* Steam consumption with automatic control valves installed is:

  425 lbs/hr. x 2435 hrs. = 1,034,875 lbs. of steam @ $7/1000 lbs. steam cost is:
  1,034,875 lbs. x $7/1000 lbs. = $7244 season steam cost on automatically controlled tracing system.

* Cost savings through use of automatic control valves on steam tracing lines:

  $12995
  − 724
  $ 5751 saved each year

Assuming an installed cost of $300 per ambient temperature sensing control valve, with each of five systems requiring its own valve, the savings would pay for the installation in two months of the first year of operation. Additional years of service would provide pure savings of at least $5751 each year.

Steam Unit Heaters Using Ogontz TL Valves

Steam unit heaters are relatively inexpensive to purchase and install and are an efficient method for heating large areas. However, they can waste large quantities of steam if they are not properly regulated.

One of the most common methods for regulating steam unit heating is to install a thermostat that senses room temperature and turns the heater fan on or off. Unfortunately, this popular practice is no longer the most energy-efficient method of heat regulation . . . because a thermostat ignores too many areas of potential steam energy savings.

A thermostat only shuts off the fan, not the steam supply. So, steam continues to flow to the heater coils and associated piping - allowing continuous radiation, convection and steam trap losses. According to the lab test data confirmed by actual field tests, such losses often exceed 20 percent of the rated heater capacity. And, thermostats are often tampered with, resulting in additional energy losses. Equipment life often suffers too, due to excessive on-off cycling of oversized heaters (research indicates that oversizing is a fairly common practice).

Reliable Control

Steam flow to unit heaters can now be reliably controlled in the same way that it is controlled in winterization steam tracing - through the use of ambient temperature sensing control valves. This method is gaining acceptance among a growing number of energy-minded companies, particularly those with great numbers of unit heaters where steam savings can be enormous.

The automatic steam control system shown installed on the unit heater in Figure II is one that has proven to be a highly reliable and energy-efficient method of regulation. Field experience with this system indicates that it will reliably control steam flow for years with little or no maintenance.

Detailed below is a temperature-actuated control valve on horizontal draft unit heater. Valve throttles steam to the heater to maintain desired air temperature.

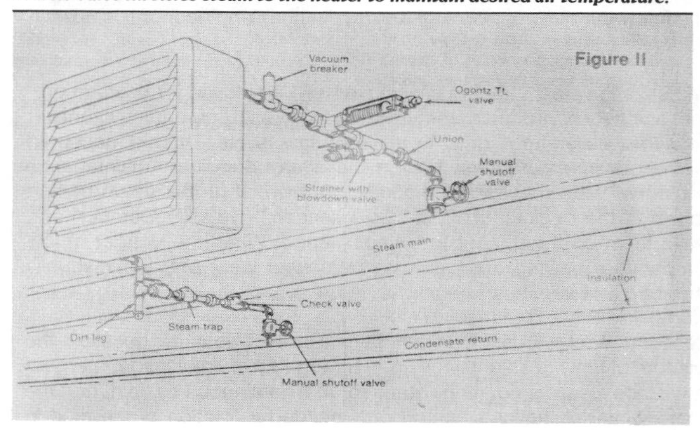

The ambient temperature sensing control valve shown installed on the unit heater's steam supply line employs the same self-contained thermostatic actuator previously discussed - providing automatic operation with no external signal or pilot valve required. As the air temperature around the valve approaches a predetermined level (e.g. 68°F), the thermal actuator closes the valve, shutting off the flow of steam. When the temperature falls below the desired level, the valve opens, allowing steam to flow as needed. With such a valve installed, convection radiation and steam trap losses at the unit heater are effectively eliminated. An electric heater can also be supplied on the actuator to "set-back" the temperature at which the valve opens and closes. This is used in those areas where personnel comfort is desired during the day and freeze protection is the concern during the night. To minimize hot air stratification, the heater fan can be set to stay on to keep air circulating to maintain a uniform temperature in the building. And, the valve's automatic operation means that maintenance crews are not needed to turn valves on or off as the seasons change.

Estimating Potential Savings

Based upon field tests and other data, 20 percent of rated capacity is a conservative estimate of typical losses when fans are off, but steam is left on.

Given: R = Unit heater rating = 750,000 Btu/hr. at 30 psig
C = Steam Cost - $8.00 /million Btu's
H = Hours fan off during heating season = 2900 (Hours in heating season = 4800 hrs.)

R x 20 percent (loss rate) x H x C = dollars saved each year

$$\frac{750{,}000 \text{ Btu/hr.} \times 0.20 \times 2900 \text{ hrs.} \times \$8.00}{1{,}000{,}000 \text{ Btu.}}$$

= $3480 saved each year per unit heater, which is nearly seven times the installed cost of 1-½" ambient temperature sensing control valve.

The days of unregulated, uncontrolled steam flow in such systems are coming to an end. Given industry's renewed emphasis on energy conservation and availability of reliable steam control technology, it is only a matter of time before this type of automatic control becomes a part of all steam systems. Old ways die hard, but major U.S. steam users are coming around. Given the cost of generating a pound of steam, and the intensely competitive market situation they all face, these users are increasingly aware of the wisdom of using only as much steam as required. The automated systems described here offer them a cost-effective way to do just that.

However, at many manufacturing and processing facilities, automating the control of steam tracing and steam unit heaters represents only a portion of the total savings potential. The self-contained thermal actuated control valve is also used to precisely control the temperature of steam traced tank heating applications involving systems process temperatures up to 200°F. The following case study describes such an application and the results achieved.

Process Temperature Application Using Ogontz ST And CTV Valves

Problem

Being a corporation that is intimately concerned with oil conservation, Petrocon sought a method to decrease steam consumption in their own plant. Petrocon Corporation is involved in collecting and re-refining waste oils and lubricants, thereby recycling existing petroleum products for use as specialty lubricants, filler materials, gear oils, coating oils, etc. The production of these products by vacuum distillation of spent oils and lubricants requires large amounts of steam for storage tank heating, freeze protection, and process heating.

Petrocon stores their finished products in a coil tank which must be maintained at a certain temperature. They would manually open the steam valves, and when the tank got too hot, close the valve. The storage tank temperature was uncontrolled, hovering between underheating and overheating causing the product to be delivered to the customer at an unacceptable temperature.

Solution

Since it is necessary to carefully control the temperature on the finish tanks, Petrocon purchased a temperature sensitive valve which is set to maintain the temperature of the tank at a fixed temperature of 120°F. Thermal cement is used to bond the valve's thermal actuator to the surface of the tank being heated. The actuator then regulates the control valve to provide the correct amount of heat necessary to maintain the predetermined temperature of the storage tank.

Also, in the steam tracer lines, Petrocon uses temperature sensitive control valves to replace steam traps. This valve controls the condensate discharge at 180°F, by forming a controlled leg of condensate which provides a liquid seal eliminating all live steam losses, which can occur with faulty steam trap operation.

Dirt is responsible for many steam trap failures. The control valve approach to steam trapping features a replaceable anti-dirt stainless steel seat and plug. This anti-dirt assembly breaks up dirt and passes it through the valve. The knife edge plug design continuously scrapes accumulated dirt away from the seat. The valve is full open at start up, purging the steam system of dirt, air, and non-condensibles; then the valve modulates to the proper opening to discharge condensate at 180°F.

Result

As applied to the storage tank heating, Petrocon found that the control valve maintained the tank at the temperature of 120°F. Previously, with no control, the steam would heat the tank to approximately 200°F which could result in degradation of the product. Petrocon calculates that 153,846 Btu/hr. are required to maintain their system at 120°F. With uncontrolled heating they were consuming 322,677 Btu/Hr. Therefore, in keeping the storage tank at an unnecessarily elevated temperature, they were wasting approximately 169,000 Btu/hr. or 52.3% of the steam.

Steam traps discharging at saturated temperatures pass quantities of live steam before they cycle off. This steam amounts to from 2 lb/hr to 10 lb/hr. Since the control valve discharges condensate between 180°F and 190°F, it completely eliminates discharging of flash steam. A standard trap will discharge condensate at about 20° below saturated steam temperature. This temperature actuated valve discharges at 180°F regardless of steam pressure within the working pressure range of 0 to 200 psi. When used in place of a steam trap, this control valve saved approximately 20% of the steam Petrocon consumed with the steam trap, by reducing sensible heat loss, plus eliminating live and flash steam loss.

Fluid Temperature Sensing Control Valve Application Using Ogontz FR Valve

Also, through the use of a thermal actuated control valve whose temperature sensor is internal to the valve, companies have been able to inexpensively freeze protect their water lines, remove condensate trapped in their steam headers and condensate return lines and by using both a direct and reverse acting valve are able to freeze and scald protect safety showers and eyewashes. In applications where cooling water is used to cool processing equipment the Ogontz FR valve is used to minimize water consumption and maintain an optimal temperature on the machinery. Injection molding equipment provides an ideal example of the benefits of automating the control of cooling water.

There are two parts of injection molding machines that require cooling water, the mold and the hydraulic oil.

The hydraulic oil maximum temperature is between 100° - 110°F. Normally the available potable water at the plant is pumped through the hydraulic oil cooler with no temperature control, and the increase in temperature of the cooling water is 4° or 5° F. The Ogontz FR valve can control the water leaving the cooler at 100°F and provide a temperature increase of approximately 50°F to the cooling water. This results in a 90 percent decrease in cooling water flow and high return on investment.

The maximum temperature of the mold end of the machine is 50° - 55°F. Refrigerated cooling water at 40°F is used to keep the molds cool. Most machines already have modulating temperature controls for the mold end. If there are no controls, an Ogontz FR valve will work well. Since the maximum allowable temperature increase of the cooling water for the mold is small, 10° or 15°, the reduction in cooling water flow will also be small. Instead of saving water, the return on investment will be based on a decrease in energy consumption of the refrigeration equipment.

A Case Study

An Ogontz project to equip the injection molding machines at Smith Corona in Cortland, NY, with FR valves illustrates the aforementioned points. All of these molding machines require water cooling for the hydraulic system. It is important to maintain an even temperature of approximately 100°F in the hydraulic fluid to maximize production speed. Higher temperatures result in thinner oil or lower viscosity that is more difficult to pump. Lower temperatures cause higher viscosity that is also difficult to pump. The existing machines come equipped with a bulb and capillary type control that sensed the hydraulic fluid reservoir and throttled the outlet water flow. Once the temperature of 100°F was attained the controller was wide open. The inlet water coming from on site wells was between 52° - 54°F. After going thru the four pass heat exchanger there was a 4° rise in cooling water temperature. The water was then discharged into the municipal sewage systems. The cost of water was as follows:

$ .08 per 1000 gallons for electricity to pump the water
$ .15 per 1000 gallons for sewage

$ .23 per 1000 gallons total

An Ogontz reverse acting self-operating temperature control valve was put on the discharge water from the heat exchanger. The type Ogontz 3/4FR95 was installed and the results were dramatic. The discharge water was throttled back until it reached 100°F now the cooling water was raised almost 50°F while going thru the heat exchanger. The hydraulic fluid was still held at at constant 100°F with just a fraction of the volume of cooling water being consumed.

A test was conducted to determine precisely how much water was saved. Water meters were installed on the cooling water sysem to various machines and recorded for a two week period with the following results:

ANNUAL WATER COSTS OF MOLDING MACHINES TESTED

	Molding Machines with Ogontz 3/4FR95CRB	Molding Machines with Conventional Bulb & Capillary Type
Qty.	Annual Cost	Annual Cost
12-300 Ton Reeds	$1,952.00	$34,805.00
8-200 Ton Reeds	$1,072.00	$14,774.00
24-100 Ton Reeds	$2,260.00	$21,537.00
5-500 Ton Cinn	$ 839.00	$ 8,395.00
2-700 Ton Cinn	$1,173.00	$11,753.00
TOTAL:	$7,296.00	$91,264.00

Savings Equals	:	$83,968.00
Ogontz Valve Cost	:	$ 8,250.00
Return on Investment	:	10.17

*Based on $ .23/1,000 gallons

The Ogontz self-contained temperature operated reverse acting control valve utilizes a wax thermal element that is an extremely reliable control. It has a 3° hysteresis that dampens its control action so that it will not hunt for a control set point but modulates to the flow requirement. The valve has a predetermined leak rate so that a small flow is always present thru the valve. This allows a sensing of the upstream temperatures at all times.

A final advantage was on increase in the time period between maintaining or descaling the heat exchanger, which usually occurred every 6 months. This was extended to at least 12 months because the reduced flow left substantially less mineral deposits on the heat exchanger.

Through the use of self-contained thermal actuated control valves you will experience significant reductions in energy consumption, repair costs, labor costs, as well as dramatically improve operations in terms of life extension, efficiency, productivity and product quality. If one is serious about competing effectively in today's global markets, then no one can afford to ignore any opportunity to reduce operating and processing costs . . . particularly in an area as significant as fluid heat transfer. For over 30 years, Ogontz Corporation has built its' reputation on providing the services and products needed to meet that need.

SECTION 3
UPGRADING EMS INSTALLATIONS

Chapter 15

RETROFITTING EXISTING BUILDING CONTROL SYSTEMS WITH COMPUTERIZED EMCS

K. Sinclair

INTRODUCTION

The evolution of Energy Management Control Systems (EMCS), Figure 1, in the last five years has turned into a revolution. Many independent companies, dealing traditionally in computer hardware, have challenged the traditional control companies. New, simpler concepts, have significantly reduced costs and greatly increased simplicity of hardware and software. Many steps in the procedure of designing and installing EMCS have been eliminated. New systems allow greater involvement of operating staff in the control strategies of buildings. Maintenance costs are greatly reduced and black box concepts removed.

New methods of procurement such as the Requests for Proposals for Computerized Energy Management Systems have been developed to keep up with the rapid evolution. New concepts in purchasing are necessary to encourage the industry to grow and evolve. Installation techniques require re-examining as the new systems resemble communication systems rather than conventional control systems. Rapid installation time allows one payment, that is, no progress payments. This concept, combined with the mutually agreed upon liquidated damages clause, insures projects are completed on time. Custom control strategy development has increased paybacks thereby making many systems viable that previously were not.

The modular approach of the stand alone panel (SAP) (Figure 1), allows for a single panel for a small building or several panels to be connected for large buildings. The approach and techniques are the same for all sizes of buildings. Building sizes that are economically viable have ranged from less than 10000 sq ft to more than 1 000 000 sq ft. The retrofit market is particularly lucrative and new micro processor based systems provide attractive paybacks based on energy savings, maintenance savings, and manpower savings.

The correct planning and the new breed of high performance EMCS can be cost effective in the retrofit of existing control systems providing greatly improved energy performance as well as significant improvements in client comfort. In addition, the system can be utilized as a monitoring system to insure achieved savings are maintained. Retraining, reorientation and infact a complete retrofit of operator, owner, consultant, and contractor is often required to provide the correct technical fit.

Figure 1

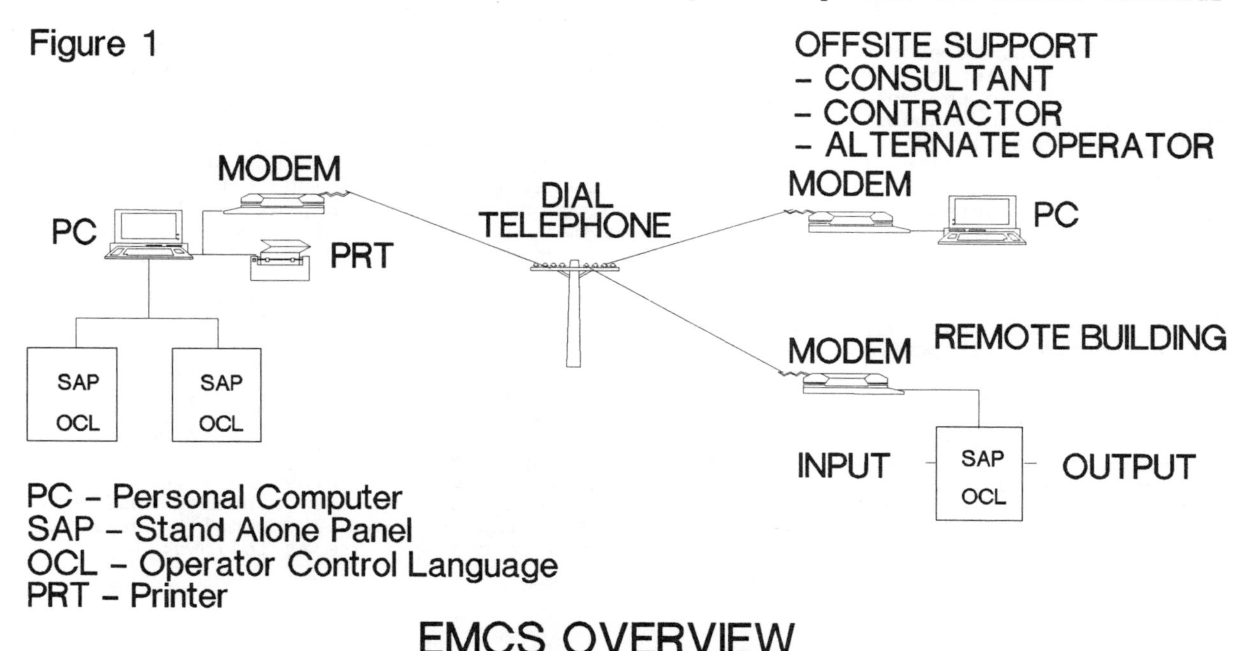

EMCS OVERVIEW

PURPOSE OF PAPER

The purpose of this paper is to make building owners, building operators, consultants and contractors aware of the new concepts utilized in retrofitting conventional and computerized control systems with the new breed of microprocessor orientated stand alone Energy Management Control System (EMCS) panels, Figure 2.

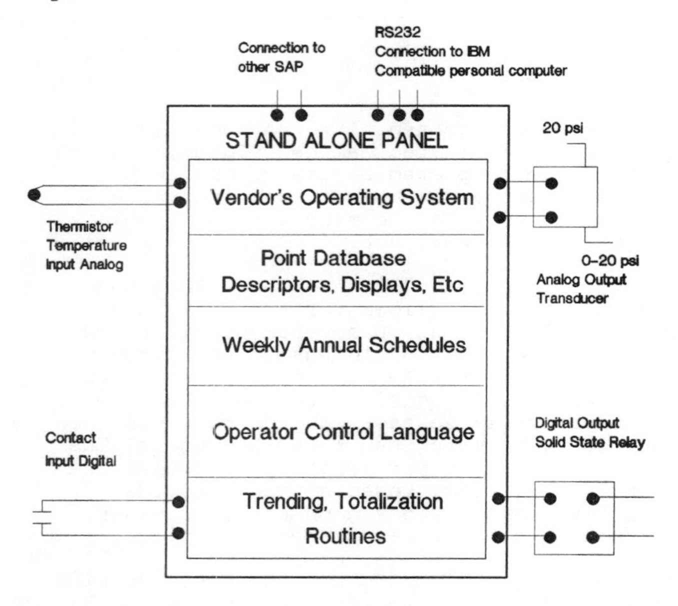

Figure 2

New concepts utilizing customized strategies is the subject of another paper entitled "Application of Dynamic Control in Retrofit of Commercial and Institutional Buildings".

JUSTIFICATION FOR EMCS RETROFIT

Every EMCS retrofit MUST be economically justifiable. Every single point added to the system must contribute to the total simple payback of the capital investment for the system. The payback can be generated by energy savings, operational/labour savings, reduced maintenance savings, etc, but every point connected to the system must be justified. A simple payback of three years is a reasonable cut off point beyond which retrofitting should not be considered.

There is a misconception that computerized direct digital control will provide significant savings. Often the closer control achieved by direct digital control can increase energy consumption by eliminating the offset or the amount of time specified control conditions were not met. If the new breed of EMCS is employed to simply replace existing controls and timeclocks, the savings will be small, if at all. The new EMCS must have changed and improved strategies to insure savings. Dynamic control, a theory of control that involves the building mass in the control equation, is one of these strategies that can provide savings from 10 - 30% over conventional control theories. In every building there are many untapped energy conservation opportunities. Identifying and quantifying these opportunities and writing control strategies in the control language of the EMCS to capitalize on them usually provides a very lucrative payback.

Maintenance savings available by utilizing the new breed of EMCS are significant. Several completed projects have provided a better than three year simple payback, based on maintenance savings, alone. Older control systems, both computerized and conventional, often had maintenance costs of 10 - 15% of the capital cost, or $100 to $150 per point. Analysis of a database of approximately 35 recent projects, ranging from 10 000 to 1 000 000 sq ft, indicates that costs to maintain the new breed of microprocessor orientated stand alone panel range between 3 - 5% of a lower capital cost, or $20 per point. Inspite of this low cost our general recommendations to clients is not to enter into a service contract as the 3 - 5% generally involves no maintenance, only a replacement risk factor. Maintenance on these products is about as exciting as a maintenance contact on a lightbulb. If the total service contract investment is used to purchase a spare panel, sensors, and relays, similar protection can be provided with better control of maintenance costs. Actual replacement of components has been very low with most failures occurring shortly after installation. After a normal stabilization period of about two months replacement of panels or sensors is rare. Normal panel replacement on most of the newer systems takes less than ten minutes and involves no tools. Every effort is made to utilize sensors that have no moving parts and require only software calibration, therefore many components of the conventional systems are eliminated.

In addition to the real maintenance savings, several unquantifiable benefits are derived from features such as allowing setpoint adjustment and control tuning from a central location. Trending of temperature, status, etc, also provides an extremely valuable tool in diagnosing problems. The ability to communicate with buildings remotely, via telephone modem, often provides the necessary payback on small buildings based on labour savings of reduced monitoring time.

NEW METHODS OF EMCS PROCUREMENT, "THE REQUEST FOR PROPOSAL"

To be able to specify an EMCS system, the system must be completely documented and provided with a solid history of performance or in other words, a system must be obsolete by today's rapid changing standards. The Request For Proposal provides a vehicle for purchasing leading edge technology while controlling the risk of a complete failure. The basic concepts are as follows:

1.0 <u>Prepare Request for Proposal "RFP" Document</u>

The document includes the following;

.1 A mandatory requirements document outlining all technical requirements.
.2 A detailed point count indicating exact number of points of each type. Schematics and reduced floor plans showing actual location within each mechanical system, plus the location within the building.
.3 An installation standard. Detail the required installation techniques for the various areas of the buildings.
.4 Description of the techniques that will be used in substantial acceptance and 7 day test of system.
.5 Legal contract documents.
.6 Liquidated damages amount. The liquidated damages is the damage cost per day that will be charged if the project is not completed by the <u>mutually</u> agreed upon date. TIME IS MONEY!

2.0 <u>Advertise Request For Proposal And Set A Preproposal Conference Date</u>

At the preproposal conference the following issues are addressed;

.1 Introduction of client and an explanation of the requirements and the nature of his business; explain how the nature of the business will affect the project.
.2 Field any questions proponents may have regarding the RFP.
.3 Explain in detail any special points that may require future clarification.
.4 Provide a site tour including all areas of the project; discuss installation techniques, etc.
.5 If questions are asked, in which a proponent provides an easier method of achieving the retrofit, or if installation techniques are modified or alternate methods approved, reassemble entire group and explain changed concept. No addenda are issued and all proponents must attend the preproposal conference.

3.0 <u>Closing Request For Proposals</u>

The amount of time allowed between the preproposal conference and the closing date for the RFP is usually two to three weeks. The RFP are accepted before or up until the stated time. No indication of the success or failure of the request is given until an actual contract is awarded.

4.0 <u>Evaluation of Request For Proposals</u>

An electronic spreadsheet that includes each of the mandatory requirements is prepared. Each proponent is rated on how their proposal meets these requirements. The substantial completion date is evaluated by multiplying the liquidated damages per day by the number of days to substantial completion. If a product is proposed that is not familiar to the evaluator, which is comparative, meets the mandatory requirements, and is cost effective, the proponent is requested to demonstrate the product to insure all questions can be answered.

The first time a client is involved with a Request For Proposal, all proposals that are close in price, and in features, should be demonstrated to allow the client insight to the comparative products available for purchase.

THE SUCCESS OF THE
REQUEST FOR PROPOSAL APPROACH

The RFP has been responsible for the creation of several independent control contractors in the Pacific Northwest EMCS market. The added competition has reduced EMCS retrofit costs from over $1000 per point to less than $400 per point. The average purchase cost, based on 35 projects, is $450 per point, and appears to be still dropping. The movement to individual room control has resulted in costs of less than $100 per point.

System functionality has increased several fold and systems have become simpler to operate. Often on a Request For Proposals the difference between the highest and the lowest costs will be more than a factor of two. The lower cost proposal invariably is the newer technology system offering more features for less cost.

Interface of several EMCS products to one central location has been achieved by making MS-DOS, and IBM compatible interface, mandatory requirements. The Request For Proposals has allowed local independent contractors to get a "toe hold" on the market. The result has been rapid growth and development of their products. The concept of no progress payments and liquid damages has brought the majority of projects to substantial completion, as agreed upon, with only a few exceeding the agreed upon date. The 7 day test has provided a clean turn over, assuring the system is complete and fully functional. The concept of having the Contractor write the EMCS software strategies so that the system responds as the conventional system did, simplifies acceptance and makes it easier to commission the complete system. When the EMCS system has passed the 7 day test more complex strategies can be provided. Operating the system as originally also removes the necessity to quickly implement new strategies. Dynamic feed forward strategies can be introduced with the assurance that all connected points are functioning properly.

EMCS RETROFIT INSTALLATION TECHNIQUES

The installation of the new breed of computerized EMCS is comparable to the installation of a communication system, rather than a conventional control system. The new distributed control system uses communication signals between field sensors and panels, as well as communication signals between panels and personal computers. The system is therefore a communication system, as well as a control system. The only areas that installation techniques should emulate traditional control technology is the point where transducers interface with activating devices and where electronic solid state relays interface with motor starters.

Opportunities often exist to utilize existing conventional control installation equipment to lower the cost of retrofitting. New electronic boards can be mounted in existing control panels; transducers can be mounted in existing control enclosures; new electronic sensors can be mounted in the same location as existing sensors utilizing existing conduit; solid state control relays can be mounted in existing terminal sections of the MCC. Utilizing these techniques significantly reduces installation costs.

Fire approved cable utilized in return air ceiling plenums significantly reduces capital costs. Conduit is normally used only in areas where mechanical damage of wiring is probable. The use of multi-conductor, or cable similar to that used by telephone companies, reduces the cost of connecting a small cluster of points to a panel several hundred feet away.

A mandatory requirement is a pressure gauge on each pneumatic transducer to quickly identify if a problem is EMCS originated or conventional hardware. Wire tagging and device tagging are important mandatory requirement used to insure quick identification of points. The Request for Proposal often calls for an airline baggage type tag in which at least the following information is put;

point descriptor name, code name, panel number that point is connected to, manufacturer's name.

THE "7 DAY TEST" AND SUBSTANTIAL ACCEPTANCE

The request to begin the "7 Day Test" is a significant milestone in the contract for both the contractor and the client. When the contractor has met the all mandatory requirements, completed point installation, setup all parameters required for the start of the test as outlined in the RFP, including the presentation of as built drawings, the test begins.

Performance of the system is monitored by a hard copy printer and personal computer on site, or remotely by modem if no on site PC is provided. All features of the system are exercised and any abnormalities noted. Check out commissioning sheets of each point, provided by the contractor, are reviewed and spot checks made to insure system integrity. An upset of the system than cannot be corrected within two hours results in the test being aborted.

When the system passes the 7 day test without upset, with all features successfully demonstrated, and all installation and electrical code requirements met, substantial completion is granted. Deficiencies such as missing sensors or features are financially assessed, and this amount held back from the agreed contract price. Total payment, less deficiencies, less Lien Holdback, is approved and the liquid damage clause is stopped. It is important that a valid claim for payment be swiftly and professionally processed.

This concept works well for retrofit and can be applied to new construction if significant modifications are made to the traditional contract relationship.

RETROFITTING THE BUILDING OPERATOR/OWNER/AND SUPPORT PEOPLE

The new breed of high performance energy management system have features and operation significantly different than conventional control systems. Updating and upgrading of control systems requires the updating and upgrading of all personnel involved with the project. The operators must be made aware of the capabilities of this new tool. The training of the building operator to read and understand the control strategies developed within the operator control language of the system is mandatory. Control languages using basic like structures utilize IF THEN logic which simplifies the training process. Most operators have mastered more complex concepts than this in the regular performance of their job. The complete operational control story that exists in the operator control language makes logic easier to comprehend than control schematics, wiring diagrams, etc. Flowcharting the strategies is not required with the new English language orientated EMCS. Flowcharting is only necessary for languages that are not English orientated and cannot easily be followed by non-programmers. Every effort should be made to modulize the coding and to keep it simple. Avoid having programs written by traditional software programmers. Control language coding should be written for and by the building operator with extra code and comments added to increase readability. The prime focus should be on readability, and functionality not programming techniques. Memory is very inexpensive in comparison to manpower.

Programming techniques are discussed in another paper I have prepared entitled "Application of Dynamic Control in Retrofit of Commercial and Institutional Buildings".

Retraining, recalibrating, reorientating, reeducating, or in other words retrofitting of people other than the operator is also necessary. Contractors are often extremely proficient at installation but do not realize the total potential of these systems.

Examples of this are utilizing remote communication capabilities for off site start up, commissioning, new methods of database building, new documentation techniques utilizing interactive database, etc.

Owners are presented with the ability of operating buildings from remote sites at no additional cost. This allows the complete operating structure of a company with several buildings to be greatly modified with little or no additional cost. Building operation can easily be contracted out, if desirable, utilizing remote access.

Control strategies can be written to constantly monitor the energy systems to insure that the intent of all logic is met. Utilization of trending of connected points is important in the evaluation of strategy performance and is another feature that is often included at no cost.

People associated with these new breed systems must realize that the scope of this new tool is only limited by their imagination.

INSURING SUCCESS OF RETROFIT

This is the most important step in the complete retrofit. The prognosis of the justification must be met or justifiable variances explained. The greatest tool to do this is the EMCS itself. Utilizing trend logs, run time totalizers, energy metering, and close monitoring the success or failure of concepts can be done by anyone understanding the system. Contractor, consultant, operator, or owner cannot hide under the electronic EMCS microscope.

Inherent problems with the building are often identified with an EMCS. A tendency to want to "shoot the messenger" is common. It is necessary to understand what information the system is relaying and to utilize the diagnostic capabilities to analyze the problem. The EMCS also becomes an excellent tool for priorizing problems; it keeps reminding you until the problems are corrected.

Insuring the success of the system requires a long term involvement by an individual to provide the concepting, justification, preparation of mandatory requirements, technical acceptance, and development of customized dynamic strategies. This responsibility is often divided among several people, resulting in no one being accountable. For success from concept to successful implementation, one person must be accountable. Obviously this person's job is to make everyone else accountable for their portion of the complete project.

MAINTAINING SUCCESS

After aggressive energy operation and improved comfort conditions are achieved, a significant challenge exists to maintain this performance. The new tool again becomes the most effective method of maintaining the achieved goals.

Generation of daily energy performance graphs quickly identify upsets that can be corrected before there are significant effects on energy consumption. The generation of a temperature comfort factor can effectively evaluate client thermal comfort. On line temperature trend data allows generation of several days of accurate temperature data to identify our success or failure. Analysis of software run time totalizer allows further identification of areas where there is deviation.

One person must be made accountable for the task of monitoring comfort and energy efficiency for success to be maintained. Performance should be judged by comfort factor and energy consumption roll ups.

SUMMARY

A summary of the concepts outlined in this paper are:

1. Justification for the EMCS is the backbone of every successful project.
2. Utilization of non traditional control strategies often provides the required payback to make the necessary justification.
3. New methods of purchasing such as the Request for Proposal "RFP" stimulate the market and increase competition lowering the cost of projects while greatly increasing functionality.
4. New construction techniques are required as the new systems resemble communication systems rather than conventional control systems.
5. The new systems are so radically different that the retrofit of all the people involved in the project must occur. New opportunities are presented.
6. English language orientated operator control languages greatly simplify documentation and project communication.
7. When project is completed all justifications must be reviewed to insure that all the objectives were met.
8. Once success with the new system is achieved a plan must be set out to maintain this success.

Times are quickly changing and computer technology is far ahead of the ways our industry is applying it. We must strive to understand more of the computer's capabilities to capture the lucrative opportunities that are presently available in the energy conservation industry.

Chapter 16

HOW TO EXPAND/UPGRADE AN EMCS

E. Ling, D. Heinz

By this time, most organizations have an energy management and control system (EMCS) installed in their physical plant. The question now is, should it be expanded and/or upgraded? And, if the system needs to be expanded or upgraded, what is the best method of contracting? Three contracting methods will then be examined.

EXPANSIONS

There are basically two reasons to expand an EMCS. One is a planned expansion due to a previous decision to not connect all existing physical plant to the system on its initial installation. The second reason for an expansion is an increase in the physical plant, such as new buildings being constructed.

If the initial installation of EMCS did not encompass all of the equipment which needed to be connected, the need to expand is clear. Likewise, if new facilities requiring connection to EMCS have been added to the inventory, the need to expand the system is clear.

CONTRACTING METHODS

The method of contracting for the expansion is the next question to be faced. Essentially three methods are potentially available: Sealed Bidding (formerly Invitation For Bids (IFB)), Competitive Negotiations (also known as Request For Proposal (RFP)), or sole source negotiation with the original EMCS vendor. For any government procurement, the correct choice is vital to a successful purchase. It is extremely important that the technical characteristics and requirements of the proposed EMCS expansion be clearly defined by the user, as this becomes a key factor in the government's determination. Each of the three contracting methods is uniquely designed to best meet the user's needs, dependent upon the required EMCS configuration. While the final decision is the Contracting Officer's (CO), the user must provide the information required for the CO to make an informed decision.

The use of sealed bidding is the preferred method, as dictated by law. However, this method of contracting for EMCS work is fraught with potential disaster. It requires a well-defined statement of work (SOW) and specifications. Not only must the EMCS requirement be so clearly stated as to minimize potential mistakes in interpretation by potential bidders, but the responses to the IFB must be subject to a fairly simple determination of "responsiveness" to the government's stated requirements. Simply specifying that the new facility must be connected to and function with the existing system is no assurance that it will occur. If the requirements are not well defined in the IFB, it becomes difficult to justify not awarding to the low bidder on the basis of "non-responsiveness." Further, using sealed bidding, it is difficult to refuse award to any bidder based on lack of past experience on similar work. After award, if a review of submittals leads to the conclusion that the system probably won't work, it would be very difficult to terminate the contract for default based, essentially, on opinion. Unless otherwise delinquent, the contractor will normally draw progress payments as the construction and installation proceeds. The user and Contracting Officer will have no way of knowing if the system being added will actually work until all construction is completed and data base generated. At that time, the contractor is likely to have been paid at least 90% of the total contract amount. If he cannot make it work, in accordance with the specifications, it is highly unlikely that the Government will recover any amounts paid if the contractor was making a good faith effort to comply with the terms of the contract. Under construction warranty provisions of the contract, the government can enforce its right to require the contractor to "remedy" the defects (non-working system), or failing this, to have another EMCS contractor make the system work and charge any additional cost to the original contractor. This is a very cumbersome and time-consuming process which, in the case of smaller companies, will most likely not succeed, at least as far as meeting the user's needs in a reasonably timely manner. As a result, additional government funds will normally have to be obtained and a fresh effort made to obtain a workable system. This has been my personal experience at least twice. The root cause of this problem is that using the sealed bidding method does not allow the user or the contracting official to review how the contractor proposes to accomplish the add-on; i.e., the award is based solely on the lowest bid found "responsive" to the IFB. Thus, the adequacy of the stated EMCS requirements becomes crucial, since sealed bidding does not allow discussions between the government and bidders to ensure vague, or complex, EMCS operational characteristics are well understood.

A better method to use is the competitive negotiation technique. This method is proper when there is more than one possible source for the work, but there are different technical approaches to achieve the desired work which need to be evaluated. This approach requires each offeror to furnish as much detail as is specified, to determine what equipment he intends to use and how he expects to perform the required connection. In addition, qualifications of the contractor can be specified, such as past experience in performing similar work. These qualifications can be as detailed as necessary to assure a satisfactory contract completion. Discussions may be held with offerors and complete verification of qualifications can be made prior to contract award. In addition, when using the competitive negotiation approach, one is not tied to the lowest price. A subjective decision can be made with respect to what is the best buy for the money, based on technical or other criteria.

The third method of procurement is that of sole source negotiation with the sole supplier of the required equipment for connection to the existing EMCS. Obviously, this is appropriate when there is only one possible source of equipment. Using this method is more difficult to justify in government contracting due to recent legislation passed by Congress, i.e., the Competition in Contracting Act (CICA). CICA mandates "full and open competition" in all government acquisitions unless certain strictly defined exemptions can be met, one of which is "only one responsible source." In the case of all newer EMCSs, it is easier to justify this type of contracting because each EMCS company retains all proprietary rights to their software and hardware. Consequently, in spite of the Corps of Engineers and others' efforts to achieve nonproprietarily expandable systems, it has not happened. Present efforts by the (ASHRAE) Committee on Standardized Protocol for EMCS may eventually result in such nonproprietarily expandable systems, but it is not expected within the next two to three years.

UPGRADES

Next, let's look at upgrades of EMCSs. Why should you consider upgrading your existing system? Certainly not just because it is not the current state-of-the-art. There are four possible economic reasons for upgrading; reduced maintenance costs, reduced expansion costs, improved operation and improved documentation of savings.

An upgrade could reduce maintenance costs in either of two ways. If the equipment has an unreasonable failure rate, it could be more economical to upgrade or even replace the entire system than to continue to maintain the unreliable equipment. An example I encountered on one of our bases where an upgrade or replacement would have made good sense was where the field equipment had an 85% failure rate per year.

Another time an upgrade might be economically justified is when the repair parts costs have increased drastically due to obsolescence of the equipment. In this case, newer equipment in the central EMCS (upgrade of system) might allow for older field equipment to be repaired by replacement with state-of-the-art equipment which is likely to be much cheaper, more efficient and more powerful than older generation equipment.

A reduction in costs of expansion can many times justify an upgrade of central station EMCS equipment. For instance, the costs of adding several buildings to an existing EMCS using outdated technology may be more than the cost o upgrading the central EMCS and adding the same buildings using current state-of-the-art equipment due to the reduced cost of the newer electronic technology. This is occurring at one of our bases now.

Improvement in operations of the EMCS might justify an upgrade of the central EMCS. Most older EMCSs are totally centralized systems and most do not perform many optimization functions nor allow much data analysis. Only a detailed analysis of the functionality of your existing system versus that provided by a potential upgrade can determine if it is cost effective to upgrade.

Another possible justification for an upgrade of the central EMCS could be improved documentation of savings attributable to EMCS. Most older systems required extensive manual inputs and compilation of data to document how much energy or dollars the EMCS was saving. Many of the newer systems can perform these tasks automatically. Thus the cost of obtaining this data could be reduced with some newer systems, making economic sense. In some cases, the need for this data might justify an upgrade, such as proof of savings to justify expanding a system to result in more savings.

If the determination is made that an upgrade is justified, or perhaps even in the process of deciding if an upgrade might be appropriate, one must consider what options are available. There may be only one possible upgrade system available. In other cases, there may be several options from which to choose. In general, I recommend commercially available upgrades versus versions that may be developed for a specific client, such as the Department of Defense. All major manufacturers constantly offer interfaces and/or upgrades from previous commercial systems to their newest equipment. Thus, one is never faced with a decision or need to totally replace an entire system due to obsolescence.

ACQUISITION OF THE UPGRADE

If an upgrade is determined appropriate, one is then faced with the same three potential contracting methods discussed for expansion of EMCSs. There is much less generalization possible in the area of upgrades. Each situation must be carefully analyzed and decisions reached based on that analysis. Generally, our experience has shown that sealed bidding is not satisfactory due to the difficulty of developing a well-defined EMCS specification that lends itself to easy interpretation, and contractual implementation, without pre-award technical discussions and evaluations. Lacking such well-defined specifications, the user must assure himself that the acquisition has a reasonable chance of being a success by using other than the sealed bidding method. In accordance with the CICA, he must be willing to document to the Contracting Officer why use of competitive negotiations would be the least restrictive procedure. In many cases, a thorough investigation of all possible options may lead to a decision to negotiate a sole source contract for upgrading.

A large physically diversified organization, such as Air Training Command, can benefit from a centralized Justification and Approval (J&A)

document, required by CICA, to support competitively negotiated and sole source acquisitions of EMCS add-ons. Since the similarities and differences of various systems are more apparent at the command level, we in engineering, in conjunction with contracting personnel, undertook the task of making those determinations for each system in the Command in accordance with the Federal Acquisition Regulation (FAR) 6.303-1(a) and (c). As a result of our efforts, model justification documents were developed for the use of both competitive and sole source negotiations by our various field activities.

The sole source J&A was made on a class basis for each type system for which it was appropriate. It should be noted that it was not appropriate for all type systems. Thus the class system (manufacturer's model) was defined to comply with FAR 6.303-2(a)(3).

The description of supplies required (FAR 6.303-2(a)(3)) was defined as:

> "acquire additions to the currently installed (manufacturer's model number) system on a sole source basis through the base level contracting office at the specific base in dollar thresholds not to exceed $100,000 per base per fiscal year for 1988 and 1989. {Note: This sample J&A is for a not to exceed amount of $100,000 which can be approved at the base level. Keep in mind that J&As over $100,000 must be approved by the Procuring Activity Competition Advocate (HQ ATC/LG)}".

The statutory authority permitting other than full and open competition was defined as "10 USC 2304(c)(1), only one responsible source."

The following is our explanation of the nature of the acquisition requiring the use of sole source authority.

> "Demonstration of Statutory Validity: EMCSs are computer driven systems consisting of many individual sensors and controls installed in buildings on a base to control and manage energy consumption in heating, ventilating, and air conditioning systems (HVAC) as well as utilities systems such as electric distribution systems, water pumping, storage and distribution systems, and sewage systems. These sensors and controls are such devices as thermostats, relay switches and two input controllers. These sensors and controls are connected to multiplexor panels (MUXs) and then to Field Interface Devices (FIDs), which are actually microcomputers. Numerous FIDs are then connected via data links (dedicated cable and fiber optics) to the master control room (MCR) containing two minicomputers.
>
> The communication protocol between the individual components (MUXs-FIDs, and FIDs-MCR) is unique and proprietary to each manufacturer; therefore, one manufacturer's equipment will not interface with another manufacturer's equipment. In addition, the associated communication protocol and wiring diagrams are also proprietary information. On a small add-on to the Manufacturer A's Model X system, the add-on must use Manufacturer A's equipment (MUXs-FIDs, and FIDs-MCR) in order for the equipment to function. The minimum requirement is for a functioning integrated system which can be provided only by Manufacturer A. Consequently, Manufacturer A is the only known source of expansion capability."

FAR 6.303-2(a)(6) requires:

> "A description of efforts made to ensure that offers are solicited from as many potential sources as is practicable, including whether a CBD notice was or will be publicized as required by Subpart 5.2 and, if not, which exception under 5.202 applies."

In response to this requirement, the following information was provided.

> "A synopsis was accomplished for the ability to purchase EMCS equipment under Basic Ordering Agreements on" a recent date. A list of companies initially responding to the synopsis was included. Some of the companies responding to the synopsis could provide only non- proprietary equipment such as sensors and controls. "Only the original manufacturers are capable of tying in compatible equipment in the proprietary arena, which comprises 30% of the cost of all EMCS acquisitions." This J&A is required for the purchase and installation of the proprietary equipment.

The following statement was made in regard to FAR 6.303-2 (a) (7) and 6.303-2 (a) (8):

> "To ensure that the government receives a fair and reasonable price, the proprietary equipment to be purchased has established catalog price lists."

FAR 6.303-2(a)(9) asks for the following:

> "Any other facts supporting the use of other than full and open competition, such as:
> (i) Explanation of why technical data packages, specifications, engineering descriptions, statements of work, or purchase descriptions suitable for full and open competition have not been developed or are not available.
> (ii) When 6.302-1 is cited for follow-on acquisitions as described in 6.302-1(a)(2)(ii), an estimate of the cost to the Government that would be duplicated and how the estimate was derived.
> (iii) When 6.302-2 is cited, data, estimated cost, or other rationale as the extent and nature of the harm to the Government.

In response, the following statement was provided:

> "Through numerous contacts with" specific Air Force Headquarters personnel, "annual trade shows such as the Association of Energy Engineers, and the American Society of Heating, Refrigerating, and Air Conditioning Engineers (ASHRAE), it is clear that the market place does not provide the services of interfacing between various manufacturer's components. Consequently, a specification for that interfacing has not been developed. In order to have open competition, the entire existing system (FIDs, MUXs, and MCR equipment) would have to be replaced at a cost of" $....

In response to FAR 6.303-2(a)(10) which asks for a list of offerors interested in the acquisition, the list of names of those who answered the synopsis were provided.

Finally, FAR 6.303-2(a)(11) asks for any actions the agency may take to remove or overcome any barriers to competition before any subsequent acquisition for the supplies or services required. We simply stated that there are no cost effective methods to overcome the barriers to competition on these purchases, as indicated in the response to 6.303-2(a)(9) above.

This constitutes the main points required to justify sole source procurement. It obviously required a great deal of work by both engineering and contracting personnel. However, it required much less manhours overall to look at all the various types of systems at one time than it would have for each base to have had to analyze their own system individually. Close cooperation between contracting and engineering personnel smoothed the road a great deal. In spite of that, it still took almost ten months to achieve the finished product.

In summary, when potential expansions or upgrades are in the offing, all aspects of the systems' operations costs and expansion/ upgrade costs should be analyzed looking at life cycle costs before proceeding. The possibilities should be examined periodically even if no obvious change is expected, since one of the possible advantages of an upgrade or replacement, as discussed earlier, is reduced operational (maintenance) costs from replacement of trouble prone equipment. While several contracting methods are available to the government in acquiring EMCS expansions and upgrades, users must perform a thorough analysis of the complexity of their requirement and provide adequate documentation to justify competitive negotiations, or sole source negotiations, when warranted.

Chapter 17

PERFORMANCE-BASED EMS SPECIFICATION

A. J. Kim

As we conclude the second decade of Intelligent Building Control, the industry still suffers from far too many failures. Despite the fact that control hardware is getting more reliable, software more powerful, and systems more usable; a high percentage of users are still less than satisfied a year after their system is started up. In this paper we shall analyze what the problems are, why it continues, and how we can change things to increase the odds for a successful installation.

THE PROBLEM

Over the years more emphasis has been placed on the product itself rather than on the application and support of that product. This has resulted in the expectation that by merely having the product your problems are solved. This does not happen. For a system to provide benefit, it must be applied in such a manner that it will perform the desired functions. Coupled with this, inadequate training and support often result in a system being reduced to its operator's level of understanding (often overridden or off).

Having seen numerous such unsuccessful projects throughout the industry, we did some probing to try to find out what the problems were. Our information concluded that in general the complaints were:

- Control sequences improper
- System too complicated
- Reliability problems
- System performance inadequate.

Other problems were indicated but these represented the majority of the complaints.

WHY IT HAPPENED

To be able to conclude why these problems occurred, we questioned numerous users as well as those who were responsible for the purchasing decision initially. While we received a wide range of opinions, when we boiled it all down we arrived at some common elements:

- Improper control strategies
- Insufficient user training
- Lack of competent support
- Inadequate system documentation.

Each of these items can be linked directly to the previously implied problems. While some specific projects had unique circumstances, the above seemed to be prevalent throughout.

HOW TO IMPROVE THE ODDS

The key to getting what you need in a Building Automation System is to properly analyze the building's needs, **accurately represent this in a performance based specification,** select a vendor that is capable of doing the job, and then require thorough acceptance testing.

Books have been written on the first item regarding energy audits and retrofit design. The second and last items regarding specification and testing shall be addressed in detail in the remainder of this paper. The third item relates to the selection process of a qualified vendor and this has also been addressed in numerous other materials.

PROPER SPECIFICATION

Far too many specifications today are being written for the benefit of the vendor rather than the user. Details are specified that serve no purpose other than to attempt to put the vendor that meets these details in an advantageous position. The end user is used rather than served in this situation. Further, attention is then given to insignificant details rather than meaningful performance.

An example of this is the paragraph that follows:

"The temperature monitoring inputs shall employ a 12-bit pulsed digital signal operating at a frequency of 2300 Hz, at 5 VDC level transmitted from the sensing/transmitting device. This signal shall be superimposed on a positive value reference voltage and shall further insure system integrity by use of an XYZ redundant error correction scheme."

This ridiculous detail serves no functional purpose and would be better replaced by:

"The temperature monitoring inputs shall be precision analogs maintaining a resolution of .1 degree F. Accuracy shall be maintained within .5 degree F through entire operating range."

This specifies what the end result should be, not how to get there.

A few basic steps need to be taken to initiate a turnaround in performance based specification:

- Educate the specifier
- Establish overall objectives
- Define sequence of control
- Get vendors out of specification writing role
- Follow through on acceptance testing.

Educating the specifier is beginning to be accomplished. Unfortunately, this is still not as widespread as it should be and seldom does a specifier have as good of a command of what the Building Automation System should be as he does of the HVAC System. The lack of competence in this industry has left the door open for vendors for years. More often than not a temperature control manufacturer will be allowed to provide the specification since "he's the expert". This normally results in the specification being written for the sole purpose of giving himself an advantageous position and total disregard for the guy who's paying the bill. It is the specifier's responsibility today to upgrade his skills in this regard, so that he may better serve his client.

Establishing the overall objectives in the specification is almost never done. Consequently the owner's objective may have been to improve energy efficiency and the vendor may feel he's providing improved temperature control. Another owner's objective may be to minimize tenant complaints through early warning functions and provide tenant accountability for usage. If the vendor feels his function is to maximize energy efficiency, he's in for trouble. These objectives can easily be stated in the introduction body of the specification.

Defining the sequence of control is probably the most critical and most crucial step in the entire specification process. While we all know that that sequence is not going to be perfect the first time and will require some fine tuning, it is imperative that this be conveyed. A sequence of control is best accomplished by taking each control point and defining <u>all</u> of the items that impact its operation.

Lastly, all of this is still for naught unless compliance is assured by thorough acceptance testing. All too often the owner nor specifier understands the system well enough to assure its compliance to the specification. Acceptance testing need be no more than just saying "show-me" on each one of the installed points or functions.

TRAINING AND ONGOING SUPPORT

In your specification you must also specify the items that live on well after acceptance of the system. These items are essential and if omitted will often allow a system to quickly revert to its previous mode of operation.

Some items to keep in mind are:

- Untrained operators <u>will</u> minimize results
- An operator <u>will</u> reduce a system to his level of understanding
- Fine-tuning is imperative for the first year and should be included in base.

We have found that there are some things with regard to training that work. First of all an operator who feels ownership will almost insure success. Also the use of video tapes in the training process are an excellent tool. This allows for "refresher training" at your convenience and also works well for training new employees.

These are the basics for writing and enforcing a performance based specification. Specify <u>what</u> is to be done and leave the <u>how</u> to the individual vendor. The following is a sample specification with all of the elements we discussed.

SAMPLE "PERFORMANCE BASED" ENERGY MONITORING AND CONTROL SYSTEM SPECIFICATION

1. <u>GENERAL</u>

1.1 The energy monitoring and control system (EMCS) as herein specified shall be provided in its entirety with all system components, engineering design, and system installation by the EMCS contractor.

All work under this section shall be done in strict accordance with laws and ordinances which may be applicable.

All equipment, apparatus and systems shall be fabricated and installed in complete accordance with the latest edition or revisions of the following applicable regulations, standards and codes.

State and Local Building Codes
National Electrical Code
Local Electrical Installation Codes
Local Power Company Regulations and Standards
Governing Telephone Company Regulations and Requirements
OSHA - Occupational Safety and Health Act

1.2 <u>Quality Assurance</u>

1.2.1 <u>Equipment</u>
All EMCS equipment installed herein shall be manufactured by firms regularly and primarily engaged in such manufacture, and whose equipment/devices shall have been in satisfactory use in similar service for not less than five years. Said equipment shall be U.L. listed.

1.2.2 <u>Application Engineering</u>
The EMCS application engineering shall be accomplished by a graduate engineer regularly and primarily employed in application and design of similar systems. Upon request, the EMCS contractor shall submit resumes of the engineer(s) and technician(s) involved with the design, installation, and support of the automation system. Said personnel shall be in the full-time employment of the EMCS contractor.

1.2.3 <u>Installation</u>
Installation of all EMCS components shall be performed under the personal supervision of the EMCS contractor. The application engineer specified in Section 1.2.2 above shall certify all work as proper and complete and shall reflect actual installation on "as-built" drawings provided under later specification. The EMCS contractor shall be responsible for the design, scheduling, coordination, and warranty of all work required under their specification. Under no circumstances shall the responsibility for any of these tasks be delegated to another contractor. All actual installation shall be accomplished by skilled tradesmen regularly and primarily engaged in the installation of EMCS systems.

1.2.4 <u>Owner's Objective</u>
It is the primary intent of this project to reduce facility energy consumption without compromising comfort conditions or equipment integrity. Secondary desire is to have an "early warning system" to alert operations people to problems.

2. SYSTEM STRUCTURE

2.1 The EMCS shall be of modular design consisting of stand alone Field Processing Units (FPU), and a Personal Computer based Host processor. FPU's shall also be accessible for data acquisition, programming changes, software override control, alarm reporting, trend/status log reporting, and program downloading remote from the facility via a 1200-baud modem and "dial-up" voice-grade telephone line. Purchaser shall be responsible for providing and maintaining this line through the warranty period. A block diagram of the system structure is shown in Figure 1 at the end of this specification.

2.2 Expansion of system shall be modular via add on I/O boards to accommodate additional monitor/control points, peripherals, or memory. Each FPU shall monitor/control no more than 120 points and additional FPU panels shall be linked by a network to allow for a minimum expansion of up to 1,000 points.

2.3 Network communications software shall provide for panel to panel exchange of information so that variables from one panel can automatically be used in another panel's program.

3. FIELD PROCESSING UNITS (FPU)

3.1 General
The system shall be at a minimum an 8-bit microprocessor-based system operating at a minimum speed of 8 mhz with full interrupt capabilities. System speed and math processing capabilities shall be sufficient to provide full direct digital control using Proportional Integral Derivative (PID) software routines, and multiple simultaneous control processes while not degrading system response time.

The FPU shall utilize a real-time operating system providing continuous communication with the I/O, supervision and polling of control program operation, and monitoring and control of peripheral devices. A multi-tasking control program shall be provided that runs under this operating system to meet the requirements listed in Sections 1706, 1707, 1708, and 1709.

3.2 Memory
The FPU shall contain a minimum of 64K of total Random Access Memory (RAM) with 32K minimum dedicated to the control program and 32K minimum for log data. This memory shall be high reliability static RAM with battery back-up. RAM memory shall be expandable to accomplish larger amounts of I/O.

In addition, Erasable Programmable Read Only Memory (EPROM) shall be provided for the firmware containing diagnostics, and basic I/O instructions.

3.3 Battery Back-Up
In the event of a power interruption, memory and time shall be maintained for a minimum of 96 hours. Upon power restoration within this time frame, reprogramming or updating shall not be necessary to return system to normal operation. System shall automatically restore itself to proper operation with no operator intervention. Restart software routines shall be provided in the control program to allow a sequenced equipment restart. Any power fail or system reset shall be detectable by the FPU and provisions shall be made to allow this information to be logged or initiate any desired sequence.

3.4 FPU Program Loading
Software entry shall be through a P.C. from a floppy disk either through the local or remote terminal port. Subsequent to program "downloading" the program variables can be updated or overridden from any of the operator interface devices described in 1708.

3.5 Diagnostics
The FPU shall contain complete diagnostic functions. When the control program is functioning, the FPU shall display a message on an Operator Interface Device when program errors occur. Additional diagnostics shall test memory, FPU, real time clock, analog and digital inputs and outputs, local keyboard and display, and display the results on an Operator Interface Device.

3.6 Communication
The FPU shall contain a minimum of two (2) communication ports for connection of peripheral equipment. An RS232 serial communication port capable of baud rates of 300 - 19,200 baud shall be available for either a local terminal or connection to a P.C. based Host system. An RS485 network communication port shall be provided for linking multiple FPU panels together on a common network and be able to share global information between panels. Baud rate for this port shall be a minimum of 1200 baud with selectable baud rates up to 9600 baud.

3.7 Environment
The system shall be capable of operating continuously in a 40 degrees - 120 degrees F environment with 20-80% RH non-condensing. An air-conditioned environment shall not be required for normal operation.

4. INPUT/OUTPUT BOARDS (I/O)

4.1 Input Capabilities
The I/O boards shall accept and interpret both digital and analog input signals. Digital inputs shall monitor dry contact status. Analog inputs shall accept signals from: 0-10VDC inputs, 0-20MA current inputs, or variable resistance. Digital inputs shall have on-board indicating lights for on/off status of each point and shall provide optical isolation on each input point. Permanent (non-volatile) memory shall store all programs necessary to convert values of analog input signals to actual engineering units for various manufacturers' sensors. This input hardware shall not be limited to sensors or other field hardware by a single manufacturer but shall be capable of interfacing to at least one other manufacturer's field devices. This shall provide the end user with at a minimum a second source for all field devices utilized.

4.1.1 Base system shall provide a minimum of 32 analog inputs and 24 digital inputs.

4.1.2 Each FPU panel shall be expandable to a minimum of 48 analog input and/or 36 digital inputs.

4.2 Output Capabilities
The I/O boards shall utilize both digital and

analog output points. The digital output points shall consist of an on-board SPDT output relay with minimum rating of 120VAC, 10A, 1/3 HP. These output points shall be provided with local hand/off/auto switches for manual override. Digital outputs shall have onboard indicating lights showing on/off status of each point. The status of local hand/off/auto switches shall be continually polled by the control program to determine if they are in the "auto" position or not. The control program can then generate an alarm condition if manually overridden. Analog output signals shall be over a range of 0-10VDC at 0-20MA.

4.2.1 Base system shall provide a minimum of 24 digital output points and a minimum of 4 analog output points.

4.2.2 Each FPU panel shall be expandable to a minimum of 48 digital outputs, and/or 32 analog outputs.

4.3 Power Supplies
Power supplies shall be included as part of the FPU panel to provide all necessary DC voltage levels to operate all EMCS components within the panel.

4.4 Enclosure
System shall be mounted in a locking metal cabinet with accessibility to optional local keyboard/display panel when cabinet is locked.

5. OPERATOR INTERFACE DEVICES

5.1 Local Interface
Local Operator Interface shall be provided and consist of a conventional full ASCII keyboard. It shall also provide a minimum of an 80-character alpha-numeric display. This device shall be mounted in the immediate vicinity of the CPU, and shall have the following keyboard commands in communicating with the CPU:

1. Operator access via multi-level access codes.
2. Start control program execution.
3. Stop control program execution.
4. View variable values in control program.
5. Display log data.
6. Enter new trend log.
7. Update an assigned variable (without editing program).
8. Display/edit present time.
9. Display/edit present date.
10. Execute hardware diagnostics.

5.2 Remote Interface
Remote terminal capabilities shall also be provided to the EMCS for support from the EMCS contractor or any other remote terminal. This communication shall be via voice-grade phone line through provision of a 1200 baud modem to be included in this proposal. Remote terminal shall be capable of the same keyboard commands as listed above for the local operator interface.

5.3 Graphics Interface Device
The optional Graphics Interface Device shall consist of a dedicated 16-bit IBM-PC-XT/AT compatible microcomputer with a minimum of 640K core memory, a minimum of 720K bytes of disk storage, high resolution color CRT, keyboard, and a minimum of 120CPS printer with full graphics capability. The following functions shall be provided by Graphics Interface Device:

5.3.1 Direct access to the FPU's to allow the same keyboard commands listed above in Section 5.1.

5.3.2 User-accessible Color Graphic Display Generation using a simplified menu-driven interface and a standard library of shapes and symbols.

Graphic CAD function shall be provided with a minimum of 16 colors, and shall also have provisions for the user to develop new shapes and add them to the standard shape library.

Any system variable shall be able to be displayed at any location on any screen, and alarm limits assigned to the variable.

5.3.3 Color Graphic Floor Plan Displays shall be provided with exterior walls, interior partitions, doorways, elevators, and temperature sensors shown in their actual location. The current value of temperature shall be indicated at the sensor locations with a keyboard command option provided to update the value without refreshing the background.

5.3.4 Graphic System Schematics of each air handling, chiller, and boiler system shall be provided, with current temperature values and status indicated in the ductwork or piping in their relative location to the other system components such as fans, dampers, coils, pumps.

5.3.5 A minimum of ten (10) Status Page Displays of preselected variables from the control program shall be user-definable for recall at any time through keyboard command. Engineering units shall also be provided for each variable.

5.3.6 Variable Update Displays shall be provided to facilitate user modification of changeable parameters such as time of day, date, cooling and heating setpoints, ambient temperature lockouts, schedule times, holidays, etc. This display shall be in tabular "spreadsheet" form for ease of operation.

5.3.7 Historical logs resident in the FPU shall be accessible through the Graphics Interface Device as a keyboard command.

5.3.8 Log graphing capability shall be provided. User selected variables shall be displayed in various user selected graph configurations (i.e., line graph, bar graph or combination of the two). The system shall have the capability of printing these displays through a printer, Plotter, or similar device.

5.3.9 Graphics Interface Device shall be capable of foreground/background operation. This will allow use of the device for non-EMCS functions without interrupting the integrity of EMCS functions such as alarm reporting and trend logging.

5.3.10 Alarm reporting shall be provided through communication between the Graphics Interface Device and the FPU. Alarms generated by the control program in the FPU shall be retrieved by the Graphics Interface Device, and display the time, date, severity and message on the CRT and printer.

5.3.11 Control Program Editing shall be provided through word processing software resident on disk of the Graphics Interface Device. A copy of the control program shall be modifiable in the memory of the Graphics Interface, stored to disk, and then downloaded to the FPU without requiring intervention to the building control program execution.

5.3.12 Optional Preventive Maintenance software package shall be available to retain complete equipment historical information and generate maintenance work orders. Preventive maintenance reports shall be runtime- or calendar-initiated to provide indication of equipment I.D., location, current accumulated runtime and instructions for maintenance. A history of each piece of equipment under maintenance shall be maintained as a file on disk. Additional information on the Preventive Maintenance package shall be provided upon request.

5.3.13 Disk-resident operator instructions shall be provided for all functions of the Graphics Interface Device and shall be accessible through a "Help" keyboard command.

6. SYSTEM SOFTWARE PACKAGE

The EMCS software shall consist of a real-time operating system as previously specified, and a building control program. The software shall be fully capable of direct digital control algorithms.

6.1 Flexibility
The control program portion of the system software shall be structured English-Language type, flexible enough for the end user to easily make modifications to all parameters. The user shall also have the flexibility of being able to generate custom control strategies not initially included.

6.2 Variable Evaluation
The control program shall evaluate both mathematical and logical expressions, and allow the operator to create new variables in the program not necessarily associated with I/O points.

6.3 Custom Written Program
Control program shall be custom-written for the installation. Standardized program modules will be acceptable only if they can be completely edited, altered, or rewritten after installation. The ability to change only certain items in program modules will not be acceptable.

6.4 Control Language Capabilities
The control program language shall allow "If", "Then", "Else" logical considerations with multiple variables and constants being evaluated by each logical test through use of "And", "Or" statements.

6.5 Multi-Processing Capability
The control language shall have the ability to perform multiple processes simultaneously.

7. CONTROL STRATEGIES

The EMCS shall have the capability of but not be limited to performing the following optimization programs:

Optimal start/stop
Programmed start/stop
Holiday scheduling
Hot deck reset
Cold deck reset
Hot water temperature reset
Fan speed/CFM control
Supply air reset
Chiller optimization condenser water temperature optimization
Demand limiting (temperature compensated)
Duty cycling (temperature compensated)
Night setback
Enthalpy economizer
Heating/cooling interlock
Boiler optimization
Lighting control
Tenant override/monitoring

These programs shall be executed automatically without the need of operator assistance, and shall be flexible enough to allow user customization. The following is a sequence of control applicable to this project.

7.1 Sequence of Control-Rooftop Units(Typical of 3)
During normal and tenant occupied modes, or during the optimal start cycle, the supply fan shall operate continuously. Outside air dampers shall remain closed if the maximum of the five space temperature on the corresponding floor is 2 degrees or more below desired cooling setpoint. If the maximum temperature exceeds 2 degrees below setpoint, the outside dampers shall be placed under control of the local discharge controller. The supply air temperature setpoint shall be reset from 62 degrees when the dampers are first enabled, to 55 degrees when the maximum temperature reaches cooling setpoint. This reset schedule shall be adjustable. Cooling setpoint shall be 75 degrees F (adj.).

When the maximum space temperature exceeds cooling setpoint, the first compressor shall be placed under control of the local discharge controller. After the first compressor is enabled, and either the maximum temperature exceeds cooling setpoint by 1 degree, or the outside air temperature exceeds 75 degrees (adj.), the second compressor shall be placed under control of the local discharge controller.

When the maximum space temperature begins to drop below cooling setpoint, the mechanical cooling and outside air dampers shall be de-energized in the reverse order as that described above. The temperature differential between equipment on and off cycles shall be a minimum of 1 degree and a maximum of 3 degrees.

Mechanical cooling shall be locked out below 55 degrees outside air temperature (adj.).

Outside air dampers shall close when the outside air enthalpy exceeds the space or return air enthalpy.

During the unoccupied mode, the supply fans, outside air dampers and mechanical cooling shall remain off, except during the optimal start cool-down cycle described below.

Provide two optimal start routines for each rooftop unit as follows:

Cool-down: If the outside air temperature is above 55 degrees (adj.) and the maximum space temperature is above cooling setpoint prior to normal occupancy start times, the equipment runtime required to achieve comfort conditions at occupancy start time shall be determined. If comfort conditions are achieved beyond 15 minutes before or after occupancy start times, the calculation shall be automatically adjusted for the next day's cool-down cycle. On Mondays, or after a holiday weekend, pre-occupancy equipment runtime shall be increased by 25 percent (adj.).

Warm-up: If the outside air temperature is below 55 degrees (adj.) and the minimum space temperature is below occupied heating setpoint (71, adj.) prior to normal occupancy start times, the equipment runtime required to achieve comfort conditions at occupancy start time shall be determined. When the calculated equipment start time is reached, the temperature control air serving the corresponding VAV boxes shall be exhausted through an electro-pneumatic solenoid valve (furnished and installed by others), the VAV boxes shall open and the electric discharge heating coil shall be energized. If comfort conditions are achieved beyond 15 minutes before or after occupancy start times, the calculation shall be automatically adjusted for the next day's warm-up cycle. On Mondays, or after a holiday weekend, pre-occupancy equipment runtime shall be increased by 25 percent (adj.). During the warm-up cycle, the outside air dampers shall remain closed. If the space temperature has not reached comfort conditions by 10:00 a.m., the electric discharge heaters shall be de-energized to prevent peak demand penalties.

7.2 Electric Baseboard Heat

During the normal occupied mode, the baseboard heaters shall be controlled on a per exposure basis (North, South, East and West). When the minimum temperature of the four exposure space temperature sensors (three each for the N, S, E, and W exposures) is below heating setpoint, the baseboard heaters on the corresponding exposure shall be enabled and under control of their individual thermostats. When the minimum temperature exceeds 2 degrees F above setpoint, the baseboard heaters shall be disabled.

During the tenant occupied mode, the baseboard heaters shall be controlled on a per floor basis. When the minimum space temperature on the occupied floor drops below heating setpoint, the baseboard heaters on the corresponding floor shall be enabled and under control of their individual thermostats. When the minimum temperature exceeds 2 degrees F above setpoint, the baseboard heaters shall be disabled.

During the warm-up cycle, the baseboard heaters shall be controlled on a per floor basis and shall be controlled as described during the tenant occupied mode.

During the unoccupied mode (except for the warm-up cycle) the baseboard heaters shall be controlled on a per floor basis to maintain a lower night heating setpoint. When the minimum space temperature drops below 55 degrees F, (adj.), the baseboard heaters on the corresponding floor shall be enabled and under control of their individual thermostats. The heaters shall be disabled when the minimum temperature rises 2 degrees.

When the outside air temperature exceeds 55 degrees F (adj.), all baseboard heaters shall be disabled.

7.3 Toilet Exhaust Fan

During the normal occupied mode, the fan shall operate continuously. The fan shall remain off during all other times.

7.4 Common Area Lighting

During normal and tenant occupied modes, the common area lighting shall be on. All other times, the lights shall remain off.

7.5 Exterior Lighting Control

Parking lot lights shall be turned on upon activation of the photocell sensor, and turned off at 11:00 p.m. (adj.). Dusk-to-Dawn exterior lights (wall washers) shall be turned on upon activation of the photocell sensor, and turned off upon de-activation of the photocell.

Parking lot lights shall only operate on Monday through Friday and shall not be turned on before 4:30 p.m. (adj.).

7.6 Auto-Dialer

The auto-dialer shall be activated when one of the alarms described below is initiated.

7.7 Alarms

Automation contractor shall provide the following alarm routines. When an alarm is activated or de-activated, the alarm status and time of occurrence shall be logged. Any alarm shall activate the auto-dialer.

7.7.1 Supply Fan Alarm shall be activated if the supply fan is commanded on, but does not indicate an ON status after a two minute time delay.

7.7.2 Supply Air Temperature Alarm shall be activated if the mechanical cooling is commanded on and, after a fifteen minute time delay, the supply air temperature exceeds 65 degrees F (adj.). The alarm shall also be activated if the supply air temperature drops below 40 degrees F.

7.7.3 Building Temperature Alarm shall be activated if, during normal occupied mode, the building maximum space temperature exceeds 78 degrees F or the building minimum space temperature drops below 68 degrees. The alarm shall also be activated if, during the unoccupied mode, the building maximum space temperature exceeds 85 degrees F or the building minimum space temperature drops below 53 degrees F.

All temperature setpoints and limits, time parameters and time delays shall be adjustable without modifying the operating program.

8. ADDITIONAL FUNCTIONS

8.1 Runtime Totalization

The EMCS system shall monitor the on-time of any connected digital point. This value may be accumulated, viewed, or reset at any time from any Operator Interface Device.

8.2 Trend, Status, or Change of State Logs

8.2.1 The user shall be able to initiate a custom log for any variable value in the control program. He shall select the type of log, number of values the log shall contain, and time interval between values.

8.2.2 Log types shall be user-selectable for either a "continuous" or a "fill and hold" type.

8.2.3 User shall also be able to store a value in a log as a result of a certain sequence of events, rather than a function of time only.

8.2.4 Logs shall be stored in system memory and shall be available for retrieval for output to any Operator Interface Device selected by the user.

8.3 Minimum, Maximum, Average Calculations
The EMCS system shall calculate minimum, maximum, or average value of a string of up to 50 values. The EMCS system shall then be capable of utilizing these calculated minimum, maximum, or average values in the control program.

8.4 Reserved Word Functions
The EMCS shall utilize reserved words as function commands that can be used in the control program to simplify complex tasks and reduce required program space. These reserved words shall function as subroutines to perform such functions as runtime accumulation, enthalpy calculations, minimum, maximum, average calculations, storing information to logs, reading input values, initiating commands, performing high level math functions, and remote alarm print functions.

9. SCOPE OF WORK

9.1 General
The EMCS contractor is expected to provide a fully integrated, complete, "turnkey" automation package. All capabilities of this system shall meet or exceed those specified herein. It is understood that upon completion of this installation, all functions of the EMCS outlined and specified hereafter shall be fully operational. The EMCS contractor shall be fully responsible, as stated previously, for all phases of the work. Specifically these phases are:

9.2 Submittals
Prior to the commencement of any work, the EMCS contractor shall submit to the owner or his designated engineer/architect the following for approval:

9.2.1 Technical Sheets
Manufacturer's technical product data for each EMCS monitor and control device to be furnished indicating dimensions, capacities, performance, and operating characteristics.

9.2.2 Operating Instructions
Operating instructions for each EMCS control device to be furnished indicating normal operation, failure detection methods, and override capabilities.

9.2.3 Shop Drawings
Preliminary shop drawings for each EMCS system specified, containing the following information: sequence of operation, label and location of EMCS devices, panel layout, and all required wiring, both inside the panel as well as between remote devices and the panel.

9.2.4 Point List
Preliminary point list showing each device to be included in EMCS sequence of control. Each point shall be labeled and defined. Each point will be designated as one of the following: analog input, analog output, digital input, or digital output. All input points shall include the sensor, contact closure, etc. which shall initiate the input signal and the type of input signal, i.e. current, resistive, voltage, or pneumatic with related transducer. All output points shall designate in detail the precise equipment or function being controlled (i.e. electric heat stages 1 and 2) and any associated equipment (i.e. contractor 4-pole 460V 30 amp). Any and all unused or spare points shall be identified as such.

9.3 Delivery, Storage and Handling
Deliver all required EMCS control devices and installation materials to the jobsite. Any control device or other such material which may be required which is not delivered in usable condition will be unconditionally replaced by the EMCS contractor.

9.4 System Installation
The EMCS contractor shall be responsible for all installation to include:

9.4.1 Materials:
- All EMCS panels, automation devices, and enclosures
- All mounting hardware
- All interconnecting wire, junction boxes, terminal strips, etc.
- All sensors, relays, contactors, and other input/output field hardware required to satisfy approved point list sequence of operations.
- Any additional materials required to constitute a complete and functional installation.

9.4.2 Installation:
- All wiring between panel and remote devices
- Mounting of panel(s)
- All electrical terminations required to satisfy approved point list and sequence of operation.
- Calibration and final check-out of all installed devices
- Installation of all devices/equipment specifically allocated to EMCS contractor.
- Provide 10% spare wire pairs on multi-conductor runs

9.4.3 On-Site Supervision:
- Continuous on-site supervision of installation personnel shall be provided by an employee of the EMCS contractor regularly and primarily responsible for such supervision.
- Sufficient on-site supervision by application engineer specified in Section 1.2.2 above shall also ensure that all work is satisfactorily completed as submitted and approved.

9.4.4 As-Built Drawings
Application engineer specified in Section 1.2.2 shall formulate "as-built" drawings showing any changes between submittal drawings and actual

installation. These "as-built" drawings shall be prepared following system start-up and submitted to architect/engineer for approval. Once approved, three sets shall be provided to owner or his designated representative.

9.4.5 Coordination With Other Trades and Work
The EMCS contractor shall be responsible for providing necessary interface to mechanical equipment, lighting, etc. as approved by architect/engineer at time of submittal. EMCS contractor shall not be responsible for providing additional interface required by significant changes to equipment/installation provided by other contractors.

9.5 Point Descriptions
Monitoring and control points herein specified shall be hard-wired to the FPU. Proposal shall include a listing of analog and digital points connected to the FPU. The input/output points to be installed for the EMCS shall include the following:

9.5.1 Rooftop Unit (each):
Supply air temperature
Power monitoring (KW)
Supply fan status
OA damper closed status
Temperature control air status
Supply air reset
Supply fan start/stop
OA damper enable/disable
Compressor 1 enable disable
Compressor 2 enable/disable
Discharge heater control
Temperature control air override

9.5.2 Baseboard Heat:
Enable/disable control per floor and per exposure and described in the Sequence of Control

9.5.3 Toilet Exhaust Fan:
On/off control

9.5.4 Common Area Lighting:
On/off control

9.5.5 Parking Lot Lighting:
On/off control

9.5.6 Building Sensors:
Fifteen (15) space temperature sensors
Space relative humidity, wet bulb or dewpoint sensor

9.5.7 Miscellaneous:
Outside air temperature sensor
Outside air relative humidity, wet bulb or dewpoint sensor
Total building power monitoring

10. SYSTEM SUPPORT

10.1 General
All materials and workmanship provided and performed under this specification shall be guaranteed by the EMCS contractor for a period of one full year commencing on date of substantial completion of system installation and being placed in useful service providing the special functions. Additionally, the EMCS contractor will provide the following services, the cost for which shall be included in the base bid.

10.2 Continuous Monitoring Capability
As defined in this specification, a voice-grade, dial-up, telephone communications capability will be maintained between the installed automation system and the EMCS contractor's office. The EMCS contractor shall be equipped with suitable equipment and personnel to provide full system monitoring, alarm reporting, program editing, program downloading and system troubleshooting. The owner shall be responsible for providing a dedicated, operational telephone line to enable this level of support.

10.3 Daily Status Report
At a minimum of once every day, including Saturdays, Sundays, and holidays, the EMCS contractor will call up the installed automation system and receive a detailed report of both current and historical information obtained and retained since the last report. The actual format and information to be polled will be submitted to the architect/engineer for approval prior to system acceptance.

The EMCS contractor will maintain copies of these reports for full warranty period. At any time during warranty period, owner or architect/engineer shall be permitted to receive copies of any or all of the daily reports at no additional cost other than basic bid price

10.4 Alarm Reporting
The EMCS shall be capable of using the aforementioned telephone link to initiate communication with personnel from the EMCS contractor. This capability will be on a 24-hr./day basis including weekends and holidays throughout the warranty period. Person(s) receiving such alarm report(s) shall be full-time employees of the EMCS contractor. These individuals shall be engineers or technicians who are fully familiar with the installed system. Upon notification of an alarm, these individuals will have the capability to communicate with the installed automation system within one hour and perform diagnostic and corrective procedures such as: program review, component status review, program modification or override, program download, and notification of building owner or his designated representative. If the problem is a failure of the installed automation system, corrective action will be initiated immediately to include, if necessary, actual travel to building by service technician. All costs for such service calls shall be included in the base bid for the initial warranty period. A permanent copy of all alarms initiated by the installed automation system will be maintained by the EMCS contractor during the warranty period.

10.5 Software Diagnostics
At a minimum of quarterly intervals during the warranty period, the application engineer specified in 1.2.2 above shall completely review the resident program loaded in the installed automation system. Adjustments shall be made to this program, as applicable, which will enhance system performance. Under no circumstances will modifications in the resident program be made which detract from the intent of this specification without prior approval of owner or owner's authorized representative. Copies of all changes made in

resident program will be provided to engineer/architect as well as owner. All charges for these services shall be included in the basic bid price.

10.6 System Documentation
At the time of system start-up, the building owner or his designated representative, as well as the architect/engineer, will be given complete system documentation which will contain as a minimum the following:

10.6.1 Building Summary
Listing of installed mechanical and lighting equipment and designation of area(s) served.

10.6.2 Sequence of Control
Text-written explanation of program and how the building is being controlled.

10.6.3 "As-Built" Drawings
Copies of all drawings showing equipment layout, sensor location, riser diagrams, wiring terminations, material list, etc.

10.6.4 Point List
Detailed listing of every point connected to automation system. Point list shall include as a minimum: point number, point name, point description, point type (analog input/output, digital input/output), sensor type, or field device.

10.6.5 Written Operating Program
Actual line-by-line listing of installed operating program annotated with comments to serve as a directory indicating the function of individual program segments.

10.6.6 Calculated and Updatable Variable Descriptions
Glossary and definition of all identifiers used in program.

10.6.7 Trend Log List
Listing of all programmed trend logs, listing type, interval, data quantity, and variable(s) name(s).

10.6.8 Update Notification
Any time a service call is made, a change is made to the program, or an alarm is retrieved, the EMCS contractor will prepare a written service or incident report and provide copies for both the building owner system handbook as well as that of the architect/engineer.

10.7 Post-Warranty Support
As a minimum the EMCS contractor shall provide, for a negotiated fee at the owner's request, all services provided under warranty. Additional services may be offered which further serve to enhance the viability of the system and/or the owner's ability to derive benefit from it. If post-warranty service is provided through an agent or distribut r other that the EMCS contractor, this fact shall be noted at time of bid.

10.8 Anti-Obsolescence
The EMCS contractor agrees to provide service and necessary system components for a minimum period of three (3) years from date of original contract.

11. TRAINING SUPPORT

11.1 System Start-Up
The EMCS contractor shall supply a complete operation and programming manual as required in Section 10.6. In conjunction with this manual, the EMCS contractor shall provide a minimum of 16 hours of formal operator training, such that all information in said manual is explained and demonstrated to owner-designated operating personnel.

11.2 Refresher and New Personnel Training
The EMCS contractor shall conduct formal operator training sessions at a minimum of semi-annually. These sessions shall cover basic system operation and familiarization. These sessions shall be held within owner's geographical area and shall be free of charge to owner-designated operating personnel.

11.3 Advanced Training
The EMCS contractor shall conduct formal specific training sessions designated to increase the competence of owner-designated personnel. These training sessions may be classroom type or hands-on. Charges for this training shall not exceed $200.00 per person per eight-hour session.

11.4 Programming Training
The EMCS contractor shall provide basic and advanced programming instruction. Basic programming shall be included in 12.1 and 12.2 above and shall be designed to familiarize the operator with program commands, structure, and day-to-day changes. Advanced programming instruction shall be designed such that the operator will be able to fully write, install and troubleshoot whole programs. Charge for advanced program shall be same as 12.3 above. Under no circumstances will the EMCS contractor withhold any programming language or routines as being proprietary. However, owner can be required to hold programming code in confidentiality such that this information is not divulged to any person not approved by EMCS contractor.

12. WARRANTY

All work performed under this specification shall be guaranteed for a period of one full year from date of installation. All necessary programming changes and updates shall be included in this warranty. Owner shall not be required to incur additional expense for the fine-tuning of the system.

13. COMPLIANCE

If the bidder's bid proposal is not in full compliance and agreement with the specification outlined herein, a written document shall accompany the proposal showing the exact item or items which do not comply.

Chapter 18

HIGH PERFORMANCE ENERGY MANAGEMENT CONTROL SYSTEMS (HP-EMCS) FOR HVAC

J. Meredith

1. HIGH PERFORMANCE EMCS (WHAT'S IN IT FOR YOU?)

High Performance Energy Management Control Systems (EMCS) have the potential to create a win-win-win scenario in the most critical area of the buildings industry.

This paper will attempt to describe to you an approach that has proven very successful for a number of building owners and consultants in British Columbia, with more and more people expressing interest in following this approach. The obvious benefit that a building owner will accrue from this philosophy, is dramatically increased energy savings, which of course translate into dollar savings. The more subtle benefits are the ones that are becoming increasingly important to progressive building owners. The two most common subtle benefits are the significant improvement in "INDOOR AIR QUALITY" (or if you prefer "Building System Performance") and the marked increase in morale of the building operators. In addition, to these benefits for the building owners/operators, there are noteworthy benefits that also accrue to the consultants. These benefits include increased fees, through an enhanced involvement in the total process, but more importantly, is the phenomenally increased job satisfaction that comes from this increased involvement in the building process.

2. CRISIS IN BUILDING MANAGEMENT

I referred to improved "INDOOR AIR QUALITY". This is an issue that is not going to go away! One of the major reasons cited by tenants for considering a move to new space is unsatisfactory interior environment! This is not totally surprising when you consider that over ninety percent (90%) of the life cycle costs of a building and its functions are consumed by the costs of the people that work within the building. Even a fraction of a percent improvement in productivity pays terrific dividends.

Weekly, if not daily, we are beseiged with articles, letters and product claims that refer to the "Sick Building Syndrome". Many people see this as an opportunity to pursue their own personal crusade to make the world a better place. This can range from ridding the workplace of smoking, to attempting to make all windows operable. Whichever way it is expressed, there are few people pleased with the interior environment of commercial buildings. This is the perception that we have to deal with in operating buildings.

Now some of you may dismiss this as the by-product of any number of things, from poor management to insufficient work. On one hand I would tend to agree with you, that the environment within most commercial office buildings, is significantly better than any industrial health standard. On the other hand, I know, from personal experience, that the traditional building process delivers a product that often has simple, but critical, elements missing or not functioning as intended. Unfortunately the resolution of these apparently simple deficiencies enters the adversarial realm of construction management where problem resolution is often less than fully effectual.

3. THE CAUSE OF THE CRISIS!

"PROBABLY THE MOST IMPORTANT MANAGEMENT FUNDAMENTAL THAT IS BEING IGNORED TODAY IS STAYING CLOSE TO THE CUSTOMER TO SATISFY HIS NEEDS AND ANTICIPATE HIS WANTS. IN TOO MANY COMPANIES, THE CUSTOMER HAS BECOME A BLOODY NUISANCE WHOSE UNPREDICTABLE BEHAVIOR DAMAGES CAREFULLY MADE STRATEGIC PLANS, WHOSE ACTIVITIES MESS UP COMPUTER OPERATIONS, AND WHO STUBBORNLY INSIST THAT PURCHASED PRODUCTS SHOULD WORK."

-Lew Young, Editor-in-Chief
Business Week

(From "In Search of Excellence" by Peters and Waterman)

Some people laugh when they read or hear that quote. They reflect upon some personal incident that has happened to them and wonder what has become of service. The sad truth is that we have a process that instills the image of apathy to the purchaser (IE. the ultimate client - the user of the space). I am not suggesting that no-one cares! I believe that we all care, a great deal! What I am suggesting is that it is not clear at all who is in charge!

Every "EXCELLENT BUILDING" that I have been involved with has had the common element of having a knowledgeable and dedicated person involved in the operation of the building systems. This point is the cornerstone of the philosophy that I want to layout for you today. The key to this approach is accountability. Someone with appropriate knowledge and desire has to be made responsible for the operation of building systems. We need good operators!

4. GOOD OPERATORS

In industry there are the production managers and the maintenance managers, both with clearly defined terms of reference as to what is expected to achieve the goals of the organization. In the buildings sector there is generally no direct analogy. Usually the operations and maintenance functions are combined into a single O&M function either given to a group or to an individual depending upon the size of the physical plant involved. The size of the group is not important. The result is typically the same. The terms of reference for the O&M group are poorly defined if defined at all. What little definition is given is towards maintaining the systems rather than operating the systems. This might work if perfectly tuned systems were handed over to "maintain", but unfortunately this is seldom done.

"GOOD OPERATORS" are very hard to come by! We have excellent designers! We have excellent installers! We have excellent maintainers! We do not have the equivalent of a production manager who is responsible for the process (IE "QUALITY CONTROL" and "COSTS"). We need someone to push the system to the limit in order to optimize the process (IE. maximize comfort and minimize energy use).

The principle tool to achieve optimum operation is the control system of the building. Prior to the advent of High Performance EMCS there were severe limitations to the degree to which an operator could "fine-tune" his building systems. This was largely due to the esoteric and proprietary nature of the systems that control companies made available to the industry. The cost and commitment required to learn how to effectively operate those systems was not cost justified. The alternative of specifying to the controls company, the performance that you want, generally meets with limited success, depending upon the degree of responsibility you give them. If you give them the total responsibility for operation you get a degree of success but it is limited to the degree of understanding of the process that the person assigned to the job has. In my experience, this is often significantly less or different than the designer. The alternate approach is to be more specific in providing directions to the controls company. This can be a frustrating experience not unlike teaching someone to paint over the telephone, and generally results in a compromised operation.

We now have the technology in High Performance EMCS to achieve dramatically improved building system performance. This is possible due to the advent of powerful and simple to use "Users Control Languages" in the "Stand Alone Panels". These allow the designers and operators of buildings to directly take control of their buildings without the need of an intermediary specialist.

What we need now are people to accept the challenge of delivering the performance that is designed into the building systems. There are a number of different groups that can provide these results. My personal view is that the people that should be able to best achieve these goals, are the design professionals who conceived and designed the systems. In fact, I feel strongly enough about this that it is an obligation of designers working on my projects to personally write the customized strategies that will make their designs function.

An integral part of the writing of the strategies is involvement in the commissioning of the systems. This approach, plus the High Performance EMCS technology, provides the designers with an opportunity to see their designs as they never have before. Through the use of modems and trend graphing designers can see how their systems really work. This often leads to significant alterations to their intended operation. Most importantly this approach aids in the communication between the designers and the operators. This has lead to a far greater understanding by both designers and operators of the systems and has greatly improved tenant comfort and energy utilization.

5. WHAT ARE OPERATIONS?

As has been stated earlier, operating is getting the most out of building systems. Maximizing client comfort and thusly the accompanying productivity, while minimizing energy and maintenance costs. In a well thought out and implemented program these can be very complimentary results. With some forethought the strategies that are developed to manage energy can also significantly increase comfort.

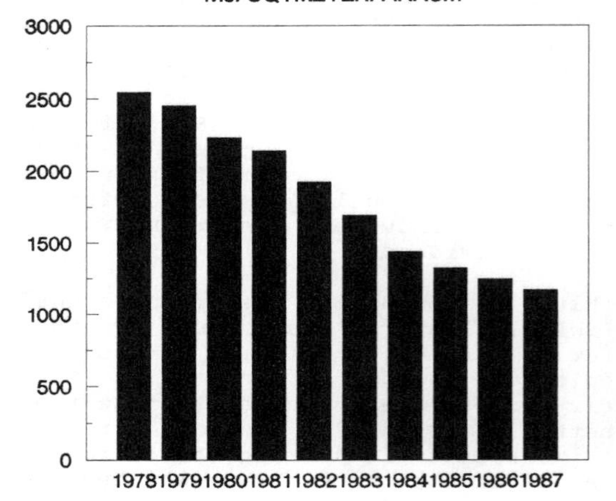

FIGURE 1.

In order to make operations accountable an objective means to evaluate performance is required! This is not very difficult to do from an energy perspective, although it is surprising how few people actually do measure energy performance. The generally accepted approach for commercial buildings is to simply measure the ratio of annual gross energy consumption per unit of gross area (Eg. MJ/sq.meter/annum see Figure 1.). There are a number of variations of Building Energy Performance Index (BEPI) calculation that take into account variables such as weather and process loads, but they should not greatly deviate from a simple ratio concept.

Client comfort has been a more difficult parameter to objectively monitor and control. The approach that we are currently using with a good deal of success is something we call a "TEMPERATURE COMFORT FACTOR". This, again, is a very simple concept (simplicity is very important for wide spread acceptance). The factor is a ratio of the percentage of time that the space temperature is within an acceptable range during the occupied hours. We have set the range of acceptability at three degrees Celsius (3C), and have allowed the operators to determine where the three degrees lay (see Figure 2.). With High Performance EMCS these factors are very easy to have automatically calculated for both individual space temperatures and for a building roll up. A similar approach should be able to be used for other comfort criteria such as humidity.

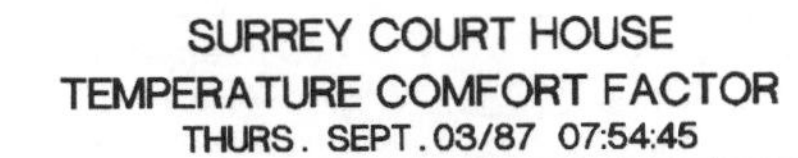

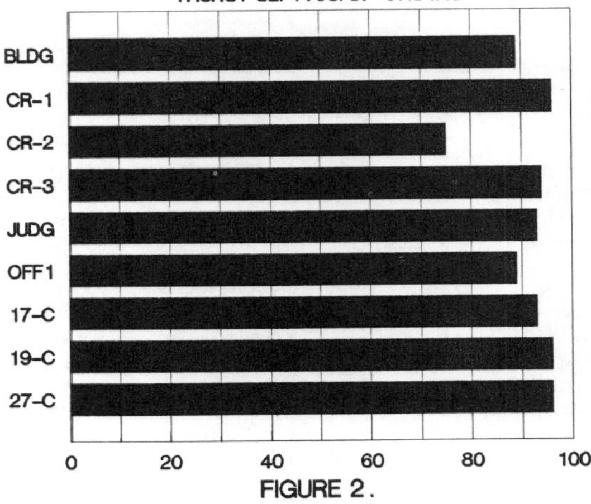

FIGURE 2.

6. CUSTOMIZED CONTROL STRATEGY THROUGH EMCS

In order to get the maximum performance out of our building systems we need to have strategies that are custom tailored to the idiosyncrasies of the buildings and the building systems. The strategies that we use have their roots in a "Dynamic Control" philosophy promoted by Mr. Tom Hartman of "The Hartman Company". "Dynamic Control" is a concept that is based upon a non-steady state approach to maintaining space temperature conditions. That is to say that there is a range of temperature acceptability. This is often mistaken for a deadband concept, which it is NOT! In classic deadband control there is no action when the temperature is within the acceptable range. With "Dynamic Control" the space temperature is manipulated within the range utilizing "free" energy in order to maximize performance.

More specifically, "Dynamic Control" utilizes a temperature predictor program to determine whether the outdoor temperature is likely to create a condition whereby either heating or cooling will be required. From this a building objective temperature will be created. This is the space temperature that the building will attempt to be driven to utilizing the "free" energy, either from the residual heat from the internal heat generation (IE. people, lights, equipment, etc.), or from the cool outside air (when available). In order to take advantage of this concept to preheat or precool a building, an optimal start program is generally developed that weighs the cost advantages, (heating and cooling energy) versus the cost disadvantages, (fan energy), in order to determine the best start time for the system.

With this approach we have found that there are a tremendous number of hours, particularly in the spring and fall, where building systems have traditionally had to heat the building in the morning and then cool it in the afternoon, where we are now able to precool the building to such a point that neither the heating nor the cooling is required for a large percentage of the year. In addition to this, surprisingly, the clients where this strategy has been applied have provided feedback indicating that they actually prefer the temperature manipulation. Remember there is a significant difference between "Dynamic Control" and deadband control (IE. with deadband control the building generally would be cooler in the winter and warmer in the summer: with "Dynamic Control" the building would be cooler on hot days and would be warmer on cold days.)

7.OPERATOR BENEFITS OF CUSTOMIZED STRATEGIES

I have spoken of the benefits of writing customized strategies for:

a) reducing the energy consumption of the building;

b) improving the client comfort and productivity, and

c) minimizing maintenance costs.

There are additional benefits that may turn out to be equal in significance to all of those I have previously mentioned. These relate to the area of operations understanding. In the past, there has been very little opportunity for us to fully appreciate how buildings and building systems operate, particularly in a non-steady state condition. The people who did have the opportunity were either in a research environment looking at a particular problem, or were involved at a maintenance level and did not have the formal training to understand and analyze the consequences of actions and reactions. We now have the opportunity, through High Performance EMCS and through having a project to write the custom strategies, to spend the time necessary to understand more completely how building and building systems work. Through this direct feedback mechanism, designers will be able to develop completely new approaches to provide comfort conditions for buildings. What has been a feedback process measured in years will now be measured in fractions of days.

This single source responsibility for strategies also has another enormous side benefit. The benefit is the increased job satisfaction that people get from being more in control of their destiny. This approach is not only fun but it also develops a terrific level of morale in the people involved in the process.

8. WHAT ARE HIGH PERFORMANCE EMCS (MANDATORY REQUIREMENTS)!

There are four fundamental requirements of an Energy Management Controls System in order for it to qualify as an High Performance Energy Management Control System (HP-EMCS):

8.1) Stand-alone Panels

HP-EMCS panels (computers) must have the ability to input, process, and output digital and analog information without the support of a central computer.

8.2) Interpanel Communication

HP-EMCS must have the ability to share information between panels. That is to say, real time data for any point must be able to be obtained and utilized in any panel, either for the operator interface, or for use in the operators' control language.

8.3) MS-DOS Pathway

HP-EMCS must provide a means to easily and quickly import real time data into a form that is useable by MS-DOS programs. This can be in the form of terminal emulation programs for systems that are able to support "dumb" terminals. This can also be accomplished through an access program that runs on a personal computer, provided that the program is able to run on a minimum configuration computer and still leave sufficient space and time to handle other tasks (IE. 512K bytes of RAM, single 360K disk, 4.77MHz).

8.4) Operators' Control Language (OCL)

HP-EMCS must have a Operators' Control Language that is easily understood by the personnel required to do custom software (either in-house or consultants, not just the contractor!).

The Operators' Control Language is necessary to convert control strategies into computer code and, therefore, is required to be able to handle logic statements and mathematical functions. It also must also have access to real time for use in control strategies. See Figure 3. from reference 1. for more specifics of an acceptable Operators' Control Language.

SECTION V: OPERATORS CONTROL LANGUAGE (OCL)

Please note that all HVAC applications programming for this project shall be done in a high level Operators Control Language provided as part of this contract. This check list is intended to determine the features of the language provided. To complete this check list, please refer to the OPERATORS CONTROL LANGUAGE (OCL) Standard, Release 1, Dated January 5, 1988. If all features as described in the referenced section of the OCL standard are included in the applications software package to be supplied, write "YES" in the Y/N column. In any features are not supplied, write "NO." If you desire to comment, enter a comment number under the COMMENT column and provide your referenced comments on a separate sheet. We recommend all "NO" responses be accompanied with an explaining comment.

Name Of HVAC Applications Programming Language Supplied with System: ____________

PROVISION	OCL STD#	Y/N COMMENT
2.1 GENERAL REQUIREMENTS		
A. Operator Overrides	2.1.A	
B. Digital Points	2.1.B	
C. Analog Point Scaling	2.1.C	
D. Variables	2.1.D	
E. System Variables	2.1.E	
F. Point Names	2.1.F	
G. Program Size	2.1.G	
H. Program Config. Options	2.1.H	
I. Preprogrammed Routines	2.1.I	
J. Program Reliability	2.1.J	
2.2 PROGRAM CONTROL		
A. Timing Functions	2.2.A	
B. Timers	2.2.B	
C. Branch Commands	2.2.C	
2.3 MATHEMATICAL EXPRESSIONS		
A. General Rules	2.3.A	
B. Standard Math Operators	2.3.B	
C. Special Math Functions	2.3.C	
D. Order Of Precedence	2.3.D	
E. Parenthesis	2.3.E	
2.4 CONDITIONAL EXPRESSIONS		
A. General Rules	2.4.A	
B. Logical Operators	2.4.B	
2.4 CONDITIONAL EXPRESSIONS (CONT)		
C. Math Operators	2.4.C	
D. Nesting Expressions	2.4.D	
E. Multiple Statememts	2.4.E	
F. Special Conditional Statemts	2.4.F	
2.5 COMMANDS		
A. Digital Commands	2.5.A	
B. Analog Commands	2.5.B	
2.6 SCHEDULES		
A. General Requiremments	2.6.A	
B. Annual Schedules	2.6.B	
C. Weekly Schedules	2.6.C	
D. Monthly Schedules	2.6.D	
E. Special Features	2.6.E	
2.7 CONTROLLERS	2.7	
2.8 INTELLIGENT TERMINAL DEVICES		
A. Global Commands	2.8.A	
B. Do Loops	2.8.B	
C. Special Commands	2.8.C	
2.9 PROGRAM SUPPORT FEATURES		
A. Program Editing	2.9.A	
B. Program Debugging	2.9.B	
C. Program Disable	2.9.C	
D. Program Backup	2.9.D	

FIGURE 3.

9. CONTRACT ADMINISTRATION OPPORTUNITIES

There are a number of approaches that can be utilized to procure HP-EMCS. In our quest for effective EMCS, we tried most of them with varying degrees of success. The following points are offered as assistance in obtaining the most cost effective and functional system possible;

9.1) Requests for Proposals

EMCS designs are evolving at an unprecedented rate. Systems are commonly obsolete within one to two years. Coupled with this fact, the various manufacturer's systems are all configured differently and often have a unique combination of features. In this situation, the production of a specification that meets a user's needs, provides a basis for competition, and allows selection of the best choice is very difficult - if not impossible. Therefore, consideration should be given to going to a proposal call philosophy, whereby the required functionality is specified and the industry is given the opportunity to respond with what it considers is its solution to the project.

9.2) Separate Contract for EMCS

Since the technology is changing so quickly relative to the conventional design/tender/build sequence, consideration should be given to pulling the EMCS out of the tender package and dealing with it at the latest possible date. In this way, there will be greater likelihood of obtaining the most up-to-date system possible.

The other side benefit of this approach is that when the system is finally put out for proposals, the operations personnel will have been identified and will take a far greater interest in both the process and the technology that is being applied.

9.3) No Progress Payments

Conventional indicators of percent completion as applied to EMCS are not valid! An EMCS may have most of its hardware on site , most of the wiring, sensors and relays installed, and most of the software debugged, but until it is all pulled together it is NOT almost finished.

Other components of buildings can provide a service that is less that 100% of design and probably "get by". With an HP-EMCS 100% completion is required to ensure "performance". For this reason, consideration should be given to NOT providing progress draws on EMCS contracts. Only pay upon substantial completion. It may cost a little more "up front", but it is generally a small amount in a competitive situation. This approach provides the additional benefits of encouraging the contractor to perform more quickly when you are holding all his money, rather than only a small percentage. CAUTION: This approach puts a lot of power into the hands of your project management people. This power will put increased pressure on your project management people to grant substantial performance earlier. Be prepared to back them up!

9.4) Liquidated Damages

If you really want to put teeth into the contract and feel the project management people can handle it, you may want to consider the inclusion of a liquidated damages clause in the contract. The accepted basis for calculating the amount of the liquidated damages is to take the amount of the projected daily savings minus the daily amortization of the estimated contract price.

10. CONCLUSIONS

Utilizing High Performance Energy Management Control Systems (HP-EMCS) and customized control strategies to enhance building system performance has the potential to create a multi-win scenario. Energy consumption and costs can be dramatically reduced. Client comfort and productivity can be significantly improved. Capital and maintenance costs for EMCS can be slashed. Everyone involved in the process will benefit from this approach. Operators will have more control over their systems than they have ever had in the past. Clients (building users) will have an environment vastly improved over anything that they have ever had before. Clients (management) will have lowered costs and increased productivity. Contractors will have an increased market once the benefits of this approach are embraced by the industry. Consultants will have an enhanced involvement by filling the void that has always existed in the area of taking "stock" building systems and customizing them into "HIGH PERFORMANCE" building systems.

The technology is now available. The philosophy has been demonstrated as sound. All that is required is for someone to take advantage of this opportunity. Someone will capture it! Will it be you?

REFERENCES

1. T. Hartman, Operators' Control Language, Release 1.0, January 5, 1988, The Hartman Company, Seattle, Washington.

Chapter 19

APPLICATION OF DYNAMIC CONTROL IN RETROFITTING COMMERCIAL AND INSTITUTIONAL BUILDINGS

K. Sinclair

INTRODUCTION

Most traditional energy management companies utilize control strategies that are simply an emulation of how the systems were controlled with conventional control hardware. The new software orientated systems provide us with a blank page to re-engineer every strategy and to optimize our new found freedom from the idiosyncrasies of conventional control hardware. This paper shares ten years experience in over 100 buildings with customized controls utilizing the Dynamic Control Theory.

What is Dynamic Control?

Dynamic control is the theory of control that acknowledges the presence of the building mass in the total building control algorithm. Traditional steady state control theories do not acknowledge the building mass, but strive to maintain constant air temperature. This often results in the mass of the building and the heating or cooling source fighting. Extra heating, cooling, capacity is required and steady state control is actually seldom achieved. Dynamic control anticipates, and preacts, by utilizing feed forward control concepts to manipulate the building mass to be on the correct side of the heating or cooling equation. The end result of applying dynamic control in the retrofitted building, is closer control than achieved by the original steady state concept, with greater client comfort.

PURPOSE OF PAPER

The purpose of this paper is to document the experiences of the author in applying dynamic control principles and custom programming techniques over the last ten years. The concept of dynamic control and custom programs are explained. The mandatory requirements of the computerized Energy Management Control System (EMCS) are outlined and also the detailed requirements of the Operator Control Language (OCL). The EMCS becomes the vehicle that carries and implements the dynamic and custom control strategies. Actual case studies are given documenting the success of this approach. The procurement and installation of Energy Management Control Systems is the subject of a paper entitled "Retrofitting Existing Building Control Systems with the New Breed of Computerized Energy Management Control System".

DYNAMIC CONTROL THEORY

Dynamic Control strategies in the Pacific Northwest, were conceived in the mid 1970s, by Tom Hartman, P E, The Hartman Company of Seattle, Washington. The strategies were the results of the development of an hour-by-hour computer energy simulation program for buildings. Tests showed that by adjusting HVAC system setpoints in anticipation of upcoming conditions, a building's mass could be employed as a thermal flywheel to reduce both the peak demands for heating and cooling and also their duration.

From the start, dynamic control introduced two new control concepts to HVAC control; first, dynamic control integrates all controlled HVAC components into a coordinated, but continually changing, control stratagem and second, dynamic control continuously anticipates upcoming weather and occupancy conditions to develop the control stratagem.

The earliest applications of dynamic control were successfully completed over twelve years ago. It has been successfully applied to over 30 million square feet of buildings of nearly every size and type. Dynamic control is a proven cost effective energy management tool.

EMCS MANDATORY REQUIREMENTS TO IMPLEMENT DYNAMIC CONTROL

The basic requirement to implement dynamic control is the software equivalent of a blank page that allows us to create a interrelationship between any of the points connected to the EMCS. The vehicle used for this task, in all suitable EMCS, is the Operator Control Language (OCL) supplied by the vendor. The mandatory requirements of this language are as follows.

The stand alone panel (SAP) shall be a standard digital computer capable of interfacing with the analog and digital points being monitored, and will have the capacity for timed start/stop on daily schedules, as well as the capacity for the Owner to develop and run custom application programs. For this, the SAP shall be provided with a proven operator control language, which shall be capable of reading the value and/or status of all system points and initiating both digital and analog control actions from any user defined combination of calculations and logical expressions which shall at a minimum include:

.1 Addition, subtraction, multiplication, and division.
.2 Square roots, powers, summations, absolute differences.
.3 Logical "and", "or", "nand", "nor", "less than", and "greater than".
.4 Time delays.
.5 Ability to imbed comments in system generated documentation.
.6 Ability to use time-of-day and day-of-year in algebraic calculators.

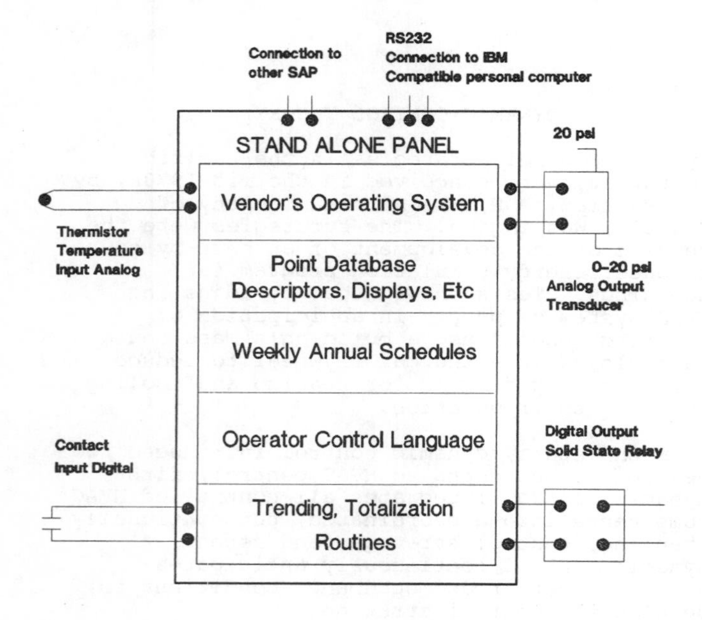

The following items on program readability and Operator Control Language (OCL) functionality, are directly from The Hartman Company Operator Control Language (OCL) Manual.

Program Readability

The purpose of the language is to provide functional and understandable programs such that operators with minimal computer or language training can read, understand, and trouble-shoot complex control sequences. It is therefore essential that the language be as readable as possible. For this, the following features are required:

1. Complex Statements: Language must permit statements constructed of multiple variables and operators without a fixed limit. Also required is a means of line continuation that allows the operator to set line breaks.

2. Mixed Math and Logic in Statements: Language must allow mixing of logic and math in statements and allow a math or logic expression in place of point name or variable.

3. No Restriction On Order of Operators in Statements: Language must allow user to construct statements in any functional order of variables, constants and operators.

4. Use of Point Names and Automatic Attribute Selection: Language must employ point names for all points and variables. Program automatically selects desired attribute of point. For example; the statement that begins IF FAN ON automatically engages the system to consider the status of that point. Starting the statement IF FAN ALARM engages the system to consider the alarm attribute of the point.

5. Minimum Parentheses Requirement: At least five levels of parentheses must be available without restrictions to force order of evaluation of math and logical statements, but parenthesis should not be mandatory in ordinary math or logical statements. Note: The use of parenthesis is mandatory for special operators.

6. Comments: Comments must be capable of being embedded anywhere in programs (including lines with expressions) through the use of special character(s) to start and stop comments.

7. Floating Point Math and Integers: Language must employ floating point math, but also permit the use of integers with automatic recognition of integers as floating point equivalence when required (5 = 5.0 in math function).

8. Line Identification: The use of line numbers in the program is discouraged as a detriment to readability. However, employing numbers in printouts and listings to identify specific lines for errors is encouraged. Automatic indents or other special characters to show a continued line or statements within BEGIN/END groups is also encouraged.

9. Program Printouts: System must include a fast and easy method of printing programs and comments, single programs or in groups.

OCL Control Function

All control decisions and calculations must be executed through OCL. The use of separate routines and enhancements such as interlock programs, special calculated points, or database enhancements to meet the requirements of OCL function is not acceptable. While many current systems can provide roughly equivalent functions by linking these special functions together, such procedure results in programs that are difficult to understand, document, or use by the operator to make control adjustments.

Hartman's 23 page OCL manual details many other features required to implement sucessful dynamic control.

Other mandatory monitoring and control features of the SAP are:

.1 Operator defined digital and analog alarms and automatic alarm condition reporting.
.2 Auto lockout of alarms when alarmed

system is shut down.

.3 Direct keyboard control of all digital and analog outputs.

.4 Addition, deletion, definition, and modification of points and point types from operator keyboard.

.5 Trend log reporting of user selected points and times.

.6 Addition, deletion, and modification of English language descriptions.

A formula to estimate the memory/storage requirements for the variety of functions which will be provided. Memory requirements for each control language function shall be provided. Points in one SAP shall be able to be utilized in another SAP's control language.

In addition to the complete unstructured blank page approach, with no preconceived features built in, the new breed EMCS must have the ability to easily interface with present computer technology. This means IBM Personal Computer compatibility which is a standard for approximately 70% of the computer industry. This link makes available an incredible base of public domain IBM compatible software for graphing, database analysis, communication, etc. It also greatly reduces the dependency on vendor developed software. The computerized EMCS built with a single, or with several modular Stand Alone Panels (SAP), simply becomes a vehicle in which control strategies can be imbedded and implemented. Another major feature of these systems is their capability to collect data. In this light the SAP, when connected to a personal computer, becomes a communication network threaded throughout a building. Space temperatures, humidity, fan status, cooling equipment, daily energy consumption, etc, is all available. Information can be stored or manipulated in any method available to the personal computer, and unlocks a flexibility never seen in conventional control systems.

IMPLEMENTING DYNAMIC AND CUSTOM CONTROL STRATEGIES

When the mandatory requirements of the OCL are met, and the system is installed and commissioned with "as found" control strategies, the implementation of custom strategies can begin.

The importance of commissioning the system with "as found" control strategies should not be under emphasized. Nothing is worse than trying to separate software problems of custom control from startup hardware problems. Contracts including dynamic control strategy development should allow for the 7 day test before substantial acceptance of the system with "as found" strategies. In new construction keep strategies as simple as possible for acceptance and commissioning. A second important feature of the "as found" or "survival" strategies concept is, that while the dynamic and custom routines are being developed, the building will operate satisfactorily on these simple strategies. This allows control strategy upgrades to occur in an organized, and unpressured atmosphere.

The weather predictor program is usually the first of the global programs to be installed. A common problem in implementing a successful predictor is to achieve reliable outside air temperature readings. In large buildings several outdoor air sensors, as selected by a software program, are analyzed. The prime components of the weather predictor are the Projected High, the Objective Temperature, and the Cold Day/Warm Day Decision. The output of the weather predictor is used throughout the EMCS system to affect cooling and heating start decisions, and setpoint calculations.

Dynamic control strategy programs for the air handling systems are usually developed next. In projects where there are several typical air handling systems, the strategies are developed on a public domain word processor and transferred by a public domain communication package. Strategies built for one system can be copied to other typical systems with a global replacement of variable points, thereby eliminating the necessity of entering duplicate coding. Building operators should be actively involved in this coding. Understanding and ownership are significantly increased by operator involvement, and the common fear of the black box approach of older systems is eliminated. The involvement is counter-productive to quick code installation, but is essential to insuring the necessary technical fit to provide overall project success. Removing the vendor from this process assists in making the operator accountable for his new system. Operators quickly learn that the vendor only supplies and maintains the equipment. The software and concepts are supplied by the operators and the examples set by their consultant. This is the start of <u>their</u> custom programs which are overlaid and entwined with the dynamic control concepts. Every control strategy ever written is described as "It works well, but!!!" <u>"But what about this condition? You forgot about....."</u>. When this problem is encountered with conventional control, more equipment and installation is required, costing more money. Who should pay? Who made the mistake? With the new breed of EMCS "forgotten, or what about conditions," can be added with only the investment of time by the operator, and or consultant. It should be noted that the next time strategy tuning is required to solve another level of conditions, it also can be handled easily.

<u>How Well Does The Normal Building Operator Cope With These New Systems?</u>

Generally systems, with the traditional approach, are engineered and installed by consultants and contractors. Operators, with very little operating training, are required to operate these systems with all their idiosyncrasies. Modifications when required, made by design consultants and contractors, are often not totally in line with the economical or aggressive operation of the building. Contractors hoping to get large service contracts limit the amount of technical transfer, and the operator learns the system with a distinct disadvantage. The blank page approach, on the other hand, with the new breed of control systems greatly reduces communication and technological transfer. The operator, consultant, and contractor can all read the same database of the OCL. The concept of dynamic control is almost always better understood by a good building operator than by the design consultant. The consultant calculates the peak loads; the operator watches the related equipment respond to time and changes in the overall loads. Ask any building operator the effect of building mass when cooling is lost for a day, and then restored; he will tell you that proper control can't be restored for several days. The effect of this mass is extremely difficult to calculate, while the real effect is obvious.

Experience shows that good operators adapt and take charge with the new breed of system, capitalizing on all the energy opportunities that occur when the mechanical system is not at peak cooling or heating capacity. Changing the traditional design is often necessary during the majority of hours when the system is at neither peak heating or cooling. Changing control strategies for various mechanical systems based on conditions with IF THEN logic allows continuous matching of the correct strategy with the correct condition. This fine tuning simply is not available with conventional control.

DYNAMIC CONTROL PERFORMANCE
ROBSON SQUARE, VANCOUVER, B C

Robson Square is a low profile development of gardens, reflecting pools, plazas and walkways, surrounding and merging with functional courtrooms and Government offices. The gross area is approximately 1 187 265 sq ft (110 297 sq M) with a conditioned space used for energy calculations of 914 716 sq ft (84 980 sq M). The complex is physically divided into three city blocks: block 51/61 and block 71. Robson Street passes over the complex between blocks 51 and 61, and the building passes over Smithe Street between blocks 61 and 71. The height of the complex varies from 67 ft (21 m) in blocks 51/61 to 148 ft (45.4 m) in block 71.

Block 51/61 is a Provincial Government office building with approximately 1300 employees and 78 departments representing a variety of Ministries. It has an open air ice skating rink, exhibition space with theater, cinema, and meeting rooms, restaurants and food fairs, garden courts with trees, flowers, waterfalls, and pools, and underground parking for 460 cars. Block 71 (the Law Courts building) encompasses the Court Registry, Provincial Courtrooms, the law library, prison holding cells, computer room, Judicial chambers, and the Great Hall Atrium.

The HVAC systems include 27 main air handlers which are variable air volume units; 56 localized supply fans which supply restaurant makeup air, car park and tunnel air, stairwell and elevator shaft pressure; and 24 exhaust fans used to exhaust contaminated washroom, kitchen, and parking air.

Three boilers provide hot water for the terminal reheat circuit, radiant slab, and domestic water heat exchanger. Two chillers and an 800 000 gallon chilled water storage tank provide water for the cooling coils. The hot water system and chilled water systems have central distribution pumps located in the central plant. These pumps distribute the water to remote circulating pumps located in the fan rooms. The central plant also provides heated and chilled water to the Vancouver Art Gallery, a complex of approximately 109 203 sq ft (10 145 sq M).

Electricity enters at 12 kV and is distributed to ten substations where it is transformed to 600 V for further distribution. Large motors are 600 V, lighting is 347 V, small motors, appliances and plugs are 120 V.

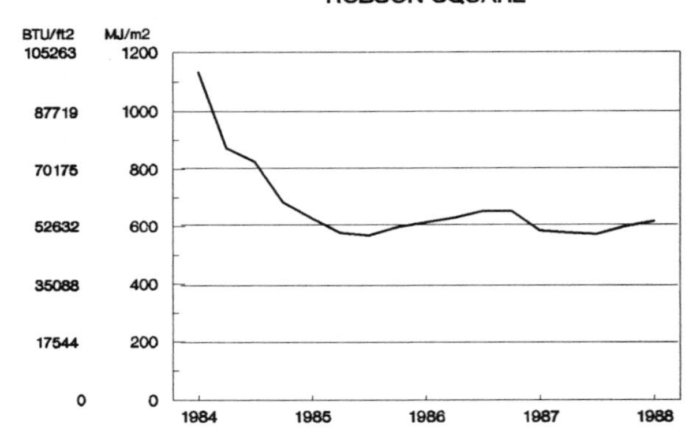

A significant energy reduction was achieved by installing a dedicated computerized energy management control system which utilizes custom control strategies which were developed onsite. The main function of this system is to manipulate the 27 air handling systems and the central plant equipment to achieve the lowest possible new energy use. Concepts such as involving the building mass and weather prediction in the control equations were used. Operating between temperature guideline limits allowed significant opportunities to both precool occupied space, and to offset anticipated losses during unoccupied hours. Other opportunities existed such as the ability to manipulate the air system to move the high internal heat load of the building to the exterior to offset transmission losses. Approximately 2000 lines of program were custom developed for this project, and reside in 18 stand alone energy management control panels.

The use of a weather predictor program allowed many energy opportunities to be exercised. Opportunities such as feed forward strategies that allowed a decision to be made on what the outdoor projected high temperature was going to be. This information often eliminated the need to heat early in the morning and then use mechanical cooling by noon. When this situation occurred, the cooler temperature was utilized to cool the building mass which was later warmed by the internal building gains, eliminating both requirements to heat and cool.

The large hot water firetube boiler is shutdown every night during heating season and an optimum start routine, based on weather projection and space objective calculations, determines the latest possible start time to achieve building conditions. After one year of this type of operation the boiler was inspected and no tube damage was noted.

All heating pumps are connected to the EMCS. This allows a significant reduction, or in many cases elimination, of the simultaneous heating and cooling by the manipulation of the heating pumps and the supply air reset of the air handling system. Over 100 space temperature sensors were installed to allow accurate feedback of space conditions.

The building architecture is very adaptive to this type of control as the mass is generally exposed; eg open concrete beams, exposed concrete walls, etc. The landscaping office concept also greatly enhances the success of this type of control.

In exploring the concept of providing a new computerized energy management control system, a maintenance opportunity was identified. The new system completely eliminated the electronic and pneumatic controls associated with 27 air handling units. This equated to a reduction of approximately $75,000 in the existing service contract. The successful EMCS contractor was selected as a result of a call for proposals, rather than the traditional equipment specification. One of the many advantages of the proposal call concept was the ability to evaluate maintenance savings potential versus the capital cost of direct digital control. A second benefit of the DDC option was a reduction of present compressed air consumption due to pneumatic component elimination. This saved money budgeted for a compressed air system upgrading.

Overall, this type of control has improved environmental conditions in the building, and has reduced the required peak capacity to the point that only one boiler is used in the coldest weather, and during the warmest weather the building can operate during the day on the chilled water storage tank, without the operation of any chillers. This provides a significant electrical demand saving over original operation.

Heating of The Atrium With Return Air From Law Courts Building

An opportunity existed to provide a major portion of the required heating of the large Atrium with the return air from the Law Courts Building. The air handling equipment for the Atrium was located in the return air plenum of the Law Courts and required modifications were minimal.

Heating of Domestic Hot Water When Boilers Shutdown

The only reason the boilers were left operating during summer months was to supply domestic hot water, and provide temperature control via terminal reheat. During warmer weather, and when EMCS control strategies were implemented, reheat was negligible. A separate source of heat was provided for domestic hot water and existing boilers were shutdown, eliminating their overhead.

The lowest capital cost heating source to install was steam, supplied by City Central Steam, because of its close proximity to both domestic hot water systems. Existing hot water heat exchangers in each DHW system were reutilized for steam operation and this greatly reduced capital costs as well as implementation time. Steam exchangers are now utilized year round, allowing more flexibility in daily shutdown of boilers, and lower overall hot water loop temperatures.

Operation Commitment

Operation commitment is an important part of any successful conservation program. The commitment, made by the operational staff of Robson Square, to the science of effective operation, played a key role in pulling together the many projects that resulted in a finely tuned building, providing both improved client comfort and energy efficiency.

Operations people quickly pointed out that not all problems could be solved with computers and reduction of lights. Mechanical deficiencies and shortcomings were identified and corrected. The focus tended to be redirected to the terminal unit rather than a previous emphasis on the central system. Fan capacities were found to be excessive, after the lighting reduction project and EMCS control was achieved. A project was identified to reduce fan speed and swap out oversized motors with correctly sized energy efficient motors.

Some areas would not fit the mold of the complex. These "exceptions", less than 2% of the total area, were addressed with localized fixes such as electric heat. This is a common problem with large buildings where less than 5% of the space determines the operation of 100% of the system, increasing both heating and cooling.

DYNAMIC CONTROL PERFORMANCE
NEW WESTMINSTER LAW COURTS

The New Westminster Law Courts provide Supreme and County Courts, Provincial Court, Probation Services, and Crown Counsel. The design concept of the building centers around a large four story atrium. The environment is open, relaxed, and planned to encourage public observation of the administration of justice.

The building's four levels total 197 653 sq ft (18 362 sq m). Level one consists of the Court Administration, Sheriff Services, Justice of the Peace, Hearing Room and a Coffee Shop. Level two is three provincial courtrooms, two heritage trial courtrooms, chambers courts, and a law library. Level three houses the Crown Counsel. Level four has one civil jury courtroom, five trial courtrooms, Registrar and Registrar's hearing room, Superior Court judiciary, and interview rooms. Additionally there is enclosed underground parking of 48 500 sq ft (4 505 sq m).

The HVAC systems include 11 variable air volume units complete with return fans, five constant volume units, three supply fans for directing air to the parking area, and 52 exhaust fans in miscellaneous areas, washrooms, and parking. Heating is distributed by a central hot water heating system of two gas fire tube boilers, radiant panels, fan coils, force flows, and unit heaters. Cooling is distributed from an electrical centrifugal chiller rated at 225 tons. Electricity enters at 13.8 kV and is distributed to substations where it is transformed to 600 V. Most motors are 600 V, lighting is 347 V and general use is 110 V with some at 208 V.

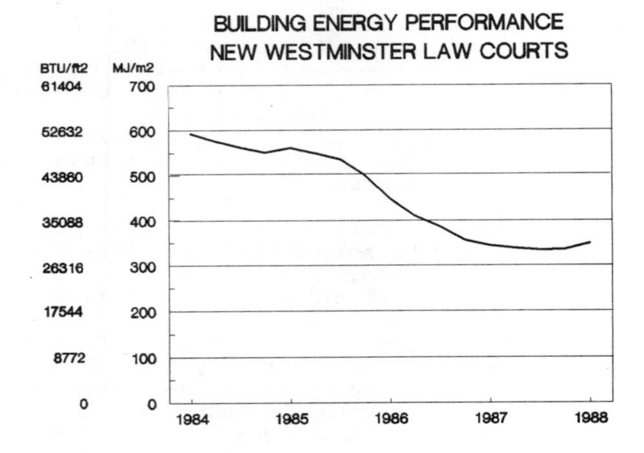

New Computerized Energy Management Control System (EMCS) - Software Strategies

A significant portion of the energy reduction was achieved by installing a dedicated computerized energy management control system which uses custom control strategies. The main function of this system is to manipulate the air handling systems, and the central plant equipment, to achieve the lowest possible energy use. Concepts involving the building mass and weather prediction in the control equations were used. Operating between temperature guideline limits allowed significant opportunities to precool occupied space, to offset anticipated heat generation by lights, people, etc, as well as preheat the space to offset transmission losses during unoccupied hours. Other opportunities existed such as the ability to manipulate the air systems to move the high internal heat load of the building to the exterior and atrium to offset transmission losses. Approximately 700 lines of programming were custom developed, and reside in six stand alone energy management control panels.

The use of a weather predictor program allowed many energy opportunities to be exercised. Opportunities such as feed forward strategies that allowed a decision to be made on what the outdoor projected high temperature was going to be. This information often eliminated the need to heat early in the morning and then use mechanical cooling by noon. When this situation occurred, the cooler temperature was offset by allowing the building mass to be warmed by the internal building gains, eliminating both requirements to heat and cool. The two 2.3 million BTU hot water firetube boilers shutdown every night during heating season. An optimum start routine, based on weather projection and space objective calculations, determined the latest possible start time to achieve comfortable building conditions. Domestic hot water is heated by a dedicated small gas boiler which shuts down each night and is programmed to start for each occupied day. Control of perimeter radiant panel heating system is by several EMCS space sensors located near the perimeter of the building. The radiant panel heating system is run as hot as possible, when required, and shutdown when not required, as sensed by the space sensors. The radiant panels, if operated at lower temperatures, have the potential of adding most of the heat to the return air rather than the space.

Close scheduling of the major air handling systems contributed greatly to the success of this project. Operation personnel and onsite security people cooperated on the close scheduling of the system operation to match the court usage. Night courts and extended trials presented unique problems, but proper communication and commitment have turned these problems into conservation opportunities. All ten statutory holidays are preprogrammed once a year; the existing time clocks were seldom reset for holidays.

The EMCS's 30 space sensors quickly identified problems with variable volume boxes, overall air delivery, zoning, and lack of total heating. As these problems were addressed the air system strategies became more effective. An example is the heating was left on in some areas because recovery was not sufficient in cold weather, and when extra heat was added, the complete heating system could be delayed from starting for several hours. Increased air volume in several areas not only allowed quick warm up, and a longer period before mechanical cooling was required, but significantly improved the air quality.

In the original concept the atrium was heated and cooled as an individual space. Modifications have been made to allow the existing air supply system to become a variable volume exhaust system, if extra air and heat is available from the rest of the courthouse. This control would have been impossible without the use of an EMCS. The software program totals the air supply of several systems to make a decision as to how much air to exhaust. If no extra air is available (air that is normally exhausted) the unit is converted by positioning dampers back to full recirculation, and heating is added. The majority of the time the atrium can be heated by the internal heat of the Law Courts.

The original heating equipment was designed with energy conservation in mind, and in extremely cold weather recovery of the building temperature becomes a problem. During these periods the amount of setback is limited, or in extreme weather (less than -5 C), heating systems are left running. The amount of hours that these conditions occur are few enough that this strategy does not significantly effect the total energy performance.

Maintenance and Operational Savings

Part of the justification for the EMCS was the $11,000 maintenance savings. These savings were real as an existing control service contract of $8,000 was eliminated. Experience todate with the system has been extremely good with almost zero repair of sensors or transducers. A few panels were replaced under warranty during the first year, and the second year of operation was trouble free.

The ability to interrogate the system remotely is its best feature since this building does not have on site support. Quick reaction to problems via the system convinced clients that on site personnel were not necessary. The building had a full time operator, later this was reduced to a part time operator, and with the help of the EMCS the onsite requirement was eliminated. The system allows the operator to be at a remote site, and it allows consultants, contractors, and other support personnel to be available via telephone modem, at anytime. When a key operator is on holidays, back up support can be provided from several sources with only a phone call and a password to arrange details.

The original project left the existing air volume control equipment operational. The high cost of maintaining this system has economically justified the transfer of control to the EMCS. In addition to reducing maintenance costs, volume control problems can be better analyzed remotely.

Operational Staff Commitment

All the energy conservation technology the world can be defeated if operations people do not care about ultimate energy efficiency. In this project the Building Superintendent not only cared about energy conservation, he was determined to have the best in energy and comfort performance in his courthouse. Commitment to achievement, married to the new EMCS technology capable of accepting custom programs that reflect the result orientated approach of the operator and consultant, is the key to this project's success. The building has no full time operating people on site, and is operated via the EMCS dial up modem from a site approximately two miles (3.2 km) away. The system consists of six stand alone panels and one modem on site. A personal computer and modem can access the EMCS by dial up phone. Daily energy monitoring of the Law Courts is done by the central steam plant Engineer at the remote site. If a problem is noted that cannot be corrected by the central operator, the Building Superintendent is called, via cellular telephone, and regardless of his locale he is able to connect to the building with a laptop computer and modem. The consultant also has access to the system if a control strategy deficiency is identified.

All functions of the system, time scheduling, control loop tuning, control strategy changes, trend log data, etc, are available remotely. In addition, the system stand alone panels can be backed up and loaded remotely. The achievements of this project, plus the introduction of high technology has greatly improved the morale of all people involved with the project.

Through this support network of people and technology the Building Superintendent has achieved his dual goals of dramatic energy management results, plus an excellent internal environment for the building occupants.

AIR QUALITY VERSUS ENERGY CONSERVATION

Concerns about the effect on the building air quality by these aggressive energy conservation measures resulted in putting in place a test measuring the carbon dioxide (CO2), carbon monoxide (CO), humidity and temperature. The measurement equipment was connected to one of the existing EMCS stand alone panels. Several software trend logs were set up on each measured test variable. Trend information was extracted from the project in Vancouver, B C via a telephone modem in Victoria, B C. Although CO2 and CO are not the only air quality concerns, CO2 appears to be a good indicator. If CO2 levels are less than 600 ppm (parts per million) air quality complaints are seldom received. The following graph is a plot of actual data collected for ten days. Note the increase in the CO2 level on the morning of April 18th as the strategies closed off the fresh air dampers on the early morning warm up routine. A drop in CO2 level is also noted around noon each day as the occupancy decreased in the sensed area. The area in which the sensors were installed was the Court Registry area; an office like area of approximately 20 to 30 people.

The following graph proves that aggressive energy conservation measures and good air quality can go hand in hand with proper planning. Often what is good for dynamic control, more air movement to transfer heating and cooling from the mass, is good for air quality.

This test has encouraged us to seek out an inexpensive, but relatively accurate, CO2 measurement device to provide feedback on building air quality conditions when manipulating for maximum energy conservation. If acceptable air quality conditions are exceeded during occupied periods by conservation routines, these routines would be overridden until proper air quality was achieved.

Air quality is simply another parameter that we must manage to successfully provide total client comfort at the lowest energy cost. To effectively manage air quality we must be able to measure it. New air quality sensors will be available in the near future that will make air quality as commonly a controlled variable as temperature and humidity.

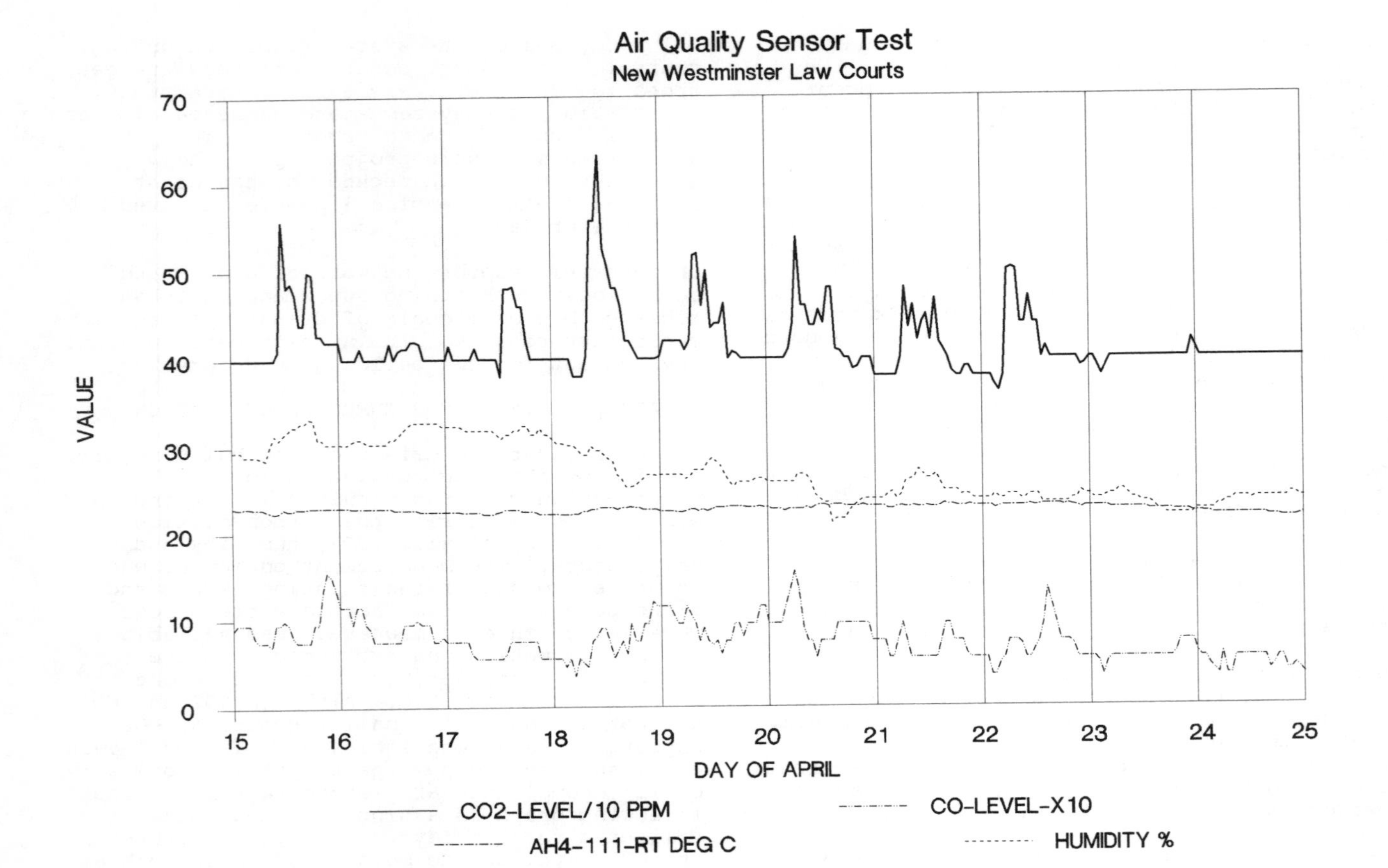

SUMMATION

In summation, Dynamic and Custom Control Strategies WORK. They require the new breed of EMCS with all mandatory requirements, including an Operator Control Language to provide a media for writing the strategies, as well as a micro processor based stand alone panel to carry out the intent of the strategies. Dynamic control concepts are the missing link that is necessary to provide the economical justification for the new breed EMCS update of conventional control. Air quality can be maintained and even improved with implementation of successful dynamic control.

REFERENCES

Operator Control Language Manual:
The Hartman Company, Seattle Washington

Chapter 20

AN ENERGY MANAGEMENT SYSTEM AT ROCKWELL INTERNATIONAL'S SATELLITE AND SPACE ELECTRONICS DIVISION

M. B. Whaley, R. V. Beck

A comprehensive energy management program began in 1987, which enhanced projects undertaken prior to 1987, on a facility that was constructed in 1966. The facility had fallen behind in state of the art technology concerning energy management. The planned projects included installing an energy management system, upgrading the lighting system in one building, upgrading the HVAC in one building, replacing equipment such as motors with energy efficient equipment, installing variable speed drives on various motors, improving the water treatment system and installing cooling tower water filters, curtailing equipment usage such as electric boilers for building reheat, and renegotiating utility contracts.

These projects resulted in annual energy cost savings of approximately $281,000 and breakeven points of less than two years. The energy consumption (KWH) was reduced by 13% and the average total demand was reduced by approximately 600KW. Several energy management projects are planned for the future.

INTRODUCTION

In 1966, energy management had very little significance as far as budgets were concerned. Electricity and natural gas rates were extremely low in 1966 when compared to 1987 rates. The major elements of the facility's budget did not include energy costs. However, in 1987, energy management has an extremely important economic significance. The current facility budget is approximately $2.5 million per year for utility payments. The incentive to reduce energy usage and therefore energy costs is very attractive.

Rockwell International's Satellite and Space Electroics Division (S&SED) began the energy management program during fiscal year 1987. The goals of the program were to reduce the amount of energy used by five percent. If the program was successful the feasibility of procuring funds for future energy management projects would be enhanced.

METHODS AND RESULTS

Energy Management System

The installation of an energy management system was one of the first goals in the energy management program. An energy management system was selected that was user friendly, user programmable, capable of expansion, and capable of monitoring and control. The system was installed in one office building of approximately 300,000 square feet. The system controls the HVAC and the lighting in the building.

The energy management system worked very well. Lighting hours of use were reduced from approximately eighteen hours to twelve and one half hours per day. The building quadrants were recircuited into nineteen subsections each. Employees and janitorial personnel can turn the lights on by dialing a numerical code after normal hours of operation.

The energy management system also controls the HVAC system including the chiller system and the air handling units. The system monitors supply and return chilled water temperatures. It controls the chilled water valves. It monitors the tempertaure and relative humidity in each quadrant of the building.

The energy management system had a breakeven period of one and a half years (Table 1) and performs very well. The number of trouble calls for hot or cold areas of the building has been reduced by ninety percent.

TABLE 1

ENERGY MANAGEMENT PROJECTS, ANNUAL SAVINGS, AND BREAKEVEN POINTS FOR SATELLITE AND SPACE ELECTRONICS DIVISION'S ENERGY MANAGEMENT PROGRAM.

PROJECT	ANNUAL SAVINGS($)	BREAKEVEN(YRS)
Energy Management System	100,000	1.5
Lighting System Upgrade	32,400	1.3
Water Treatment Filters	9,525	1.8
HVAC Upgrade	90,000	1.7
Energy Efficient Equipment	12,705	0.6
Variable Speed Drives	60,000	1.3

Lighting System Upgrade

In the facility office building, a major project to replace fluorescent ballasts and tubes with energy efficient ballasts and tubes was undertaken. The existing fixtures are three tube fixtures. Three ballasts are required for every two fixtures. One tube was eliminated from each fixture and one ballast was eliminated from every two fixtures. The fixture wattage was reduced from approximately 140 watts to 81 watts. Several fixtures in hallways were disconnected. The lighting levels at the workstations were maintained using existing task lighting.

The lighting retrofit had a breakeven period of approxmately sixteen months (Table 1). There have been no complaints from employees. The failure rate for the new ballasts and tubes has not changed from the previous rate for the existing ballasts and tubes.

Effective Water Treatment

Several of the chiller tubes were plugged and corroded at the start of this project. The water treatment was inadequate to keep the cooling tower water system clean. Filters were installed at the cooling towers which reduced the reliance on water treatment chemicals considerably. The water treatment is monitored several times per week by an outside vendor to insure that the treatment chemical concentrations are within acceptable tolerances.

The cooling tower water filter installations had a breakeven point of twenty-two months (Table 1). The filters are relatively inexpensive and their use results in a reduced need for water treatment chemicals. Overall this project improves the efficiency of the cooling tower water portion of the HVAC system tremendously as well as extending the life of the equipment.

HVAC Upgrade

In the office building chiller plant, two 670 ton chillers were used to maintain chilled water to the air handlers. It was determined that the chillers were never fully loaded and the entire building required only 350 tons of cooling on a summer-peak day. One chiller was excessed and replaced with a more efficient 350 ton machine. The existing chiller when operating at forty to fifty percent load required approximately 2KW/ton while the new chiller requires only 0.6KW/ton at eighty to ninety percent load.

The reduction in required chiller tonnage was accomplished by allowing the chilled water temperature to float five degrees. Then the energy management system selectively cycles the air handlers on and off as required.

The chiller installation had a breakeven point of twenty months (Table 1). The chiller is also capable of soft start and stop. It has a variable speed drive which maintains the amperage by varying the frequency and voltage. The drive reduces the kilowatts required to operate, increases the power factor, and serves as a power conditioner.

The existing air handling system consisted of a double duct system with hot and cold ducts and a common return air plenum. Electric boilers were used to reheat the supply air. It was determined that the electric boilers were not needed. Enough reheat was gained from the lighting system to satisfy comfort HVAC requirements. The two 750 KVA boilers were permanently curtailed.

Energy Efficient Equipment

An assessment of the motor loads for twenty-four air handlers was undertaken. The loads were measured on a peak-summer day. It was found that the motor design loads were twenty to fifty percent higher than required. Twelve of the motors were changed for energy efficient smaller sized motors. A typical motor was reduced from fifty horsepower to thirty horsepower.

The motor replacement installation had a breakeven point of approximately seven months (Table 1). The motor downsizing is very cost effective and it was easily handled by plant maintenance personnel.

Variable Speed Drives

The twelve motors downsized were also matched with variable speed drives. These drives are capable of maintaining a constant amperage by varying the frequency and voltage. The drives are also tied into the energy management system. The variable speed drives are anticipated to breakeven in approximately fifteen months (Table 1). These drives also have soft start and stop capabilities.

With variable speed drives on the equipment, especially pump and air handler motors, the proper media volumes can be supplied. This results in an unloading of the chillers and increases the efficiency of the HVAC system. These drives also increase the power factor as well as serving as power conditioners.

Renegotiate Utility Contracts

After a careful review of the existing natural gas contract it was determined that the division's gas rate was extremely high when compared with diesel fuel rates. The local natural gas utility was contacted, after two months of operations using diesel fuel instead of natural gas, and a more favorable contract was negotiated.

CONCLUSION

The primary goal of the energy management program was to realize a reduction in kilowatt-hours and therefore

FIGURE 1
ELECTRICAL ENERGY CONSUMPTION FOR S&SED

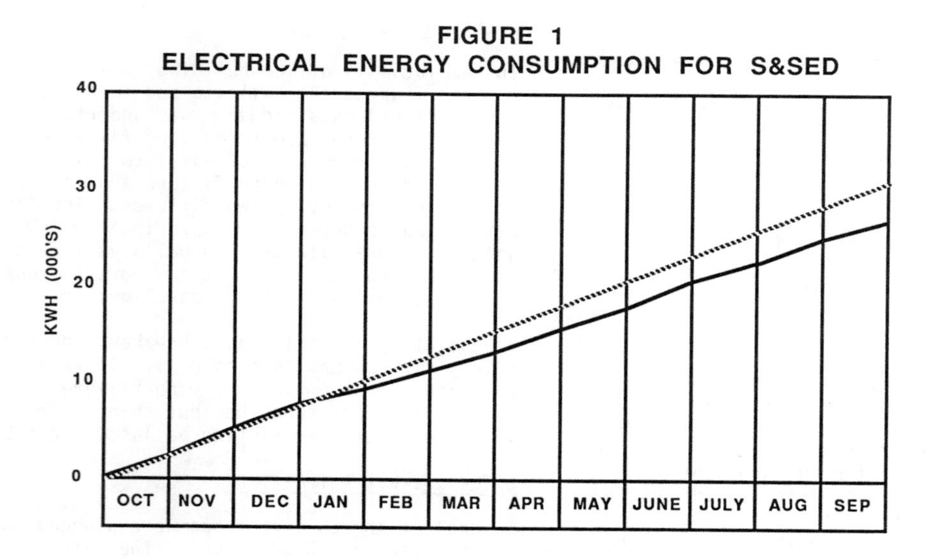

FY 86 ACTUAL: 31,170,000 KWH
FY 87 GOAL : 29,690,000 KWH (5% REDUCTION)
FY 87 ACTUAL: 27,069,000 KWH (13% REDUCTION)

FY 86 ACTUAL (dotted line)
FY 87 ACTUAL (solid line)

a reduction in energy costs. The numerical goal was to reduce the total kilowatt-hours from the 1986 actual usage of 31,170,000 by 5% to 29,690,000 (Figure 1). The result of our program's first year was to reduce the kilowatt-hours to 27,069,000. This reduction translates to a 13% decrease in energy usage for the year.

The program was more successful than anticipated and it has opened the door to many future projects. The energy cost savings realized by comparing FY 1986 electrical energy costs versus FY 1987 electrical energy costs totaled $281,000. This number would have been even greater if the projects had a full year to operate.

REFERENCES

1. McGowan, J. J. 1987. Automation and control system program management. Energy Engineering 84 No. 6:33-39.

2. Lazar, I. A. 1987. Electrical demand control - how to attain it. Consulting-Specifying Engineer 2:76-81.

3. Energy management systems, specifiers guide. 1987. Consulting-Specifying Engineer 2:31-38.

4. Dulanski, G. 1987. Managing energy with lighting controls. Light Design Applications 17:44-46.

5. Jones, D. H. and Hall, D. N. 1988. Dual duct HVAC system at veterans nursing homes. Energy Engineering 85 No. 2:36-41.

Chapter 21

REPLACING AND UPDATING YOUR EMCS

C. E. Neal, Jr.

ABSTRACT

Various types of Energy Management Systems have been in operation for some time. As time goes on, it becomes evident the system should be upgraded and/or replaced. The normal method is to program a replacement in the O & M program and await its turn for contract action. Due to the current less than enthusiastic support for energy projects and other funding constraints, approval and funding for such a project will be delayed at best. This paper will discuss another approach that has proven successful; however, there were many problems to overcome.

INTRODUCTION

The Energy Management Control System for Sheppard Air Force Base was started in 1976. It consists of a Honeywell, Inc., Delta 2500/Level 6 system with approximately 8000 points in 100 facilities. The system is Class C, which means it provides maximum energy conservation through a centralized computer running various optimization routines while also providing the capability to troubleshoot HVAC and various other utility systems. Operation through the years has revealed short comings with the early EMCS software. The CRT did not have graphic capability. While the computer could still be maintained, it was becoming more difficult. The FID's were no longer a standard off-the-shelf item but could be obtained through special order thus increasing the cost and raising the question of supportability. It became increasingly obvious something had to be done. This paper describes the thought process and methods used which proved successful. Hopefully, some new or enhanced idea will be presented which will be useful to your program.

LONG RANGE PLANNING

Where Do You Want To Go: By now everyone with an active EMS should have a master plan. This should cover more immediate items such as adding buildings or points that are not needed. This in itself can be no mean task and it should be done prior to any upgrade action. Due to the many field alterations, authorized and unauthorized, what you see on the screen is not what you have. Once the floor plans and AHU zones are corrected, the CRT display, I/O summary and card slot identification are matched. You may be surprised to see the number of points on the I/O that don't exist, or the card may be in the FID but in a different slot or all of that is correct but the sensor does not exist, or is sensing something entirely different. All of these type problems must be corrected prior to new hardware installation as the new systems may not be as forgiving of errors as the old work horse and just not operate.

Along with the new front end, you should look at your data transmission medium. At Sheppard AFB we have our own system to maintain. This system is like any other utility line that should be programmed for replacement. We have the added problem of numerous spring storms that play havoc. The decision was made to go to a fiber optic system. The natural extension is to put our fire alarm, security alarm, intercom, and any other computer data links that may be needed on to the same line. The first phase of this is scheduled for FY88.

With the data transmission system planned out, your attention can be focused on the hardware. Knowing that the future lies in DDC, selections should be made in that light. Also, funding constraints will not allow for a complete front-end/FID replacement so some type of phasing plan needs to be developed. After reviewing the many options we decided, due to cost/phasing constraints and technical problems, to stay with Honeywell so the system could be changed out in stages and finally end up with a DDC system when authorized by Air Force.

This decision process should not be done lightly. Doing your homework early on will pay off during the procurement phase.

How To Get There: The most obvious and cleanest method is to program phased projects in the O & M program. This has been very successful in the past. However, since energy is not a top issue, it is difficult to get sufficient priority from the Facilities Board, so your projects, even the first phase, may be two or three years off. Some work, due to its magnitude, must go the O & M contract route. This work can be broken down so the phases are not unreasonable in cost. Also, if the phases are delayed, the overall operation is not affected.

There is another source to use and that is through your squadron's funds and that being the supplies account. This is the method that Sheppard AFB selected and will be described below.

PROCUREMENT PROCEDURE

Using Supply Funds: HQ Air Training Command had been working with Honeywell on various methods of computer procurement. The particular unit Sheppard AFB wanted was available through a Honeywell GSA contract. The computer was broken down into small components which could be purchased as money became available. The thinking was that it could take more than a year to get the computer in this fashion but it would certainly be faster than the Contract Program route. This appeared the direction to go.

A problem did arise over the fact that the new

computer would not talk to the existing FIDs through the existing data transmission medium; however, it could through fiber optics. As the F/O replacement was doubtful any time soon, it was decided two interface modules would be purchased so two channels could be connected. By the time all the bugs were ironed out on the first two channels, the other projects should have been completed allowing complete changeover.

The first step was to prepare the supply documents. The same procedure was used as would be if a replacement pump was ordered. After everything was prepared, it was placed in an awaiting funding file. Constant contact was maintained and as funds became available, items covering that amount were released to Material Control. When the first request hit Supply the first hurdle was encountered.

Hurdle 1: This problem is one that anyone with an EMS will face yearly. When Base Supply sees your request and realizes it is computer-related, then they immediately want to send the request through the Data Processing approval route. Your battle is to show your EMS is real property. There is clear guidance that Base Supply has accountability for PCs, etc., to a certain value and Communications community for other ADP equipment. Unfortunately, the EMS systems were not addressed. The fact is, the last time I checked, Headquarters didn't know how to handle it. The burden then is on the requester to prove EMS is Real Property.

While it usually takes several letters, two things have proven successful. The first is found in Air Force Manual 172-1, Vol I (c3) para 8-15. This deals with Base Procured Equipment. You must show this item cannot be classed as equipment. The paragraph describes what type funds are used for items including RPIE. The key is in paragraph b.(1) and (2) which reads as follows:

> b. The following are not financed with 57*3080 BPAC 84xxxx funds.
>
> (1) Equipment costing less than $5,000.
>
> (2) Spares, components, assemblies, repair parts, and other material (expense or investment) that do not qualify as end items of equipment.

The key is the part I underlined. You must prove the item is not an end item.

The second document found helpful is Air Force Manual 86-2.(c8) which gives Category Codes to Real Property items. Paragraph 23-5 cover EMCS nicely. For added strength, I include Air Force Manual 300-4, Vol IV, which gives the Civil Engineering Cost Account codes.

Aggreement is reached and you are home free. However, remember that the system is being purchased in pieces over some period of time so with personnel changes, you may have to go through this time and again. A clear cut policy from Headquarters is surely in order and hopefully, in time, will be provided. Now comes the biggest hurdle of all--Procurement.

Hurdle 2: When the request hits procurement the same questions are asked and you have to go through all the steps of Hurdle 1 again! This does not take too long as anyone they would contact for answers you have convinced in Hurdle 1.

The next question is why bits and pieces are being purchased? Are you trying some funny business to get out of approval requirements or funding limitations? These are valid questions. The key here is that the purchases will take place over a long period of time as funds are provided. You would prefer to do it at one time, but the system needs replacement and can't wait the three years or more it looks like it would be before it could be replaced through contract action. Be prepared to back up your claim of urgency for replacement with downtime reports, increased maintenance problems, etc.

The next problem was to fight the brand name procurement. Procurement wanted to go out for bids from any supplier. This could lead to disaster. Great effort must be expended to carefully prove that direction is not the way to go. It is difficult for a buyer to be an expert in all areas, so great care must be exercised in helping them understand the situation. This process included many letters, phone calls, and face-to-face dialogue and can take a month or more.

With this problem resolved then, since using a GSA contract, procurement would proceed. This may not be so. There are two types of GSA contracts--mandatory and not mandatory. Honeywell's was the latter. This means Procurement could use it or not. Due to Procurement's attitude through the whole thing, they chose to use the Contract as a bid and went on the street for other quotes.

At first there was real concern going this direction, but it worked out fine and was even cheaper than the GSA contract. One surprise was the purchase of a rebuilt CPU. There is a Federal Acquisition Regulation 10.010 that covers acquiring used or reconditioned material. It gives the authority to the Procurement officer to do this if they choose.

One other point to remember that if the item costs over $25,000, then a whole new set of Procurement rules come into play. Avoid that if at all possible. All the parts are in and you are ready for Hurdle 3.

Hurdle 3: Funds are needed for the installation of the system which includes the software, training, programming, etc. This was identified as a Sheppard project and the Air Training Command Inspector General added support to the need for replacement. Local funds were made available by Budgeting. At this writing, the contract has not been released. Our position is that it should be a sole source procurement to Honeywell, Inc. Supporting documents have been provided by Headquarters, Air Training Command. The package is at Base Legal. Preliminary words are that Legal agrees it should be sole source. The final decision, however, will be with the contracting officer.

CONCLUSION

There are many hurdles to overcome following this procedure. However, if you do your homework well, it works. It has taken about one year to get to this point which is much quicker than the normal action. A key player is the Contracting Officer who can be a real help or a hindrance. We must recognize their position and work within their system for success.

Think things through, get all the facts possible, and develop a clear, long range plan, then start working it. Before long you will have a state-of-the-art EMCS.

Chapter 22

IPDFACS: A STANDARD FOR EMS

J. P. Cilia

ABSTRACT

The trend toward lower cost of new technologies in building automation and energy management in the 80's has forced the controls companies to produce individual microbased Distributed Direct Digital Controls (DDDC) systems. The impact of new technologies unfortunately raises a tremendous number of problems for large facilities, such as incompatibility between different manufacturers' data communication protocols, configuration networks, message formats, sensors, and source program languages. This paper describes the IBM Poughkeepsie Distributed Facilities Automation Control System (IPDFACS) standards used in the bidding process, the philosophy behind them, the type of instrumentation specified to be installed, and how to maintain this complex DDDC system.

INTRODUCTION

The lower cost of chips has been one of the principal forces in the development of new technology in the building automation and instrumentation fields. Fast changes are causing manufacturers to engage in research and development rather than investing in their existing hardware. These new directions are complicated by the fact that commercial grade hardware and instruments do not comply with any standard due to the lack of such guidelines in this trade.

This condition impacts our future strategies for new and retrofitted builddings. Since no guidelines are currently available,we have prepared for our site a set of standards that will assist facilities engineers, outside consulting engineers, and maintenance departmetns.

The purpose of IPDFACS is to standardize:

- Basic instrumentation
- Systems
- Specifications
- Technician training programs

IPDFACS does not represent the only way to design, install and maintain a building automation and energy management system, but it does reduce some of the problems caused by new technologies.

INSTRUMENTATION STANDARDS

The first and most important part of the instrument standardization process was to select the quality of instrument to be used. The selection of instrumentation was a choice between the commercial and industrial grades. It was determined that the cost of installation of any similar instrument in both grades was the same. The actual instrument cost was about 250% higher for an industrial grade than the cost of the commercial grade. The approach to evaluate the cost effectiveness of this extra burden encompassed the precision of the instruments, the coverage under warranty, and the convenience of replacement and repair.

Precision

Precision of an industrial grade instrument can reach .1% accuracy, however, this level of accuracy may not be required. A commercial grade instrument is accurate to 1 to 2%, but it may not retain its level of accuracy as long an industrial grade instrument. Practically, it is required to calibrate a commercial instrument twice a year versus once a year for its industrial counterpart. Savings which may be generated by selecting the industrial grade instrument are:

- Calibration maintenance cost reduction by 50%.
- Energy savings due to higher accuracy.

Warranty

The warranty of a commercial grade instrument is one year, as compared to two or three years for the industrial grade instrument. Assuming that the life expectancy of any of these instruments is longer than the warranty period, a defective part will be detected during the first few months of its operation in both types. An additional warranty, over a year in this case, is used to cover the construction and installation time required to complete the project. In general, a small job can take a minimum of six months, while a larger project may take over a year or two.

Any contractor can install a "new" instrument with its warranty period already expired, but we have generally stated in previous specifications that parts and materials should be warranted for one year after being approved by the site. For example, if a defective instrument is detected during the first few months of operation and its warranty is expired, this problem would be handled by the contractor. The contractor will undoubtedly involve facilities engineers and/or maintenance departments if the instrument manufacturer requires a repair fee for the defective instrument. With the extended warranty of an industrial grade instrument, reduction of time and money lost by Facilities Services will be considerable.

Replacement and Repair

For this example, assume that both types of instruments are "out-of-warranty." The change in the field differs, depending on the grade of the instrument. Again, due to the lack of standards, one of the pitfalls of the commercial grade instrument is the possibility of a different type of instrument being provided. Thus, incompatibility will result in an additional cost for labor and material by adding and/or modifying the wiring and piping of the instrument. Additional time is required to understand, maintain, and train personnel for this new product.

In the case of trying to repair the instrument, the chances of having parts stocked by the commercial instrument manufacturer are greatly reduced, due to the small difference in the cost of the instrument versus the cost of the parts and the labor involved to repair it. Therefore, an additional premium must be paid to replace this instrument. The reliability of the industrial instrument is greater than the commercial one and so it is well worth its extra cost.

Advanced Manufacturing

The complexity of advanced manufacturing facilities relies heavily on the proper functioning of the building automation. The need for reliability and the consequences of a shut-down due to an inferior quality instrument is more evidence to promote the use of an industrial grade instrument.

Selection

Selection of specific industrial instruments in design and application is based on our experience and on feedback from the field. The selection of the "IPDFACS Instrumentation Standards" was reviewed by our facilities engineers and by our maintenance departments to modify our selected list of instruments before being adopted for our site last June. Each facilites engineering and maintenance department was provided with 600 pages of specifications, installation maintenance bulletins, and price sheets for instruments included in our two volumes.

The "IPDFACS Instrumentation Standards" Volumes I and II facilitate the comprehension, design, installation and maintenance of each instrument. The result has been time saved for each trade whenever we design, purchase, calibrate or replace and instrument, thus promoting cost effectiveness.

IPDFACS System Standards

The purpose of this section was to provide "templates" of each type of heating, ventilation and air conditioning (HVAC) system. Because automatic control of HVAC systems requires mechanical, electrical and electronic control devices, the "templates" have been prepared with the following sections.

HVAC Schematics - Each template is made of an HVAC Schematic showing the duct work, fans, and the location of instruments required to perform as per the description of operation. All instrumentation symbols and identification used in our standards are from the Instrument Society of America Standard ISA-S5.1.

Description of Operation - Each type of HVAC system to be used and operated has a complete descriptive sequence of operation with all strategies included to provide maximum energy conservation. The engineer selects the strategies he/she desires to solve a specific application. The solution of more exotic and complex problems can be easily implemented directly on CADAM, without totally rewriting the description of operation of the system. While designing these systems, we included the following strategies for potential energy conservation:

- Time of Day (TOD) and Day of Week (DOW) Schedule
  This will permit starting and stopping equipment only when required.
- Night Set Back and Set Up
  This strategy will maintain the lowest and/or highest temperature and humidity permissible in the building without the use of fresh air intake. This method will override capabilities of each controllable point until the desired night temperature and humidity have been reached.
- Optimum Start Up
  This is provided to delay equipment start-up as late as possible before building occupancy and without the use of fresh air intake. It is based on

the inside and outside building temperature and humidity.

- Cycling Control
  Controllable equipment can be turned off if desired after the TOD programming mode is started.

- Demand Control
  Each controllable device can be selected to be turned off or the set point may be adjusted to reduce the electrical demand in the building after the TOD programming is started.

- Economy Cycle
  To maintain a mixed air temperature in the range of 55 to 60°F (adjustable) in the air handling unit (AHU) system by modulating the outside and return air dampers.

- Enthalpy Control
  Uses 100% cool outisde air if the total heat content of the return air (enthalpy) is greater than the total heat content of the outside air while the system is running during the cooling mode.

- Dead-Band between Heating and Cooling
  This strategy prevents the AHU from supplying simultaneous hot and cold air at the discharge of the AHU.

- Dead-Band between Humidification and Dehumidification
  This strategy prevents simultaneous humidification and dehumidification at the AHU.

Control Loop Diagrams

Each application described above has individual wiring and piping loop diagrams of each instrument specified in our Instrument Standards. The "templates" provide detailed hookup information for all loop components from the field device to the Local Microprocessor Control Panel (LMCP).

The loop may be selected by the engineer to solve his/her special application by simply modifying the template already stored on CADAM. The solution of a more complex system or for special applications can easily be implemented directly on the template without redesigning a total system.

Field I/O Devices

Each "template" has an I/O device list with a cross reference between the tag number, the item, and the area served. In addition, the manufacturer's name, product number, range of the instrument, and the cross reference to our IPDFACS Instrumentation Standards volumes has been included to simplify the search for the type of instrument to be purchased.

Control Available to Operator

In this case, we provided one section for the points, reports, graphics and point names that will be used during the data base generation (DBG) by the controls contractor at the PC/AT to be included on the DBG at the host computer. This prevented the contractor from incorporating a point name that was different from the one on our host computer.

These prepackaged IPDFACS systems standards templates stored on our CADAM system are available to all our facilities services departments. In addition to keeping future projects more homogeneous in our site, engineering and design time to all parties concerned is reduced considerably.

IPDFACS SPECIFICATIONS STANDARDS

In the past, the mechanical contractor subcontracted the instrumentation to a controls contractor. At the same time, the electrical contractor had to wire these instruments mounted by the mechanical contractor. This approach is more complicated when you have a general contractor that coordinates the work of all its subcontractors. When problems occured, it was often difficult to pinpoint the responsible party.

Single Source Responsibility

We simply combined these tasks under a single source of responsibilty in section 17000 of our Specifications Standard. The controls contractor is fully responsible for the complete installation and proper operation of the IPDFACS system. This includes, but is not limited to interface of the PC/AT to the S/1, and from the PC/AT to the LMCPs, memory units, peripheral devices, communication links, sensors and controls. The contractor is reponsible for the debugging and calibration of the IPDFACS system, including all software.

Manufacturer Training

Instruction of IBM personnel is provided under our specification by the controls contractor to provide competent personnel to give full instruction in the adjustment, operation, and maintenance of the requirements of all equipment specified.

IPDFACS TECHNICIAN TRAINING PROGRAMS

The challenge of today's new technologies in the maintenance of the building automation and energy management system, relies heavily on the training of technicians in the correct instrument maintenance and repair procedures. This is the program that will make or break any similar organization in the corporation. After extensive research, the ISA - Instrument Technician Training Program (ITTP) was selected for our site. A seminar was prepared by the ISA training center in Raleigh, North Carolina to add specific IBM training requirements and to set up a course schedule with different instructors from ISA at the Poughkeepsie site. Preparation of the classroom, laboratories and equipment requirements were negociated for a successful program. The description of our first training program last spring was as follows:

- Instrumentation basics
- Instrumentation maintenance
- Pneumatic instruments
- Microprocessors and digital systems

Each section covered troubleshooting, adjustment, calibration, replacement and repair with emphasis on safety. Approximately 60 hours of videotape were included with over 220 hours of classroom instruction and hands-on experience with equipment. Over 220 hours of out-of-class studies and applications were included in this program which will be repeated twice a year for the next two years.

A special agreement regarding ITTP was initiated by Poughkeepsie to provide better prices to the IBM sites. This agreement was given to RECD last February for corporate evaluation.

CONCLUSION

IPDFACS standards are by no means the total solution to all problems we encounter because of the continuous changes in the technologies of building automation and energy management systems. However, IPDFACS standards serve as a tool to improve and reduce the cost of engineering, installation and maintenance. Further, they serve to improve the reliability of the site operation supported by trained technicians and, therefore, reduce production costs and increase plant efficiency.

Chapter 23

AN INTEGRATED ENERGY MANAGEMENT SYSTEM FOR PLANT AND HVAC FACILITIES: A STEP AHEAD

G. D. Debban, A. Kaya

ABSTRACT

This paper describes a totally integrated energy management system throughout the plant and facilities. The equipment involves: the power plant (boilers, cogeneration turbines, chillers); HVAC of buildings; air handling and other distribution systems; and the equipment for the electric load demand management. The implementation is done by the distributed control system suitable for the total spectrum of equipment involved.

The energy management system has the architecture for data transmission and processing including multilevel control and optimization through the plant and buildings. The key step in this work is to interface the power plant control and management system with the rest, under a unified energy management concept.

The main advantages of the system has been: (1) the ability to optimize and coordinate the overall operation under the conflicting optimum goals of each subsystem at the highest level, (ii) to control the feedback loops for their optimum operation at the lowest level, (iii) to communicate between the levels of control hierarchy, all from the same operator station, and (iv) provide the high reliability and availability of the system.

The communication throughout the plant is provided by the plant loop bus to which operator interface units, other process control cabinets, as well as HVAC and air handling control units are connected.

It has been demonstrated that the additional benefits such as energy savings, and the reductions on equipment, human resources, maintenance, and training expenses are realized. The operational displays of various subsystems including the live data on plant status are shown by the same operator interface unit.

INTRODUCTION

The recent trends in control systems hardware present an increasing uniformity and controlling a broader plant equipment. Specifically, the process control suppliers introduced the controls for industrial power plant and so did the power plant suppliers, conversely. Also, the same pattern is seen in commercial and industrial buildings where the same equipment from a single supplier can control and manage the complete facilities.

This trend is due to the flexibility and control power of new distributed control systems creating a reduction in number of plant operators and a change in their role. The advantages exist in using a uniform operator console for easy interaction. Furthermore the uniform hardware reduces spare part needs.

This work describes the combination of powerful energy management functions for power plant and for HVAC facilities under one uniform system. Such system performs the necessary functions throughout and coordinates the total plant operation for overall maximized efficiency.

Background on energy management

Increasing energy and accordingly the cost of electricity has brought the attention to the control of HVAC systems for minimized energy use. The problem has been considered in buildings which include the power plant and the HVAC spaces where energy is consumed. Overall schematic of the system is shown in Figure 1. However, energy management is presented and discussed for both power plant and for HVAC spaces separately but not totally coordinated.

There are numerous energy savings measures exercised by means of common sense operations. Some of them are: control of lighting; shedding the loads on increased electrical demand; operating the air moving fans intermittently to save energy. The interest here is on control and optimization methods to improve energy efficiency and still to provide the necessary comfort. This has been the main driving force of energy management.

Opportunities

The energy management in Power house (boilers, turbines, and chillers) have been extensively studied, designed, and implemented with proven performance records [1-5]. No details on the power house equipment will be given as they are described in References 1-5. There may be multiple chillers and boilers supplying the same chilled water and steam line respectively. Figure 2 describes one zone of HVAC system and air handling unit (AHU). Note that the climatizer contain heating/cooling coils and humidifier. Obviously there are many AHU's within the system. Energy savings for each area of the plant are given in the following.

Energy saving opportunities in Power plant through control and optimization are given in Table 1 [1-5]. Note, that, the load allocation functions are based upon multiple units serving a common supply.

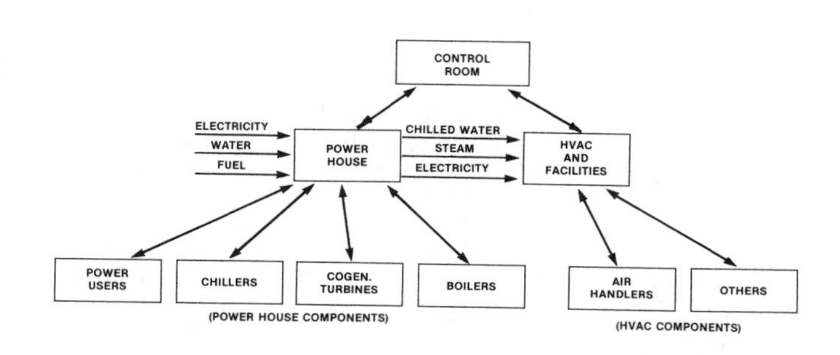

FIGURE 1 — Total Plant Schematic

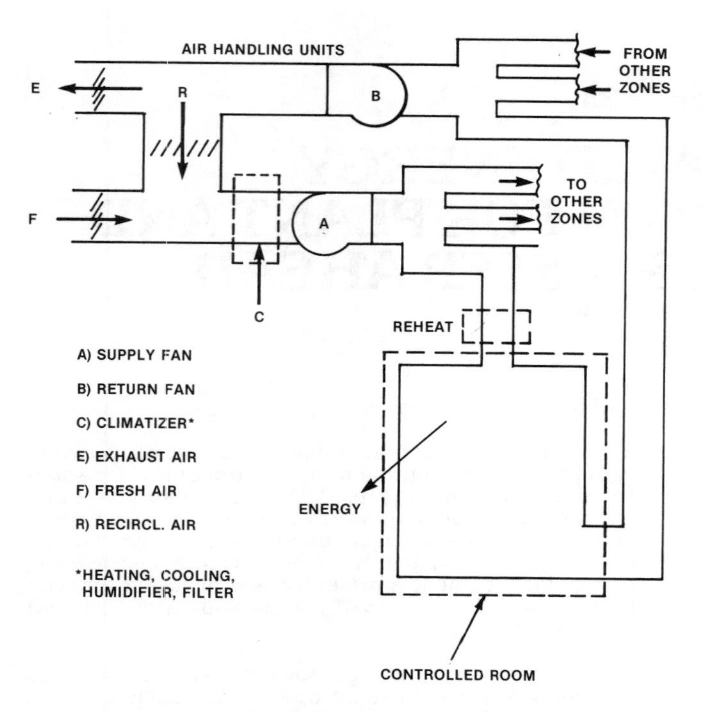

FIGURE 2 — HVAC Space and Air Handling Unit (Terminal Reheat)

TABLE 1. ENERGY CONSERVATION OPPORTUNITIES IN POWERHOUSE

ENERGY SAVING OPPORTUNITY

UNIT OPTIMIZATIONS

1. Optimization of chilled water supply temperature
2. Optimization of condenser water temperature leaving cooling tower
3. Optimization of heat recovery of condenser discharge

OPTIMUM LOAD ALLOCATION

4a. Optimum pumping load allocation

4b. Optimum boiler load allocation

4c. Optimum chiller load allocation

5. Optimum Turbine Cogeneration.

The driving force behind the energy conservation in a HVAC space is based on the enthalpy measurement which is the function of two variables, dry bulb temperature and the relative humidity. The comfort conditions in a HVAC space are affected by three variables; dry bulb temperature, relative humidity, and air velocity. Figure 3 describes the control region in psychrometric chart (area ABCDE) within which the variables of comfort for the HVAC space are kept. If the properties of outdoor air fall within this region, the introduction of 100% outside air into HVAC space is justified. If the outdoor air properties are outside of this region, either minimum outdoor air is introduced into the HVAC space or the outdoor air is mixed with the return air. This philosophy is commonly practiced. Also "on-off" control of air handling units are still exercised.

The comfort region per ASHRAE standards is also shown in Figure 3. However, the HVAC space must be within this region to satisfy the comfort requirements. Based on this information and on common sense ideals, the energy conservation opportunities are given in Table 2. Furthermore, the opportunities at the highest plant energy management level exist as given in Table 3.

TABLE 2. ENERGY CONSERVATION OPPORTUNITIES IN HVAC AREA

ENERGY SAVING OPPORTUNITY

FIRST LEVEL FEEDBACK CONTROLS

1. Modulating control of AHU supply temperature and humidity for accuracy.

SUPERVISORY AND OPTIMIZATION

2. Supervisory set point adjustment of supply air properties.
3. Enthalpy-based control of dampers for minimized cost of supply air.
4. Optimum start/stop strategy of equipment for energy savings.
5. Power demand management to reduce the electric cost
6. Fixed-time scheduled start/stop operations and control-point settings.

TABLE 3. ENERGY CONSERVATION OPPORTUNITIES BY OVERALL MANAGEMENT

ENERGY SAVING OPPORTUNITY

1. Supervisory pumping load and cooling load coordination
2. Optimum load scheduling
3. Tie Line Management

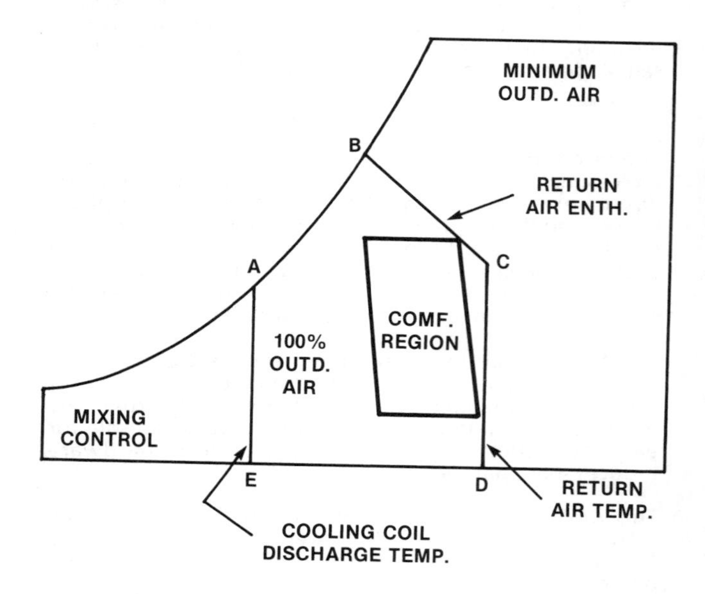

FIGURE 3 — Comparison of the Control Methods in Psychrometric Chart.

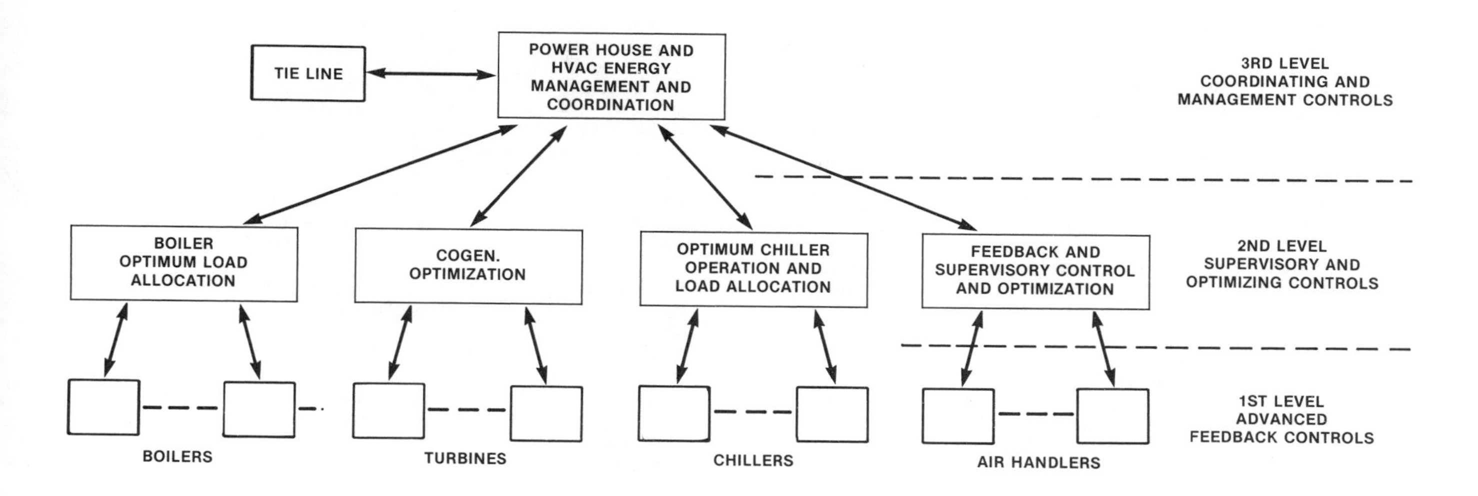

FIGURE 4 — Plant Control and Management Structure and Functions

II. ENERGY MANAGEMENT STRUCTURE AND FUNCTIONS

Energy management is based on plant information that is the measurements. Energy management system is structured by considering the plant equipment, the functions to be implemented, and the hardware to be used. Here, the energy management structure is shown in Figure 4 along with the functions implemented in each block. The opportunities listed in Tables 1-3 are implemented within this structure.

Energy Management Structure

Figure 4 is structured by proper partitioning of the equipment as well as by delegating the control functions at various levels in a practical and economical manner. The dedicated feedback controls are used to control a specified variable considering the dynamics of process, and the controls must be responsive. At the higher levels, the dynamics are not significant and usually ignored; but, the process scope is larger and many variables are reduced to a few by aggregation, and dealt with in a slower time frame. A simple example to that is the feedback control of pressure or level in a boiler, versus the boiler efficiency which includes many variables but no significant dynamics concern. Next, the functions in this structure are discussed.

Energy Management Functions

Referring to Figure 4, the energy management and control functions related to Tables 1-3 will be discussed. The efficiency improvements through tight control of process variables about their set points in single units have been reported on extensively. However, the adjustment of control set points for maximized unit efficiency is left out frequently. supervisory and optimizing controls for chillers are given below. Referring to Table 1 and the second level in Figure 4, the chilled supply water and condenser water temperatures are optimized by adjusting the set points for a maximized chiller efficiency under partial cooling loads.As the chilled water temperature increases and as the condenser water temperature decreases, the efficiency of the refrigeration cycle increases. The incremental change in efficiency is about 1% per 1F in both cases. The details of the concept and implementation are given in References 3 and 4.

For the optimum heat recovery of condenser discharge, the similar concept holds. Normal industrial plant hot water control systems supply hot water at constant temperature. If the hot water supply temperature is allowed to vary while still meeting hot water heating demand, then the cost of both refrigeration and steam can be reduced. The energy saving strategy is based on utilizing the minimum hot water temperature necessary to meet hot water heating demand. The reduction of the hot water supply temperature means the reduction of the hot water temperature at the split condenser, which in turn reduces the discharge temperature and pressure of the compressor (reducing the power input to the compressor). Cooling tower loads are reduced since condenser heat is partially dissipated by reheating the plant hot water supply. Steam use is also reduced since some of the hot water heating energy is recovered from the condenser. The control policy on this saving is given in References 2 and 3.

Continuing on Table 1, the next is optimum load allocation for overall maximized efficiency of boilers, chillers, and pumps. Although each unit is operating at its highest efficiency (lowest cost), additional energy savings are possible with multiple units operating in parallel. This is possible because of differences in unit performance and variations of energy cost.

The optimum load allocation algorithm is configurable in distributed control instrumentation function blocks without using a high level computer program [5, 6]. The implementation is given in Reference 4. These functions are handled in the second level in Figure 4. The last item in Table 1 is optimum turbine cogeneration which is handled in cogeneration optimization block in Figure 4. The goal of cogeneration optimization is to minimize the cost of cogeneration and satisfy the steam load demand by utilizing the most efficient extraction flows and/or condensing flows. The decision of using extraction or condensing flows (or both) is made at the higher level (3rd level, Figure 4) by tie line management. This topic has been studied and reported in References 7 and 8 and implementation is shown in Reference 4.

At this point, the first and second level controls for air handlers are discussed (Figure 4). Now, the opportunities in Table 2 are explored. The first level feedback controls of AHU's should be controlled by modulating the valve positions via PID and more advanced control functions as needed. This approach provides more accurate control than the selenoid-driven two-position controls. Figure 5 describes the instruments, valves and other actuators etc., which are interfaced with and controlled by distributed control system. Accurate control of variables about the set points minimized the deviations from optimum values to proved energy savings.

Continuing on Table 2, the supervisory set point adjustments for feedback controls provide a minimized cost of energy by still keeping the HVAC space within comfort region (Figure 4). A point within comfort region in Figure 4 corresponds to a minimized energy cost. Then, the temperature, relative humidity, and air flow control set points are adjusted to keep the HVAC space at the optimum values corresponding to that point. This function is performed by the second level control of air handlers (Figure 4).

Optimization routine calculates the set points of variables (temperature, RH, velocity) corresponding to minimum energy use for given outside and return air conditions. If the outside air is in the comfort region, outside air and room air can be mixed to maintain the room air within the comfort region with no energy use. Since outside air is usually out of the comfort region, the comfort variables corresponding to minimum energy use, fall on the boundary of the comfort region. Based on that, optimum set point values are found on the comfort boundary corresponding a minimum energy use of air handlers. The optimum may well correspond to minimized fresh air in space.

At the same time another enthalpy-based supervisory function (Table 2, item 3) operates the dampers to introduce the appropriate ratio of outside and return air into the system to minimize the energy use. The control region in Figure 4 is indicative of the damper positions. Multiple enthalpy measurements and calculations are performed to make different damper control decisions for different zones.

The optimum start/stop strategy in (Table 2; item 4) avoids waste of energy during the change of the comfort requirements of HVAC space. The function optimizes the start time of HVAC equipment and air handling units so that the space is within the comfort zone as the occupancy begins. This is done by the set point adjustments. Also, this function provides gradual loading (soft start) of the equipment for demand limiting. The same strategy occurs as the space occupancy ends.

The power demand management function in Table 2 minimizes the electric energy consumption at peak load times as signaled by the energy management level (3rd level, Figure 4). This occurs by cutting back the energy use or turning off the equipment (AHU's) based on the priority, rotation, and the measurements in HVAC spaces.

The fixed-time scheduled start/stop operations and control set-point shiftings are provided based on the schedule of occupancy (holidays, weekends etc.). The calender can be converted into decimal values and the equipment is scheduled by simple instructions.

Table 3 introduces opportunities at the energy management level. The supervisory pumping and cooling load coordination minimizes the summation of these two energy use. Consider the coordination between chilled water temperature optimization and pumping system optimization. Chilled water optimization maximizes the chilled water supply temperature while meeting cooling demand by increasing the chilled water supply flow (and consequently the pumping energy). In contrast, pumping system optimization minimizes the power required to deliver chilled water. Pumping cost depends on pumping load demand, and this demand is lower with decreased chilled water supply. The goal is to minimize the sum of these two costs. The trade-off between pumping energy and compressor energy is coordinated at a supervisory level. While chillers and pumps operate at their optimum, a performance measure determines the total energy cost versus chilled water temperature and the incremental cost, accordingly. Whenever the incremental cost is zero, the optimum is reached. The control algorithms to implement it is worked out in References 1 and 2.

The supervisory load scheduling is a future policy. So far, the control decisions have been a single-decision policy based on the current operation of system. There are further cost savings if the control decisions are based on a multi-decision policy, considering a time span. Future estimations and projections are made and a control policy is followed to minimize the cost over a time cycle (hour, day, week, etc.). The main factors to consider are the weather, peak electric demand, and equipment availability.

As an example, consider a day with a high cooling load. Plant is operating at its minimum cost with an optimum chilled water temperature based on a single decision policy. As the cooling demand increases, the chilled water supply temperature must be decreased (or pumping must be increased) to respond to the demand. This possibly means an increased electric use beyond the limit of peak demand. The result may be high energy cost. If there is a maximum limit set on electric peak demand, insufficient cooling will result.

Multidecision policy provides sufficient cooling by still staying within the limits of electric peak demand. Chilled water supply temperature is scheduled to drop to a lower value than the optimum before the cooling demand increases. During the high cooling load the electric use stays within the limit of electric peak demand and chilled water temperature does not increase beyond the threshold value of discomfort. The capacity of chilled water within the system plays an important role in implementing this concept.

Supervisory scheduling is provided by an optimization routine such as dynamic programming or linear programming. The data such as cost versus load from load allocation hierarchy (2nd level) is received by the optimization routine, along with other information (weather, condition, electric demand limit, unit capacity and availability).

The necessary computer program is within the capability of a distributed control system. A host computer can also be used and distributed control system communicate with host computer.

The last is tie line management. Electrical power requirements can be satisfied from three sources of power:

Purchased power
Power generated by extraction flow
Power generated by condenser flow

The power generated by extraction flow is fixed by steam requirements. Power generated by condenser flow is controllable, but expensive. In some cases plants can purchase power at a lower cost than they can produce it by condensing. This is usually during the "off-peak" demand period. The situation may reverse during the "on-peak" period. In order to determine how a plant may reduce its cost of electric power, the utility's rate structure must be examined. Utility rate structures and methods of computing monthly bills vary considerably from utility to utility. However almost universally the cost of power is based on the "billing demand". Some utilities make a direct charge for the "billing demand" while others step their charge for KWH usage downward as the monthly average KW approaches the "billing demand".

Analysis at several plants have shown that utilizing power generated by condenser flow only during the on peak periods will reduce the "billing demand" and the cost to supply the plant's electrical requirements. The target on peak demand setpoint must be established by a review of the plant's electrical usage history and operating experience. A tie line management problem of determining the purchased and generated power is worked out in Reference 9 under one sort of rate schedule.

The usual philosophy is that the increased plant demand must be delivered by the source with minimum incremental cost $/kwh. In case the demand decreases, the reduction should be made from the generating source with the maximum incremental cost. There are cases when increase in generation or in purchase is not feasible. Then, the least significant loads should be shed.

Tie line management decides on the source of electric generation. The cogeneration optimization at the lower level generates electricity optimumly from the specified source. Further details on tie line managements is given in Reference 8.

IMPLEMENTATION

The implementation of energy management is done by using the measurements which are already needed for the first level feedback controls. Using these measurements, the performance of each unit or component is calculated for monitoring optimizing etc.

Energy Management Application

It is well known that accurate measurement and control can improve the boiler efficiency for energy conservation. Boiler efficiency is calculated in real time based on the transmitter readings, boiler design data, fuel analysis, etc. The boiler efficiency information is the basis for minimizing the steam cost when multiple number of boilers serve a common header. The implementation is done by generating the optimum load values of each boiler coming out of optimization module which operates in real time as well. The load of each boiler is changed through the boilermaster. The specific details of this implementation as well as for other units such as chillers and pumps exist in application guides [4].

The methods and implementations of energy conservation and management for power house which includes the boilers, turbines, and chillers are well covered by previous literature with proven performance records [3, 5, 6, 7, 8].

The emphasis is placed on discussing the implementation of energy, conservation and management functions for HVAC area including their integration into one uniform energy management system. The implementations of these functions listed in Table 2 will be treated in the respective order.

The lowest control level for HVAC area is the feedback control of an air handling unit (AHU). Modulating control provides closed loop control for continuously varying quantities, such as air flow, temperature and humidity; compares sensed or calculated values to either fixed or calculated setpoints, then sends appropriate control signals to the device actuator. The instrumentation of AHU is shown in Figure 5.

A typical air handling unit (AHU) needs three control loops for heating coil, and cooling coil, and humidifier, plus accessory control devices to make them all work in harmony. How these control loops operate has a major effect on the amount of energy used to condition the air. All these automatic control and accessory devices determine how much energy is used to maintain desired environmental conditions in a building.

The PID control is commonly known. The set points of these controls are adjusted at supervisory level for minimized energy use. The transmitters and actuators (valves) for the variables are respectively: TT_1 and V_1 for cold deck temperature control; TT_2 and V_2 for space temperature control; RH_1 and V_3 for humidity control.

As for the safe and harmonical operation, the TT_3 measurement is used for low limit to prevent freeze up. The PT_1 measurement monitors the fan operation and filter status. The interruptions are provided for each actuator for safety and energy conservation as well. The PID control functions are primarily implemented by local control modules (LCM) with supervisory adjustments from multifunction controller (MFC) and/or from engineering work station (EWS).

PID control provides fast and responsive performance; decreases the differential of controlled value; and eliminates hunting, overshoot, and steady state error.

PID control generates pulsed electrical control signals or voltage and current levels needed by the actuator to position the controlled device, bringing the controlled variable to within established limits. In applications requiring pneumatic actuator control, pulse width modulated (PWM) control signal is sent to device controller, typically a Pneumatic Interface which translates the control signal into the bleed/feed action moves the actuator. The pneumatic actuator controls the device (such as valve) that directly affects the controlled variable.

The supervisory set point adjustments of control points to modify the supply air properties and eventually the HVAC space are calculated based on outdoor and return air. For example, the setpoint of a hot water supply can be raised (according to the reset schedule) as the outside air temperature decreases. This permits the HVAC system to automatically compensate for the increased heat loss caused by the falling temperature of the outside air.

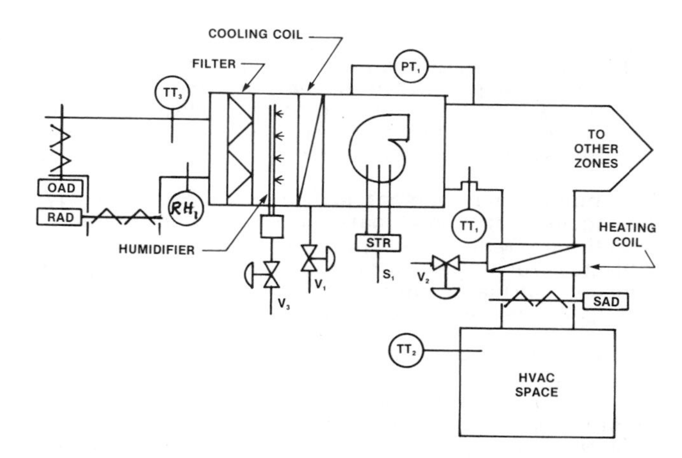

FIGURE 5 — AHU Instrumentation (Terminal Reheat)

Also, a reset schedule defined by the operator may determine the amount of change permitted in the controlled variable. The return air temperature and enthalpy is measured to adjust the supply air temperature, relative humidity and air flow to provide the comfort in HVAC space and minimize the energy use. Also, the outdoor and return air temperature readings are used to calculate the mixed air temperature corresponding to the desired amount of outdoor air admitted at full flow. The outside and return air dampers are adjusted for variable air volume (VAV) systems according to actual flow to provide a minimum specified level of ventilation.

The economizer and mechanical cooling bands control the operation of the economizer dampers and the cooling coil. Overlapping these two bands allows the most efficient use of economizer cooling. The heating band setpoints determine the heating coil operation. Temperatures in this band can be compensated by zone or return air temperature. Other setpoints, in addition to those shown, can be defined by the operator. The enthalpy based supervisory control of dampens provides minimized energy use. This function calculates the enthalpy content of outdoor and return air. Based on the comparison of these enthalpy values, the mechanical cooling, minimized outdoor air conditions are implemented by changing the damper positions. This function provides an improvement over standard economizer control by comparing the total heat contents, rather than just dry bulb temperatures of outdoor and return air.

A binary command is sent to all local control modules LCM that control AHU's and other equipment under start/stop control. If return air enthalpy is higher, normal economizer control occurs on AHU's. If outdoor air enthalpy is higher, the command overrides the economizer control and causes the dampers to maintain the minimum percent outdoor air setting. Equipment under start/stop control will be either started or stopped according to system requirements, based on the enthalpy comparison. Multiple enthalpy comparisons can be made for facilities that have varying interior conditions. The optimum start/stop strategy optimizes the start time of HVAC equipment so that the temperature of a particular area will be brought to the correct level by the time the occupied period actually begins. It makes up for the lag in HVAC system response by advancing the normal start time of heating or cooling equipment. It calculates the system response time and brings the temperature to the desired level.

When the temperature sensed by the controlling sensor is below the heating setpoint, the optimized start occurs to perform warmup; when it exceed the cooling setpoint, optimized start occurs to perform cooldown. The optimized start does not occur when the desired conditions are within the deadband.

The power management provides complete energy management functions that limit energy consumption, significantly reducing energy costs. Using information collected from other energy conservation functions, this function calculates energy usage trends necessary for energy-efficient HVAC equipment management.

By controlling equipment operation through load shedding, this function ensures the most efficient operation of the HVAC system possible without affecting comfort levels. Automatic failsafe back-ups assure energy-efficient operation by providing predetermined default operating strategies.

Based on demand limiting signal coming from tie-line management the selective load shedding is used. Temperature-based demand limiting is used in shedding AHU loads. AHU's closest to the desired setpoint are shed first. Those furthest from the setpoint are allowed to run as long as possible. Rotation-based demand limiting alternates operation of equipment according to minimum on and off and maximum off time settings. Priority-based demand limiting permits the user to select equipment for shedding according to preference. This permits equipment needed for critical processes to remain on, while others are shed in order of their priority status. Duty Cycling repeatedly cycles equipment on and off for designated time periods to reduce energy consumption without affecting occupant comfort. Each piece of equipment can be cycled individually. Or, groups of equipment can be cycled on a rotating basis. Each piece of equipment has user-definable entries for minimum on, minimum off, and maximum off time to avoid short-cycling (hunting).

Soft Start prevents excessive power consumption by staggering the starting of major HVAC loads after power has been restored following an outage or at the morning start.

A time scheduled start/stop operation of equipment and set point changes of controls can be implemented for energy savings to match the occupancy and operation of facilities. Clock/calendar schedules for controlling binary devices are used. Weekday, weekend and holiday schedules provide automatic control of each device. The special holiday schedule automatically cancels after one use, and any schedule can be overridden to permit controlled devices to be manually swtiched on or off. Scheduled Control can be used alone or in combination with other control functions for most flexible control priorities possible.

Modulated devices are interlocked to shutdown positions when the AHU is not running. Two fixed setpoints (occupied and setback) can be selected; plus, actuators can be forced open or closed based on time, any sensed value or a switch input.

All the above functions are implemented by LCM, MFC, and OIU or PCU. The tasks are distributed among these units in parallel to hierarchy of the control functions, as shown in Figure 6. LCM handles the tasks directly dealing with AHU. MFC handles calculations for supervisory and optimizing functions. OIU has displays for monitoring, and operator-interface for implementing the energy management. Also all these are communicated through a local area loop. This local area network is linked to NETWORK 90[R] plant loop so that the monitoring of the HVAC system as well as communicating the decisions of overall energy management (Table 3). It is also possible to integrate the HVAC management functions into the Process Control Units (PCU) as an option. However, the hardware features of HVAC equipment justify the existence of the proposed structure in Figure 6 to be practical and economical.

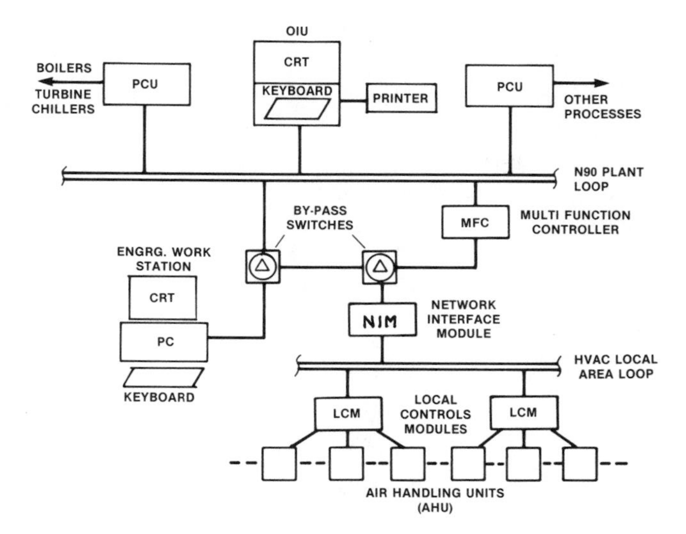

FIGURE 6 — Control System Architecture for HVAC Area

SUMMARY AND CONCLUSIONS

Benefits of energy management by distributed controls are indisputable. It has been demonstrated that energy management of HVAC systems can be achieved by multi-level control techniques and implemented by distributed control systems. Energy savings are substantial and additional side benefits are significant. The integration of HVAC energy management with the other power house control system provides a uniform system and additional benefits. The control system provides the distinct features [3]. The CRT operator interface unit replaces panel board instrumentation. It provides color graphics, alarming, loop control, and trend displays, along with tuning and configuration functions. furthermore, alarm management and instructions, energy management, procedural instructions are transmitted through CRT. Main source of tangible savings are in energy. In addition, a smooth operation of the plant and a good grasp of its operation are realized.

The control system has been improved to a point that the variables were steady with a small deviation. This means maintaining an efficient operation. It has been shown by the working cases that, as the energy management is turned "on", the plant productivity (ton of cooling per kw-hr) is increased [3]. The savings in percent can be combined with the electric load to arrive at the energy and cost saving figures. Furthermore, the swings in the plant chilled water temperature was decreased when the energy management was "on". [3].

The supplier and user can work together to develop customized configurations on each step of the project. In some cases, the energy management system provided benefits beyond the expectations of the user. The flexibility and the expandability of NETWORK 90[R] with a minimal additional cost proves easy addition to integrate the HVAC management. As the facility requirements change or expand, the upgraded modules can be installed to provide new capabilities. Spare parts stocking is reduced dramatically plus the maintenance of control system had been simply replacing the modules. The entire control system has self diagnosing capabilities to lead the operator for easy intervention.

In summary, the real-world projects with demonstrated energy savings and other benefits are examples of customer and user cooperation to achieve the expected results. The technological developments in microprocessor hardware has made these achievements possible and must be given credit.

Also, through this integrated system, the methods and procedures used for developing the industrial control systems are implemented in HVAC controls as well. Industrial controls have more rigid testing, quality assurance, and start-up procedures. As a result, this system has superior features over the conventional HVAC control systems.

REFERENCES

1. Kaya, A. and Sommer, A.C., "Energy Management of Chillers by Multilevel Control and Optimization", ASME Transactions Journal, Vol. , No. , pp .

2. Enterline, L.L., Sommer, A.C., and Kaya, A., "Chiller Optimization by Distributed Control To Save Energy", ISA Transactions Vol. 23, No. 2, pp. 27-37, 1984.

3. Rice, L.S., et al, "An Integrated Chiller Energy Management System at Georgia Tech". Proceedings of World Energy Engineering Congress, Atlanta, GA., October 1987.

4. "Application Guides on Energy Management by Network 90", Bailey Controls Co., Wickliffe, Ohio, U.S.A.

5. Matski, T.N., et al, "Optimal Boiler Load Allocation in Distributed Control", Proc. of 1982 ACC, Arlington, VA, USA.

6. Kaya, A., et al, "Load Control for Energy Converters", U.S. Patents 4, 412, 136 (Oct. 25, 1983) and 4, 430, 573 (Feb. 7, 1984).

7. Matski, T.N. and Dziubakowski, D.J., "Energy Management for Cogeneration, A Case Study" ASME 103 Winter Annual Meetings, Phoenix, Arizona, Nov., 1982.

8. Derreberry, B.C., et al, "Instrumentation and Control for Energy Efficient Cogeneration" ISA Power Industries Division Symposium, May 1985.

9. Keyes, M.A. and Kaya, A., "Plant Energy Management by Multilevel Coordination" Proc. of Fifth IFAC International Symposium, PRP/5-Automation, Antwerp Belgium, 1983.

SECTION 4
ENERGY MANAGEMENT SOFTWARE

Chapter 24

ENERGY MANAGEMENT SOFTWARE DEVELOPMENT—STANDARDIZED PROTOCOL

C. Rayburn

The automatic controls industry has gone through some rapid development over the past 20 years. Considering energy prices in the 70's and the advent of increasingly sophisticated microprocessors the picture becomes more focused. Industry need has been great for a higher level of facility and energy management as well as control systems. This demand has created a flood of pneumatic, electronic, direct digital control(DDC) and more recently unitary control products and vendors into the market. Because many manufacturer's designers have different ideas about the best approach to use for system architectures, the end result has been numerous vendor-specific protocols.(1)

Accordingly, the lack of compatibility between manufacturers makes communication between systems unlikely if not impossible. This situation has brought many facility owners to the viewpoint that some type of standardization is desirable. Until recently vendors have been less than enthusiastic about supporting the development of a standardized communication set, if for no other reason than preserving their competitive advantage. However, substantial improvments have been made in the standardization of the types of input signals acceptable by field panels, for example a 4-20 ma standard input signal can be used by virtually all manufacturer's system. But continued owner pressure has resulted in industry, agencies, and manufacturers to start to recognize some potential benefit in the development of an all-encompassing standard. Hence, the following information is presented on protocol standards.

DEFINITIONS OF STANDARDIZED PROTOCOL

1. Defines the content and format of messages communicated between computer equipment used for the monitor and control of building HVAC/R systems.(2)
2. An agreement by which communication can begin and can be terminated.(3)
3. Rules that define the nature and meaning of the signals that govern the exchange of data used in a communication network.(4)
4. An exact specification of the allowable sequence of events in time.(4)

Some key phrases/words are "agreement," "exchange of data," "content and format of massage," and "specification". Basically there are three types of protocols, data transmission, network control and network management.(5) Data transmission protocols involve actual transmission of useful user data from one node to another. (A node is a logical connection to the communication channel, i.e a system point.) An example would be flow control.

Network control protocols involve negotiation among nodes prior to useful user data transmission to allow data to ultimately be transmitted. A physical link activation mechanism is an example. Network management protocols involve observation of a running network. Performance tracking, and error logging are two examples.

Factors in open communication systems that determine the selection of protocols include: amount of main memory, availability of secondary storage and the internal archi tecture of the machine.(5) The internal performance of the machine needs to be considered in regards to buffer management, time management and terminal I/O interface. (An open communication system is a system that can exchange information with another by using common protocols.)

VENDORS AND USER : PROS AND CONS

The future of the protocol development is dependent on users and vendors attitudes and perceptions toward it. This is not simply to state that users are for and vendors are against protocol evolution. Arguments can be made for both groups that support and detract from the creation of an industry standard on communication systems protocols.

Affirmation by users for protocol development include the view that in the past an Owner possessing equipment which he judged to be outdated or inadequate for the tasks involved, would be faced with the choice of either eliminating the existing system entirely or upgrading to a more sophisticated system by the same manufacturer. With a standardized protocol, he would have the option of retaining and upgrading the existing system with peripheral equipment and products from different manufacturers. Similarly, in the case of a facility expansion the Owner would have the option of utilizing a different vendors' equipment for the expansion, and linking it with the existing system in a potentially standard way. Given the building trend towards sophisticated, factory-packaged, unitary control of individual HVAC equipment,(6) there would appear to be a developing need for some manner of integrating these controls. A standard protocol would allow these control systems to be linked to a central automation system, which in turn would allow a more global scheme of domain control to evolve. The job of operating a facility utilizing a standard protocol would be greatly simplified. The need for extensive employee training would be negated since generally only one type of system wculd exist. A standard protocol would lead to a more competitive environment for proposed control projects, virtually eliminating sole-sourcing and ultimately resulting in lower project costs.

On the negative side for a user a protocol may be the lowest common denominator between different systems, thereby limiting the level or complexity of control possible. Customizing software to fit a particular need or priority would no longer be feasible because the standard would place a limit on potential technological advances. Given a situation in which the protocol would allow little more than data exchange between heterogeneous systems, the Owner would be required to possess multi-system knowledge, leading to a greater required knowledge base than if all systems followed a single architecture. The use of multiple vender systems might hamper troubleshooting efforts with the vendors each claiming that the other's system was at fault. Although enhancing the competitive bidding process, specifying a standard protocol system could lead to an overall increase in project cost. For example, a standard protocol system might possess more stringent wiring requirements, or eliminate the possibility of employing power line carrier for data transmission.

Favorable support for the protocol from the vender point of view includes the option of competing with an established OEM, effectively increasing the size of their potential market. The need for in-house software developmental and associated staff might be decreased, thus leading to lower overhead costs. In the area of service contracts, which have been of increasing interest to vendors, protocol could significantly expand their market share. With a standard protocol it is possible that one vendor could service a competitors system with spare parts and technical training.

However, most vendors would probably consider it undesirable that system expansions or upgrades would not necessarily require the use of their products. Being an OEM would no longer be preceived as a competitive advantage. Also, the possibility of signals being confused or adversely affected is greatly increased.(7) There could be significant expense associated with changeover of a manufacturer's system to a standardized protocol, such as required reconfigurations of chips, additional wiring, etc.

CURRENT GUIDELINES AND DEVELOPMENTS

Several protocol guidelines currently developed include the Open System Interconnection (OSI) by the International Standards Organization (ISO); the Public Host Protocol (PHP) and the Public Unitary Protocol (PUP) both by American Auto- Matrix. Two companies have developed and implemented protocols. Manufacturing Automation Protocol (MAP) and Technical and Office Protocol (TOP) both by General Motors; and IBM's Facilities Automation Communication Network (FACN) Protocol running a program called General Purpose Automation Executive-Distributive (GPAX-D). These protocols all attempt to allow interaction between multi-vendor systems.

The Open System Interconnection (OSI) was developed by ISO (International Standardization Organization) to serve as a framework for future standard protocols for networks of heterogeneous systems.(5) (FIGURE 1) It allows maximum interoperability among different network architectures. This is achieved by a seven-layer model which breaks the problem of inter-systems communication into manageable pieces. The application layer is the highest and is characterized by human users, applications programs and services. The presentation layer is next and represents the uniqueness of exchanged data. The fifth layer, session, provides the mechanisms for organizing and structuring the interactions between applications processes. Fourth, transport, is the transparent transfer of data between field devices. The network layer, third, is characterized by network access functions. The data link layer is second and provides detection and correction of errors which may occur in the physical layer which is first. This layer controls the physical medium.(4) This model ensures that gateways can be built to the standard reference architecture. This model will be used by the ASHRAE Technical Committee SPC 135p and the Intelligent Building Institute. (IBI) Also, General Motors used this model in the development of their Manufacturing Automation Protocol (MAP) and Technical and Office Protocol. (TOP)

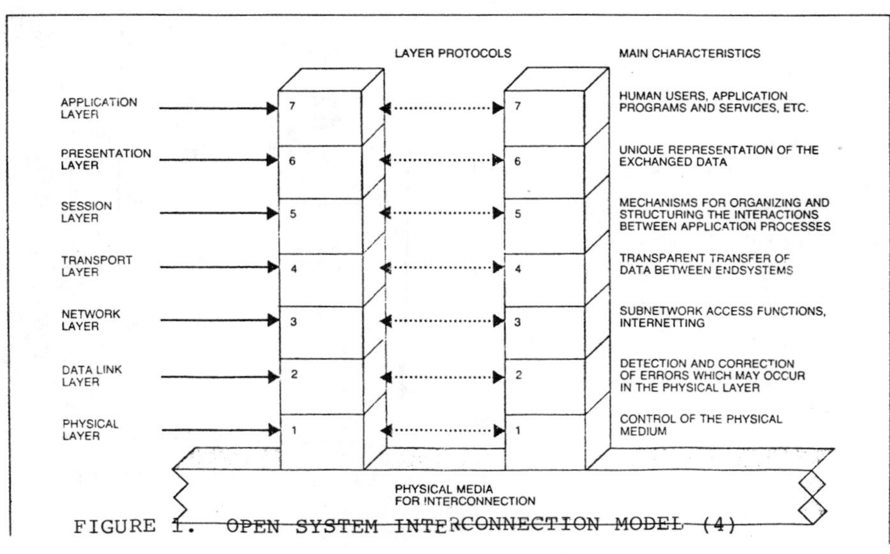

FIGURE 1. OPEN SYSTEM INTERCONNECTION MODEL (4)

Public Host Protocol (PHP) and Public Unitary Protocol(PUP) both reject the seven-layer protocol in favor of using a single-layer functional architecture. Created by American Auto-Matrix, the purpose of PHP and PUP is to "promote a standard protocol which can be used effectively by multiple vendors in both dedicated wiring and dial-up systems."(1) Since there are a wide variety of manufacturers the guidelines have "attempted to deemphasize the number of low level operations which are available to the public."(1) Several things PHP and PUP do not attempt to do is: 1)require any manufacturer to disclose any secrets. 2) "universalize" proprietary protocols and communications media.. 3) force vendors to utilize the same man/machine interface. 4) upgrade an unintelligent system into a computer-based system. PHP and PUP use standard asynchronous ASCII communications sets allowing for local area network (LAN) applications . These sets consist of one start bit, eight data bits, no parity bit, and one stop bit. Both allow low-level access to data. PHP uses hosts to communicate with field equipment to provide query/response messages and to act as virtual/dumb terminals. Other features of PHP include exception message handling, storage medium and multidrop operation (network sharing). PUP uses the slave/master scenario where each unitary controller accesses the network for it's own purposes and then relinquishes control to the next unit. Other features of PUP include the use of virtual terminals, half-duplex multidrop networks, error detection and data exchange.

Manufactured Automation Protocol (MAP) and Technical and Office Protocol (TOP) were both created by General Motors and seven other participants. MAP and TOP are applicable to multi-vendor environments in manufacturing, technical and office settings.(8) As stated, both protocols are based on the OSI seven-layer model. MAP and TOP encourages vendor development of MAP and TOP products and participation with standards groups. MAP specifies functional network protocols for information processing on the factory floor. The cost per node connection of a MAP-based system is quite high, over $2000.00 and may have too many capabilities for building automation control applications.(1) TOP specifies the same for business offices and technical environments.

Facilities Automation Communication Network (FACN)is based on the use of IBM PC's as translators for connecting non-IBM microprocessor-based DDC systems to a host which is running the General Purpose Automation Executive-Distributive (GPAX-D) licensed program. (FIGURE 2) GPAX-D operates on an IBM Series/1 and "serves as a communications center for gathering, monitoring and controlling data through the facilities automation system."(9) GPAX-D can communicate with Remote Intelligent Subsystems (RIS) via Binary Synchronous Communication or Asychronous Communication provided by Original Equipment Manufacturer

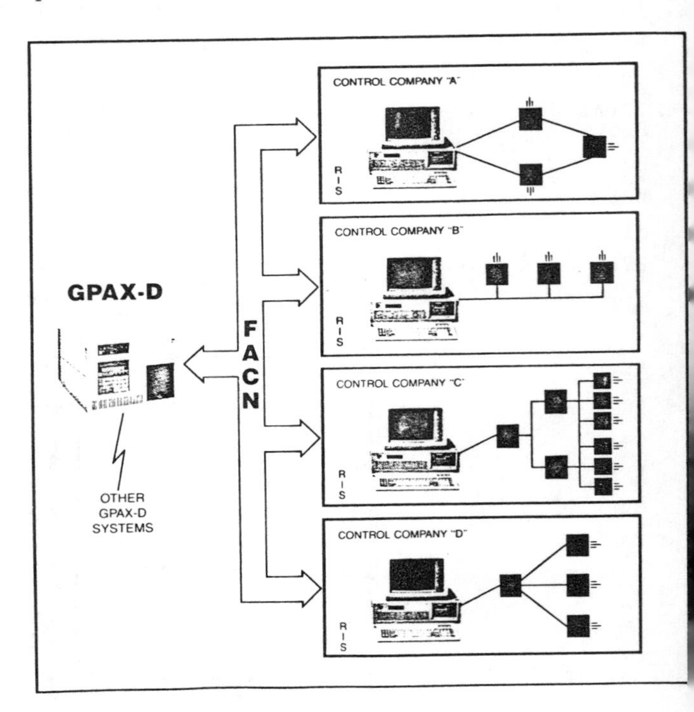

FIGURE 2. GPAX-D SYSTEM CONFIGURATION (9)

(OEM). The RIS typically includes an IBM or equivalent that serves as a gateway to the GPAX-D system. GPAX-D must communicate through a program which translates generic data from itself into device-specific protocols. This "front end" system also provides common energy and facilities management control functions.

One item that should be mentioned here is a software package that supports multi-vendors - The Interpretor 2000 by Universal Controls. The two major components of it's architecture are the operator interface and the vendor translator. A translator is a software component that is EMS vendor dependent(i2000).(10) Basically it contains protocol information for communication to specific vendors' equipment. The operator interface at the host provides the same monitor/control screens for each EMS allowing for generic programming and scheduling.

DEVELOPMENT BY INTERESTED PROFESSIONAL GROUPS

American Society of Heating, Refrigeration and Air Conditioning Engineers (ASHRAE):

ASHRAE Technical Committee SPC 135p - Energy Monitoring Control Systems Message Protocol whose purpose is to "define the content and format of messages communicated between computer equipment used for the digital monitoring and control of building HVAC & R systems, thereby facilitating the application and use of this technology."

The scope of the committee is to "provide a comprehensive set of messages for conveying binary, analog, and alphanumeric data between devices." Each basic message type will also require the capacity of supplying ancillary information such as reliability, priority, real-time and other related data. This scope will also provide for the format of each data element.

Desirable elements of a good protocol, as defined by the committee include:

a) interoperability - independence of any particular manufacturer's hardware
b) efficiency
c) low overhead
d) seek highest common multiplier rather than lowest common denominator
e) compatibility with other applications and networks
f) layered architecture along the lines of the ISO (International Standards Organization) model OSI (Open Systems Interconnection)
g) flexibility - i. not limited to present hardware technology ii. can be implemented in parts iii. permits the use of multiple kinds of physical mediums
h) extensibility
i) cost effectiveness
j) transmission reliability
k) must apply to real-time processes
l) maximum simplicity
m) should allow priority schemes
n) fairness with respect to medium access
o) stability under realistic loads

Three working groups have been formed to provide recommendations and information to the full committee pertaining to their subject area. Each group has the responsibility to clarify nomenclature and review existing protocols.

1. Object Type and Properties Working Group - Determine the types of data and the properties associated with each object type. A simple object type would be a binary status input point. A complex object type would be a control loop. System object types could be made up of both simple and complex object type, for example a VAV box. The goal of this group is to come up with a comprehensive set of object types which is "both reasonable in number and is general enough to cover the majority of physical and logical systems and subsystems commonly encountered in industry."

Attributes will be identified for each object type so they can be implemented and supported easily by host systems. Unitary controllers will also be considered.

2. Primitive Data Format Working Group - Binary, analog, character string, time and schedule are some primitive data categories. The purpose of this group is to define an acceptable binary format for presentation for each of these type of data. This format would be used in the data blocks and messages determined by other working groups. Since Energy Management Systems operate in a real-time environment, how time is represented in the protocol will be evaluted.

3. Application Services Working Group - The scope of this group will cover the entire range of communications in energy management and control systems. However, other types of building functions and services and their interaction will be considered. "There is a need to define the message types that the protocol should be able to support from a functional perspective." The messages need to be generically organized and independent of system layout. The protocol should provide message transmission, change of state report mechanisms, point data request/response, parameter modification, point command, alarm processing, block read/write functions, terminal support and other application functions.

Highlights of the SPC 135p working groups to date include the research of other protocols currently developed or being developed in the United States, Canada, and overseas, a document defining control terms, and the decision to use the OSI Reference Model as a framework for development of the protocol. A Building Management Control Systems Communication Protocol Laboratory at the Center for Building Technology at The National Bureau of Standards was established to carry out research, evaluation and testing of the protocol. (11)

Intelligent Buildings Institute (IBI):

Another group currently working toward the development of a protocol standard is the Intelligent Building Institute (IBI). They have sponsored four "Executive Forums on Open Protocols." Their concept of applications for open protocols includes voice and data communications, building controls, security, fire-life safety, and related building functions. Four levels conforming to the OSI reference model have been outlined:

1) Level I would permit point to point or sensor to sensor communication. 2) Level II would permit remote intelligent microprocessors (ddc modules), to which sensors are connected, to communicate with other intelligent microprocessors.

3) Level III would permit host to host communications (These hosts could be either PC's or minicomputers.)

4) Level IV would permit communication at the Information Management Network.

IBI feels that Levels I and IV can utilize existing protocols. Level II and III are estimated to require five or six years to develop, depending on vendor participation.(12)

SUMMARY

As you can see, the development of protocol standards for HVAC controls is rapidly becoming a perceived priority by concerned parties. To date, efforts by various commitees and manufacturers have resulted in some standard protocol guideline development that seeks to ultimately result in a situation beneficial to vendors and users. The technology is challenging and the process difficult, but the need is there and industry is responding.

References

1. National Reserch Council, Committe on Controls for Heating, Ventilating and Air Conditioning Systems. "Controls for Heating, Ventilating and Air Conditioning Systems" 1988.

2. David Fisher, "Public Host Protocol Guidelines" American Auto-Matrix, Inc. 1987.
Lawrence Gelburd, David Fisher, "Public Unitary Protocol Guidelines" American Auto-Matrix, Inc. 1987

3. Center for Advanced Technology, "Some Insights Into Office and Factory Communications Systems" March 1986.

4. A. Fleischman, S. Chin, D. Effelsberg, "Specification and Implementation of an ISO Session Layer" _IBM System Journal_ Vol 26, No. 3 1987.

5. B.C. Goldstein, J.M. Jaffe, "Data Communications, Implications of Communications Systems For Protocol Design" _IBM System Journal_ vol 26, No. 1 1987.

6. Richard Mullin, "Beginning the End of the Beginning" _Energy Users News_ Feb 8, 1988, p.1.

7. Richard Mullin, "Conflict on Protocols' Value Still Evident at Committee" _Energy Users News_, Feb 8, 1988, p.1.

8. "Maufacturer Automation Protocol and Technical And Office Protocol" SME.

9. IBM, "General Purpose Automation Executive-Distributed Program" Executive Summary.

10. Universal Controls and Engineering, Inc. "Interpreter 2000 Energy Management Front End Software"

11. American Society of Heating, Refridgeration and Air Conditioning Engineers (ASHRAE) Minutes to date, Technical Committee SPC-135p.

12. R.H. Geissler, "News Release" Intelligent Building Institute (IBI), March 8, 1988.

Chapter 25

SIMPLE MODELING AND ANALYSIS OF HVAC SYSTEMS WITH YOUR EMCS

A. F. Stewart

The operations and maintenance engineer is challenged to use facility energy resources in the most efficient way. Computer based energy management systems (EMS) are the tools used in large facilities to implement efficient operation. In this paper I will discuss some techniques for automating a measurement of performance for your heating, ventilation, and air-conditioning (HVAC) components and systems using the EMS. By using simple models to quantify actual energy consumption, the choices for equipment replacement, maintenance, or facility modification can be based on your actual costs. At the end of the paper, I include an example of a contract specification requiring energy analysis models. If you are installing or upgrading your EMS, consider including the analysis models in the specification.

Computer models enjoy extensive use in engineering design. The kinds of models we operations and maintenance (O&M) engineers need just are not available as part of the standard software in energy management systems. Design models include many assumptions about how a facility system should work. O&M models must reflect how our HVAC systems actually work; the model must be verifiable when viewing the actual building performance today. Many of us are responsible for the management of facility plants with annual energy consumptions of millions of dollars. The question we must be able to answer is "How effective are we at the task of saving energy and dollars?" Answer the following questions for yourself:

a. Can you document how successful your energy management program is working? How many dollars are you saving?

b. What is the most efficient portion of your plant? (HVAC or facility envelope) Is your answer based on results from data gathered continually or estimates based on original design?

c. What is the least efficient portion of your plant? (HVAC or facility envelope) Do you know why? (The systems are old is not an acceptable answer.)

d. Do you know the actual operating efficiency for all your chillers and boilers?

Before we jump into a discussion about the use of models for energy management there are two points I think should be addressed. First, "What is a model?" And second, "What do we want to model?"

WHAT IS A MODEL?

A model for our purposes is a simple simulation of a real event. The model is a mathematical function, we choose the independent variables and the model finds the solution. Using a chiller as an example, we choose the ambient conditions and the performance data from the manufacturer will indicate the rated output for the chiller. Here the vendor provides the model. The model is useful when it adds to our understanding of the cause and effect relationship of any process. Conversely, if the model is so comprehensive that the process is obscured we have failed. Many point with pride to the design analysis programs (BLAST or TRACE for example) but my experience is that the data run in these programs is off the "blue line drawings" and the results are difficult to reconcile to actual values. Given those boundary conditions for our model we can look at what it is we want to model.

O&M ENGINEER APPROACH

As the O&M engineer, we want to be able to model chillers, boilers, and the facility. By modeling the devices supplying the heating and cooling we can determine operating efficiency. By modeling the facility we can determine if the load imposed on chillers and boilers is correct based on known conditions. Most of our models will be a composite of the individual components; the chiller, piping systems, and cooling towers for example.

DESIGN ENGINEER APPROACH

The approach to energy conservation has been very much from the perspective of the design engineer. The design engineer knows we can make the facility envelope more resistant to the heat gain or heat loss. This approach has spawned many insulation and window replacement projects. We do have less of a cooling load/heating load from the external forces on the envelope. But when we compute the air conditioning loads for a typical administrative facility, the interior electrical loads (including lighting) and personnel loads (including ventilation) are more significant than the transmission losses through the envelope. Today's higher quality envelopes make a situation where the internal heat loads become the driving force. To achieve the efficiency we need for today we must insure our facilities are actually operating as designed.

COMPUTER MODELING

The concept of computer modeling, while not new to any engineering discipline, has not been aggressively applied to the HVAC arena. There has been plenty of modeling done for new facilities and for retrofits in terms of additional insulation or changes in the HVAC system. The analysis starts with the design of the building as given. These are table top evaluations and do not actually look at the operation of the facility; these are open loop evaluations. Modeling as it is used by the O&M engineer relates to actual performance of the facility. The power of the model is the ability to go back and compare the actual consumption with values predicted by the computer model; closed loop evaluation. The computer models are required for the following reasons:

a. The computer model provides the ability to establish an expectation, that is we can establish what we expect to save in terms of dollars or energy for a given facility. Then we have the ability to reconcile our expectation with actual consumption for the facility.

b. The modeling allows us to document those savings which we have trouble quantifying. Those would be savings from many of the reset type programs using outside air for free cooling or changing deck temperatures to optimize the energy consumption.

c. We have the ability to establish a meaningful history on the operation of our facilities.

When we are trying to evaluate the performance of some particular HVAC system we must have a method to quantify the performance. It has been estimated (no source) that perhaps 70% of the savings of an energy management system exists in the start-stop function, basically those functions of the time clock. I would submit to you that perhaps on the order of 50% of the savings to be attributed to an energy management system come from the start-stop. (I will defend the 50% hypothesis in an example later) The other 50% would be available through proper operation of the facility during equipment "on" time and proper maintenance of the plant providing the heating and cooling. We have two time related functions, either the equipment is off (when its off we are certainly saving energy) or on. Many buildings are in use 50% of the hours in the week. Other facilities, hospitals and other 24 hour a day facilities have no savings from start-stop but tremendous savings from analysis of the equipment, analysis of airflows and just a wide range of diagnostic features. The key here is to keep our minds receptive to the savings to be attained. If we assume from the onset that start-stop functions are the only savings to be realized then we shall be blinded to other opportunities.

The concept of the modeling of devices, predominantly chillers and boilers, can be divided into two methods. The first method would evaluate the device over some specific time period and analyze the total energy input and the total product produced (average performance). The second method applies to devices which can be analyzed by taking a "snap shot" of the actual device performance (instantaneous performance).

COEFFICIENT OF PERFORMANCE

The analysis over time, six or eight hours, allows us to analyze the results of the average performance. This actual performance is compared to the manufacturer's stated performance. A report should be generated to document the results. The actual performance can be stated as a percent of rated output. I prefer to think of this as a coefficient of performance (COP). The key here is to have your EMS monitor the energy in and the product produced and document the results.

REAL TIME EVALUATION

The second method of evaluation we have is a real time evaluation (instantaneous). Our EMS can display the results of the product (chilled or heated water) produced through a calculated point in our EMS software. The energy input can also be displayed through the use of a calculated point in our EMS. With current EMS software we have difficulty setting up a calculated point for a family of curves, such as that required for a chiller. However, choosing the chilled water supply at a nominal temperature and typical outside air temperatures will provide a first cut for evaluating efficiency. This type of model is good for a chiller or boiler under steady state conditions. We currently measure, on some of our chillers, the actual energy or cooling produced. (For a chiller the output is a function of water flow and temperature differential.) Additionally, with the real time value we can actually demonstrate through the calculated point again what percent of performance the device is actually achieving. The results with this approach have been very encouraging. We have identified chiller performance as low as fifty percent of rated output.

The problem with this approach is that some devices are not well suited to real time evaluation. Centrifugal chillers certainly are; boilers may or may not be. The conventional sensors we have on a boiler are inadequate to measure the performance of the heat transfer when the boiler has cycled to make up for losses in the thermal jacket. Also we could measure the stack temperature but I do not support over instrumenting the devices. As I indicated in the beginning of this paper my intent is to develop simple models -- energy in (electricity and gas) against the product produced (hot water or chilled water). We seek to avoid situations where we are not in equilibrium conditions.

The evaluation over time is much more useful to the O&M engineer as it provides two distinct benefits. First, we show the actual transfer of energy through the device and second you see the efficiency over time. A device may have good COP (transfer ratio) but may be oversized for the application. For example a very large boiler may work well but the losses may cause the percentage of usable heat to be low.

For those readers that have not had an opportunity to evaluate chillers or boilers and actually look at component performance, it is very difficult to understand that the device could be installed and actually be functioning significantly outside of design performance. Only after you actually make a number of site inspections do you begin to see the magnitude of the problem. Our mechanical design has sufficient safety factor to tolerate a large reduction in actual output. In fact output may degrade to 70-60-50% of capacity. However, if the original sizing of the device was in fact 200% of actual load then we will not know there is a problem until we reach only 50% of the output. (If the chiller is sized for 200% of the existing load the O&M engineer would not know there is a problem until the occupant finally cannot be cooled.) Even when we have technicians on duty, the attitude exists that says run the plant with more units on line (large plants with multiple units).

THE EXAMPLE

In this example, we have an administrative facility with approximately 30,000 square feet of office area. The three floors of this facility are served by a multi-zone air handler with eleven zones. The air-conditioning is provided by a 70 ton reciprocating air cooled water chiller. The building HVAC controls were converted to direct digital control (DDC) in 1987. As part of the new control scheme we installed a flow meter on the chilled water. The chilled water supply and return temperatures were also monitored. A power meter was installed on the chiller. (Because of problems with our power meter, manual readings were made on the power input to the chiller during our load tests.)

On the first warm day after the new sensors were installed we set about running a load test. With the vendor's performance data in hand and outside air temperature at 90 degrees F., we increased the load on the chiller with outside air by direct operation of the outside air dampers. (The economizer cycle for the air handler is also under DDC.) We had started our test with the building internal temperatures in equilibrium. The chilled water was stable at 45 degrees F. We increased the percentage of outside air until the chiller could no longer maintain the 45 degree supply temperature.

The chiller actually developed 37 tons of cooling. A check of the part numbers on the two compressors indicated they were down sized by 5 horsepower on the last replacement. Some interpolation of the vendor's data indicated we could still expect to produce approximately 65 tons. The refrigerant gas pressure was low on the suction side. Since this unit was nearly ten years old we suspected scaling in the chiller barrel.

After the refrigeration shop had chemically treated the chiller barrel with a mild acid we again performed a load test. This time we were able to get 45 tons. The chiller is now scheduled for replacement as 45 tons is only working at 70 percent capacity.

The insight to be gained from this example is this:

a. As the output from this chiller decreased the hours of operation for the HVAC system were extended. Shop personnel incorrectly attributed the longer hours (over the years) to the addition of numerous computer systems in the building.

b. Approximately 40 percent of the HVAC load is made up of pumps and the air handler. These are fixed loads and the related energy consumption increases in direct proportion to the extended hours of operation. The remaining 60 percent is the compressor. When the output from the compressor is one half, our cost of operation is twice the proper amount.

c. To satisfy the temperature requirements for this building we have extended the hours of operation for the HVAC equipment by approximately 40 percent. These extended hours allow the building to start under required temperature control in the morning and eventually drift out of control by late afternoon under the hot Texas sun. The additional annual cost for this mode of operation is approximately $3,000/year.

TEST INSTRUMENTS ON WHEELS

One of the questions that surely will be asked is how can I afford all the hardware to monitor all those diagnostic points on the HVAC equipment? The decision to monitor certain HVAC systems is not without cost. However, the current state of EMS hardware facilitates a very cost effective approach. Current EMS vendors allow the field panels to be dialed up by phone. A properly outfitted smart field panel can be configured with all the type of sensors you want to gather data. These sensors can include clamp on current transducers and volt meters (to measure electrical load), temperature sensors (to temperature differential), flow meters, solar energy meters or any measurement device you want to install for a temporary period. With this approach you have the ability to record all your data into your EMS. Also, you can decide which sensors provide the characteristic information you need to model any particular type of facility and then permanently install only those sensors.

MAINTENANCE PROJECTS

The most over looked area in existing facilities for energy conservation is in the retrofit option. When we analyze a large military base approximately 70 percent of the buildings are good candidates for many of our reduction schemes. Some buildings have a requirement for 24 hour a day operation, those buildings are disqualified from some of the savings from start-stop or temperature resets. Those are the facilities that receive the most benefit from schemes which look at the actual operational efficiency of the boiler, chiller plant, air handlers, water distribution systems or facility envelope.

RETROFIT PROJECTS

The model can help us replace systems too. By establishing a COP for each of our devices we can develop a prioritized list for replacing systems. This list allows our mechanical devices and facilities to be rank ordered from efficient to inefficient. Imagine having the ability to assign your limited personnel assets against your verifiable worst system. This approach also provides us the data to defend the life cycle cost for our action. When we calculate return on investment for projects that re-insulate or change windows the term of investment is long, perhaps 15 years. Certainly one of the key ingredients to any energy management system is recognizing what the return on investment is in terms of the actual operation.

THE MOVING ENERGY TARGET

Every day more and more computer equipment is being added to our buildings and the energy intensity per square foot for some facilities is actually increasing. The Department of Defense tried to relate the amount of energy used at the military bases to the amount of facility area (total metered base consumption / total heated or cooled base area). The energy plan required a reduction in the energy use per square foot from the baseline year. This seemed like a method to establish a baseline for consumption, however, this has turned out to be not a very valid measure of energy reduction efforts. Certainly mission change is going to reduce or increase energy consumption. These changes are outside the scope of what can be accomplished by the energy management system.

THE COMMITMENT

We are required to provide a level of service to our occupants, efficiently. The current tools that are available to energy management systems are insufficient to affect those savings. Modeling is the only way that we can identify real energy waste. A very large portion of the potential savings that energy management systems can deliver will be identified through modeling techniques.

SAMPLE SPECIFICATION FOR ENERGY MODELS

XX. ANALYSIS PROGRAMS:

XX.1 General: The purpose of this section is to take the EMS data and create an EMS information system through analysis of the data. The analysis programs evaluate the EMS operation and indicate how well, quantitatively, we are achieving our goals. Also, the analysis programs evaluate how well the various end items controlled by EMS are functioning. This analysis can be accomplished by comparing actual performance to manufacturers performance data, for individual items like chillers and boilers. In addition an interface will be required for a commercial spreadsheet program to obtain data telemetered by the EMS. The spreadsheet program shall be resident in an IBM personal computer (PC) compatible machine. The data may be transferred to the PC by diskette or by direct computer connection.

XX.2 Program Inputs: This program shall use historical or current trend files for actual data inputs. The data for all programs covered under this section will be gathered under the following methods: All data telemetered by the EMS computer shall be automatically available for use as inputs to the analysis programs without having to be reentered by the operator. Where data is not available or unreliable the programs shall default to normal engineering values for the process being modeled. The operator may override any values of telemetered or default data.

XX.2.1 The EMS computer will retain historical information on all start-stop devices. This information will consist of the previous 31 days of equipment run time retained by day. And also the previous 13 months by month. Any data that is unreliable should be tagged to indicate the error. Each piece of equipment with start-stop capability will have a record of 31 individual daily run times and 13 previous monthly totals for run time. This information will be automatically available to all programs covered in this section.

XX.2.2 Spreadsheet Inputs: A EMS program shall be provided to facilitate the conversion of EMS telemetered data into a format compatible with commercial spreadsheet programs. (this section will include any specific spreadsheet program requirements, for example, if you use LOTUS 123 then require the program interface)

XX.3 Program Outputs: The output shall be available in tabular form and in bar chart/plot format. The results shall include the actual measured values, the computed/theoretical values and the difference between the actual and computed values.

XX.4 Energy/Dollar Saver Program: This program shall totalize the actual energy and dollars saved by the EMS. The program shall accumulate all savings resulting from the shut down of any equipment connected to EMS. The program shall be implemented by the contractor for all on/off type devices connected to the EMS. The operator shall be able to define a single device or any number of devices as a subset for this program. The savings shall be defined for any subset or the total as specified by the operator. The program shall keep track of two separate types of loads; fixed loads (such as pumps) and variable loads (such as chillers). The cost saving will be the product of the kwh saved times the kwh cost. The provision shall be made to show savings from a reduction of total kw demand. This program shall be available to run at specific times as required by the operator and/or automatically run on a weekly basis.

XX.4.1 Program Inputs:

a. Device identification (building,unit,point)

b. kw fixed (normal running load for constant consuming devices)

c. kw variable (minimum load for variable load devices)

d. kwh cost

e. kw cost (the incremental cost per kw from the power company)

XX.4.2 Program Outputs: MBtuh and dollars actually saved.

XX.5 Chiller/Boiler Evaluation Program: This program shall allow the operator to enter the manufacturers performance curve for discrete devices. The performance data shall be entered in increments of five percent of load values. The manufacturers performance data shall be compared to actual run time samples from current data or from historical files. The program shall compute the actual performance from actual temperature differentials and flows.

XX.5.1 Program Inputs:

a. Type of device

b. Logical system name (building,unit).

c. Operator entered performance data (Btu/kw, Btu/Kcf, Btu/gal...etc.) from zero to 100% of load from the manufacturers performance data.

d. Chilled water flow rate.

e. Chilled water entering temperature.

f. Chilled water leaving temperature.

g. Condenser water entering temperature.

h. Condenser water leaving temperature.

i. Condenser water flow rate.

j. Boiler water flow rate.

k. Boiler fuel flow rates.

l. outside air temperature.

m. Boiler pressure.

n. Boiler supply water temperature.

o. Boiler return water temperature.

XX.5.2 Program Outputs: Actual Btuh delivered and operating efficiency.

XX.6 Building Load Analysis Program: This program shall allow HVAC modeling of a facility for comparison to actual data. Primary emphasis is placed on proper sizing of the A/C plant and the heating plant. The program shall determine the following loads: ventilation, personnel, equipment, lighting, and transmission load through the facility envelope. Solar loads will be measured by sensors or entered by the operator. The load program should be modeled along the lines of the ASHRAE cooling and heating load method. Walls and roof sections shall be chosen by the operator from a selection of typical sections, similar to the ASHRAE program. The program will be used to simulate actual facility performance during some chosen day and for a specified period of at least 8 hours (after the internal temperatures have reached equilibrium). During the same period the total energy consumed by the facility will be recorded for comparison against the simulation. The result of this analysis shall be reported as the coefficient of performance for the facility, a value of 100% would indicate the facility HVAC load equals the computed HVAC load.

XX.6.1 Program Inputs:

a. All physical parameters required in the ASHRAE cooling and heating load calculation method.

b. Trend file of the actual inside and outside temperatures for the test period by quarter hour.

c. Period for test (in hours)

d. Solar load values (may be operator entered if not metered)

e. Chiller load values (Btu/hr)

f. Boiler load value (Btu/hr)

XX.6.2 Program Outputs:

a. Actual Btu load/hour.

b. computed Btu load/hour.

c. Coefficient of Performance (COP) for the facility during the test period. COP will be defined as (COMPUTED LOAD/ ACTUAL LOAD) X 100 percent.

Chapter 26

SELECTION OF ENERGY MONITORING SOFTWARE—MEETING YOUR NEEDS

M. A. Friedle

Abstract

The advent and wide distribution of reliable and inexpensive personal computer hardware has been accompanied by a cornucopia of software, even in such specific fields as energy monitoring. Like software products in other specialties, some are good, many are bad, and all are designed to meet certain needs. Selecting the proper software involves matching your needs with the benefits provided. This paper provides a down-to-earth look at some of the problems associated with selecting energy monitoring software, and discusses workable selection techniques.

Introduction

We know from life experience and from basic economics that items are selected based on their perceived values. When buying an apple, we select between the apple and our money. When we pay for an apple, we trade something we value less than the apple (money) for something the grocer values less than money (the apple). Both parties come out true winners. When we select between red and yellow apples, the analysis is only slightly more involved, as we select between three items, the red apples, the yellow apples, and our money.

In our personal lives as consumers, decisions are often made intuitively and are based largely on what appeals or "tastes" the best. We generally do not take the time to estimate the benefits provided by one type of apple verses another. Consumer advertisers, aware of our behavior in this area, generally strive to appeal more to our senses than to our intelligence, and, as the saying goes, sell us the sizzle instead of the steak.

In the industrial environment attempts are made to quantify the selection process. Many mathematical techniques have been invented to help us with this task. Return-on-investment, for example, could help us decide what size boiler to install; payback could help us decide which computer system to purchase; discounted-cash-flow could help us decide which apartment complex to renovate; return-on-capital-employed could help us decide which plant expansion is given approval. Attempts are also sometimes made to force such non-quantitative items as employee morale and customer satisfaction into the above calculations [1].

Many people are employed across industry performing these types of calculations every day. This quantified selection process gives people security, lets companies maximize their profits, and helps the economic world go around. Adam Smith would be proud of us.

Cost/Benefit Analysis

The quantified selection process described above is based on three major assumptions. First, it assumes we know the true cost of apples (which includes such things as the prices of delivering the apple, storing the apple, preparing the apple for consumption, consuming the apple, cleaning up after consuming the apple, disposing of unused apples and parts, product liability to cover misuse of the apple, and other costs). Second, the quantified selection process assumes we know the true benefits the apple provides (which includes the apple's health value to us, the apple's energy value, the apple's esthetic value, the apple's investment value, etc.). Third, it assumes all the costs and benefits can be quantified in the same common set of units, usually current dollars (this requires knowing, among other things, the project schedule, the time-dependence of all costs and benefits, future financial data such as inflation rate, future wage rate, etc.).

It is easy to imagine a situation in which the above assumptions are close to being valid. A piece of property, purchased with fixed-rate financing, is let out to a stable tenant under a long term lease. This ideal situation does not strain any of the assumptions mentioned above. The requirement that we know the true cost, the requirement that we know the true benefit, and the analysis requirement of common units are all met quite nicely. The most we could argue about perhaps, is lost opportunity costs.

Most industrial selection decision, however, are not nearly as ideal. We find that in actual practice the above assumptions are difficult to meet, and that the selection decision often bogs down under the efforts involved in making projections and quantifying costs [2]. Estimates also have a way of becoming political "hot potatoes," with varying interests groups making conflicting projections. The end result is that the process often break down under its own requirements. The decision maker is left in the uncomfortable position of making selections based on what "appeals" to him the best or on political factors. The method which is supposed to help him actually leaves him no better off that the consumer selecting between the two apples. The selection process is invalidated.

The two main villains responsible for this break down are our imperfect knowledge and our imperfect ability to predict the future. Most of us will admit our knowledge is somewhat imperfect and have had a experience or two to prove this to us. Perhaps we specified a valve, only to realize later that it was not suitable for the application, resulting in additional costs for replacing the valve. Most of us have also had experiences where making a prediction of the future failed us. An example would be a project with a picture-perfect gant chart which runs into considerable delays, perhaps due to something as simple as the weather disrupting the concrete pouring schedule, increasing overtime pay and disrupting production schedules. It is easy to visualize these problems in larger selection decisions, especially when related to the military projects [3], but they are present to some degree in all projects.

The tendency for the quantified selection process to break down is particularly common when selecting energy monitoring software. There are several reasons for this, starting with a common lack of experience. People have been selecting boilers for a long time and many people have devoted their careers to becoming boiler experts. Text books are available on boiler technology, and we can usually find someone in our organization with enough background and experience to help us. Selecting energy monitoring software, however, is new to us. We have very few points of reference in our work experience to relate it to.

This lack of familiarity, together with the "mystique" of computers, leaves some of us reluctant to subject it to the same type of scrutiny that other selections receive. This psychological "computer resistance" is difficult to overcome, and can greatly impact the selection process [4].

The rate of technological change impacts our ability to make proper selections as well. Boiler manufactures may take exception, but the technology involved in designing a boiler is static when compared to the pace of change in computers and software. The average product life expectancy of personal computer hardware is two or three years. Software products have correspondingly short life cycles. It is difficult to make accurate projections within this environment.

Perhaps the greatest difficulty specific to software is its conceptual nature. A boiler design is worked out in iron, and is there for everyone to touch and see. The differences between "good" and "bad" boiler designs are often all too obvious to the trained observer. Software, however, is more like that book which you can't judge by its cover. It is difficult to tell the good from the bad until after it has been purchased and is in use. Even experienced computer users have difficult accurately judging the merits of one software product over another until after it is in use.

In summary, there are many ways for the quantified selection process to break down. This tendency is pronounced with software in general and with energy monitoring software in particular. Even if we had an army of cost accountants and engineers available to us, pinpointing the best selection to make is often difficult or impossible due to uncertain costs and benefits, and difficulty in predicting the future. When all the detailed analysis and estimations are done, we may find ourselves still making decisions based on what "feels" the best to us.

Needs-Fulfillment Analysis

Where does this leave us in practice? If selecting energy monitoring software based on quantified selection techniques is a futile effort, what's a poor manager to do? Those looking for an easy answer to these questions will be disappointed to learn that there is no easy answer. There is, however, a technique available which often produces better results than cost/benefit analysis. The technique is called needs-fulfillment analysis.

Start with the Right Question

When asked "where do I start," a good answer would be "the best place to start is at the beginning." When selecting energy monitoring software, starting at "the beginning" means starting with YOUR NEEDS [5]. Do not start with the technology, or with the schedules, or with the hardware, or with the software, but your NEEDS.

With energy monitoring software, it is all too easy to start with the technology. We look at the product, learn what it can do, and then try to figure out where to use it. But starting with the technology is actually a trap. It puts the cart before the horse and forces us to play the vendor's game of "my technology can beat up your technology." We end up examining the sizzle instead of the steak. The reason vendors like to talk in terms of speed, chips, power, memory size, bytes, and bits, is that such items are eye-catching, and modern technology makes it EASY. It is EASY to a build a state-of-the-art system that has lots of memory, visual appeal, and a bit of flash. What's HARD is meeting the customer's needs.

The most commonly asked question about energy monitoring software is "what can this do for me?" Turn this question around. Don't start by asking the vendor what this software can do for you; ask yourself what do you need from this software. This admonition may sound simple and straight forward enough, but it must be kept before us if we are to avoid putting technology in the driver's seat.

Determining your Needs

The needs-fulfillment technique will only work if you take the time to determine what your needs truly are. Buying energy monitoring software simply because you have money in your budget, to "automate" some procedure, or to "be modern" is a sure prescription for disappointment unless you also determine your needs and how they will be met.

There are many ways to do this. To list them even a fraction of them is outside the scope of this paper. You might have a brain storming session, hire a consultant, or scribble ideas on a napkin. What ever your method, it is important to itemize your needs in some sort of logical list. As an example, imagine a utilities department which produces steam, potable water, and chilled water, and distributes electrical power for a variety of users. The utilities superintendent may list out his needs as follows:

Problem: Accurate and timely records of daily utility production and distribution are not being kept.

Current Method: Employees walk and drive around to all the metering sites at or around the end of the month. They write down the meter readings in the meter log book. Other employees transfer the readings from the meter log book to an accounting form. Other employees calculate the differences between meter readings, multiply by the scaling factor, and write up a monthly production and usage report. Other employees distribute the report to all production and building supervisors.

Needs: Automate monthly meter readings, flow calculations, and report generation and distribution.

The needs-list may be as detailed as required, and more than one item may be listed. Listing the current method of accomplishing the task, as in the above example, is also usually helpful. There is much room for flexibility in the scope and format of the list. The important thing is to take the time to write it out. It will serve as a reference point as the process continues.

Head towards the Right Destination

Alice (of Wonderland) asked the Cheshire Cat "Would you tell me, please, which way I ought to go from here?" The Cat answered, "That depends a good deal on where you want to get to." "I don't much care where" said Alice. "Then it doesn't matter which way you go," said the Cat. [6]. When selecting energy monitoring software, we DO care where we want to go. We want to head towards selecting the software which best meets our needs.

Determining if the Software will Meet your Needs

List the method each software system would use to meet your needs. To continue the above example, such a list could look like this:

First Need: Automate monthly meter readings.
How: Personal computer and printer are installed in manager's office. Electronic devices are installed at each meter. Both personal computer and electronic devices are connected to the existing telephone system. Once a day, the personal computer, using the phone system, takes the meter readings through the electronic devices. The personal computer prints out the daily readings each morning, and the monthly readings on the first day of each month.
Additional Requirements: Install personal computer, printer and electronic devices mentioned above. Administrative assistant to supply printer paper and ribbons. Buy computer desk. Minimal power requirements.

Second Need: Automate monthly total flow calculations.
How: Included in above.
Additional Requirements: None.

Third Need: Automate report printing.
How: Included in above.
Additional Requirements: None.

Fourth Need: Automate report distribution.
How: Not provided for.
Additional Requirements: Administrative assistant makes copies of automatic reports and distributes them to the appropriate people.

Listings like the example above simplifies greatly the comparisons between energy monitoring software. For example, the above list is for a fairly automated system. Many systems, however, are not. For non-automated systems, the "additional requirements" section in the above example would grow rapidly with such items as who will operate it, what skills are needed, who passes what information to whom, etc.

In summary, making a list of your needs and listing the extent to which each energy monitoring software package meets them is a useful exercise which will help you to select the best software for you application.

Hiring-an-Employee Analogy

While the above mentioned lists will work for you, it suffers in that it is a somewhat "dry" method. A more enjoyable method is to pretending you are hiring an employee.

Some tools, like a fork lift or pipe wrench, provide productivity improvement but not automation. Other tools, including most good software, provide productivity improvements AND some level of automation. Automation is almost like getting another employee, one that can perform the specific tasks assigned to it. Selecting energy monitoring software can be made more enjoyable and be brought somewhat "down-to-earth" by pretending you are hiring an employee. Complete a help-wanted ad. For example:

WANTED: Person to read steam, gas, chilled water, and electric meters once per day. Will calculate totalizer differences, and compute total flow. Will present raw readings to production and to accounting. Will produce and distribute management summaries based on these readings.
QUALIFICATIONS: Must understand and be familiar with a wide variety of meters and their use.
PAY: 110 AC outlet provided. Surge protection coverage provided.

Selection would then be made based on how closely the software could fill this want ad.

A Final Word

The proper technique, in summary, is to start with your needs, and then work towards the benefits provided. The above two examples (list your needs explicitly and filling out a classified ad) are just two of the many methods available to visualize selecting energy monitoring software. One may work for you, or you may find an analogy that works better.

Needs-fulfillment analysis does not eliminate the need for cost/benefit analysis. Both are useful techniques for making selection decisions. Needs-fulfillment, however, has several distinct advantages over cost/benefit which make it more applicable for selecting energy monitoring software. The primary advantage is a lack of dependence on assumptions. (A careful analysis of this technique shows that assumptions are not eliminated, but are replaced with assumptions which are much more likely to be met. The twin assumptions of knowing true costs and knowing true benefits are replaced with the assumption of knowing your needs and knowing if the software meets those needs. This transformation itself is based on the assumption that the software that best meets your needs WILL provide the best benefit for the least relative cost.)

Summary

When buying any tool, be it a torque wrench, blow torch, or computer software, issues such as "who will be using it," "who will be installing it," and "who will be maintaining it" need to be addressed. A poor match between the tool and the user can be as disastrous as giving a right-hander's baseball glove to a south-paw. For example, some software requires extensive use of the keyboard. If the user can't type, frustration with the software will likely result regardless of how nifty the software is.

Perhaps the most important question to ask yourself is "what do I want out of the software'? If the primary need is to automate monthly meter readings, the software might only be used once a month. This mandates the utmost simplicity for the operator who may not remember from month to month exactly how to operate it. The needs of other people, such as your boss, your subordinates, and other departments, also must be considered.

Remember always to think of energy monitoring software as a TOOL to make your life easier, nothing more. The whole idea is not to be overwhelmed by the new technology, but to USE it to make your life easier. This is only possible when there is a good match between your needs and the benefits provided by the software.

References

1. "How the New Math of Productivity Adds Up," *BusinessWeek* , June 6, 1988, p103.

2. "Uncovering the Hidden Costs," *Wall Street Journal*, June 12, 1987, p14D.

3. Augustine's Laws, Normon R. Augustine, 1987, Penguin.

4. "How to cure computer resistance," *Gas Industries*, May 1988, p18.

5. "Tracking Utility Costs by Microcomputer," *State of California Publication #P400-86-007*, August 1986, p6.

6. The Annotated Alice, Lewis Carrol, 1960, Bramhall House, p.88.

Chapter 27

MICRO VS. MAINFRAME: PLANNING AN ENERGY MANAGEMENT DATABASE

V. Filler

In energy management, everyone can agree on the need for a good computer database. In order to perceive needs, analyze alternatives, and evaluate results, there must be adequate data. This said, however, the arguments begin. In today's rapidly changing world of computers, anyone planning a database must ask several critical questions: What kind of computer hardware and software to choose? What kind of database to set up? And how to manage the database?

Planning is itself a word that should signal caution. With databases it is well to look ahead, but too much planning can bring problems. Computer hardware costs money, and computer programming can cost even more. Too much planning can be an expensive burden if plans change, as they tend to do. The best course, then, is to let the database grow organically from the needs of energy management, and avoid grandiose planning.

Think small might be a useful maxim in this discussion, in more ways than one. The question of size arises right at the start, when one thinks of what sort of computer to use.

Until just a few years ago, a large database required a large computer. Nothing but a mainframe or minicomputer had either the memory capacity or the speed to process large quantities of data. (Old-timers who can recall ancient history--say twelve years ago--like to recount how a big minicomputer once could not handle what a child's toy can rip through today.)

Big computers, however, have inherent drawbacks. They are expensive and cumbersome. And perhaps more importantly, they are not user-friendly. They require a computer programmer at every turn, for installation, modification, and maintenance. This is both costly and inflexible. Even after a system has been set up, it cannot be altered without further programming assistance.

THE MICROCOMPUTER REVOLUTION

All this has now changed with advances in microcomputer technology. (The terms microcomputer, desktop computer, personal computer, and PC are commonly interchangeable.) Today, much of what was once conceivable only in terms of a mainframe computer is now well within the capability of a microcomputer using commonly available software and the skills of the non-programmer. This means databases can be cheaper and easier to maintain. And they can work better.

A BETTER WAY

This last point may be surprising, and bears repeating. A database on a PC can be better than one on a big computer. How can this be so?

The answer lies not in the intrinsic characteristics of the computer, but in how it can be used. A PC is user-friendly in the extreme. This is an incalculable advantage over a mainframe computer. Those who work with the database can easily learn how to use the computer at every step of operation. As a side benefit, all who must work with the energy management database acquire in the process computer skills of great value in other aspects of their work.

With a personal computer, unlike a mainframe, those who use the database also understand how to maintain the database and tailor it to their work. They do not have to call on a computer programmer for every problem that arises. They can modify or manipulate the database according to the task at hand. Thus a PC database does not provide a quick, cheap substitute for a real database; it is actually better.

Of course, a PC does not have the memory or speed of a large, expensive machine. There are trade-offs for being cheap and user-friendly. But as we shall see, the losses do not offset the gains. Recent advances in computers, if used imaginatively, can make the PC the "appropriate technology" for a large database.

AN ENERGY DATABASE

The meter database at Stanford University's Operations and Maintenance department illustrates all of these points. Stanford uses over 100 million kilowatt-hours of electricity a year, distributed to over 400 buildings and monitored by over 250 meters. Over 800 million pounds of steam for heating and 20 million ton-hours of chilled water for cooling are also distributed, monitored by several dozen more meters. The meter data serve various functions: billing and accounting, management planning and forecasting, budgeting, energy conservation, and building retrofitting. The data are distributed in a variety of regular and special reports, as well as being available for immediate inquiry.

Until several years ago, these records were collected in handwritten meter books, extended on a hand calculator, and typed on a typewriter. It was a slow, laborious, and unreliable process.

Mainframe Problems

In an attempt to streamline the operation, a computer database was set up using a mainframe program called Mark IV. After many months of study and programming effort, and great expense, the Mark IV program was put to work, but it never really achieved what was intended. It had implacably rigid user-interface screens, presumably to prevent data input mistakes by inexperienced office workers. Unfortunately, it was easy to enter erroneous figures, and then difficult and inconvenient to make corrections. It was a major task to enter a new meter into the system or do other routine chores. Often, a programmer had to be called for minor matters.

Once set up, the system could not be modified without a major reprogramming effort. Any ad hoc reports and analyses had to be done on the hand calculator. And no single meter or facility in the database could be handled individually: either a giant report had to be run, or nothing at all.

After several years of frustration with this system, the microcomputer came on the scene. As hardware and software developed and as computer skills improved, the entire database has gradually been put on a microcomputer. The improvement has been satisfying in all respects.

A SPREADSHEET-BASED SYSTEM

The hardware we use is a personal computer with a 20-megabyte hard disk. Our software is well known as the most popular PC spreadsheet program, which we have found to be better than specifically designated "database" software. (To confound the discussion slightly, much of the so-called "spreadsheet" software is also designed to provide database functions.)

The spreadsheet is simple to use, yet highly complex operations are possible. All data are arrayed in easy-to-read columns and rows, and can be entered or modified by the merest beginner as the need arises. Once entered, the data can be moved about or manipulated with ease. Thus one is never locked into a particular structure. The database develops in modular fashion, and can be tailored to the particular energy management requirements that arise.

Storage and retrieval of the data present no problem. An ordinary data file, on a single large spreadsheet of up to 400,000 bytes, can easily hold six or seven years of electric meter data. For larger masses of data more than one file can be used. On a fast hard disk, tens of megabytes can now be stored and rapidly retrieved. Thus there is no need to insert and remove floppy disks in order to find, assemble, combine, or reorganize data.

It must be conceded, of course, that a desktop computer cannot at present handle the data in the same fashion as a giant computer. (This may be just a few years down the road, however, when more powerful computer chips and programs are expected to be available.) A big computer is unmatched for speed and memory with massive amounts of data.

But our experience is that the pluses of the PC far outweigh the minuses. Indeed, a supposed shortcoming of the PC is sometimes not the handicap it seems. For example, certain operations that are normally done manually on a PC would commonly be automated on a mainframe. (If more than one file must be retrieved on a PC, this could be accomplished automatically with "macro" commands, but might better be done manually, though it might require a few minutes of work at the keyboard.) But such tasks are generally minimal and tolerable, and often we have been grateful for manual steps. They afford a sense of control over the program and the data, and virtually eliminate the need to call the programmer if something goes wrong.

Database Maintenance

To return to the Stanford electricity database, the basic data reside on one spreadsheet that contains the current year's meter data. (We have separate files for each utility.) Each month we update the database by entering the current month's meter readings. Only the meter readings have to be entered--the usage for the month is calculated automatically.

Among database programs, spreadsheets are unexcelled for the ability to apply calculations to large masses of numbers. On the meter spreadsheet, using formulas, one can quickly check the numbers for obvious errors and unusually high or low readings.

Once the database has been made current, the data can be manipulated in many ways. The data can be sorted, extracted, or combined with other data. Thus the data for the current year can be analyzed by itself, or combined with other spreadsheets to produce multiyear analyses. One of the most important features of the PC, to repeat, is its flexibility: one is not locked into preprogrammed operations, but can use the data quickly and effectively for the purpose at hand. At the same time, routine tasks can be automated and effortlessly accomplished.

At Stanford we produce reports showing energy consumption and costs for the University as a whole, and for different sectors, e.g., housing, athletic department, medical school. We produce reports showing month-to-month comparisons and year-to-year. Some reports deal with only one utility--say electricity--some bring together data for a given facility on all utilities. We are able to extract the data on the 50 biggest electricity users for several years and track changes in consumption. Graphs are used to follow usage trends and check for metering errors.

A Success Story

Perhaps the most telling improvement to come out of the desktop database is in our building energy reports. These are periodic reports sent one or more "building energy managers" in each of the campus facilities as part of a continuing energy conservation campaign. The report profiles the electricity consumption of the building in the current and previous years. When the reports were generated by the aforementioned mainframe program, it was a major computing job, and fraught with problems.

Now the entire system, once thought too big for any computer but a mainframe, has been duplicated on a PC-based spreadsheet. It is not, as might be supposed, a barely adequate stand-in for the mainframe program. It is better in all respects. This is a surprising new capability of the PC, formerly thought suitable only for small jobs. The building energy reports consist of 250 pages of printout, at one time a huge and unwieldy job for the mainframe. But the PC makes short work of it.

THE BROADER VIEW

I mentioned trade-offs--what is it that we give up by using a PC rather than a mainframe-based system? One possible disadvantage is that the PC can handle only files of a limited size. A single file can contain seven years of data for 200-300 electric meters. This is quite sufficient capacity for most tasks we have encountered so far.

For larger jobs, say to combine information on a given facility from different spreadsheets treating different utilities, one needs to use the ability of the program to combine data from separate files. This, as with other data handling tasks, can be automated, using the "macro" method of automated keystrokes and programming commands. For some operations, however, it is better to do it manually--it is not especially burdensome, and maintains better contact between the operator and the database. This, we have found, works better than attempts to simulate the total automation of mainframe programming, which gains little and introduces troublesome complication.

Size, then, is not a barrier. What about speed? With some machines, using floppy disks, a large database can be maddeningly slow. But with a machine of the IBM-AT class and a hard disk, the speed is quite acceptable. In fact, overall the system works faster than a mainframe. Control is local. Printing is done on one's own printer, not queued at a remote site. The PC seldom is "down," and if it is, one can use another PC. And there is seldom the need to call in a programmer to find out what went wrong.

A final consideration is security. One supposed advantage of a mainframe computer is that the database is impenetrable except by secret passwords and tamperproof against all but the trained hand of the programmer. These attributes are doubtless essential to airlines, banks, and other massive and sensitive databases. With an energy management database, however, it must again be stressed that user-friendliness is an advantage. The flexibility to use, innovate, and improve is an asset.

One complaint sometimes heard about PC-based spreadsheets is that they are too easy to work with: everyone who touches the worksheet is tempted and able to make "improvements" according to his judgment. Thus one never knows what has been done to the data or the spreadsheet. This raises a false issue. Increased computer knowledge and user-friendliness are factors in thinking about security, but they do not pose an insuperable danger. Files can be backed up and audited as much as necessary. And the easier it is to understand the computer, the easier it is to train users in correct procedures.

As for protection against deliberate mischief with the database, routine security measures seem adequate under normal circumstances. Locking the office door or desk drawer and the careful, frequent backing up of files can usually assure the accuracy and integrity of the database. For suspected sabotage, of course, there can be special measures, and if the proverbial "disgruntled ex-employee" is on the loose, even more special ones. If someone is determined to corrupt the database, this is more than an energy management problem. In this connection, of course, it should be noted that even a mainframe database is not impregnable.

THE PSYCHOLOGICAL BARRIER

In the light of the foregoing, why would anyone do otherwise than use a PC for an energy management database? I have the feeling that there remains a large psychological barrier to be surmounted. Because the desktop computer has developed so rapidly in recent years, it is sometimes difficult to conceive of a tiny computer doing jobs that were once the exclusive preserve of the giant mainframe computer and its programmers. Many still associate the PC, even using highly sophisticated and well-advertised software, with relatively simple spreadsheets and data listings.

Experience shows that one should not approach a database management project with a feeling of limitations. The personal computer and a good spreadsheet program have proven themselves tools equal to the most challenging energy management database project.

SECTION 5
DIRECT DIGITAL CONTROLS

Chapter 28

DDC PROBLEMS AND SOLUTIONS

J. R. Sosoka, K. W. Peterson

ABSTRACT

Since the advent of the energy crisis the status of controls in the HVAC industry has never been the same. The initial thrust of this change was to use central computers to monitor the performance of the control systems. When the inadequacies of the control systems were discovered, the role of the central computer was expanded by adding outputs in order to reset the pneumatic controls. The next attempt at upgrading was to go to direct digital control and eliminate the pneumatic controls.

Currently there is a lot of concern in the industry because the new controls are often performing very poorly. At the present time the Army and Air Force will not allow DDC systems to be installed because of the poor performance they have experienced. Commercial and institutional users have similar concerns. Another basic concern is the problem of obtaining competitive bids when a proprietary control system needs to be expanded.

This paper will discuss the changes that must be made if HVAC control systems are to operate properly. These changes involve the owner, system designer and the manufacturer. Basically, the owner must provide adequate funding, the designer must become more knowledgeable and the manufacturer must be willing to provide more information on their products.

The paper will also explain the new directions that control system design is taking in the HVAC industry. A major reason for the change is the development of the new "smart" control modules. Increasingly the unitary controller version of these modules is being installed in products at the factory. These same modules also provide a basis for easy retrofit of existing systems because they can be integrated into a variety of central computer systems. Finally, this paper will present an example of a system mixing unitary controllers and industrial microcomputer based controllers with a proprietary control system.

INTRODUCTION

A major problem that the HVAC industry is currently facing is that many large control systems do not operate properly. While this is not a new phenomenon, attention is currently being focused on the situation because it is now practical to monitor the problem with central computers. Also occupants are demanding better control of their environment. In many cases, computer based control systems have made the control problem even worse because of the general lack of understanding of the new technologies involved.

Unfortunately, the industry cannot quickly improve the controls problem because of a combination of many factors. First, owners have not been prepared to pay for a properly designed and operating system. Secondly, HVAC system designers do not generally understand control systems and could not design a properly operating system even if the funding was available. If the HVAC system designers did understand controls it would still be difficult to design the system because of the lack of information available from manufacturers regarding their proprietary control systems. Finally, even if a properly designed system is installed, many building owners do not believe that it is necessary to provide properly trained operating personnel that can operate and maintain the system.

HISTORICAL BACKGROUND

Historically, properly operating control systems have not been important to the HVAC industry because people are able to adapt to their environment. The emphasis for the design of the buildings was on the architectural aspects and cost factors rather than human comfort. Also, air conditioned building are, in a sense, relatively new and people were pleased just to have the building cooled. As a result, commercial HVAC controls are relatively inexpensive compared to their industrial process control counterparts; where the failure of the control system to operate properly was not acceptable.

Another factor contributing to poor HVAC controls has been the industry's practice of basing the design fee on a percentage of the construction cost. Control system costs are relatively low and can involve a considerable amount of work to obtain a properly operating system. This is especially true if system commissioning is included. For this reason it was more profitable for the HVAC system designer to concentrate on the basic heating and cooling systems and leave the design of the control system to the controls manufacturer. Servicing of the system was also often given to the controls manufacturer. With the introduction of proprietary computer based control systems it was generally mandatory to use the control manufacturers service organization.

The problem with this approach is that after a while the only member of the construction team that understood controls was the controls manufacturer. It eliminated the checks and balances provided by the various team members during construction. Under the pressure of trying to reduce the cost of controls and without anyone else actually reviewing the system, it is understandable that contractors might limit their design and startup expenses. Especially since the scope of the original design was probably not clearly defined.

The seriousness of the control problem really started to become apparent when the early energy management computers started providing data on the systems performance. What was observed as periods of discomfort and high operating costs by the occupants and owners, respectively, could now be verified with hard data. In a rush to try and correct the problem there was an attempt to improve the systems operation by resetting the controls from the central monitoring computer system. In most cases, a lack of understanding of these systems, the complex software and the multitude of communication activity led to the failure of these systems. While the computer based systems continued to improve by distributing the control functions to lower levels in the system, the lack of involvement by owners and designers continued to prevent the systems from achieving a successful level of performance.

In addition to excessive energy usage, another factor is increasing the need for better controls. This factor is the growing concern of the occupants regarding their environment. The publicity regarding poor air quality in the work space has increased their insistence on better temperature and air quality control.

CURRENT SITUATION

A major change that has occurred in the area of HVAC controls has been the development of application specific modules or unitary controllers. They allow microprocessor based controls to be installed at the factory. This approach substantially reduces the installed cost of microprocessor based controls. It also simplifies design and installation because the application software in the module can operate independently of the central computer system.

From the standpoint of the owner, designer and HVAC contractor, these factory installed controls offer the potential for increased reliability and reduced cost of installation. The problem is that if equipment is purchased from several different manufacturer's the modules will probably have different proprietary protocols depending on the module manufacturer. The necessity of the manufacturer having to use a module that matches the proprietary control system being used for a particular construction project greatly increases the complexity of the module concept.

Now that owners are beginning to realize that they should be able to obtain properly operating control systems, they are asking the HVAC designers why they can't make the control systems operate properly. As a result, HVAC designers are beginning to realize that their lack of involvement in the control system design and installation process has left them without the training required to address the problem. Furthermore, they realize there was never any money in their fee structure to provide the design services and commissioning effort required to implement a good control system.

While manufacturers are discussing common protocols, they still see proprietary protocols as a way of protecting their share of the controls market as well as the service market. This is particularly true of the large control companies. Many smaller control companies have demonstrated that they are more willing to provide some measure of open or common protocol.

SOLUTIONS

Changes will be required in order to resolve some of the control problems facing the HVAC industry. The building owners need to change their method of operation. System designers must be more involved with the control system and manufacturers will need to provide more information regarding their products.

Owners must revise there thinking regarding controls. Control systems are no longer simply thermostats on the wall that occupants and building engineers can play around with to try and obtain acceptable conditions in the space. The hardware is available that can provide occupant comfort and operating efficiencies. However, there are changes that must be made in the owners approach to controls.

To begin with, the owner must be willing to pay for both the design of the system and the commissioning of the system. The commissioning phase is extremely important because it involves not only the tuning of the system to obtain optimum performance, but also the training of the operating personnel to insure that it stays that way. Having decided to properly fund the controls effort, the owner must then make certain that the controls system designer is really qualified for the task. With the present state of the art this may well require that the design team have a separate controls consultant with a proven record of performance.

A major change in the HVAC industry must take place in the role of the control system designers. They must understand the HVAC system, the control hardware, control theory and programming. Design of the system has to be more than just listing inputs and outputs and providing a general description of what should happen. The control system design should be approached with the same detail as any other portion of the system. The HVAC designer would not normally provide design documents that simply tell the contractor how much air is to be supplied to each room and rely on the contractor to select the air handler, size and route the ductwork, and select the diffuser. The same design detail must apply to the control system design. Design of the control system must include the selection of specific actuators, valves and control devices. Wire types and sizes should be specified. Control algorithms should be specified and approximate tuning constants should be provided by the control system designer. Since the real control is in the software programs the designer must be able to understand and provide the necessary applications programming when required. Simply furnishing a sequence of operations is transferring the real controls design to the contractor.

While this extra effort will require additional design fees it should be partially offset by reduced construction costs. The additional savings from reduced operating cost should more than offset the remaining extra cost.

The major problem will be to develop control system designers that are really capable of providing the necessary services. Until now, the only real source of training for the industry has been ASHRAE. As a percentage of ASHRAE's total effort, the controls training represents a minimal effort. Controls and instrumentation represent 2 or 3 technical committees out of a total of over 80 technical committees. By comparison on the industrial side, the controls function is served by the Instrument Society of America (ISA) which only addresses the design of control systems. This is further supplemented by smaller special purpose organizations such as the programmable controller groups and by the controls committees of larger organizations such as IEEE.

It is obvious that if HVAC control system designers are to develop to the required level of expertise, more support from professional organizations is required. This would provide a forum where the subject of controls could be treated in depth rather than as an ancillary subject. The best approach might be a specialty organization developed under the sponsorship of a larger organization. This would allow for the development of separate conferences that could focus solely on control problems and training.

CURRENT SYSTEM ALTERNATIVES

Currently, there is no common protocol within the industry, making it difficult to integrate unitary controllers from different manufacturers. There is an effort taking place within the industry to develop a common protocol for energy management systems and the unitary controller should be a logical starting point for this work. Since the communication protocol in the unitary controller can be changed relatively easily, the controller can be offered with a choice of protocols. One choice might be a common protocol developed by ASHRAE. In the meantime, some manufacturers are providing documentation for their protocol to assist with the integration of their controllers to other manufacturers.

Currently, building owners and system designers have various alternatives in integrating more than one manufacturer in the control system. Which alternative to implement depends on the capability of the system designer and the type of project. One alternative would be for the manufacturer or a third party, usually a consulting firm, to provide an interface between the unitary controllers of one manufacturer and an existing front end of another manufacturer. In systems built around field panels this approach would require the interface to look like a field panel. See Figure 1 for a diagram of an upper end interface. The software development required for this approach can be costly due to the complexity of the communication protocol used in the upper level of control systems.

Another alternative would be to use a unitary controller protocol interface to another manufacturer's field panel. Figure 2 shows an example of this type of interface. Once again the interface would most likely be provided by a third party. This approach would need to be connected to a field panel in the existing system. If an existing field panel is not close or has no more expansion capability, a new field panel would be required. The software development for this type of interface would cost substantially less and be easier to maintain.

The last alternative would be to specify a non-proprietary or public domain protocol to be used in the unitary controllers. This approach can first be used in new systems where substantial future expansion is expected. Specifying a protocol with which other manufacturers can emulate would allow competitive pricing in the future expansion by multiple manufacturers. While the industry has not yet developed a standard protocol, different manufacturer's controllers can and are being integrated in a single control system.

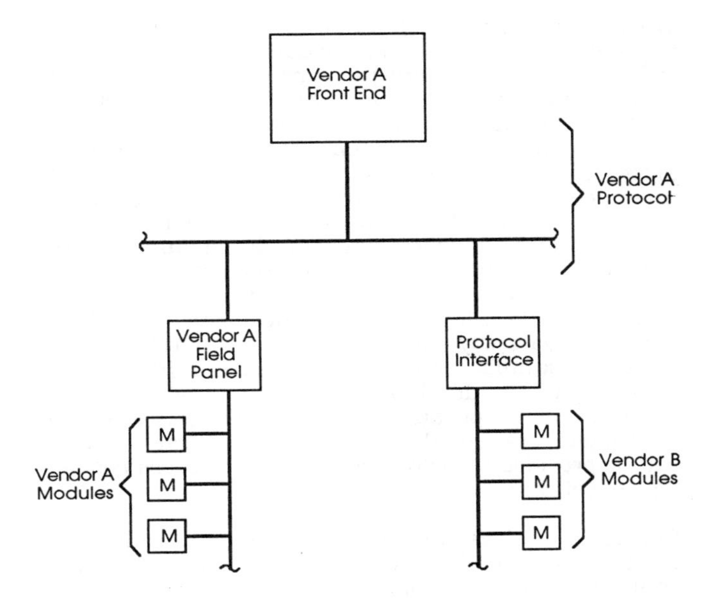

FIGURE 1. UPPER END INTERFACE FOR MULTIPLE MANUFACTURERS

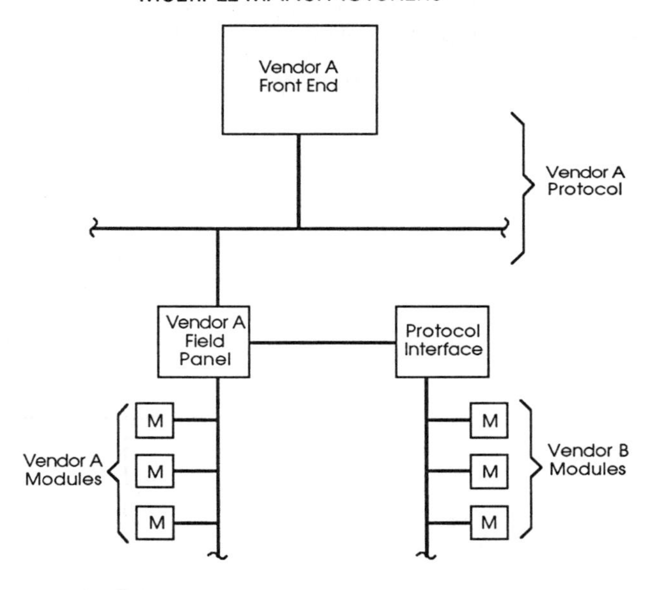

FIGURE 2. MODULE INTERFACE FOR MULTIPLE MANUFACTURERS

EXAMPLE

A large university needed to upgrade the control system in an old 22,000 square feet office/classroom building. The building was served by a dual-duct system with approximately fifty mixing boxes. All of the cooling is provided by a centrifugal chiller and the heating is supplied by a hot water boiler. The existing controls were pneumatic. The University wanted to completely replace the pneumatic controls with a new DDC system which could interface with their existing energy management system.

The following example describes an approach that can be used to insure competitive bidding on future system expansion even when the primary system is a proprietary control system. As a condition of the specification for the primary system there was a requirement that the system manufacturer provide protocol information that would allow the University to interface the system with other manufacturer's devices. Specifically, the University wanted to receive permission for a independent controls consultant to be able to develop interfaces for various application level devices. These included industrial micro's and application specific or unitary controllers. While this approach generally does not appeal to manufacturers, there are manufacturers that are willing to try it. It is also possible to develop these systems without any involvement of products from proprietary system manufacturers. In these cases the central software can be provided by using an industrial process control type of software.

Control of the system was accomplished through the use of an industrial micro in combination with unitary controllers used as smart I/O devices. Very detailed documentation was supplied and the system was installed by an electrical contractor. Refer to Figure 3 for an example of the wiring instruction. The University's personnel were responsible for the insertion of the micro's and modules into the control boxes as well as loading the application software and point addresses. The controls consultant was responsible for the applications programs, systems interface and the training of the University's personnel.

It was decided that an industrial single board computer acting as a local controller along with unitary controllers would provide the university with the most cost-effective solution of retrofitting this system. The unitary controllers were programmed with the same protocol as the local controller and were used as smart input/output modules.

The interface between the new control system and the existing front end was handled by using a protocol converter between the local controller and an existing field panel tied to the front end. The protocol converter looked like a series of smart modules to the field panel. A partial diagram of the control system is shown in Figure 4.

This system required detailed drawings showing the system components, panel layouts, complete wiring schedules and device connections. Figure 3 shows an example of a portion of a wiring schedule. The contractor was responsible for the proper installation of the control panels, controlled devices, sensors and wiring. The control system designer was responsible for the proper operation of the control system including the control sequences and software. Protocol interfaces were also coordinated and verified

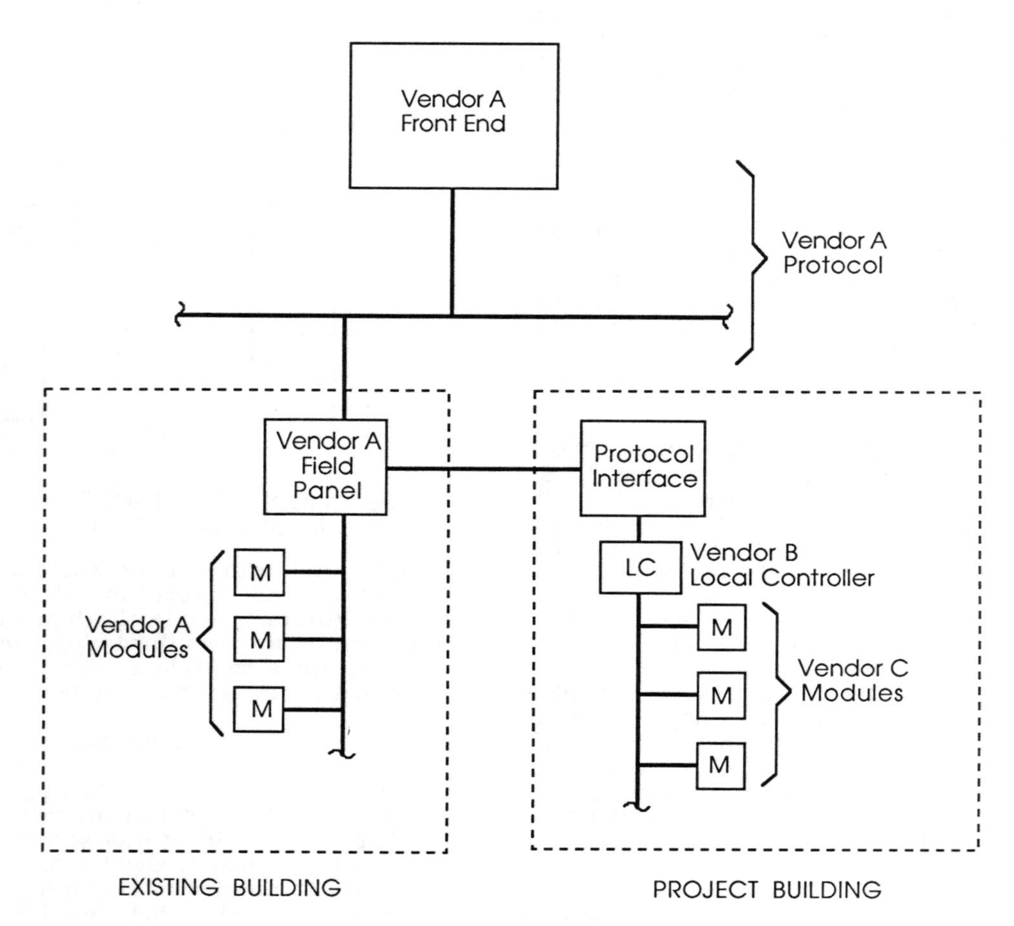

FIGURE 3. PARTIAL SCHEMATIC OF EXAMPLE CONTROL SYSTEM

MODULE:	M1	LOCATION:	Mech Rm					
CONNECTED DEVICE				MODULE/BOARD		CABLE/WIRE		
DESCRIPTION	LOCATION	TERMINATION		TERMINATION				
		POINT	DETAIL	POINT	DETAIL	SIZE	COLOR	TAG
Boiler	Mech Rm		6/B	(R1)T-2/1	6/A	2#14	BK	1-1-1
HW Pump P-5 Starter	Mech Rm		6/B	(R2)T-2/2	6/A	2#14	BK	1-1-2
HW Pump P-6 Starter	Mech Rm		6/B	(R3)T-2/3	6/A	2#14	BK	1-1-3
HW Valve V-2 Actuator	Mech Rm	1	6/G	24VAC	6/D	TP	W	1-1-4
		2		T-3/4			BK	
		3		T-1/16A		TP	W	1-1-16
		4		T-1/16B			BK	
		5		T-2/4		TP	W	1-1-5
		6		T-2/5			BK	
Chiller Starter	Mech Rm		6/B	(R4)T-2/6	6/A	2#14	BK	1-1-6
CHW Pump P-2 Starter	Mech Rm		6/B	(R5)T-2/7	6/A	2#14	BK	1-1-7
CHW Pump P-4 Starter	Mech Rm		6/B	(R6)T-2/8	6/A	2#14	BK	1-1-8
HWS Temp Sensor T-6	Mech Rm			T-1/9A	6/D	TP	W	1-1-9
				T-1/9B			BK	

FIGURE 4. WIRING SCHEDULE

by the designer.

CONCLUSION

If properly operating HVAC control systems are to be achieved a major change must occur in the industry's approach to the problem. Manufacturers must relinguish their insistence on proprietary systems. This approach is simply not in the best interest of the users because it limits cost effective choices and minimizes the involvement of the owner and designer. The increasing use of unitary controls in HVAC equipment will intensify the need for a common protocol at this level.

Owners must be willing to restructure their thinking regarding the design of control systems. These specific design function should be funded with the understanding that the control systems cost is highly leverage with respect to operating costs. A poorly designed control system can substantially negate the performance of an otherwise well designed system. The cost of control system design should include system commissioning as it relates to tuning. The HVAC design then must develop the skills necessary to design properly operating and cost effective control systems. The engineer must take over responsibility to see that the total system is operating effectively. Because of the magnitude of this training problem it will require substantial support from some professional organization to provide a forum for the distribution of technical information.

Even now it is possible to design very cost effective non-proprietary control systems using industrial micro's and/or unitary control modules. By providing very thorough documentation these systems can be installed by an electrician without any knowledge of controls. The installation of the software and the tuning of the system is a joint effort of the designer and the owners personnel. The joint effort also doubles as a training experience for the owner's personnel. All of this will, however, require that the control system designer have a good understanding of control algorithms, sensor selection, programming, etc.

Chapter 29

DEVELOPING A SPECIFICATION FOR A DDC EM SYSTEM FOR A COMMERCIAL OFFICE BUILDING

M. Heis

OBJECTIVES

A specification that results in:

1. State-of-the-art DDC system at a reasonable cost.

2. A user friendly and simple to operate system.

3. Reasonable time schedule for installation and start-up.

4. Adequate training for operators and maintenance personnel.

5. A smooth acceptance with both parties understanding conditions for acceptance.

6. A good payback in energy and labor savings

7. Improvement in comfort conditions for building occupants.

8. Easily serviced and maintained system.

9. A dependable and reliable system.

10. A system that is expandable.

DDC systems may be installed to upgrade existing controls and/or reduce energy consumption. If an old system is to be replaced, list problem areas that the new system should solve. Often new control systems are installed with little thought to the objectives. Old systems may be replaced with new systems having the same design problems. Define what needs to be controlled and how. This is true whether the system is in a new building or existing building. One objective of a DDC system should be improved control and increased occupant comfort level. When a DDC is installed strictly for energy management, analyze potential savings and set realistic paybacks. Payback should consider labor savings because the new system should reduce occupant complaints and operation time. I look at DDC systems as termperature control systems. DDC systems are very efficient temperature control systems. They save energy because they provide much closer control of HVAC equipment.

DDC systems should not be thought of as "black boxes". If a system is too complicated and mysterious, stay clear of it. A DDC system should be easily understood and operated. Maintenance and HVAC operators should be able to service and maintain a DDC system. Expensive service contracts should not be necessary with modern DDC systems. Vendors should provide all manuals and training necessary to make a workable, easily understood, and user-friendly system.

GENERAL SECTION

Section 1 of my specification defines what work is to be done and by whom. It should clearly specify what, if any, work is to be performed by the owner. The scope of work is defined in this section. The following areas are covered in Section 1.

1. Scope of work--general description
2. General system description
3. Contractor/subcontractor responsibilities
4. Equipment to be controlled/monitored
5. Training requirements--type, number of hours
6. Existing conditions, reuse of devices
7. Delivery and storage of equipment
8. Testing and acceptance of system
9. Submittals and shop drawings
10. Design review by owner/engineer
11. Work schedule
12. Documentation requirements including manuals
13. Warranty
14. Electrical and other local codes, insurance
15. Bid evaluation method

Section 1 is from an existing 17 page specification. Figure 1 is only the first page of section 1.

HARDWARE

Points List: A points list must be provided to enable the vendor to determine the number of panels, digital inputs (DI), digital outputs (DO), analog inputs (AI), analog outputs (AO), relay, sensor and other field hardware requirements. An input/output summary form can be used or the points can be listed. An example of both is given and I have used both methods. The input/output summary form was provided by United Technologies Carrier Building Services Division. (see figure 2 and 3) It is a good idea to provide notes to go along with the points list. The example (figure 4, 5) provided further clarifies the hardware requirements. It establishes the number of field panels and their location. Any special requirements should be included in the special notes. In the example, section 4 is the special notes section. Section 5 of the specification is the points list. Notes for Section 5 (figure 6) provide further information for each point in the points list. Also as part of Section 5 is the control sequences required for the points. (figure 7). Notes provide a good method of specifying exactly what is required for the points listed on the points list. They can make the difference between a vendor understanding or not understanding what is required.

OPERATOR INTERFACE

The specification should specify whether a personal computer or a host computer is required. Most systems today use a personal computer that is IBM compatible. In the example provided, Section 3 specifys the personal computer requirements (figure 8, 9).

DDC FIELD PANELS

I like to use Section 2 of the specification for the DDC panel requirements. One page from this section is given as an example (figure 10). When specifying a DDC system, it is very important to specify how the field panels are to communicate with each other and the operator terminals. If the field panels are not stand-alone control units, then they will not function without the host computer. This can be a problem when the host is down for service. Therefore, I make sure that the vendor is supplying stand-alone units capable of functioning without a personal computer or Host computer.

SUMMARY

A specification should be written with specific objectives in mind. The writer must do his homework. He should understand how DDC systems work and how they differ from each other and other types of systems. The specification must set minimum standards for hardware, host/personal computers, software, control programs and training. Time spent on the specification can save a great deal of time and headaches later on.

DIRECT DIGITAL CONTROL SYSTEM

SECTION 1

1. SCOPE OF WORK

The work covered by this specification consists of providing all engineering, technical supervision, software, programming, submittals, shop drawings, installation drawings, computer hardware, installation supervision, training, documentation and transportation as required to furnish and install a fully operational environmental control system to monitor and control the equipment listed in the points list in strict accordance with these specifications, subject to the terms and conditions of the contract.

The owner will furnish the necessary labor to install the field DDC processor, sensors, control devices, conduit, control tubing, and wiring in the facilities in this specification.

Conduit and power wiring to both the central computer, and associated peripherals and all field DDC processors shall be furnished and installed by the owner.

General construction work including labor and material necessary to remove and replace ceilings, cut and patch walls and floors shall be the responsibility of the owner.

2. GENERAL SYSTEM DESCRIPTION

The DDC units shall perform direct digital computer control of the various HVAC systems and equipment identified in specifications.

The central IBM PC/AT or compatible PC computer shall perform data consolidation, reporting, maintain the data base configuration of each DDC processor, provide centralized alarm reporting, operator interaction, system supervision and control. The software shall include an operating system and a variety of application programs to accomplish the requirements detailed in this specification.

The field direct digital controllers shall be capable of stand alone operation and for complete control of their associated inputs and outputs, performing direct computer control from software algorithms maintained in local memory of the DDC control units. All units shall continue to function without disruption of full HVAC control function in the event of the failure on any portion of the communications network, central personal computer or other DDC panels.

(Figure 1)

Location K, 27th floor Mechanical equipment room, S-1 fan system

DESCRIPTION	NOTES	TYPE
1. SUPPLY FANS (2)	6,10,22	DO, DI
2. RETURN FAN	6	DO, DI
3. SUPPLY FANS AIR FLOW MEASURING STATIONS	7,10,23	AI
4. RETURN FAN AIR FLOW MEASURING STATIONS	7	AI
5. MIXED AIR TEMPERATURE	3	AI
6. RETURN AIR TEMPERATURE	3	AI
7. DISCHARGE AIR TEMPERATURE	3	AI
8. PRE-HEAT AIR TEMPERATURE	3	AI
9. MINIMUM AIR DAMPERS	8	DO
10. OUTSIDE AIR DAMPERS	9	AO
11. RETURN AIR DAMPERS	9	AO
12. RELIEF AIR DAMPERS	9	AO
13. SUPPLY FANS INLET VANES	10, 11	AO
14. RETURN FAN INLET VANES	11	AO
15. CHILLED WATER COIL VALVE	11	AO
16. FAN STATIC PRESSURE	13	AI
17. EF - 13 TOILET EXHAUST	6	DI, DO
18. OUTSIDE AIR TEMPERATURE	3	AI
19. OUTSIDE AIR HUMIDITY	15	AI
20. WARM-UP CHANGEOVER	8	DO

(Figure 2)

201M84-008

Input/Output Summary

System Apparatus, or Area Point Description	Analog In										Discrete In								Outputs							System Features																		A/I	A/O	D/I	D/O
	Measured						Calc.												D/O				A/O			Alarms						Programs															
	Supply Air Temp.	Return Air Temp.	Return Air Humidity	Cool. Coil Dis. Air Temp.	Temperature Ind.	Humidity Ind.	KWH	Enthalpy	Run Time	Efficiency	Status	Filter	Smoke	Freeze	Off/Normal Alarm	Hi-Lo	High Temp.		Off-On	Open/Close	Enable/Disable			Damper Pos.	Valve Pos.	Hi Analog	Low Analog	Hi Discrete	Low Discrete	Status		Time Scheduling	Demand Limiting	Duty Cycle	Start Opt.	Enthalpy Opt.	Reset	Runtime Total	Nighttime Free Cooling	Damper Opt.	Power/Fail Restart	Discharge Air Opt.	Night Setback				

Notes:

Page ______
Of ______

BSF1083-047

(Figure 3)

SPECIAL NOTES

SECTION 4

1. In the event of a failure of the system or a stand-alone DDC panel, provisions must be made to manually or automatically switch back to the existing control system. Describe in the technical proposal how this is to be done. All devices needed (including pneumatic, electric and/or electronic) to accomplish the above, must be included.

2. A set of drawings is provided showing the existing control system. Contractor must indicate what devices will be re-used and how the switch back to the existing control system will be accomplished in an emergency.

3. One complete spare panel must be provided. This DDC stand-alone panel must be complete and include enclosure and any necessary software for operation. This panel will be used for training and for spare parts. All interface boards will be included for hook-up to system and PC. This panel is to be delivered within 30 days of contract award.

4. All temperature sensors provided in this contract must have an accuracy of plus/minus one deg.F. Relative humidity sensors will have an accuracy of plus/minus five percent. System shall have a resolution of .1 deg.F. and 1% RH.

5. All equipment provided (i.e. DDC panels, auxillary panels, remote relays, sensors, interfacing devices, modems, etc.) must include enclosures, any required mounting hardware, power supplies and other devices required for proper installation and operation. The owner will provide wiring, conduit and miscellaneous small hardware (i.e. anchors, bolts, screws and the like). Strap-on kits should be provided for liquid immersion sensors (i.e. hot water, chilled water, condenser water, etc.).

6. Space and duct sensors may be either thermistors or 4-20 MA. The proposal should clearly identify the type of sensor proposed and the price difference for using the other type. Space sensors must be complete with decorative cover and suitable for mounting over a standard electrical box. All mounting hardware must be included. Duct sensors must be a minimum of 18 inches long and include all mounting hardware.

7. The outside air temperature sensors must be 4-20 MA sensors. Provide sun shields and weatherproof assemblies for mounting.

8. Liquid immersion temperature sensors may be either thermistors or 4-20 MA. The proposal should clearly identify the type of sensor proposed and the price difference for providing the other type. Probe

(Figure 4)

and sensors heads shall be removable without breaking fluid seal. Sensors must include enclosure and be suitable for well or strap-on applications.

9. Space relative humidity sensors shall be wall mounted devices with cover and all mounting hardware required for installation. The sensor will have a linear output for the range of 20-90% RH.

10. Duct mounted relative humidity sensor shall be designed for duct mounting complete with enclosure and mounting hardware. The sensor must have a linear output for the 20-90% RH range.

11. Proposal should identify components provided to make the air flow measuring stations operational. Specification sheets must be included and must specify accuracy and range of operation.

12. Proposal should identify components provided for building static pressure and duct static pressure measurements. Specification sheets must be included and must specify accuracy and range of operation.

13. Proposal should identify number and type of air flow switches being provided.

14. Proposal should identify number and type of contactors and relays provided.

15. Electrical demand metering shall be from pulsing dry contacts provided by the owner and installed by the utility company.

16. Location A thru O are DDC panels. Master-slave and/or fewer DDC panels is permitted. The following rules must be followed in setting up master-slave arrangements or combining points into fewer panels:

a. Panel A must be a master; panels B and C may be slaves to A.

b. Panel D must be a master; panels F and G may be slaves to D or the points included in Panel D.

c. Panel E must be a master; panels H and I may be slaves to E or the points may be included in E.

d. Panel J can be a master, slave or the points can be included in any other panel on the 12th floor.

e. Panel K must be a master; panel M can be a slave or the points can be included in panel K.

f. Panel L must be a master; panel N can be a slave or the points can be included in panel L.

g. Panel O may be a master, slave or the points combined with K or L.

17. Any pneumatic devices or interfaces provided must be of the non-bleed type.

(Figure 5)

Notes For Section 5

1. MONITOR AND TOTALIZE ELECTRIC PULSES FROM CG&E'S METER; FOR ELECTRIC DEMAND CONTROL.

2. PROVIDE CONTACTOR AND ON/OFF/AUTO SWITCH.

3. PROVIDE SENSOR.

4. EXISTING CONTACTOR.

5. PROVIDE ALL NECESSARY HARDWARE INCLUDING PROBES, TRANSDUCERS AND HARDWARE; REFERENCE POINT TO BE OUTSIDE.

6. PROVIDE AIR FLOW SWITCH, EXISTING STARTER.

7. EXISTING AIR FLOW MEASURING STATION, PROVIDE TRANSDUCER AND ANY HARDWARE NECESSARY TO MAKE IT OPERATIONAL.

8. PROVIDE EP SWITCH.

9. EXISTING PNEUMATIC OPERATORS, PROVIDE PNEUMATIC AO FOR OPERATION.

10. TWO SUPPLY FANS, SEPARATE CONTROL POINTS FOR START/STOP AND STATUS (2-DO, 2-DI).

11. EXISTING PNEUMATIC OPERATORS.

12. PROVIDE HUMIDITY SENSOR-DUCT.

13. PROVIDE TRANSDUCER, REUSE EXISTING STATIC PRESSURE LOCATION.

14. PROVIDE HUMIDITY SENSOR-WALL.

15. PROVIDE OUTSIDE AIR HUMIDITY SENSOR.

16. EXISTING STARTER, STATUS FROM STARTER.

17. PROVIDE ANOLOG OUTPUT (PNEUMATIC) FOR BOILER CONTROL, STATUS POINT TO INDICATE "OFF/ON", PROVIDE "DO" FOR ON/OFF CONTROL

18. STATUS FROM CHILLER PANEL.

19. CONTROL 60, 100% POINTS.

20. PNEUMATIC AO.

21. EXISTING CONTACTOR, PROVIDE ON/OFF/AUTO SWITCH.

22. TWO DO AND TWO DI POINTS.

23. TWO POINTS.

24. PROVIDE VALVE AND OPERATOR OR ADD OPERATOR TO EXISTING VALVE.

(Figure 6)

SECTION 5

CONTROL SEQUENCES

The following software and programming is to be included in the proposal. Owner must approve all programs before contractor enters programming into DDC panels.

1. Start/stop, time-of-day, and optimum start/stop for all DO points.

2. Mixed air control for all supply fans with reset control (from 55 deg.F to 65 deg.F).

3. Supply air control for all supply fans with reset control (from 55 deg.F to 65 deg.F).

4. Chiller optimization and demand control.

5. Boiler optimization and demand control.

6. Peak demand control using digital and analog points.

7. Air volume control for supply and return fans using air measuring stations.

8. Fan tracking control.

9. Static pressure control.

10. Outdoor air reset programs for boilers, chillers, mixed air and supply air points.

11. Enthalpy changeover using return and outside air RH sensors.

12. Warm-up changeover programming.

13. Alarm and event initiated programs.

14. All programming required to allow for proper operation of systems.

15. Chilled water and hot water by-pass control.

16. Chiller efficiency calculations (kw/ton).

17. Power failure motor restart programs.

18. Holiday schedules.

19. Summer/winter switchover programs.

(Figure 7)

DIRECT DIGITAL CONTROL SYSTEM

SECTION 3

1. PERSONAL COMPUTER/HOST COMPUTER

Provide an IBM-PC/AT Computer or equivalent. The operating system untilized shall be single-user, multi-tasking. This software shall provide for concurrency which means that the user can accomplish several tasks at the same time. One task need not be complete before another starts. The operator should be able to run other application programs in the "fore-ground" while the DDC program continues to run in the "back-ground". The operator will be alerted to alarms by the sounding of an audio tone.

Describe operating system and all software packages in technical proposal.

If an IBM-PC/AT Personal Computer is not being provided, please **describe the Personal Computer and its capabilities.**

2. PERSONAL COMPUTER HARDWARE CONFIGURATION

Provide necessary hardware configuration and describe in proposal.

a. IBM-PC/AT with 512K memory, keyboard and one dual-sided, floppy disk drive.

b. Quadram board with 256K, real-time clock, Asynchronous communications port, and parallel printer port.

c. IBM color/graphics display adapter board.

d. Color monitor.

e. Printer with cable (IBM or equivalent).

f. Twenty (20) megabyte hard disk drive.

3. SYSTEM CAPABILITIES

a. Two-way proprietary twisted pair communications with multiple DDC panels.

b. Two-way leased-line communications with multiple DDC panels.

c. Head-end receives immediate off-normal condition reports from DDC panels (alarms and change-of state reports).

(Figure 8)

d. Data base down loading to DDC panels.

e. Logs and summaries for DDC system analysis.

f. English language descriptions for DDC points.

g. User friendly operation- menu driven.

h. Dynamic color graphics.

4. ACCEPTABLE SYSTEMS

This section is not intended to eliminate control companies from bidding because of differences in personal computers or software. However, this section outlines the basis for evaluation of the IBM or other PC proposed.

5. DESCRIPTION

All hardware and software provided must be fully described in the technical proposal.

Figure 9

The system, as specified, shall independently control the building's HVAC equipment to maintain a comfortable environment in an energy efficient manner. The building operator shall communicate with the system and control the sequence of operation within the building.

2.1 SYSTEM ARCHITECTURE

The building control system shall consist of a network of independent, stand-alone control units. Each stand alone control unit shall be capable of performing all specified control functions in a completely independent manner. All operator communication with the system shall be via operator terminals provided as required. It shall be possible for each control unit to have a dedicated local display or for a collection of several control units to share a single operator terminal.

2.2 STAND-ALONE CONTROL UNIT

Each control unit shall be capable of full operation either as a completely independent unit or as a part of the building-wide control system. All units shall contain the necessary equipment for direct interface to the sensors and actuators connected to it. No more than two (2) auxillary panels are allowable per stand-alone control unit. (Master-slave arrangement)

It shall be possible to define control strategies at each control unit, and for the control units in the system from any operator terminal in the system. Each control unit shall provide the ability to support its own operator terminal if so desired.

Each stand-alone control unit shall include its own micro-computer controller, power supply, input/output modules, and battery. The battery shall be self-charging and be capable of supporting all memory within the control unit if the commercial power to the unit is interrupted or lost for a minimum of eight (8) hours. The stand-alone control unit shall be listed by Underwriters Laboratory (UL) against fire and shock hazard as a signal system appliance unit.

2.3 SENSORS/INPUT SIGNALS

Each stand alone control unit must be capable of direct interface to a variety of industry standard sensors and input devices.

It shall be possible for each stand alone control unit to monitor sensors of various types as follows:

- analog inputs

Figure 10

Chapter 30

WHY THE NAVY SHOULD LIFT THE BAN ON, AND ALLOW THE USE OF, DDC SYSTEMS IN ITS FACILITIES

W. G. Hahn

ABSTRACT

Direct Digital Control (DDC) systems have been commercially available for approximately 10 years. When introduced, the Navy allowed the use of DDC systems where they proved economically feasible and when control and operational sequences could be met. However, due to what seemed an inordinate amount of problems with DDC systems for HVAC control, the Navy instituted a ban on the use of DDC on HVAC equipment in their facilities, until such time as the problems could be studied and solutions provided.

Some of the problems included: poor installation of many systems; inoperable or failure prone hardware; awkward operators consoles (when provided); confusing operating software instructions; poorly written or nonexistent documentation; failed maintenance requirements; and nonprofessional, extemporaneous and haphazard training. These problems led to HVAC systems that were out of control, and which could not be operated or maintained by base personnel. Further, of the DDC systems that seem to operate properly, there is no method to insure that another manufacturer can add to the DDC system now installed, thereby depending upon only one vendor.

During the ban, the Navy has recognized many of the advantages of DDC systems are being lost to the Navy. These advantages include: system accuracy allowing for superior occupant comfort and system operating economy; inherent energy conservation programs and routines allowing for continuous energy savings; in many cases, lower first costs (with DDC system costs decreasing); lower life cycle costs; increased control system flexibility and versatility, allowing for increased and enhanced HVAC and building management capabilities; and control system speed which can provide for application of advanced and improved facility automation systems operational capabilities.

This paper discusses the steps taken, through the new Navy Guide Specification for DDC Systems, that will help to eliminate or minimize the problems the Navy previously encountered, other methods that the Navy will also use to reduce or eliminate these problems, and various additional advantages available to the Navy in using DDC systems in their facilities.

PRESENT REQUIREMENTS

Overview

The control systems that will be controlling the HVAC Systems, Life Safety Systems, Security Systems and other processes in Navy buildings in the year 2000 are being designed now and will be installed during the next five years.

These systems in the past have gone by many names. Automatic Temperature Controls (ATC), Building Automation Systems (BAS), Facility Management Systems (FMS), Facility Management and Control Systems (FMCS), Building Management Systems (BMS), Energy Management Systems (EMS) and Energy Monitoring and Control Systems (EMCS), just to name a few.

In general, in the past, these have mostly been pneumatic control systems, with some electric or electronic control systems. Where energy management has been applied to HVAC systems, the controls have usually been central computer systems. In some cases, the control systems have been integrated or interfaced with fire alarm systems (Life Safety Systems), lighting controls and security systems. When this was done, the proliferation of the above names came about.

Navy Automatic Temperature Control (ATC) System Design Criteria For Architects/Engineers (A/Es)

Present Design Criteria:

1. Navy Design Manual (DM) 3.03 (Jan 87 3.23.4) prohibits DDC systems, except on pilot projects.

2. Control systems are required by DM 3.03, Section 6, to be designed for "future EMCS".

3. Southern Division Naval Facility Engineering Command design manual, Section 15000 requires provision for separate zone control for building areas with different uses and different loads and also requires occupied-unoccupied control.

4. MILITARY HANDBOOK, FACILITY PLANNING AND DESIGN MANUAL (MIL-HDBK-1190, formerly DOD 4270) Chapter 8, (Energy Conservation Criteria) has no control requirements. "Energy Conservation Criteria" in Chapter 10, (Air Conditioning, Dehumidification, Evaporative Cooling, Heating, Mechanical Ventilation, and Refrigeration) simply requires compatibility with future EMCS. Chapter 11 (Energy Source Selection and Central Heating Criteria) has no control requirements under "Energy Source Selection".

5. DM 4.9, EMCS Manual, Sept 1983, (Army TM 5-815-2, Air Force AFM 8-36) has only EMCS criteria, no closed loop control requirements.

6. There are no control systems that are presently standardized with another, or standardized with an existing control system for a facility addition or alteration. Sequences of operation, documentation, installation criteria, execution and user interface also are not standardized.

DDC Guide Specification: The Navy, through review by all of its design divisions and many vendors, has recently developed NFGS-15972 Control System Guide Specification. It will be available mid 88. It is based upon Direct Digital Control using commercially available single loop or multiple loop controllers with good quality commercial components. This guide specification does not include specific sequences of operation for various HVAC equipment, but does allow for development of almost any sequence. Therefore, it is flexible in that it allows the Architect/Engineer (A/E) to specify a sequence of operation that fits the building HVAC system design. It has reasonable cost and vendor competitiveness with good quality commercially available equipment. The specification also includes requirements for more adequate documentation, warranty, installation, testing and training.

PAST EXPERIENCE

Energy Monitoring And Control Systems (EMCS) In The Navy

Approximately 10 to 12 years ago, an attempt was made to provide a design manual/guide specification that would provide all of the desirable features of all of the manufacturers of EMCS equipment and at the same time "standardize" EMC systems. This evolved into the "Tri Service EMCS Specification". It did not meet expectations. Even though the Tri Service specification was designed to meld all of the desirable features of all of the manufacturers, it instead caused most manufacturers to modify their standard hardware and/or software in order to allow them to compete for this large market. Hence, almost all systems installed were either "one of a kind" systems or nonstandard "bastard" systems of the control company's standard package. While the specification tried to standardize the features required, it could not standardize the interface protocol, hence, very few systems could be expanded except with the original vendor's equipment. EMCS additions became very expensive. High service costs and lack of experience with the modified systems caused maintenance to either be very haphazard or ignored. Many EMC systems installed during this era are presently not operating. Only some EMC systems operated up to specifications and expectations.

Because EMC systems of this generation depended upon one main Central Processing Unit (CPU) minicomputer they, by their very nature, developed two serious problems that eventually led to their demise. And, because they operated from one CPU through "dumb" field panels, if the communication cables between the CPU and field panels would fail, or if the CPU would fail, the entire system failed. This, in effect, sometimes stopped an entire facility HVAC system! Secondly, because of the large point count of these systems and because they operated through one CPU, the large amount of activity in the system would slow down speed of response to several minutes thereby making monitoring and energy management ineffectual. (Later EMCS hardware reduced these problems through "Smart" field panels.)

Therefore, if the systems were expensive to maintain, stopped the HVAC upon CPU or communication failure and were not dependable for monitoring and energy management, it is not difficult to understand why their outputs were bypassed and the systems abandoned!

Early DDC Systems' Questionable (Bad) Reputation

Many of the early DDC systems were designed, engineered, installed and serviced by the personnel and companies who provided EMC systems. Therefore many of the errors were carried over. Also, the earlier DDC systems were usually of high point count, hence when they failed, not only did they affect the EMCS items, but also the closed loop control operation. While rushing their products to market, many manufacturers had not perfected their systems, hence, failure prone hardware and many software bugs were encountered. Further, because of the veritable explosion in numbers of systems being marketed, there were not enough qualified control contractor or branch personnel to properly design, engineer, install, warrant and service the systems. Poorly written or nonexistent documentation was equally predominant with these early systems.

Most user operators of the early DDC systems were drafted from previously operating pneumatic control systems and were generally not CRT or computer oriented, hence they not only did not understand the DDC system, but sometimes actually feared these systems! Many of the early systems had confusing software operating instructions and awkward or nonexistent operator's consoles, further leading to the alienation of the non retrained system operator. Most systems were turned over to the user without any test and verification as to whether the system was operating in accordance with the design criteria and specifications. Operator, warranty and service training was either nonprofessional, extemporaneous, haphazard or nonexistent. Again, with all of the above stacked against the DDC system, many were overridden, turned off, never to run again.

CONCERNS WITH DIRECT DIGITAL CONTROL TODAY

Nature of The Business

Heretofore mediocre design specifications, schematics, point lists and sequences of operation have led to mediocre or poor installations. With a DDC system being new to most user/operators, it is important to have the DDC system that is turned over to them by the contractor be operating at 100% so that they can concentrate on operating and learning the system, rather than fixing the system or making it work. Most vendors and contractors have the capability of installing a good operating system, but with the nature of competition being what it is, "what you specify is what you get"! And even though you may have an excellent technical specification, with all of the necessary hardware and software required, if it isn't checked, tested and verified, you probably won't get that 100% system.

Areas of DDC System Specification Concern Are:

1. A two tier DDC system design process should occur. The first should be a conceptual outline of the control company's interpretation of the engineers design with a meeting to review and insure that the outlined system covers the design's salient points. Second, the final design should incorporate review process considerations as well as all specification submittal requirements.

2. In many cases, adequate equipment data is not being required and/or submitted.

3. In some cases the DDC schematic design is not included in the contract documents.

4. Often the sequence of operation is not detailed enough or it does not cover the operation of all equipment.

5. Software flow charts are seldom required, yet they aid immensely in analyzing and trouble shooting a system.

6. Room/sensor, valve/operator, and damper/operator schedules are not provided and/or specified.

7. Approximately 50% of DDC system designs do not include point lists which are invaluable in detailing what is to be monitored and controlled, and in outlining the parameter operational requirements of these points.

8. Panel wiring and tubing schematics and layouts, as well as panel details should be required in a submittal, but are seldom specified or checked.

9. Wiring and tubing overlays on floor plans should always be required with the "as-builts" to aid the user in maintenance and system additions and modifications. Also, as-builts should include corrected and updated submittal data.

10. Interfaced equipment (chillers, fire alarm panels, boilers, etc.) details, schedules and wiring/tubing diagrams should be required of the DDC system contractor, but generally are not specified or supplied.

11. Software program listings and attributes, referenced to the software flow diagrams and sequence of operation can aid immensely in the analysis of a DDC system, but are usually not provided by the DDC system contractor.

12. In order that proper check out, verification and testing can be accomplished on a system, and that proper training can be provided, a performance testing and verification outline and procedures manual should be submitted 60 to 90 days prior to the expected test.

13. Also, a training manual, schedule, plan and outline should be provided 60 to 90 days before the training is expected to start.

14. Further, at this time, a DDC system failure procedures manual should be submitted.

Note: Each of these will have standard elements. However, in that the manufacturer of a given DDC system will be different than another manufacturer, and in that each HVAC system controlled is unique, each of the above must be tailored to the manufacturer and the building HVAC system being controlled.

Areas Of System Inspection Concern Are:

At this critical stage in the development of DDC systems through out Navy and other facilities it is incumbent that they operate at 100% when turned over to the user so that the HVAC system can "hit the ground running." To do this, the owner/user or the consulting engineer must test and verify the system before the contractor has left the project.

Note: As of November 87, the Air Force has a detailed test and acceptance procedure, and field team, to aid in training their contract field personnel in the test and verification of their electronic control systems. This program is being expanded in 88 with their contract field inspection personnel instructed to request assistance of the test and acceptance team for projects being completed. A similar program is being carried out by Southern Division and the Navy Energy and Environmental Support Activity (NEESA) in the verification and test of a DDC system at a Navy facility in Florida.

Either the Navy will provide for this test and verification procedure, or it will be contracted, as a separate responsibility, to the consulting design HVAC engineer or qualified controls system engineer. However, in this event, a qualification program will have to be developed.

Areas Of Acceptance Concern Are:

1. Once the test and verification has been successfully concluded, an endurance test, with acceptance parameters, is completed before system acceptance. System acceptance includes an integrated turnover to the user, with all the "players" (control contractor, mechanical contractor, operating and maintenance personnel, and the HVAC design engineer).

2. Also, at this time, certified documentation is turned over to the user.

3. The integrated turnover also commences the training program and system demonstration. Extended training is required, into the next heating/cooling season. Factory training schedules are also made available to the user.

Concerns During System Warranty Are:

1. A DDC system failure procedural manual, along with DDC hardware and software backup, is provided and demonstrated at the integrated turnover.

2. Designated spare parts are provided by the control contractor during the warranty period (stored on site), with an option to the user to purchase the spare parts upon warranty completion.

DDC System User Responsibilities Are:

As with any new technology, trained personnel are a scarcity. More computer trained and computer literate (along with HVAC system trained) personnel will be brought on board for facility and station maintenance staffs.

Other System Improvements and Concerns:

1. As with EMC systems, very few manufacturers' DDC systems can interface or operate with another manufacturer's system. A common communication protocol is needed. Although most DDC systems can and do operate through the same P.C. as with that of another system or systems, they do not communicate with each other through this common P.C.

2. Much of the hardware being installed in systems today has been available for up to five years, and as such, is time tested. Obviously, manufacturers are continually enhancing or updating their microprocessor products, but these too are being tested more thoroughly, and are "burned in" (powered up and operated on line) for several days before shipping.

3. Even as much of the hardware is keeping abreast of modern microprocessor technology, so to has much of the software for closed loop control, operator interface and system manipulation been developed and been operating for several years. Therefore most "software bugs" have been corrected and adaptation to new hardware is relatively easy. Most new software is being developed for artificial intelligence and for networking systems. Standard HVAC system operation and sequence software has generally been forged in the "real world" crucible and has become more reliable.

4. Because of the fact that most hardware and software has been in a period of stable development for several years, manufacturers have had the chance to enhance and improve their documentation so it better fits the users needs.

5. Most systems have been developed to be P.C. compatible, hence their usefulness for operation and management is expanded. Almost all control companies have MODEMS at their facilities to aid the user in operation, maintenance, trouble shooting and service of their DDC systems.

6. Both system installers and users will require more education; become computer literate; and have technical associate degrees or other experience in computer science and/or electronics.

7. Both local control companies and branches, and manufacturers have improved their central and local training capabilities for engineering, installation, operation and maintenance.

8. Because of these and other considerations, design consulting engineers will more often than not specify DDC for their HVAC systems.

ADVANTAGES OF DIRECT DIGITAL CONTROL

First Cost

1. First cost of Direct Digital Control systems is equal, or less than, the cost of comparable pneumatic or electronic systems as applied to the same main HVAC equipment. (Chilled water system, AHUs, heating system, etc.)

2. Presently, the first cost is about double that of conventional controls for terminal units. (VAV) boxes, fan coil units, reheat coils, etc.)

Operating Costs

However, because of the accuracy, the built in energy management routines, and the sequence programming capability of DDC, it will always significantly reduce HVAC operating costs.

Speed

DDC is extremely fast reacting for both control and monitoring. Obviously, this is important for emergency alarms, (high temperature, low water level, water leakage on computer room sub floor, etc.) but it is equally important for fast response in certain control sequences. (Alternating or sequencing chillers based upon flow, differential static pressure control for lab areas, etc.)

Interface

Because DDC receives digital inputs and provides digital outputs, it can easily be interfaced to all Facility Automation Management and Control Systems (FAMACS) via supervised wiring for fire alarm and life safety, security and access control, computer room monitoring, sprinkler system monitoring, elevator operations, lighting control, energy storage systems and of course HVAC systems. Through event and time preprogrammed sequences, the "Smart Building" is a reality.

Reliability

In spite of the failure of some early DDC systems, they are, never the less, much more reliable than pneumatic and electronic systems, and they are self reporting through preset high and low operating parameters.

Versatility

Because DDC can accept any variable or discrete input, or provide any variable or discrete output, it is versatile enough to be applied to almost any building or process function (i.e. monitor and control a building's exterior lighting) and therefore is an invaluable tool for the innovative user and operator.

Flexibility

All of the sequences and routines required for an HVAC system and its interfaced systems is accomplished in software. With pneumatic or electronic systems, for instance, if a set of dampers were required to operate in a given manner from one set of fire alarm system contacts and then were to be changed to operate in another manner from another set of fire alarm contacts, the wiring and the tubing would have to be modified. With a DDC system, these same changes could be accomplished through the keyboard, thereby providing a great deal more flexibility for modifications and additions. Once the hardware is attached as a DDC input or output, what it does, is simply a matter of software and keyboard entry.

Expandability

Because today's DDC systems are modular in increments of eight or sixteen, they are designed to be extended. Hence, there are no "closed-end systems", as may be the case with electronic or pneumatic systems. Cases in point are the many pneumatic and electronic panels that have been modified to include additional gages, switches and pilot lights added in a haphazard manner to the bottom of the panel, or to a non-matching panel extension.

Maintenance And Service

While the new user of a DDC system may have to learn how to operate and maintain a product that is relatively foreign to him, the actual maintenance and operation of a DDC system is less demanding and easier to carry out than that of a pneumatic or electronic system. With the pneumatic or electronic system, each component (high pressure selector, averaging relay, etc.) would have to be checked for proper operation and/or wiring/tubing connections, in order to check proper system performance. In some systems there may be hundreds of these items. With DDC, most of these functions are carried out in software and therefore can be checked through the CRT and can be programmed to alarm or alert upon abnormal condition. Actual maintenance then applies to the sensors or controlled devices, the same as with pneumatic and electronic systems (except that sensors with DDC systems require less service and maintenance, and have longer lives). DDC maintenance applies to the microprocessor circuit card and/or power supply, which is easy to trouble shoot (it is self reporting) and replace. The DDC circuit card is probably easier to trouble shoot and replace than its corresponding pneumatic or electronic receiver controller.

MODEM Support

Additionally, maintenance and service is greatly aided through the remote CRT modem communicating off site to the DDC system. Many corrections, changes and adjustments to a system can be made simply and easily by maintenance personnel or a service organization from a remote CRT through a modem without the additional cost or time delay that would be required by a site visit. Control service contractors can handle up to 50% of their service calls on DDC systems without leaving their office!

Total HVAC System Control

By use of compatible unitary microprocessors at the various controlled HVAC pieces of equipment, the HVAC system DDC controls can be networked to form a homogeneous integrated single building system with all aspects of operation and management of the HVAC system coordinated for maximum efficiency. To do this with a pneumatic or electronic system would, or course, require a great deal of wiring/tubing and interface devices. With networking, functions such as whether to operate two chillers at 40%, or one chiller at 80% can be analyzed with all aspects of the HVAC system taken into consideration.

Accuracy

Accuracy of pneumatic systems is known to be notoriously poor. While the accuracy of electronic systems is much better, each control loop must have an indicating sensor, making this and its wheat stone bridge its weak point. DDC systems will maintain the accuracy of the chosen sensor, with an added accuracy burden in the analog to digital conversion of 0.25% or less. With thermistor room sensors, for instance, (low cost) end-to-end accuracy is +/- 0.5F. Again, these accuracies can be monitored through high and low alarm and alert limits. Obviously, greater accuracy provides greater comfort and greater economy of operation. In laboratory or computer room situations, greater accuracy can provide an improved environment and greater protection and longer life to operating equipment.

Energy Management

1. Although the reputation of the previous generation of EMC systems may have been dubious, the development and implementation of these systems did provide a very solid foundation of EMCS software routines and sequences that are now inherently available in the DDC microprocessor. Therefore, when a DDC system is installed, the full retinue of EMCS is automatically included simply by initiating the data base for a given procedure or routine.

2. If all of the HVAC equipment in a facility is operated and controlled by networked DDC, the HVAC system operating economy can be greatly improved. For instance, if all of the lighting and terminal unit operation of an office or classroom can be monitored and controlled, it stands to reason that this equipment will be operated and used only when these spaces are occupied. If it is a multi-tenant facility, with each tenant paying for its share of the facility operation, greater watchfulness on the part of the tenant will be forth coming if the tenant know that he is being billed for the actual HVAC and lighting usage, instead of on a square foot basis, which is all that would be available without networked DDC.

Comfort

As stated above, greater accuracy leads to greater comfort. For instance, DDC allows for the HVAC system to be preprogrammed to over cool a conference room or banquet room prior to a large gathering so as to prevent over heating later as the instantaneous cooling capacity becomes overloaded.

Also, on those "heat in the morning, cool in the afternoon" days, the DDC system can be preprogrammed for a warmer "toasty" temperature in the early morning, to a slightly cooler set point temperature in the warm afternoon, while still remaining within energy conservation guidelines.

Facility Management

For medium and larger buildings, the Facility Automation Management and Control System (FAMACS) is applicable. As stated above, this includes the integration or interface of all building functions and systems into one homogeneous system. An integrated system is one that operates on one single microprocessor network with the same communication protocol for all system microprocessors being coordinated through a central computer. The advantage of an integrated system is that it is software oriented and is therefore very flexible. The disadvantage is that, by nature, it effects life safety, it must be UL listed or otherwise certified, and usually this process takes one to two years, hence, new innovations are slow in coming. Also, with an integrated system, if the central computer fails, the system fails. On the other hand, an interfaced system, uses separate microprocessor based systems for each of the building functions. These microprocessors control their own functions, independent of the other, usually with separate communication protocol, and each with its own operators device. Operations with each other are interfaced through hardware digital or analog inputs and outputs. The disadvantage of an interfaced system is that it is not very flexible. However, if any one microprocessor should fail (i.e. fire alarm, security, lighting control, etc.) it will not affect any of the other systems or functions. Also, because only some of the functional systems must be certified or UL listed (i.e. fire alarm, security), those not requiring this type of certification are usually on the leading edge of technology. In either case, DDC lends itself very well to FAMACS whether as an integrated or interfaced system.

Intelligent Building

With all of the information being sensed and monitored through a microprocessor, it is but a simple step to store this information for a short time period, then download it to a central CPU or P.C. where it can be manipulated, massaged and, in simple words, managed. By managing this information, we can manage our facility. The smart or intelligent building becomes a reality. Especially with an integrated or interfaced FAMACS, a facility can be available to a user through the use of his touch telephone and voice instructions from the DDC computer or be available to him through a card access system in communication with the DDC computer. Total building operation can be analyzed and dissected by massaging the sensor information through P.C. programs similar to debase or LOTUS.

DIRECT DIGITAL CONTROL FUTURE DIRECTION

Protocol

The problem of communication protocol between DDC panels of different manufacturers continues to plague the controls industry. Over a year ago, ASHRAE established a special industry wide committee to address communications protocol for energy management and controls systems. (SPC 135.) The committee and its three subcommittees have had several meetings, but it will be a very slow and difficult process to obtain a consensus among users, manufacturers and designers for the protocol. It does appear that a standard will be available in 1991 or 1992.

Product

1. Each sensor or sensor package will be a self contained micro processor with an indication window and set point window. For instance, a room or space sensor, will have a small LED or LCD window which can indicate the actual space temperature, and indicate the space set point. Communications between the terminal unit and/or network microprocessor will be over a single pair of shielded wires acting as a "data highway" to each sensor package and network microprocessor.

2. HVAC equipment and AHU microprocessors will be approximately 16 points and be designed with an indicator window and keyboard to allow direct access (with proper security access code) and will handle a single piece of HVAC equipment. (AHU, chiller, boiler, door control etc.)

Market

Controls are now approximately 50% DDC and 50% pneumatic. By 1995 it is expected they will be 80% to 90% DDC and 20% to 10% Pneumatic. Also, most HVAC equipment (VAV boxes, fan coil units, AHUs, chillers, etc.) will be supplied with self contained unitary DDC.

NAVY ENERGY OBJECTIVES AND GOALS

The Navy's goals for reduction in energy consumption are:

To Reduce Energy Consumption In Existing Buildings By:

*6% by 1990
*12% by 1995
*15% by 2000

To Reduce Energy Consumption In New Buildings By:

*1.0% per year up to 10% by 1995.

Note: These are based upon 1985 energy consumption figure

Present Navy Schedule

The present Navy schedule calls for 1500 existing buildings to be provided with Energy Monitoring and Control Systems (EMCS) by using DDC single building controllers. Per building costs are estimated at $25,000.00 with savings estimated at $12,500.00 per year per building. (Two year pay back)

FUTURE NAVY DIRECTION

Assessment

1. Because of the recent development, implementation and acceptance of the microprocessor for HVAC control throughout the industry, the Navy is at a pivotal point in implementing DDC technology to improve its facilities. Technology must not over take the Navy, hereafter requiring the Navy to keep grappling with even newer technology.

2. In order that Direct Digital Control provide all of the benefits available, with as few of the problems that could occur, attention will be directed to three areas of concern: the specification; the test and verification procedure; and operation and maintenance.

3. To properly employ Direct Digital Control in Navy facilities, the following procedures or programs will be implemented:

DDC System Design Phase

1. The Navy is preparing NFGS-15972 which is a control system specification that includes and allows for pneumatic and electronic controls, but also strongly covers direct digital control.

2. The NFGS-15972 controls guide specification covers the below specification items. However, as systems are installed, operated and maintained, there may be areas of the specifications that will require rework and/or improvement.

a. Sensor and controller accuracies.

b. Provide for commercially available DDC equipment.

c. Allow multi closed loop system controllers.

d. Allow single closed loop unitary controllers.

e. Require networking between all DDC panels and devices.

f. Allow no more than two HVAC pieces of equipment (AHUs, Chillers, etc.) to be controlled from one DDC panel.

g. Provide for a building operators console and keyboard.

h. Require that the equipment be "burned in" at the factory.

i. Require that the software and hardware be already field operational. Prohibit prototype installations.

j. Require that the software protocol of all DDC panels in a system be P.C. compatible.

k. Require an on site MODEM to communicate with:
   1. Control company service group MODEM.
   2. Base, activity, Engineering Field Division or station MODEM.

l. Provide for an experience clause for the control contractor and his personnel.

m. Require two tier submittal procedure.

n. Specifications and drawings include:
   1. Sequence of operation.
   2. Schematics.
   3. Point list.

o. Documentation includes as a minimum:
   1. Software flow diagrams.
   2. Software program listings referencing sequence.
   3. Schedules-room/sensor, valve/operator, damper/operator.
   4. Panel schematics, layouts, details.
   5. Wiring/tubing plan layout and as-builts.
   6. Interfaced equipment details, schedules, wiring and tubing requirements.
   7. Performance test and verification plan and procedures outline.
   8. Training manual schedule, plan and outline.
   9. Failure procedures manual.
   10. Operators manual.

p. Requirement for test and verification.

q. Requirement for endurance test.

r. Integrated system turnover.

s. Training for:
   1. Engineering and design modification.
   2. Operators.
   3. Maintenance and service.

t. Training to include:
   1. Local on site over a period of two months.
   2. Local refresher at next change of season (winter or summer).
   3. Factory training schedules and availability.

u. Warranty designated spare parts with purchase option.

v. Warranty failure procedure and replacement

<u>DDC System Inspection; Test And Verification; And Endurance Test - Construction Phase</u>

1. It is of the utmost importance that the DDC system be turned over to the user in 100% operating condition. Without proper test and verification by <u>qualified</u> personnel, this is just not possible.

2. Even the most conscientious control companies do not have check out, calibration and adjustment procedures that tend to be thorough enough without outside supervision.

3. Test and verification, as well as supervision of the endurance test can be implemented in one or both methods:

   a. Develop a Navy test and verification training team to educate and train inspectors from the construction offices to become familiar with, and gradually take on the responsibility of, testing and verification of DDC systems.

   <u>Note:</u>This is the approach that the Air Force has taken on their analog electronic control systems.

   b. Allow for a service, negotiated in the A/E's fee, to be provided by a qualified A/E or his contracted qualified controls engineer, to inspect, test and validate, and supervise the endurance test of the DDC system.

<u>Activity Operator and Maintenance Personnel - Maintenance Phase</u>

1. Even assuming that the DDC system installed is the best available, has an abundance of excellent documentation and is operating perfectly when turned over to the user, it is still very possible that the DDC system will not be used to its fullest extent unless the operator and maintenance personnel are qualified and are willing to accept the system.

2. Hence a recruiting and/or training program will be instituted for the system operational and maintenance personnel. The program will encourage existing activity personnel to receive Navy or outside training in computer process control and HVAC systems.

3. Simply stated, personnel must become computer literate in order to operate and maintain DDC systems. If this cannot be done with present personnel, then personnel will be required who have associate degrees in computer science and heating ventilating and air conditioning.

SUMMARY

The success of DDC systems in Navy facilities is directly the responsibility of the troika of Design, Construction and Maintenance. Success starts with design. While the new DDC systems are being designed and installed, the programs for the training of inspection and maintenance personnel should be implemented. Hence, when the DDC systems now in the initial stages of design reach the field, the final step of their success will be assured.

SECTION 6

NEW TRENDS IN BUILDING AUTOMATION SYSTEMS

Chapter 31

THE INTEGRATED BUILDING AUTOMATION SYSTEM

T. R. Weaver

Integration may well be the most overused, misused, and maligned buzzword in the Building Automation marketplace in the late 1980's. In its most broadly defined sense it sometimes means as little as utilizing a single proposal binder to convey cut sheets of functionally and physically discrete systems. This end of the spectrum can be described as "Marketing Integration" or "Brochureware Integration." While there may, indeed, be some purchasing or performance leverage gained by the client in such a packaging scheme, this type of practice has tended to confuse rather than clarify the performance benefits and design objectives of Integrated Building Automation Systems.

At the other end of the spectrum is the fully Integrated Building Automation System, wherein related systems are both physically connective and fully capable of sharing data on both a real-time event basis and a shared data base or historical file transfer basis. Availability of totally integrated systems at this writing ranges from rare to nonexistent. The Building Automation Systems industry has two decades of experience in interfacing dissimilar systems for the purpose of controlling, monitoring, or overriding specific functions from one to the other. These interfaces, however, have typically been performed in external hardware and have been done on a project-specific, custom basis. Although a digital interface between subsystems would be highly desirable and far more flexible over a system's lifetime, few interfaces are accomplished in this manner. In this chapter we will explore some of the driving forces behind this apparent contradiction--market forces and technical performance opportunities that would demand integration, and the yet unanswered questions of how it should be accomplished or by whom.

First, let's examine the market forces or the potential end-user benefits of a fully Integrated Building Automation System, if such a product were to exist. By definition, a fully Integrated System would include physical and functional access to building operations data from any system or subsystem within the building, effectively interfacing all controllers, processors, or data acquisition paths. Current technology would suggest that most, if not all, of these devices are digital in nature and would interface on some sort of data communications link. The list of systems potentially falling under this umbrella is extensive, but would include at a minimum the following:

1. Engine Generator or Standby Power Systems
2. Uninterruptible Power Supply (UPS) Systems
3. Emergency Lighting Systems
4. Lighting Control Systems, either discrete or integrated into fixtures
5. PBX or Telecommunications Systems
6. Office Automation or Management Information Systems
7. Control Systems on packaged equipment, including chillers, boilers, computer room HVAC, kitchen equipment, laboratory equipment, etc.
8. Fire Management Systems, including detection and smoke control devices
9. Security Management and/or Access Control Systems Time and Attendance Recording Systems
10. Maintenance Management Systems
11. Miscellaneous Building Systems, including booster pumps, sump pumps, etc.
12. Elevator Control Systems

Assuming that both physical connectivity and appropriate protocol conversion or standard protocols existed between all of the above systems, a truly holistic approach to building operations could be applied. Experience in Building Automation to date has shown that the more fully connective and integrated these types of subsystems are, the more effectively and efficiently a building can be operated. Some examples of this are included in the Case Studies section of this book.

Take, for example, the potential overlay of an electric load control scheme on such a system. Given the necessity to reduce electrical usage during either a peak demand or time-of-day billing situation, a system with high level access could accomplish the load reduction without significant service disruption or occupant discomfort. A few percent of lighting load could be reduced by selective reduction of lighting levels at the perimeter through either dimming ballasts or split wattage fixtures, assuming sufficient daylight is available. Elevator speeds could be reduced, or a portion of cars temporarily parked. HVAC system load could be temporarily reduced by incrementing set points rather than wholesale shutdown of equipment. Engine Generator or UPS equipment could be exercised and at the same time contribute to load reduction at the incoming service. Occupancy schedules, either programmed or monitored by the Access Control System, would automatically alter the strategy.

Fire Management functions could also benefit significantly from this type of integration. Readily available real-time access to the control or override of all building systems would allow many functions to take place from a fire fighter's control panel. In addition to receiving alarm and event information from the full spectrum of building systems, fire fighters could readily issue override commands to smoke control systems, elevator systems, electrically locked doors, and auxiliary power and pumping systems.

While these types of Integrated functions--either independently or in combinations--have been implemented in current systems, they have typically required a great deal of interfacing and/or overriding of existing controls by the Building Automation or Energy Management System. Typically, then, a subsequent change to the original strategy requires significant changes to either hardware or software.

EXPERT SYSTEMS

Expert Systems can be defined as the capturing of expert logic or operations knowledge in a delivered or installed system. Expert Systems technology has made its way into many businesses and industries, and in fact it is present on a limited basis in existing Building Automation Systems. Optimal HVAC start-up prediction, chiller plant automation, and adaptive tuning algorithms are examples of lower level implementations of Expert Systems currently in use today. As Expert Systems are more widely understood and accepted, software tools will become more readily available for the application of them to real-time problems. Imagine the productivity and performance of a building-wide network whose operations could be directed with the multifaceted knowledge of experts from throughout the user's company or the building operations industry in general. Current real-time computing technology applied to a fully Integrated Building Automation System would yield many real and justifiable enhancements to building operation. Add, as a likely future development, the evolution of real-time Expert Systems into the Building Automation designer's tool kit, and the case for full-scale integration of Building Automation Systems makes practical and potentially economic sense.

PROTOCOL TRANSLATORS

In the scenario described above, the key element is the ability of systems and subsystems to intelligently interact the way a human would perform the same operations. In other words, anything an operator can do to a packaged HVAC unit, a chiller, an Engine Generator, or an elevator controller from a control or service panel could be accomplished through a remote digital control link. This requires, of course, some method by which digital processors can talk to each other in mutually acceptable, real-time languages or protocols. At this writing, such a function requires protocol translation between a Building Automation System and the controlled subsystem. This protocol translation can be done in a computer with firmware, simply converting and/or buffering bits and bytes between the systems, or in a software-driven protocol translator most commonly built around microcomputer hardware.

Protocol translation is somewhat more myth than reality. First, it is expensive. One is typically confronted with the challenge of any custom software development project--limited resources, limited time, and loosely defined performance criteria. Since in many cases the ultimate functionality of the integrated system is yet to be discovered, it is somewhat difficult to put boundaries on the capabilities or performance criteria of the protocol translator.

Again, opposing forces come into play. The ultimate functionality is usually directly proportional to the amount of effort and expense involved in designing and building the translator. Real-time performance, specifically accuracy and throughput, are very difficult to determine in the early stages of designing and building a protocol translator and may, in fact, not be verifiable until the project is complete. If a protocol translator is built on a custom, one-of-a-kind basis (as many are), it is almost assured that long-term support and maintenance of the software will be both challenging and costly. It is important to understand the role of protocol translators, however, because they are the essence of what most "brochure-ware" integration is made of. They are cloaked in a variety of depictions and buzzwords, but usually appear as a box on a network diagram called a "gateway," a "network interface," or some other suitably cloudy term. The function and intent, however, is always the same--to make two dissimilar systems interact as if they were the same, with a device in the middle doing the conversions to make either side of the network comfortable.

If protocol translation is an undesirable solution, why do we see it required and/or proposed? Very simply because this solution, although expensive and cumbersome, provides an answer of sorts to the basic need for dissimilar systems to operate in an integrated fashion. The fact that their shortcomings are accepted as a solution gives credence to the assertion that both market wants and performance capabilities of fully Integrated Systems are recognized and accepted.

SHARED COMMUNICATIONS MEDIA

For a while, a considerable mystique developed around the idea of a single network backbone performing all data communications in a building. Again, marketing hype, rather than functionality, tended to be the driving force. While it is true that building-wide communications on a single medium--broadband or baseband cable, fiber optics, or even twisted-pair wire--has a very elegant ring to it, the economics of converting several existing communication interfaces to these types of networks rarely proves favorable. Again, the question "Just what functionality does this integration provide?" generally goes begging. Integration for the sole purpose of sharing communications media is sometimes an economic benefit. It is certainly a more complicated technical solution.

If there is marketplace want and user benefit from fully Integrated Building Automation, then why isn't it more readily available? Well, it's not easy to do. The current state of standards for real-time interchange of data between devices provides a system designer with no logical starting point. Although some basic standards for the physical link between devices such as RS232, RS422, RS485, etc., exist, the protocols or communications data structures between devices have not captured the attention of recognized standards bodies or vendors each pursuing independently their own "best" structures for their own needs.

This is, of course, a historical problem with newly emerging technologies, and the digital communications industry is no different, except that developments and technical progress has occurred at such a rapid pace that standardization is almost doomed to be after the fact. In essence, the designer of a fully Integrated System of any type currently faces a moving target as newer, faster, and more exciting data interchange techniques become available almost monthly.

Integrated Building Automation is not cheap. Many users fantasize about the idea of interconnectability between systems of different manufacture. This is generally thought to be in the user's interest, since one could presumably shop for the best price/value combination in adding a feature or function to an existing system. Only rarely is the idea that such flexibility might have a heavy initial cost impact openly discussed. In fact, one reason that standards are so difficult to arrive at in any industry is that each designer or vendor truly and ardently believes that his approach is the most efficient, the most cost effective, or the most desirable for the user. Standardization of interfaces and protocols would probably, therefore, impose some limitations on designers who would find their abilities to optimize cost/function relationships limited by the prescribed communications media and protocol standardization efforts. On the other hand, the need for standards to reduce life cycle costs should inevitably create a minimal restriction for designers in these areas.

This line of reasoning does not oppose nor rationalize the industry's lack of development and adherence to standard interfaces or protocols, but simply illustrates some of the restraining forces that have kept widespread standards from becoming a reality.

When and how will higher levels of integration occur within Building Automation Systems? Both "when" and "how" are difficult questions. One thing is for sure. It will not be by some sort of megamerger or "we do it all" strategy by a major vendor. Technologies involved in control systems, telecommunication systems, data communications networks, etc., are each diverse, complex, and rapidly moving in their own industries. It is beyond rational thinking to suppose that one organization is going to be able to provide the most competitive and state-of-the-art offering in all of the electronic system disciplines required for a given facility. Any user should be skeptical of an organization which proclaims to be a "one-source" solution to the entire gamut of technologies involved in building and operating a building inhabited by a modern business. No one is going to have the best solution across the board at any one time, and most building owners recognize this fact. On the other hand, the emergence of interface standards between systems and devices in the commercial building would provide a responsive answer to the marketplace, while at the same time preserving the necessary technical and marketing specialization within each of the subsystem disciplines. Thus, a marketplace and technical balance is achieved very similar to that existing today in consumer audio and video equipment, where specialty niche vendors co-exist and prosper alongside major system vendors, each complementing the other.

Clearly, this discussion centers around level interfaces for the purpose of digitally sharing data and control signals between major subsystems. It is most likely that these types of standard interfaces would evolve on a "top-down" basis. It is unlikely and probably only marginally desirable that individual sensors, actuators, or terminal devices would become standardized or interchangeable. In other words, a significant amount of performance and functional differentiation would still exist between competitors in the subsystems arenas.

How might standard interfaces or protocols between building systems take place? It is likely that efforts on the order of those being put forth to standardize Manufacturing Automation Protocol (MAP) and Technical Office Protocol (TOP) will be necessary. As a further assumption, then, this charter must be accepted by an existing standards body or must emerge from within the Building Systems industry. It is probably safe to say that a disciplined protocol structure, such as the Open Systems Interconnect (OSI) model as developed by the International Standards Organization (ISO), will form the basis for a successful standard. Nearly all protocols currently in existence for use within Building Systems have been developed by and for vendors considering their use only in proprietary systems. A few vendors have clouded the issue somewhat by publishing their protocols, thereby claiming them to be "open." Most, if not all, of these existing protocols bridge multiple layers of the OSI model and, therefore, likely preclude their use by the broad spectrum of systems listed earlier in this chapter.

A likely ingredient to such a standardization effort would be the commitment by one or more large users to involvement in and support of a standardization body. General Motors has become a critical ingredient in Factory Automation integration efforts through MAP, and has acted as a catalyst to drive the continuing work on MAP standards.

So much for the future. What about the present? Users desire and are currently specifying functions which require a level of interface between multiple electronic systems within buildings. Uniform interface standards do not exist and probably will not for some time. The essential ingredient in a successful project is, therefore, the ability to integrate, on an engineered system basis, the necessary and essential functions required to make a building work efficiently and effectively. This requirement for effective integration capability is an organizational rather than a product attribute. To be successful in this endeavor, an organization must be chosen that has the necessary engineering and technical skills to design, manufacture, install, and commission combinations of standard and custom hardware or software necessary to meet the design objectives of a project. It is essential also for such an organization to have a great deal of experience in this type of work. In the final analysis, that organization must also have a measure of flexibility and not be committed to a single vendor solution, but rather an "open interconnection" posture. Although high tech systems and skills are being offered as solutions, the basic objective is an efficient, safe, and comfortable building.

Chapter 32

ELECTRICAL WIRING MANAGEMENT FOR THE INTELLIGENT BUILDING

A. B. Abramson

In the past, communications requirements for voice, data and control systems were met through the use of dedicated cables individually wired between their respective points of interconnection. As communications requirements have grown, designers have typically addressed the expanding cabling needs of buildings via the provision of flexible, typically expensive, raceway systems. The rapidly expanding need for new and varying point to point communications links has led to the situation where buildings contain a maze of unidentified, often abandoned, cable. This occurs because new cable is added and old cable is left to remain because of the high cost of cable removal. Therefore, the apparently large and flexible raceway system may become filled with such cable and, in fact, size and flexibility has helped to exacerbate the same problem which it was intended to solve.

A new approach to this problem has been developed, which is best called Cable Management, that addresses the needs of both a facility and a tenant in a manner that significantly reduces the life cycle cost of the provision of signal level communication services. The key to the concept of cable management as presented in this chapter is the understanding of the need to plan requirements, recognize commonality, design flexibly, and manage communications for the life of the facility. This new approach has resulted from an examination of future directions in communication technology as they relate to the much slower changes that occur in the building process.

DEFINITIONS

The term "Cable Management" as used in this chapter relates to a process and not to a physically defined system. A detailed presentation of this process appears later in the chapter.

Further definition must also be given to the item being managed. For the purpose of this chapter, the term "Cable Management" will be used to identify all types of signal level transmission media; this includes copper media of all sorts, such as coaxial cable or twisted pairs, fiber optic cables, and other links such as microwave or satellite channels. It is, of course, implicit in such a discussion that other distribution equipment related to the transmission medium such as interconnection equipment be considered in the overall management process.

It is also important to note that one generically accepted definition of the term "Cable Management" relates solely to the provision of a flexible raceway system. Acceptance of this narrow definition will inevitably lead to a facility where cable will be unmanaged to the point where the raceways become full as described earlier. At that point in time, gaining control of the situation will be extremely difficult because the magnitude of the unmanaged cables will be significant due to the size and flexibility originally provided. The provision of a flexible raceway system as the sole solution to cable management is a poor approach and, in fact, represents an approach which may be termed "cable management avoidance." "Cable management avoidance" results from a lack of consideration for the transmission media and an intent on the part of the designer to postpone its consideration to the point in time where responsibility for cable becomes someone else's problem.

SUB-SYSTEM COMMUNALITY

Cable management is essentially a communications problem. The basic intelligent building systems for which communication is required have been described in previous chapters. The commonality among these systems relates to the transmission of the following signal types:

- Voice
- Data
- Images
- Video
- Other discrete signals (sensors, controls, etc.)

Ideally, the degree of communications commonality among these systems encourages the consideration of integration and uniformity with regard to planning, design and continuous management.

Prior to presentation of this formal process, three critical points of caution that are common to all intelligent building approaches must be expressed. They are as follows:

1. Integration of systems should never be attempted merely for integration's sake. One must always look to achieve a functional synergy when integrating. Examples of the functional synergy that should be sought in an integrated cable management approach include items such as reduced first cost, reduced life cycle cost, increased flexibility or expandability, or increased ease of maintenance. Practicality must be stressed, and solutions whose only benefits are aesthetic do not belong in the design and building process.

2. Cable management cannot be successful if related only to signal level transmission media. A consistent approach must be taken to the distribution of electrical power and cooling media as well. In addition, equipment related to the interconnection of the network or cables must also be considered in the context of current and future space allocation.

3. There are inherent roadblocks in the design and building process and in the respective tenants' organization to integration. For example, in a facility being designed for a developer/landlord, the HVAC control systems would likely be part of the base building design. Typically, installation of cables for tenant communication requirements would be part of the tenant space design process. Therefore, integration of those systems for cable management purposes would likely be impossible. A further example exists in the organization where responsibility for different intelligent building systems is separated among different departmental managers. While it is becoming less common, there are still many organizations who separate the managerial responsibilities for voice communications, data communications, and EDP facility management. These organizational barriers must be addressed prior to the consideration of an integrated cable management approach.

THE CABLE MANAGEMENT PROCESS

The cable management process has three distinct phases. These phases are as follows:

- Strategic Planning
- Distribution System Design
- Data Base Management

Strategic Planning

As previously noted, the rapid expansion of cabling requirements within an organization has typically led to a crisis management atmosphere with regard to the provision of these services. Therefore, the typical telecommunications manager, regardless of his or her career path in the organization, has likely never had the opportunity to plan for growth. Furthermore, in the multi-vendor environment created by the recent regulatory and anti-trust rulings, it is difficult to find a single source of responsibility for the coordinated planning of transmission media requirements. It is, therefore, imperative that prior to the consideration of any facility engineering solutions, the transmission requirements for the present and future be studied and presented in a strategic plan for the user.

The strategic plan must address the use of communications as a tool to accomplish the mission of the user in the short and long terms. To accomplish this process, interviews must be held with those responsible for the direction of the respective organization from a business point of view; those in the responsible administrative and facilities operation areas; and those to whom responsibility has been entrusted for support of communications as a business tool.

Every business and/or organization has a different character with regard to its perception of communications as a tool. In fact, the strategic planning process often becomes a time for rethinking the business mission and for education of senior management regarding the opportunities that modern technology offers.

Corporate culture must be recognized and incorporated into the communications strategic plan. A strategic plan will ultimately result in a fairly definitive projection of the functions and/or systems directions related to voice, data, administrative services and facilities as they fit the organization's mission.

A critical part of the strategic plan will be the provision of a migration plan which relates tactical approaches to a schedule of implementation for the acceptance and integration of new technologies. It must be noted that this is not an engineering-oriented task. It is a management and business planning task which may be lengthy and painful, but is also absolutely necessary. Without such a plan, the risks associated with any given engineering solution may be significant.

Distribution System Design

A distribution system is the cables, adapters and other supporting equipment that connects telephones, data processing devices, control panels and other communication devices, permitting them to "talk" to one another. (See Figure 14-1.) The distribution system is represented by the assemblage of hardware in addition to the method of arranging this hardware within a building or campus in a coherent fashion. The intent of this interconnection is to future-proof the facility to permit the economical and flexible expansion of the networked devices without the need to significantly disrupt the facility physically.

The goal of a well-defined and designed distribution system is to permit integration of new requirements over time without disturbing floors, walls, and ceilings in the process. In addition, such a system should permit quick, inexpensive interconnection of all devices in a manner that suits the aesthetic requirements of the facility. The ingredients of a distribution system are the following sub-systems:

- Work Station Wiring
- Horizontal Distribution
- Riser or Backbone Wiring
- Interconnection and/or Network Equipment

The work station wiring connects the piece of equipment to a receptacle or jack on the wall or in the furniture. This cable is similar to a power line cord, is usually flexible and may be provided with the equipment.

The jack in the wall or furniture is connected through cabling that is permanently installed in raceways via a horizontal distribution sub-system. The form of this cable, in terms of the type and quantity of transmission media, will be determined based upon an engineering analysis related to the requirements stated in the strategic plan. Selection of the appropriate cable requirements is a critical part of a distribution system. Typical in modern distribution systems is the selection of multiple media which are distributed uniformly to all jacks in a facility. When jacks are placed at regular intervals and prewired as an in-place horizontal sub-system, the layout is typically called a uniform cabling plan. The essence of such a plan is the anticipation of short and long term media requirements and the slightly greater investment in providing these anticipated requirements prior to their actual need. The horizontal distribution system typically radiates from an interconnection point located in a communications room or closet.

The multiple closets in a facility are interconnected by a cable riser or backbone. The horizontal distribution system is interfaced to the backbone via interconnection hardware and/or network equipment. The interconnection hardware may be as simple as terminal blocks with patch cords or may be as complex as a rack of sophisticated local area network communications management equipment.

Media Selection

Media selection must be related to the migration path presented in the strategic plan. For example, if an organization intends to have a specific vendor's equipment for the long term, it should select a uniform cabling system supported by that vendor. If multiple vendors require support, then a hybrid approach may be appropriate. If high speed, high volume requirements can be anticipated, then a fiber optic approach may be appropriate.

It is also advisable to consider multiple uniform cables in the same facility if required. For example, in a health care facility, one type of uniform cable may be proper for the patient care areas while a different approach may work better for support areas. Typical media include twisted pairs (shielded and unshielded), coaxial cables, and fiber optic cables. Selection is based on an evaluation of the risk of not having cable available for a future vendor or tenant requirement versus the first cost of covering all contingencies. Therein lies the importance of the strategic plan and the criticality of an "artful" interpretation of the migration path.

User Sophistication

The beauty of a uniform cabling distribution system is its flexibility in dealing with multiple levels of sophistication and its adaptability to grow with a user over time. It may be as simple as a hard wired, point to point interconnection network within a limited location via "jumpered" connections within a closet. It may also permit point to point connections via cross connection on and off of a multiconductor backbone on different floors. When the user becomes more sophisticated, local area network equipment may be introduced where required by installing LAN equipment at the interconnection points. Sophisticated environmental control may be achieved by utilizing distributed direct digital control panels or fire alarm (smoke control) panels at these interconnection points.

Building Issues

Closets or communication rooms must be properly sized and/or designed for expandability to permit equipment requirements to grow.

With such a cabling system, flexibility of the raceways becomes a less important criteria, even though the requirement for its existence does not disappear. First cost for raceways will likely be reduced, even though initial cable costs are higher. Concurrent power and cooling media provisions must be made to make the plan work.

Data Base Management

Hopefully, it has become obvious by the previous discussions that the flexibility of a well-designed distribution system, combined with the multiple varieties of interconnected equipment, provides an extraordinarily large and changing mass of data with which management must keep current.

A Data Base management approach is absolutely necessary to avoid the problems noted in the beginning of this chapter.

While such an approach may be manual (by hand), a computer-based approach is recommended. Furthermore, there is a great advantage to utilizing computer-aided design technologies during the design period to create the data base for eventual management use.

CONCLUSION

Cable management is a multi-phased process that is evolutionary in its application. While good engineering is an integral part of the process, planning and continuous management are equally as critical to success and life cycle cost containment.

Chapter 33

HVAC, LIGHTING AND OTHER DESIGN TRENDS IN INTELLIGENT BUILDINGS

F. S. Dubin

The intelligent building is normally defined in terms of currently available building automation, information processing and telecommunications services.

However, most of the buildings which are defined as intelligent are not really so smart because they are deficient in two broad areas. One of these areas is the building envelope and plan which are not as smart as the electronic systems that they enclose.

The other area is mechanical, electrical and structural systems which are not smart because they do not provide the environmental quality required by an intelligent building. Conversely, the mechanical and electrical systems do not adequately serve the special needs of information technology and telecommunications equipment and systems in an intelligent building.

Existing technology and design knowledge are available to address these deficiencies and can be employed by a smart, interdisciplinary, building design team in a cost-effective manner. This interdisciplinary or, better yet, trans-disciplinary team must consist of an architect, mechanical engineer, electrical engineer, structural engineer, acoustical engineer, interior designer, strategic planner, communications consultant, physiologist, epidemiologist, and members of other related disciplines.

The members of this building design team should be people whose professional abilities cross over normal disciplinary lines. They should be people who understand and feel comfortable with each other's expertise.

Overall, the truly intelligent building which this team will design requires:

* A building envelope which reduces heat loss and gain through insulation and thermal mass;

* Moisture control which is achieved by properly locating vapor barriers and external shading devices to reduce glare and solar heat gain;

* Properly-located windows equipped with insulating glass, heat rejection glazing, and light shelves or other devices capable of simultaneously reducing glare and reflecting daylighting deeper into the building's space. This interior use of ambient light is needed to minimize contrast between the building's immediate inside perimeter and occupied spaces 20 to 30 feet from the perimeter.

* Properly-located windows equipped with insulating glass, heat rejection glazing, and light shelves or other devices capable of simultaneously reducing glare and reflecting daylighting deeper into the building's space. This interior use of ambient light is needed to minimize contrast between the building's immediate inside perimeter and occupied spaces 20 to 30 feet from the perimeter.

* Double or triple panes of glazing which are necessary to prevent condensation on windows due to the higher humidity levels required for proper operation of electronic equipment.

* Vertical shafts for duct work and power and communication cables which are larger than normal from the standpoint of past practices and strategically located to reduce horizontal distribution runs.

* Telephone and electric closets on each floor which are adequately sized and ventilated or cooled to protect equipment.

Building design must meet these requirements in an intelligent building because it is the first line of defense in terms of reducing energy requirements. In addition, building configuration, materials and construction influence the type, size and performance of mechanical and electrical systems.

In effect, the building is the catalyst in achieving desirable inter-relationships between thermal, visual and acoustical environments and systems. At the same time, building structure and materials affect worker productivity, aesthetics, maintainability and life cycle costs.

In addition to these building design requirements, there are a number of additional HVAC, lighting and other needs which a structure can and must be designed to cost-effectively satisfy if it is to truly qualify as being an intelligent building.

AIR QUALITY MONITORING AND CONTROL

One of the major requirements for an intelligent building which is not presently being met is an air quality monitoring and control capability. Air quality in office and other types of buildings today is oftentimes very poor, resulting in health problems, reduced productivity and increased absenteeism.

There are many different kinds of pollutants in indoor air, including formaldehyde which is outgassed from building materials, furnishings and glue in carpets; chlorine, sulphur and ammonia products; carbon dioxide and monoxide; contaminants resulting from copying and reproduction operations; and cigarette smoke.

These pollution problems have always been present; but they are currently growing in severity due to the increasing use of new, synthetic materials in buildings and the expanding implementation of energy conservation measures. These measures make buildings tighter to reduce infiltration while minimizing the amount of outside air used in ventilation.

In addition, the problem is exacerbated by the development of the intelligent building since the insulation on cabling used to connect computers and other electronic equipment can be a significant source of outgassing. The level of one toxic material may be low enough to cause no health problem. But the same contaminant in combination with one or more other toxic materials can produce a mixture of compound injurious to health.

Ways To Improve Air Quality

The best way to improve air quality is to minimize if not eliminate pollution at the source wherever possible by specifying building materials, furnishings and equipment which do not outgas contaminants.

Another effective approach is to group together copying and printing processes and rooms where cleaning materials are stored so that contaminants can be efficiently exhausted to the outside. Smoking areas with special exhaust capabilities can also be established as necessary. Non-smoking areas are also very desirable and smoking is being banned in more and more buildings.

Outside air intakes should be located away from obvious pollution sources such as exhaust discharge ducts and truck loading docks. Contaminants coming into the building with outside air should be monitored and controlled through filtration, absorption, sprays or electrostatic precipitation, depending on the type of contaminant. In addition, indoor sensors should be used to constantly monitor air quality in terms of the chemical composition and parts per million of each contaminant. However, if indoor air pollution still exceeds predetermined limits, the next step is to control air quality through increased ventilation and improved air distribution patterns.

Use Ventilation To Control Air Quality

The most cost-effective way of providing building air circulation for heating, cooling and ventilation today is through the use of a variable air volume (VAV) system. A VAV system conserves energy by reducing the volume of air delivered to a space based on its thermal requirements, thus reducing power required for fans. A number of different types of VAV systems can be used.

One is a VAV dump system which closes dampers and recirculates air from supply back to return through a ceiling plenum without delivering it to a space when temperature requirements of the space are being met. Since less air is circulated through the duct system, the system's static pressure is reduced and less supply fan horsepower is required.

Another system uses damper supply ducts to build up pressure as a way of reducing air volume. Operating a little more efficiently, a vaned inlet system reduces air intake to the supply fans to achieve the same objective. The return air fan is keyed to the supply fans so that it also reduces the amount of exhaust air, thus saving power. Another VAV system approach involves using a three-speed fan motor to vary air volume. The most efficient system consists of a variable frequency drive which can provide an infinite number of fan speed settings based on space temperature requirements.

All of these VAV systems conserve energy by reducing the amount of ventilation or outside air delivered to a space when permitted by temperature conditions. However, this reduction in air volume has undesirable effects on air distribution patterns.

For example, air may short-circuit from supply diffusers to return air grills. In addition, air does not circulate properly through personnel workstations, particularly if they are built with walls to the floor and fitted with acoustically- absorbent materials or baffles necessary to reduce noise but that further obstruct air movement. As a result, work station air quality suffers because contaminants are neither diluted with outside air nor flushed by overall air flow.

One method of solving this problem is to install power or induction units at outlets of VAV mixing boxes to induce additional room air into the primary air stream. This makes it possible to maintain proper air distribution volumes and patterns around workstations, even though a VAV system may be operating with reduced air volume to conserve energy.

Another approach is to use mixing dampers to deliver an increasing percentage of outside air when air quality sensors indicate contamination buildup while recapturing sensible and latent heat from the exhaust air stream through the use of a heat exchanger. Applicable to cooling as well as heating operations, this method retains the energy-conserving benefits of VAV operations while providing contamination control.

BAS Provides Ultimate Control

Air quality can be controlled using multiple, small air handling units, each of which varies the amount of fresh air delivered to a space based on input from pollution sensors installed at floor, zone, office and even individual workstation levels.

But what if indoor pollution rises to an unacceptable level, even though outside air dampers are opened to provide 100-percent outside air? In this case, VAV system fan speed would have to be increased to bring in even more outside air. At the same time, it would be necessary to have mixing boxes equipped with terminal heating and cooling coils to respectively prevent overheating and overcooling. Furthermore, if outdoor air quality were subnormal, additional air treatment would have to be provided rather than bringing in more outdoor air.

Ultimately, the intelligent building must therefore have a Building Automation System (BAS) capable of sensing and optimizing all of the environmental conditions or factors involved in achieving required air quality. These factors include not only temperature, humidity and energy consumption but new inputs such as air quality, air movement and number of building occupants. Equipment and controls required to implement this building automation system are available today.

The building automation system will optimize all of these factors to achieve the best possible results while insuring that required air quality is maintained. This will result in a better work environment while improving personnel health, safety and productivity. In addition, it will provide an improved operating environment for the proliferating use of electronic equipment in the intelligent building.

DESIGN FOR INCREASING LOADS

The intelligent building's use of electronic equipment such as workstations is expanding rapidly. Some intelligent buildings currently provide workstations in the form of video display terminals, personal computers and other input/output devices for as many as two out of every three personnel. Eventually, this workstation/personnel ratio will become one-to-one.

As a result, intelligent buildings must be designed to handle much greater electrical loads. Five years ago, for example, conventional buildings were designed to provide 1-1/2 watts/sq. ft. Today, similar buildings are being designed with 2 to 3 watts/sq. ft. But an intelligent building requires 6, 8 or even 10 watts/sq. ft. in addition to lighting needs and 20 to 30 watts/sq. ft. at workstation locations.

Workstation growth is also resulting in fast-growing heat loads because each workstation is equivalent to a heat load of four to eight people. Conventional buildings are normally designed with one ton of refrigeration for every 600 sq. ft. But, based on a one-to-one workstation per person ratio, the intelligent building needs one ton of refrigeration for every 200 sq. ft. an increase of 1.2 CFM (cubic feet per minute) per sq. ft.

In addition, it is anticipated that the "energy crisis" will re-emerge in future years because the current oil glut and stable prices are only temporary. Prices will rise as world economies recover, fossil fuels will once again be in short supply and the future of nuclear power will remain uncertain.

Consequently, due to growing heat loads and rising energy prices, it will become increasingly more important to use a building automation system to increase the efficiency of the building envelope and HVAC system operations in intelligent buildings. At the same time, other improvements will be made through the use of more efficient alternative energy and power sources.

Increased HVAC System Efficiency

Increased HVAC system efficiency can be achieved through the use of more sophisticated building automation system control programs. Currently, for example, cooling operations are typically controlled on the basis of chilled water temperature and heat load. The control objective is to use warmer chilled water for cooling as loads decrease because this saves energy.

However, when chilled water is warmer, more air must be used to achieve the same cooling level. This results in increased energy consumption in terms of fan horsepower and duct losses which normally are not considered in HVAC system control operations.

As a result, there is a need for a more sophisticated control program which will further lower energy consumption by optimizing chilled water temperature, fan horsepower and duct losses. Other variables which should also be optimized by this program include chilled water pump horsepower and piping losses. In addition, the program should balance these variables with air quality and ventilation requirements. Expert systems are now being developed which will be capable of preparing this program.

The use of cooling storage is another way in which significant improvements can be made in HVAC system efficiency. Cooling storage is accomplished by economically operating a chilled water or ice-making plant at night during off-peak hours and with lower condenser temperatures. The following day, the stored chilled water or ice is circulated through cooling coils in air handling systems or fan-coil units without running chillers, resulting in reductions in operating costs and peak electricity demand.

The HVAC system for an intelligent building must provide greater humidification to reduce the possibility of static electricity which can cause electronic equipment to malfunction or even fail. At the same time, cooling efficiency can be increased by as much as 20 percent by the use of dessicant when dehumidification is required. Absorbing rather than condensing moisture, dessicant use reduces the amount of chilled water required and permits the use of higher chilled water temperatures since it eliminates the need to achieve dehumidification through cooling.

The multiple use of equipment is an additional way of achieving increased HVAC system efficiency. An example of this is the use of a heat exchanger as an indirect evaporative cooling system.

Alternative Energy Sources

The use of solar energy for heating new intelligent buildings will increase rapidly in the future. Conserving energy while reducing heating costs, solar energy uses major passive techniques such as direct gain, trombe walls and sun spaces including greenhouses and atria. Building envelopes will also be designed to do a better job of controlling heat loss or gain. In this regard, building materials can provide both shelter and energy storage, each of which can be controlled as necessary.

Photovoltaic cells are being used for direct conversion of solar energy into electricity primarily in remote and specialized installations. However, conversion efficiency will continue to increase while costs come down from the current $6 to $2 per peak watt. This will make photovoltaic electricity generation competitive with other electrical energy sources, triggering a huge increase in installations.

Intelligent buildings should therefore be designed with a built-in capability for installing photovoltaic cells in terms of orientation and slope of outside surfaces. In addition, electrical systems must be designed for compatibility with alternative energy sources such as photovoltaics, and HVAC systems must be designed for compatibility with solar thermal collectors. Low-temperature media in the heating system permit the use of low-cost solar collectors which provide reduced hot water temperatures or air for heating.

Power Sources

Co-generation or on-site electric power generation will be a growing factor in intelligent buildings because it makes it possible to more efficiently handle increasing electrical and heat loads and generate different-voltage power as needed. It also has the capability to produce cleaner power without spikes, resulting in smoother operations and fewer malfunctions in electronic equipment. Furthermore, more efficient absorption cooling units that use lower temperature fluids for the generator will permit the use of less costly solar collectors.

It is also important to correct the power factor if it is below .95 since this would indicate that the system is paying for power that is not being utilized. This adversely affects performance and operating costs of motors and equipment.

Technology is now available to automatically correct the power factor when it changes in response to load and the addition of equipment and motors with inherently poor power factor. In addition, an uninterrupted power source is necessary so that computer operation and storage integrity are maintained. Batteries are not only the most reliable source of uninterrupted power but eliminate the delays which are experienced with diesel generators used for emergency power. Adequate ventilation must also be provided as required.

DESIGN FOR MOVING LOADS

Electrical and heating loads in intelligent buildings are not only increasing; they are also constantly moving. This is because organizational changes result in continual personnel relocations. In some cases, changes may affect as many as half of all employees per year. As these people move, their workstations and related loads go with them.

Intelligent buildings therefore must be designed with the adaptability necessary to accommodate these moving loads. Based on this requirement, there is a trend to handling base loads centrally while using decentralized, unitary equipment to meet needs created by moving loads. Installed as an integral part of workstations, this equipment consists of miniaturized, local HVAC units with cooling, heating, humidification, exhaust, ventilation and control capabilities.

These HVAC units can be served by duct work and power and communications cabling which is installed either above a hung ceiling or below a raised floor. In the past, the hung ceiling has primarily been used for duct and cabling installation. But, because of increasing load movements, there is a trend today toward use of the raised floor.

Use of Raised Floor

The raised floor has been used for many years to accommodate cables and cool air distribution systems in computer rooms. More recently, raised floors have been installed in large office areas. The floors are initially used to carry cables and then to transport warm or cold air supply and return distribution.

Raised floors provide an adaptability to moving loads which is far beyond that of hung ceilings. The floors can be designed with a "tap-in" capability so that workstation duct and cable connections can be simply plugged in at desired floor locations, eliminating the need for ceiling drops or power poles.

The complete availability of all services at floor level directly under workstations provides a number of benefits. HVAC system efficiency and effectiveness are improved, for example, because air does not have to be as cool when it comes out of the floor and air movement is located exactly where it is most needed. In addition, installing a raised floor in place of a hung ceiling can be more cost-effective overall.

The construction cost of a raised floor is $6 to $8/sq. ft. in comparison to $2/sq. ft. for a hung ceiling. Use of a raised floor therefore results in a net addition to costs of $6/sq. ft.

However, a raised floor can be built with less depth than a hung ceiling, making it possible to reduce floor-to-floor building heights. Moreover, the installation of an HVAC system in a raised floor is easier and less costly.

These savings can pay for the added construction cost of a raised floor, resulting in the same or even lower first costs. Most importantly, costs involved in making changes to accommodate moving loads can be substantially reduced since a raised floor permits changes to be accomplished faster, easier and at less cost.

GREATER INDIVIDUAL CONTROL

Greater individual control of HVAC system operations is another trend which will become increasingly more evident in intelligent buildings equipped with unitary workstations. This will enable building occupants to completely control heating, cooling, air movement, filtration and even humidification, if desired, at the workstation level. Individual control will be further enhanced through the use of personnel sensors and other control devices capable of automatically activating HVAC system operations.

Individual building occupants could also possibly program their own environment through the use of direct digital control capabilities which are increasingly being implemented in intelligent buildings. Providing simpler operations and faster response times, direct digital controllers are able to sense changes in environmental conditions and immediately and directly operate HVAC devices such as dampers or valves, eliminating the need to send signals back to a central computer to achieve device activation. Manufacturers are now producing packaged air handling units with integral direct digital control capabilities.

MORE EFFICIENT, EFFECTIVE LIGHTING

Intelligent buildings will make growing use of natural daylight illumination to improve lighting quality while reducing energy costs. Other trends will include increasing implementation of task/ambient lighting and higher-efficiency lamps and other devices.

Use of Daylighting

Daylighting can provide a higher-quality, more attractive ambiance while significantly cutting energy usage, particularly in areas which are primarily used from 8 a.m. to 6 p.m. An entire top floor of a building, for example, can be top-lighted or skylighted, eliminating the need for any other ambient lighting. Light shafts extending from the roof down to lower levels can be used to bring light in and distribute it within a building.

Another daylighting technique is to install light shelves outside windows. Reducing glare at the perimeter, these shelves reflect even, ambient illumination up to 25 or 30 feet into interior areas. Atriums will also be increasingly used to bring in multi-directional, natural lighting as well as solar energy and natural ventilation to interior spaces.

Photocells will play a growing role in reducing electric lighting to predetermined levels when daylight is available. Personnel sensors to control lighting and heating will be increasingly used for control and energy savings. In addition, increasing use will undoubtedly be made of a currently-emerging coating technology which makes it possible to use an electric current to vary light and heat transmission through windows.

Task/Ambient Lighting

To insure maximum personnel productivity, lighting must provide not only the correct amount of illumination but good quality in terms of minimizing direct glare, reflected light and lighting contrasts. This requirement is accentuated in the intelligent building because of the accelerating use of workstations with video display screens.

Consisting of vertical rather than horizontal work surfaces, these screens are internally illuminated so the need for ambient or ceiling lighting is minimized. In fact, high levels of ambient lighting can and, in many cases do, cause visual problems due to direct glare, light reflected from screens, and uncomfortable contrasts between screen and background illumination levels.

Consequently, there is a marked trend in intelligent buildings towards task/ambient rather than uniform ceiling lighting. Capable of being varied in terms of intensity and color based on specific job requirements, task lighting consists of either specially-located, fluorescent fixtures with low-brightness lenses and directional louvers or desk-type lamps that are built into workstations and office furniture. The lamps may be a new type of fluorescent tube which is much shorter than conventional tubes -- only 9 inches long -- and can be plugged into power outlets.

Ambient lighting can be provided in a number of different ways. One is in the form of daylighting. Another is indirect lighting produced by floor-mounted, high-intensity discharge lights or other types of lamps which can be used to reflect light off ceilings. In addition, uniformly-spaced, pendant-mounted, surface-mounted or recessed-ceiling lighting can be used.

Task/ambient lighting provides high-quality illumination that eliminates undesirable glare and reflection problems on video display screens. The lighting also conserves energy and can be installed at a lower capital cost in new intelligent buildings.

Higher-Efficiency Lamps, Devices

A number of higher-efficiency lamps which produce more lumens per watt are now becoming available, including argon-filled tubes and high-intensity discharge and sodium lights. New electronic ballasts efficiently provide the same amount of light with 10 to 20 fewer watts. Programmable ballasts make it possible to use electric current to vary light output.

SOLVE ACOUSTIC PROBLEMS

An intelligent building may have greater acoustic problems than other types of buildings because of the growing use of workstations which require the operation of impact printers. These problems can be solved through the use of vertical, movable acoustic baffles which are hung above workstations. Four feet long, one foot high and one inch thick, these baffles can be moved in any direction to absorb printer noises. Acoustically-treated furniture and stub partitions are effective but care must be exercised to assure proper air circulation through workstations.

ACHIEVE TOTAL BUILDING PERFORMANCE

The advent of the intelligent building is spurring a total re-evaluation of how buildings are designed, constructed and operated or managed from initial concept throughout useful life. Buildings can offer the ultimate in high tech services, but fail to respond to individual occupant needs for adequate flexibility, power, health and comfort. Consequently, the building can no longer be viewed as a passive setting for office work. It is now, in fact, a holistic system critical to the operational and economic performance of its occupants.

Based on serving people rather than creating technology for its own sake, this development of intelligent buildings from this overall point of view will make it possible to maximize total building performance in terms of HVAC system, lighting, acoustical and energy management capabilities.

ABOUT THE AUTHOR

A registered P.E. in 24 states and five foreign countries with degrees from Carnegie Institute of Technology (BSME, 1935) and Pratt Institute of Technology (M. Architecture, 1978), Fred S. Dubin is president of Dubin-Bloome Associates, P.C., New York, N.Y., and Fred S. Dubin Associates International, Rome, Italy.

Dubin is also Andrew Mellon Professor in the School of Architecture of Carnegie Mellon University, Pittsburgh. He also served as a core member of the Orbit 2 team which conducted the Office Research Buildings Information Technology project. In addition, he is a member of two National Academy of Sciences committees which are respectively concerned with "Technologically Advanced Buildings" and "Office Planning."

SECTION 7
HVAC RETROFIT

Chapter 34

HVAC SYSTEMS REPLACEMENT SAVES 44% ON ENERGY FOR A SHORT PAYBACK

A. J. Partridge

Lumen Christi High School, located in Jackson, Michigan, was constructed in 1967. The gross building area is 155,000 square feet divided into the following functional use areas: classroom - 65%; gymnasium - 20%; kitchen/ cafeteria - 10%; and administration - 5%. A student population of approximately 1100 plus 40 teachers occupy the building during normal school hours. Evening and weekend activities include adult education, dances, bingo, meetings and intramural events. The kitchen is used to prepare school lunches and meals for special events.

The classroom, cafeteria, administration portion of the building was heated and air conditioned using rooftop multizone, gas-fired air handling units. The vestibules are heated with electric resistance cabinet unit heaters. The domestic hot water is provided by gas-fired storage type water heaters. The gymnasium was heated and ventilated using Rooftop Gas-Fired Air Handling Units. The locker rooms were heated and ventilated with Rooftop Gas-Fired Air Handling Units providing 100% outside air.

Beset with escalating operating energy costs, excessive maintenance, numerous roof leaks, limited system reliability and lack of proper environmental control necessitated an HVAC systems analysis. The evaluation of eight alternative HVAC systems was based on the following criteria:

a) capital expenditure, maintenance, and operating costs

b) life expectancy, noise generation, simplicity of operation and maintenance, and impact on the building during the construction phase

c) current condition of rooftop units, ductwork and temperature control system

d) past maintenance costs

e) impact of current applicable codes and regulations

f) effect on architectural, electrical and other mechanical elements

In addition an Energy Analysis Report was prepared that contained a list of possible energy cost avoidance measures. These energy cost reduction ideas ranged from installing storm windows to the utilization of solar domestic water heating. Five of these conservation measures were implemented because of their short estimated simple payback period.

RECOMMENDATIONS

Based on the HVAC systems analysis a closed loop water source heat pump system was recommended to replace the current rooftop multizone units. The primary advantages of the heat pump system was:

a) it's ability to redistribute heat from the core area of the school

b) eliminate the rooftop equipment

c) eliminate roof leaks directly attributable to the current HVAC equipment

d) reuse of the existing ductwork system

e) individual space temperature control

f) ease of adaptability to an occupied building

This system would also accommodate the extended evening use of the facility.

Five energy conservation measures identified were implemented because of their short payback period:

*energy management system

*sealing fiberglass ductwork joints

*gymnasium overhead circulating fans

*"time of day" scheduling of domestic hot water circulating pumps

*domestic water heater flue dampers

MODIFICATIONS

Outside air intake penthouses were installed at the multizone unit discharges. Relief hoods were installed at the return air openings to the multizone unit. The old supply air ductwork then became outside air ductwork to the classrooms. This ductwork was interrupted at the classroom and a heat pump unit installed. The discharge air ductwork and diffusers were not modified. A return air duct from the conditioned space plenum was added and connected to the heat pump return. While this arrangement allowed for air economizer utilization, capital costs and projected operational cost avoidance precluded its installation.

A microprocessor based energy management system was installed to control the individual heat pumps and ancillary equipment. The system is also utilized to limit the building electrical demand peak by shedding various mechanical loads in a prioritized sequence to maintain a comfortable facility.

Two gymnasium heating and ventilating units replaced the roof mounted units serving the gymnasium and locker rooms. A ventilation concept allowed the elimination of 90% of the outside air supplied directly to the locker rooms. Ventilation air supplied directly to the locker rooms. Ventilation air supplied to the gymnasium was introduced to the locker rooms using transfer fans. Supplemental heating coils in the transfer air ductwork maintained locker room temperatures.

Four additional energy cost avoidance modifications were implemented. They were (1) sealing the fiberglass ductwork joints (2) overhead circulation fans in the gymnasium (3) "time of day" scheduling for domestic hot water circulating pumps, and (4) domestic water heater fuel dampers.

The existing fiberglass ductwork transverse and longitudinal joints were separated requiring cleaning and taping to ensure the supply air entered the classroom. This work allowed the new heat pump system to operate less frequently, and more efficiently.

The gymnasium is circular with the dome apex 47 feet above the floor. The circulating fans eliminated stratification during both occupied and unoccupied modes of operation.

The domestic hot water circulating pumps ran constantly. By scheduling the pumps operation only when the gymnasium and laundry were in use, the piping heat losses and pump power consumption were minimized.

CONCLUSION

The implementation of the closed loop water source heat pump system along with other energy conservation recommendations resulted in a first year 44% metered decrease in the high school's energy consumption. Figure 1 indicates the 15 year energy history for the facility. Fiscal year 1985 had an energy consumption of 73,756 BTU's/sq.ft./year and 77,902 BTU's/sq.ft./year for fiscal 1986.

Figure 2 projects a ten year energy cost avoidance using current rates and consumption. The average energy cost increase for the years 1971 through 1982 has been assumed to continue for the next ten years. The upper datum line is based on the "old" rooftop multizone system. The bottom datum line reflects utility cost with the new environmental system and energy conservation measures implemented. The aggregate savings for the ten year period will approach $1,332,823.

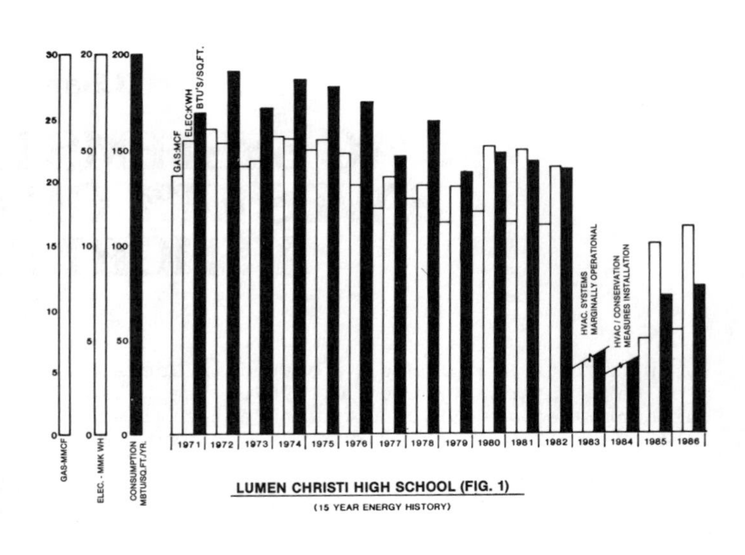

LUMEN CHRISTI HIGH SCHOOL (FIG. 1)

(15 YEAR ENERGY HISTORY)

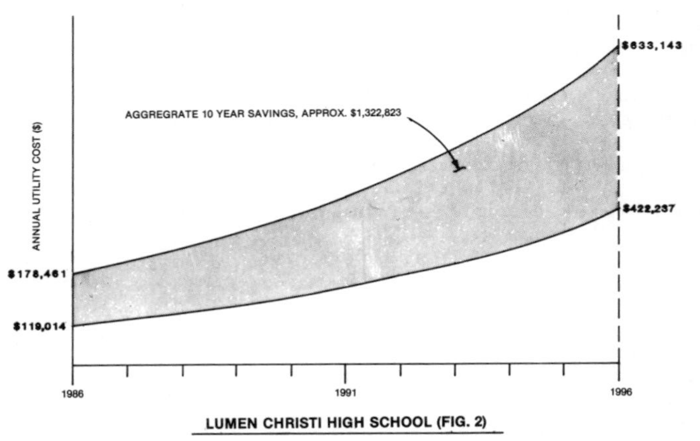

LUMEN CHRISTI HIGH SCHOOL (FIG. 2)

BASED ON AVERAGE ANNUAL UTILITY INCREASE OF 13.5% FOR PREVIOUS 15 YEAR ENERGY HISTORY

Chapter 35

ENERGY MANAGEMENT CONCEPTS IN SMALL DX APPLICATIONS

K. S. Harmon

There are many recommendations being disseminated on what consumers who use Direct Expansion (DX) air conditioning equipment should do to maximize their energy savings. Too often too much emphasis is placed on some of the least effective energy conserving measures. The prime example is the emphasis placed on changing evaporator filters while totally neglecting both the condition of the condenser and the freon charge. Under certain conditions one might use more energy cleaning the filter than is saved by the cleaned filters. This could well be true if the cooling load is predominately latent and the dry bulb set point of the space thermostat would otherwise be raised to take advantage of the reduced humidity. On the other hand, either a dirty condenser coil or undercharged condenser could literally double the cooling energy consumption. This paper addresses these specific items and other commonly employed solutions to energy savings in small DX HVAC systems applications.

Evaporator Filter Cleaning

The evaporator filters in typical residential and small commercial DX unit applications do not load up quickly enough to need cleaning every month. In fact, filtering is better (more efficient) when the filter is dirty. A dirty filter decreases the air flow (CFM) and, depending on the fan curve relationships, a decrease in air flow may decrease the fan horsepower energy needs. This, however, causes a slight decrease in refrigeration efficiency, and thus causes a shift in the ratio of latent versus sensible cooling. In many cases there is no payback in cleaning filters more often than every four to six months. This is even more evident if the system is off at times and if the fan automatically cycles on to satisfy the load when the system is on, minimizing the fan run time even more.

A dirty evaporator filter or one dirty enough to have a slightly increased static pressure drop performs like an inlet damper on a VAV system: It saves fan horsepower and does not significantly decrease the total cooling efficiency of the system, provided air flow is sufficient to both prevent coil freezing and do the cooling job. A manometer type draft gauge installed to give pressure drop across the filter allows one to know when a filter should be cleaned or changed, and in many cases proves to be a good investment.

A test showed that a filter severely clogged (i.e. to the point of reducing air flow by 39%) reduced the total cooling capacity by 14%, the electrical load by 7%, and the total cooling efficiency by no more than 7%. However, the sensible cooling percentage dropped from 68% to 54%, and compressor flooding began to occur. But remember, under certain conditions where the sensible load is light and humidity is high, this increased latent cooling can allow thermal comfort at a higher dry bulb temperature[1] which can offset the effect of efficiency loss.

An evaporator filter clogged with dirt "inducted into the unit" is not likely to cause compressor failure but can reduce compressor life. If the air flow across the evaporator coil were sufficiently restricted, then the compressor will probably cycle off-and-on due to low suction pressure. Too much cycling can cause premature compressor failure and accelerated degradation of starter contacts. For most environments such clogging would require serious neglect; i.e. no cleaning over a period of several years. A frozen coil, on the other hand, would contribute to flooding of compressor which would cause more immediate failure. Under no circumstances is compressor floodback acceptable.

There are a number of compressor failure modes that are directly traceable to evaporator air flow and heat transfer problems. Not all equipment of this type is designed to the same parameters and the variable ambient conditions have a considerable effect. For example, an evaporator air flow reduction coupled with low outside ambient temperature will result in problems more quickly than it would with high ambient temperatures.

One should be concerned about the overall ability of the evaporator section (coil, drain pan and condensate line) to dispose of condensate. Condensate backed up from dust, dirt and debris or from improper drainage grading of the drain pan and condensate line can result in the growth of slime mold or mildew. Spores from fungi can contaminate the air harming allergic occupants and attacking organic articles such as cotton clothing and leathergoods. Allowing the fan to cycle off at the end of each compressor cycle allows condensate to drain more freely and reduces humidity by minimizing re-evaporation of condensate. Indoor air quality[2] is important to occupant health and may need cleaning beyond that which can be accomplished with ordinary filters. Where the intake of outside air is reduced to save energy, activated charcoal filters may be needed to reduce the level of harmful pollutants of indoor origin.

Condenser Cleaning

The refrigeration efficiency can be significantly decreased by a dirty condenser coil. Regular cleaning here, such as every two to three months, will really pay off. Spray cleaning fluid under pressure through the coil from inside to outside, opposite the direction of air flow.

There is an energy related area where the residential user might be misled. Residential users have been receiving cash rebates based on tons installed from several electric utilities, for installing high efficiency air conditioners and heat pumps. This tends to cause some new units to be deliberately oversized in order to increase their rebate.

Despite the nameplate rating and rebate incentive intent, the excess capacity, plus the increased condenser surface area which is often attained through more fins-per-inch, combine to create a potentially inefficient operating system. The condenser coils will, over time, clog considerably before effecting the capacity enough to be noticed. The resulting higher head pressure draws more compressor horsepower per ton reducing efficiency but still providing cooling. This condition will persist until the coils are cleaned or the compressor fails due to exceeding the head pressure limits. Through the final months of operation the unit may well be less efficient than a standard air conditioning system would have been. Under such circumstances the utility company who paid the rebate will have failed to achieve the anticipated system load reduction. Tests have shown a 44% loss in efficiency with an increase in head pressure of 58%. Oversized units will cycle more frequently, will operate more at the inefficient partially loaded conditions and will experience the decrease in efficiency over the entire operating life. The occupants will experience discomfort from thermostat overshoot (excessive cooling each cycle) attendant with oversized equipment.

For maximum efficiency units should be properly sized and properly maintained to handle the normal design load. Design load calculations should be performed in accordance with the good engineering practice for every project.

Refrigerant Charging

"Accurate, safe and quick charging of split DX system air conditioning and refrigeration equipment utilizing capillary tube or other fixed-orifice metering devices at other than optimum conditions has long been a serious problem. With the advent of high efficiency cooling equipment this problem has become even more difficult. High efficiency equipment has also increased the problem of refrigerant floodback and slugging." The foregoing statements were made by Mr. Richard J. Avery, Jr., the inventor of a visual accumulator-charger device he calls the "Accu-Charger" that appears to be highly accurate, and a more sophisticated method of refrigerant charging.

The recurring theme in numerous HVAC publications, manufacturers' service bulletins, trade journal articles and application engineering bulletins, is the fact that refrigerant overcharge, through the effect of floodback and slugging, is a major contributor to premature compressor failure. The following statements are from the Copeland Application Engineering Bulletin entitled "Liquid Refrigerant Control in Refrigeration and Air Conditioning Systems."

"One of the major causes of compressor failure is damage caused by liquid refrigerant entering the compressor crankcase in excessive quantities. Regardless of design there are limits to the amount of liquid a compressor can handle.

The potential hazard increases with the size of the refrigerant charge and usually the cause of damage can be traced to one or more of the following:

1. Excessive refrigerant charge.
2. Frosted evaporator.
3. Dirty or plugged evaporator filters.
4. Failure of evaporator fan or fan motor.
5. Incorrect capillary tubes.
6. Incorrect selection or adjustment of expansion valve.
7. Refrigerant migration.

If an expansion valve should malfunction, or in the event of an evaporator fan failure or clogged air filters, liquid refrigerant may flood through the suction line to the compressor as liquid rather than vapor."[3]

Equal in importance to the adverse effects of refrigerant overcharge on compressor life is the effect of refrigerant overcharge or undercharge on the operating efficiency and economy of a split DX air conditioner system. An undercharged system will cause a loss in capacity with the addition of a disproportionate reduction in EER. An overcharged system will cause a slight increase in system capacity over the nominal, but possibly at a higher cost in terms of electrical demand (kW). In addition, reduced equipment life expectancy is a direct result of refrigerant overcharge or undercharge.

In 1983, Texas Power and Light Company conducted a controlled test on a 1.5 ton packaged DX unit. A loss in efficiency of 52% was experienced from a 23% undercharged system with no comfort changes that were expected to be noticeable to the user, partially because of oversizing. Refer to Figure 1. Overcharging was found to improve efficiency, but can decrease compressor life too drastically to be considered a viable option. Refer to Figure 2.

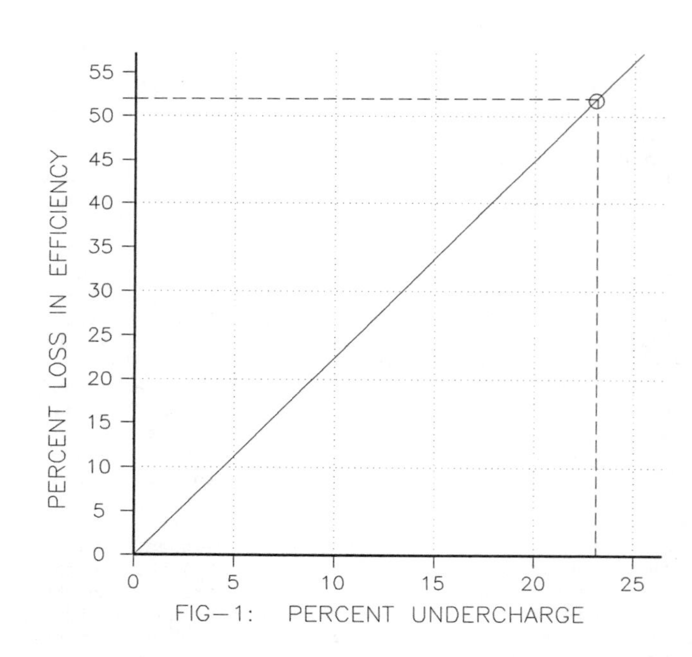

FIG-1: PERCENT UNDERCHARGE

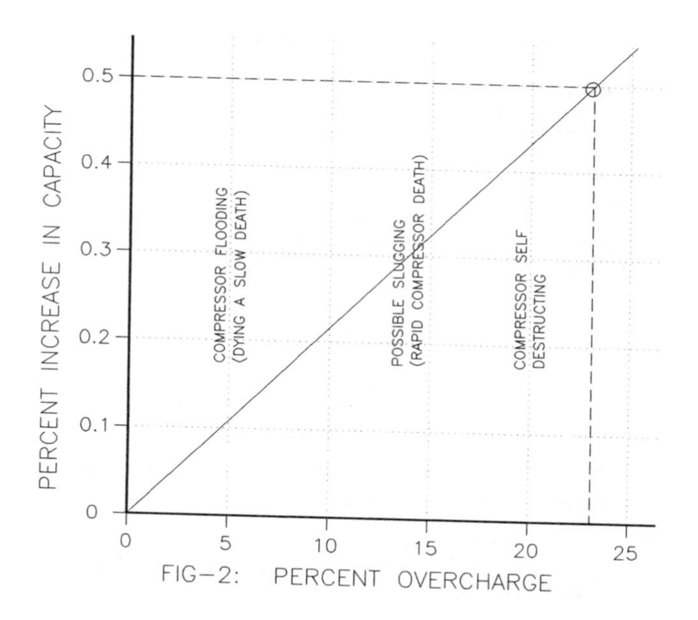

FIG-2: PERCENT OVERCHARGE

Duty Cycling

In recent years past, a number of vendors established a market for residential and small commercial duty cyclers at a price range of $200 to $800 each. If a duty cycler works well on the system, so would a smaller capacity system.

The Air-Conditioning and Refrigeration Institute (ARI) developed general guidelines for the application of energy management control systems which cycle air conditioning and refrigeration systems. ARI's "1987 Guideline for Energy Management Systems and Load Management through Duty Cycling"; Guideline A states:

"ARI recognizes the desire of many consumers, users and other building owners to install some sort of energy management system (EMS) device on heating, ventilating, air conditioning and refrigeration equipment. It is also recognized that some power suppliers feel the need to effect "load management through duty cycling", a program designed to reduce the peak load on a power distribution system and hence delay or eliminate the need for additional generating capacity. ARI offers these guidelines without stipulating that either energy savings, user comfort or equipment performance will be achieved.

"The product scope of ARI encompasses a wide variety of products. The availability of various types of EMS devices is very broad and the effect of such devices on equipment warranties may vary product-by-product and manufacturer-to-manufacturer. Therefore, ARI urges that the equipment manufacturer be contacted for specific recommendation concerning that equipment.

"The general guidelines are as follows:

1. Safety

   Do not alter, disable or bypass any of the safety controls.

2. Control Circuits

   Control the unit operation through the control wiring. An auxiliary power supply may be required to carry the load of any additional field supplied controls. Additional load on the equipment transformer can cause voltage drop, chattering contactors, and ultimate failure of motor-compressor or other components.

3. Fail-Safe Requirement

   In the event of failure of an add-on control device(s), the normal operation of the equipment being controlled should not be jeopardized.

4. Cycle Rate

   Do not short cycle motor controllers, motors, or motor-compressors. The compressor off cycle must be five (5) minutes or longer. If more than four (4) cycles per hour are anticipated, contact the equipment manufacturer for specific recommendations.

5. Fossil Fuel Heating Equipment

   Do not short cycle or underfire fossil fuel heating equipment. Adequate burner operating time and temperature is necessary to prevent condensation damage to heat exchanger and/or flue.

"In the event of any conflict between the manufacturers' specific instructions and these Guidelines, such instructions should prevail over these Guidelines.

"The information on these Guidelines is current as of the date of publication. These Guidelines are only guidelines and should not be referred to or construed as a standard, certification or warranty. The appropriate steps to be taken with respect to duty cycling may vary with particular circumstances. All work with respect to wiring of cycling devices should be done by and under the supervision of qualified and experienced personnel to insure proper installation, and should be properly inspected. However, no changes in these Guidelines (when identified as ARI guidelines) shall be made without the approval of ARI.

"Released for publication by the ARI General Standards Committee on June 12, 1985."

Excessive Heating and Cooling

The amount of heating or cooling called for by traditional thermostats, as well as the heating or cooling provided by most HVAC systems, have been found by test to be excessive. This is due to insufficient occupant interaction and the occupant's inability to judge and request what he actually needs. This causes any system to perform below achievable efficiencies. Utilizing Thermal Comfort Dynamic (TCD) control strategies can literally cause systems with normal efficiency to out perform the high efficiency systems discussed elsewhere in this article. Energy is used to meet the dynamic occupant needs rather than calculated building needs.[4]

Refrigerant Subcoolers

Heat exchange type subcoolers are being installed on some existing split DX systems to save energy. The subcooler is installed ahead of the expansion valve or capillary tube to cool the refrigerant liquid to reduce flash gas bubbles. This not only improves capacity of refrigerant flow to evaporator but also reduces compressor head pressure which thereby reduces compressor motor energy draw. Subcoolers have also been installed ahead of condenser coils for energy consumption reduction. Subcoolers installed ahead of the condenser does reduce energy consumption and compressor head pressure but in the process reduces cooling capacity and efficiency. A simple technique which has long been used for affecting subcooling is by soldering a section of the liquid line to the suction line and covering that combined line section with insulation.

Condenser Air Precooler

Evaporator precoolers are being used on air-cooled condensers in hot, dry climates to improve heat rejection, reduce compressor head pressure and save energy. Condenser fans are not generally designed to accommodate this added pressure drop. It is possible to negate the effect of the reduced air temperature with the restricted air flow. These types of water saturated precoolers must be generously sized and properly sealed against short circuit air flow to realize optimum benefit. The precooler media panels should be removed during heating season if the condenser is on a heat pump system.

Conclusion

When attempting to reduce energy consumed, or even more importantly peak demand required, the nameplate efficiency rating may be almost meaningless unless:

1. The equipment is matched to the load.
2. The controls system can emulate the human body's ability to sense and make adaptations to operative temperature drifts or ramps.
3. The indoor air flows are proper.
4. And, the heat transfer surfaces are maintained in such a way as to allow the system to perform as intended.

Otherwise, neither model codes nor utility peak reduction programs can be expected to achieve their goals. Instead, efforts will be better rewarded if concentrated on the thermal integrity of the structure involved.

REFERENCES

1. "Thermal Environmental Conditions for Human Occupancy" ASHRAE Standard ANSI/ASHRAE 55-1981.
2. "Ventilation for Acceptable Indoor Air Quality," ASHRAE Standard 62-1981.
3. "Liquid Refrigerant Control in Refrigeration and Air Conditioning," Copeland Application Engineering Bulletin.
4. "Thermal Comfort Dynamics-An Advanced Control Strategy," by Bruce K. Colburn, Ph.D., P.E. and Kermit S. Harmon, P.E., C.E.M. Presented at the AEE 6th World Energy Engineering Congress, Atlanta, Georgia, November 1983.

Chapter 36

FREE COOLING—HOW AND WHY

M. J. Dwyer, Jr.

BACKGROUND

The University of Arkansas for Medical Sciences campus is located in a residential section in the City of Little Rock in what is considered the Sunbelt region of the Southcentral United States. The main campus is situated on a 26 acre block and consists of 11 major buildings totalling over 1.25 million square feet.

Steam and chilled water are generated in a central power plant and distributed to all but one building on campus. The power plant contains one 750 ton steam driven turban chiller, one 900 ton centrifugal chiller, and one 2,500 ton centrifugal chiller. The campus was constructed in 1956 and initially consisted of a 350 bed hospital, a nine story educational building, and a ten story student union building. Additional educational buildings, research buildings, and outpatient clinic facilities have been added.

Over the years numerous renovation projects have taken place, primarily to convert former teaching space into research laboratories and animal holding areas. As research became more complex and more heat generating equipment was installed (both in the hospital and research buildings), it became necessary to provide chilled water to many locations on the campus year round. This was true even though the outside air temperatures were below freezing for a number of days at a time. To produce chilled water under these conditions modifications to the chiller were required in order to make sure it would not unload itself to the point where it would shut down.

In 1975 the 2,500 ton centrifugal chiller was installed and all steam absorption chillers were phased out. It was thought at that time that over the next few years the cost of electrically generated chilled water would be cheaper than gas. During the summer months (especially when the temperature is above 95 degrees) it is necessary to maintain 41 to 42 degree chilled water in order to provide adequate cooling to the operating room and labor and delivery. As the outside temperature decreases, the chilled water temperature is allowed to rise to approximately 45 degrees during the day and up to 48 to 49 degrees at night and on weekends, when many of the critical areas are not in use (See Table 1).

Temperature (Degrees Fahrenheit)	Nov.	Dec.	Jan.	Feb.
Average Daily	49.7	42.4	40.2	47.1
Average Maximum	56.7	50.0	49.3	55.9
Average Minimum	42.6	34.7	31.0	38.3
Degree Days				
Heating	454	694	762	496
Cooling	0	0	0	0

TABLE 1 TEMPERATURE AVERAGES FOR A FOUR MONTH PERIOD.

FREE COOLING

Although there is a clear cut definition of free cooling, to me free cooling means cooling without the use of mechanical refrigeration. In reality there is no such thing as free cooling. By that I mean we cannot cool without some cost. Because even if we are not using refrigeration, we will still have to move air some way and probably some type of cooling liquid, both of which cost money. However, if you are doing the cooling without the use of mechanical refrigeration you have taken the biggest energy user out of the system (compressor, or chilled water pump, or solution pump, or boiler in the case of an absorption machine).

Let's talk about some of the different methods of what I consider free cooling. You've all heard of the kiss method, to keep it simple. The earliest method known to man as free cooling was simple - open the window. Well as we know, many buildings now days don't even have windows that will open and if they did, most administrations don't want people opening the windows at random. In health care institutions it may even be against code to open windows in certain areas.

Probably the most used method of free cooling is the economizer cycle. This cycle is accomplished through the use of some sort of air handling unit or air handling equipment and is quite common. Here an air handling unit is used that is delivering air to the space. The equipment needed for this system would be some sort of outside air mixing box or a way to bring in outside air. The other requirement is having economizer controls.

Another type of system that you may have is a unit ventilator system which also may have an economizer cycle. Quite often, a variation is an economizer controllable lockout of there refrigeration or chilled water coil so that while your doing your free cooling you won't also be doing mechanical cooling.

Other types of free cooling involve the use of a previously installed chilled water system in one form or another. I think there really are three different types of free cooling (using the existing system) two of which I will describe briefly and the third is the subject of this paper. Basically, all three use the principle of heat exchange in that one side of the heat exchanger is connected to the existing cooling tower system and the other side to the chilled water system. For these systems to operate, the outside air must be low enough so the cooling tower water (which usually operates in the 75 to 85 degree range) can be lowered to approximately 45 degrees or a fixed temperature that has a direct relationship to the required chilled water temperature. At this time, the cooling tower water becomes colder than the building system's chilled water and, therefore, heat will be exchanged from the building to the cooling tower water.

The simplest system to install uses an existing chiller as a heat exchanger and its refrigerant as the actual heat transfer medium. There is minimum renovation required to the chiller and none required to the existing piping system, either in the chilled water side or the cooling tower side. Thus, it is also by far the cheapest to implement. However, there are certain disadvantages: the first being that this system will only generate 35-40% of the tonnage available from the chiller when it is operated manually, i.e., when it is a 1,000 ton chiller it will provide 350 to 400 tons of cooling. The other disadvantage and probably most significant, depending on the area of the country in which the facility is located, is that the chilled water can only be cooled to about 10 degrees above the temperature of the cooling tower. Of course, in northern climates where temperatures remain below 40 degrees for long periods of time, this is not a disadvantage.

The second free cooling system is called the strainer cycle (see Figures 1 & 2 below). This cycle once again works with a cooling tower and chilled water systems. In effect, it allows the cooling tower water to enter the chilled water system through a strainer. Obviously, when you inject cooling tower water, which is relatively dirty, directly into the chilled water you can sludge up a cooling system very quickly. For this reason, strainers and filters are added to prevent dirt from entering the chilled water system. This method also has some disadvantages. The strainers take up a large amount of room in the boiler house and need to be located somewhere close to the existing chiller. Another problem is that you have to clean the strainer on a regular basis. The newer systems have an automatic backwash system that would operate approximately every hour during the initial system clean-up. But once this was completed, the backwash would occur only once or twice a day. When you are backwashing strainers, you have to dispose of the water down the drain and as a result your building's water usage goes up. A major advantage of this system over the use of the chiller is that if you have 40 degree water coming from the tower, you have 40 degree water in your chiller water system. Therefore, this system can be operated many more hours per year than the previously described system.

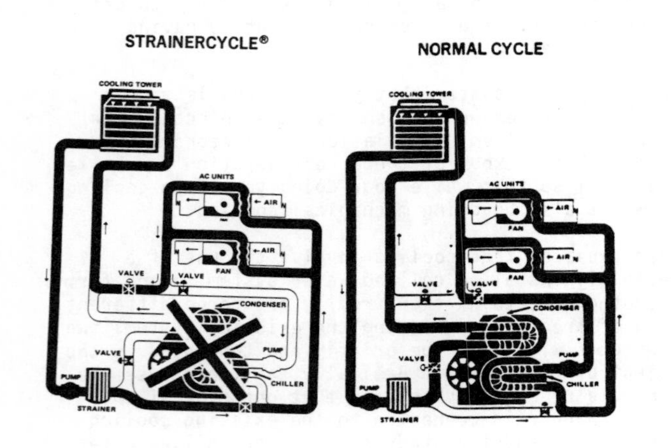

FIGURE 1. WATER FLOW IN STRAINERCYCLE.

FIGURE 2. WATER FLOW IN NORMAL CYCLE.

The third system is a combination of the first two and involves a flat heat exchanger that allows heat to transfer directly form the chilled water system to the cooling tower system. Thus, it is possible to generate chilled water that is only one to two degrees warmer than the cooling tower, but does not run the risk of contaminating the system and actually takes up less space than the equipment needed for the strainer cycle. However, it does require the additional piping and valves that the strainer requires so its initial cost is about the same as the strainer but a lot more than the chiller (about $100,000). The actual energy savings and payback period will vary greatly, depending on the system used as well as the temperature variations experienced in the area.

OPERATION

During the first cycle of the Hospital and Colleges Energy Grant program, the University of Arkansas for Medical Sciences modified its 2,500 ton York centrifugal chiller for free cooling. The total cost for this modification was approximately $35,000. Thus, it cost the University half this amount under the matching funds grant program. As predicted, we are only able to get about one-third the rated tonnage of the chiller (750 tons) during the free cooling mode. However, tonnage-wise this was enough to meet the needs of the campus during the cold winter days. More significant was the fact that there was more than a 10 degree loss between the cooling tower water and the chilled water during the free cooling process as shown in Table 2.

Free Cooling Chiller Temperatures

Time	Outside	Cooling Tower	Chilled Water
0600	38	40	51
1200	39	36	47
1800	36	35	47
2400	33		
0600	31	37	48
1200	35	39	50
1800	35	40	50
2400	33	40	51
0600	34	38	48

TABLE 2 TEMPERATURES DURING FREE COOLING PROCESS.

Because of the winter temperatures found in the Little Rock area, this 10 degree change significantly reduced the number of hours that free cooling could be used and in many cases limited free cooling to evening hours. A greater problem, however, was that of freezing of the cooling tower itself. The cooling tower fan was temperature controlled and automatically shut off when the cooling tower water got below 45 degrees, however, at about 30 degrees ice started to form on outer edges of the tower. It was felt that major damage could occur if this ice was allowed to build up for any significant length of time. Strange as it may sound, as the temperatures decreased we were forced to turn the chiller back on in order to heat up the condenser water. Although successful during hours of operation, this type of free cooling was limited by both an upper and lower temperature range. A review of the daily temperature ranges in the Little Rock area indicated that free cooling would be extremely beneficial if it was not limited by lower temperatures and a method could be devised by which there is a minimum Δ T between the cooling tower water and the chilled water. An efficient heat exchanger would meet these requirements if at the same time modifications could be made to the cooling tower to allow it to operate at lower outside temperatures with out the problem of freezing. A flat plate heat exchanger can be installed in a manner very similar to that of a strainer cycle as far as the amount of additional piping and valves is concerned; the difference being the heat transfer takes place in the heat exchanger rather than in the building itself (see Figure 3).

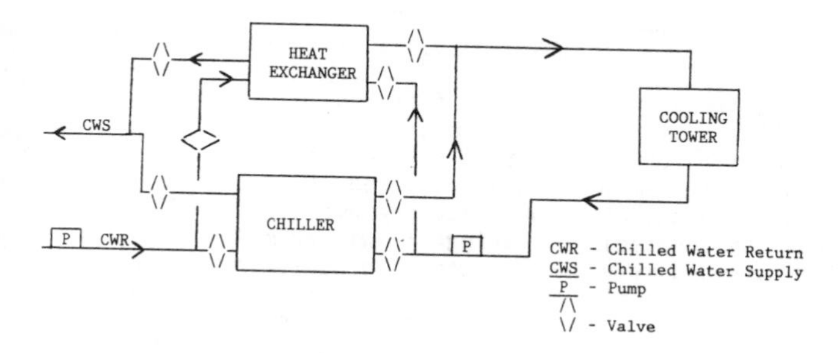

FIGURE 3. HEAT TRANSFER IN HEAT EXCHANGER.

Because there was no other flat plate heat exchanger system in the state of Arkansas, the State Energy Department determined that such a system could be installed on an experimental basis and energy funds could pay for the whole project, in contrast to the energy grant program which is normally a 50/50 match. A Traenter Model UX-896-UP-335 Flat Plate Heat Exchanger was installed in the central power plant adjacent to the 900 centrifugal chiller. It was recommended that the flow rate of the chilled water be as close as possible to the condenser water flow for the system to achieve maximum efficiency. In our case, the system was designated for 4,490 gallons per minute of chilled water and 4,400 per minute of condenser water flow. It was estimated that this system could achieve chilled water temperatures range between 44 degrees and 49 degrees with condenser water temperatures of 42 degrees and 46 degrees. Table 3 shown below is an example of what was actually achieved on two separate days.

Free Cooling Flat Plate			
Time	Outside	Cooling Tower	Chilled Water
0600	38	46	47
1200	40	45	48
1800	41	45	48
2400	38	46	45
0600	33	44	43
1200	48	44	49

TABLE 3 TEMPERATURE RANGE WITH CONDENSER WATER TEMPERATURES OF 42 AND 46 DEGREES.

In order to solve the cooling tower freezing problem, heat was added to the tower basin and the cooling tower fan, motor and drive were modified so that the fan could reverse direction when necessary. When ice begins to form the fan is reversed (run backwards) until the tower thaws out.

It was hoped that the heat exchanger installation would be completed in time to use it during a complete winter season. Late equipment deliveries delayed completion of the project until late January, so the first year there was only one month of actual use.

Table 4 indicates the number of days in which free cooling was used during the four coldest months (November, December, January and February). The old free cooling system (the 2,500 ton chiller) was used in the first three months and the flat plate heat exchanger was only used in February. December was the coldest month of the year with January and February temperatures being way above the norm.

	Free Cooling			
	Chiller		Flat Plate Heat Exchanger	
Month	Days	Hours	Days	Hours
Nov	2	28		
Dec	15	246		
Jan	12	118		
Feb			8	128

TABLE 4 DAYS IN WHICH FREE COOLING WAS USED.

Table 5 indicates the outside temperatures and chilled water temperatures during four days in which the flat heat exchanger system was used continually. These days were all week days and it is obvious from the maximum temperatures that the chiller free cooling system could not have been used during a major portion of the daylight hours. A chiller was turned on at about 10:00 a.m. on the 21st when the outside temperature finally reached 51 degrees and the chilled water temperature was at 48 degrees. For every hour the flat plate heat exchanger is on, the medical sciences center realizes a savings of $30.93. Therefore, during the period described above, there was a total of 128 hours of free cooling for a savings of $3,959.20. Using the actual experience of late February and the climatic data for the months of November, December, January and February, it appears that it would have been possible to use the flat plate heat exchanger for a total of 1,435 hours at a savings of $30.90 an hour, which would equate to an annual savings of $44,384. In energy this would amount to 986,366 KWH. The cost figures above are based on an average cost of electricity or .045 per KWH. It is anticipated that once a full year of data is available and we can show the number of hours which we will not have to operate one of our chillers, we can also realize additional savings by obtaining a reduction in the cost of our chiller maintenance contracts

	Temperatures					
	Outside		Cooling Tower		Chilled Water	
Date	Max.	Min.	Max.	Min.	Max.	Min.
17	36	30	43	41	46	44
18	45	32				
19	42	37				
20	42	37	46	43	49	43

TABLE 5 TEMPERATURES DURING CONTINUAL USE OF FLAT PLATE HEAT EXCHANGE SYSTEM.

Now for the bad news. There was one serious problem incurred in operation of the free cooling system that has since been corrected. Operation of the type chiller installed on the UAMS campus was limited to cooling tower water of no less than 65 degrees. Therefore, prior to putting the free cooling system into effect, there was a period of time after the chiller was shut down before the cooling tower water was cold enough to implement free cooling. This period of time ranged from 10 minutes to 30 minutes, depending on the outside temperature. Of course, since free cooling was implemented at the highest outside temperature possible, it normally took 30 minutes for the cooling tower to reach the mid 40 degree range necessary to switch to free cooling. This period of time without any cooling at all resulted in unacceptable temperatures in many areas of the facility. The problem was solved by using the free cooling of the 2,500 ton chiller, starting approximately 15 minutes before the switch to the flat plate heat exchanger and carrying over 15 minutes afterward. This was enough to keep the chilled water at an acceptable level. Modifications have since been made to the 900 ton chiller (normally run in the winter prior to use of free cooling) that will allow passing a controlled amount of the tower water, therefore allowing the chiller to operate at a much lower water temperature. We now have an additional year of operational experience with our flat plate heat exchanger and are extremely pleased with the results. Table 6 indicates the results of operation from November 1987 to March 1988. The average outside temperatures for this year have been 2.8 degrees greater than the last year. Table 6 does not indicate any free cooling on our large York chiller although it was used as an aid in starting our flat plate heat exchanger operation.

Free Cooling

Month	Chiller		Average Outside Temperature	Flat Plate Heat Exchanger	
	Days	Hours		Days	Hours
Nov			55.45		
Dec			46.62	22.21	533
Jan			38.53	22.54	541
Feb			45.15	9.08	218
Mar			54.57	.33	8
	0	0		54.17	1,300

TABLE 6 FREE COOLING OPERATION RESULTS.

The 1,300 hours of operation of the heat exchanger equate to an annual savings of $40,265 based on the average cost of electricity at .046 per KWH. This is actually less than last year's prediction of $44,480 but this can be attributed to the fact that we did not have as cold a winter.

With additional hours of operation, we did uncover another problem that we are in the process of solving. The cooling tower we are using with the flat plate heat exchanger has two cells. When we installed the flat plate heat exchanger we also installed not only base heat but the capicity to reverse the fan in one cell of the cooling tower. There was enough water carried over to the other cell to cause an icing problem so we decided to have the second cell fan designed to have reverse operation.

Again, there is no doubt that the installation of the flat plate heat exchanger has realized significant monetary savings for the University of Arkansas for Medical Sciences campus. We also have experienced continual free cooling operation using a greater tower water temperature than anticipated. On cloudy days we have been able to maintain free cooling with tower water temperatures as high as 48 to 50 degrees.

We are currently installing an additional chiller in the central power plant in order to handle the increased demand resulting from the construction of the Arkansas Cancer Research buildng. As a result of the experiences we have had in the past, we are also including a flat plate heat exchanger in the installation of this new chiller. We are extremely anxious to try out our new chiller and flat plate heat exchanger and to develope new procedures that will allow us to use combinations of chillers and flat plate heat exchangers depending on the weather and the demands on the system.

CONCLUSION

In most parts of the country free cooling can achieve significant savings in both energy and dollars. Of course, the actual amount of savings will depend on not only the temperature encountered at a particular location installation but also the amount of cooling needed by the facility during the winter months. With more and more complex computer equipment being installed in hospitals and medical centers, the need for year-round cooling is on the increase. Although not absolutely free, free cooling should be investigated by all plant engineers as a means to reduce escalating costs in both operating and utility budgets.

Chapter 37

INDUCED AIR DISTRIBUTION SYSTEM

P. L. Kogan, J. R. Tuckner

INTRODUCTION

An air distribution system greatly affects the performance of a building's HVAC system. Designed to provide a comfortable level of occupancy, it can sometimes become a stumbling block in the way of proper system operation . This discussion provides some insights into how the air distribution system within a large administration building, located in Northern California, effects the HVAC systems present and future performance.

In recent years, there have been numerous complaints from the building occupants regarding the uncomfortable space temperatures. These complaints occurred most frequently in the hot summer (temperatures rise to 80°F) and cool winter months (temperatures drop to 65°F). The building also experienced uneven temperatures throughout the central core portion of the building at the various floors. The space conditions got even worse due to the lack of any discernible air motion at the working levels.

The concerns of the building's management over the inadequate HVAC system capacity are exaggerated by the fact that new office automation equipment is to be installed in the near future that would generate approximately 1,000,000 Btuh of additional heat. Therefore, an extensive investigation into the design and performance of the several mechanical systems, including the air distribution system, was made.

BUILDING DESCRIPTION

The building houses approximately 2,000 people and has a total area of almost 600,000 square feet, with 75 to 80% being inhabitable space. The major portion of the building area is open clerical offices with some space allocated to the computer center, cafeteria, auditorium and mechanical facilities. The building structure consists of six floors above grade and a basement level. All major mechanical and electrical equipment is located in the basements' mechanical rooms. Designed between 1973-1974, the building was put into operation in 1975.

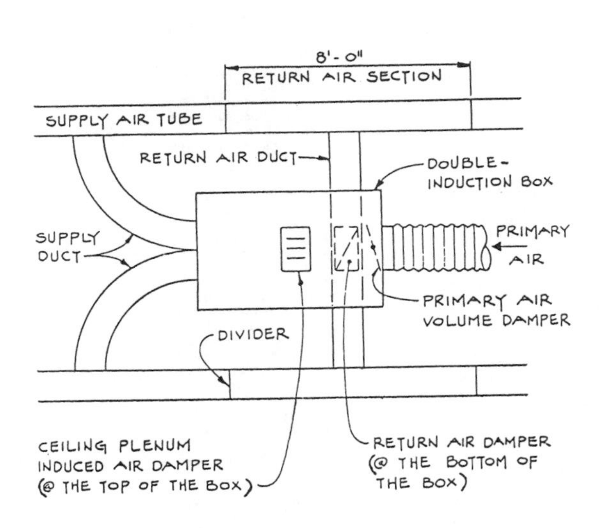

PLAN VIEW

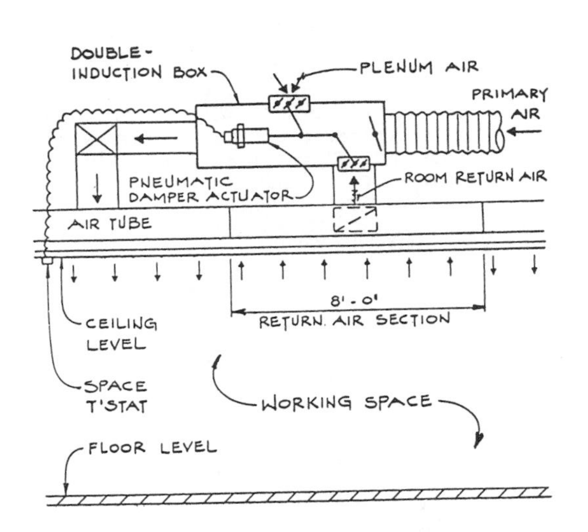

ELEVATION VIEW

DOUBLE - INDUCTION BOX

FIGURE 1

AIR DISTRIBUTION SYSTEM

There are several air distribution systems in the building. The one discussed in this paper is the system which serves the interior building zones on floors 2 thru 6. The interior zones occupy the largest portion of the building's inhabitable space and is where a major portion of the cooling load is concentrated. The design of the interior HVAC system was considered innovative at the time it was installed. Incentives were given to engineers to develop an integrated design that would minimize life cycle cost, size of equipment and ductwork and annual energy consumption, when compared to a traditional building. It was designed and built to meet the design conditions listed in Table 2, Column 2. However, the system performance has significantly declined, which was confirmed by the field testings and measurements, performed under the direction of the authors of this article and described in the following paragraph "Field Tests".

The interior zones, which encompass 234,750 sq. ft., are laid out on a modular basis, with each floor divided into four quadrants. Each quadrant is served by a medium pressure, primary air handling unit located in the basement mechanical room. Each air handling unit supplies a constant volume of primary air to a series of double-induction boxes located above the ceiling of each floor. The induction box, as shown on Figure 1, uses the primary air supply to induce air flow from either the space or ceiling plenum or both. The primary air is mixed with the induced air and then is supplied to the space through an air bar and air tube assembly as shown on Figure 2.

The primary air is cooled by the chilled water coils located at each air handling unit and then the tandem damper on the induction box controls where the induced air comes from, based on a signal from the space thermostat. The temperature of the primary air is constant and can be changed only by the manual setting adjustment on the discharge temperature controller.

The average sized double-induction box is designed to serve 1800 square feet with an air flow rate of 0.6 cfm/sq.ft. (0.65 cfm/sq.ft. for the 6th floor). The design induction ratio (total air flow to primary air flow) is 1.5 and therefore, the design air flow of 1080 cfm (1800 x 0.6 = 1080) consists of 720 cfm of primary air and 360 cfm of induced air.

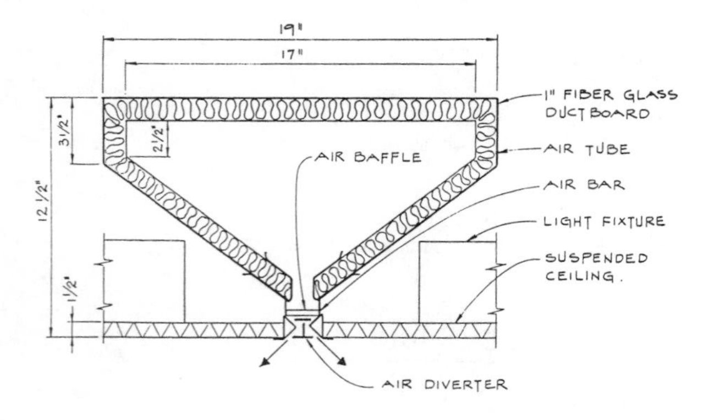

AIR TUBE & AIR BAR DETAIL

FIGURE 2

FIELD TESTS

Field tests were done to determine the performance characteristics of the air distribution system. Four double-induction boxes were tested which best represented each of the four air handling systems (Unit AC-1 thru AC-4), one box per system. All of the boxes tested were located on the 6th floor, where the cooling load is at a maximum and where the boxes are furthest from the primary air fan and have the lowest available static pressure.

A summary of the test data for the double-induction boxes is shown in Table 1 and a comparison to the design data is presented in Table 2.

TABLE 1

Air Flow at Air Bars with Double-Induction Boxes

Test No.	Test Description	Data Recorded (R) or Calculated (C)	Cooling Mode
1.	Induction Box 6th Floor, "In-System", Unit AC-1	(R) Total Air Flow, cfm	1100
		(R) Primary Air, cfm	720
		(C) Induction Ratio	1.53
		(C) cfm/sq. ft.	0.61
		(R) Inlet Pressure at Box, in. w.g.	2.30
		(R) Outlet Pressure at Box, in. w.g.	0.18
		(C) Pressure Drop at Box, in. w.g.	2.12
		(R) Inlet Pressure at Air Tube, in. w.g.	0.05
		(R) Primary Air Temperature, °F	57
		(R) Outlet Air Temperature, °F	62
		(R) Room Temperature, °F (11 a.m.)	75
		(R) Plenum Temperature, °F	76
2.	Induction Box 6th Floor, "In-System", Unit AC-2	(R) Total Air Flow, cfm	960
		(R) Primary Air, cfm	640
		(C) Induction Ratio	1.50
		(C) cfm/sq. ft.	0.53
		(R) Inlet Pressure at Box, in. w.g.	3.00
		(R) Outlet Pressure at Box, in. w.g.	0.18
		(C) Pressure Drop at Box, in. w.g.	2.82
		(R) Inlet Pressure at Air Tube, in. w.g.	0.02
		(R) Primary Air Temperature, °F	57
		(R) Outlet Air Temperature, °F	64
		(R) Room Temperature, °F (10 a.m.)	74
		(R) Plenum Temperature, °F	76
3.	Induction Box 6th Floor, "In-System", Unit AC-3	(R) Total Air Flow, cfm	995
		(R) Primary Air, cfm	520
		(C) Induction Ratio	1.91
		(C) cfm/sq. ft.	0.55
		(R) Inlet Pressure at Box, in. w.g.	1.30
		(R) Outlet Pressure at Box, in. w.g.	0.15
		(C) Pressure Drop at Box, in w.g.	1.15
		(R) Inlet Pressure at Air Tube, in. w.g.	0.02
		(R) Primary Air Temperature, °F	54
		(R) Outlet Air Temperature, °F	62
		(R) Room Temperature, °F (1:30 p.m.)	75.50
		(R) Plenum Temperature, °F	76
4.	Induction Box 6th Floor, "In-System", Unit AC-4	(R) Total Air Flow, cfm	1485
		(R) Primary Air, cfm	825
		(C) Induction Ratio	1.80
		(C) cfm/sq. ft.	0.825
		(R) Inlet Pressure at Box, in. w.g.	2.00
		(R) Outlet Pressure at Box, in. w.g.	0.23
		(C) Pressure Drop at Box, in. w.g.	1.77
		(R) Inlet Pressure at Air Tube, in. w.g.	0.04
		(R) Primary Air Temperature, °F	62
		(R) Outlet Air Temperature, °F	68
		(R) Room Temperature, °F (2:30 p.m.)	77
		(R) Plenum Temperature, °F	78

The main air handling units were tested at the same time as the double-induction boxes in order to coordinate the data obtained at the units with that from the induction boxes and a summary of the test data is presented in Table 3.

In addition, the following conditions were noticed:

* The pressure drop through the air tube/air bar assembly was significantly higher than anticipated in the original design.

* The static pressure drops rapidly in the air tubes. Only a few hundredths of an inch of water gauge pressure was available at the first one-quarter of air bar length.

* Air movement was measured by the rotating vane anemometer at various locations at desk height and at 5 feet above the finished floor. The test revealed that there was no discernible air motion at either elevation at the end of the first one-quarter of the air bar length.

* The air flow volumes were checked at various points of floors 2, 4 and 6, using the 4-feet long air flow hood. Minimal air flow, very often close to zero CFM, was obtained in some of the tested points located away from the induction boxes. This indicates that most of the air is "dumped" in the beginning of the air bar.

* Physical location of air bars being tested and relative to the adjacent return air light fixtures is only 6-1/8 inches. The design also incorporates an 8 foot return air tube in the middle of the supply air tube (Figure 1). This arrangement creates a situation where a significant amount of the supply air could be short-circuiting to the return air plenum, rather than entering the occupied space.

TABLE 2

FLOW CHARACTERISTICS COMPARISON CHART
DOUBLE-INDUCTION BOXES

1	2	3	4	5	6	7
Flow Characteristics	Design Conditions	Test No. 1	Test No. 2	Test No. 3	Test No. 4	Test * Conditions
Primary Air, cfm	720	720	640	520	825	680
Total Air, cfm	1080	1100	960	995	1485	1135
Air Flow Rate, cfm/sq. ft.	0.6 - 0.65	0.61	0.53	0.55	0.825	0.63
Induction Ratio	1.50	1.53	1.50	1.91	1.800	1.69
Inlet Pressure, in. w.g.	1.4 - 1.80	2.30	3.00	1.30	2.000	2.15
Outlet Air Temperature, °F	59.2	62	64	62	68	64
Primary Air Temperature, °F	50.8	57	57	54	62	57.50
Room Temperature, °F	76.0	75 (11:00am)	74 (10:00am)	75.50 (1:30pm)	77 (2:30pm)	---

* Test conditions are an average data based on field tests.

LOAD CALCULATIONS

Computer load calculations were performed using data which represents the building structure, occupancy, equipment load (present and future), lighting, etc. The building was modeled with U-values for the roof, walls, and windows, and shading coefficients for windows as applicable. The temperature difference was calculated as a difference between the design room temperature of 76 °F DB and space supply temperature, which is changing correspondingly to a change in temperature leaving the cooling coil (primary air temperature). Since it's a constant volume system, the primary air temperature was the variable used to evaluate different conditions, as steps to approach the desirable design parameters.

It is important to note that the current net available primary air flow to the building is 285,475 cfm, which includes the 97,905 cfm for interior zones (total air flow of units AC-1 through AC-4, Table 3) and 16,510 cfm for exterior zones. The remaining air flow of 171,060 cfm (285,475 - 97,905 - 16,510=171,060) is serving the rest of the building. Only two building systems, interior and exterior, are using the principle of the induced air. Therefore, 171,060 cfm + 1.5x16,510 cfm =195,825 cfm - available air flow for the entire building excluding the interior system. This air volume is an important figure because it will have to be subtracted out of the flows for the present and future block load calculations, when determining any additional load to the interior system. A summary of different computer calculation runs is presented in Table 4. Each particular case will be discussed further in the text.

TABLE 3

FAN TEST DATA CHART

1	2	3	4	5	6	7	8	9
UNIT NO.	RPM (D)	RPM (T)	S.P. (D)	S.P. (T)	CFM (D)	CFM (T)	HP (D)	HP (T)
AC-1	1125/1238	1145	6.5	5.40	23,110	22,663	40	26
AC-2	1100/1210	1200	6.5	5.85	22,810	21,675	40	30
AC-3	1149/1264	1265	6.5	5.50	26,315	23,944	40	32
AC-4	1140/1254	1310	6.5	5.90	25,670	24,584	40	36
AVERAGE:		1230		5.70		23,200		

(D) Design
(T) Tested

TABLE 4

BUILDING LOAD COMPUTER CALCULATIONS

1	2	3	4	5	6
ITEM DESCRIPTION	DESIGN DATA	PRESENT LOAD ΔT = 14.9°F	FUTURE LOAD ΔT = 14.9°F	FUTURE LOAD ΔT = 13.1°F	FUTURE LOAD ΔT = 17.3°F
COOLING LOAD (MBH)	N/A	8,982	9,969	9,969	9,969
HEATING LOAD (MBH)	N/A	6,409	6,409	6,409	6,409
TONS OF REGRIGERATION	N/A	748.5	830.8	830.8	830.8
AIR FLOW (CFM)	285,475	368,655	417,360	489,006	370,287
CFM/SQ.FT. (BLDG.)	N/A	0.71	0.80	0.94	0.71
SQ.FT./TON	N/A	695	628	626	626
PRIMARY AIR TEMPERATURE	50.8	53.7	53.7	56.3	50.0
AIR TEMPERATURE (DELIVERED TO SPACE)	59.2	61.1	61.1	62.9	58.7
CFM/SQ.FT. (INTERIOR)	0.6-0.65	0.74	0.94	1.25	0.75

N/A - NOT AVAILABLE

SYSTEM EVALUATION AND CALCULATIONS

1. **Building Load and Air Flows:**

The first step in our evaluation was to review the capacity of the existing air handling units. For the purpose of this analysis, we will use an average fan air flow of 23,200 cfm (Table 3) and the existing chilled water supply temperature to the cooling coils of 42° F.

A computer simulation of the coil was then performed using existing coil data. The primary air temperature leaving the coil was found to be 53.7° F.

This temperature represent a theoretical approach and, of course, somewhat different from the actual field measured temperatures. The supply temperature to the space, "developed" as a mixture of primary air and induced air, was calculated by performing a heat balance around the induction box. The temperature of the air entering the space can be approximated by:

$$\frac{(720\ \text{cfm})\ (53.7°\text{F}) + (360\ \text{cfm})\ (76°\text{F})}{1080\ \text{cfm}} = 61.1\ °\text{F}$$

This temperature produces a temperature difference of 14.9 °F (76-61.1=14.9) which was used in the building block load calculation. The computer calculations for this case are presented in Table 4, column 3 and 4 for present and future conditions (future condition accounts for additional heat generated by new equipment). The air flow rates for the interior zones, which were derived from the computer data are as follows :

Present Load:

$$\frac{368{,}655\ \text{cfm} - 195{,}825\ \text{cfm}}{234{,}750\ \text{sq. ft.}} = 0.74\ \text{cfm/sq.ft.}$$

Future Load:

$$\frac{417{,}360\ \text{cfm} - 195{,}825\ \text{cfm}}{234{,}750\ \text{sq. ft.}} = 0.94\ \text{cfm/sq.ft.}$$

It can be seen from these results that the existing 0.6 cfm/sq.ft. is not sufficient to meet the present or future cooling needs of the interior spaces.

When considering the alternatives, the owner requested that it would be preferable to avoid modifications to the air distribution system so it will not disturb the office workers.

Consequently, all modifications were limited to work within the mechanical rooms. The owner of the building had previously tested the induction boxes and found that the maximum achievable air flow rate was 0.75 cfm/sq.ft. Therefore, the initial option to system improvement was to increase the air flow rate to a level of 0.75 cfm/sq.ft. To achieve this air flow rate, new supply fans would have to be installed with a rated air flow of 29,300 cfm, as shown below:

$$\frac{(0.75\ \text{cfm/sq. ft.}) \times (234{,}750\ \text{sq. ft.})}{1.5\ (\text{induction ratio}) \times 4\ (\text{fans})}$$

This increase in air flow would cause an increase in primary air temperature from the existing coil and therefore we looked at ways to reduce the primary air temperature as required.

2. **Lower Primary Air Temperature:**

We first determined that the primary air could not be adequately cooled using the present chilled water temperature. Our next step was to consider utilizing the existing coil and providing chilled water at a colder temperature. A cooling coil computer simulation was run utilizing 32°F chilled water and the new air flow of 29,300 cfm.

It calculated that the coil leaving temperature or the primary air temperature would be 56.3 °F (temperature increase is due to a larger air volume) and that by utilizing the previous procedures, the supply air temperature would be 62.9 °F and temperature difference would be 13.1 °F.

Computer calculations for the future load case with this temperature difference is presented in Table 4, Column 5.

The air flow rate for the interior zones was then calculated to be 1.25 cfm/sq. ft. which is much higher than the recommended limit of 0.75 cfm/sq. ft. A word of caution, however : one should not confuse the calculated air flow rates (based on computer generated air flow, accounting for lights, people, building structure loads, etc.) with air flow rate of 0.75 cfm/sq.ft., which represent maximum achievable air flow induced through induction box.

In addition to the air flow rate exceeding allowable limit (1.25 cfm/sq.ft. vs 0.75 cfm/sq.ft.), the primary air temperature and consequently the temperature of air supplied to the occupied space, is still higher than required one (56.3° F vs 50.8° F).

We then determined what primary air temperature, at a supply air flow rate of 0.75 cfm/sq. ft, would be required to meet the cooling load. Through trial and error, the required primary air temperature was found to be 50.0 °F. It should be noted that the air handling unit is a draw-thru configuration and therefore, to determine the required air temperature leaving the coil, one must subtract the heat gain from the fan. For this system, the heat gain was 5°F. Consequently, the air temperature leaving the coil would need to be 45°F (50°F - 5°F).

Using the 50°F primary temperature, the supply temperature was calculated to be :

$$\frac{(720\ \text{cfm})\ (50.0°\text{F}) + (360\ \text{cfm})\ (76°\text{F})}{1080\ \text{cfm}} = 58.7\ °\text{F}$$

The temperature difference would then be: 76 - 58.7 = 17.3 °F.

Evaluation of computer calculations, which were run with a temperature difference 17.3 °F (Table 3, Column 6) reveals that major cooling load requirements will be satisfied now.

The calculated air flow rate for the Interior Zones will be 0.75 cfm/sq ft, based on computer output.

CONCLUSIONS AND RECOMMENDATIONS

To begin with, we will consider several alternatives which offer problem resolutions by means of the modifications made mostly in the vicinity of mechanical rooms, air shafts, etc., all outside the occupied space.

The discussion presented above provided a basis for laying out one of the alternatives in solving the problem with the restrictive induced air distribution system. It's apparent that multiple solutions must be considered: increased air flow, lower primary air temperature, etc. Brief conclusions on different aspects of these solutions are indicated below.

o ALTERNATIVE NO. 1

* General: The major aspects of this approach are larger air volumes and new supply of low temperature water (32°F) to additional cooling coils installed at four major air handling units.

* Air Handling Units: As we have shown, a higher air flow rates will be required. We recommend to install new supply and return air fans to accomplish this task. Also, due to the large pressure drop through an air tube/air bar assembly, the static pressure capacity of the new fans should be increased.

* Low Temperature Chillers: We recommend the installation of the additional low temperature chillers to provide a lower chilled water temperature and, consequently, lower supply air temperature. This goal can be accomplished with new low temperature chillers, pumps and controls, and creating a separate loop circuit with a new set of cooling coils, installed in series with existing ones. One may question: why low temperature chillers, which are not yet common in the mechanical rooms and more complicated to maintain? We have attempted to use the conventional 42°F supply temperature, but it would require far larger coils, more GPM of chilled water, more power consumption, etc. We found this solution impractical and not economical.

* Controls: At the present time the supply air temperature is varied based on outside air temperature plus modulation of the return/exhaust dampers. In addition, the space thermostat governs the position of tandem damper at ceiling induction boxes, alternating the return air induction directly from space or from ceiling plenum.

  We propose to use a series of space temperature sensors (per floor basis), average the signals per each quadrant and send the signal (high signal controller) representing worst conditions to the air handling unit controls. These controls should operate the chilled water control valve at the existing (42°F) cooling coil to furnish primary air temperature at 56.3° F.

  The controls at the new (32°F) cooling coil should govern the chilled water valve only to recool the air leaving the first coil from 56.3° F to 50.0° F, when conditions are calling for.

* Balancing: The balancing of the entire HVAC systems, air and water sides, is considered very important. During the length of building life, almost thirteen years, the HVAC systems have not been balanced, adjusted or corrected. In addition, new fan systems would require all air distribution systems to be balanced to provide new increased air flow rates.

* Calibration: It's recommended to calibrate all HVAC instruments, and controls. This operation will ensure the owner that the systems are functioning properly, according to new set points.

o ALTERNATIVE NO. 2

* General: The major features of this alternative are larger air volumes, larger water volume supply to new cooling coils at main air handling units and new booster cooling coils at 6th floor with low temperature water supply.

* Air Handling Units: New air handling units are proposed as in Alternative No. 1.

* Low Temperature Chillers: Small size low temperature chiller is proposed only to feed the new booster recool coils at the top 6th floor supply ductwork, where load is at the maximum. This will also greatly decrease the possibility of overcooling on the "interior" floors without roof load. Existing coils with 42°F supply will be replaced with a larger size to keep up with increased load demand.

* Controls: Controls, in general, can utilize the scenario described for Alternative No. 1, with addition of a separate control circuit for the new booster coils and low temperature chiller.

* Balancing & Calibration: Same procedures are recommended as for Alternative No. 1.

The alternatives presented above are based on the improvements to major mechanical equipment. We consider it prudent to mention several possible solutions carried out by modifications to the air distribution system by itself.

o ALTERNATIVE NO. 3

* Install new supply and return fans to increase available air flow and static pressure.

* Remove all ceiling mounted air tubes and air bars. Install new linear slot diffusers with low static pressure drop.

* Balance, test, adjust and calibrate all air and water sides of the system.

* Clean thoroughly all ductwork.

This solution will improve the air distribution system by allowing more air supply to the space due to increased air flow from air handling units and lower resistance of the air terminals.

o ALTERNATIVE NO. 4

* Remove all ceiling mounted double-induction boxes.

* Install new ceiling mounted fan-powered VAV boxes.

* Balance, test, adjust and calibrate all air and water sides of the system.

* Clean thoroughly all ductwork, air tubes and bars.

This solution can enhance the air distribution system by increasing the rate of induced air and the total volume of air supplied to the occupied space.

The optimum system composition is still in design at the present time. No doubt, a positive outcome will emerge and the performance of the air distribution system will be enhanced. It's most exciting and challenging (and most difficult) for an engineer to see the outcome of major modification to a large HVAC system.

Chapter 38

INTEGRATED COMFORT™ SYSTEMS: THE NEW FRONTIER

E. L. Smithart

Most building industry disciplines today agree that we're living in an age of rapidly changing technology. John Naisbett, in his book **Megatrends,** dubbed it "the information age." It's an age where computer literacy has virtually become mandatory in order to survive. It's an age where the only way of really discovering the limits of the "possible" is to venture a little way past them into the "impossible."

Naisbett also says that the only way to understand the future is to first understand the past and the present. A rather profound statement. The last 15 years have brought about dramatic change in the way we live and conduct our business. In 1972, we had an Arab oil embargo that resulted in an energy cost crisis. And, since then, our energy costs have been doubling on a seven-year cycle.

In 1975, we saw building energy conservation standards based on ASHRAE Standard 90 at both Federal and State levels, which created a new "mind set" for energy efficient design of buildings.

Beyond this, we've seen massive government funding for renewable energy research. . . primarily solar energy. We've seen massive govern ment deregulation and we're now seeing tax reform which is changing building investment strategies for building owners and developers. Short-term owners are now becoming longer term owners with strong focus on efficient building management and increased cash flow generation. We've also seen a great deal more interest on the part of corporate owners in the "quality of comfort" in their buildings for **health and other reasons.**

Beyond this, we have gone from a national economy to a global economy with the ensuing balance of trade and currency devaluation that we are seeing today. All this has created a strong need for a concentrated focus on increased productivity and building market ability to remain competitive in the world marketplace.

The "productivity crunch" translates directly into a need of building owners and developers to differentiate their buildings from their competitors by providing safer, healthier and more productive working environments to maintain acceptable tenant occupancy rates and cash flow which translates into building resale value.

We must never lose sight of the fact that the resale value of a building is a direct function of the building's ability to generate cash flow.

About ten years ago, as a reaction to this need for better building environmental comfort, our industry focused on advanced development work to attempt to utilize space age microelectronic technology for better control and management of building comfort systems. In the process, it established a strong emphasis on the marriage of in-depth experience in automatic temperature control with microelectronic technology.

As the development efforts progressed, it became evident that there were a couple of major shifts in technology taking place in the building industry:

1. A shift from electromechanical and pneumatic to microbase equipment and system control.
2. A shift from field-engineered and installed automatic temperature controls to pre-engineered and installed "smart" microbase system and equipment controls.

Many exploratory interviews were conducted with a large number of owners and system designers that provided interesting insight into a few more industry trends.

In the past, building automation and energy management were pretty much limited in application to buildings of 100,000 square feet or larger. And now, because of the cost of energy, some form of energy management is being applied to virtually every building constructed today. . . fast food, branch banks, etc.

In building automation systems being furnished today, there is a strong trend toward **distributive processing and control.** Said another way, the intelligence in the system is moving closer and closer to the end device being controlled. There is also a trend toward the use of **direct digital** control where the control is in the software rather than the hardware. It was also the general opinion of the designers interviewed that a natural separation **exists** for control software location, where that necessary for standalone equipment operation must reside at the equipment level and that required for global functions on the building should reside at the host or central computer in building automation systems; global functions being those things that affect the total building. . . demand limiting, etc.

The reasons for placing equipment control algorithms at the equipment level were given as:

- **The reliability increase with factory pre-engineered and installed specific control matched to specific equipment involved.**
- Standalone control capability if for some reason communication was lost to the host computer.
- Standalone control capability where automation does not exist.

In the general building market, several characteristics were commonly listed by **owners** for the ideal comfort control and management system:

1. Quality environmental comfort control (including noise).
2. Reliability.
3. Low owning and operating costs.
4. Flexibility to accept changing tenant needs.
5. Simple to operate and easy to maintain.
6. Competitive first cost.
7. Availability of local service and **training** from a single source.

On the other hand, the **system designers** interviewed had a **little different perspective.** They listed these characteristics for a good comfort control system:

1. System reliability.
2. System cost effectiveness.
3. A system that leads to optimization.
4. Integrated but does not cross code boundaries and a system that combines only areas where a common design discipline exists.

5. A system that has application flexibility.
6. A system which is upward expandable.
7. A system that can interface with other building systems where necessary and desirable, i.e. fire and smoke control system interface with air handling equipment control.
8. A system from a supplier with good local ongoing support.

Given the significant building industry trends and these extensive customer criteria, the industry, in general, is currently embracing an interesting system concept which they call the Integrated Comfort system. Basically, this new concept integrates the heating, ventilating and air conditioning equipment with automatic temperature control and the building comfort and energy management system, all within a common design discipline and preferably from a single source whenever possible. But certainly under single responsibility.

The system architecture is simple to understand and yet possesses a degree of application flexibility that allows it to be utilized on virtually any building type or size. It is based on the fundamental principle that there are three general levels of comfort control:

1. Building level control.
2. System level control.
3. Equipment level control.

At the equipment control level, the thing that sets microbase control apart from pneumatic or analog electrical is the processing of the input data and the development of the output control logic which takes place in accordance with a program within the circuitry of the microprocessor itself. Because of this, control logic changes are possible by simply changing the software program within the microprocessor.

A good example of a microbase controller is a standalone device that is dedicated to control the functions of an air handler.

It is a **programmable, unit mounted, standalone unit level equipment controller** with upward communication capability to a BAS. As such, it is referred to as a "unit control module" or UCM.

Each control loop within the equipment controller is then an assembly of equipment associated with the task of maintaining a controlled variable at setpoint. And each control loop includes a sensor, a controller with setpoint and a controlled device such as an actuator. The function of the **equipment control system** is to hold the controlled variable, such as discharge air temperature, at a value that corresponds to the control setpoint.

In the case of an air handler used in a variable air volume system, the unit mounted control will typically have several control loops contained within it and whenever the digital controller is located within the closed control loops, **where it exercises direct control,** the configuration is then termed direct digital control (DDC). Said another way, the control logic is contained within the software in the digital controller.

In a **VAV air handler,** the three closed control loops contained within the unit mounted control would be:

1. The supply air temperature control loop.
2. The discharge static pressure loop.
3. The inlet mixed air temperature loop.

Here is the **real thing!** Factory **prewired** and **mounted.** However, while this establishes the air handler operating modes and controls the air temperatures and duct static pressure in accordance with the program setpoints and algorithms, it can do nothing to integrate the air handler into the remainder of the system other than provide a communication interface to a higher level of authority such as a system level controller.

It's a fact that many systems, of necessity, contain multiple pieces of equipment of the same general type, each having its own unit control module for individual control. Good examples of this are the variable air volume system which must, by necessity, contain several variable air volume control terminals or air valves.

Another example is the application requiring multiple chilled water machines on the same chilled water loop. In installations of this type, it is generally cost effective to consider an intermediate controller which can be classified as a system level control. In the case of the variable air volume system, the system level controller would control several individual variable air volume unit control terminals with their individual unit control modules. It essentially functions as a data gathering station to free up the data link and CPU.

In the case of a multiple chiller installation, a system level controller is required to control the sequencing of the multiple chilled water machines in an optimum manner to match the system loads. These system level controllers will generally have standalone control capability and will be field programmable.

When we consider an air conditioning system that consists of several kinds of units, each with its own individual unit control module, **with or without system level controllers,** a higher level of authority is required to draw these units together to participate in a programmed building control strategy. In this case, the higher level of authority is a building automating computer. This controller is then connected via a twisted wire pair, or serial communications data link to the various system level or unit control modules for the various pieces of equipment. In the case of a chilled water variable air volume system, the building level controller would be connected via twisted wire pair, or communication data link, to the chiller system control, the air handler unit control module and the variable air volume system level controller.

The communication data link would then provide two-way serial communication, enabling the computer to issue commands **to** and receive data **from** the unit control modules or system level controllers **within** each of the three independent control loops.

The function, then, of the building level controller is to act as a **central command post** for all of the system level and unit level controllers and to provide programmed instructions covering, for example, start-up sequences and time keeping plus other global functions for the overall building control. As such, it keeps track of the days of the week, including holidays, and the programmed occupied and unoccupied hours for each. In addition, it stores zone morning cool-down time requirements averaged over the past several days. Then, knowing the occupied setpoint for the subject zone, the start-up time is advanced to achieve that temperature at time of occupancy in the most optimum manner.

Some of the other functions performed by the building level controller are global functions. These include:

- Night setback.
- Morning warm-up.
- Night setup.
- Night economizing.
- Override of schedule.
- Demand limiting.
- Diagnostics and alarm.
- Duty cycling.

Of all of the capabilities of the building level controller, perhaps one of the greatest is that of communication. Through the use of a personal computer and printer, system temperatures, pressures, equipment operating status and any other sensed conditions communicated through the unit control modules within the remote control loops to the building level controller, can be displayed on the CRT and printed out as reports.

This not only gives the building operator a chance to review the operating status of the system and equipment, but also enables him to communicate through the personal computer. He can also key setpoint or algorithm changes to change equipment and system operation for optimum building performance.

In addition to those functions already mentioned, it's important to note that the building automating computer is not limited to the control of air conditioning functions exclusively. It can also be programmed to manage things such as building lighting, building access, report writing and maintenance scheduling. When assigned these kinds of tasks, the computer literally becomes the heart of a building management system.

A name commonly used for building level controllers is a building automation system. As such, it integrates or merges the air conditioning and other building functions into an overall system. This performance is then managed automatically in accordance with the customized programming instructions.

In earlier discussions, it was mentioned that integrated comfort systems are being applied in virtually any size of building across a broad spectrum of building types or classifications. As such, they come in a variety of sizes and types.

Integrated Comfort systems utilizing small BAS panels provide an inexpensive standalone building management system with anywhere from 5 to 50 points of control. This system is typically suitable for building sizes ranging from 3,000 to 30,000 square feet. It provides an ideal system for multiple location chain store applications, branch banks or fast food establishments.

A good one may offer a self-prompting, easy-to-use operator interface with an integral operator keypad. It may typically feature twisted wire communication capability with products such as mid-range rooftop air conditioning units and may also feature remote telephone communication ability and auto-dial alarm capability. A good system of this size will also generate building management reports and feature building management capabilities such as:

- Time of day scheduling.
- Tenant timed override.
- Demand limiting.
- Duty cycling.
- Optimum start/stop.
- Trend log.

The next incremental size of building level controller or building management system will typically handle 50 to 100 points of control and may allow several controllers to be linked together with unit-to-unit communication. This system is ideal for buildings ranging in size from 30,000 to 300,000 square feet. It, too, will usually have a self-prompting easy-to-use operator interface.

As with the smaller building controllers, it will feature twisted wire communication capability with HVAC equipment, but because of its size and expandability, it will normally need to communicate with a larger array of equipment, including but not limited to:

- Centrifugal chillers and smaller 20-200 ton air and water-cooled recip, scroll and screw compressor water chillers.
- Air handling units.
- Variable air volume terminals, fan coils and unit ventilators.
- Packaged equipment such as large 20-105 ton rooftops, 20-60 ton vertical self-contained equipment, 5-20 ton rooftop equipment and water-source heat pump equipment.

It will also typically feature remote telephone communication and alarm auto-dial-out capability. Its building management capabilities will usually include as a minimum:

- Time of day scheduling.
- Tenant timed override.
- Demand limiting.
- Duty cycling.
- Optimum start/stop.
- Reports and logs including energy usage, after-hour usage, preventive maintenance, zone temperature and system status.
- Remove monitoring.
- May have color graphics.

The top end of the building controller family is usually a large central station building automation system which includes intelligent remote panels that communicate with powerful central processing equip ment. This system usually has an easy-to-use operator interface which is menu driven and normally features expanded color graphics. It typically can handle up to 10,000 or more points of control and will provide precise comfort monitoring and control for individual zones through DDC/VAV. It will usually have several communication channels and auto-dial alarm capabilities. It will also have extensive equation processing capabilities for flexibility and expandability.

Beyond this, it will usually include the ability to provide comprehensive building management information such as scheduling and tracking of after-hour tenant use and daily, monthly and yearly reports. In addition, it must, of course, have the Integrated Comfort link with a large variety of heating, ventilating and air conditioning equipment types to greatly simplify installation and provide increased cost effectiveness and reliability.

With the smaller and intermediate sized building control systems, we are also seeing a trend toward use of what we call a building management network which features networking of software for building controllers. It is typically an IBM personal computer base system which uses nondedicated lines and telephone modems. It normally should feature a self-prompting, easy-to-use operator interface and automatic dial out of alarms, advance remote monitoring and diagnostics, preventive maintenance, nighttime dial up of alarms and may feature capability for direct connection to a large centrally located mainframe like an IBM 3033 for:

- Electronic wires.
- Electronic bulletins.
- Guaranteed updates downloaded electronically from the large mainframe located remotely.

In summary, then, these products form the key top end building controller component of an integrated comfort system.

Thus far, we have explained essentially what an integrated comfort system is and the various levels of control associated with it.

Its real claim to fame is the fact that not only is it a more reliable building management and control system when compared to conventional, pneumatic or electromechanical control systems with building automation, but it's also very cost competitive when compared against those same types of conventional systems. So, in the true sense, the owner is getting more capability with more features in a more reliable system for equal or lower installed cost.

We all know the old saying, you get what you pay for. But with the ICS concept, on a comparative basis, you usually get much more for equal or less installed cost dollars. Let's look at some of the reasoning behind this.

- Factory pre-engineered and installed unit control modules to protect and operate the equipment on a standalone basis. You've essentially eliminated a great deal of automatic control field engineering and installation costs. It's done at the factory!
- The use of the daisy chained single twisted pair communication data link in lieu of individual home run wiring from the building control panels to the individual control points on equipment greatly reduces field wiring and installation costs.
- You get factory matched pre-engineered automatic microprocessor based electronic controls which are more reliable and from a single source, which eliminates costly field delays in trying to coordinate the installation process with various field disciplines.
- You get the possibility that field balancing and tuning of the HVAC system can essentially be done from a central location over the serial communication data link which substantially reduces the jobsite project engineering and checkout labor.
- You get remote monitoring diagnostic capability that allows problem solving from your office rather than at the jobsite, generally in a manner that is much more organized and cost effective. This greatly reduces call back labor costs.

To put the installed cost of building management in proper perspective, let's look at general office building construction costs. The cost of an office building today will probably run somewhere between $80 and $100 per square foot. The HVAC system installed cost will be somewhere between $6 and $10 per square foot, depending upon the system complexity. The temperature controls on this type of building typically run somewhere between 30 cents and a dollar per square foot, again depending upon the complexity of the controls. The building management system will typically cost between 30 cents and 50 cents per square foot on an installed cost basis.

Typical utility costs on a building per year are approximately $1.85 per square foot. The energy cost savings normally associated with the building management system would be 10 to 15 percent or somewhere between 19 and 28 cents per square foot per year which, in and by itself, would pay for the building management system in an approximate two-year time frame.

In the past, this would be the typical economics utilized to justify the use of a building management system. While these figures are still considered important, they by no means outline the total true advantages associated with a state-of-the-art integrated system.

I previously stated that we are living in a global economy where we are currently undergoing a **"productivity crunch" in order to remain competitive in the global market.** Recent BOSTI studies have indicated that poor environmental control in office buildings can result in a typical productivity loss of between two and four percent. If we look at an average occupant salary of $15,000 per year and an average occupant space of 150 square feet, we arrive at a cost of $100 per square foot per year. This translates to a productivity gain of $2 to $4 per square foot per year in office buildings with a system that provides clearly superior comfort control capabilities.

Beyond this, in recent owner surveys, the lack of comfort was a factor in approximately 50 percent of all tenant moves. Toan owner, the loss of a tenant is a major factor in the profitability of his building. Let's examine tenant retention and its effect on a 100,000 square-foot office building. The average tenant size in a building of this type is approximately 7,000 square feet and a typical lease rate for that space in a large city would be $20 a square foot per year.

The effect of a lost tenant will typically translate to a vacancy for that space of approximately six months and necessary modifications to accommodate a new tenant of approximately $5 a square foot. In addition, the brokerage fees would typically be $4-5 a square foot and probably, in this economy, rent concessions to get a new tenant to sign a lease will typically be about six months. This translates to an average loss on the building of $210,000 or $2.10 a square foot in a 100,000 square-foot building that good comfort control may have avoided.

Many more astute owners are finding today that the ability to account and bill for after-hour usage of their building beyond the hours agreed to on the leasing agreement can substantially increase the cash flow of their buildings. With an Integrated Comfort system utilizing building management networking, after-hour tenant tracing is relatively easy to accomplish.

On a 100,000 square-foot building, again using an average tenant size of 7,000 square feet, with a rate of $25 per hour for after-hour use, an additional billing per year would come to something like $91,000, or 91 cents per square foot per year. This assumes that 50 percent of the tenants will use the building for an additional two hours per day, five days a week and 52 weeks per year, and will require comfort control during that time frame.

In summary, here are some of the key benefits, as we see them, on the Integrated Comfort system concept:

- HVAC experience: HVAC equipment suppliers have HVAC equipment and system control experience. **We've been doing it for years!**
- ICS systems are simple systems: To install. To start up. To operate.
- With the ICS, the owner, engineer and contractor get single source responsibility for HVAC, ATC and BAS.
- They are assembled on the bench, so to speak. With this comes improved quality/reliability: Factory-engineered controls. Factory installation. Factory checkout. Simple field integration.
- Additional features/functions.
- Guaranteed system performance from the single source.
- Integral building management features with PC management networking.
- Lower installed cost.

And last, but not least, is the ability of the owner to utilize this new technology and resulting superior building control to market the building services to prospective tenants. If you're visiting Chicago and are in the vicinity of O'Hare Field, you'll see a billboard advertising Madison Plaza that states "You'll admire her beauty. You'll love her brains." Here is the real building!

This is Integrated Comfort system technology in action as a building marketing tool and I think clearly illustrates building differentiation in a competitive marketplace.

I'd like to leave you with a thought for the day. "There is nothing in this world today more expensive to own and operate than a building with a cheap, inefficient environmental comfort control system." More astute owners are finding that for a relatively minor additional frontend investment in a better controlled system, the returns on that investment are most gratifying.

Chapter 39

HVAC EQUIPMENT REPLACEMENT STUDY—ENERGY SAVINGS THREE WAYS

D. S. Teji

ABSTRACT

For many years much of the Energy Conservation efforts have been concentrated towards no cost/ low cost Energy Conservation measures. Also, in accordance with the availability of funds, major Energy Conservation retrofit projects are also being implemented. Looking at other areas of Energy Conservation, the City of Phoenix thought of replacement of HVAC equipment with higher efficiency equipment. Equipment manufacturers, in line with energy saving trends and through research and development, have improved the efficiencies of their products.

This study on the equipment replacement was conducted under the Federally funded Urban Consortium Energy Task Force program. As a result of the study, it was found after extensive testing that replacement of HVAC equipment could benefit the City three ways. The primary objective of this study was to survey all the old HVAC equipment in 82 City owned buildings for their performance standard and compare with the currently available higher efficiency equipment.

Secondly, make an assessment of the aging effect on the equipment as to the extent the efficiency has further deteriorated. At the time of original system design, because of very cheap utility rate, energy cost was not a consideration. Consequently, liberal allowances and safety factors were provided resulting in much over sized equipment. Thirdly this fact provided scope for energy savings by capacity optimization in accordance with the building cooling and heating loads. In evaluation of the economics of equipment replacement, the use of equipment run hours by assumption method was an error. On actual testing through lapse meters the results were astonishingly favorable for replacement justification.

As a side benefit, during testing operation, certain areas of poor maintenance came to light. While considering the replacement equipment some of the new energy saving strategies such as thermal energy storage were considered. On the whole the equipment replacement program turned out to be much more beneficial than it was originally considered. A detailed description of the study is presented in this paper.

BACKGROUND

The City of Phoenix owns and maintains approximately 250 buildings occupying 2.5 million square feet of air conditioned area. The older buildings age range between 10 to 25 years approximately.

The age of some of the equipment falls under this range. The major energy consumption in these buildings is from lighting and air conditioning system equipment. The type of equipment used for heating, ventilating and air conditioning system is a wide spectrum. The equipment considered in this study included chillers, boilers, pumps, fans, DX air conditioning units and heat pumps and heating units. The City of Phoenix had completed the energy audit of approximately 100 buildings under the U.S. Department of Energy Technical Assistance program. This study provided wide range opportunities for energy savings through no cost/low cost energy conservation projects. The City of Phoenix has also to its credit successful completion of retrofit projects such as variable air volume system, capacity optimization and Thermal Energy Storage system. These projects were implemented through the Urban Consortium Energy Task force (UCETF) programs. Replacement of HVAC equipment with more efficient ones is the current project in hand through the UCETF and is scheduled for completion by December 31, 1988.

Urban Consortium is both a network of the nation's 43 largest cities and counties and a partnership of major players within these jurisdictions, focusing collective talents and resources on group research efforts and solutions transfer initiatives. The City of Phoenix has so far completed seven UCETF projects and has the eighth one currently in progress.

THE PROJECT PURPOSE - SAVINGS THREE WAYS

This project was proposed with the objective of analyzing the HVAC equipment for replacement with the ones at higher efficiency with a reasonable payback. There are three anticipated factors associated with the equipment replacement program.

Replacement of the equipment with higher efficiency

The HVAC equipment components (compressors, boilers, pumps, fans and motors) typically have long lifetimes, some lasting for more than twenty five years. Many technological advancements have been achieved in the area of HVAC equipment efficiency, increasing it quite substantially from that of the equipment available ten to twenty five years ago. The need for higher efficiency arose, of course, from the effort to conserve energy. Much of this equipment was manufactured at a time when energy cost to operate this equipment was of secondary importance. Since then the energy rate has increased enough to receive management's attention.

Capacity Optimization

The standard engineering practice was to size the compressors, pumps and fans and piping to satisfy the maximum heating and cooling loads. The average heating, ventilating, and air conditioning (HVAC) system generally operates at full capacity less than three percent and at a minimum capacity less than five percent of the heating and cooling seasons. The systems usually operate between 50 and 60 percent of their rated capacity 50 percent of the time. Also there are several allowances inherent in the original design. In the current analysis the designer could design the capacity of the equipment using the newly calculated loads, selecting the lower instantaneous loads. Moreover the oversized equipment that operates below its design capacity can be very inefficient.

Aging Effect

The efficiency printed on the label is the efficiency of the equipment when new. As the years of operation pass by the efficiency deteriorates to even lower ratings. This factor provides greater opportunity for equipment replacement, also affecting the payback. To determine if HVAC equipment does in fact operate less efficient with time, a sampling of buildings were tested for efficiency. A description of the testing method is described elsewhere in the report.

ACTUAL STUDY PLAN

The preliminary survey and major study was contracted out to a consulting Engineering company. The study plan consisted of the following elements:

- Provide equipment data for all HVAC equipment including chillers, boilers, unitary air conditioning units, heat pumps, fans, pumps and cooling towers in 82 City owned buildings.

- Provide a capacity optimization analysis, comparing the size of the existing equipment with the actual (analyzed) capacity required for chillers, fans, pumps and boilers.

- Provide an efficiency analysis, comparing the state-of-the-art higher efficiency equipment currently available in the market.

- Provide an economic analysis for all equipment considered as candidates for replacement because of their age. The economical analysis includes the installed cost of the replacement equipment, with the highest efficiency of an optimized size, estimated annual savings and the resulting payback period.

THE SURVEY

The data gathered for the 82 buildings included: conditioned square feet area, estimated occupation in hours per year, air conditioning capacity in refrigeration tons, type of HVAC equipment, type of energy management system (if any) and type of controls.

Because of space limitations, tables containing building and equipment date, optimized capacity and recommended replacement equipment for all buildings and equipment is not included in this paper. However, the type of information contained in these tables is stated as follows. These lists are repeated for each type of equipment such as remote units, chillers, boilers, pumps, fans, other heating units and cooling towers. A typical table for a remote unit is reproduced for easy comprehension.

1. Building No.
2. Building Name
3. Make
4. Model No.
5. Equipment Start Date
6. R. Tons
7. Unit EER/SEER
8. General Condition
9. Estimated Run Hours
10. Recommended Unit Tons
11. Replacement Make/Model
12. Replacement SEER/EER
13. Estimated Installed Cost
14. Energy Savings KWH/Year
15. Payback in Years
16. Unit Type
17. Area Served
18. Unit Serial No.
19. Building Location
20. Equipment Location
21. No. of Compressors
22. Compressor Make/Model
23. Compressor Volts/Ph.
24. Compressor Amps
25. Refrigerant
26. No. of Evap. Fans
27. E. Fans Amps
28. E. Fans H.P.
29. E. Fans RPM
30. No. Of Condensor Fans
31. Condensor Fans Amps
32. Condensor Fans H.P.
33. Condensor Fans RPM
34. Control Volts
35. Precooler
36. Comments

ANALYSIS/RECOMMENDATIONS

Chillers

The survey and analysis of the existing chillers included those ten years and older. The study concluded that the best payback would be accomplished by reusing the shell of one of the existing 400 ton chillers at the Municipal Building and replacing the compressor and motor with a 200 ton compressor and motor. This can be done with a payback of about four years. The reciprocating chillers would not be cost effective to replace because the efficiency of these chillers has not changed significantly.

Boilers

Due to low annual heating degree days in Phoenix area there are very few boilers used in City buildings. The capacity optimization was performed by using Trakload computer simulations where necessary. One typical boiler in the Municipal Building was recommended for replacement with five 150,000 BTUH gas-fired pulse combustion boilers.

Pumps

All HVAC pumps with motors two horsepower and above were analyzed. Pumps include chilled water, hot water, condenser water and combination pumps. Actual brake horse power (BHP) was calculated with the formula:

$$BHP = \frac{3xExIx\ P.F.xEm}{746} \quad \text{(for three phase)}$$

Where E = line voltage
I = measured current flow
PF = power factor (.95 for VFD motors .8 for others)
Em = efficiency of the motor

Estimated pump efficiencies were calculated using the formula:

$$E = \frac{Q \times H}{3960 \times BHP} \times 100$$

Where
Q = measured flow rate (gpm)
H = measured total head (feet)
BHP = break horse power

The brake horse power savings that can be achieved by connecting to variable frequency drives (VFD) were calculated using the formula:

$$BHP2 = \left(\frac{GPM2}{GPM1}\right)^3 \times BHP1$$

Where suffix 1 and 2 represent actual and design values.

Fans

All fans with motors two horsepower or greater were analyzed in a similar fashion to the pumps. Similar to the pumps, the energy savings from the fans using VFD's were calculated using the formula:

$$HP2 = \left(\frac{CFM1}{CFM2}\right)^3 \times HP1$$

Where suffix 1 and 2 refer to optimal and design values.

Energy savings were calculated using the equation:

$$\text{Savings KWH/yr} = \frac{\text{HP saved x fan run hours/yr}}{\text{motor efficiency}}$$

The study showed that most of the HVAC supply and return fans have already been optimized with the VFD's which automatically vary the air flow corresponding to the actual building load.

Other Heating Units

It was recommended that Natural Gas furnaces be replaced with new gas-fired pulse combustion furnaces at the end of their useful life.

Dx Air Conditioning Units and Heat Pumps

The evaluation was done on the units of two ton and larger capacity. The study indicated good potential for replacement on all units five years and older. In the analysis the estimated compressor equivalent full load run hours in the cooling mode were obtained from one of two methods.

In method one the cooling kilowatt hour as recorded during the Technical Assistance (Energy Audit) process was used in calculating the full load hours.

$$\text{Unit equivalent F.L. hours} = \text{Bldg cooling kwh/yr} \times \frac{\text{Unit EER or SEER}}{\text{total bldg tons x 12}}$$

In method two where no Technical Assistance reports were available, a computer simulation program TRAKLOAD was used to estimate equivalent full load run hours.

The evaluation of energy savings (kwh/year) by replacement of the existing unit with the higher efficiency was done using the equation:

$$S = H \times T \times 12 \quad \frac{E2 - E1}{E1 \times E2}$$

Where S = Energy savings (kwh/yr)
H = Unit run hours/year
T = Existing unit nominal Tons
E1, E2 = Existing unit EER and Rpl. unit EER respectively.

Electric rate of $.07 for large downtown buildings $.09 for medium size and $.11 per kwh for smaller buildings was used.

The resulting payback and savings on typical small building is shown in the table.

EQUIPMENT RUN TIME ERROR

Looking at the estimated run time used by the consultant the Energy Conservation Office felt that these were under estimated and which resulted in longer payback for replacement units. Consequently an actual run time testing program was initiated. This testing program was designed to produce actual documented data to replace the consultants estimated data. Elapsed run time meters were installed on eighteen air conditioning units throughout the City.

Readings were taken on a monthly basis. The run time hours derived from the elapsed meters were rightfully assumed as full load run hours. As is apparent from the table 1 the measured run hours were much higher than the estimated figures. Consequently greater number of air conditioning units became potential candidates for replacement on the basis of greater savings with reasonable payback periods.

ACTUAL VS. NAMEPLATE EFFICIENCY

Another aspect of the Consultant's data that caused concern was the Energy Efficiency Ratings. The Energy Efficiency Ratings used were on the basis of new equipment when installed ten to twenty five years ago. It is reasonable to assume, however, that the equipment has undergone some degree of degradation during operation for a number of years. It was therefore, decided to determine the effect of this degradation on the efficiency. A simple formula is used to define the efficiency of the equipment.

$$\text{Energy Efficiency Ratio} = \frac{\text{Net Cooling Capacity (BTU/hr}}{\text{Electric Input (watts)}}$$

Published values are calculated at designated operating conditions, 95 degrees F Dry Bulb air on the condensor, 80 degrees F Dry Bulb and 67 degrees F Wet Bulb on the cooling coil (evaporator), and 40 degrees F evaporating temperature. Net cooling capacity is calculated using the formula

$$Q = F \times T1 - T2 \times M$$

Where Q = Net cooling capacity
F = Air or water flow (CFM or GPM)
T1 = Inlet temperature Deg. F
T2 = Outlet temperature Deg. F
M = Conversion factor

CONSULTANT STUDY							ACTUAL TEST			
BLDG	TONS	PUB EER	REPLC EER	RUN TIME	KWH/YR SAV	PYBK YRS	PUB EER	RUN TIME	KWH/YR SAV	PYBK YRS
A	10.0	8.2	9.0	630	1261	26.5	4.75	2136	25,482	2.6
A	7.5	8.7	9.3	622	541	61.6	7.00	2187	6,954	7.0
B	5.0	5.6	9.0	669	2709	12.8	4.80	2307	13,458	2.6
C	7.5	7.6	9.2	1651	3400	16.4	3.95	2516	32,714	1.0
D	4.0	7.7	13.2	2468	6410	3.7	6.90	3074	10,206	2.3
D	2.5	7.7	11.9	1437	1961	7.3	7.80	2498	3,284	4.4
H	4.0	7.0	13.2	1636	5268	4.5	6.00	2598	11,337	2.1
H	2.5	6.9	11.9	1612	2928	4.9	5.90	2446	6,245	2.3
E	4.0	7.0	13.2	2044	6584	3.6	5.60	2044	10,087	2.3
F	5.0	6.3	8.6	1972	5024	3.1	4.50	1972	12,535	1.2
F	5.0	6.3	8.6	1972	5024	3.1	5.80	1972	6,642	2.3
G	5.0	6.3	8.6	1508	3841	4.0	5.40	2000	8,269	1.9
G	5.0	6.3	8.6	1508	3841	4.0	4.40	2000	13,319	1.2
J	4.0	7.7	13.2	1632	4240	5.6	7.00	2076	6,686	3.5
J	2.5	7.7	11.9	1632	2227	6.4	7.00	2770	4,859	2.9
K	10.0	6.6	9.0	710	3443	19.3	5.60	1939	15,697	4.2

Table 1 Comparison of Consultant and Test Results

A two weeks testing period was selected. Testing was conducted as follows. Air flow rates were obtained using the duct flowhood. An ultrasound flow measuring instrument was used to record the water flow rate in chilled water systems. Supply and return air temperatures were taken with an electronic analyzer. A handheld kilowatt meter was used to record kilowatt usage. The result of these tests and the new run hours produced astonishing results and brought several units of equipment within the acceptable range of payback for replacement. Table 1 shows the difference between the Consultants study and the results after the tests. A more extensive testing program has since, begun for all city buildings.

CONCLUSION

The limited test results indicate that energy savings are substantially increased when accurate run times and actual Energy Efficiency ratings are used.

The testing program can also prove valuable for equipment maintenance purposes. During the testing process in a few buildings several maintenance problems, such as pinched closed flexible duct and clogged filter were found. It is the intention of the Energy Conservation Office to expand the testing program to include testing of all 82 buildings and beyond.

ACKNOWLEDGEMENTS

The author is thankful to the U.S. Department of Energy for providing the grant which made this project possible. Thanks are also due to the chairman of the Urban Consortium Energy Task Force for selection of City of Phoenix for the grant award to do this project. The author acknowledges the splendid work done by the Energy Conservation staff with utmost professionalism and enthusiasm.

0813e

Chapter 40

VARIABLE AIR VOLUME SYSTEMS—THE BUILDING OWNER'S PERSPECTIVE

M. Heis

The purpose of a variable air volume (VAV) system is to provide the correct volume of air to building occupants at a constant temperature. Changing conditions in a space require varying amounts of air conditioning. With constant volume systems, the temperature of the air is changed. Often this requires the air to be cooled and then reheated because the volume to the space remains constant. Simultaneous heating and cooling is common with constant volume systems. Normally all of the air is cooled at the air handling units and reheated or mixed with warm air in the individual zones. This is true for all seasons.

Constant volume requires the fan to operate at the same horsepower continuously. VAV systems allow fans to run at reduced volumes when the cooling load is reduced. In addition, diversity is normally used to design the original VAV fan system. Diversity becomes a factor in VAV because it is assumed that less than 100% of the VAV boxes will ever be open at the same time. Not only is the supply fan sized smaller in VAV; the supply fan will run at reduced loads for long hours.

VAV systems are very good selections for office buildings and classroom application. The volume of air will automatically adjust in response to the changing loads from lights, people and the solar heat gain. Varying occupancy rates and times provide for long periods of reduced cooling loads. Classroom schedules and class size provide excellent savings opportunities using VAV systems. Loads in offices will normally vary to a lesser degree but still provide significant opportunities. Vacant rooms with the lights out require much lower volumes than crowded meeting rooms. Perimeter areas will have varying amounts of solar gain during the course of a day. Varying loads demand variable volumes of constant temperature air.

ADVANTAGES OF VAV SYSTEMS

1. Load diversity of 20-30% allows for smaller air volumes, smaller supply fans, smaller heating and cooling equipment.

2. Lower installed cost for HVAC equipment because of diversity factor.

3. Significant energy savings for HVAC equipment.

4. Close temperature control in zones at a lower cost.

5. Tolerant of minor air balancing problems.

6. Lower cost for the addition of zones to an existing system.

7. Lower maintenance costs including fewer air filter changes.

8. Quieter operation at reduced loads than constant volume systems.

9. Quick and easy change-over from winter to summer operation. VAV boxes close for heat.

10. Fewer draft problems and increased occupant comfort.

BASIC TYPES OF VAV

Cooling only VAV with perimeter heating: The perimeter heating would normally be radiation heat. In this system, the VAV system provides cooling during all seasons of the year. When cooling is not required, the VAV box closes to a minimum. When enough boxes close, the supply air static pressure raises and the supply fan backs off reducing fan horsepower and static. During winter months, it would be common for the interior VAV boxes to be open and the exterior boxes closed.

Interior cooling only VAV with perimeter VAV reheats: In this system, the interior zones are cooling only. The perimeter zones are both heated and cooled by the VAV system. In the cooling mode, the VAV boxes open for cooling and close off to a minimum for heating. When heat is required, the perimeter VAV reheat units heat the discharge air at the VAV box. Perimeter reheat boxes have a minimum designed air flow when heating to provide adequate air flow to deliver the heat to the space.

Interior VAV cooling with constant volume perimeter: In this system, a VAV system provides cooling for the interior zones. The perimeter is served by a constant volume, variable temperature.

Interior VAV system with dual duct VAV perimeter: The interior is served by a cooling only VAV system. The perimeter is served by a dual duct VAV system. VAV is desirable on both the hot and cold decks to obtain fan savings on both systems. Separate fans for the hot and cold decks also allows for the use of return and outside air. During summer operation, return air can be used for the hot deck. During the winter months, outside air can be used for the cold deck.

Interior cooling only VAV with terminal air blender perimeter: Once again, the interior is served by a cooling only VAV. The perimeter uses a fan powered air blender unit. When the cooling load drops off, the blender reduces cooled air and introduces return air from the ceiling to blend with the cooled air. The central fan load is reduced while the volume of air to the space is constant. When additional cooling is called for, more cool air is used from the central fan and less return air is mixed with it.

Interior cooling only VAV with change-over VAV perimeter: The perimeter VAV boxes are either provided hot or cold air by the single duct system. Heating and cooling is done at the central fan. The VAV boxes must be switched from the summer to the winter mode. The interior is served by a cooling only VAV system. With this system, only heating or only cooling is available to the perimeter areas.

Bypass VAV systems using one of the above perimeter heating systems: The system incorporates a pressure-dependent damper which allows supply air to bypass to the return air plenum when the space is not calling for cooling. On a call for cooling, the damper opens to the space and closes to the plenum. There is no savings on fan horsepower with this type of system. The savings is in heating and cooling costs.

HVAC drawings for VAV systems should clearly indicate CFM ratings for all VAV boxes and diffusers. The final air balance report should be checked against design. Any discrepancy should be resolved before the system is accepted. Proper air balancing should provide for maximum human comfort and efficient system operation. VAV systems are somewhat tolerant of air balancing problems. To some extent, VAV systems are self-balancing if proper static pressure can be maintained.

Newly installed systems should be carefully inspected for poor workmanship and defective components. Each thermostat should be checked for proper operation. Each box should have a maximum and minimum CFM setting and this should be checked. (The design engineer should be specific as to CFM settings for all boxes and diffusers). All ductwork, including ductwork above the ceiling, should be inspected for proper installation and air leaks. Static pressure should be checked against design specifications. Supply fans should be checked for proper operation and their ability to maintain proper static pressure under varying loads.

VAV systems are easy to maintain and service when properly designed and installed. Each building will have multiple units that are identical. A stock of spare parts should be kept. At least one maintenance person should be trained to test and service VAV Boxes and terminals. Most problems with VAV boxes involve either the thermostat or some other part of the control system. In addition to the VAV boxes, the supply fan and controls must be maintained. Static pressure control is critical in VAV systems. Only trained maintenance people should work on the static pressure and safety controls. Building static pressures will vary if the supply and return air system volumes vary. When a return air-exhaust air fan is used, its volume must be balanced with that of the supply air fan. Maintenance personnel must be aware of the importance of proper fan tracking and static pressure control.

Poor static pressure control can result in excessive duct pressures damaging the ductwork or fan. Excessive pressures result in excessive duct leakage and poor VAV box terminal performance. Higher than design pressures can also result in excessive noise levels. Fan instability can also result from poor static pressure control. The end result can be erratic terminal operation, damaged duct and lost system energy savings.

Fan safeties should be in place to protect against excessive pressures. The fan should be started unloaded. The duct pressure should be limited for several minutes after start-up to allow the controls to sense valid pressures. Overloads in the starter should be properly sized to protect the motor. Safeties must be working and checked. This will provide protection for the fan motor, ductwork, terminals and other components.

Fan energy savings depends to a large extent on the type of fan volume control used. Variable frequency/variable speed drives provide maximum savings at low fan BHP operation. Outlet damper control is the least efficient of the methods of fan air volume control. Variable pitch is more energy efficient than inlet vane control. Variable speed and variable fan pitch are very common in modern VAV systems. Both offer very good energy savings. With changing loads, the fan operates continuously at varying air flow conditions. It is extremely important that the appropriate fan and fan control be selected to insure stable and efficient system operation.

Air diffusers are often overlooked in VAV systems. Proper selection of air distributing diffusers is critical. The diffuser must provide acceptable room air motion. A common complaint with VAV systems is that the floors are "stuffy". Another consideration with air diffusers is sound level. They must provide acceptable air motion with an acceptable sound level. Air distribution is the third consideration. The diffusers must have the ability to provide acceptable air distribution over the variable volume range provided by the system.

Successful VAV performance depends on proper design, installation and maintenance of the system. The first step is a good understanding of VAV systems and their selection. Most vendors of VAV equipment and controls can provide you with good printed matter to read on the theory and application of VAV. Maintenance personnel should have manufacturer engineering and maintenance information on the equipment to be maintained. Every component is important in a VAV system.

Chapter 41

ENERGY CONSERVATION VS. COMFORT AT MILITARY INSTALLATIONS: MUST THERE BE A CONFLICT?

T. R. Todd, M. E. Pate, III

INTRODUCTION

This paper is based upon the authors' experience, and that of their firm, in performing energy conservation studies, designs, construction management, testing and balancing and follow-up up services in over 2500 buildings and industrial facilities for a variety of public and private sector clients.

It has been observed in the course of this work that energy conservation is approached on two fundamentally different levels. The first, and lower level of the two, concentrates on "raise the thermostat setting" type recommendations. These almost always result in a sacrifice of occupant comfort. The second, and higher level of the two, concentrates more on making the buildings work as they should. In doing this, one almost always finds that energy conservation is achieved at the same time that comfort is being improved. This paper attempts to illuminate the difference between these two approaches by giving examples.

It must also be recognized that, although utility costs are important, they generally are much smaller than those costs associated with personnel. In the case of a typical administrative building utility costs may be on the order of $1 to $3/SF-YR. Personnel costs, on the other hand will likely be more in the $200 to $300/SF-YR range. It's clear that a "raise the thermostat" recommendation that reduces utility cost even 50%, but in the process results in an even 1% decrease in productivity (e.g., a few extra trips to the water fountain each day), would be a very poor investment indeed. Those responsible for controlling utility costs must be careful to consider <u>all</u> of the effects of their actions. A major goal of this paper is to point out that the best recommendations will almost invariably <u>improve</u> worker productivity (because they improve comfort conditions in the space).

THE DESIGN PROCESS

Design is, by its very nature, a problem in optimization in which the engineer's principal task is to reach an acceptable compromise between various conflicting factors. The three principal factors at play are the client's desire for a building that (1) is a comfortable and pleasant place in which the occupants can work or live, (2) within the first costs that budget has placed upon the project, and (3) inexpensive to operate and maintain. Energy efficiency, and the design decisions pertaining to it, is clearly one of the important elements in both first cost and operating cost. It is generally felt, in a negative sense, that it is also an important element in the comfort factor. In this paper the authors agree but contend that it can and should be viewed in a more positive manner.

Clearly there are tradeoffs that cannot be avoided. For instance, if an owner wishes to have the maximum comfort for his occupants, it will be necessary to zone the building extensively. Without question this will result in a higher first cost and, probably, a higher energy cost. This is unavoidable and is not the central issue of this paper. What this paper will contend, however, is that there are numerous ways that this zoning can be done. Of these, one tends to find that those which give the best comfort also tend to be the more energy efficient. This is the level of thought that experience has shown to frequently be missing in designs.

This paper concentrates on heating, ventilating and air conditioning (HVAC) issues. The reason for this is simple. These are almost always the most attractive retrofit opportunities available to the person doing an "energy study" of an existing building. They tend to have much better paybacks than building envelope options as an example. In addition, at the time of the original design, they tend to offer the greatest opportunity for improving the comfort and efficiency of a building with the smallest incremental first cost investment (often with no additional investment or even a savings). It is the authors' contention that, by ignoring some of the most basic principles of good design, these opportunities are often lost.

Without attempting to be all inclusive, the paper next gives examples of typical design decisions facing the new building or retrofit designer which emphasize the cases where the three design factors (i.e., comfort, first cost and operating/maintenance costs) are, and are not, in conflict.

"RAISE THE THERMOSTAT SETTING" MODIFICATIONS

First, consider the type of recommendations that normally come to mind when someone mentions energy conservation. Most people immediately think about being either too cold or too hot and not having enough light for their work. This level of recommendation misses the whole point and neglects the importance of comfort. In these recommendations energy conservation is achieved at the expense of comfort and, hence, productivity.

It is the authors' opinion that this negative approach is generally found when the persons performing the services are not sufficiently experienced in the technical issues to realize that far better options exist. These engineering intensive options tend to require technical skills beyond those of the analyst. No further consideration will be given to recommendations based upon this negative philosophy.

(Using "raise the thermostat setting" as symbolic of this category of negative options might cause confusion with the desirability of repairing controls to insure that whatever temperature is desired is, in fact, maintained. Often a control system will be in such a poor state of repair that it gives low cooling season, and high heating season, temperatures. When the system is repaired to control temperature to its set point, the occupants will often complain. This complaint arises from what the set point is rather than from the fact that the system is now controlling to it. Policy must be adjusted to reflect reality. One may think that he has been holding cooling season temperatures to 75 °F only to find that they were, in fact, 72 °F. When the controls are repaired the occupants may well be unhappy at 75 °F. A management decision must be made to determine what temperatures should actually be maintained.)

SYSTEM SIZING

Perhaps no other issue makes the interplay of comfort, first cost and operating and maintenance (O&M) costs as clear as that of sizing the HVAC equipment. To discuss it properly it is necessary to set the stage by examining in some detail the process through which the design engineer typically goes in selecting equipment for an HVAC system.

The first step is to calculate the heating and cooling loads that are expected for the facility. To do this the engineer must deal with some rather complex thermal questions. To illustrate this consider, as an example only, the cooling load calculation for the heat gain through walls and roof. (The situation is similar for other elements of the cooling load. The accuracy of infiltration estimates is even worse. While ASHRAE leads the designer to use methods based upon the tightness of the envelope, e.g., the "crack method," research has shown that as much as 80% of a building's infiltration is attributable to door openings.) The envelope is, in fact, a very complex thermal system. Since the information needed is the instantaneous load at some particular time and since the driving forces for heat transfer (i.e., ambient dry bulb temperature, solar load, etc.) vary in time, there is no avoiding the transient problem (in all of its messiness). ASHRAE (1) provides the standard for making these calculations. The methods most often used are the Transfer Function Method (usually with the aid of a computer) or the Cooling Load Temperature Differential, CLTD, Method and the associated Cooling Load Factor, CLF, Method which deals with fenestrations and internal loads (often done manually). The CLTD and CLF methods are, in fact, nothing more than a collection of results obtained by the Transfer Function Method. (These have replaced the older Total Equivalent Temperature Differential Method of the 1972 ASHRAE Fundamentals Volume.) To put things mildly, there are problems. Cases are frequently found in which the results of the calculations are inaccurate. This has led to the 1985 ASHRAE Fundamentals Volume (1) stating that RP-359, which had been sponsored to extend the range of cases for which CLTDs and CLFs were calculated, has reported that they have found an unexpected sensitivity to the weighting factors which are inherent in the method. The test then goes on to state: "These new developments, identified in 1984 and clearly beyond the scope of RP-359, eliminated the inclusion of recalculated data in this volume and made necessary the following caveats concerning appropriate use of this or alternate methods". The caveats then listed place a heavy requirement on the "engineering judgement" of the designer. The point in all of this is that the cooling load that a designer might calculate, using the best available methods, are estimates only. Designers have understandably responded by including a healthy safety factor onto a value that is, in all probability, very conservative.

Next, starting with a load that is conservative, the designer must determine how much water and air he wants to move. Having completed the psychrometric necessities he can then specify CFM and GPM, usually with another safety factor. He is then faced with estimating the pressure drops of the systems (air and water) at the chosen flows. This also is difficult to do with any degree of certainty. Predictablily, the designer again opts for a healthy safety factor. Finally, the designer must select equipment to do all of this. He calls a vendor (or looks at equipment cut sheets) and determines which pump or fan can safely be expected to do the job (another safety factor).

The designer next must select a heating or cooling coil to serve the load. The same process takes place. The designer takes his load, water and air flow values and asks a vendor to help him select a coil. The vendor looks at his coils and selects the next larger unit. Again, rounding up introduces another safety factor.

Set this in a time in which most AEs live in fear of professional liability claims and it is understandable that equipment gets oversized. Add to this the fact that the Department of Defense's current emphasis on "quality" has been interpreted as a need to pursue AE errors and omissions vigorously (e.g., recent orders by the Navy to their local OICCs) and it becomes even clearer. (Instructing AEs, as DoD does to design without safety factors does not do much to counter all of the other signals the AE receives daily from his client. It is an unrealistic instruction in the context of the design environment within which the engineer must work.) It should, then, certainly come as no surprise that when one evaluates 2500 buildings, as the authors have, he finds that HVAC equipment is on the average about twice the size that it needs to be! This is sad but true.

How does this tendency to "play it safe" affect comfort, first cost and energy consumption? First, addressing comfort, one can see that high water flows should have no appreciable negative impact on comfort. (With certain types of capacity control it may even improve dehumidification by keeping the coils cooler.) High air flows tend to make a space drafty and lead to occupant discomfort. Most observations of discomfort are associated with high, rather than low, air flows.

The real comfort concern associated with overcapacity is the system's ability to keep the desired dry bulb temperature in the space and, at the same time, give a room humidity ratio in a reasonable range. (Much of ESI's experience has been with the Navy. Many Naval facilities are in humid climates. This makes humidity control a major concern.) In the interest of conserving energy, the military design guidance (2,3) effectively preclude the use of any form of active humidity control. The authors have written extensively (4,5,6,7) concerning these issues and have coined the term "passive humidity control" to describe how well a system maintains the humidity in a space as the load varies over its normal range of part load conditions, without any active means of sensing and controlling humidity. FIGURE 1 shows that the sensible heat ratio (SHR) of the cooling provided by the unit generally increases as the system unloads. The SHR actually required by the space, on the other hand, almost always decreases at lower likely part loads. Room humidity, floating as a free variable, drifts upward whenever the amount of dehumidification provided is less than that needed. The room's humidity will reach an equilibrium at some elevated value. Now, think about the effect of oversizing a system. Supposed the system was twice the size it needed to be. This has the effect of moving everything to the left in FIGURE 1. The humidity at full load would be high. The humidity at a more normal part load (say 50% of actual design or 25% of installed capacity) would be higher still. These effects have been found to be significant and one of the major reasons that buildings are found to be uncomfortable. In Okinawa, admittedly a severe case, almost half of the large buildings have room relative humidities in excess of 70%.

While this discussion has presupposed chilled water, the same observations are made for direct expansion equipment. When a coil is first energized it is warm and begins to cool down. At first very little dehumidification occurs. FIGURE 2 shows the extreme sensitivity of a coil's latent capacity to the temperature of the refrigerant. Khattar, Ramanan and Swami (8) have investigated the start up performance of a typical DX system. FIGURE 3 is taken from their work and shows how sensible and latent cooling vary with time. If the system has been oversized the sensible load may become satisfied before any real latent cooling can be done, allowing the room humidity level to rise to an elevated equilibrium, and often unacceptable, value.

With respect to first cost, it is clear that big equipment costs more than small equipment. Oversizing, therefore, results in a high equipment first cost. Big equipment also takes up more space than small equipment and raises the first cost of the building by requiring a larger envelope. For example, oversizing air flows can often result in a greater floor to ceiling height than that which would be required if they had been selected properly. Large chillers and air handling units require additional mechanical room space.

Finally, consider the energy implications of oversizing equipment. The operating penalty on most primary equipment is not so bad. Chillers and DX machines unload rather efficiently. (Boilers are not as good.) Pumps and fans, however, are a different situation. If, based upon an overly conservative estimate of the peak required, a system is sized to pump an excessive GPM throughout a chilled water loop, it does so, not only at peak load, but also at all part loads. This adds up to a considerable waste. The same is true of the air moved. With the exception of variable air volume (VAV) systems, air handling units will move the design air flow any time the system is on. One generally finds that pumps and fans consume more energy in a building than all of the primary equipment (i.e., chillers and boilers) combined.

The disconcerting aspect of this is that it is often so unnecessary. While it may be difficult to get engineers to accept the risk of eliminating their safety factors, it is often very easy to make modifications after the equipment is installed. This can even be done in a new design as a part of the testing and balancing procedure spelled out in the specifications. Pump impellers can be trimmed and fan shieves changes to efficiently reduce capacity. Even sadder, the authors have seen innumerable cases where old, oversized equipment has been replaced "in kind" with new, oversized equipment. Whereas the original designer may have been faced with uncertainty about the loads, the replacement designer has the advantage of knowing that, for instance, the chiller never loads beyond 50%. By replacing "in kind" an opportunity for correcting the original overdesign is lost. The new equipment has all of the negative features of the original equipment. It costs more than necessary to buy and operate and it gives poor comfort.

Clearly, properly sizing equipment results in lower first cost, lower operating cost and better comfort and, hence, productivity. There is no optimization or trade off required. It is the opinion of the authors that the Department of Defense should make an more serious effort at getting designers to eliminate the excessive safety factors that are so common in DoD designs. To be effective, however, this must be accompanied by a more understanding attitude when a design is actually short of capacity. If this were done, a tremendous first and operating cost savings would be realized and the buildings' comfort conditions (and, hence, the workers' productivity) would be improved. This would be much more than needed to cover the costs of those cases where underdesign actually leads to a change order.

METHOD OF CAPACITY CONTROL

Most buildings perform well at design load. The same cannot generally be said at part load. Since a typical building will spend over 75% of its time at less than half load, this is an important failing. A key aspect in determining the performance at part load (i.e., comfort and energy efficiency) is the method of capacity control that has been selected. Somewhat surprisingly the authors' have found that very little thought goes into this crucial element of the design and have written a number of papers concerning the subject (4,5,6,7). As mentioned earlier, much of the work cited has been done for the Navy. With many Naval activities located in humid climates, an emphasis has been given to room humidity control. The importance of the concept is, however, more general and has been demonstrated for a number of climates which are not technically classified as "humid" (e.g., Memphis).

Most systems are controlled by sensing the dry bulb temperature in the space being conditioned and taking some sort of action to regulate the amount of cooling that is done so that the space dry bulb stays within some acceptable range. (The fact that this might involve sensing some intermediate variable and cascading a control does not change the principle.) In the simplest systems this may be nothing more than taking the action of turning the system on and off. More often large systems modulate the amount of cooling.

Before beginning to evaluate the comfort conditions provided by the various control system options it is appropriate to mention an item which has been observed to significantly affect maintenance costs. In the field one frequently observes condensation and associated property damage. Again the method of capacity control is usually responsible.

As one might imagine, this is a very geometry-specific phenomenon, but the depression of supply-air dry-bulb temperature below the room dew point which is the critical variable. Thus, good design frequently requires paying the fan energy penalty and moving more air to raise the supply air temperature (an economic optimization). For this same reason, any system which produces low supply air dry bulb temperatures in the course of its operation is subject to problems. This is particularly true of those that produce low temperatures at problem part load conditions when the room dew point is higher than normal. The temperature of supply air at part load is, therefore, another constraint that the designer must consider in his selection of a capacity control method.

Modulating Control Systems: To explore the behavior of the room/coil system during quasi-equilibrium operation a computer program has been written by the authors (7). At the heart of the analysis is the model used to predict the behavior of the cooling coil under the variety of conditions it will see. Goodman (9) and later McElgin and Wiley (10) developed an approach for treating wet cooling coils as heat exchangers in which the driving potential is the difference between the enthalpy of the entering air and that of saturated air at the temperature of the entering water or refrigerant. Nussbaum (11) later extended this using the notion of heat exchanger "effectiveness" and the "NTU" approach showing that most coils can be modeled very accurately as counterflow heat exchangers. The authors have used this as a basis to develop a coil model which uses one known operating point (i.e., entering and leaving dry-bulb and wet-bulb, airflow, water flow and entering water temperature) to determine constants which represent the physical characteristics of the coil in the effectiveness calculations. These constants are then used to calculate the effectiveness and, hence, performance of the coil under other conditions of interest. The coil model can then be coupled to various types of control system models and a room model to explore the part load performance of a load and the equipment serving it. For instance, the chilled water valve system iterates the GPM through the coil until the room's sensible load and the coil's sensible cooling are in equilibrium. A variety of pertinent variables (e.g., coil entering and leaving dry-bulb and wet-bulb, sensible, latent and total cooling, etc.) are recorded and output.

Referring to FIGURE 4, there are only a limited number of variables which can be controlled. Entering dry-bulb (EDB) and entering wet-bulb (EWB) are set by room and outside air conditions and, hence, uncontrollable. This leaves the three variables entering water temperature (EWT), water flow (GPM) and airflow (CFM). All methods of capacity control involve manipulation of one or more of these variables.

Using the computer model described above, the effect of varying each of these three parameters can be examined. In each case, the key criteria of coil sensible heat ratio (CSHR) and leaving dry-bulb temperature (LDB) are shown.

FIGURE 2 shows that varying coil temperature is a very poor way of controlling capacity. This has been excluded from further consideration.

This leaves only GPM and CFM. FIGURE 5 shows the effect of varying GPM while holding the other parameters constant. (It should be pointed out that it is immaterial to the coil whether this is done with a two-or three-way valve.) Clearly, a similar problem exists. The latent cooling falls off much faster than the sensible (i.e., CSHR increases rapidly). The result is that the room humidity ratio will increase dramatically under the more probable part-load conditions. Note that leaving dry-bulb rises on a fall in load. This is desirable in the sense that it minimizes the chance of condensation during quasi-equilibrium operation. Its variability can, as pointed out above, lead to condensation problems when a sudden change of setpoint occurs.

FIGURE 6 shows that decreasing CFM in response to a reduction in load while holding all other variables constant (i.e., VAV with a "wild coil") gives considerably different results. Sensible cooling actually falls faster than latent in this case as one would want. Note, however, that the leaving dry bulb drops rapidly with a fall in load. This can (and often does) lead to condensation problems.

A variation of the VAV concept is the "face and bypass" system. Here the air being sent through the coil is varied in response to load in the same manner as with VAV, but the remaining air is mixed back after the coil, keeping total discharge flow constant. FIGURE 7 shows that the SHR of the system behaves exactly as the VAV did but that the leaving dry bulb rises on a fall in load. For these reasons, this system is desirable in that it will tend to provide good passive humidity control without risk of condensation during quasi-equilibrium operation. (Note, however, that the variable leaving dry bulb still admits the possibility of condensation on an upset of setpoint.)

FIGURES 8 and 9 summarize the situation for quasi-equilibrium operation. Systems which vary chilled water flow (GPM) are unsuitable because their SHRs increase dramatically with a fall in sensible load. Variation of CFM alone fails because its leaving dry bulb falls rapidly. Of the methods considered, only the face and bypass meets both of the quasi-equilibrium performance criteria.

Thinking about the problem as a whole, the choice of a modulating control system can significantly affect first cost, operating and maintenance costs and occupant comfort. It deserves careful attention.

On/Off Control Systems: An example of on/off control might be a smaller DX system where the room thermostat closes a solenoid valve in the refrigerant line to the coil. The designer is immediately faced with an important decision. What should be done about the supply air fan? If it is allowed to continue running while the coil is de-energized, warm moist air is blown across surfaces that only moments before were cooled by the supply air. If these surfaces are sufficiently below the dew point of the air, condensation can occur. This has been documented and studied previously by Spielvogel (12). This control method can also be shown to lower the amount of latent cooling done by the system and introduce an additional latent load on the space resulting in a high room humidity. This has been investigated by Khattar, Mukesh, Ramanan and Swami (8).

Clearly, cycling the fan on and off with the solenoid will interrupt the supply of ventilation air. This may or may not be acceptable. The authors have found that, in many cases, the amount of ventilation air required and the amount provided work out as one would want because both the cooling load and required ventilation are closely related to the number of people in the space. This is particularly true of buildings such as mess halls. In any event there will be a significant energy savings from the reduced fan operation and the reduction in the total amount of ventilation air brought in over a period of time. Claims have been made that the cycling of fans will result in additional maintenance and repair costs. There is certainly some truth in this, but every case that the authors have analyzed has shown that the savings far outweigh the additional costs. With an overall improvement in comfort, the scales would seem to tip in favor of fan cycling.

The point in this discussion about capacity control is that there are major consequences in terms of comfort and operating costs. The Navy specifically instructs the designer (13) to evaluate the performance of his design at likely part load conditions. It has been the authors' experience that this is almost never done. Without recommending any particular method of capacity control, it can be stated without reservation that, if designers were to do this, they would immediately come to realize how important the capacity control design decision is and take it more seriously.

VARIABLE PUMPING

This final example again illustrates a case where efforts to save energy (i.e., reduce operating costs) may have comfort implications. As discussed earlier in the section on equipment sizing, most chilled water pumping systems operate at a constant volume regardless of the cooling load. The better designs today try to reduce pump energy by pumping only the water that is really needed in the system. FIGURE 10 shows a typical large system and is an example of how this is often attempted (14).

In this control strategy pumps P1 and P2 are constant volume and run anytime the associated chiller is on. Pump P3 is equipped with a variable frequency drive controlled from a pressure sensor strategically placed out in the system with the idea of only pumping the water that is really needed. The general notion is to run one or two chillers as necessary to keep a small flow from right to left in the bypass line. If two chillers are on and the flow in the bypass becomes equal to the flow of a whole chiller (indicating that the load has become small) one of the chillers is turned off. If the load then increases to the point where the flow becomes left to right in the bypass, the second chiller is started.

Unfortunately this strategy has a problem. It has been the authors' experience that these, and similar systems (though very common) do not work as intended and are often found to be operated in a manual mode and that they often lead to comfort problems. To address this it is necessary to first discuss some basic but little recognized facts about chilled water flow.

FIGURE 11 shows the basic relationship between total cooling and water flow for a typical cooling coil. (The straight line represents a linear relationship between cooling and flow and is used in the discussion which follows.) Note that, for a typical coil, a flow of 50% does cooling of 75% and a flow of 25% does cooling of 50%. Think now about how the system of FIGURE 10 would respond to various loadings. At full design load the flows would be as shown (in arbitrary units).

Next, consider the loading case shown in FIGURE 12 where both Load 1 and Load 2 have been reduced to 75 units each (a total of 150 units). The flows, however, would be 50 units each (a total of 100 units). Any slight decrease beyond this point would call for one of the chillers to be turned off. But this will cause trouble because, although the flow is down to 100 units (within the range of a single chiller) the load is still 150 units (too much for one chiller). A similar problem arises when one chiller is on and the load is rising. When the flow through the loads increases beyond 100 units the flow in the bypass would begin to go from left to right and the logic would turn on the second chiller. This would be too late because the cooling required would already be 150 units (more than the single chiller can do).

Now, consider the loading shown in FIGURE 13. Here the total cooling load is 100 units. The flow is now 100 units as in the case of FIGURE 11. A further decrease in load would force the bypass flow above 100 units and say that it is time to turn a chiller off. This time there would be no problem. Similarly a rise in load with one chiller on would start the second chiller before cooling capacity was beyond the capability of one chiller.

Think next about how these two extremes might be approximated. The first case, where all coils are operating at the same part load ratio, might occur on a mild night. The case where some coils are fully loaded and others fully unloaded might occur on a cold day when those zones on the west side are under a big solar load but the others have no load. Of course, one would never find these limits, but there could be cases that get close. The point is that there is no one-to-one relationship between cooling load and system flow. There can, therefore, be no one-to-one relationship between bypass flow and load and the bypass flow cannot serve as an indication of what the load is and how many chillers are needed. The best that can be said is that the actual load/flow points observed will be found somewhere within the zone bounded above by the linear relationship and below by the typical coil relationship. Translating this shows that the water temperature rise will remain the same or increase as the load falls off. This might help explain some of the systems which have plenty of chiller capacity but are not able to satisfy even relatively low part loads.

(As a complication to this the designer must consider that all of the above has been based upon the mixed air entering the coils remaining the same. This is not the case. Units with a lot of ventilation air will tend to have a lower entering air temperature at the lower part loads. This may actually result in the water temperature rise decreasing at some part loads. It is impossible to generalize.)

From the viewpoint of energy use, the pumping energy that can be saved by stretching the water temperature rise will usually far outweigh any negative effects on chiller energy. Since this is equivalent to saying that less water will be pumped, it can also mean that smaller pipe sizes will be allowable, saving first cost. As discussed earlier, just going to variable flow will allow smaller pipes to be used and save even more in first cost. The net effect on first cost of less piping but more controls can go either way.

There is, however, a comfort consideration. As water temperature rise increases, holding leaving water temperature constant, the coils operate warmer. As seen in the earlier discussions, this means a loss of latent capacity and will result in higher room humidities. The effect of this on comfort must be assessed and the result weighed against the first cost and operating cost savings.

As before, the point is that the engineer must realize that he is faced with an important design decision and that his choice can affect all three major design considerations (i.e., comfort, first cost and operating costs). The evidence seems to be that this is not realized.

CONCLUSION

In some cases energy conservation and comfort are very definitely in conflict and the design engineer must be careful to consider all aspects of the optimizations with which he is faced. More often, the negative comfort connotations brought to mind by the words "energy conservation" are entirely unnecessary and avoidable. Fifteen years after the first oil embargo, energy efficient design is nothing more than good design. Good design always takes into consideration the various factors that are important to the client. Energy, comfort and first cost are all important.

References

1. American Society of Heating, Refrigerating, and Air Conditioning Engineers 1985. ASHRAE Handbook: Fundamentals, Atlanta, GA.

2. Naval Facilities Engineering Command 1987. Heating, Ventilating, Air Conditioning & Dehumidifying Systems (Design Manual 3.03), Alexandria, VA.

3. Department of Defense 1987. Policy Guidelines for Installation, Planning, Design, Construction and Upkeep, Washington, DC.

4. Todd, Tom R. 1985. "Energy Conservation Experiences with HVAC Systems in the High Humidity Climate, A Case History," Second Annual Symposium on Improving Building Energy Efficiency in Hot and Humid Climates, College Station, Texas.

5. Todd, Tom R. 1986. "Start-up of Air Conditioning Systems after Periods of Shutdown (Humidity Considerations)," Third Annual Symposium on Improving Building Energy Efficiency in Hot and Humid Climates, Arlington, Texas.

6. Pate, Marvin E., III and Todd, Tom R. 1986. "Innovative HVAC Cycles For Severe Part Load Conditions in the Humid Climate," Fourth Annual Symposium on Improving Building Energy Efficiency in Hot and Humid Climates, Houston, Texas.

7. Todd, Tom R. and Pate, Marvin E., III 1986. "Passive Humidity Control: A Comparison of Air-Conditioning Capacity Control Methods For the Humid Climate," ASHRAE Far East Conference on Air Conditioning in Hot Climates, Singapore.

8. Khattar, Mukesh K., Ramanan, N., Swami, M. 1985. "Fan Cycling Effects of Air Conditioner Moisture Removal Performance in Warm, Humid Climates," Florida Solar Energy Center, Cape Canaveral, FL, ISA.

9. Goodman, William November 1938 - May 1939. "Performance of Coils for Dehumidifying Air," Heating/Piping/Air Conditioning.

10. McElgin, John & Wiley, D. C. 1940. "Calculation of Coil Surface Areas for Air Cooling and Dehumidification," Heating/Piping/Air Conditioning.

11. Nussbaum, Otto J. December 1983. "Calculate Counterflow Coil Performance with a Programmable Calculator", Heating/Piping/Air Conditioning.

12. Spielvogel, Lawrence G. April 1980. "Air Conditioned Buildings in Humid Climates, Guidelines for Design, Operation and Maintenance", ARMM Consultants, Gloucester, New Jersey.

13. Naval Facilities Engineering Command 1980. Tropical Engineering (Design Manual - 11.1), Alexandria, VA.

14. The Trane Company 1983. "Variable Flow Chilled Water Systems," Engineers Newsletter, La Crosse, WI.

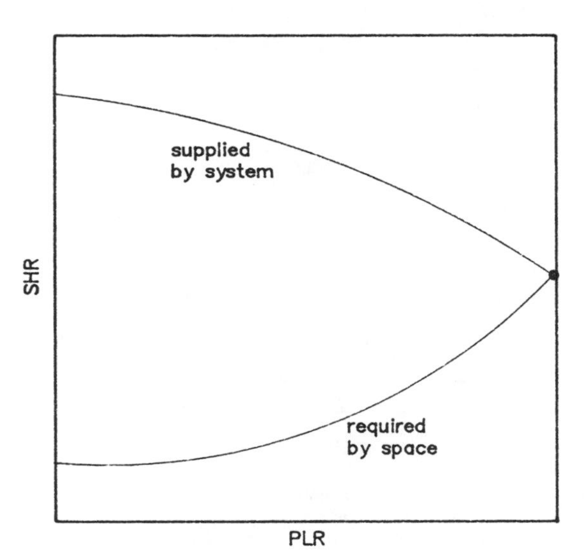

FIGURE 1. Sensible heat ratios required by a typical space and provided by a typical system at various part load conditions.

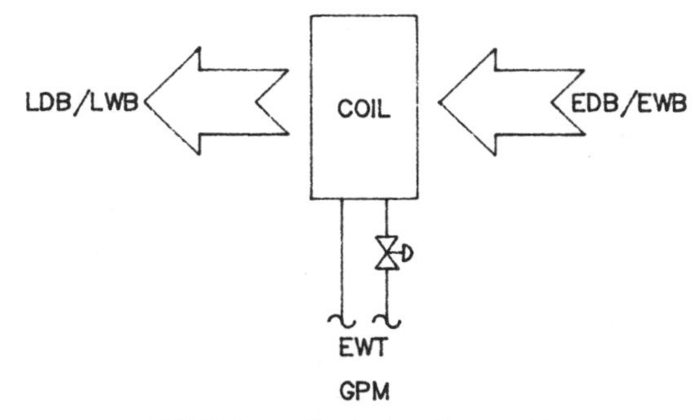

FIGURE 4. Typical coil variables

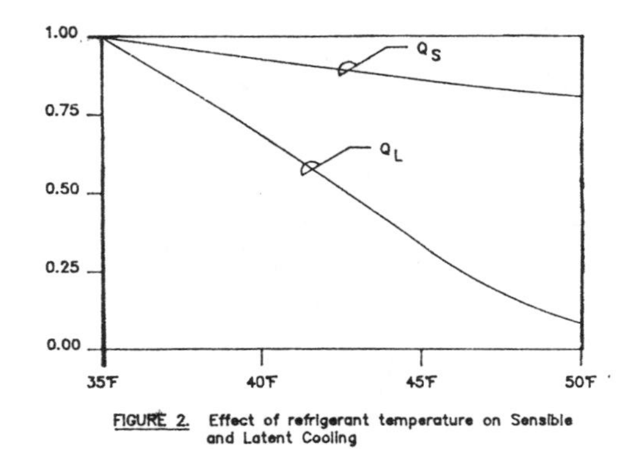

FIGURE 2. Effect of refrigerant temperature on Sensible and Latent Cooling

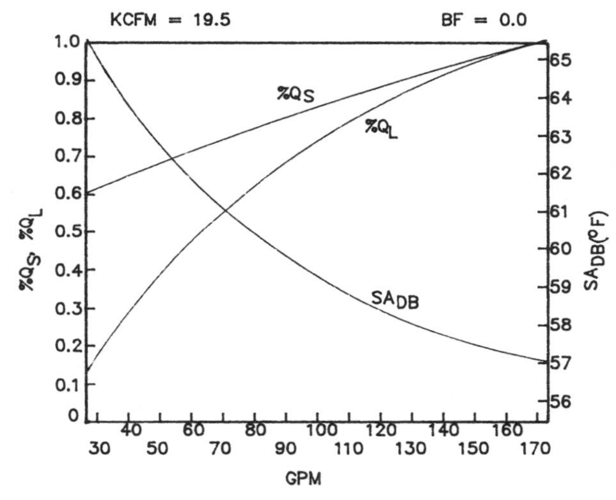

FIGURE 5. Effect of water flow.

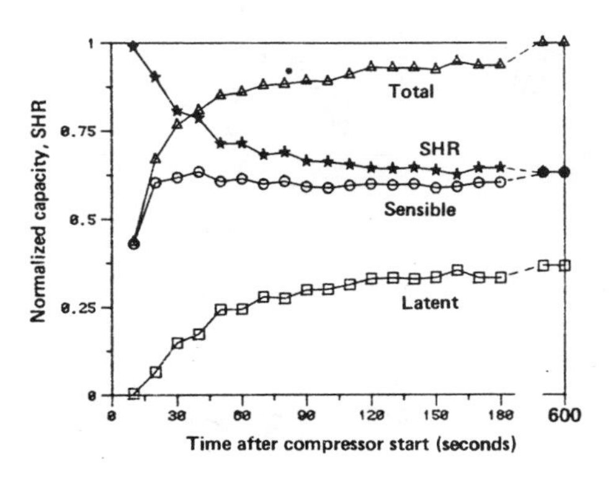

FIGURE 3. Air Conditioner Transient Performance

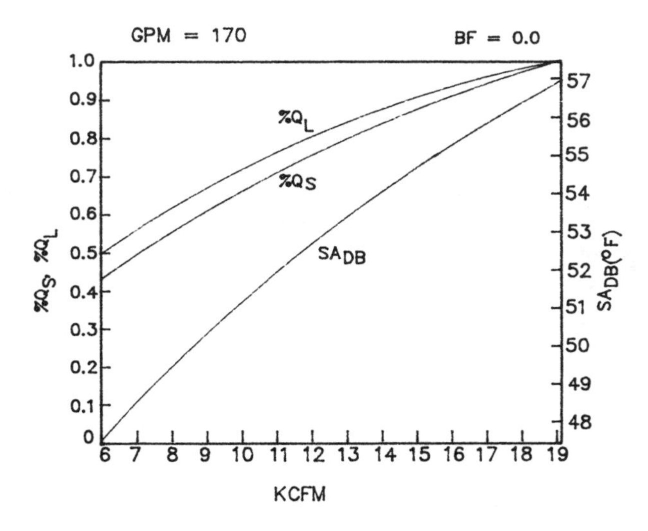

FIGURE 6. Effect of air flow.

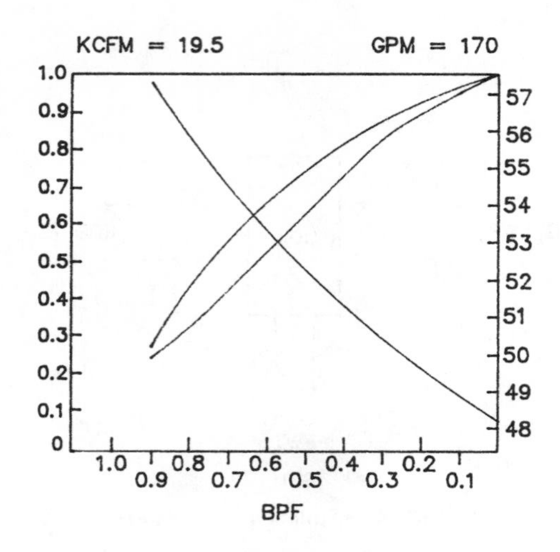

FIGURE 7. Effect of bypassing.

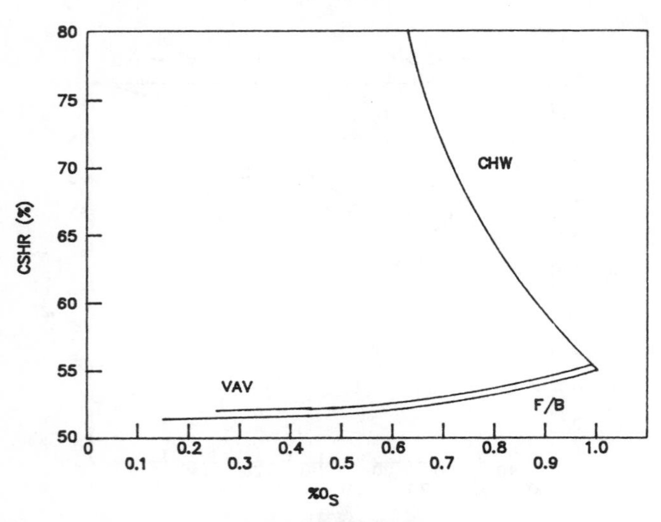

FIGURE 8 COIL SENSIBLE HEAT RATIO VS.% SENSIBLE COOLING CAPACITY

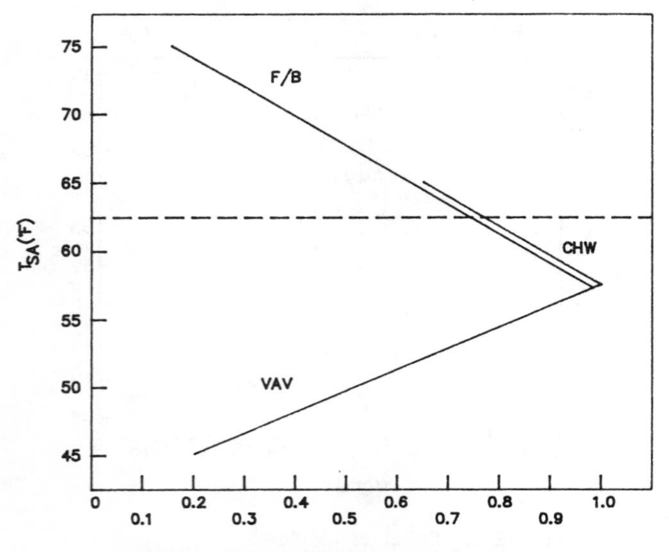

FIGURE 9 SUPPLY AIR DRY BULB TEMPERATURE VS.% SENSIBLE COOLING CAPACITY

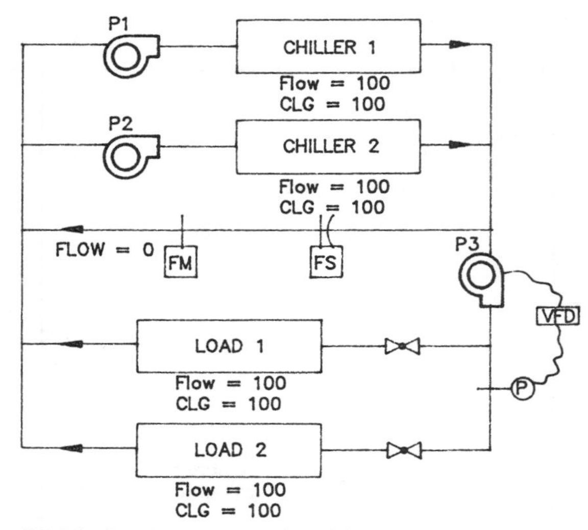

FIGURE 10. Common variable chilled water pumping system at full load.

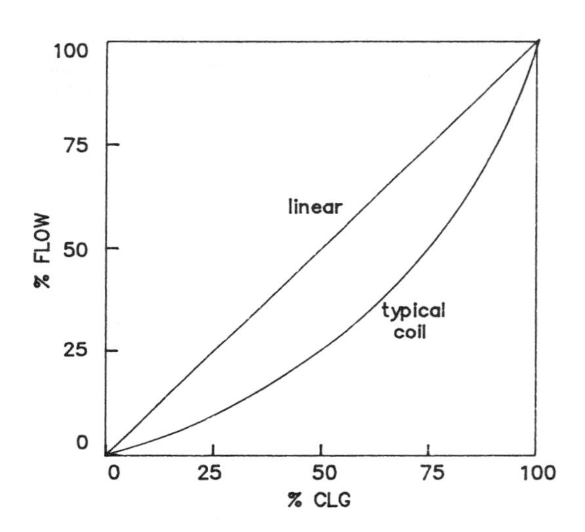

FIGURE 11. Relationship of cooling and flow.

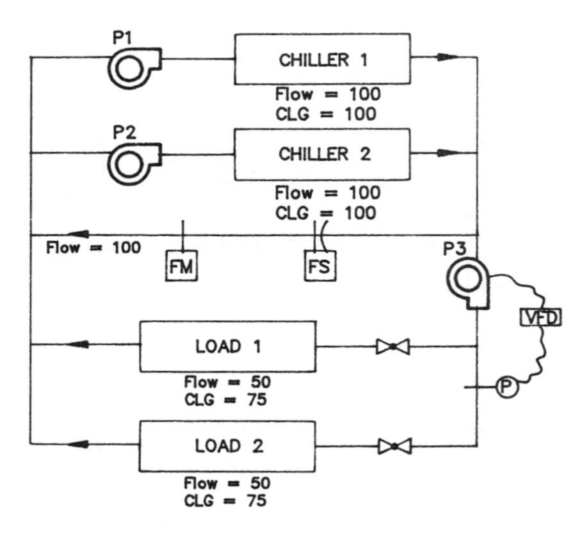

FIGURE 12. Common variable chilled water pumping system at part load.

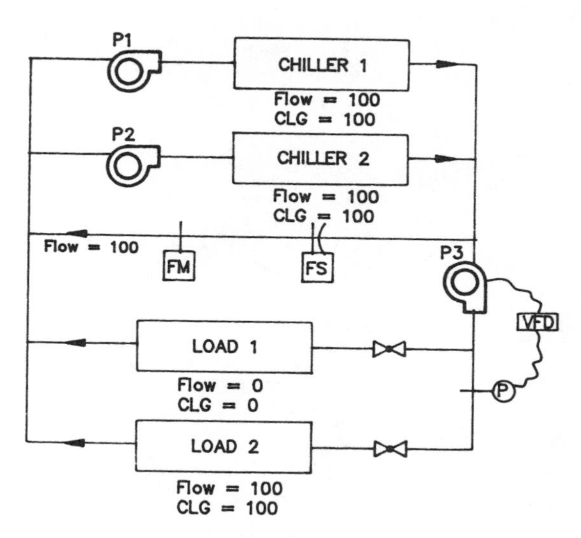

FIGURE 13. Common variable chilled water pumping system at part load.

Chapter 42

USING ADJUSTABLE SPEED DRIVES AND MICROPROCESSOR CONTROL FOR HVAC SAVINGS

W. L. Stebbins

Abstract

When chilled water refrigeration loops and air handling units are operated manually, the resulting inefficiencies cause higher electrical power requirements than necessary.

Installation of an automatic air conditioning control system maximizes chiller efficiency, minimizes the number of chilled water pumps operating while maintaining adequate capacity, and controls the air volume by varying fan speed to move only the air required to maintain desired room conditions.

Project payback may be less than three years, depending on energy costs, with the space conditions actually improving by providing a more even control of temperature and humidity. Details provided include selection of control equipment, calculation of savings, motor load reading techniques, and adjustable speed drive fundamentals involving selection of equipment.

Introduction

All dollar values are approximate, and are based on 1989 equipment and energy costs in the Southeast. Specific brand names and equipment suppliers are included to present a cross section of all equipment available; the listing is not intended to be all inclusive.

There are a variety of firms with the capabilities to supply satisfactory equipment, hardware, and software. United Technologies Building Systems [1] and Southcon [2] were chosen for this system based on their previous experience on similar applications.

Application of this equipment, hardware, and software for these specific requirements should not be construed as a general endorsement by either Hoechst Celanese or by the Author. It is important to note that a variety of brands and suppliers should be evaluated by anyone considering a similar application.

Values given for flows, temperatures, pressures, and unit energy costs approximate actual numerical quantities and are used to illustrate the methodology employed.

Background

Since 1973, the Celanese plants of Hoechst Celanese Textile Fibers have had an extensive energy management program in place. Results of this program have yielded, within the Celanese plants, a reduction over 30% in BTU/NGP. Since the majority of the easier projects have now been implemented, significant investments in time and money are now required to achieve continued energy savings. Improvement of an HVAC system serves as a good example of this continued investment.

A typical industrial building, Figure 1, built in the late 40's may consist of several floors serving a variety of production functions, Figures 2 and 3. For example, a building of 10 to 15 acres of conditioned space for processing requirements and creature comfort will have an air change of every 5 to 8 minutes. The heat load may be predominantly ambient in nature, with a smaller effect from process machines and lighting.

HVAC Equipment

The building may be served by several chillers. For example, three 500 ton chillers and one 800 ton chiller, connected in parallel to a common chilled water sump. The chilled water is typically pumped to a number of air handling units (AHUs), Figure 4, on various floors and the roof, an example listing is shown in Table 1.

Table 1 includes typical data on fan air flow, fan motor HP, and kW at 60 Hz and 45 Hz.

Motor Load Survey [3]

An initial survey of AHU fan burden may indicate that a number of motors are significantly under loaded, and thus operating in the 75-85% efficiency range. When induction motors are operating in the 75-85 percent efficiency range, the reduction in power achieved for each percentage point improvement is approximately 1 kW/100 hp of actual running load.

Thus if a 100 hp motor runs continuously for one year, the savings from improving its efficiency one percent point would be

$$8760\ \frac{hr}{yr} \times \frac{1}{100}\ \frac{kW}{hp} \times \frac{\$0.05}{kWh} = \$438/yr.$$

If a suitable smaller hp motor is already available, the one time change out costs are estimated at $500-$600, material and labor. To identify the actual motor candidates, a motor survey should be carried out.

The detection and change out of large under-loaded induction motors to smaller and/or higher efficiency induction motors contributes greatly to improved system efficiency and power factor. Such under-loaded motors can be detected by the use of devices such as a stroboscope digital readout tachometer (4).

Note that a digital-style tachometer is required to

enable reading the speed to the nearest single digit. Thus the commonly used stroboscope tachometer, with the speed determined by reading the markings on a 4-in plastic top dial, is not suitable for detecting motor slip r/min.

Slip as a Measurement of Torque

The synchronous speed of an AC motor is determined by the line frequency (assumed to be 60 Hz in the USA) and the number of motor poles. The relationship of 60 Hz is as follows:

Poles	Synchronous r/min
2	3600
4	1800
6	1200
8	900
10	720
12	600

The actual running speed of an induction motor depends on the motor design and the load on the output shaft. The speed at full load is indicated on the motor nameplate, accurate to ± 10 percent. Full-load slip r/min is the difference between the synchronous r/min and the full-load r/min:

slip r/min = synchronous r/min - running r/min.

For a four-pole motor with a full-load speed of 1700 it would be

1800 - 1700 = 100 r/min slip.

Measurement of motor output is based on the principle that slip r/min is linear from ten-percent load to 110 percent load. For example, the four-pole motor discussed earlier was found to be running at 1760 r/min. If the nameplate output is 10 hp, what was the output at the time

Full-load slip = 1800 - 1700 100 r/min. = 100-percent load.
Running slip = 1800 - 1760 = 40 r/min.

$$\text{Running load} = \frac{\text{running slip}}{\text{full-load slip}} = \frac{40}{100} \quad 0.40.$$

Output load = 10 hp x 0.40 4 hp.

Note that below about 50 percent of full-load amperes, the motor current is not a valid measurement of motor load. Due to the decrease in motor power factor, more and more of the apparent current is reactive amperes as the motor load decreases.

The greatest benefits from using this technique with optical and mechanical digital tachometers include the following:

1. ability to locate under loaded induction motors quickly;
2. ability to watch loading of equipment versus other conditions such as throughput, filter conditions, temperature, and pressure;
3. ability to determine approximate motor efficiency at any load when used in conjunction with conventional kW monitoring on the motor input leads;
4. ability to assist in sizing future motor requirements based on actual load data;
5. ability to assist maintenance mechanics in periodically checking motor loading, which would help detect worn bearings, clogged filters, etc.

Motor Efficiency

To determine motor efficiency, an additional piece of equipment is required to provide data on kW power input to the motor. The following list outlines several suitable devices:

Device	Vendor	Approx. Cost
1. PG 191 Analyzer	Westinghouse (5)	$2000
2. IDM-3 Wattmeter	EPIC (6)	$ 600
3. Model 2000 A	TIF (7)	$ 400
4. Power Master III	Esterline Angus (8)	$1500
5. Model 2433	YEW (9)	$1000
6. Model 808	Dranetz (10)	$3500

If the motor has an individual kWh meter, the instantaneous power can be determined by timing the disk and using the formula (11):

$$kW = \frac{(PTR)\ (CTR)\ (K_h)\ (3600}{(1000)\ (s/rev)} = \frac{(PK_h)\ (3.6)}{s/rev}$$

where

PTR potential transformer ratio,
CTR current transformer ratio,
K_h meter constant found on the face of the meter,
PK_h primary meter constant = K_h (PTR) (CTR).

Other Considerations

More complex methods than those described can be used if special accuracy is desired, but they are usually unnecessary in a general survey. Moreover, the more complex the procedure, the more costly the equipment and the higher the operator-skills requirements. Higher precision may be desired if the survey uncovers areas where motor change outs could save significant amounts of energy by sizing motors to run closer to their rated loads.

It is possible to be lulled into a false sense of security by assuming that everything is functioning as planned. For example, some meters may be recording properly while others are either running high, low, or slow. Therefore, it is wise to interrogate the system and people. For example, are you sure that the wiring is correct How did you verify it

Generally, kilowatt-hour meters run well when they are connected correctly. Faulty readings are caused by 1) current transformers and voltage transformers with reversed polarity connections, 2) voltage transformers with blown fuses, and 3) shorting blocks with shorting screws not removed.

The over sized fan motors indicated in Table 1 should be changed out prior to installation of the adjustable speed drives.

Scope Of Project

The typical project strategy is detailed in Appendix A. In summary, the manual controls for air flow and chiller operation are replaced with adjustable speed inverters, Figures 5 and 6, connected to microprocessor-based controllers, Figures 7 and 8, yielding a potential annual savings of several hundred thousands of dollars.

An example system configuration is shown in Figure 9, and a typical system operating description specification is detailed in Appendix B.

In Figure 9, various chillers, cooling tower fans, and AHU fans are operated by the unit controllers, (UC). These unit controllers are connected to gateways (GW) and ultimately the CPU host computer, Figures 10 and 11, by a four conductor 20 AWG cable.

The UCs in use feature up to 16 analog or discrete input channels and 12 analog or discrete output channels. A more recent UC now offers up to 60 analog or discrete I/O in modules of 4 or 8.

The UCs can stand alone and if necessary control conditions without communications from the host CPU. The host serves to log trends, generate reports and alarms, change set points, and display system status.

Appendix C, (12) provides a brief review of the types of adjustable speed drives available. In general, when using existing induction motors, the choice falls to either VVI or PWM inverters.

As shown in Figure 5, the power is connected from the original motor starter, through an isolation transformer, to the inverter.

Few subjects cause as much and as vehement debates as isolation transformers. The benefits of an isolation transformer are many, including that of limiting fault current to a solid-state control. Most isolation transformers limit fault currents to about 30 times their full load rating, while also providing electrical isolation from the power lines to ground for the safety of personnel and inherent ground fault protection against a motor or wiring ground. With an isolation transformer, one ground developing will not cause a problem.

A delta-primary/wye-secondary is most often recommended in the isolation function, and this unit has an additional benefit in that it will cancel the third order harmonics which might be a feedback to the power line. It also acts as a noise filter and as a buffer against severe line disturbances. Users of PWM type controls will also find that the isolation transformer will prevent a clipping of the line at the level of the d-c bus that occurs when the PWM inverter is connected across a service (13).

Project Challenges

The 100 hp fan motor on AHU Number 10, Figure 12, was an Allis Chalmers 23D8 frame unit approximately 40 years old. (See Figure 13 for nameplate date). As illustrated in the following table the motor would not run at 60 Hz on the new six step inverter, and was subsequently replaced with a new unit. Investigation into the suitability of the old motor for use on an inverter resulted in the letter shown in Figure 14.

Measured Parameter	Test Sequence			
	1	2	3	4
Hz	60	48	58.5	45
V	470	368	450	347
Nameplate AMPS	124	124	116	116
Running AMPS	106	136	116	76
kW	76.8	72	73	34

Test Sequence:
1 - Old motor on conventional 3 phase 60 Hz power.
2 - Old motor on 6 step inverter.
3 - New motor on 6 step inverter at FLA.
4 - New motor on 6 step inverter at 45 Hz.

Other challenges involved the following areas:

1. Asbestos insulation used prior to 1968 on chilled water lines. Installation of eight control valves required portions of asbestos to be removed in accordance with all applicable guidelines and regulations, Figure 15.
2. Significant maintenance and overhaul of existing pneumatic operated control valves.
3. Installation of a dedicated instrument grade air supply line and individual filter regulators for each of 11 unit controllers and eight new control valves, Figure 16.
4. Fabrication of new mounting bases for the seven downsized motors and one replace motor noted in Table 1. Installation required millwright crafts to dial indicate for run out and check for shaft and pully vibrations, Figure 17.
5. The usual system hardware and software debugging associated with a complex microprocessor-based control scheme connected to a CPU data collection host. This required the dedicated support of the utility HVAC operators, Figure 18.

Summary

The typical reduction in fan kW, 735 at 60 Hz to 319 at 45 Hz from Table 1, produces an annual savings of several hundred thousand dollars depending on energy costs. In addition, a reduction can be expected in the number of chillers, cooling water and chilled water pumps, and cooling tower fans required to operate the building.

Anytime a project of this magnitude is installed, a formal economic audit should be carried out to document the total project savings and calculated payback. Based on this final analysis, other plant areas may then be considered for similar HVAC system control upgrades.

References

[1] United Technologies Building Systems
Building D Suite 400
5775 Peachtree Dunwoody Road
Atlanta, GA 30342.

[2] Southern Industrial Controls
10901 Downs Road
P. O. Box 410328
Charlotte, NC 28241-0328.

[3] W. L. Stebbins, "New Concepts in Electrical Metering for Energy Management", IEEE Transactions on Industrial Applications, Vol IA-22, No. 2, March/April 1986, pp 382 - 388.

[4] Speed Measuring Instruments
Biddle Instruments
510 Township Line Road
Blue Bell, PA 19422
Bulletin 35i.

[5] Westinghouse Electric Corp.
Relay Instrument Division
4300 Coral Ridge Drive
Coral Springs, FL 33065.

[6] Epic, Inc.
150 Nassau Street
New York, NY 10038.

[7] TIF Instruments, Inc.
9101 NW 7th Avenue
Miami, FL 33150.

[8] Esterline Angus Instrument Corp.
P.O. Box 24000
Indianapolis, IN 46224.

[9] Yokogawa Corp. of America
2 Dart Road
Shenandoah, GA 30265.

[10] Dranetz Engineering Lab., Inc.
1000 New Durham Road
Edison, NJ 08818.

[11] ANSI/IEEE Standard 141-1986
The Red Book, P.463.

[12] Adapted from Allen-Bradley Publication DGI-2.1 and Southcon AC Drive Comparison Booklet.

[13] James R. Bradley, VP of Engineering for Southcon, in Article "Optional Equipment", Published in International Fiber Journal, June 1987, Pages 39-41.

TABLE 1 - EXAMPLE EQUIPMENT LISTING

UNIT NO.	FAN - MCFM[2]	FAN MOTOR HP ORIGINAL	FAN MOTOR HP FINAL	kW @ 60Hz	kW @ 45Hz
1	55	-	50	30	12
2	75	-	50	40	16
3	100	-	60	40	17
4	160	-	100	80	35
5	160	-	100	75	32
6	160	-	100	85	37
7	80	75	50	40	17
8	85	100	60	45	20
9	130	-	75	50	22
10[1]	190	-	100	80	36
11	60	100	50	35	15
12	75	125	50	35	16
13	100	100	60	40	18
14	70	100	50	40	17
15	100	125	60	20	9
TOTALS	1,600		1,015	735	319

NOTES:

1. Motor was originally 100 hp, 23D8 Frame @ 1170 RPM.
2. MCFM - 1,000 Cubic Feet of Air per Minute.

FIGURE 1
TYPICAL INDUSTRIAL BUILDING

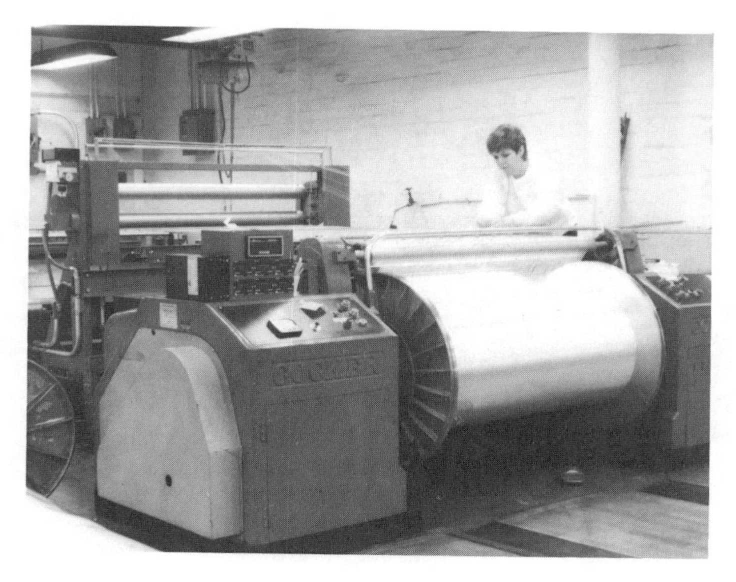

FIGURE 2
REWIND PRODUCTION EQUIPMENT

FIGURE 3
HIGH SPEED BEAMING EQUIPMENT

FIGURE 4
TYPICAL AIR HANDLING UNIT

FIGURE 5
INVERTER AND ISOLATION TRANSFORMER

FIGURE 6
INVERTER INPUT AND OUTPUT CONNECTIONS

FIGURE 7
CHILLER MICROPROCESSOR
BASED CONTROLLER

FIGURE 8
BUILDING AREA MICROPROCESSOR
BASED UNIT CONTROLLER

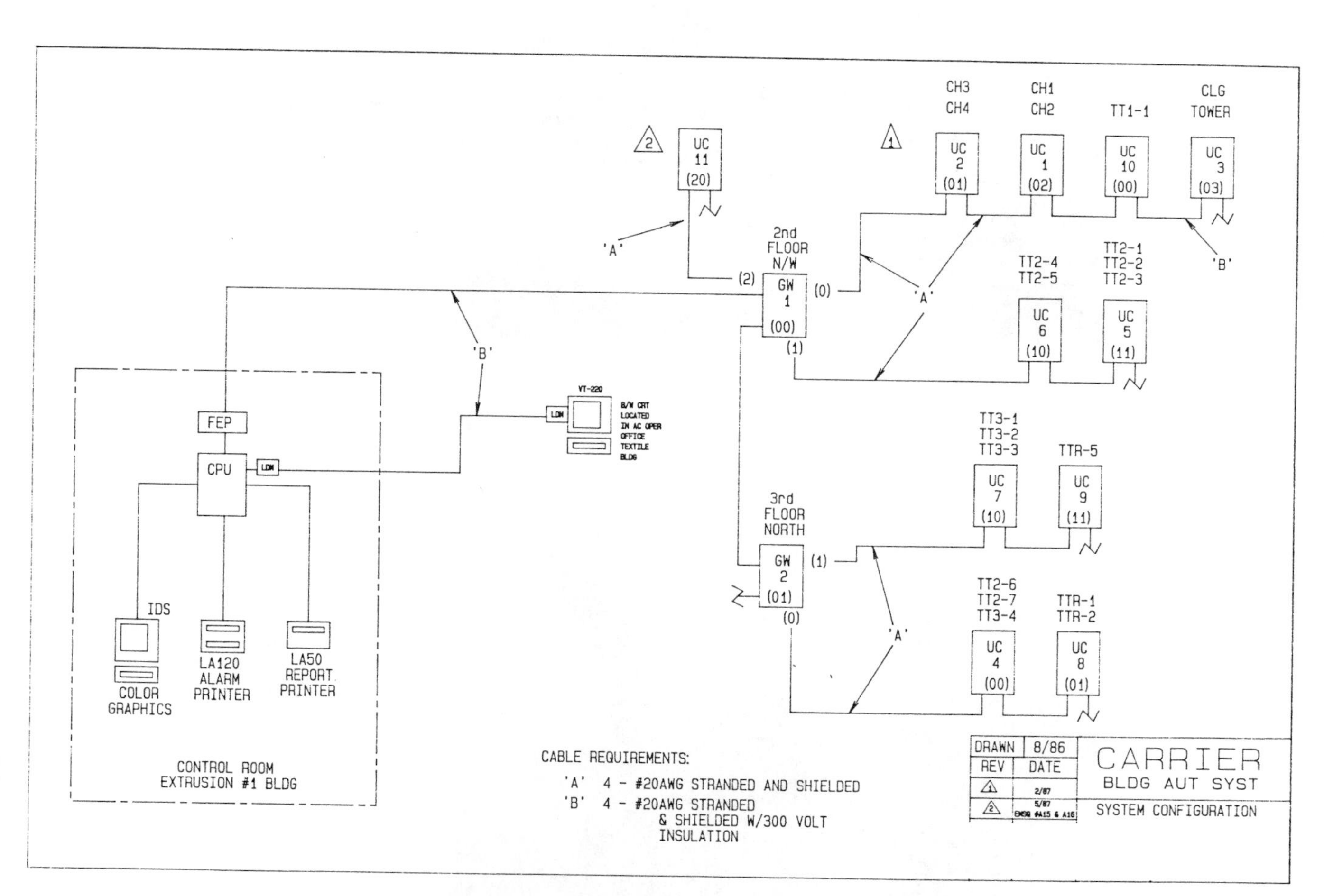

FIGURE 9: SYSTEM CONFIGURATION LAYOUT

FIGURE 10
CPU HOST COMPUTER

FIGURE 11
OPERATOR CONSOLE AND ALARM PRINTER

FIGURE 12
FORTY YEAR OLD 100HP FAN MOTOR

CR-195

ELECT. DEPT. MOTOR RECORD — MOTOR SHOP NO. 3329

H. P.	100
MAKE	Allis Chalmers (Norwood, Ohio)
TYPE	AW
SPEED	1170
AMPS	124
VOLTAGE	440
PHASE	3
CYCLES	60
ENCLOSURE	
FRAME NO.	23D8
KVA CODE RATING	F
INSULATION	
SERIAL NO.	23D8-51-645-659-39
MODEL AND STYLE NO.	
TORQUE DESIGNATION	NEMA B
SERVICE-FACTOR	1.15
TEMP. RISE CONT.	40 C

REQ. NO.	DO-26047 Roc
ORDER NO.	36142 RH
APPLICATION	
BLDG.	Textile
DEPT	Air Cond. 3rd. fl.
DRIVE	Supply Fan
BEARINGS	
MAKE	
NUMBER P.E.	SKF 6319J
NUMBER F.E.	FAFNIR 315K

MAINTENANCE RECORD

DATE	M1C	M2CL	LUB	REMARKS
7-31-80	✓			

REPAIR RECORD AND REMARKS: Replaced Wedges, dipped + Baked – E-S 7-31-80 c 11-10-83 MIC REPLACED BEARINGS NO LOAD 35A. Scrapped 5-13-87 (JC)

Shipping Weight 1880 LBS.

FIGURE 13: DATA CARD ON FAN MOTOR FOR AHU 10

SIEMENS

January 6, 1987

RECEIVED
JAN 12 1987
W. L. STEBBINS

Wayne Stebbins
Senior Electrical Project Engineer
Celanese Fibers Operations
2850 Cherry Road
Rock Hill, S.C. 29730

Subject: 100 HP Motor

Dear Wayne:

At your request, I have investigated the suitability of the old 23D8 frame, 100 HP motor you have, for use on a 6 step inverter. According to several of our senior engineers, the designs used when that motor was manufactured (40-50 years ago) are not suitable for use on an inverter. The use of rectangular wire, form coils, and the internal connections used then would not allow the motor to operate satisfactorily on an inverter. The motor would run extremely hot and draw excessive current.

I hope this answers your questions, but if not, feel free to call.

Regards,

Russ Pettit
Russ Pettit
Senior Project Application Engineer
Industrial Motor Division
Siemens Energy & Automation

RP/rv

cc: Harrell Irby
President
Energy Conversion Corp.
P.O. Box 7169
Charlotte, N.C. 28217

Siemens Energy & Automation, Inc. Industrial Motor Division
14000 Dineen Drive • Little Rock, Arkansas 72206 • Tel (501) 897-4905 • TWX: 910-722-7320

FIGURE 14: LETTER FROM MANUFACTURER OF ORIGINAL FAN MOTOR

FIGURE 15
REMOVAL OF ASBESTOS INSULATION

FIGURE 16
DEDICATED INSTRUMENT AIR SUPPLY
TO UNIT CONTROLLERS

FIGURE 17
NEW MOUNTING BASE FOR DOWNSIZED
FAN MOTOR

FIGURE 18: A WELL TRAINED HVAC OPERATOR IS THE KEY TO A PROPERLY RUN SYSTEM

APPENDIX A
Project Scope

Strategy

The water chillers and the air handler units throughout the plant are presently operated with minimal controls. Chilled water temperature is set so as to not unfavorably impact the room conditions. However, there is no automatic adjustment of chill water temperature. Operators occasionally monitor room temperature and humidity charts and make adjustments to the chilled water supply temperature. If insufficient chillers are on line to supply the required capacity, an additional unit will be started manually. Chilled water pumps and cooling tower fans are similarly started when necessary to assure conditions are satisfied. Since room conditions are not monitored continuously, reserve capacity must be running at all times to assure that normal process and atmospheric swings can be handled.

Each air handler presently controls the local area conditions by regulating the chilled water flow. The air flow is not adjusted, due to lack of any control devices. This required that full airflow be used year-round even though the tonnage required in the winter is less than half that of the summer.

The new control system will monitor the room conditions and continuously adjust the chilled water temperature to be as warm as possible while still maintaining control. Cooling tower fans will also be started and stopped as needed to provide the coldest possible condenser water. Since chiller efficiency is better at the higher chilled water temperature and low condenser water temperature, the electricity usage will decrease. Additionally, the chilled water pumps and the condenser water pumps will be automatically started and stopped, assuring that only required capacity is running at any given time.

The strategy for controlling the air handling units will be modified to manipulate air flow as well as chilled water flow. Large electrical inverters will be installed on the air handler fans to achieve adjustable speed. Airflow reductions will be limited to 50% to assure that distribution patterns are not disturbed. However, since centrifugal fan energy in existing duct work varies as the cube of airflow, the electricity requirements reduce disproportionately. For example, 50% airflow only requires 12.5% of the full flow horsepower.

Location

All chillers and pumps are located in the equipment room on the first floor. The air handlers are located throughout the building. The new control system will be connected to the field equipment and the CRT-based operator interface will be located in a control room on the first floor of an adjacent building.

Equipment

Each of the four chillers will be equipped with the following control equipment:

1. Condenser and evaporator freon temperature sensors.
2. Chilled water and cooling water inlet and outlet temperature sensors.
3. Control valves on chilled and cooling water.
4. Inlet vane position actuator.
5. Motor start-stop control and amperage sensor.

Common chiller system controls will include:

1. Chilled water pump start-stop controls.
2. Chilled water pressure controls.
3. Cooling water pump start-stop controls.
4. Cooling tower fan 2-speed controls.

Each of the 15 air handlers will be equipped with:

1. Inlet air temperature and humidity.
2. Leaving air temperature.
3. Isolation transformer and inverter for fan speed control.
4. Chilled water valve positioner.

A distributed programmable control system will monitor all sensors and manipulate all outlets. A terminal will be installed in the control room for all routine operator interface. Menu-driven software will provide easy access to all information. The system will routinely log all important data and will produce alarms when abnormal conditions occur.

Savings

The savings detailed below fall into two major categories; that attributed to implementing variable air volume control, and that caused by upgrading the chiller plant.

Variable Air Volume

The largest area of savings results from modifying the air handler control strategy to include Variable Air Volume (VAV). The current system simultaneously controls space temperature and humidity by manipulating chilled water and reheat steam valves. The new strategy will manipulate chilled water and air volume, essentially eliminating reheat.

Historical operating logs show that the total building chiller load reached a maximum in the summer at 1355 Tons. Eliminating reheat will reduce this maximum to 1355-208 - 1147 Tons or 85% of the current maximum.

The minimum cooling load in the winter was 680 tons, and will be reduced to 680-208 = 472 Tons or 35% of the current maximum. With the new control strategy, airflow can be reduced proportionately to tonnage requirements, within the limitation that good air distribution must be maintained. At 35% of design airflow, distribution problems would be expected, therefore during the coldest winter months the airflow cannot be varied proportionally with tonnage. A reduction to 50% of design is generally considered the lowest allowable airflow to assure proper distribution, and since extensive testing is not practical it is assumed that the minimum airflow that will be achieved will at 75% of the current value. The year-round average airflow will then be 75% + 85% / 2 = 80% of the current fixed flow.

Fan laws state, and a trial confirms, that the power required in an existing fan varies as the cube (third power) of the air flow. This causes the relatively small volume reduction to cause a large power savings as follows:

Fraction of original power = (fraction of original flow)^3
= $.80^3$ = .51 = 51% of original power

Fan energy savings = 1 - .51 = .49 = 49% of original power

However, the inverters themselves have electrical losses which amount to an estimated 5.4% of the running load. Using the average motor load after inverter installation, the net impact on electricity usage becomes:

(0.49 x Original kW)) x 0.054 = A kW (increase from inverters)
- B kW (0.49 x Original kW)

- C kW (net reduction)

The annual savings then become:

C kW x 8760 hr/yr = D mkWh/yr

Chiller Plant Controls

Based on operating logs of the chiller tonnage on line during the year and weather data, it is estimated that an extra chiller is on line for 2190 hours during the hear. Automatic start/stop of chillers will eliminate the rotating overhead of 150 kW associated with a typical 500 Ton chiller. Savings are:

150 kW x 2190 Hr = 328 mkWh/yr

With the addition of water block valves for off line chillers, a chilled water pump can be shut down half of the times and excess a chiller is shut down. The typical electric power consumed by a 125 HP chilled water pumps is 100 kW. The savings are:

$$100 \text{ kW} \times \frac{2190 \text{ Hr}}{2} = 109 \text{ mkWh/yr}$$

Use of colder cooling tower water will cause improved efficiency in the chillers as they have to compress freon against a lower head pressure. Typical efficiency improvements range from 1.75% - 2.00% per degree of water temperature. Weather data indicates that the cooling water temperature can be reduced by an average of 6 degrees during the months that chillers are operated. Using the average tonnage expected after VAV implementation of 900 Tons, the current chiller efficiency of .82 kW/ton, the hours of chiller operation and an estimated fan energy increase of 20 kWh, the savings are:

[(6 degF x .0175/degF x .82 kW/Ton x 900 Ton) - 20 kW x 5784 hr = 333 mkWh

APPENDIX B
Operating Description Specification

General Description

The Building Management System (BMS) shall provide both monitoring and control of the chilled water equipment, and the electrical system for energy conservation and facility management.

The system shall use fully distributed processing and closed loop direct digital control. This control shall provide stable operation while maintaining precise control of any set point. The set point to be controlled shall be settable over the full range of control.

Standard components shall be used for the built up systems. The system shall offer the flexibility to accept the field sensors and control equipment from different manufacturers.

The local Direct Digital Control Unit (DDCU) shall be capable of operating independently to perform building control and energy management. The DDCU shall have the ability to automatically switch the equipment that it is controlling to a fail safe position in the event of a malfunction or power failures.

The system shall be designed for continuous operation in a manned or unmanned mode.

The system shall offer flexibility with regards to:

- System layout - Placement of field sensors and control devices.
- Field equipment locations - CPU Arrangement.

The system response time shall be as follows:

- Maximum time from command initiation to device output: 30 seconds.
- Maximum scan time for DDCU: 6 seconds.
- Typical point annunication (single point alarm): 15 seconds.

The construction of the equipment, both physically and electronically, shall be modular in nature.

The system shall be made up of the following components:

- Central Processing Unit (CPU), peripherals and telephone modems.
- Direct Digital Control Unit (DDCU).
- Multiplexing Panels (MXP).
- Software Programs.
- Field Sensing and Control Devices (FSC).
- Interconnecting conduits, wiring, and control piping between system components.
- Portable Keyboard Terminal (PKT).

Central Processing Unit (CPU)

The CPU shall consist of a general purpose, solid state minicomputer.

The software system shall be a disk based, multi user, real-time, multi-tasking operating system designed for interactive program development of on-line processing of application programs.

The application programs data, base, and the operating system shall be provided on a 32 megabyte bulk storage disk.

The CPU shall have modular architecture allowing programmable asynchronous serial line multiplexers to interface with up to a total of (8) peripheral devices, operating on EIA RS232C channels up to 9600 baud.

The CPU shall contain self-test diagnostic routines which shall be automatically executed every time the processor is powered up, the console emulator routine is initiated, or the bootstrap routine is indicated.

The CPU shall contain an operator front panel and a built-in console emulator to allow control from any ASCII terminal without the need for a front panel with display lights and switches.

The unit shall contain an automatic bootstrap loader to allow system restart from a variety of peripheral devices without the use of manual toggling or key pad operations.

A real time clock with a seven day battery back up shall be provided. This clock shall allow for both calendar and interval time keeping by the system software.

The CPU shall be modular so that additions or deletions can be easily accomplished.

Economic Methodology

Fan Control

The hours per year of the outdoor air ambient conditions for the various temperatures during the occupied period take into account perimeter ambiguities.

The reduced fan kW is calculated by the reduced load on the fan as the CFM requirements are reduced. These CFM requirements are reduced as the space requires less cooling. Rather than cool a constant volume of air and then reheat, air flow will be reduced to avoid reheat. During occupancy, the main load variable is ambient conditions. A direct relationship exists between ambient conditions and if quantity required.

It is important to note that in this strategy there may be some problems which have to be addressed. One is the possibility of dead spots at the end of long duct runs. A possible solution is to reduce the air capacity of all diffusers by a certain percentage. Another possible problem is water carry over through the eliminators on the air washers. In order to avoid this problem, air velocity must remain at a level of 300 to 600 feet per second.

The summer outside air reduction savings are also based on lower air volume requirements. As the discharge CFM is reduced, so will the minimum amount of outdoor air required to be cooled be reduced.

The following will be the control strategy for the 15 air handlers:

1. Each fan system will have an associated space temperature sensor. During unoccupied hours the system will monitor space conditions and automatically bring fan system to a higher speed if conditions get out of range.
2. When the fan system is operating, the existing pnuematic controls will maintain selected discharge conditions.

3. Volume of air discharged from fan will be varied according to the space conditions through the fan motor adjustable speed controller.

Chiller Control

1. EMS will keep chillers off during silent hours as long as space conditions remain within acceptable range.

2. When the cooling start signal is given, the EMS will start one of the 17M 500 ton chillers, necessary pumps and cooling tower fan. After a time delay, if the chilled water return temperature exceeds a set point, the EMS will shut off the 17M chiller and start the 19C 800 ton chiller. After a time delay, if chilled water return temperature exceeds set points, the system will start a 17M 500 ton chiller after partially unloading the 19C machine. Both chillers will then be ramped up to full load. Chilled water and condenser water pumps will be run as required to provide water at chillers design requirement.

   This procedure will continue until chiller plant capacity is fully utilized if required.

   As cooling demand decreases as determined by chilled water return temperature, the system will reverse the procedure described above.

3. Condenser water will be controlled to maintain a minimum temperature entering the condenser to minimize lift required by the impeller. Condenser water set point will be reset by outside air dew point. EMS will be set up to maintain minimum differential between condenser water entering condenser and chilled water supply temperature.

4. If outdoor conditions will not permit condenser water to be controlled below 85 deg F, the EMS will reset chilled water set point on each chiller one degree at a time as long as chilled water return temperature does not exceed set point.

APPENDIX C
Adjustable Speed Drives

There are several types of adjustable speed drives that can be used with pumps and fans. Each has desirable features along with disadvantages, as outlined below.

I. Variable Pitch

Uses a mechanical means of belts and variable pitch sheaves or pulleys to change speed.

A. Features are low first cost and simplicity.
B. Disadvantages include lack of remote control, belt and sheave wear, problems with high inertia loads, limited range of 5 to 50 HP.

II. Eddy Current

Uses a constant speed induction motor coupled through a eddy current clutch to the load.

A. Features are low first cost, simple electronics, and the ability to handle high inertia loads.
B. Disadvantages include the lack of repair shops, the special clutch can not be bypassed, the efficiency decreases with speed, and tach feedback is required for speed control.

III. Direct Current

Uses a phase controlled bridge rectifier delivering variable DC voltage to a DC motor armature and field.

A. Features are simple technology, high efficiency over the entire speed range, and smaller controllers.
B. Disadvantages include a rather large and expensive DC motor which requires more maintenance than an induction motor, a tach generator is required for good speed regulation, power factor decreases with speed, and bypass is not possible.

IV. Current Source Inverters

Uses output SCRs to "steer" the current to the motor at the desired frequency.

A. Features are the use of standard motors, good efficiency, the ability to bypass, and inherent fault current protection.
B. Disadvantages include the need for tach feedback, motor matching to the inverter electrical characteristics, the inverter will not run at all without the motor connected, and limited to usually only one motor.

V. Variable Voltage Inverters

Uses a separate section to control output voltage, with frequency controlled by an output bridge which switches the variable voltage to the motor at the desired frequency.

A. Features are the use of standard motors, good efficiency, the ability to bypass, testing of the inverter with no motor connected, and multimotor applications limited only by the inverter full load rating.
B. Disadvantages include a high initial cost, high power components in the inverter, and low input power factor at light loads.

VI. Pulse Width Moduated Inverters

Uses a constant DC voltage that is chopped in a manner to produce the desired average voltage at a particular output frequency.

A. Features are the use of standard motors, good efficiency, the ability to bypass, testing of the inverter with no motor connected, multi-motor applications, and high power factor.
B. Disadvantages include a high initial cost, high power components, more complex regulator, high frequency switches required, relatively high audible motor noise, and potentially higher motor heating.

Chapter 43

WEATHER AND REFRIGERANT PROBLEMS AND EVAPORATIVE COOLING REMEDIES

J. R. Watt, A. A. Lincoln

ABSTRACT

"Greenhouse effect" and "ozone-layer" decay are already affecting world weather, with 1980s global average temperatures topping recorded history. Meanwhile, "sick buildings" and increased Indoor Air Quality ventilating codes further boost cooling loads.

Thus, world cooling demands are rising while climbing ambient temperatures and energy costs respectively erode air conditioner condensing efficiencies and overall economics.

Perhaps worse, the beginning phase-outs of R-11, R-12, and other refrigerants to protect the ozone layer suggest coming scarcities, the forced use of less-efficient substitute refrigerants, reduced equipment capacities, and an all-out cooling crisis.

Fortunately, most existing air conditioning equipment can recover lost capacity and economy by adding evaporative precooling stages: indirect (dry air) evaporative panels to precool return air from rooms and fresh outdoor makeup ventilating air, and direct (adiabatic) evaporative precoolers for air-cooled condenser air.

Five well-established trends threaten present comfort-cooling and refrigeration patterns. They are explained below, followed by evaporative cooling remedies.

TREND ONE

Progressive "greenhouse effect," from centuries of burning carbonaceous fuels which raise global atmospheric CO_2 content, is trapping more solar heat on earth and increasingly warming and drying world climates.

The historic 0.029% CO_2 fraction has become a still-rising 0.034% today[1], caused by both growing fuel consumptions for heating, power, and transportation, and decreasing areas of green vegetation which convert CO_2 to O_2. Not only is population pressure slashing more forests and jungles today, but related over-cropping and over-grazing is turning more arable land into deserts. Yearly permanent forest losses alone may exceed 27 million acres, much by "acid rain" from burning coal, and no end is in sight.[2]

Also trapping solar heat is atmospheric methane, believed increasing 1% yearly. The hydrogen content of natural gas, it is also released by hydrocarbon oxidation and decay and animal and human excreta. Gas leaks, incomplete combustion, garbage from cans to landfills, and sewers, sewage plants, animal housing and barnyards generate it. Burning garbage merely creates CO_2 instead.

However, many landfills and sewage plants trap methane to burn, as do bio-gas pits fed food waste and manure on some farms and cattle-feeding stations and in Third World villages.

Already, global average 1980s temperatures are the highest in recorded history.[3] For untold centuries despite slowly increasing solar absorption, global temperatures scarcely reacted, ballasted by our enormous weight of seawater. Now, some authorities expect a 2-8 deg F (1-4 deg C) rise in the next 60 years[4]. Others, at least 0.54 to 1.44 deg F (0.3-0.8 deg C) per decade. Either way, polar ice will melt, flooding many coastal cities[5]. Our central states may thus become a dust-bowl, new deserts and famines will appear everywhere, and Canada and Siberia may have to feed the world.

Even if this warming trend is merely a 30-year cycle starting, as some meteorologists believe, or a 100-year cycle, effects will still be devastating.

TREND TWO

World populations are steadily growing, fastest in hot areas. Particularly in the West, but increasing elsewhere, more people are demanding and can afford greater summer comfort.

U.S. fuel prices, retirement policies, and declining northern industry are sending new millions into our Sun Belt states. These transplants' cooling usage likely exceeds the natives', and will tax local utilities' summer capacities.

Thus, human demography and subjective needs further increase the overall demand for cooling.

TREND THREE

The decay of the eons-old stratospheric ozone (O_3) layer is already increasing ultraviolet radiation on earth, increasing skin cancer and affecting animal and plant life in yet unknown ways. The Antarctic ozone losses of 4-5% in eight years are warnings. Current estimates suggest losses already of 1.7% over south and central U.S.A., ranging up to 3% over Canada, and a global average of 2.6%[6].

The apparent cause is over 40 years of escaped chlorinated flourocarbon (CFC) gases like refrigerants R-11, R-12, and related Rs 113, 114, and 115, drifting up and attacking the ozone. Today, about 35% of it leaks from refrigeration and air conditioning systems, another 35% from plastic-foam cells in thermal insulation and products like egg packages and drinking cups, about 18% from solvent-cleaning refrigerative and electronic items, and 5% for sterilants.

Formerly, large quantities came from pressurized aerosol spray cans of insecticides, polishes, paints, and hairsprays, etc. The U.S., Canada, and Sweden have banned CFCs in such items, but other nations still make and use them.

Last year, 31 United Nations member-nations signed a CFC-control treaty which the U.S. Senate ratified this March, leading to a 20% U.S. CFC production cut next year, and a 30% cut in 1994. Dupont, the world's largest manufacturer is phasing-out R-11 and R-12 much earlier.[8]

Meanwhile, several years leakage of CFC's is still drifting upwards, so ozone damage and increasing ultraviolet radiation will continue unabated for some time. Ozone from lightning, Man's electric devices and incomplete combustion is too little and short-lived to replenish the losses.

One dire effect may be the killing of microscopic sea-life upon which larger organisms, and ultimately our whole seafood chain, depends.

Phasing-out key refrigerants will affect billions of tons of installed refrigeration and air conditioning built especially for them. Refrigerant scarcities and prices likely will threaten auto air conditioners most because of their frequent leaks, followed by most small systems ill-adapted to substitute gases. Window air conditioners seem at great risk.

Merely changing refrigerants is expensive. Once, old ones were merely discharged into the air; now, removal must involve safe confinement akin to nuclear-waste disposal.

Some ozone-inert substitutions are possible. R-22 and experimental R-141b can replace R-11 in foam products, other solvents can substitute for R-113 in delicate cleaning, and R-22 and new R-134a may replace R-12 in some refrigeration and cooling service; however, adequate supplies seem years hence.[9]

Replacement refrigerant CFC 123 is also under test,[10] but may be little available before 1992. All such substitutes' operating efficiencies and prices are conjectural today.

In much cooling, substituting safe R-22 for R-12 overloads motors up to 60% at equal speeds, so reducing capacity is mandatory, easy with belted compressors, not with modern hermetic drives. Other required changes include starting relays and windings, gas pressure safety valves, expansion valves and devices, overload and high- and low-pressure cut-out switches, and even condensers and evaporators. Even then, oil-return problems may occur.

R-22 in automotive cooling may create 650 psi (46 kg/cm²) pressures when radiator fans idle on hot days. Most large centrifugal compressors designed for R-11 may possibly use R-123 when available, but sacrifice about 10% capacity.

In general, the coming refrigerants crisis suggests severe shortages and high prices ahead, mitigated by substitutes and reduced cooling capacities.

Fortunately, most such lost output can be recovered by combining two types of evaporative cooling with the refrigerative systems, as detailed later. Scrapping and wholesale rebuilding of equipment may not be needed.

TREND FOUR

The advent of well-designed but "sick" buildings demanding more ventilating makeup air necessarily increases demands on installed equipment.

Thousands of buildings designed after the OPEC 1970s oil crisis cut heating and cooling costs by minimizing natural infiltration and exfiltration. The planned mechanical ventilation then couldn't remove the formaldehyde and other emanations from new materials, fabrics, pressed woods, paints, etc., fumes from built-in garages and utilities, and even radon gas and asbestos.

Occupant morale, health, and productivity suffered, and tentative standards were raised to require 15 cfm (.43 m³/min) of fresh outdoor makeup air per person.[11]

Because makeup air must often be cooled 50-60 deg F (29-33 deg C), and heated even more in winter, raised standards demand larger equipment with greater investment, power costs, maintenance, and leakage, and larger utility plant CO_2 and acid rain emissions.

As noted later, the cheapest solution to increased makeup ventilating air is to use indirect evaporative precoolers.

TREND FIVE

The steadily rising cost of power will affect air conditioning and refrigeration for years ahead. Causes include climbing fuel prices, generating plant overhead and operating costs, and the amortization of frequent expansion, together with cost-overruns for nuclear plants, even abandoned ones, and their indifferent performance to date.

Falling oil prices aid only indirectly, most power being generated from coal and gas. However, long-term dwindling oil reserves and conservation measures must raise all fuel prices. In March, U.S. Geologic Survey data cut U.S. reserves 40%; OPEC's inner divisiveness seems ending, so major price drops are unlikely hereafter.

More significant than fuel prices are the top-heavy excesses of summer over winter power demands. The four Trends above promise to exacerbate this imbalance for the forseen future.

To finance idle winter capacity, utilities exact punitive summer demand and "ratchet" charges. In defense, large users cut summer peak demands with auxiliary gas-fired absorption systems, ice-bank thermal-storage units, or seek price advantages via winter heat pumps.

<u>Trend Summary</u>

To summarize the five Trends discussed above, three greatly increase present and future demands for cooling, but the refrigerants problem and rising power costs create enormous obstacles.

Fortunately, two types of evaporative cooling can assist refrigerative plants here, particularly in increasing outputs at little extra power consumption. They are detailed below.

EVAPORATIVE COOLING[12]

Indirect evaporative cooling appeared in Arizona about 1930, and employed indoor fan-equipped, finned water-coils connected through pumps to oversize natural draft cooling towers outdoors. Their principle applies with variations to all indirect evaporative coolers today: evaporating water isolated from air can cool it dry.

Dry outdoor air blowing through the cooling towers evaporatively cooled water spraying down in them, rendering it sometimes 30-35 deg F (17-19 deg C) below original air temperature.

This water circulated through the indoor coils to cool indoor (sometimes outdoor) air moved through by the fans. The rooms were often kept 20-25 deg F (11-14 deg C) below outdoors, with air as dry as from refrigerated coils.

Thousands of these systems cooled the Southwest in the 1930's, the world's first consumer air conditioning. Ultra- simple, they were also expensive, unsightly, and prone to clogging from lime in the regional hard water. In California, they often precooled return or makeup air for early refrigerated cooling systems.

Evaporative Cooling Theory.

The cooling towers, like indirect evaporative cooling today used a modified adiabatic saturation process. When relatively dry outdoor air at outdoor dry-bulb (ordinary thermometer) temperature freely contacts water, some of its sensible heat evaporates water and becomes latent heat in the vapor created. The reduced sensible heat cools the air toward its original wet-bulb temperature, the lowest temperature that air can reach by the evaporation process, governed by its original humidity.

The difference between the air's original dry-bulb and wet-bulb temperatures is called the wet-bulb depression. It is the maximum number of degrees that air can be cooled by perfect air-water contact when isolated from external heat flows.

The better the contact, the closer the air can approach the wet-bulb temperature. The effectiveness of most evaporative cooling equipment is thus measured by the percent its output air's dry-bulb temperature is reduced, compared to the wet-bulb depression.

Thus, indirect evaporative coolers can cool dry primary air 40 to 80 percent of wet-bulb depression, and direct evaporative coolers (see later), which use adiabatic saturation exactly, can cool their output air up to 95%, but make it very humid.

Indirect Evaporative Cooling Equipment.

With better water and cooling tower management, the original system is still employed where needed. However, most manufacturers today use heat exchangers which combine cooling tower and air-cooling functions compactly for greater economy.

Two U.S. manufacturers cool the dry air, hereafter called primary air, by drawing it through banks of parallel hollow horizontal tubes, dry inside but kept wet outside and blown with fresh outdoor air, termed secondary air, which helps evaporate water on the tubes' outer surfaces and cool the dry primary air flowing inside them.

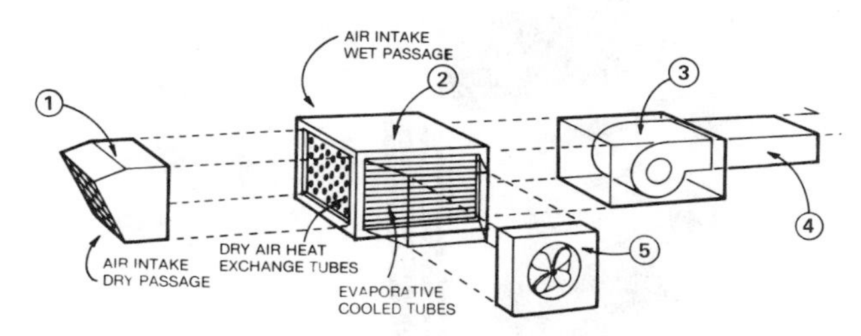

Fig. 1. Schematic view of modern tube-type indirect evaporative coolers. Fresh outside primary air enters at 1, is cooled dry passing through bank 2 of plastic tubes with absorptive outer sleeves kept wet by a pump. Meanwhile, fan 5 draws other outside secondary air across and between them to hasten evaporation. Blower 5 propels the primary air and delivers it into duct 4.

These coolers use multiple round plastic tubes about one inch (2.5 cm) in diameter with thin absorbent cloth exteriors which ensure better wetting. The primary air enters through filters at one end, drawn by centrifugal fans at the other, which also deliver it for cooling use.

The fresh outside secondary air is drawn across and between the tubes by small propeller fans while dripping water wets the tubes and falls in a sump below for recirculation.

Fig. 2. A small DiPeri tube-type indirect cooler with intake filters removed. Primary air enters the open tube ends at right center, drawn by an enclosed blower at far left. Secondary air enters between the wet gauze-covered tubes at left center, drawn by a fan at far right.

A third manufacturer makes very compact and self-contained indirect coolers with closely spaced vertical one inch (2.5 cm) square tubes. Here, primary air blows horizontally between and around them, while water sprays down inside them against rising secondary air. Absorbent coatings are not used.

Three other U.S. firms and one in Australia use hollow flat plates as tubes. These are essentially broad flat, vertical tubes closely spaced together so that primary air is drawn horizontally between them. Meanwhile, water films or sprays descend inside the plates, counter-flow to secondary air. When the plates are under ¼ inch (6 mm) thick and apart, as many are, the mass of primary air is only millimeters from the evaporating water and cools easily.

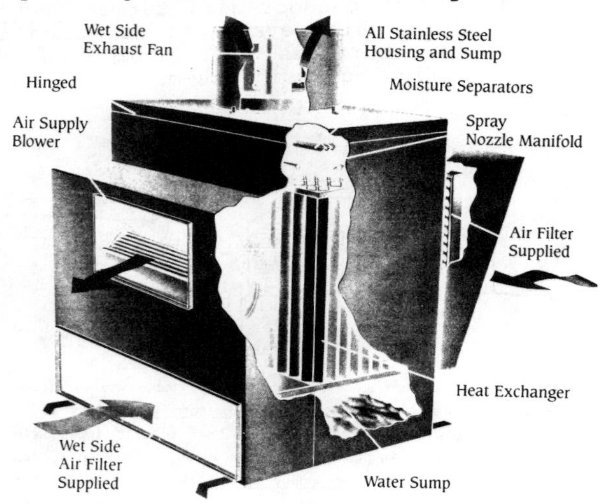

Fig. 3. Schematic of a compact, self-contained Energy Labs Ltd. indirect cooler with vertical square tubes. Primary air is first filtered at far right, then blower, left, draws it horizontally and zig-zag between the tubes and delivers it at far left. Secondary air enters at lower left, and the fan above draws it up inside the tubes, counterflow to descending water spray and films.

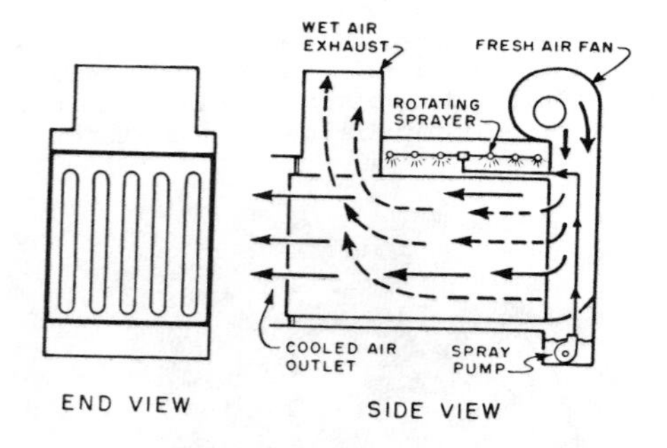

Fig. 4. Schematic of a single-fan indirect cooler with horizontal hollow plates, designed for cheap precooling of makeup ventilating air. Air entering the fan divides, part entering the plates and becoming primary air delivered left. Part becomes secondary air in the spray space around and between the tubes and exhausts upward.

Some designs use plates of corrugated or dimpled sheet aluminum; others, thin sheet plastic. Embossing creates turbulence in the primary air stream and thus increases heat flow to the evaporating water.

One indirect cooler brand with aluminum plates uses absorbent coatings on the wet sides, as does another using plastic plates.

A very different new type of indirect evaporative cooler is composed of banks of externally finned, closely spaced, parallel and horizontal round aluminum tubes called "heat-pipes". These are lined end-to-end with fabric wicking partly soaked with liquid refrigerant, and are permanently sealed at both ends.

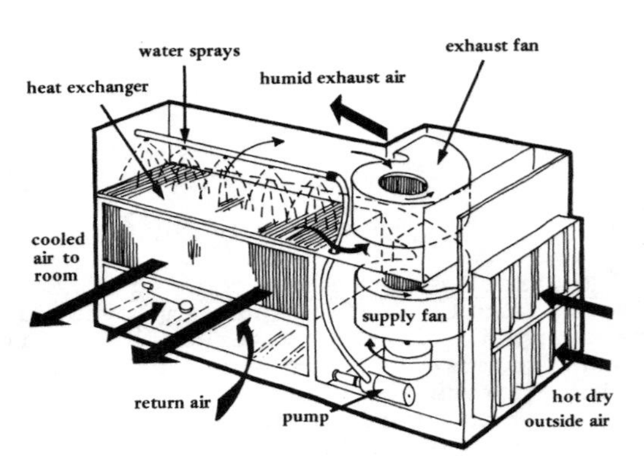

Fig. 5. An Australian indirect cooler with very closely spaced thin plates of embossed sheet plastic. Primary air enters, right, and the lower fan blows it horizontally and leftward between the hollow plates. The upper fan draws secondary air up inside the plates, counterflow to the downward moving water films.

Using cool return air from rooms as secondary air gives a regenerative performance boost.

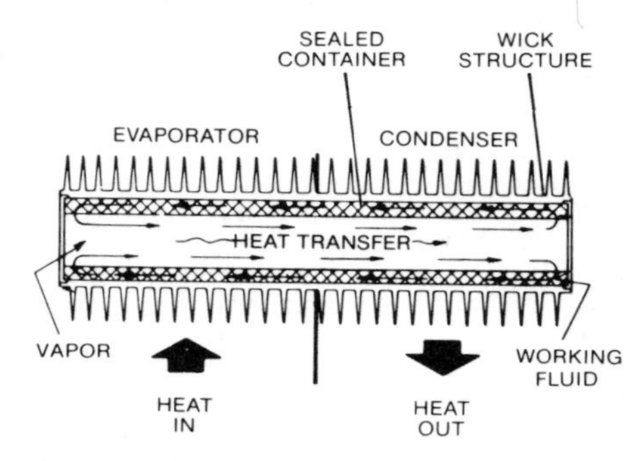

Fig. 6. Schematic of an Industrial Energy Corp. indirect-cooler heat pipe. Flowing primary air gives heat to the finned tube's left half, evaporating sealed-in liquid refrigerant. Its vapor flashes into the right half where spray and secondary air remove its heat and condense it. The wicking continuously returns the liquid to the left half again.

One half of each bank acts as cooling coil, and primary air is drawn horizontally between and across its pipes. The other half serves as cooling tower and is sprayed with water and blown with secondary air. A partition separates the two halves and the primary and secondary air streams.

Operation is simple. Primary air transfers heat to the cooling-coil halves of the tubes. This evaporates refrigerant within; its vapor flows into the cooling-tower halves where the evaporating water outside cools and condenses it. The wicking continuously returns the liquid refrigerant to the coil end to repeat the process.

The system is simple, foolproof and efficient, with nothing to wear and little to clog or corrode. It should insert easily into refrigerative cooling systems, as noted below.

Indirect Evaporative Coolers and the Coming Crisis.

All the above indirect cooler types cool air much more cheaply than does refrigeration. Accordingly, they can be inserted into existing refrigerative air conditioning systems to boost capacity and save power. Often their bare heat-exchangers will suffice, the original equipment fans moving the air through them.

Net performance, of course, varies with average secondary air wet-bulb depression, high in the Southwest, low in the lower Mississippi Valley and along shorelines, and moderate elsewhere. The indirect coolers can be staged accordingly: single-stage in arid areas, two in series for average climates, and three-stage for humid zones.

Two different opportunities exist. First, such units can precool return air from the rooms, reducing its temperature 5-15 deg F (3-8 deg C) at little cost, ahead of the conventional evaporator coils. This may well restore the capacity lost by substitute refrigerants.

Second, where buildings need significant makeup ventilating air, as under revised ventilation codes, the indirect coolers can precool it very cheaply.

Where fresh outside makeup air must be cooled from 100 F to 50 F (38 C to 10 C), compressor power is wasted, for indirect evaporative cooling could be installed to precool it 10-20 deg F (6-11 deg C) and cut the compressor's load. Altogether, this helps restore lost capacity, saves power, and reduces compressor strain and wear.

The performance of indirect precoolers can be regeneratively increased two ways. Used ventilating air being discharged from buildings should be used as precooler secondary air. Its lower temperature and humidity than fresh outdoor air will deliver cooler primary air.

Some large indirect evaporative coolers already contain ducts and dampers for such usage. Others can have them easily added.

Then, condensate from refrigerative evaporator coils should be piped into the indirect precooler sump pans, for two gains: the colder water will help cool the primary air and, as essentially distilled water, the condensate will reduce scale deposits in the system while saving water and eliminating disposal problems.

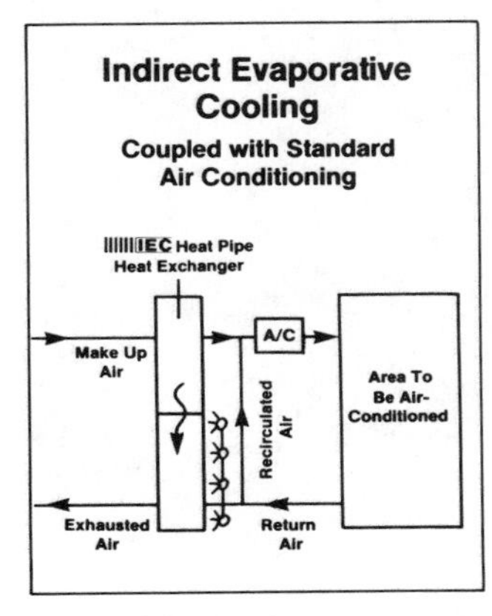

Fig. 7. A heat pipe indirect cooler precooling fresh outside makeup ventilating air for a refrigerative air conditioner. Cool dry return air from the rooms becomes secondary air for increased performance and economy.

Indirect Precoolers as Winter Fuel Savers.

Most indirect precoolers are ready-made winter fuel savers where ventilating loads are significant. Specifically, with water shut off, all but cooling-tower-and-coil types can also preheat needed fresh winter makeup air.

The incoming outside air passes through the coolers exactly like summer makeup air. However, warm, used ventilating air being discharged is ducted out through the heat-exchangers' secondary air passages and transfers heat to the much colder entering air.

Often, the latter is warmed 60% toward indoor temperatures. In cold climates, where buildings require much ventilation, fuel savings may exceed those of summer power.

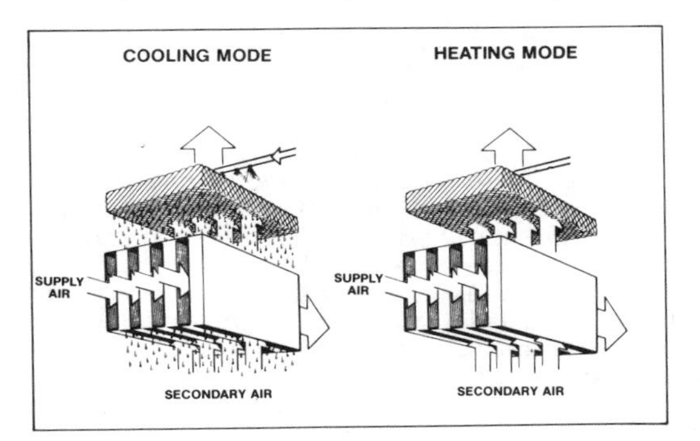

Fig. 8. How Norsaire plate-type indirect coolers save winter fuel. Here, fresh outside makeup air flows horizontally between the plates all year. In summer, water sprays and secondary air cool it. In winter, the sprays stop and warm used air discharged from the rooms becomes secondary air and preheats the cold new air.

With large ventilating loads in cold climates, the winter savings may exceed summer ones.

Prudent building owners should install such indirect precooling systems before the refrigerants crisis predicted here makes such coolers scarce. The two-season usage should bring two-year paybacks or less.

Direct Evaporative Cooling.

This superseded the indirect systems of the 1930s, as cheaper, more compact and sightly, and giving colder air for less power.

However, it cooled air by humidifying it, so gave comfort only in arid states. There, millions of homes, schools, churches, and businesses have used such cooling at savings up to 80% in first, maintenance and power costs compared to refrigerative cooling. Hundreds of thousands still do.

This evaporative cooling operates by direct contact between evaporating water and the air being cooled. Specifically, fans draw fresh outside air through wet porous pads, wet wire-mesh wheels, or dense water sprays. This creates adiabatic saturation; sensible heat in the air evaporates water into it, the reduction of sensible heat reducing its temperature instantly, sometimes 80 to 95% of wet-bulb depression.

Such cooled air, termed washed air, cools rooms by circulating vigorously through them once and discharging quickly outside. In this minute or two, it absorbs heat from walls, windows, ceilings, floors, lights, furnishings, and occupants, and carries it outdoors through provided open windows, doors, ventilators, or exhaust fans.

Though warmer than refrigerated supply air, washed air absorbs equivalent heat through greater volume flow and faster indoor circulation. Usually it also removes indoor smoke, steam, odors, and some dust.

Washed air overcomes its humidity and creates human comfort largely by circulatory velocity around persons. Here, it feels like much colder air, giving a comfort slightly different from refrigerated air but acceptable to most people.

Direct Evaporative Cooling Equipment.

The simplest, commonest, and least expensive direct coolers are called drip coolers. They consist of usually cubical boxes with louvered walls and squarish water pans for bottoms. These contain (usually) centrifugal fans which draw outside air in through the louvers and discharge it out one side or the bottom. Lining the louvered walls are porous pads about 2 in. (5 cm) thick of (usually) loosely packed aspenwood excelsior. Perforated troughs above them, supplied by small recirculating pumps in the bottom pans, drip water into them, wetting the fibers. Surplus water returns to the pans.

Most drip coolers mount outside windows and blow directly indoors without ducts. Most others mount on roofs and deliver through duct systems. Either way, air drawn through their pads is cooled perhaps 80% of wet-bulb depression, and cools factories, animal barns, greenhouses, power plants, etc. besides residences, schools, and small businesses. Unfortunately, most pads need replacement yearly.

Fig. 9. An Essick drip cooler with right and rear pad-holders removed, but showing the left aspenwood excelsior saturation pad. The water pan below shows its strainer-enclosed pump, left, and float valve, right. Above is the centrifugal fan which draws fresh outside air in through the pads and delivers cool washed air at top right.

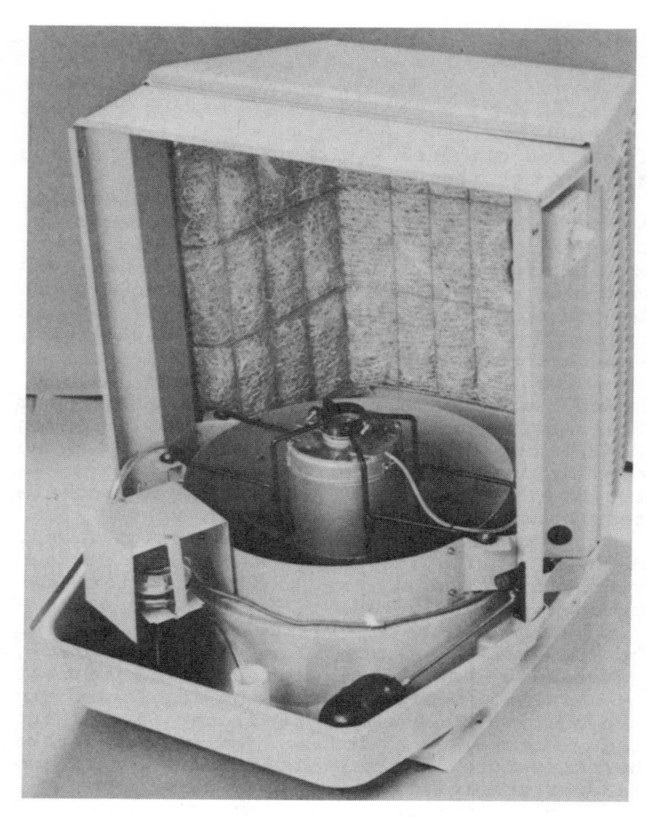

Fig. 10. A Bonaire (Australia) six-sided down-draft drip cooler with frontal pads removed. Its propeller-type fan blows down at center; its pump and float valve are at left and right respectively.

For equal air flow, the larger the pad area, the cooler the washed air and the lower the power consumption.

Fig. 11. This Bessam-Aire commercial-size sprayed-pad direct cooler has long-lasting "filters" wet by a centrifuging water slinger wheel in the water pan below. A centrifugal fan, not shown, draws air through for cooling and dust removal.

Fig. 12. This Airfan rotary direct cooler has a thick permanent wheel of corrugated wire mesh one quarter submerged in a water tank below. As it revolves slowly, the wet metal carries water films through the path of air drawn through by centrifugal fans.

Larger direct coolers usually have single square vertical pads with water sprays on one side and centrifugal fans drawing air through them on the other. A competing type has thick vertical wheel-shaped pads of heavy wire screening. Standing one-quarter deep in water and slowly revolving, these expose wet mesh to air drawn through them. Both types are long-lived and maintenance free.

The largest coolers once had spray chambers through which air was drawn to be cooled and washed. Today they have single pads perhaps 10 ft (3.05 m) tall and 20 ft (6.09 m) wide and 12 in. (30.5 cm) thick, composed of corrugated, very absorbent sheets bonded together to form thousands of pencil-size horizontal tunnels. Wet from above, these can cool air over 95% of wet-bulb depression extremely cheaply. Smaller models now compete with drip coolers.

Fig. 13. This Buffalo Forge direct cooler typifies man's cheapest artificial cooling. Pads of Munters-patent long-lasting absorbent rigid media can be almost any height and length. Usually a foot thick, they contain thousands of pencil-size horizontal tunnels with wet walls which both cool and cleanse air drawn through.

Similar single-pad direct coolers now compete with drip coolers.

Direct Evaporative Coolers and the Refrigerants Problem.

As noted, direct evaporative cooling provides comfort only in low-humidity areas. Thus, it can replace refrigerated air conditioners if necessary in only about 40% of the basic U.S., namely central Texas to Southern California and Arizona to Wyoming.

Even then, it requires much larger ducts and registers, so cannot instantly substitute for conventional systems if such cannot change refrigerants.

However, throughout the Southwest and in areas bordering it are many "piggyback" systems combining evaporative and refrigerative cooling, using each according to the weather. These can be expected to multiply as hedges against refrigerants problems and to employ the evaporative phase more as power costs rise.

In the same area, as refrigerative window units fail and cannot accept available refrigerants, some return to window-mounting drip coolers seems likely.

In the about 40% of moderate humidity U.S.A., indirect evaporative coolers with direct evaporative second stages, though little tested there, should deliver adequate comfort. Several firms make them, and power savings over refrigerated cooling should be 30-60%. However, their easiest use likely will be in new construction, not as replacements.

Fig. 14. This Vari-Cool indirect-direct staged cooler has an indirect stage at right center and a direct vaporative one at far left. Primary air entering at the right is cooled dry in the fabric-covered white tubes, then in wet rigid media. The resulting cold washed air is suitable for comfort cooling in many areas of moderate humidity.

Evaporative Precooling Refrigerative Condensers.

The greatest number and total tonnage of world refrigeration systems have air-cooled condensers. Their simplicity, reliability, and lower cost have displaced many cooling towers and evaporative condensers almost everywhere, despite higher compressor power consumption.

However, global temperatures are seemingly on a permanently upward trend particularly penalizing air-cooled refrigeration. Not only must power consumptions rise inordinately and cooling capacities fall, but climbing head pressures increase breakdowns and shorten compressor lives.

Few such systems can be economically rebuilt for cooling towers or given evaporative condensers, but direct evaporative precoolers for condenser air should suffice.

Most are thick saturation pads formed into panels easily mounted on the entering-air faces of existing air- cooled condensers. Made in both stock and custom sizes, they fit almost all tonnages of cooling equipment, using the condenser fans to draw outside air through both. Some may contain water circulating pumps, but more have automatic valves to wet the pads in "pulses" according to air temperatures.

Most begin precooling when outdoor air exceeds 80 or 85 F (27-30 C). Lower range precooling may lower condenser pressures so refrigerant flow through evaporators becomes inadequate, both for the cooling loads and oil return to compressors.

Today, precoolers are less limited; new controls feed evaporators regardless of condenser conditions. Precooling power savings and cooling capacity increases thus occur whenever the system operates.

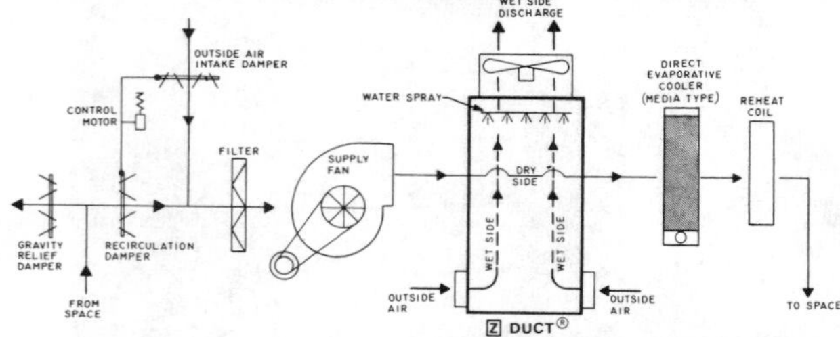

Fig. 15. This large rooftop Des Champs Laboratories indirect-direct cooler serves all year. Its schematic, below, shows dampering at left to meet ventilating needs, then the supply blower pushing primary air horizontally between the indirect heat-exchanger plates, center, and through a rigid media direct evaporative pad. The ultra-cool washed air costs 30-60% less than refrigerated supply air.

In winter, with sprays stopped, warm air exhausted from the rooms is drawn through the hollow plates to preheat mixed return and makeup air, saving much fuel. Final heating is by a steam or hot-water coil, right, shut off in summer.

Precooled Condenser Savings.

One manufacturer cites power savings of 5-10% in the East, 20% in the Southwest. But capacities also increase. A 40-ton rooftop system operating in 95 F (35 C) 40% relative humidity rooftop air was given a precooler. Its EER (Btuh/Watt) rose from 9.0 to 10.3, its capacity from 450,000 Btuh to 482,000 Btuh, or 7.1%, and its power draw fell from 50 kW to 47 kW, or 6%.[13]

Its head pressure fell 13%, probably cutting service calls and increasing compressor life at least as much.

Engineers at the University of Alabama at Huntsville, under contract to Tennessee Valley Authority, made very precise tests on a 8.25 EER, 90,000 Btuh split-system air conditioner consuming 10.9 kW at standard conditions.

Four precoolers were installed, two on each side of the condensing unit installed in a laboratory "enthalpy loop" allowing close control of operating conditions.

Air entering the precoolers was varied from 84 F to 120 F (29-49 C) at a humidity range representing the TVA region.

Test results showed that in the Memphis area, seasonal power savings of 7-10%, would result, with water costs approximating 20-33% of power cost savings. (TVA power prices are unusually low, so water costs elsewhere would be proportionately much less).[14]

A related use of precoolers is to cool the compressor or "motor" rooms of supermarkets, hotels, motels and restaurants, where all possible small air-cooled condensing units are grouped together for convenience and noise isolation.

Most are on rooftops and are cooled only by exhaust fans drawing air through screened openings. Temperatures inside may exceed 100 F (38 C) on cool summer days, so the small compressors inside run at marginal efficiency at best.

Installation of precooling panels, or where no exhaust fans exist, simple drip coolers, might payoff in a single season.

Fig. 16. Five Research Products condenser-air direct evaporative precoolers mounted on a large condensing unit. Their thick, low air-flow resistance pads are wet at timed intervals depending upon outdoor temperatures, so use little water.

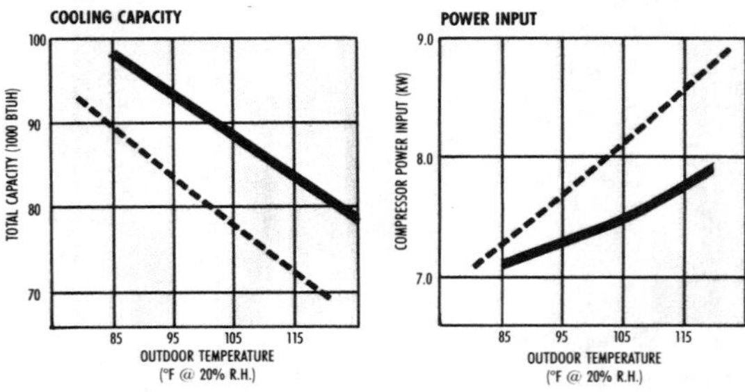

Fig. 17. These Research Products curves suggest the savings from direct evaporatively precooled condenser air. The broken lines represent operation without precoolers; the heavy lines, operation with.

The power input savings suggest compressor head-pressures and resulting maintenance reductions. The cooling capacity gains should counteract many losses due to substitute refrigerants in the difficult years ahead.

CONCLUSION

All comparisons cited here are transitory. Ambient temperatures are steadily rising and humidities slowly falling. Note three successive drought years in many U.S. states, including very dry summer 1988.

Accordingly, even without refrigerants problems, most refrigeration systems face not only rising loads but long-term declining efficiencies and rising power consumptions, together with increasing maintenance costs and shorter lives.

Simultaneously, evaporative cooling devices face long-term growing wet-bulb depressions and efficiencies. Thus, in the chaotic times ahead for refrigerated cooling, using the former to help the latter seems the least of engineering common sense.

* *

1. Time, October 19, 1987, pg. 63.
2. Atlanta Journal Constitution, March 3, 1988, pg. 4-HE.
3. International Solar Energy Society News, May, 1988, pg 2.
4. Time, op.cit.
5. Atlanta Constitution June 7, 1988, pg. 26-A
6. Wall Street Journal, March 16, 1988, pg. 38.
7. Air Conditioning, Heating, and Refrigeration News, April 18, 1988, pg. 8.
8. ACHR News, April 25, 1988, pg. 18.
9. ACHR News, February 1, 1988, pg. 3., and May 16, 1988, pg. 2.
10. ACHR News, February 8, 1988, pg. 2.
11. Journal of the American Society of Heating, Refrigerating and Air Conditioning Engineers, July, 1987, pgs. 17-25, 49.
12. John R. Watt, Loren Crow, and Alfred Greenberg, Evaporative Air Conditioning Handbook, Second Edition, Chapman and Hall, New York and London, 1986.
13. ACHR News, April 11, 1988, pg. 6.
14. Experimental Evaluation and Simulation of Air Conditioning Precoolers, Report to TVA, by T. K. Spain, G. R. Guinn, and J. A. Hall, supplied by Research Products Corporation, manufacturer.

Chapter 44

THE BILLION-DOLLAR CLEAN ROOM HUMIDITY CONTROL RACE

R. J. Morris

The American semiconductor industry is in a billion dollar race with the Japanese. New markets are springing up everywhere. More memory is needed for local area networks, text scanners, bar code readers, voice cards, artificial intelligence, digital automobile mapping and others. Although the race is complex with hundreds of variables, the number one deciding factor will be quality control. Quality control experts agree that the clean room environment must have stable humidity in the range of + or - 2% to beat the Japanese in providing microprocessors at a lower price. The stability is necessary to produce microprocessors with smaller and smaller circuits. One percent improvement in production could generate as much as $170 million dollars a year.

Millions of dollars have been spent in research and development to find ways to stabilize humidity in clean rooms. The first assumption was that standard commercial controls were not accurate enough and the solution to the problem was to buy expensive industrial controls. Industrial controls were installed and in most cases humidity was controlled at + or - 10%. The industry still finds itself in and out of tighter and tighter specifications. Some American companies have tried expensive methods of humidity control such as direct expansion units or desiccant cooling. Even though small scale results are encouraging, the large scale reliable solution has still eluded the industry.

The problem is getting worse because production departments add more and more equipment and therefore more exhaust requirements to the system. More exhaust means more outside air therefore bigger humidity problems.

What is the solution?

Before we explain the solution lets study the problem in more detail. Understanding humidity control in clean rooms requires a working knowledge of psychrometrics. The easiest way to describe psychrometrics is with the psychrometric chart depicted in Figure 1. Psychorometrics is the science of the properties of air. The properties of air described are dry bulb temperature, wet bulb temperature, dew point temperature, enthalpy (BTU/lb), relative humidity, specific volume, vapor pressure and humidity ratio. Any two of these properties will pinpoint air on the chart.

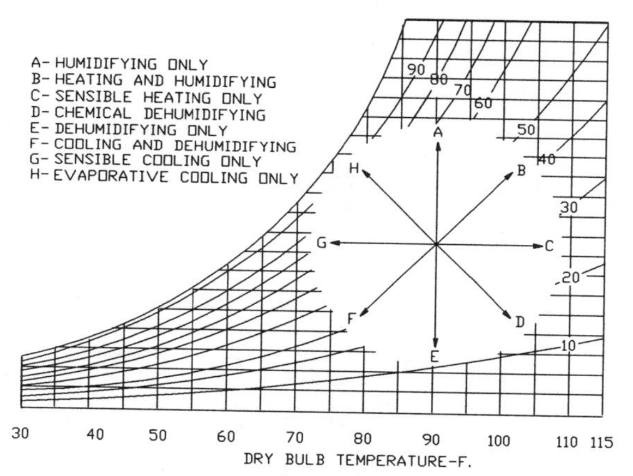

FIGURE 1. ALL OF THE AIR IN A CLEAN ROOM WILL BE DESCRIBED SOMEWHERE ON THIS CHART. MECHANICAL COOLING AND/OR HEATING CHANGES THE PROPERTIES OF AIR AND MOVES THE DESCRIPTION OF AIR ON THE CHART.

Any air involved with a clean room will fall somewhere on the chart. As you will note the properties of air vary with changes in temperature or relative humidity. Figure 2 shows the change in the properties of outside air as it passes through a precool coil.

The problems of precise humidity control are numerous:

1. Some mechanical engineers specify the mechanical systems, then depend on vendors to provide control design.

Solution: Mechanical engineers must learn the control systems at the grass roots level. They must know the limits of the sensors, transducers and controllers. They must be able to program at machine language level. Without a working knowledge of the system, the designer cannot get acceptable results.

Fill-in-the-blank loop control is not programming! Also a series of control loops does not constitute system automation!

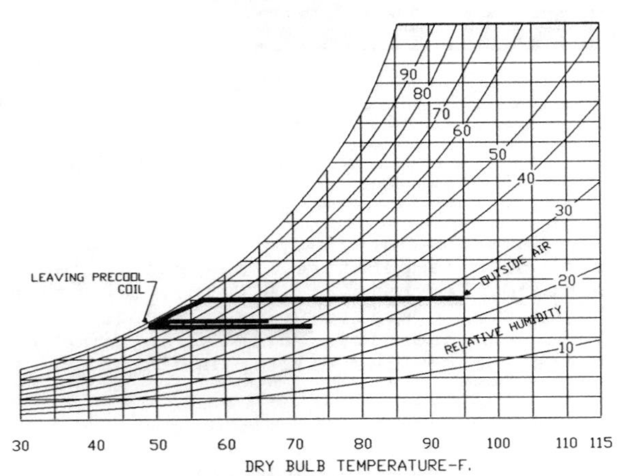

FIGURE 2. FRESH OUTSIDE AIR PASSES THRU A PRECOOL COIL AFTER FILTERING. THIS IS THE FIRST STEP OF CONDITIONING THE MAKE-UP AIR. A LARGE VOLUME OF MAKE-UP AIR IS NEEDED TO REPLACE CLEAN ROOM EXHAUST AIR.

A well written automation program plots all of the air in a clean room in memory so that all systems work together. We have seen humidifiers pumping steam into a duct when the air was already saturated. Ten feet down the duct water was dripping out. This would not happen if the programming was written as a system instead of a series of control loops. The control loop is a carry-over from the days of pneumatic controllers. A control loop consist of a sensor, a controller and an output device. Normally the controller has a setpoint and the controller modulates the output to maintain control. It is going to take years to deprogram engineers, vendors, technicians and servicemen who look at building automation as a series of independent control loops. The power of automation is harnessed using system control. For example: Setpoints for the coils in the make up air handler should be calculated from actual data in the clean rooms not from a loop control reading a mixture of return air. Figure 3 shows how changes in the clean room must be accounted for in the make up unit.

2. Matching a sensor with an actuator for loop control using a setpoint is not adequate. The result of a series of loop controllers is wasted effort and wasted energy. The clean room will be out of spec more than it is in spec.

Solution: The air handling equipment must be automated as a system not a series of loop controllers tuned to individual setpoints. The preheat, precool, secondary heat, secondary cooling coils must work in conjunction with humidifiers and reheat coils to provide stable results. The system must work in harmony like a fine tuned orchestra. What is in place today is analogous to a ten piece band with ten different sheets of music. For example: a makeup cooling coil overcools dry air, then a reheat coil warms it back up pushing the relative humidity in the wrong direction. Good system automation instead of loop control would eliminate this situation. All heating and cooling must work together to get desired results.

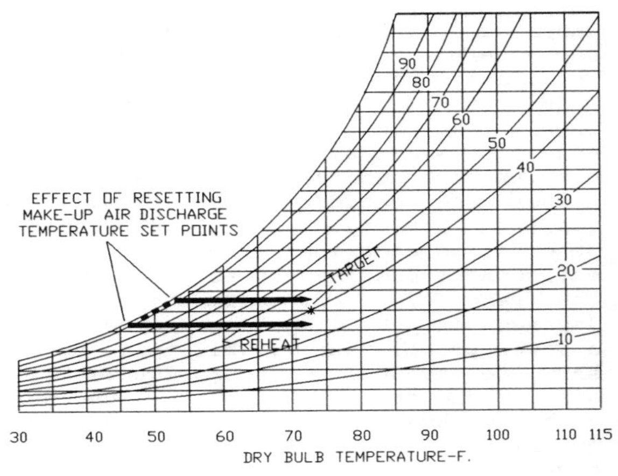

FIGURE 3. THE DISCHARGE TEMPERATURE OF THE MAKE-UP AIR CAN BE RESET UP OR DOWN TO REACH A TARGET OF 40% RELATIVE HUMIDITY THIS APPROACH REQUIRES SYSTEM AUTOMATION NOT JUST PNEUMATIC LOOP CONTROL.

3. Humidity sensors in use in clean rooms are obsolete. Some clean rooms are using lithium chloride sensors or even worse, fiber sensors to measure dewpoint or relative humidity. These old sensors are slow, hard to calibrate and unreliable.

Solution: New polymer film technology has yielded better sensors. These sensors read relative humidity and produce + or - 2% accuracy. They are reliable and repeatable. For accuracy in the 1% range, the chilled mirror sensor is available. It is much more expensive and requires more maintenance but in some areas the accuracy could be worth it. Quality control could better answer specification questions if they are applied with accurate data.

4. Temperature sensors are obsolete. Designers are still using RTDs. To obtain .1 degree F accuracy with a RTD requires special hand made sensors of exotic materials. Field calibration is required.

Solution: Use precision thermistors. A thermistor with .1 degree F accuracy cost less than $30.00 versus $300.00+ for an equal precision RTD. Unlike RTDs thermistors need no calibration.

5. Many sensors are not located properly. Sometimes temperature sensors are located in chases where air blends from two separate clean rooms.

Solution: Install sensors as close to the work area as possible. Avoid areas where blocking and blending occurs. Use several sensors per clean room so that hot spots can be located and corrected. Best results will be obtained if a temperature sensor is located at each work station.

6. Technical maintenance people are underpaid, undertrained and undermotivated. Operators and maintenance personnel must be technically qualified to keep equipment operating at optimum performance levels.

Solution: Hire technicians that know psychrometrics, electronics, pneumatics, computers, air balance, water balance and refrigeration. Pay salaries that compare with the scientist, chemist and engineer. The maintenance decisions regarding a clean room are worth millions of dollars in quality product output. Hire maintenance management that compares in quality with other departments such as quality control or research and development.

7. A mountain of data is collected in a semiconductor plant. This data is a resource that is wasted.

Solution: Manage data as carefully as electricity or water is managed. Building automation can make all departments run more efficiently if the data is compiled into useful pertinent reports. Quality control, production, research and maintenance all need accurate data to make good sound business decisions.

8. The clean room must maintain a slight positive pressure even though 30% - 40% of the air is exhausted. This requires vast amounts of outside make up air. This air must be cleaned, cooled/heated and either dehumidified or humidified. Even if most clean rooms could be tuned into specifications, big problems still occur when production departments add exhaust equipment.

Solution: Put a ceiling on exhaust levels. Maintenance managers should provide costs figures for additional equipment required to meet new higher levels of make up air. These costs should be calculated into the production budget not in the maintenance budget. Also find new ways to reduce exhaust when it is not needed. New variable speed fan technology and low pressure static control is available to cut exhaust requirements drastically. Here again system automation is the key element that is missing.

The problems associated with maintaining humidity in the clean rooms of today are diverse. These problems can be solved but it will require commitment from the highest levels of management. Executives must attack the problems with a new perspective. The old ways of doing business will not work. Designers must be held accountable for their work. The days of writing a specification around vendor equipment are over. Performance contracting is the answer. If the designer is required by contract to stay in spec for one year (or be penalized) he/she will put together a system that works.

Other forms of performance contracts will evolve. In the future, manufacturers will pay service companies or engineering firms to maintain conditions in clean rooms for a monthly price. This will put the burden on the technicians and designers where it should be. If performance contracting is unacceptable then quality maintenance personnel must be added or existing personnel must receive extensive training. Maintenance is not insignificant any more. Highly trained technical people are needed and they will cost more.

New hardware and technology must be put in place. Automation has evolved just as fast as the computer industry, therefore three year old control systems are obsolete.

We can beat the Japanese but it will require bold new steps and fresh new ideas. The entire design and maintenance process needs an overhaul. The companies that take quick action will save millions through improved production and quality control.

SECTION 8
VENTILATION

Chapter 45

VENTILATION, HEALTH AND ENERGY CONSERVATION— A WORKABLE COMPROMISE

G. Robertson

INTRODUCTION

The outcome of two recent legal claims by employees -- one in Florida and another in California -- involving indoor air pollution has received the prompt attention of many office and institution managers across the country.

In both instances, employers were forced to financially reimburse their employees for illnesses that the workers blamed on the indoor air environment of their offices. Considering the widespread nature of indoor air problems nationwide, clearly these cases constitute only the tip of the iceberg.

To appreciate the potential scale of the problem, consider the April 7, 1988 news release of the American Medical Association. This concerned the results of a study completed at the Walter Reed Army Institute of approximately 400,000 recruits during basic training over a four year period. This study examined the incidence of acute respiratory disease at four U.S. Army training centers and it was determined that trainees housed in modern, energy efficient barracks were about 50% more likely to contract a respiratory infection than were trainees in older, less air tight buildings.

Couple this data with the review of Garibaldi and Dixon published in the American Journal of Medicine (1985) who estimated that respiratory tract infections annually account for $15 billion of direct medical care costs, approximately 150 million lost work days, and at least $59 billion of indirect costs such as lost income from work absenteeism.

So whether the driving force is the threat of litigation or simply a move to improve productivity and reduce operating costs, it is hardly surprising to see an increasing focus on indoor air pollution. We believe that at ACVA we are at the forefront of such studies.

Since we established ACVA in 1981, we have pioneered a multi-disciplined approach to the investigation of internal pollution. Investigators include chemists, microbiologists, and air conditioning engineers -- three disciplines unused to working as a team. Our client list includes numerous government agencies; multi-national companies in insurance, finance, industry, banking, and property management; colleges, schools, and numerous hospitals. Most of our clients now not only ask us to examine other buildings that they own, but also enter into long term contracts of regular monitoring and preventive maintenance. In fact, as of March 1988, we have now studied the indoor air quality of over 42 million square feet of property.

INDOOR POLLUTANTS - THE SOURCES

Virtually everything we use in the interior sheds some particulates and/or gases. When a building is new, some compounds are given off quickly and soon disappear. Others continue "off-gassing" at a slow pace for years. Common office supplies and equipment have been found to release dangerous chemicals-especially duplicators and copiers and we have even found formaldehyde being released from bulk paper stores.

People themselves are a major contributor since each person sheds literally millions of particles, primarily skin scales, per minute. Many of these scales carry microbes but fortunately the vast bulk of these microbes are short lived and harmless.

Clothing, furnishings, draperies, carpets, etc. contribute fibers and other fragments. Cleaning processes, sweeping, vacuuming, dusting, etc. normally remove the larger particles, but often increase the airborne concentrations of the smaller particles. Cooking, broiling, grilling, gas and oil burning, smoking, coal and wood fires also generated vast numbers of airborne particulates, vapors, and gases. If the windows and doors are closed all of these can only accumulate in that internal environment.

INDOOR POLLUTANTS -- THE TYPES

There are many types of indoor pollutants, gases, vapors, dusts, fibers, and viable and non-viable microorganisms. Some of the more common ones are described below.

Organic Chemicals

There are arguably the widest range of pollutants with literally thousands of specific types fortunately occurring in very dilute concentrations which are usually expressed as parts per million or per billion. Most of these are presumed to be safe at the very low levels encountered, although some synergism between different organics or some incidences of organics "sensitizing" people to other pollutants cannot be ruled out. Usually the organics are more a problem in the typical home than the office and concentrations in the home are usually higher than the office mainly due to lower air exchange rates.

Radon Gas

Radon, a decay product of uranium, is present in variable quantities in soils. It moves from the soil by diffusion into the soil's air pockets or into soil water. Then the radon can migrate from the soil air through unvented crawl spaces, building foundation cracks, etc. into the indoor space. Some building aggregates, cinder block, etc. also contain radon and out-gassing from these materials add to the indoor air levels. In other cases radon enters a building via the water supply. Some of this radon is released when there is turbulence of the water such as a running tap. It has been estimated by some researchers that anywhere from 10 to 15% of the average radon we are exposed to comes from such water. However, the general consensus is that the principal source of radon in buildings undoubtedly is the soil gas. Pollution by radon is far more prevalent in homes than in offices, again mainly due to the lower air exchange rates in homes plus the fact that homes have a larger area of exposure to soil relative to building volume and soil leakage area.

Inorganic Oxides

Carbon dioxide is produced by respiration and combustion, oxides of nitrogen and sulphur are combustion products associated with gas stoves, wood, coal fires, and kerosene heaters. Carbon monoxide is emitted from unvented kerosene heaters or wood stoves and it frequently diffuses into buildings from automobile exhaust fumes generated in adjacent garages. Small to trace quantities of each of these gases and other organics are present in cigarette smoke.

Ozone is another gas that is generated, usually in very small quantities, by miscellaneous copying machines and by certain electrostatic precipitators that are used to clean up the air. In one specific case that we studied, the maintenance staff of a building switched off the main air supply fans over the weekend, but omitted to switch off the central electrostatic precipitators. Thus, ozone accumulated inside the air handlers and was subsequently delivered to the staff first thing each Monday morning. When the fans were switched on this caused a severe, though temporary, period of discomfort to the people working in the areas involved.

FIBERS

Asbestos

Prior to 1973, asbestos was the material of choice for fire-proofing, thermal insulation, and sound insulation. It was used as a spray-on insulation of ceilings and steel girders; as a thermal insulation of boilers, pipes, ducts, air conditioning units, etc.; as an abrasion resistant filler in floor tiles, vinyl sheet floor coverings, roofing, and siding shingles; as a flexible, though resistant, joining compound and filler of textured paints and gaskets; as a bulking material with the best wear characteristics for automobile brake shoes, and in countless domestic appliances such as toasters, broilers, dishwashers, refrigerators, ovens, clothes dryers, electric blankets, hair dryers, etc. In fact, the EPA has estimated that approximately 733,000 or 20% of all government, residential, and private non-residential buildings in the U.S. contain some type of friable asbestos-containing material.

The fact is that many asbestos bearing materials or products are of no health risk whatsoever when used in the normal course of events. However, if for any reason of wear, abrasion, friability, water damage, etc., any of the asbestos fibers are released into the air and inhaled into people's lungs, there is a health hazard. The scientific evaluation of all available human data provides no evidence for a "safe" level of airborne asbestos exposure, thus any quantity should be considered potentially dangerous.

Glass Fibers and Other Man-made Fibers

The glass fiber (usually referred to as fiberglass) industry is in its infancy compared with asbestos and since asbestos related illnesses only manifest themselves tens of years after exposure, there are some schools of thought that suggest glass fiber fragments will also accumulate in the lungs and cause later problems. This may be so, but it is unlikely to be anywhere near as severe. The fibers of glass are not shed in such large quantities as asbestos and most of the resins, etc. bonding the fibers together appear to be extremely effective and long lasting. However, some fragmentation does occur and this is especially noticeable when the loose fiberglass insulation, popularly used in attics and ceiling voids, is disturbed. Most of us have experienced itching on contact with fiberglass and dermatitis-type reactions are not infrequent due to airborne fiberglass particles.

MICROBES

In our reviews of the literature, the one area of indoor pollution that has received least study or research has been contamination due to microbes. Nine percent of the first 223 major buildings studied by ACVA have exhibited high levels of potentially pathogenic or allergy causing bacteria, including Actinomyces and Flavobacterium species. In addition, Legionella pneumophila, the cause of the dreaded Legionnaires' disease has frequently been isolated from inside air conditioning systems. Perhaps more significantly, we have found over twenty-eight different species of fungus contaminating air handling systems (see TABLE 1).

Alternaria sp.	Aspergillus sp.
Aureobasidium sp.	Candida sp.
Cephalosporium sp.	Chaetomium sp.
Chrysosporium sp.	Cladosporium sp.
Curvularia sp.	Diplosporium sp.
Fusarium sp.	Helminthosporium p.
Monilia sitophila	Monosporium sp.
Mucor sp.	Mycelia sterila
Oospora sp.	Paecilomyces sp.
Penicillium sp.	Phoma sp.
Rhizopus sp.	Rhodotorula sp.
Saccharomyces sp.	Scopulariopsis p.
Streptomyces sp.	Tricothecium sp.
Verticillium sp.	Yeasts

TABLE 1. FUNGI ISOLATED FROM AIR CONDITIONING SYSTEMS BY ACVA SYSTEMS -- 1981 TO 1987

Of the 223 buildings studied by ACVA between 1981 and 1987, thirty-four percent have been found to contain high levels of potentially pathogenic or allergy causing fungi, including Alternaria, Aspergillus, Cladosporium, Fusarium, and Penicillium species. In many buildings with excessive staff complaints, either Aspergillus and/or Cladosporium species of fungus were found growing to excess in the air conditioning ductwork systems. In some investigations, epidemiological tests run by various doctors have confirmed severe allergic reactions to the spores of these fungi in all affected staff. Subsequent cleaning and removal of the sources of these fungal contaminants have resulted in a complete abatement of complaints.

DIRT IN DUCTWORK

HVAC systems also have been found to be poorly designed and negligently maintained. Excessive dirt accumulations are common in ductwork, even in hospitals. Frequently dirt is built into the systems during construction since the ducts are installed long before the windows, etc. and construction dusts from the site, plus wood shavings, lunch packets, coke and beer cans, etc. find themselves brushed into the vents then "out of sight -- out of mind." Thereafter over the life of the building, more dirt enters with the supply and return air. Good filters reduce the rate of this accumulation, but the only perfect filter would be a brick wall. All filters, even the ultra-efficient HEPA filters used in hospital operating rooms allow fine particles through. Many of these fine particles coalesce, sticking to each other by adhesion or electrostatic attraction and larger particles simply grow with time. In commercial buildings, much cheaper and far less efficient filters are common. Many will stop birds and moths, but that is about all. Occasionally we find that the filters have been omitted and very frequently we find they are undersize, resulting in large air gaps that allow massive volumes of air bypass to occur. Then, there are the large electrostatic precipitators that theoretically provide ultra-efficient air. In one major building we found 16 out of their 18 precipitators were inoperative due to broken parts, many had not worked for over a year. In a major hospital, we found the power pack was missing from one of these units. When inoperative electrostatic precipitators provide zero filtration.

Dirty ductwork is a prefect breeding ground for germs. It provides an enclosed space, constant temperature, humidity, and food-- which is the dirt. No germ could wish for more!

The extent of this potential problem is huge and it is very surprising what we have found in ducts. Dead insects, molds, fungi, dead birds and rodents are common. In 1984 we found two dead snakes in air supply ducts. We have also found rotting food, builders rubble, rags, and newspapers. All of these contaminate the air we breathe. It is the dirt that encourages germs to breed -- germs which cause infections.

The dirt and dusts also may be allergenic, in fact most of the dusts are, by definition, household dusts which are notorious for causing allergies in many people.

In a survey of a 750,000 square foot hospital in Virginia, we found 14 miles of ductwork. Here are a few examples of the problems we encountered in that maze of ducts. Smoke detectors blocked by dirt and inoperative; fire dampers jammed open by dirt -- they were unable to close; reheat coils completely blocked by dirt sealing off the fresh air supply; turning vanes and even the exhaust grilles completely sealed with dirt accumulations -- in the operating suite the exhaust fan was still working against these duct blockages causing such immense negative pressure in the ducts that the ducts were bowing inward almost to the point of collapse; huge excesses of bacteria and fungi were present inside the air handling chambers and throughout the ductwork; cross infection rates were high and nurses, doctors, and patients complained about poor air quality. We have since cleaned all the air handlers and the 14 miles of ducts and have overseen the installation of more efficient filter systems. That hospital has been dramatically improved and its air quality is now well above average.

SYMPTOMOLOGY OF INDOOR AIR POLLUTANTS

In general, when one hears of a polluted building or a so-called "sick building," one hears familiar symptoms from occupants including eye and nose irritation, fatigue, coughing, rhinitis, nausea, headaches, sore throats, and general respiratory problems. Without doubt, the pollutant most often blamed for these symptoms by the public is environmental tobacco smoke (ETS). However, there are usually confounding variables presented by a number of potential contaminants that precludes a quick analysis establishing a single source of contamination. The main problem being the incredible similarity between symptoms from widely different irritants or even environmental conditions. For example, identical symptoms have been reported for individuals exposed to formaldehyde, ammonia, oxides of nitrogen, and ozone. In addition, similar symptoms are reported by those individuals suffering allergic type reactions to numerous dusts and to microbial spores such as Aspergillus, Penicillium, and Cladosporium fungi, among others. Similar symptoms have been reported from exposure to cotton dust and fiberglass fragments and an ever increasing and similar problem is encountered due to low relative humidities. The latter is well known to frequent flyers of airliners where relative humidity levels are frequently as low as 10%, compared to a normal lower comfort level of say 40%.

This similarity of symptoms is usually unappreciated by the public and in part it accounts for a bias against tobacco smoke, which happens to be the sole visible air pollutant. Furthermore, due to their unreliability, we, as a policy, refuse to rely upon or otherwise use the information generated by subjective building occupant questionnaires. Only upon careful investigation of the entire indoor environment and ventilation system of a building can be draw informed conclusions about the various causes of poor indoor air quality. As a result, we have made it our business to perform precisely such investigations. Despite being the main suspect of the occupants in many of the buildings we have examined, we have determined high levels of environmental tobacco smoke to be immediate cause of indoor air problems in only 4% of the 223 major buildings investigated by ACVA between 1981 and 1987 (see TABLE 2).

Total building studies	223
Number of square feet	39,000,000
Est. number of occupants	225,000

Summary of most significant pollutants found:

Major Pollutants in Air	% Buildings
Allergenic Fungi	34
Allergenic or pathogenic bacteria	9
Glass fiber particles	7
Tobacco smoke	4
Carbon monoxide (vehicles)	3
Miscellaneous gases	2

TABLE 2. ACVA SYSTEMS EXPERIENCE -- 1981 TO 1987

Significantly, in those few cases where high accumulations of ETS have been found, ACVA also has discovered an excess of fungi and bacteria in the HVAC system. These microorganisms usually are found to be the primary cause of complaints and acute adverse health effects reported by building occupants.

VENTILATION AND INDOOR POLLUTION

The fact is that the accumulation of many pollutants is itself a symptom of a more serious problem -- a problem of inadequate ventilation. Medicine teaches us that treating the symptoms simply does not work, one has to go after the cause of the problem.

Improper ventilation can sometimes be carried to extremes. The fresh air dampers were closed completely in over 35% of those buildings studied by ACVA (see TABLE 3).

Sample Buildings:		223
totalling 39,000,000 square feet		
Period:		1981-1987
(1)	Poor Ventilation	
	No fresh air	35%
	Inadequate fresh air	64%
	Poor distribution of air	46%
(2)	Poor Filtration	
	Low filter efficiency	57%
	Poor design	44%
	Poor installation	13%
(3)	Contaminated Systems	
	Excessively dirty ductwork	38%
	Condensate trays	63%
	Humidifiers	16%

TABLE 3. SICK BUILDING SYNDROME CAUSES -- ACVA EXPERIENCE

Three years ago we found a building where the "maintenance engineer" had bricked up the fresh air vents to save energy. In Washington State, one NIOSH investigator of a sick building found heavy duty polyethylene sheets sealing off the fresh air intakes. It turned out that these had been installed two years earlier to reduce the levels of silica dust being carried into the building from Mount St. Helens. There are also numerous incidences of inadequate ventilation due to hidden blockages inside ducts. Using fiber-optic technology, we have found many classical examples of such where turning vanes, dampers, and reheat coils inside ducts have been totally sealed with massive accumulations of dirt, loose insulation, etc.

Perhaps the most serious problem of ventilation is that there is no effective legislation mandating the uniform use of minimum fresh air requirements. Certainly some authorities do specify ventilation rates at the design stage -- most of these are based on ASHRAE or BOCA standards. However, the major problem is that there is no legislative structure, nor is there a practical policing methodology to ensure that the operators of buildings run their ventilation systems according to such designs.

THE EFFECT OF ENERGY CONSERVATION

Some of these examples of inadequate ventilation were due to ignorance or accidents, however, the complex of symptoms that I have mentioned -- the "sick building syndrome" -- may result primarily from energy conservation efforts to seal buildings and reduce the infiltration/exfiltration of air. Such efforts have reduced the natural infiltration of fresh air that previously existed in many buildings, exacerbating the often undiscovered problem of a poorly designed or maintained HVAC system.

In addition to tightening buildings and sealing windows, building managers have shut down air conditioning systems at night and on weekends in an effort to lower energy costs. When the air conditioning is shut down in humid climates, condensation builds up and settles inside the ductwork. If dirt is present in damp ductwork, spores and microbes can flourish, only to be spread throughout the building once the HVAC system is turned on the next morning. This often results in Monday morning complaints of building odors or building sickness that disappear during the week, only to recur the following Monday morning. To save more energy, automatic temperature controllers are used to cycle fans on and off during the day. Vibrations from the start-up of these fans can cause dirt and microbes trapped inside ductwork to be dislodged and carried into occupied areas.

Another energy conservation effort that may contribute to sick building syndrome is the recirculation of indoor air, at the expense of fresh outdoor air. The 35% of the buildings mentioned above were saving energy by shutting off all the fresh air.

Extremely bad distribution of air throughout the building is common, especially in those systems using multiples of fan coil units mounted throughout the various floors of the building. Local thermostats switch off individual units independently of others and micro-environments are set up. Often it is necessary to ensure that when the heating or cooling is not required, all the fans should be left running to aid circulation throughout the areas concerned.

Variable air volume systems (VAV) using VAV mixing boxes mounted in the ceiling void frequently have louvers opening into the void. When certain temperature conditions are met, the louvers open and return or exhaust air from the void can be induced into the supply air, bypassing the filtration system. We have found fiberglass, asbestos, fungi, and ETS to be recycled throughout an office due to this design.

More and more frequently one finds the following design condition, exhaust fans rated at say 70 to 80% of the supply fans. The supply fans are often automatically throttled back for energy savings, say to 25% of their rated capacity. If the exhaust fan is not adjusted at the same rate the exhaust fan can overpower the supply fan and no fresh air gets into the building. The open fresh air louvers now act as addition exhausts and the whole building runs at negative pressure. When this occurs, unfiltered outside air infiltrates into the building or, worse still, exhaust fumes are sucked up from underground garages.

In addition, as described above, the substitution of low cost, low efficiency filters to reduce pressure drops and save energy seriously reduces the efficiency of building filtration systems, and can lead to serious indoor air quality problems.

COSTS OF DEALING WITH INDOOR POLLUTION

Ventilation Costs

Without doubt, the major resistance to increasing ventilation rates has been the cost of such increases. Most companies have incorporated energy management programs and new operating budgets based on saving every energy dollar possible. However, if one reviews the actual increased energy costs and equate them to the costs of maintaining healthy staff, it is obvious that many energy conservation measures have been misguided.

Consider a 100,000 square foot building typically housing 667 staff, based on national averages. Using an average annual cost of electricity of $0.09 per kilowatt hour, the costs of providing different levels of ventilation are summarized here:

CFM of fresh air per person	Kwt/hour	Costs per annum
5	3.51	$31,500
10	3.95	$35,500
15	4.68	$42,000
20	5.56	$50,000

(CFM = cubic feet per minute, Kwt = kilowatt)

Note: Worst case of moving from 5 CFM to 20 CFM costs an extra $18,500 per annum. This equates to $28 per employee per year or less than one half day of absenteeism. In a building of 100,000 square feet housing 667 staff, the minimum cost of 1% absenteeism would be $100,000 per annum.

Filtration Costs

Consider the same 100,000 square foot building with 667 staff. Assume two main air handling units. Three classes of filter are considered, all in common use in major buildings.

Filter	Fibrous glass pad	Pleated panel	Bag filters
Efficiency*	8%	32%	80%
Unit cost	$150	$450	$2,500
Life	6 weeks	15 weeks	52 weeks
Annual cost	$1,300	$1,560	$2,500

*(ASHRAE Standard 52-68)

Note: Doubling the price yields a tenfold increase in efficiency and all staff breathe cleaner, healthier air. Note that the extra cost per employee is less than $2 per annum!

Air Handling System Cleaning

Most of the dirt problems are within the air handling units themselves. Here cleaning costs range from $200 to $1,000 per main unit. Ductwork systems may need cleaning but usually this only applies to specific sections of ductwork. However, even in the worst possible case of a heavily contaminated system throughout a building where the costs could run as high as $0.50 per square foot of building, we only need to improve the attendance by 0.5% per year (by reducing absenteeism) for this expense to be recouped in one year.

CONCLUSIONS

Over the last two to three years there has been a dramatic increase in public awareness on matters of indoor pollution. Some of the well intentioned steps to address this problem have backfired. For example, we see a proliferation of so-called "clean indoor air acts" throughout this country. On examination, these acts are limited to bans on smoking. This indicates a single-minded focus on environmental tobacco smoke -- the only visible indoor pollutant. However, simply removing tobacco smoke is certainly no way of guaranteeing clean air.

If smoke accumulates in an environment, you have a marker of poor ventilation. If the visible pollutant is trapped, it is inevitable that the invisible is also trapped. Many of these invisible and odorless pollutants are potentially far more dangerous than the smoke.

Persistent indoor air quality complaints can be resolved if the building managers and operators are prepared to focus on their ventilation systems in an appropriate manner. Instead of a single-minded focus on specific pollutants, we believe very strongly in a generic engineering approach to deal with all pollutants at the same time. We have pioneered a proactive approach to monitoring the ventilation systems in the firm belief that prevention is far more efficient than cure.

Chapter 46

LEVELS AND CONTROL OF ENVIRONMENTAL TOBACCO SMOKE IN MECHANICALLY-VENTILATED OFFICE BUILDINGS

C. W. Collett, E. M. Sterling, T. D. Sterling

ABSTRACT

Increasing complaints about the air quality by occupants in offices and other mechanically ventilated buildings have focused on one of the more easily identified elements of indoor pollution, environmental tobacco smoke (ETS). The first part of the paper reviews measured concentrations of ETS-related constituents (nicotine, respirable particulates and carbon dioxide) in mechanically ventilated buildings where smoking is permitted and also where smoking is restricted. The measurement data has been extracted from a Building Performance Database, containing 3,003 measurements of 189 different substances from 389 published or official reports. Results show minimal differences in average substance levels in smoking and nonsmoking offices.

The focus of attention on ETS as an indoor pollutant has lead to a dramatic increase in smoking regulation policies in both public and private sector organizations in North America. A common approach to smoking regulations has been the designation of smoking areas within buildings. The second part of this paper examines the need to provide separate mechanical ventilation to a designated area. ETS-related constituents were measured in smoking lounges and nonsmoking office areas which share a common ventilation system in several office buildings. Results indicate that ETS-related constituents are not recirculated throughout the nonsmoking offices by the mechanical ventilation system under typical operating conditions. These findings agree with previous research conducted by Health and Welfare Canada in offices in Eastern Canada.

INTRODUCTION

Increasing complaints by occupants in offices and other mechanically ventilated buildings about the air quality have focused on one of the easily identified elements of indoor pollution, environmental tobacco smoke (ETS).

The focus of attention on ETS as an indoor pollutant has lead to a dramatic increase in smoking regulation policies in both public and private sector organizations in North America. A common approach to smoking regulation has been the designation of smoking areas within buildings. In a minority of offices, smoking may be completely prohibited, or the workspace may be divided into smoking and nonsmoking areas. However, in most offices, smoking is prohibited in the actual workspace (with minor exceptions for private offices) while, at the same time, special areas are being set aside for smokers.

In some buildings, designated smoking areas may be separately ventilated, and so provide a workspace that is completely free from environmental tobacco smoke (ETS) [1] However, in many buildings, provisions for separately ventilating designated smoking areas cannot be made, either because of cost, occupancy patterns, or because the design of the building does not permit installation of additional ventilation. In these instances, air containing ETS from designated smoking areas may be circulated throughout the nonsmoking workspace. An important question therefore arises. To what extent does recirculated air from smoking areas effect the quality of the air in nonsmoking office areas? The effect can be estimated through comparing ETS levels in offices where smoking is unrestricted, where smoking takes place in designated and separately ventilated areas, and where smoking takes place in designated but not separately ventilated areas.

The first part of the paper compares measured concentrations of ETS-related constituents in office and other mechanically ventilated buildings where smoking is permitted and also where smoking is restricted.

The second part of this paper examines the need to provide separate mechanical ventilation to designated smoking areas by presenting the results of two separate field studies.

PART ONE

Methods

Environmental conditions in a large number of buildings have been investigated by both private and agencies, frequently in response to complaints about poor indoor air quality. As of April 1988, the results of 395 studies of workplaces such as offices, hospitals, schools and other public buildings, have been abstracted and entered into a computerized Building Performance Database (BPD).[2] The database currently contains 3,003 measurements of 189 substances. The investigations were conducted between 1974 and 1987. The BPD contains, for each building where possible, architectural and engineering data, ventilation and thermal characteristics, materials and machinery used in the buildings, information about the occupants and the building use, and levels of gaseous and particulate compounds in the ambient air.

A search of BPD was conducted to identify all measurements of carbon monoxide, particulates and nicotine in buildings where smoking was allowed and where it was restricted. Measurements of carbon dioxide were also identified where smoking was allowed and where it was restricted as a general indicator of the performance of the mechanical ventilation system to remove other occupant generated substances.

BPD contains 458 data records for carbon monoxide, (265 measurements in areas where smoking was allowed "smoking" and 193 where smoking was restricted "nonsmoking"); 346 data records for particulates, (110 "smoking" and 136 "nonsmoking"); 46 data records for nicotine, (18 "smoking" and 28 "nonsmoking") and 354 data records for carbon dioxide, (158 "smoking" and 196 "nonsmoking").

Table One presents the number of measurements, the range, and the median value for each of these substances measured in smoking and nonsmoking areas. Figures One through Four provide histograms of the distribution of values for these four substances in smoking and nonsmoking areas.

TABLE ONE

COMPARISON OF SMOKING AND NONSMOKING MEASUREMENTS
EXTRACTED FROM THE BUILDING PERFORMANCE DATABASE

	SMOKING			NONSMOKING		
	NUMBER OF MEASUREMENTS	MEDIAN	RANGE	NUMBER OF MEASUREMENTS	MEDIAN	RANGE
Carbon Dioxide	158	550 ppm	ND - 1613 ppm	196	600 ppm	300 - 2075 ppm
Carbon Monoxide	265	2.8 ppm	ND - 268.7 ppm	193	2.1 ppm	ND - 58.6 ppm
Particulates	110	0.038 mg/m^3	ND - 2.7 mg/m^3	136	0.012 mg/m^3	0.004 - 0.323 mg/m^3
Nicotine	18	1.8 $\mu g/m^3$	ND - 43.7 $\mu g/m^3$	28	ND	ND - 10.9 $\mu g/m^3$

FIGURE ONE

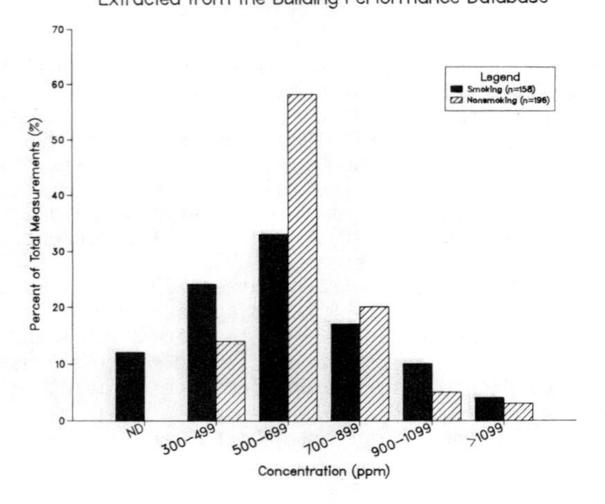

FIGURE TWO

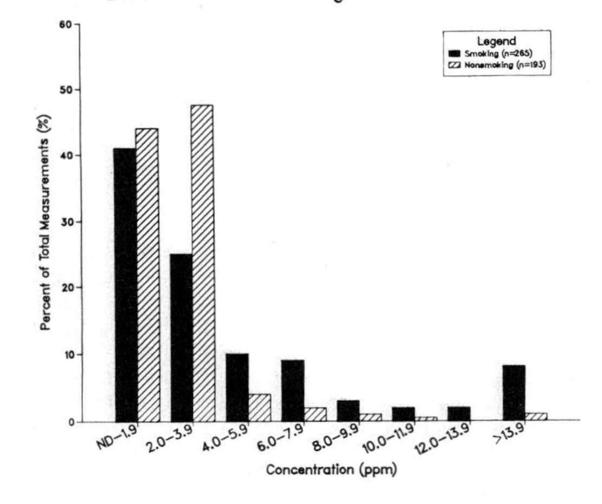

FIGURE THREE

Distribution of Particulate Measurements Extracted from the Building Performance Database

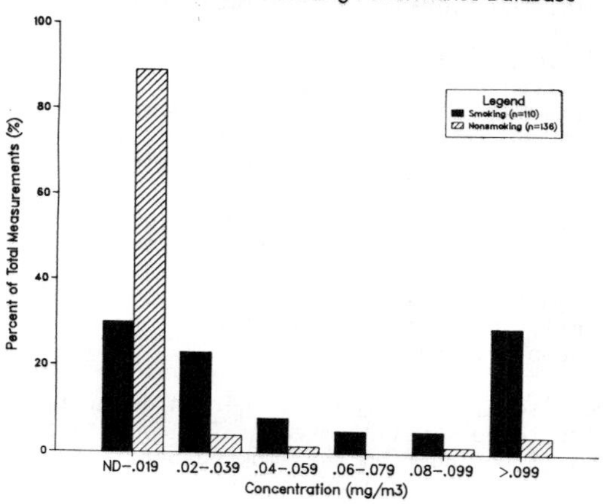

FIGURE FOUR

Distribution of Nicotine Measurements Extracted from the Building Performance Database

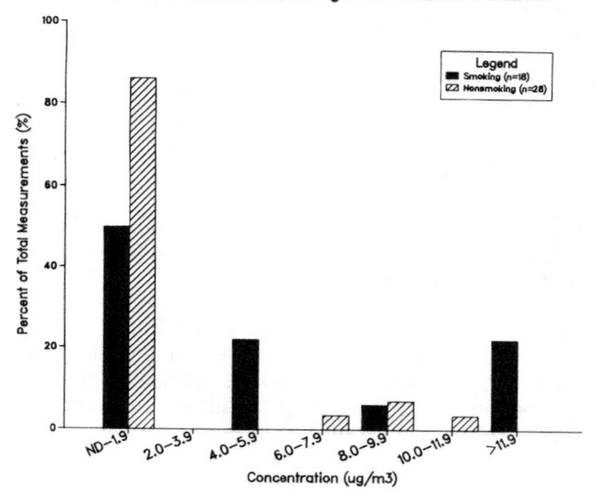

Results

The median values and ranges for the indoor substances related to ETS and for carbon dioxide listed in Table One and illustrated in Figures One through Four provide benchmark data on the pollutant levels to be expected in office workplaces, both where smoking is allowed and where it is restricted.

Table One shows that in typical office buildings whether or not smoking is allowed, concentrations of ETS-related constituents are low and of little concern. However, concentrations of both particulates and nicotine were noticeably higher in areas where smoking was allowed. Carbon dioxide, suggested by many air quality authorities as an index of ventilation adequacy, was slightly higher in nonsmoking areas. This is not surprising considering that ventilation rates may be reduced where smoking is prohibited.

Figures One through Four illustrate the similarity between measures of ETS-related constituents in offices where smoking is allowed and where it is restricted. For carbon monoxide, the majority of measurements for both smoking and nonsmoking areas were below 4 ppm, however for both cases levels above 13 ppm were found. For particulates, the majority of measurements were below 0.04 mg/m³. However while virtually all levels in nonsmoking areas were below 0.1 mg/m³, there were levels above 0.1 mg/m³ in smoking areas. For nicotine, as would be expected, concentrations in areas where smoking was allowed were substantially higher than where it was restricted. Although carbon dioxide concentrations were similar in both smoking and nonsmoking areas, more measurements occurred at higher concentrations in nonsmoking areas.

PART TWO

Methods

Part Two presents the results of two separate field studies of ETS levels in Canadian offices under conditions of normal occupancy, smoking and ventilation. The two studies were designed to provide data on the following question:

How does restricting smoking to specially designated, but not separately ventilated areas, affect the levels of these four substances in nonsmoking offices?

Field Study One: In the first study, nicotine, CO and CO_2 concentrations were measured on the 7th and 11th floors of a government office building. (Measurements of RSP were also taken but, unfortunately, were to be inappropriately analyzed.) Smoking was permitted freely on the 11th floor but was restricted on the 7th floor to a 22.5 m² (242 ft²) coffee/smoking lounge where smoking was permitted at all times. The layout and design of the two floors were almost identical, with an open-area office of approximately 780 m² (8,400 ft²) surrounding a 114 m² (1,230 ft²) mechanical/service core.

Both floors had independent ventilation system which recirculated between 80% to 85% of the air returned from each floor and provided at least 20 cfm (cubic feet per minute) of fresh air per person. Except for leakage through elevator shafts and stairwells, no mixing or recirculation of air between floors occurred. The coffee/smoking lounge on the 7th floor was on the same ventilation system as the rest of the 7th floor.

Field Study Two: In the second study, RSP, nicotine, CO and CO_2 concentrations were measured in two adjacent buildings (A and B) containing a mixture of open-area offices, private offices and public waiting/service areas.

Building A was a sealed, mechanically ventilated four storey office building with two levels of underground parking. Smoking was prohibited in all areas of the building except for a smoking section, in the fourth floor cafeteria, which was not separately ventilated. The ventilation system mixed indoor air from all parts of the building before recirculation.

Building B was a 12-storey unsealed office building where most areas were passively ventilated by building leakage. Few areas had a separate ventilation system, and these systems were not connected to other ventilation systems. Consequently, there was no mechanical mixing of air from different floors or offices. Smoking was prohibited in all areas of the building except in the smoking section of a basement cafeteria. Heated/cooled air was supplied separately to the cafeteria and exhausted through windows.

Sampling Protocol for Field Studies

Nicotine: In Field Study One, eight one-hour nicotine samples were collected in the designated smoking room, ten one-hour samples in the nonsmoking offices on the 7th floor, and ten one-hour samples in the smoking-permitted offices on the 11th floor.

In Field Study Two, six one-hour nicotine samples were collected in each of the smoking and nonsmoking sections of the cafeteria of Buildings A and B; two samples on each of the four floors in the nonsmoking offices of Building A and two samples in nonsmoking offices of Building B. Of the samples obtained in the nonsmoking offices of Buildings A and B, six collected air for two hours, three collected for four hours and one collected air for eight hours.

Ambient nicotine was sampled with a portable air sampling pump housed inside a briefcase. The sampling apparatus was designed to collect samples unobtrusively becuase of the previously noted effect of observation on occupant behaviour.[3] Nicotine samples were collected by pumping air at one litre per minute through sorbent tubes containing XAD-4 resin. Samples were analyzed using a gas chromatograph equipped with a nitrogen-phosphorous detector.

RSP: RSP's were measured in Field Study Two during the entire period of nicotine sampling and were averaged over each sampling period. RSP (particles less than 5 um diameter) levels were measured with a Shibata Scientific Technology P-5H digital dust indicator which senses light side-scattered by suspended particles.

CO and CO_2: CO and CO_2 levels were measured over three to four minute periods approximately midway into the one- and two-hour nicotine sampling periods and at least twice during the four- and eight-hour nicotine sampling periods. CO was measured using a direct reading electrochemical analyzer (Nova 310L). CO_2 was measured using extra low range CO_2 Gastec detector tubes and a manual sampling pump.

Other Observations: During each sampling period, the number of occupants and the number of cigarettes smoked in each predefined observation area were recorded. The observation areas were defined by the ability to survey the area. For purposes of comparison, the average number of persons per 10 m^2 and cigarettes smoked per hour per 10 m^2 were calculated.

Results

Field Study One: Table Two presents the average concentrations of nicotine, CO and CO_2, the average number of persons per 10 m^2, and the number of cigarettes smoked per hour per 10 m^2 (where applicable) for the smoking-permitted and nonsmoking floors and the designated smoking area.

Although the three areas differed substantially in the number of cigarettes smoked per hour per 10 m^2, only ambient nicotine levels responded in a similar fashion. For example, the average smoking intensity (cigarettes per hour per 10 m^2) was 10.8 times greater in the smoking designated area than in the smoking-permitted floor. Similarly, the average ambient nicotine level was 15.6 times greater in the smoking designated than the smoking-permitted areas. Corresponding ratios for CO_2 and CO were only 1.3 and 1.7 respectively.

Smoking was not observed on the nonsmoking floor. The average ambient nicotine concentrations were below the limit of detection (i.e. less than 1.6 ug/m^3) for the one hour sampling periods used. CO and CO_2 levels were slightly lower on the nonsmoking floor than on the smoking-permitted floor.

Table Two also lists the average person densities for each of the three sampling areas. The average person density was 30% higher on the smoking-permitted floor and 60% higher in designated smoking area than on the nonsmoking floor. Corresponding ratios calculated for CO and CO_2 show a similar pattern and range.

Field Study Two: The data for the smoking and nonsmoking cafeteria sections of Buildings A and B were combined because there were only very small differences in the results for each building.

Figures Five through Eight summarize the CO_2, CO, RSP and nicotine concentrations measured in four areas under different smoking and ventilation conditions: the smoking sections of the two cafeterias, the nonsmoking sections of the two cafeterias, the nonsmoking offices of Building A which received recirculated air from other areas of the building, and the separately ventilated nonsmoking offices of Building B which did not receive recirculated air from other areas of the building. In each figure, the height of the bars give the average concentrations, n gives the total number of samples on which the average is based. The range of observed values is given by the vertical lines.

Figure Five shows that there was little difference between the CO_2 levels in the smoking prohibited offices of Building A and Building B, whether or not they received recirculated air. The smoking and nonsmoking sections of the cafeterias had slightly elevated, but not statistically significant, CO_2 concentrations when compared to the nonsmoking offices. These small increases in CO_2 levels could have been

TABLE TWO

Comparison of ETS Related Air Quality Variables (Averages) in Smoking Prohibited and Permitted Work Areas and in Designated Smoking Areas, Site 1

	Nicotine ($\mu g/m^3$)	CO (ppm)	CO_2 (ppm)	Persons per $10m^2$	Cigarettes/ hour/$10m^2$
Smoking Permitted	4.8	2.5	720	0.79	0.36
Smoking Prohibited	<1.6	2.1	680	0.61	NA
Designated Smoking	75	4.2	960	0.97	3.9

partly due to occupant density which was twice as high in the cafeteria sections than in the nonsmoking offices.

FIGURE FIVE

Comparison of Averages and Ranges of CO_2 Concentrations for Different Smoking Regulations

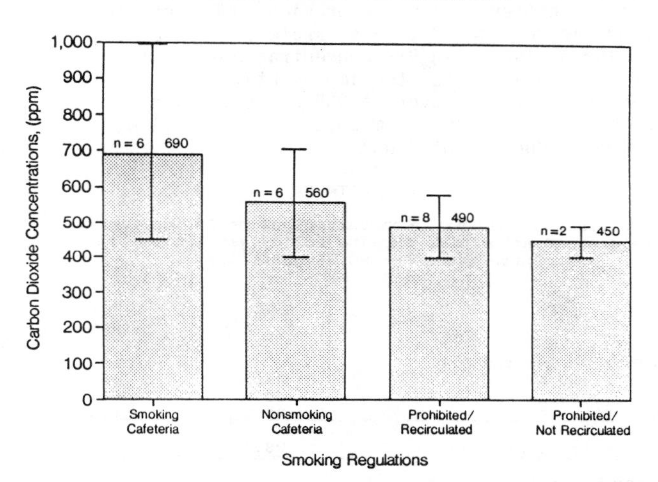

The distribution of mean CO levels is similar to that for CO_2, as shown in Figure Six. Although mean CO levels were higher in both cafeteria sections than in nonsmoking offices, none of the differences were statistically significant.

FIGURE SIX

Comparison of Averages and Ranges of CO Concentrations for Different Smoking Regulations

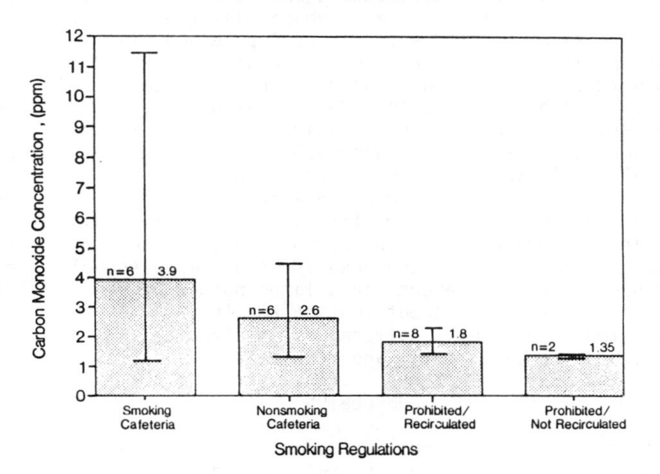

There were no differences in RSP concentrations between the nonsmoking offices with and without recirculated air, as shown in Figure Seven. However, the mean RSP level in the smoking section of the cafeterias was 2.6 times that in the nonsmoking offices while the mean level in the nonsmoking section of the cafeterias was about 1.7 times that found in nonsmoking offices.

FIGURE SEVEN

Comparisons of Averages and Ranges of RSP Concentrations for Different Smoking Regulations

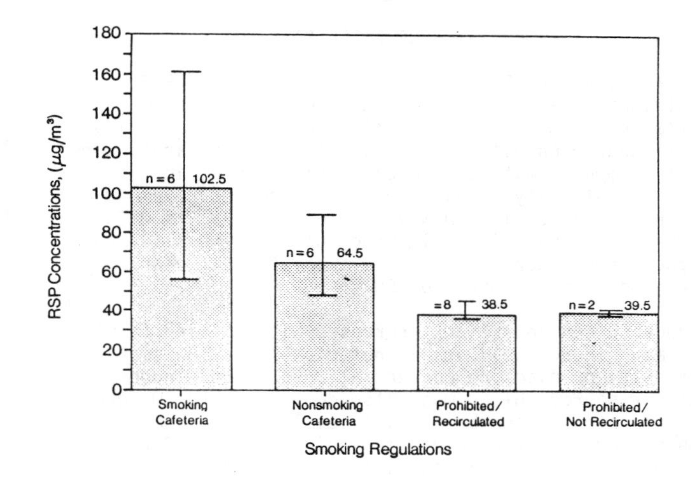

Figure Eight gives average ambient nicotine concentrations. In some instances the nicotine levels in the cafeteria sections were below the detection limit of 1.6 $\mu g/m^3$. The mean nicotine concentration measured in the smoking sections of the cafeterias was more than twice that in the nonsmoking sections of the cafeterias and at least 15 times that in the nonsmoking offices which received recirculated air.

FIGURE EIGHT

Comparison of Averages and Ranges of Nicotine Concentrations for Different Smoking Regulations

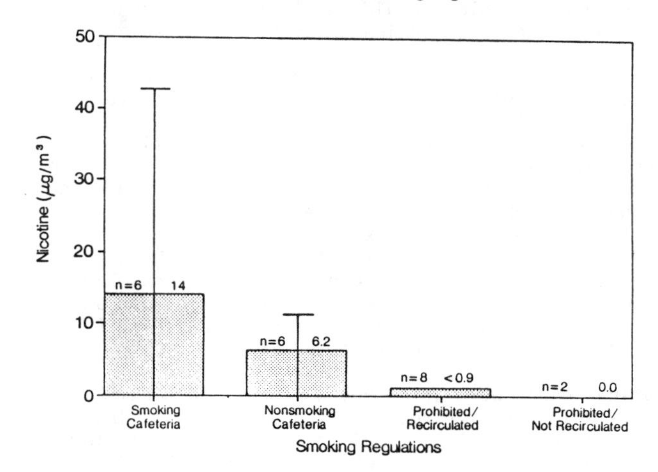

The lower detection limit for nicotine is dependent upon the amount of air that is sampled. For the method used in this study, the lower detection limit was 0.8 $\mu g/m^3$ for a two-hour sample. 0.4 $\mu g/m^3$ for a four-hour sample, and 0.2 $\mu g/m^3$ for an eight-hour sample. Two-, four-, and eight-hour samples were taken in nonsmoking offices in Building A which received recirculated air. None of four two-hour samples were above the detection limit; one of three four-hour samples were above the detection limit and gave a determination of 1.0 $\mu g/m^3$; the single eight-hour sample was also above the detection limit and yielded a determination of 0.8 $\mu g/m^3$. These results indicate that the ambient nicotine concentration in these nonsmoking offices was not larger than the maximum positive result of 1.0 $\mu g/m^3$.

Two two-hour nicotine samples were taken in nonsmoking offices in Building B which did not receive recirculated air. Nicotine was below the detection limit of 0.8 μg/m³ in both samples.

DICUSSION OF FIELD STUDIES

The results from Field Study One suggest that ETS contributes little to indoor CO_2 levels. The differences in CO_2 concentrations between the smoking-permitted, prohibited and designated smoking areas were small compared with the observed differences in smoking intensity (cigarettes smoked per hour per 10 m²) and nicotine levels. However, the sample sites with higher CO_2 levels in Field Study One and Two also had higher person densities. This suggests that people were the primary source of CO_2.

Indoor CO concentrations increased with smoking (Table One adn Figure Two) but did not closely sollow smoking intensity or nicotine concentrations. Other indoor and outdoor sources of CO must also contribute to CO levels, as indicated by the background level of 1.35 ppm in the nonsmoking office without air recirculation. Part of the higher CO level in the cafeteria could also be due to cooking activities.

RSP are produced both by smoking and by many other processes. The background RSP level, as indicated by the results for the nonsmoking office without recirculated air, are about 39 μg/m³.

Not surprisingly, of the four substances measured, nicotine shows the strongest association with smoking. There are few, if any, significant sources of nicotine in the non-industrial indoor environment other than smoking and it follows that nicotine is an accurate marker of ETS exposure. Improvements in nicotine measurement technology could result in the widespread use of nicotine as an indicator of ETS exposure.[4]

Restricting smoking to specially designated areas which are not separately ventilated appears to effectively prevent high ETS concentrations of components in adjacent nonsmoking areas. Both RSP and nicotine concentrations declined sharply from the smoking to the nonsmoking sections of the cafeterias in Building A and B (Figures Seven and Eight). The recirculation of air from the smoking and nonsmoking sections of the Building A cafeteria further diluted ETS to the extent that the levels of CO, CO_2, and RSP in the nonsmoking offices of Building A were approximately the same as those levels in the nonsmoking office in Building B, which did not receive recirculated air.

Nicotine levels were at or below 1 μg/m³ in nonsmoking offices which receive recirculated air from smoking designated areas. This level of exposure is very low. For example, breathing air which contains 1.0 ug/m³ of nicotine for one hour at an average respriation rate for office activity of 0.48 m³/hr[4] is approximately equal to 1/1,900thof the 900 μg of nicotine inhaled by a smoker from the mainstream smoke of one cigarette.[6]

There are few published studies of nicotine levels in nonsmoking offices. The nicotine levels observed by Bayer and Black[7] are, unfortunately, not comparable with the results given here because their results are given in ng/m²/min. Nevertheless, they did not detect nicotine in two of three nonsmoking offices.

Other office studies have also found low RSP levels in nonsmoking areas close to smoking designated areas that are not separately ventilated.[8] These findings are reinforced by Health and Welfare Canada who measured RSP, CO and CO_2 concentrations before and after the implementation of a no smoking policy in offices that received recirculated air from a designated smoking area. After the no smoking policy was implemented, mean RSP levels decreased by 8 μg/m³, from 26 μg/m³ to 18 μg/m³.[3]

Another Health and Welfare Canada study measured RSP and CO_2 on three floors of an office building before and after the implementation of a no smoking policy.[9] This study differed from the previous HWC study in that smoking was restricted to a separately ventilated area. Table Three gives mean RSP concentrations for the three floors before and after the implementation of the no smoking policy. The average RSP concentration on the three floors decreased from 28.1 μg/m³ to 21.1 μg/m³, for a net reduction of 7 μg/m³.

TABLE THREE

Comparison of RSP Mean Concentrations on Office Floors Before and After a No SMoking Policy was Implemented (Extracted from HWC, August, 1985)

	RSP (μg/m³)	CO_2 (ppm)
Before No Smoking Policy		
Floor A	29.3	663
Floor B	30.0	614
Floor C	25.0	606
Overall Average	28.1	627
After No Smoking Policy		
Floor A	22.7	591
Floor B	19.8	551
Floor C	20.8	503
Overall Average	21.2	551

The results of Field Study Two and the HWC studies indicate that recirculated air from designated smoking areas contributes less than 8 μg/m³ of RSP to nonsmoking offices which receive such air. HWC further suggests that ad lib smoking in offices under normal ventilation and occupancy conditions contributes about 7 μg/m³.[9] These findings are substantially different from the widely quoted estimate by Repace and Lowrey[10] that smoking increases RSP levels in offices by 170 to 200 μg/m³. This estimate, aslo accepted by Health and Welfare Canada,is based on modelling exposures. However, Repace and Lowrey's estimate is up to 21 to 29 times greater than the observed effect of smoking on RSP levels in the two HWC studies and, in fact, is 5.3 to 6.3 times larger than the average RSP concentration of 31.8 μg/m³ from all sources found in smoking designated areas in the HWC study.[9] These results further verify the results of a comparison of air substance measurements in a large number of buildings with and without smoking regulations.[11] The study found only small differences in airborne particles (combined total and respirable particles).

CONCLUSION

While smoking regulations are here to stay and will affect most offices under federal, provincial or municipal control in the United States and Canada, the haste to regulate smoking may have been based on unrealistic modelled estimates or on "worst Case" measurements in poorly ventilated workplaces, instead of on actual measurements of ETS levels in typical offices. In a majority of buildings where ETS levels were measured, levels in smoking and nonsmoking areas did not differ substantially. In fact, even in smoking areas ETS levels were at such low levels that concern for health of occupants due to ETS exposure was not warranted.

Although the smoking regulations being implemented may not be warranted as protection for nonsmokers, the provisions of a designated smoking area appears to effectively reduce ETS constituent levels in nonsmoking offices. Also, ETS levels measured in buildings with smoking restrictions clearly demonstrates that designated smoking areas need not be separately ventilated to reduce ETS levels. However, we should caution that an exclusive reliance on regulating smoking while ignoring other sources of indoor pollution in the office environment will not provide a solution to indoor air quality problems, expecially in so-called "Sick Buildings".[11]

REFERENCES

1. C.W. Collett and T.D. Sterling, The Regulation of Smoking in the Canadian Workplace (Submitted to Canadian Journal of Public Health), 1988.

2. E.M. Sterling, J.F. Steeves, C.D. Wrigley, T.D. Sterling and J.J. Weinkam, Building Performance Database. Los Angeles, California (Proceedings of the International Conference on Building Use and Safety Technology), 1985.

3. Health and Welfare Canada, Investigation on the Impact of a New Smoking Policy on Office Air Quality of the National Library of Canada Hull, Quebec. Occupational Health Unit Medical Services Branch, 1985.

4. Health and Welfare Canada, Investigation on the Impact of a New Smoking Policy on Office Air Quality of the Auditor General, Ottawa, Ontario. Occupational Health Unit Medical Services Branch, 1985.

5. ASHRAE, ASHRAE Standard 61-1981R, Ventilation for acceptable indoor air quality. American Society of Heating, Refrigerating and Air-Conditioning Engineers, Atlanta, Georgia, 1986.

6. M. Muramatsu, S. Umemura, T. Okada and H. Tomita, Estimation of personal exposure to tobacco smoke with a newly developed nicotine person monitor. Environ Res, (35): 218-227, 1984.

7. C.W. Bayer and M.S. Black, Passive smoking: Survey analysis of office smoking areas vs. environmental chamber studies. Atlanta, Georgia (Proceedings, ASHRAE IAQ'86, pp. 281-291), 1986.

8. H.K. Lee, Investigation of the impact of a new smoking policy on office air quality in the National Library of Canada in Hull, Quebec. Ottawa (Proceedings of the 78th Annual Meeting of the Air pollution Control Association), 1985.

9. Health add Welfare Canada, Involuntary Exposure to Tobacco Smoke, Report of the Advisory Committee, Environmental Health Directorate, Ottawa, 1987.

10. J.L. Repace and A.H. Lowrey, Tobacco smoke, ventilation, and indoor health. Science (208): 464-472, 1980.

11. T.D. Sterling, C.W. Collett and E.M. Sterling, Environmental tobacco smoke and indoor air quality in modern office work environments. J Occup Med, 29: 57-62, 1987.

Chapter 47

DESIGN AND APPLICATION OF AIR FILTRATION SYSTEMS FOR SUPPLYING VENTILATION TO COMMERCIAL BUILDINGS

J. H. Giles

ABSTRACT

Problems associated with indoor air pollution have led many international, national and local organizations to search for a better understanding of its nature as well as practical solutions for its control. ASHRAE, in its STANDARD 62-1981R, has comprehensively addressed proper ventilation with both outside air as well as recirculated and filtered air as one of the most important factors leading to acceptable indoor air quality.

The proper design, application, operation, and maintenance of air filtration systems is presented in this paper as a definitive criteria for utilization of energy efficient recirculating air systems for supplying ventilation in lieu of outside air in commercial structures.

The conclusion of this paper is that properly designed, operated and maintained air filtration systems are in many cases superior to outside air systems for achieving quality indoor air, and that HVAC systems employing filtration are more energy efficient, cost effective, less complex and easier to manage than are those making use of large quantities of outside air.

INTRODUCTION

During the 1970's the world experienced several energy crises which ultimately resulted in an all out effort to reduce its consumption. Of the many ways to achieve energy conservation inside commercial buildings, reductions in ventilation were among the most popular, since lowering outside air intake achieved reductions in both required air conditioning capacity and operating costs associated with conditioning ventilation air. Other energy conservation efforts were achieved by reducing air circulation, installing inoperable windows, sealing the building envelope against the outside elements and other construction and occupancy related efforts.

Shortly after these new energy efficient buildings were occupied, a new problem called "Tight Building Syndrome" began to appear." Initially unexplained, the sickness associated with this phenomenon ultimately was explained as the manifestation of exposure to air pollutants trapped inside these new, tight structures.

Today, improved knowledge about the indoor air pollution problem and its causes seemingly presents a dilemma for builders and building occupants alike, for conventional wisdom pits indoor air quality and energy conservation against one another. However, this does not have to be the case.

VENTILATION

It is now known that if commercial buildings are to be free of indoor pollution at levels where the health and safety of workers is unaffected, large amounts of ventilation air must be employed to dilute these contaminants. Ventilation is traditionally thought of as the use of outside air to replenish oxygen and to dilute air contaminants inside buildings. To achieve this stated purpose, the outside air must be relatively free of air pollutants, since using polluted outside air to dilute polluted indoor air can potentially exacerbate the indoor pollution problem. Even when outside air is of relatively high quality, the difficulty of bringing it into a building's HVAC system and distributing it to areas where it's needed is formidable. These problems make use of outside air for ventilation difficult, and combined with the energy consumed in its temperature and humidity control, its use oftentimes becomes expensive and impractical.

A practical, energy efficient alternative to outside air for ventilation is readily available in filtered, recirculated air. However, since conventional filtration does not produce oxygen, ventilation with filtered air is confined to that portion of the ventilation component which is required to dilute indoor contaminants. Under normal conditions, five (5) CFM of outside air is sufficient to replenish oxygen, so ventilation air in excess of 5 CFM can potentially be provided by filtered, recirculated air from which objectionable air contaminants have been removed.

ENERGY SAVINGS

ASHRAE, in its proposed Standard 1981R, suggests that a minimum of 20 CFM of ventilation air be used inside an office building. This represents a four fold increase over the presently accepted 5 CFM of required ventilation. If this additional 15 CFM of air comes from the outside, it must be heated and/or cooled, humidified and/or dehumidified, cleaned (if it does not meet EPA ambient outside air quality standards) and then distributed throughout the structure. This is a difficult, energy inefficient and expensive process which can dramatically increase the heating and/or cooling load on a building's HVAC system and create a need for larger, more expensive equipment with extended operating hours.

The added expenses associated with these energy inefficiencies can be reduced by a large degree by employing filtration systems to remove contaminants from the air in lieu of diluting their concentration with less polluted outside air. By employing filtration systems which meet the criteria discussed below, air from which pollutants have been removed can be recirculated and the energy which would have normally been used to condition a like amount of outside air can be conserved.

FILTRATION DESIGN CRITERIA

In order for filtration to achieve its desired goal of mass contaminant reduction resulting in high air quality for ventilation, the filtration system must be designed, located, applied and maintained properly. This implies that a proper system of filters must be employed which is comprised of high efficiency particle filters and gas sorbers, located to ensure that contaminants are circulated through them at constant high rates and applied in such a manner that air distribution and mixing are enhanced. Furthermore, it is critical that the filters in these systems be inspected and changed on a regular schedule to ensure their ability to remove contaminants of concern.

Since particles in the respirable size range of .2 to 5 micrometers are those with which there is the most concern in indoor air quality health issues, particle filters must be of sufficient efficiency and capacity to remove these ultra small particles with a high degree of reliability. Therefore, filters with a minimum ASHRAE 52-1976 efficiency of 90% and a minimum depth of 6 inches (for desired six month minimum life cycle) are required. In addition, these filters must be used in conjunction with gas sorbers capable of removing a broad range of gases present or anticipated in the airstream. In addition, they must be of sufficient capacity to remain active for a full service cycle (recommended as six months at 24 hours per day, seven days per week operation) and designed so that velocity through the collection bed provides a residence time approximating .1 seconds, yielding high efficiency collection of many commonly found gases. The entire system of filters must also be designed to operate with a minimum energy penalty for static pressure compensation.

Since low circulation rates are a characteristic of modern central air-handling systems, and since poor distribution efficiency often creates poorly mixed air, filtering at a central air handler is usually not effective in removing enough indoor contamination to produce a high state of indoor air quality. Therefore, filtration systems must be utilized in locations where high circulation rates can be obtained with careful attention paid to distribution in order to ensure proper mixing. A filtration system should therefore contain its own fan system and operate either in conjunction with the HVAC system or independently of it. Air should be circulated through these systems at rates of between 6 and 10 times per hour with distribution of cleaned air being ducted to discharge diffusers that direct air in a plug, or horizontal laminar flow. The air can then circulate in a sweeping pattern across a space to return air intakes on the opposite side of an occupied zone, such intakes either being ducted to the filtration system or open to a conditioned return plenum. Filtration units should be located in as small a zone as possible to enhance efficient circulation and distribution.

When integrating high efficiency particle and gas filtration systems with a building's HVAC system, air-handling equipment with the capability of circulating air continuously at 6 to 10 circulations per hour in small zones is required. Equipment such as fan coil units or series type fan powered mixing boxes which operate on continuous fan, meet this criteria. In such systems, these air handlers must employ the necessary combinations of high efficiency particle filters and gas sorbers operating against total static pressures in the range of 2 inches W.G. produced by the filters. In addition, constant-volume central air-handling systems, employing proper distribution systems and high circulation rates (6 to 10 times per hour), could be designed to integrate the required filtration into the system for maximum cost effectiveness. A word of caution, however: Whenever air handling and filtration are integrated with equipment containing cooling coils, the filters must be located downstream of the coils and drain pan in order to ensure that microbiological contaminants living on those wet surfaces are removed before the air is distributed to the occupied zones.

Air filtration systems operating independently of HVAC systems should be employed in situations where integrated air handling/filtration systems are either not possible or are undesirable. In such cases, these systems should be designed to employ the same filter combinations, circulation rates, and distribution effectiveness as integrated systems. Air should be circulated in its own closed-loop system with contaminated air being circulated out of the room into return air grilles, ducted to the filtration unit, cleaned, and then discharged into the occupied zone through high velocity diffusers at the opposite end of the zone, ensuring a plug flow across the space. These systems should operate with continuous flow at all times when the zones are occupied.

Maintenance of filtration systems to ensure that filters, are changed on a regular basis is paramount to the systems performance. Particle filters and gas sorption media should be inspected on a regular basis (no longer than six months) and replaced when necesary to ensure their effectiveness in removing contaminants.

CONCLUSION

In summary, recirculated air, filtered for removal of indoor pollutants, provides designers and users an opportunity to properly ventilate commercial spaces without dramatically increasing a building's HVAC capacity, energy consumption, and operating costs. When properly designed, located, applied and maintained air filtration systems are employed inside a building and used in combination with minimum outside air (5 cfm/person), the result is more efficient air distribution, reduced energy consumption, enhanced thermal control, and, in general, acceptable air quality free of contaminants at harmful or objectional concentrations.

REFERENCES

ASHRAE. 1981. ASHRAE Standard 62-1981, "Ventilation for Acceptable Indoor Air Quality." Atlanta: American Society of Heating, Refrigerating, and Air-Conditioning Engineers. 19pp.

ASHRAE. 1987. ASHRAE Standard 62-1981R, "Ventilation for Acceptable Indoor Air Quality." Atlanta: American Society of Heating, Refrigerating, and Air-Conditioning Engineers. 63pp.

ASHRAE. 1982. Position Statement on Indoor Air Quality. Atlanta: American Society of Heating, Refrigerating, and Air-Conditioning Engineers. 7pp.

Green & Associates. 1984. Indoor Air Quality. An Architectural Primer. Washington, D.C., 20pp.

Janssen, J.E. 1983. "Air Mixing Efficiency." Proceedings of An Engineering Foundation Conference on Management of Atmospheres in Tightly Enclosed Spaces. Atlanta: ASHRAE, 4pp.

National Research Council Committee on Indoor Pollutants. 1981. Indoor Pollutants. Washington, D.C.: National Academy Press. 537pp.

Turk, A. 1983. "Air Cleaning with Granular Adsorbents." Proceedings of An Engineering Foundation Conference on Management of Atmospheres in Tightly Enclosed Spaces. Atlanta: ASHRAE, 4pp.

Chapter 48

OPTIMIZING ENERGY EFFICIENCY AND INDOOR AIR QUALITY

S. J. Hansen, R. P. Bartley

Sick Building Syndrome (SBS), is rapidly becoming part of the energy engineer's and facility manager's lexicon. The complaints are mounting: sinus congestion, scratchy eyes, chronic headaches, excessive fatigue. And the cause? By inference, implication or direct attribution, energy efficiency seems to be the culprit. Linda Moore, Managing Editor of Buildings, writing in the April 1988 issue, "The Closed (Sick) Building Syndrome, a by-product of the 70's energy-efficient construction methods..." seems to summarize the problem as viewed by many.

In today's litigious society, the specter of legal action becomes very real, as evidenced by the February 18, 1988 Wall Street Journal report:

> 'SICK BUILDINGS' LEAVE BUILDERS AND OTHERS FACING A WAVE OF LAWSUITS
>
> More office workers are filing lawsuits, claiming they were made ill by indoor air pollution from such things as insect sprays, cigarette smoke, industrial cleaners, and fumes from new carpeting, furniture, draperies and copiers. Most cases involve energy-efficient buildings. [underscore supplied]

In addition to implying that energy efficiency is to blame, litigation costs loom large against potential energy savings.

Financial losses due to absenteeism and lower productivity attributed to indoor air pollutants are increasingly costed against energy cost savings. Implying a straight line relationship, a federal official recently observed, "5 percent reduced energy consumption equals two minutes per day lost management time."

The stage is set for indoor air quality (IAQ) to deal a sweeping blow to energy efficiency before people learn to discern which energy efficiency changes, if any, are needed to provide quality indoor air. Or, to learn there are ways to optimize IAQ and energy efficiency.

Indoor Air Quality is being heralded as "The Issue of the 90's." This growing concern stems from the following factors:

1. People spend up to 90 percent of their time indoors.
2. Pollutant levels indoors frequently exceed those outdoors and may exceed outdoor pollution standards.
3. Air contaminants consist of numerous particulates, fibers, mists, fumes, bioaerosols, and gases or vapors that can impair human performance and as such present a full range of implications from mild irritation of the upper respiratory system to serious health threats. Some indoor pollutants are clearly carcinogens or chemical sensitizers.
4. Increased awareness through press coverage of specific pollutants, such as radon and formaldehyde.
5. The economic and legal implications for building owners and operators.

IAQ concerns are real and are not going to go away. A recent U.S. Army report estimated health problems related to indoor air quality has cost $15 billion in direct medical costs and about 150 million lost workdays. A twenty-two story office building in Northern Virginia labeled "SBS" stood empty for months until an IAQ team found the problem. Under the Superfund Bill (P.L. 99-499), Congress authorized the Environmental Protection Agency to research radon and indoor air quality. Predictably this will lead to regulations that are expected to dwarf its previous asbestos efforts.

Indoor air contaminants, or pollutants, can be divided into particles (solids or liquid droplets) and gases or vapors.

Particles of specific interest include:
1. Respirable particulates as a group;
2. Tobacco smoke (solids and liquid droplets) [tobacco smoke also contains many vapors and gases];
3. Asbestos fibers;
4. Allergens (pollen, fungi, mold spores, insect feces and parts); and
5. Pathogens (bacteria and viruses), almost always contained in or on other particulate matter.

Vapors and gases of particular interest include:
1. Carbon monoxide (CO), carbon dioxide (CO_2)
2. Radon (decay products become attached to solids);
3. Formaldehyde (HCHO);

4. Other volatile organic compounds (VOC); and
5. Oxides of nitrogen (NO and NO_2).

Energy efficiency is not on the list of contaminants. Misguided attempts to reduce energy consumption may exacerbate conditions, but well coordinated efforts to improve IAQ and maintain an energy efficient building are not incompatible.

NIOSH's Health Hazard Evaluation Team investigated over 400 "Tight Building Syndrome" facilities and found 53 percent of the complaints were directed at poor ventilation, thermal comfort or humidity. The quality of indoor air is influenced by thermal conditions, humidity, air movement, etc.; so energy efficiency can actually have a beneficial effect on these aspects of IAQ. Even in the realm of controlling contaminants, it is possible at times to optimize both energy efficiency and IAQ.

To put the relationship between energy efficiency and IAQ in context, it is desirable to examine the factors which influence, not necessarily degrade, air quality as shown in Figure 1.

FIGURE 1

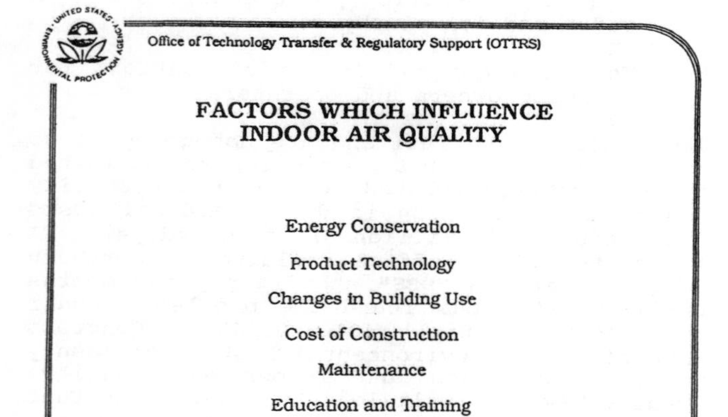

Office of Technology Transfer & Regulatory Support (OTTRS)
FACTORS WHICH INFLUENCE INDOOR AIR QUALITY
Energy Conservation
Product Technology
Changes in Building Use
Cost of Construction
Maintenance
Education and Training

It is worth noting that energy conservation is only one of the influencing factors listed by EPA. Contaminants can be controlled through: a) control at the source -- removal or substitution; b) filtration and purification; c) encapsulation -- coating, interfering with material's ability to off-gas or release particles; d) time of use; e) education and training -- mitigating action resulting from knowledge and sensitivity; and f) dilution-- increased ventilation. In a great many instances, removing or controlling contaminants at the source is more cost-effective than diluting them through increased ventilation.

Measures to improve IAQ may: (1) be energy neutral; (2) cause an increase in energy consumption; or (3) improve both energy efficiency and IAQ.

ENERGY NEUTRAL REMEDIES

Many of the ways to improve the quality of indoor air focus on maintenance. The growing cost of poor quality indoor air gives added importance to preventive maintenance (PM). In fact, today PM carries increasing overtones of the other PM, preventive medicine. Some PM measures representative of low cost source control and training are:

- o checking duct linings, which are breeding grounds, especially in areas of high humidity;
- o exercising care that steam for humidification has no contact with boiler additives;
- o removing partitions impeding critical air flow;
- o checking condensate pans in unit ventilators, fertile sources for biological growth of fungi and bacterial organisms; treating with algaecide as directed on container;
- o exercising great caution in the use and storage of toxic chemicals in operations, maintenance, pest control, kitchens, etc. and using less toxic chemicals wherever possible;
- o checking common reservoirs for microbial contamination; i.e., flush toilets, humidifiers and ice machines;
- o sealing cracks and openings around basement drains and openings, especially in radon affected areas; or
- o positioning air intake grills so that the re-entry of fumes from the building's own exhausts or other avoidable contaminants such as fumes from delivery alleys and loading docks are not a factor.

This list of measures for IAQ control argues well for an effective preventive maintenance program before costly increases in energy consumption are implemented.

IAQ MEASURES THAT INCREASE ENERGY CONSUMPTION

Several measures can be taken to improve the quality of indoor air that will have only modest effect on energy consumption. Others, such as increased ventilation, will have a far greater impact.

Typical measures with modest effect on energy consumption are:

- o scheduling painting, coil cleaning or other projects which involve the use of volatile organic chemicals during minimum occupancy and then purging, or "airing out," the facility; and

- avoiding backdrafting by supplying fuel-fired heating appliances with more fresh air.

Through design modification, some measures are practicable from both the cost of modification and the efficient use of energy. A good example is the use of an air curtain in open design restaurants that are required to provide non-smoking areas, which would otherwise probably be faced with providing separate systems.

After measure, which have a relatively slight effect on energy consumption are taken, it may be necessary to dilute remaining contaminants through ventilation. However, before dilution is implemented air flow patterns need to be carefully examined. Increased ventilation may prove to be a "simplistically" complex problem. Three examples illustrate that more ventilation may not be the answer. First, more is not necessarily better. Increasing ventilation in the upper stories of a building can cause a negative pressure in the basement, which could increase radon flow into the building. Second, the quality of outside air can be as critical as the quantity. Bringing in "bad" air is no solution at all. Third, increased fresh air near the ceiling, as shown in Figure 2 will not have the same beneficial effect as the increased ventilation in the other two air flow patterns.

FIGURE 2

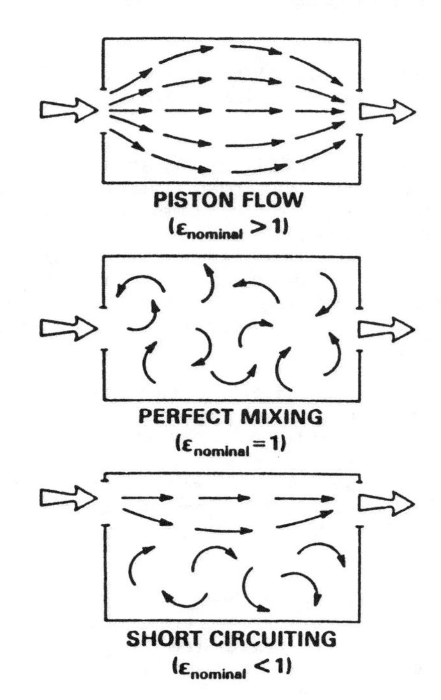

Current ventilation standards prescribe quantities of outside air. They do not address the quantity in relation to contaminants, nor do they require air in a room to be effectively distributed within the occupied space. The Proposed ASHRAE Standard 62-1981R, "Ventilation for Acceptable Indoor Air," is moving toward further integrating air exchange needs with the concentration of contaminants as well as supply considerations.

OPTIMIZING ENERGY EFFICIENCY AND INDOOR AIR QUALITY

Many factors that have a negative impact on energy efficiency also have an adverse effect on IAQ. Correcting such situations can improve energy usage and enhance the quality of the air. A few examples will illustrate the point:

- poor maintenance of pulleys, belts, bearings, dampers, heating and cooling coils and other mechanical systems can increase resistance causing a decrease in air supply. Good maintenance improves both energy efficiency and IAQ;
- water damaged insulation, ceiling tiles, rugs and internal walls support biological growth. Wet materials nullify insulation properties. Replacement increases energy efficency while removing grounds that breed such growth;
- the routine replacement of filter media facilitates the filtering of particles and avoids the resistance to air flow caused by clogged filters, which left unattended reduce energy efficiency;
- humidity control can reduce the likelihood of mold while contributing to comfort and energy efficiency; and
- frequent causes of combustion contaminants are defective central heating systems in which the exhaust is not vented properly, or there are cracks or leaks in equipment. Defective or poorly maintained systems are less efficient and, at the same time, polluters.

NEW CONSTRUCTION

'Engineering out' materials and conditions known to contribute to indoor air pollution in new construction or renovation work may prove to be the most "energy efficient" measure of all.

Siting of a facility should consider potentially negative influences on IAQ, such as the quality of the soil particularly in radon affected areas, the proximity of roadways and vegetation which may, depending on the type and placement, filter out contaminants or serve as a source of allergens.

Some building materials may produce varying amounts of air contaminants. The careful selection of construction materials and furnishings can help alleviate indoor air problems. Among the things to watch for are formaldehyde emission from pressed wood products and volatile organic compounds (VOC). Standards can be specified that meet Federal regulations (CFR 24, Part 3280) or industry association guidelines (NPA 8-86 and HPMA FE-86). Caulks, sealants and adhesives have various levels of VOC emissions depending on the compound. Building, pipe and duct insulation with urea-formaldehyde insulation or asbestos-containing acoustical, fireproof and thermal insulation can have an adverse effect on IAQ. Paints have highly variable mixtures of VOCs and have high release rates during the short-term curing process. Paint should be applied and cured in well-ventilated conditions.

Care should be taken during construction to insure that IAQ is not compromised before the building is used. Materials and furniture should have had sufficient "curing" before the building is occupied. This means timing construction so it is completed before being brought into use. This will reduce off-gassing and other noxious effects of new materials or equipment.

In addition to the consideration given to the thermal environment and humidity, air flow patterns need careful HVAC designer attention. Ventilation effectiveness, the ratio of outside air reaching occupants compared to total outside air supplied to the space, determines the capability of the supply air to limit the concentration of the contaminants.

Filter access for replacement is an important consideration. Filters used should be rated for efficiency; i.e., tested in accordance with ASHRAE Standard 52-76. Higher efficiency filters (above 30 percent) can be effective at reducing levels of respirable particles. Filters can also be specified based upon their removal efficiency for particles of a specific particle size.

Operating the building and its equipment to meet environmental needs requires good maintenance and conscientious custodial practices carried out according to manufacturers' schedules and directions.

Codes and standards are useful references in new construction and may be mandated by state or local laws. In view of the growing concern regarding IAQ, codes and standards are apt to be revised and regulations promulgated.

RULES, REGULATIONS, STANDARDS AND GUIDELINES

Indoor air quality is receiving attention at all levels of government. Lead agencies at the federal level are the Environmental Protection Agency (EPA), the Occupational Safety and Health Administration (OSHA) and the Consumer Product Safety Commission (CPSC). Other federal agencies actively involved are the National Institute of Occupational Safety and Health (NIOSH), the Center for Disease Control, and the Department of Energy.

As a sign of the potential explosion in exposure standards, agencies and associations are rushing to coin labels: OSHA uses Permissible Exposure Limits (PELs), the American Conference of Governmental Industrial Hygienists uses Threshold Limit Values (TLVs), and the American Industrial Hygiene Association has developed Workplace Environmental Exposure Limit (WEEL).

The National Governor's Association has expressed considerable interest in the IAQ issue. NGA issued a policy statement, Indoor Air Pollution, adopted in February 1986 and revised in February of 1988. The Governors recommended that the federal agencies continue national research efforts to: a) characterize exposure of the general public to indoor air pollutants, b) provide comprehensive information for consumers, industry and state and local governments and develope national programs and product standards designed to protect public health. The governors also urged that regional concerns be addressed by federal and relevant state agencies so that working together they can assure that the public health is adequately protected.

Several state governments are actively involved in IAQ. The Maryland State Department of Education has developed a monograph with federal assistance, held workshops for school administrators and is now working with districts to develop a protocol. Public awareness programs are dependent upon the adequacy of information and the soundness of underlying research. States do not have the resources to conduct such research and must rely on the federal sector, private consultants and industry sources. Given the appropriate support, the states can provide the conduit for informing local governments and individual citizens.

ENERGY EFFICIENCY AND INDOOR AIR QUALITY: WE CAN HAVE BOTH!

Increasing energy efficiency as we improve IAQ has merit and is feasible. Rather than layer IAQ requirements over energy systems, we should design them to be an integral part of the product that building operators deliver: a comfortable, productive environment. Rather than competing goals, IAQ and energy efficiency can and should share a commonality of purpose.

The knee jerk reaction of increasing ventilation flow rates and settling for that and that alone must be avoided. Masking potentially dangerous materials that may have long term hazards by making the environment bearable through increased ventilation will leave us far short of what can be done to improve the quality of indoor air.

The key to quality indoor air is the building operator. We all know from sad experience that standards and codes are only as effective as the regulated community makes them. Without this support, the IAQ problem will never be fully resolved. Natural bridges to the building operators are state energy offices, energy engineers and consultants. They have built their networks and facility managers and operators look to them for information. By utilizing federal government research and development resources and these existing networks, information to guide and assist facility managers and building operators in attaining a quality indoor environment in an energy efficient manner can happen. This complete system approach can result in true Indoor Air Quality.

Just turning up the fan will not do it!

REFERENCES

U.S. Environmental Protection Agency, EPA Indoor Air Quality Implementation Plan. Report to Congress. Washington, D.C.

Milton Mechkler, "Impact of New Ventilation Standards on Indoor Air Quality & Energy Efficiency," AEE Product Showcase. Atlanta: Association of Energy Engineers, 1988.

U.S. Department of Energy, Indoor Air Quality Environmental Information Handbook: Radon. Washington, D.C.: January, 1986.

National Governors' Association, D-45. Indoor Air Pollution. Washington, D.C.: National Governors' Association, 1988.

Lawrence Berkeley Laboratory, The Concentrations of Indoor Pollutants Date Base: A Status Report. Berkeley: University of California: July, 1984.

American Lung Association, Indoor Air Pollution in the Office. and Lung Hazards on the Job series; Welding, Solvents, Auto Repair.

California Department of Education, Program Advisory: Toxic Art Supplies Legislation. Sacramento: State of California, 1987.

Maryland Department of Education, Indoor Air Quality: Maryland Public Schools. Baltimore: State of Maryland, 1987.

Teichman, Kevin, Indoor Air Quality Overview. Washington, D.C.: U.S. Environmental Protection Agency, 1988.

SECTION 9
HEAT PUMPS

Chapter 49

HEAT PUMP PERFORMANCE EVALUATION

D. P. Mehta, D. C. Zietlow

I. INTRODUCTION

A. Background

In September, 1984, a utility company initiated a research project in collaboration with Bradley University with the objective of examining the marketability of Earth Coupled Heat Pumps (ECHP). For this purpose five different heating, ventilating, and air-conditioning (HVAC) systems were instrumented for monitoring. These systems included: an air to air heat pump (AAHP), and add on heat pump AOHP), natural gas furnace with an air-conditioner (NGAC), electric heat with air-conditioner,(EHAC) and an earth coupled heat pump (ECHP). The experimental data for at least one cooling and one heating season were collected at each site and used in the current report to evaluate the marketability of an ECHP compared to the other HVAC systems. Although the measured performance of an ECHP deviated a little bit from the manufacturer's specifications and only a limited amount of experimental data were available, it was found that for the residential sector an Earth Coupled Heat Pump was not marketable at the current prices of fossil fuels and electricity.

B. System Monitored and Test Procedures

The heating/cooling system at the AAHP site consisted of a Rheem Heat Pump (Model 048JA) of three ton capacity with an Electric Furnace for auxiliary heating. The system at the natural gas furnace site consisted of a HEIL Natural gas furnace (Model No. NOGF100AF) and a Westinghouse Central Air-Conditioner (condensing Unit Model SL030C). The ECHP used in this study was a GEOSYSTEM (Model # SWPR351) with an electric furnace by Lennox (Model # E1203-20). The ECHP had an outdoor coil which was buried in the ground outside the house.

The method of obtaining and analyzing test data for the systems evaluation consisted of several test sensors, programmable data acquisition systems and analysis software developed for use with an IBM Personal Computer.

II. DATA ANALYSIS

Two software packages were developed and used for the technical analysis of data--one for data reduction and the other for performance simulation. These computer programs calculated the energy transfers, energy demands, and performance parameters.

Prices for electricity and natural gas were taken from the current rate structure of the Utility. Prices for the oil and propane gas were obtained for the local suppliers. Measured and manufacturer's data were used to simulate the performance using bin method. Measured values of heating/cooling requirements adjusted for a typical weather year and measured performance parameters were used for the ECHP, for NGAC, for the AAHP and for the ELAC Performance parameters for the remaining systems viz: Oil Furnace with Air-Conditioner (OIAC) and Liquid Propane Furnace with Air-Conditioner (LPAC) were taken from the manufacturer's specifications.

The data on the acquisition costs, maintenance costs, and salvage values for all the HVAC systems were obtained from the local contractors. The economic analysis for the home owner was based on marginal costs and revenues. In 1984, detailed marginal costs studies were performed by the Utility for both the gas and electric systems. These studies were adjusted to reflect the current load growth projections and were used in the analysis.

III. RESULTS

A. Results for the Home Owner

The data used and the results obtained from an economic analysis made for the home owner are shown in Table 1. The annual operating costs for seven different types of heating/cooling systems viz: Electric Furnace with Air-Conditioner (ELAC). Air to Air Heat Pump (AAHP), Liquid Propane Furnace with Air-Conditioner (LPAC), Natural Gas Furnace with Air-Conditioner (NGAC), Oil Furnace with Air-Conditioner (OIAC); Add-On Heat Pump (AOHP); and for the Earth Coupled Heat Pump (ECHP) are shown in Figure 1. As seen from this figure, Earth Coupled Heat Pump has the lowest annual operating cost. The annual savings in the operating costs which are possible by installing an ECHP compared to the rest of the six heating/cooling systems are shown in Figure 2 and the corresponding payback periods (P.B.) and the rates of return (R.O.R.) are shown in Figure 3. It can be seen from Table 1 and Figure 3 that an investment in an ECHP by the home owner has a payback period of 9.5 years and a rate of return of about 6 percent when compared to an AAHP.

Similarly an investment in an ECHP was not found to be economically attractive for the home owner when compared to the other heating/cooling systems with the exception of ELAC. So a sensitivity analysis was done. In this analysis, required variations in the first cost, heating load, fuel prices, and in performance were determined which will assure at least a 10% rate of return for the home owner. These variations are summarized in Table 2 A-E. As shown in Table

Table 1. Input Data and the Results for the Home Owner Analysis

	ECHP	AOHP	AAHP	Natural Gas with A/C	Propane Gas with A/C	Oil Heat with A/C	Electrical Heat with A/C
Life	15 yrs	15 yrs	15 yrs	15 yrs	15 yrs	15 yrs	15 yrs
Acquisition Cost	$7075	$3193	$2700	$2586	$2886	$3194	2796
Annual Heating Load (Btu)	6.7×10^7	6.7×10^7	6.7×10^7	6.7×10^7	6.7×10^7	6.7×10^7	6.7×10^7
Annual Cooling Load (Btu)	1.6×10^7	1.6×10^7	1.6×10^7	1.6×10^7	1.6×10^7	1.6×10^7	1.6×10^7
Heating Cost	$263.43	$413.15	$601.40	$464.65	$554.10	$431.27	$816.64
Cooling Cost	$171.29	$211.32	$291.17	$263.29	$263.29	$263.29	$263.29
Operating Cost	$434.72	$624.47	$892.57	$727.94	$817.39	$694.57	$1079.93
Savings*	--	$180.75	$457.85	$293.32	$382.67	$259.85	$645.20
Savings**	--	--	$268.10	$103.47	$70.01	$192.92	$455.46
Savings***	--	--	--	--	--	--	$187.36
Payback Period*	--	20.4 yrs	9.5 yrs	15.3 yrs	10.9 yrs	14.9 yrs	6.63 yrs
Rate of Return*	--	< 1%	6%	< 1%	4.25%	< 1%	13%
Payback Period**	--	--	1.84 yrs	5.8 yrs	1.6 yrs	1 year	1 year
Rate of Return**	--	--	> 5%	15%	> 50%	> 50%	> 50%
Payback Period***	--	--					
Rate of Return***	--	--					

* Vs ECHP
** Vs AOHP
*** Vs AAHP

Table 2A. Sensitivity Analysis for a 10% R.O.R. from an Investment in an Earth Coupled Heat Pump Vs Air to Air Heat Pump

Required Reduction in First Cost of ECHP	Increase in Natural Gas Price	Required Increase in Space Heating Requirement	Required Increase in C.O.P. of ECHP
$895.42	---	0.34 times	2.51

Table 2B. Sensitivity Analysis for a 10% R.O.R. from an Investment in an Earth Coupled Heat Pump Vs Add-On Heat Pump

Required Reduction in First Cost of ECHP	Increase in Natural Gas Price	Required Increase in Space Heating Requirement	Required Increase in C.O.P. of ECHP
$2440	$3.09/Therm	3.14 times	--

Table 2C. Sensitivity Analysis for a 10% R.O.R. from an Investment in an Earth Coupled Heat Pump Vs Natural Gas Heating with Air Conditioning

Required Reduction in First Cost of ECHP	Increase in Natural Gas Price	Required Increase in Space Heating Requirement	Required Increase in C.O.P. of ECHP
$2262	$0.34/Therm	2.48 times	--

Table 2D. Sensitivity Analysis for a 10% R.O.R. from an Investment in an Earth Coupled Heat Pump Vs Propane Gas Heating with Air Conditioning

Required Reduction in First Cost of ECHP	Increase in Natural Gas Price	Required Increase in Space Heating Requirement	Required Increase in C.O.P. of ECHP
$1278.20	$.175/Gal	1.58 times	4.8

Table 2E. Sensitivity Analysis for a 10% R.O.R. from an Investment in an Earth Coupled Heat Pump Vs Oil Heating with Air Conditioning

Required Reduction in First Cost of ECHP	Increase in Natural Gas Price	Required Increase in Space Heating Requirement	Required Increase in C.O.P. of ECHP
$1905	$0.39/Gal	2.49 times	--

2-A, the first cost of the ECHP must be reduced by $895 before the home owner can expect a R.O.R. of 10% from this investment compared to the purchase of an AAHP.

It is shown in the next section on the "Results for the Utility" that the Utility has the highest before tax profit from an Air to Air Heat Pump installation and so the Utility cannot afford to initiate a marketing program in favor of ECHP over an AAHP. On the contrary, the Utility can implement marketing efforts to promote the installation of Add-On-Heat Pumps and Air to Air Heat Pumps.

B. Results for the Utility

Marginal revenues, costs, and before tax profits for the Utility for all the seven systems viz: AAHP, ECHP, AOHP, NGAC, LPAC, OIAC, and ELAC are shown in Table 3. It can be seen from Table 3 that the ECHP ranks fourth in the order of producing before tax profits for the Utility. So the Utility cannot be advised to market this product over the frist three ranking products viz: AAHP, ELAC, and AOHP.

The ECHP does produce net before tax profits for the Utility when compared to NGAC, LPAC, and OIAC. However, the rates of return from an investment in ECHP by the home owner compared to these systems (see Table 1) are not attractive. So efforts by the Utility to market ECHP over these systems are not likely to succeed.

It can also be seen from Table 3 that in the residential sector. The Utility is benefited the most from the use of AAHP. However this is not the best choice for the home owner. Same is true about the ELAC. So the Utility should concentrate on the next profitable system viz: Add-On Heat Pump. The net before tax profits from the use of AOHP over ECHP, NGAL, LPAC, and OIAC were converted into after tax profits at a tax rate of 49.46%. The present values of these after tax profits were calculated for each system, which are the maximum rebates that the Utility can offer to market AOHP, and are shown in Table 4. In fact, it can be seen from Table 1 that the rates of return for the home owner from an investment in an AOHP are very attractive even without a rebate program. So the amounts shown in Table 4 may be used to allocate a marketing budget to implement efforts for educating the potential buyers of heating/cooling equipment for use in the residential sector.

Table 3. Impact on Utility's Profits

Type of Residential Heating and Cooling Equipment	Marginal Revenues ($/year)	Marginal Costs ($/year)	Before Tax Profit ($/year)
Air to Air Heat Pump	$669.36	$125.00	$544.35
Electric Furnace and Air Conditioner	$808.93	$388.42	$420.51
Add-On Heat Pump	$434.40	$89.01	$345.40
Earth Coupled Heat Pump	$326.31	$78.48	$247.83
Nat. Gas Furnace and Air-Conditioner	$293.30	$106.30	$187.00
LPG. Furnace and Air Conditioner	$199.30	$106.35	$92.95
Oil Furnace and Air Conditioner	$199.30	$106.35	$92.95

Table 4. Maximum Possible Rebates to Market Add-On Heat Pump

System	Rebate
ECHP	$391
NGAC	$525
LPAC	$785
OIAC	$785

IV. CONCLUSIONS

Seven types of heating/cooling systems for the residential sector have been evaluated with the objective of examining the marketability of Earth Coupled Heat Pumps. The ECHP ranks fourth in the order of producing before tax profits for the Utility. An investment in an ECHP (compared to the first two profit producing systems for the Utility viz: ELAC and AAHP) is not economically attractive to the home owner at the current level of the acquisition and operating costs of the system.

The ECHP does produce net before tax profits for the Utility when compared to NGAC, LPAC, and OIAC. The first cost of the ECHP must drop by $1278, for the potential buyers of LPAC, $1905 for OIAC buyers, and $2262 for NGAC buyers before the R.O.R. can be expected to be 10% for them from an investment in ECHP. Alternatively, the price of natural gas must increase $0.34/Therm, or the price of LP must increase by $0.18/gallon, or the price of oil must increase by $0.39/gallon, before an investment in an ECHP by the home owner can result in a 10% R.O.R. Thus the Utility may consider initiating marketing efforts for ECHP when the above said changes take place. At present the Add-On Heat Pump seems to be a good candidate for promotion by the Utility.

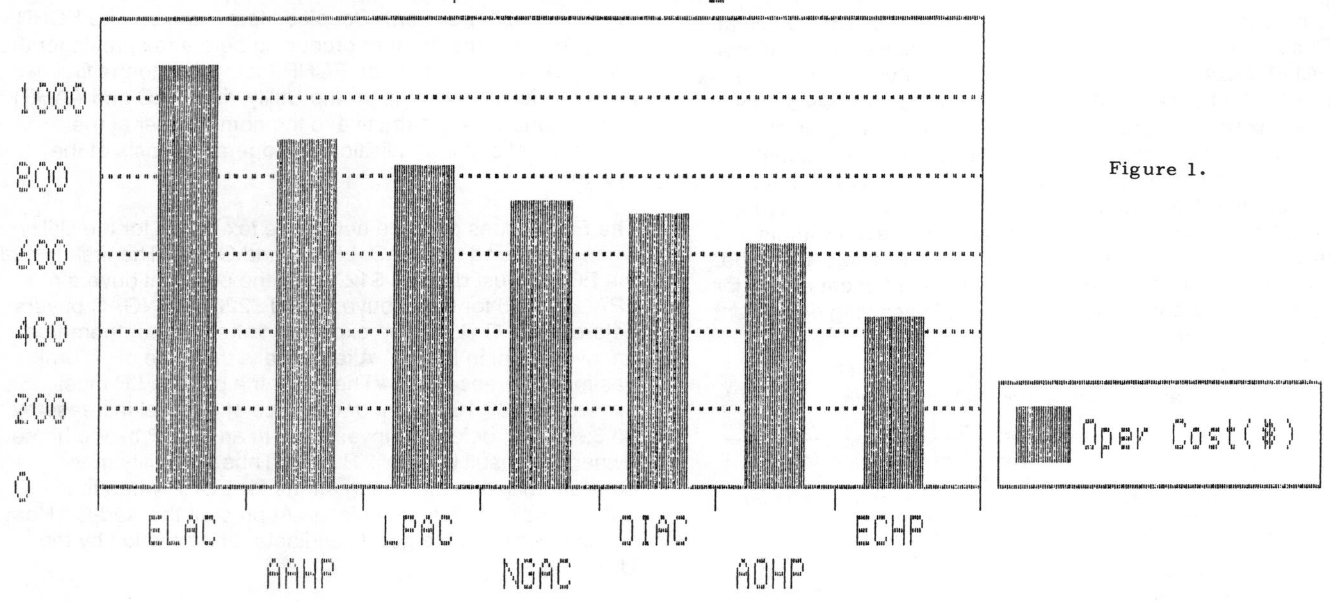

Figure 1.

Heating and Cooling System

ELAC -- Electric Furnace and Air-Conditioner
AAHP -- Air to Air Heat Pump
LPAC -- Liquid Propane Furnace and Air-Conditioner
NGAC -- Natural Gas and Air-Conditioner
OIAC -- Oil Furnace and Air Conditioner
AOHP -- Add-On Heat Pump
ECHP -- Earth Coupled Heat Pump

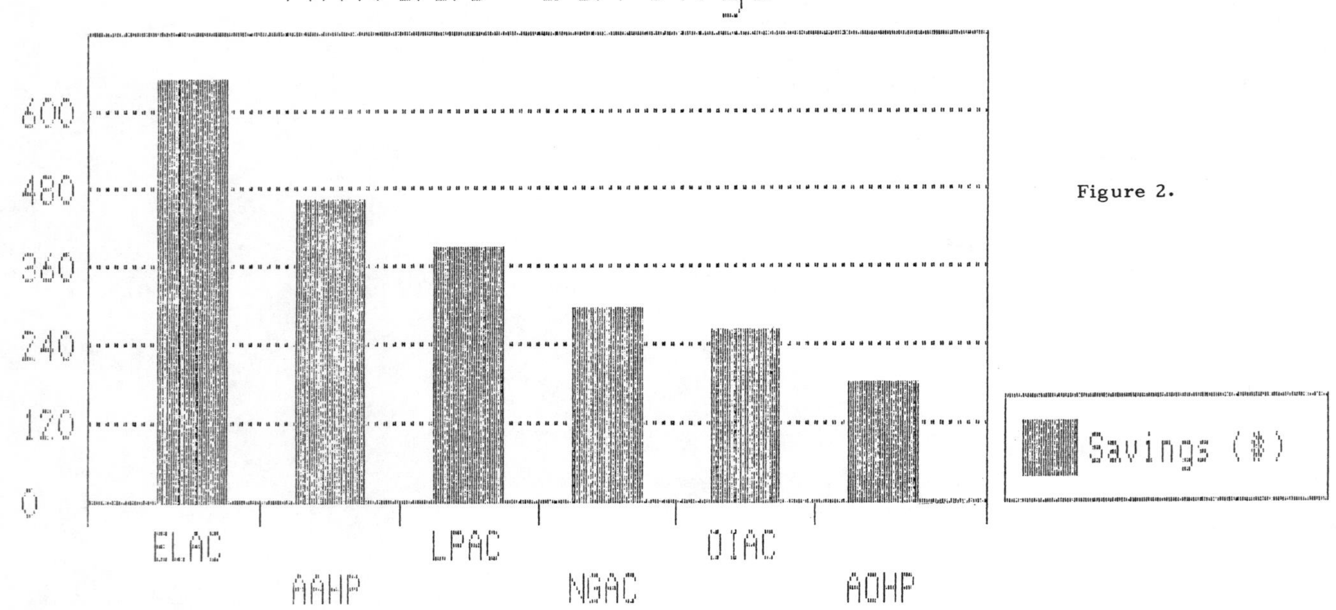

Figure 2.

Heating and Cooling System

ELAC -- Electric Furnace and Air-Conditioner
AAHP -- Air to Air Heat Pump
LPAC -- Liquid Propane Furnace and Air-Conditioner
NGAC -- Natural Gas and Air-Conditioner
OIAC -- Oil Furnace and Air Conditioner
AOHP -- Add-On Heat Pump
ECHP -- Earth Coupled Heat Pump

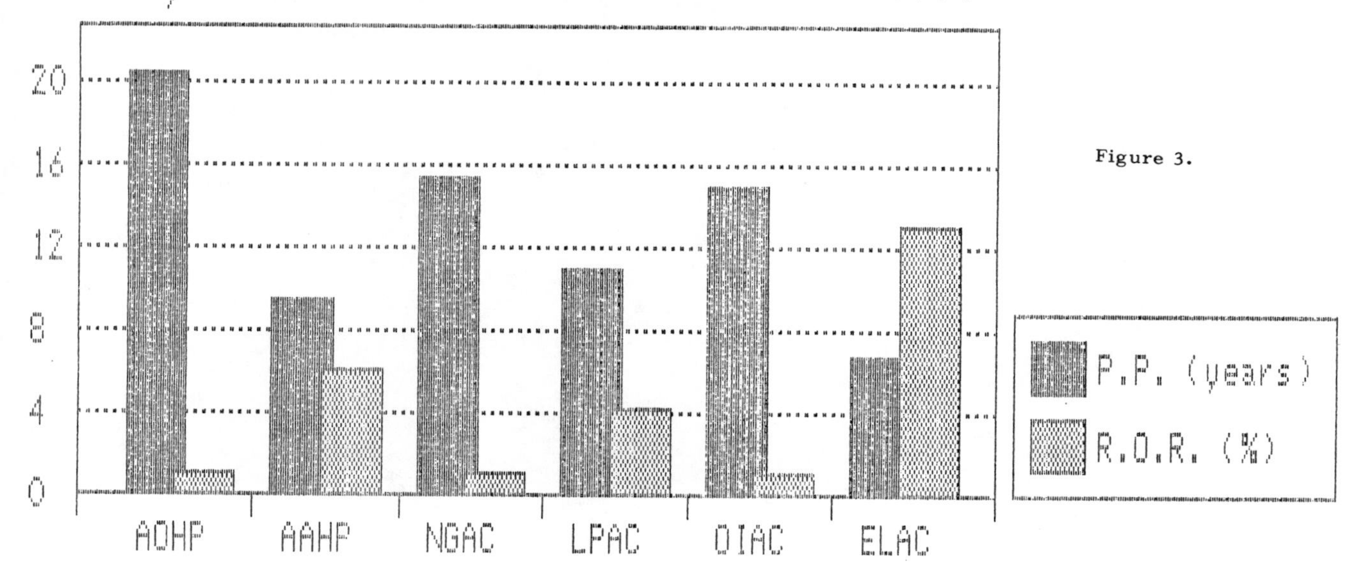

Figure 3.

Heating and Cooling System

ELAC -- Electric Furnace and Air-Conditioner
AAHP -- Air to Air Heat Pump
LPAC -- Liquid Propane Furnace and Air-Conditioner
NGAC -- Natural Gas and Air-Conditioner
OIAC -- Oil Furnace and Air Conditioner
AOHP -- Add-On Heat Pump
ECHP -- Earth Coupled Heat Pump

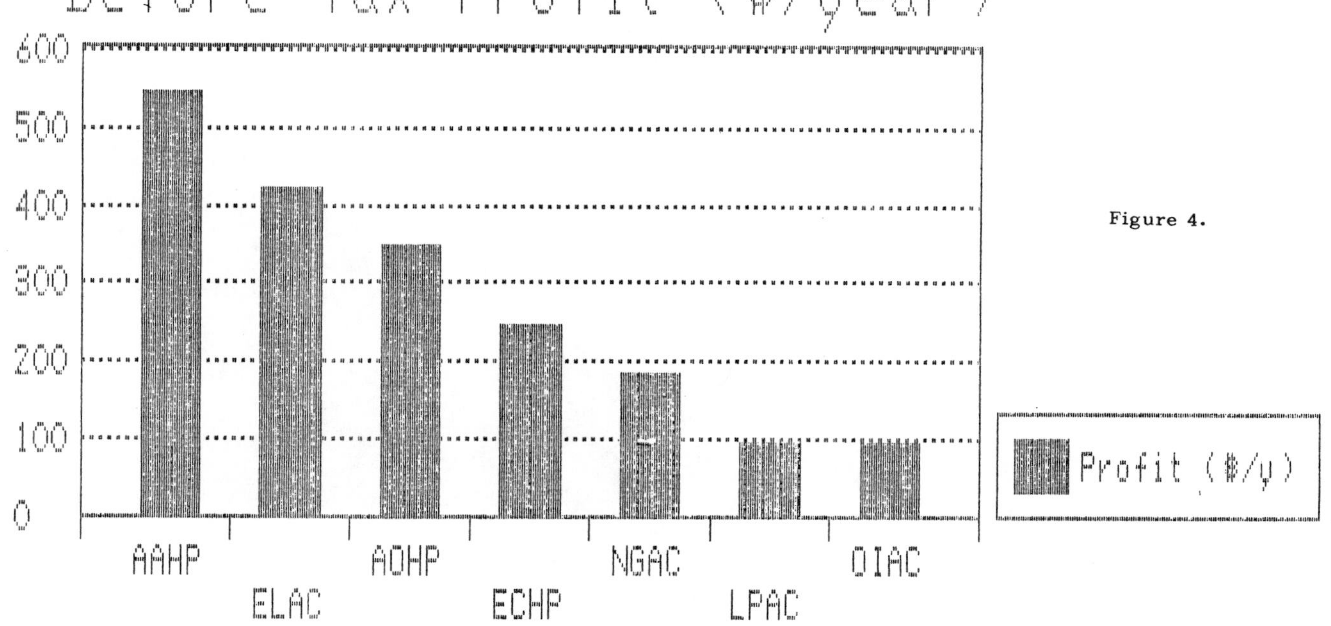

Figure 4.

Heating and Cooling System

ELAC -- Electric Furnace and Air-Conditioner
AAHP -- Air to Air Heat Pump
LPAC -- Liquid Propane Furnace and Air-Conditioner
NGAC -- Natural Gas and Air-Conditioner
OIAC -- Oil Furnace and Air Conditioner
AOHP -- Add-On Heat Pump
ECHP -- Earth Coupled Heat Pump

Chapter 50

HEAT PUMP PROMOTION IN THE TVA POWER SERVICE AREA

R. L. Culpepper, W. S. Morrow, K. J. Ward

In 1933, the Tennessee Valley Authority was created for "planning of proper use, conservation, and development of the natural resources of the Tennessee River drainage basin and it's adjoining territory for the general social and economic welfare of the nation..." As the only government organization with the flexibility and initiative of a private enterprise, TVA works for the people of the Tennessee Valley region. Throughout the years of its existence, TVA has developed and implemented energy use programs for the people's benefit.

Immediately after its creation, TVA undertook an aggressive effort that brought electricity to a predominately rural area of the country, the majority of which had never before been served. TVA created demonstrations which showed how electricity could help both the farmer and his wife through the purchase of basic appliances. These early demonstrations set a TVA tradition for programs designed to help people use electricity to save money and labor, while increasing productivity and improving living conditions.

As the needs of the Valley ratepayer changed through the years, TVA broadened the scope of its energy use programs to include not only agricultural applications but also commercial and industrial lighting and space conditioning demonstrations and more specifically the promotion of residential electric space conditioning. Due in part to the low cost of TVA residential electricity (as low as 89 mills/kWh in 1966), residential electric resistance heat became very popular in the TVA power service area. This high saturation of electric resistance heat contributed to TVA being a winter peaking utility and lead TVA planners to become interested in heat pumps, which provided a summer air-conditioning load, as a means of helping to balance annual electric loads. As early as 1956, many of the 160 power distributors of TVA power were promoting heat pumps to furnish both heating and cooling requirements for existing and new buildings.

However, the era of cheap electricity ended in the early 1970's with the onset of double digit inflation, the energy crisis, and the oil embargo. Almost overnight the emphasis on promoting heat pumps changed to one of conserving electricity as Valley ratepayers struggled with rapidly escalating electric bills. The first concerted region-wide effort to promote heat pumps as a reliable, cost-effective form of electric space conditioning culminated in the Certified Heat Pump Installation Program.

The TVA Certified Heat Pump Installation Program, similar to the Alabama Power Company heat pump program in effect at the time, offered the following services to residential consumers interested in heat pumps:

- Customized heat pump equipment application and sizing information.
- Heat pump operating cost estimates.
- Certified heat pump dealer list.
- Free post-installation inspection for compliance with the TVA Standards for Application and Installation of Heat Pump Systems.

The certification program was intended to build consumer acceptance of heat pumps by assuring high quality, reliable installations of heat pump equipment. In 1973, heat pump schools for dealers and servicemen taught courses which included sessions on the proper installation and servicing of heat pumps. These dealer training schools were part of TVA's initial program to certify heat pump dealers and to promote energy conservation through the widespread use of the heat pump. Soon after the creation of these schools, power distributors began inspecting heat pumps for certification in 1974.

The Certified Heat Pump Installation Program was designed to help assure consumers that a quality heat pump was being installed properly by a dealer who was certified, i.e., the dealer agreed to install heat pumps according to TVA's Standards for Application and Installation of Heat Pump Systems and had been certified as having the expertise to properly install heat pumps. Those who participated in this program appreciated the assurance of a properly installed heat pump, but this assurance was apparently not enough incentive to attract large scale participation in the Certified Heat Pump Installation Program. For example, over the years of its operation (1973-1979), less than 1,000 heat pumps were installed and certified through the Certified Heat Pump Installation Program. This low number was rather disappointing to TVA planners. Other possible incentives needed to be explored if larger scale consumer participation was to be achieved.

Approximately one year after the TVA Home Insulation Program was started in 1977, TVA planners recognized its rapid expansion in popularity among consumers. In contrast to low participation in the Certified Heat Pump Installation Program, the TVA Home Insulation Program had weatherized 13,310 homes in its first year. Why? While both programs offered comparable levels of technical assistance to consumers, the Home Insulation Program also offered financing to consumers desiring to install attic and floor insulation, caulking, weather stripping, and storm windows. Financing, being the only major difference between the two programs, indicated that it was *the* incentive causing greater participation. Therefore, in 1979, TVA developed and introduced heat pump financing through the Heat Pump Program. All of the services available under the preexisting Certified Heat Pump Installation Program were retained, but in addition the TVA Heat Pump Program offered financing to program participants to cover the following items:

- o Heat pump equipment.
- o Ductwork, piping, and materials.
- o Wiring including electric service upgrades as required.
- o Manufacturer's extended warranty.

The financing feature of the Heat Pump Program has been the vehicle by which the program has penetrated the residential heating market. The attractive financing terms (10 years to repay and low interest rate) plus the ease and convenience of acquiring the loan created an attractive incentive for consumers to participate. In addition, the monthly payment amount was applied to a participant's monthly electric bill which eliminated the inconvenience of a separate bill associated with a loan from a conventional financial institution.

The consumer demand for Heat Pump Program financing caused heat pump dealers to begin participating in the program in greater numbers. However, heat pump dealers also began to recognize that they also benefited from the financing program since it reduced their credit risks. The local power distributor paid the dealer directly after the heat pump installation passed inspection. Therefore, with the Heat Pump Program, the dealer's major concern was that their heat pump be installed correctly in order to pass the TVA heat pump inspection. In addition, since all jobs had to meet the same installation standard, competition among dealers was more equitable and profit margins were more predictable. Today, approximately 800 heat pump dealers participate in the Heat Pump Program in the TVA power service area.

The financing of the Heat Pump Program has proved to be the major incentive needed to boost heat pump participation. As indicated in Figure 1, 460 heat pumps were installed through the Heat Pump Program (HPP) (with financing) for the fiscal year ending September 1979, as opposed to the average of less than 200 units per year certified through the Certified Heat Pump Installation Program (CHPIP) (no financing) during the five years ending in fiscal year 1978. This is particularly noteworthy since the Heat Pump Program was in operation for only the last five months in 1979. The Heat Pump Program's participation rate has grown tremendously, interrupted only by the effects of the recession during years 1980-1982. After recovery from the recession period, participation in the financing program has grown approximately 25 percent per year indicating strong gains in consumer acceptance of the program and of heat pumps. In 1988 9,792 units were installed during fiscal year 1987 ending in September.

Figure 2 illustrates the change in saturations of heat pumps, electric resistance heaters (all types), and natural gas equipment (all types) as a percentage of the total number of residential customers receiving TVA power. As of January, 1983, there were approximately 2.5 million residential customers in the TVA service area. Similar to the previous figures, the saturation of heat pumps was relatively flat until 1982. Subsequent to 1982 there appears to be a steady growth in heat pumps. Heat pump saturation has grown from 7.9 percent in 1982 to 12.6 percent in 1986. Conversely, the saturations of electric resistance heat and natural gas heating appliances appear to have had a slight decline over the same period of time after 1982 (-1 percent for electric resistance and -2.2 percent for natural gas appliances). Figure 3 tends to corroborate the viability of electric space heating by illustrating the changes of electric and non-electric space heating appliance saturations in the TVA service area over the same time period. The electric category includes electric resistance heat of all types plus heat pumps while the non-electric category includes all space heating appliances fueled by natural gas, LP gas, oil, and wood. Since 1982 the saturation of electric space heating appliances has increased approximately 3.7 percent at the expense of the non-electric space heating appliances. Again, TVA planners believe the TVA Heat Pump Program has contributed significantly to these favorable trends in electric heating saturations.

In conclusion, the promotion of heat pumps in the TVA service area slowly evolved over thirty years before reaching its present form, the TVA Heat Pump Program which included financing. The steps in this evolution included simple information campaigns in the 1950's; equipment improvements in the 1960's and 1970's; the Certified Heat Pump Installation Program in the 1970's, which improved installation quality and heat pump dealer expertise; and the financing incentive which was added in 1979. It is the belief of TVA staff that the successful penetration of the residential heating market by the TVA Heat Pump Program is primarily attributable to the program's availability of convenient heat pump financing. This belief tends to be corroborated by a 1984 report by Westat, Inc. "Residential Energy Conservation Attitude Survey." This survey of 1,600 respondents investigated consumer attitudes about the TVA energy conservation program which included home weatherization, heat pumps, and heat pump water heaters. Their findings indicated that financing was "a critical aspect of TVA's conservation effort" and that without financing "over 40 percent of the installations done might have been cancelled." This belief in the value of financing in promoting heat pumps is shared by two other utilities, the Alabama Power Company and Ohio Edison, which have implemented heat pump financing programs similar to TVA's. After two years of operation, an Alabama Power Company official reported that participation in their program is still escalating rapidly and stated his belief that financing is one if not the most important heat pump marketing tools available to them. Likewise, as TVA continues to modify its energy use programs to meet changing customer needs, the TVA staff remains convinced that utility financing is a key factor in demonstrating that heat pumps are a reliable and economical form of space conditioning.

Figure 1

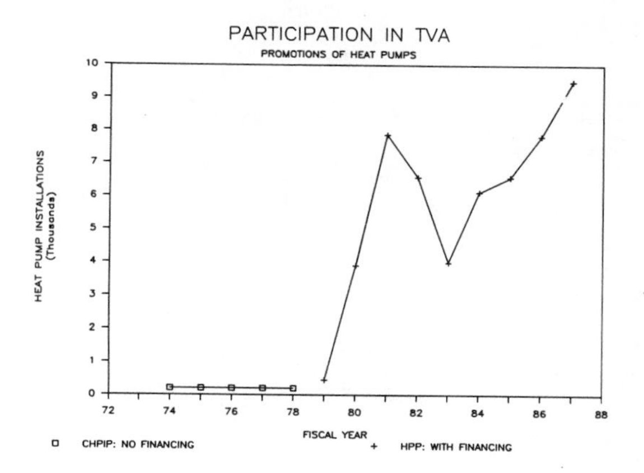

Figure 2

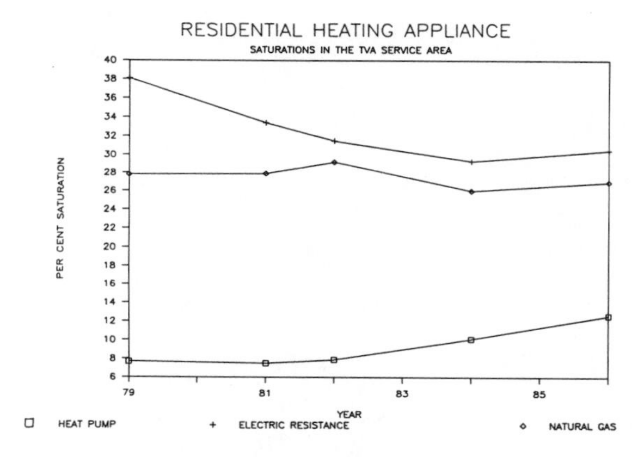

Figure 3

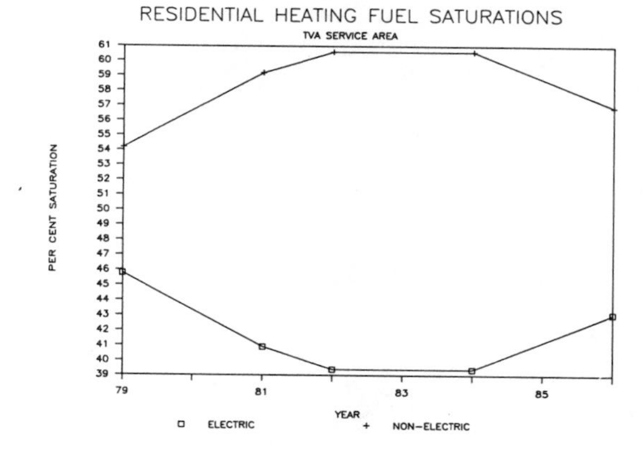

References

1. Residential Energy Conservation Attitude Survey, Final Report, Rockville, Maryland: Westat, Inc., November 15, 1984.

2. 1986 Residential Survey, Customers of Municipal and Cooperative Distributors of TVA Power, Chattanooga, Tennessee: Distributor Services Branch, Division of Energy Use and Distributor Relations, Office of Power, Tennessee Valley Authority, December 1986.

Chapter 51

DISTRICT HEATING AND COOLING USING NORWAY'S LARGEST HEAT PUMP

P. Skjaeggestad, U. Rivenaes

INTRODUCTION

The energy situation in Norway.

Norway's population is 4,2 million, and the electricity requirement is generated by 650 hydro electric power stations with a total, combined capacity of 23 000 MW. These power stations produce a total of between 95 and 120 TWh/year depending on the downfall.

Approximately 100 TWh is bound in contracts. The rest of the power is either exported to other Scandinavian contries through special agreements or it is sold as occational power. In years when Norway is short of power, electricity is imported from other Scandinavian countries. Since hydro electric power has been relatively cheap, about 70 % of all residential homes in Norway are heated by electricity, at a consumer cost of 30-40 øre/kWh (5-6 US cents/kWh) (1988).

There are no large thermal or nuclear power plants in Norway.

Norway's hydro potential is not by far fully used, however environmental aspects have limited the potential considerably.

The Norwegian government want to limit the use of electricity for heating/cooling purposes, and large heat pumps and cogeneration plants (natural gas fired/district heating) will now supplement the hydro electric power plant system.

Location of Norway's largest heat pump

The largest district heating and cooling plant in Europe is located 5 miles west of Oslo in Norway (Figure 1), in the county of Bærum.

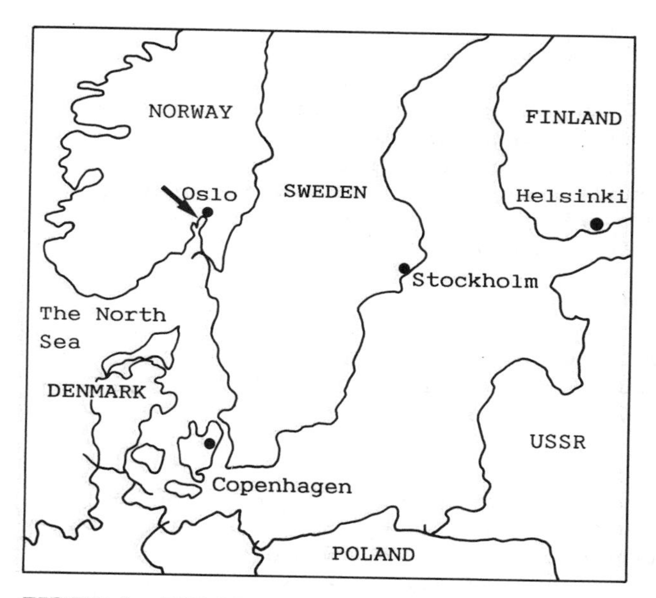

FIGURE 1. THE LARGEST DISTRICT HEATING AND COOLING PLANT IN EUROPE IS LOCATED IN NORWAY.

Time schedule

The preliminary design started in 1986. Construction started in March 1987, with commissioning for phase 1 during late summer 1988. A large number of new buildings will be connected to the plant during the years to come. The area is expected to be completely developed with new buildings by the year 2000.

The owner of the plant

The plant is owned by Bærum energiverk, the local energy utility serving the entire county (population 85 000) with electricity.

Parts of the county will now be served with district heating and district cooling as well. Bærum energiverk will also run and mantain the plant.

PROJECT BACKGROUND

Energy for new buildings

Sandvika is located by the sea in the county of Bærum. This region is on the threshold of an extensive exploitation. A total of 300 000 square meters of floor space is being planned, for commercial, business and residential purposes.

This large concentration of buildings require energy for heating and cooling. The local energy utility, Bærum energiverk, is supplying the consumers with energy at the lowest possible consumer price. In this case it was found cheapest for Bærum energiverk to build its own heat pump and to supply the new buildings with hot and cold water for heating and cooling purposes.

Existing buildings

A relatively large number of existing buildings in Sandvika is heated by individual oil-fired boilers and radiator systems.

These buildings are offered connection to the district heating system for a price competitive to oil-prices. Even with the low oil- prices of today, connection will be profitable for the consumer.

Energy requirements in existing and new buildings

District heating: In existing buildings, the energy requirements which can be converted to district heating are approximately 35 GWh/year. The capacity requirement by consumers is approximately 17 MW, and the temperature demand is between 80 - 85 ^{0}C at design temperature.

The district heating potential in the new buildings in this region has been the subject of rigorous evaluation. In domestic dwellings, school, hospitals etc. the district heating potential is significant. Office/commercial building with large internal loads and heat recovery from cooling machines are less suitable for a district heating system. The fact that there are a significant number of such buildings planned in the area, was the background for the initiation of the district cooling project.

Based on the assumption that heat recovery from cooling machines within the building is neglected, the district heating potential for new buildings within the area is approximately 25 GWh/year with a capacity demand of approximately 13 MW. This gives a total district heating potential of 60 GWh/year with a total capacity requirement of approximately 24 MW.

District cooling: The cooling requirement will be large in the type of buildings planned in Sandvika. More than 210 000 m^2 will be office/business premises with a total cooling requirement of approximately 12 GWh/year and a capacity demand of approximately 9 MW.

In the evaluation of the profitability of connecting these buildings to a district heating system, a heat pump plant was considered. The costs related to the district cooling plant are therefore evaluated as marginal costs.

Cooling will be obtained from the same machine which supplies heat to the district heating grid - a combined heat pump/cooling plant.

District cooling as a necessary supplement to district heating in new office/commercial buildings.

Buildings with their own systems for producing cooling will also produce heat. This heat can cover all or part of the heat requirements of the building, or may not be utilized at all. Today it is regarded as prudent to invest in equipment which can utilize this heat in the majority of new office/commercial building.

By connecting these buildings to a district cooling plant, heat recovery will be realized and the total heating/cooling reguirement can be covered by one energy production plant.

Sensivity analyses for the Sandvika project show that the district cooling plant gives a positive contribution to the profiability of the project, even with very low district cooling prices. The reason for this is that an increase in the sale of heat will be achieved. In other words, modern office buildings do not need large quantities of district heat if they are equipped with individual cooling system with 100 % heat recovery. In order to sell district heat, the energy utility must also take care of the buildings cooling requirements.

ENERGY PRODUCTION PLANT

Location

The energy production plant is located in a specially made rock cavern, blasted close to a 25 mile long sewage tunnel. This tunnel takes all the sewage from 400 000 people in the region and delivers it to a very large sewage treatment plant 20 miles down stream.

The production of heat in the energy production plant is divided into base production and peak and emergency heat production. The base load will be covered by a heat pump plant with an output of approximately 60 % of the maximum capacity demand. The peak and emergency load will be covered by oil-fired boilers with a capacity equal to maximum demand of the heating system.

Heat source

Mechanically threated sewage water from the sewage tunnel is the heat source for the heat pump plant. The temperature of the sewage water varies over the year (Figure 3).

The same sewage water is being used for another heat-pump located up-stream in Oslo. This heat-pump is serving a district heating net-work (no district cooling) (Figure 4).

Energy production plant - base load

The energi production plant is an underground plant (see Figure 5).

The sewage screening part consists of gates (1), two parallel filters (2), and dry displayed sewage pumpe (3) with pipe to the heat pumpt units. The screens have an aperture of 1 mm. The flow of sewage is kept constant at 300 kg per second per heat pump unit.

There are two heat pump units (4). Each unit is equipped with a valve system (5) for

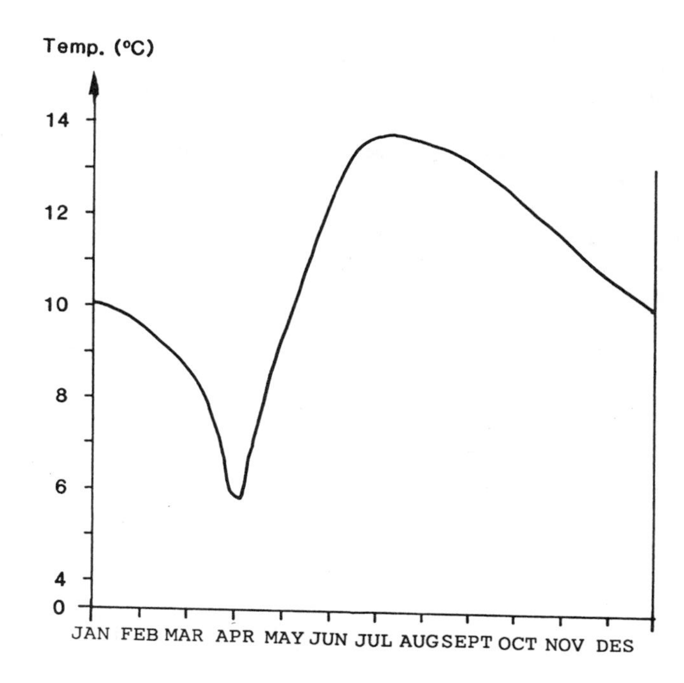

FIGURE 3. SEWAGE WATER TEMPERATURE. THE TEMPERATURE IS LOWEST IN MARCH/APRIL WHEN THE SNOW SMELTS

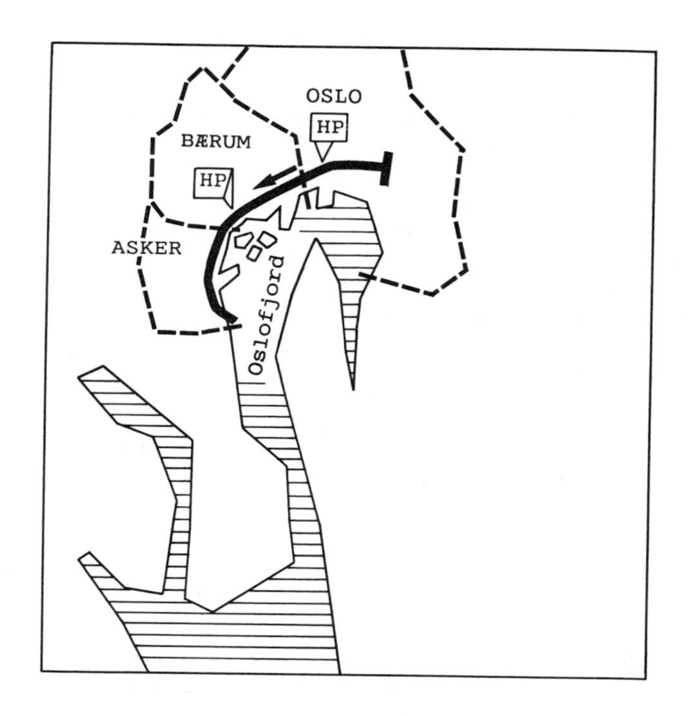

FIGURE 4. THERE ARE TWO HEAT-PUMPS WHICH UTILIZE THE SEWAGE WATER IN THE LONG SEWAGE WATER TUNNEL BUILT TO PROTECT THE OSLO-FJORD FROM POLUTION.
THE ONE IN BÆRUM IS DESCRIBED IN THIS PAPER.

reversing the flow of sewage in order to avoid clogging in the intake to the evaporators.

Circulating pumps (6), pressure retaining equipment and water treatment vessels (7) for the district heating and cooling grids are located in a separate room. High voltage units and ventilation equipment (8) are located at the entrance.

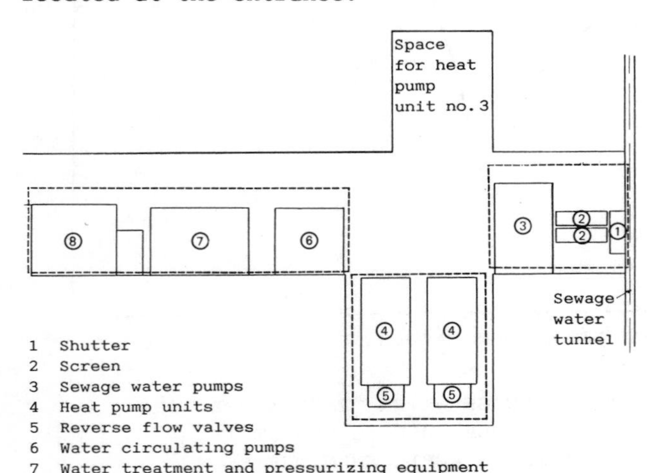

FIGURE 5. ROCK CAVERN. PLAN - GENERAL ARRANGEMENT.

Each heat pump unit has the following main components (Figure 6):

- Condenser with sub-cooler (9) (for district heating)
- High pressure (HP) compressor (11)
- Low pressure (LP) compressor (12) (these two compressors can be capacity regulated independently of each other)
- Intermediate pressure vessel (13)
- Combined evaporator/condenser (14) (for heat exchange with the mechanically treated sewage water)
- Evaporator (15) (for district cooling)

Heat pump operating conditions

The operation conditions of the heat pumps will vary according to the district heating and district cooling demand (see the heat and cooling duration curves in Figure 7).

Heat requirement dominates, low cooling requirement: The capacity of the HP-compressor is regulated according to the outgoing temperature requirement of the district heating water.

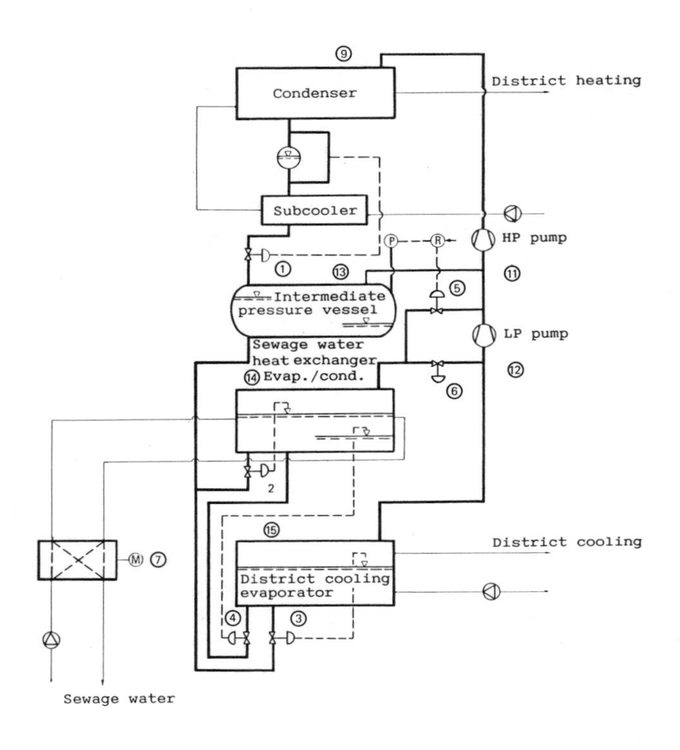

FIGURE 6. HEAT PUMP UNIT - GENERAL DIAGRAM

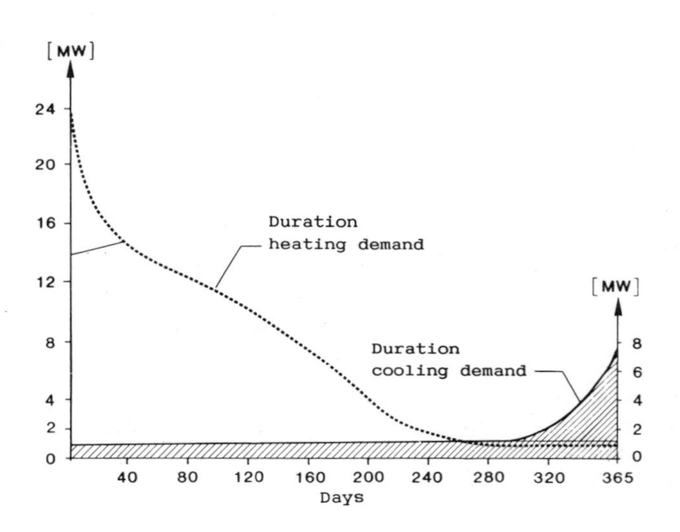

FIGURE 7. HEAT AND COOLING DURATION CURVES

The capacity of the LP-compressor is regulated to optimize the intermediate pressure. The outgoing temperature of the district cooling water is permitted to fluctuate freely as long as it is below 5^{0}C.

Cooling requirements dominates: The capacity of the HP-compressor is regulated according to outgoing temperature requirement of the district heating water.

The capacity of the LP-compressor is regulated according to the outgoing temperature requirements of the district cooling water. The sewage heat exchanger is working as a condenser.

Selection of sewage heat exchanger

The sewage treatment system consist of two parallel screens (with an aperture of 1 mm).

The flow of sewage to each heat pump unit is kept constant at 300 kg/s.

A tube heat exchanger was selected. To avoid clogging or growth within the tubes, several measures are taken:

- One-pass type shell and tube heat exchanger

- Four-way valve for recirculation to avoid clogging due to fibre, (7) in Figure 6

- Copper-based materials is used in the tubes to avoid growth

In addition, there are space and possibilities to include equipment for further filtering and cleaning with detergents.

Peak- and emergency load

The heat pumps are not designed to cover the entire heat demand during winter. On very cold days, additional heat is provided by oil-fired boilers. Three oil-fired boilers (2,5 MW, 9 MW and 10 MW) are installed in a separate building 400 meters from the heat pumps (Figure 8). The oil-fired boilers will supply about 3 - 5 % of the annual energy requirement.

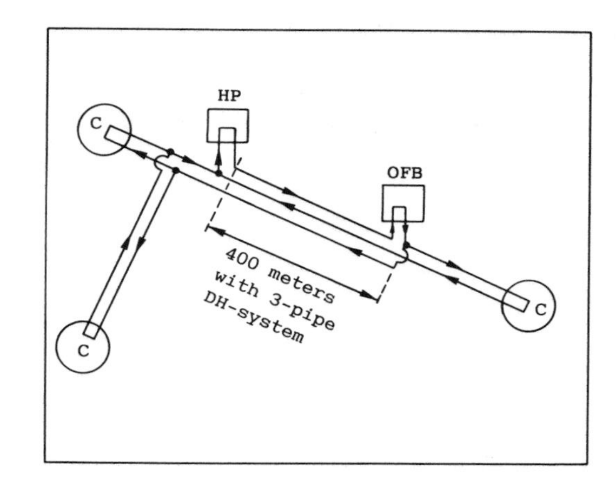

FIGURE 8. THE OIL-FIRED PLANT FOR PEAK AND EMERGENCY LOAD IS LOCATED 400 FROM THE HEAT PUMPS.
HP=HEAT PUMP
OFB=OIL-FIRED BOILERS
C=CONSUMERS

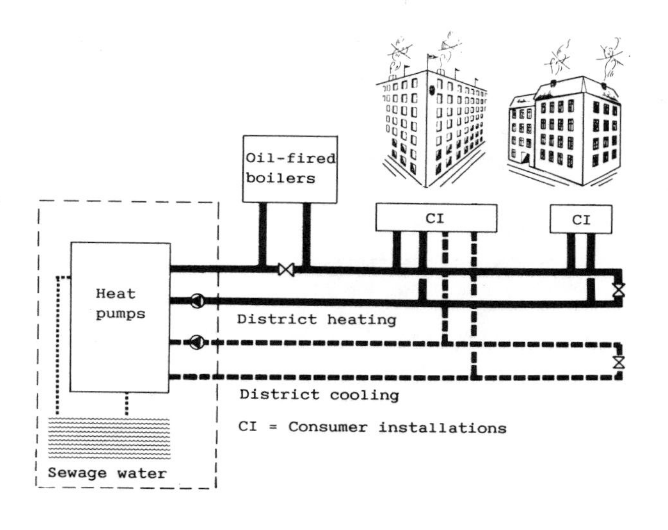

FIGURE 9. ENERGY DISTRIBUTION SYSTEM

THE DISTRIBUTION SYSTEM

The distribution system

The energy distribution system is shown in Figure 9.

District heating network

The hot water is distributed to each consumer through preinsulated underground district heating pipes of steel, with a built-in alarm system to detect leakages.

District cooling network

The chilled water is distributed to each consumer through plastic pipes.
In tunnels, wall mounted steel pipes are being used. The district cooling pipes are not insulated.

Temperature levels

The temperature levels of the district heating and cooling water are shown in Figure 10.

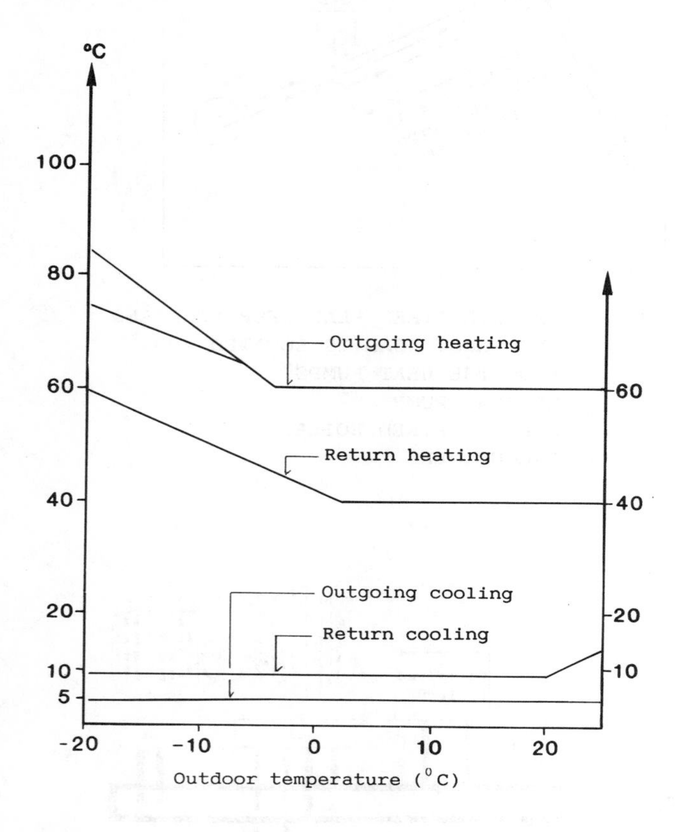

FIGURE 10. HEATING AND COOLING DISTRIBUTION TEMPERATURES

Heat exchanger for consumer connection

Each consumer (building) is connected to the district heating network by a separate heat exchanger. This is also the case for the consumer who wish to connect the building to the district cooling network.

CONTROL STRATEGY

General

The complete project includes a new and separate building from which the system is operated and run.

In order to optimize the complete system each consumer installation related to both the district heating and the district cooling system are connected to the operating room via a communication network, consisting of a separate cable network installed together with the district heating and cooling pipes.

Each consumer installation is equipped with a programmable process computer for process controll and energy consumption measurement and connected to the main computer in the operating room. Similar process computers are connected to the heat pump units, the oil-fired boilers and to other equipment (for water treatment, pressure retaining, water circulation, the sewage water screening plant etc.).

Consumer service

The local energy utility can with this system monitor and control the energy consumed by each individual consumer.

The master control system will generate reference values for the outgoing temperature demand for the district heating and cooling water, based on meteorological values and registered changes in consmuer demand.

Thus the local energy utility Bærum energiverk, can minotor and in cooperation with the consumer, even control the energy requirement of the heating and cooling installations in each building connected to the system (about 50 buldings when completed). This unique cooperation between the energy utility and its consumers makes it possible for the utility to optimize temperature levels and thus the energy consumed by the heat pumps and the three oil-fired burners. In addition, each consumer can obtain from the utility a complete energy report, showing the heating and cooling demand for any period of the year, of the month, of the week or of the day, analyzed every 30 seconds if required.

The programmable logic control units in the energy measuring units installed for each consumer, will every month automatically bill the consumer for energy consumed.

RESEARCH AND DEVELOPMENT

The system design and the operating conditions of the entire plant will be investigated/followed closely by a separate 200 000 USD/year research programme, starting in 1988.

The research programmes are funded by the Government, the local utility Bærum energiverk and the main consultants involved in the system design. Other local electric utility companies are invited to participate in the programme.

The programme is also open for financial support from foreign companies, with tailor-made results and research reports in English. Interested companies can write to: Hafslund Engineering a.s, Box 5012, 0301 Oslo 3, Norway.

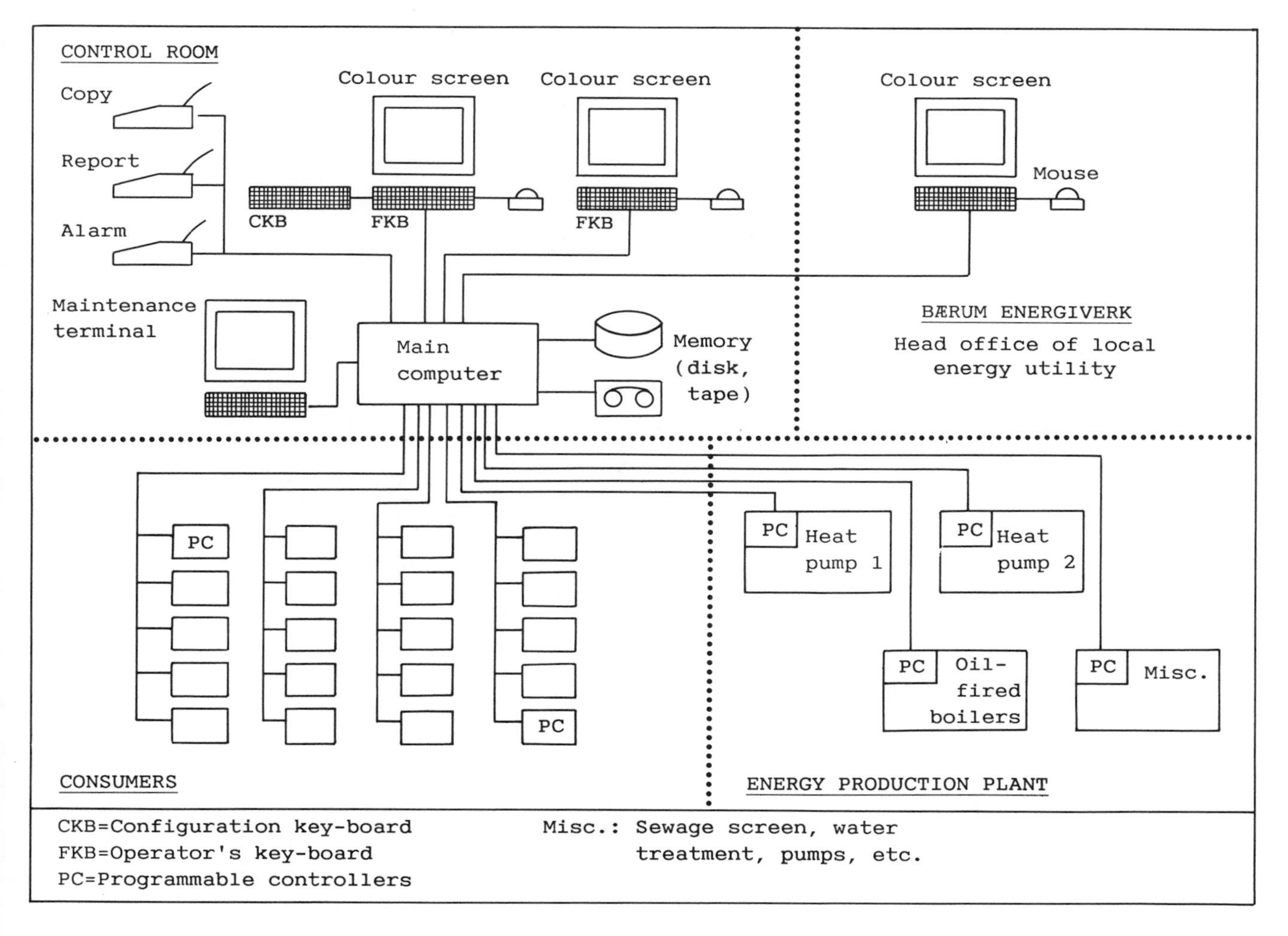

FIGUR 11. COMPUTER SYSTEM FOR OPTIMAL OPERATION AND ENERGY MANAGEMENT

SECTION 10
CHILLER SYSTEMS

Chapter 52

ABSORPTION CHILLERS AND CENTRIFUGAL CHILLERS IN A COMBINED PLANT—OPTIMIZING THE MIX

J. Poulos

INTRODUCTION

The selection of absorption or centrifugal chillers in a central plant varies based on building location (utility rates, climate) and building type (operating hours). This decision becomes even more complex for buildings requiring more than one chiller. In multiple-chiller configurations, the optimum size and type of each chiller must be determined. This requires that, in addition to the data above, the following information be known (or approximated):

- o Equipment Efficiencies
- o Equipment Run-Hours
- o Parasitic Loads
- o Peripheral Loads
- o Building Energy Consumption
- o Peak Building Demand
- o Equipment Control Sequence
- o Owner's Economic Criteria

It is evident from this list that the optimum chiller plant mix will vary with each building analyzed. Therefore, those readers expecting an absolute all-encompassing answer will be disillusioned by the following discussion. There is no single, correct answer. However, there is a correct methodology, which is presented herein.

It can also be surmised that manual calculation of the above components would be tedious. Therefore, Jones, Nall and Davis developed an hourly simulation computer program to perform parametric analyses of multiple-chiller plants. This paper is not intended as a description of the computer program. Instead, this paper presents all the factors which should be considered in a mixed plant evaluation and, subsequently, which are included in the computer program. This information should permit the reader to modify the methodology to suit his or her specific utility rate and owner-mandated conditions.

BASIS OF ANALYSIS

Many chiller analyses are performed using typical chiller load profiles or equivalent full load hours as presented in ASHRAE Handbook of Fundamentals and other publications. However, assumptions required by these estimation methods limit several aspects of the analysis. For example, the cost per KWH in the Georgia Power region can vary from $0.02 to $0.10 depending on where in the rate schedule a building falls. The average cost per KWH is actually a combination of this KWH cost and a demand charge which can be as high as $20.00 per month per peak KW. These costs cannot be defined using an equivalent full load hour approach. The entire building's energy requirements must be determined to identify the applicable incremental KW and KWH costs.

In order to provide each individual building owner with an optimum design for his building and criteria, the analysis should be based on hourly energy requirements, not "rules of thumb" such as typical chiller load profiles. Unfortunately, the decision as to which chiller type(s) to use and the approximate sizes of each chiller must be derived early in the design process when much of the building design information is unknown. To permit an educated decision with incomplete information, the JND-developed hourly simulation program allows the user to describe a given building at whatever level of detail is available. The program uses default values and accepted engineering practices to calculate probable values for any inputs not provided by the user.

Another analysis shortcut that should be avoided is the analysis of chillers without regard to the peripheral equipment required by each chiller. Peripheral and parasitic loads vary by chiller type. Therefore, an accurate analysis must compensate for the difference in these loads among the chillers. For example, centrifugal chillers use an oil pump, an oil sump heater and a control panel which total from about 1.5 KW to 3.5 KW for chiller sizes of 200 tons to 1,500 tons. Conversely, direct fired absorption chillers use a solution pump, a refrigerant pump, a purge pump, a control panel and burner which can total from 5.7 KW for a 100 ton machine to 30.8 KW for a 1,500 ton machine. These are not accounted for in manufacturer's efficiency claims. Absorption chillers also have greater heat rejection which requires a larger cooling tower and/or a larger condenser water pump than a centrifugal chiller.

Finally, actual equipment costs must be used instead of average costs per ton if the owner requires an economic analysis. Based on actual manufacturer's quoted prices, a 3 stage centrifugal chiller can range in price from about $150/ton to $450/ton while a direct-fired absorption chiller can range from $320/ton to $760/ton. In general, the larger the chiller the smaller the cost per ton in a given model number. Other equipment costs which vary with chiller type must also be considered such as cooling towers and condensing water pumps.

The methodology developed by JND for use in the computer model, and as outlined below, accounts for the numerous, interactive components in a chilled water plant. This will produce a more accurate determination of the optimum mix of electric centrifugal and gas absorption chillers in a combined plant. The added benefit of a parametric computer analysis permits refinement of the mix as the design progresses.

METHODOLOGY

The reader may utilize any of dozens of analysis tools available. However, the analysis must address the factors discussed below.

A typical mixed chilled water plant consists of six components. These are:

- Electrical Centrifugal Chiller(s)
- Gas Absorption Chiller(s)
- Chilled Water Pump
- Centrifugal Chiller Condenser Water Pump
- Absorption Chiller Condenser Water Pump
- Cooling Tower Fan

The energy use of these pieces of equipment depend on the size of those components. Therefore, the size(s) of all equipment must be determined. This requires that the building peak heating and cooling loads be calculated. Any acceptable engineering load calculation can be used. To simplify the analysis process, the JND-developed computer program includes a peak load calculation module which uses the ASHRAE GRP 158 method.

The equipment listed above must then be sized based on the peak loads. Again, acceptable engineering practices should be used. And again, the JND developed program includes a sizing and selection module which parallels manufacturer's recommended selection processes. Once the equipment is sized and selected, the analysis itself, can begin. Throughout this discussion, both peak demand and energy consumption calculations are shown. In areas where demand charges are not applicable only the consumption calculations need be used. The variables included in the equations are defined in Table 1.

Electrical Centrifugal Chiller

The electric centrifugal chiller(s)' energy consumption varies with its operating load. The units' rated efficiency (COP or KW/ton) is applicable only at peak capacity. Its efficiency varies at partial capacities along a curve known as the part-load-ratio. Part-load ratios are equipment specific. Figure 1 illustrates the part-load-ratio for a Trane Centravac chiller. The curve denotes the energy consumption at part loads as a percentage of the unit's energy consumption at peak conditions. These curves are listed in manufacturers' literature. Peak efficiencies can range from 0.56 KW/ton to 0.8 KW/ton for current technology centrifugal chillers.

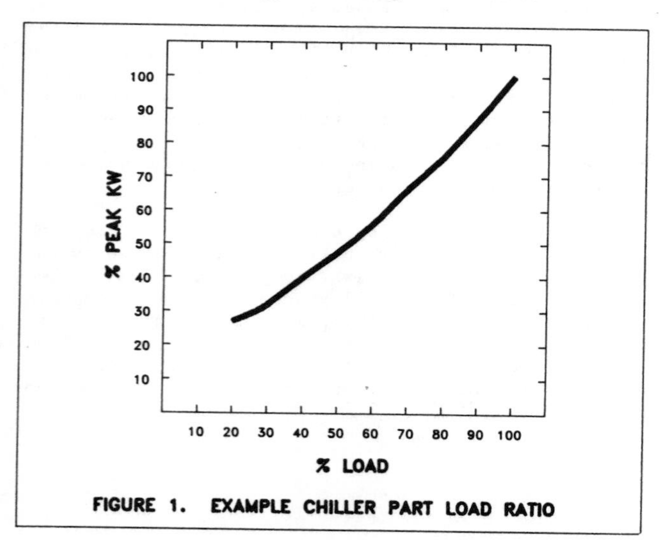

FIGURE 1. EXAMPLE CHILLER PART LOAD RATIO

The monthly chiller KW demand is calculated using Equation 1.

$$\text{CHILLER KW} = \text{KW/TON} * \text{MAX TON} \quad (1)$$

(See Table 1 for definition of variables.)

The monthly chiller KWH consumption is calculated using Equation 2.

$$\text{CHILLER KWH} = \text{KW/TON} * \text{CENT. CHLR. TON-HRS.} \quad (2)$$

The chiller also has additional energy consumption due to an oil pump, an oil sump heater and a control panel. These parasitic loads are relatively constant in a centrifugal chiller; usually about 2.5 KW.

The monthly parasitic load in KWH is calculated using Equation 3.

$$\text{PARASITIC KWH} = 2.5\ \text{KW} * \text{CENT. CHLR. RUN-HRS.} \quad (3)$$

It should be noted that the monthly chiller run hours can be different from the monthly chilled water plant run hours when a waterside economizer is specified.

Gas Absorption Chiller

There are two heat sources for gas absorption chillers; direct fired and steam. The efficiency of a direct fired absorption chiller shown in the manufacturer's catalog is usually given in COP. However, the COP given is based on the lower heating value of natural gas. Since the gas utility company charges its customers in therms based on the higher heating value of natural gas, it is necessary to adjust the lower heating value COP such that the therm of gas used can be calculated correctly. This is shown in Equation 4.

$$\text{ADJUSTED COP} = \frac{\text{LHV}}{\text{HHV}} * \text{COP(LHV)} \quad (4)$$

The annual energy consumption of the direct-fired absorption chiller can then be calculated using Equation 5.

$$\text{CHILLER THERMS} = \frac{\text{ABSORBER TON-HRS.} * 0.12\ \text{THERM/TON-HR.}}{\text{ADJUSTED COP}} \quad (5)$$

Parasitic loads of the direct fired absorption chiller consist of the energy used by the solution pump, refrigerant pump, purge pump, control panel and burner. The total of these loads varies with the size of the absorption chiller. The parasitic loads for a range of sizes of a Hitachi direct fired absorption chiller are shown in Table 2.

The annual load is then calculated using Equation 7.

$$\text{PARASITIC KWH} = \text{PARASITIC KW} * \text{ABSORBER RUN-HRS} \quad (7)$$

The efficiency of the steam absorption chiller shown in the manufacturer's catalog is usually given in lbs. of steam/ton-hour. This can be translated into COP using Equation 8.

$$\text{COP} = \frac{\text{BOILER EFF.} * \text{DIST. EFF.} * 12{,}000\ \text{BTU/TON-HR.}}{\text{\#/TON-HR.} * \text{BTU/\#}} \quad (8)$$

The energy consumption of the absorption chiller can now be calculated using Equation 9.

$$\text{Chiller Therms} = \frac{\text{Absorber Ton-Hrs.} * 0.12\ \text{Therm/Ton-Hr.}}{\text{COP}} \quad (9)$$

Steam absorption chillers do not have a burner. The heat generation process of the absorption cycle is provided by the steam. The parasitic loads for three popular steam absorption chillers are shown in Table 3. The annual parasitic energy consumption is again calculated using Equation 7.

Chilled Water Pump

The chilled water pump size and energy consumption for a given cooling load, chilled water delta T and flow resistance are constant regardless of the chilled water plant configuration. Therefore, the chilled water pump size and energy consumption only needs to be calculated once. The reason it is included in chiller analyses is to account for its electrical usage when determining the building's rate schedule ratchet. The calculations to determine the chilled water pump size and energy consumption are shown in Equations 10 through 13 below.

$$\text{CHW GPM} = \frac{\text{PEAK COOLING LOAD}}{500 * \text{CHW dT}} \quad (10)$$

$$\text{CHW BHP} = \frac{\text{CHW GPM} * \text{CHW HEAD}}{3960 * \text{PUMP EFF}} \quad (11)$$

$$\text{CHW KW} = \frac{\text{CHW BHP} * 0.745\ \text{KW/HP}}{\text{Motor Eff.}} \quad (12)$$

$$\text{CHW KWH} = \text{CHW KW} * \text{COOLING HRS.} \quad (13)$$

Centrifugal Chiller Condenser Water Pump

The size and energy consumption of the condenser water pump(s) is dependent upon the heat rejection rate of the chiller(s). The formulas for calculating the size and energy consumption of the centrifugal chiller condenser water pump are presented in Equations 14 through 18 below.

$$\text{HEAT REJECTION} = \text{CENT. SIZE} * (12000 + 3413 * \text{KW/TON}) \quad (14)$$

$$\text{CW GPM} = \frac{\text{HEAT REJECTION}}{500 * \text{CW dT}} \quad (15)$$

$$\text{CW BHP} = \frac{\text{CW GPM} * \text{CW HEAD}}{3960 * \text{PUMP EFF}} \quad (16)$$

$$\text{CW KW} = \frac{\text{CW BHP} * 0.745\ \text{KW/HP}}{\text{MOTOR EFF}} \quad (17)$$

$$\text{CW KWH} = \text{CW KW} * \text{CHILLER COOLING HRS.} \quad (18)$$

Absorption Chiller Condenser Water Pump

The size and energy consumption of the absorption chiller condenser water pump is similar to that of the centrifugal chiller condenser water pump described in the last section. Heat rejection from the absorption chiller is much higher than from a centrifugal chiller. The heat rejection rates for the two types of absorption chillers can be calculated using Equations 19 and 20, respectively. The energy use calculations for the pump are the same as listed above in Equations 14 through 18.

For Direct Fired Absorption Chiller:

$$\text{HEAT REJECTION} = \text{ABS. SIZE} * 12{,}000 * (1 + \text{BURN EFF/ADJUSTED COP}) \quad (19)$$

For Steam Absorption Chiller:

$$\text{HEAT REJECTION} = \text{ABS. SIZE} * (12{,}000 + \text{\#/TON-HR.} * \text{BTU/\#}) \quad (20)$$

Cooling Tower

The analyst can usually assume there is only one cooling tower for the entire chilled water plant if more than one is not specifically required. That is, both the centrifugal and absorption chillers use the same cooling tower to reject heat. However, there may be more than one cell in the cooling tower.

Cooling tower selection is a function of the design entering tower water temperature, design leaving tower water temperature, and design O.A. wet bulb temperature. This selection procedure is described in manufacturers' catalogs. The JND-developed computer simulation includes a selection routine based on the Marley Series 220 cooling tower. Since both the centrifugal and absorption chillers are assumed to use the same cooling tower, the mixed entering tower water temperature can be calculated using Equation 21.

$$\text{ENT. TEMP} = \frac{(\text{CW GPM} * \text{CW DT})_{cent} + (\text{CW GPM} * \text{CW DT})_{abs}}{\text{TWR GPM}} \quad (21)$$

The cooling tower fan BHP is a function of two variables; the tower selected and the tower flow rates. Table 4 lists the fan horsepower and range of GPM for various sizes of the Marley Series 220 cooling tower.

When the calculated tower fan BHP is greater than 50 BHP, the tower GPM should be divided by the smallest integer which will result in a tower fan BHP smaller than 50 BHP. This integer will be the number of cells in the cooling tower.

Estimating cooling tower fan run-hours is not a precise process. However, it can be assumed for analysis purposes that the fan will operate whenever cooling is needed and the outside air wet-bulb temperature is greater than the supply air dewpoint. In addition, the fan will run some of those hours when cooling is needed but the outside air wet bulb is less than the supply air dewpoint. The JND-develped program estimates the cooling tower fan run-hours during normal operation according to the following conditions.

Condition @ Each Hour	Hour Counted
No cooling load	0
Cooling load and O.A. WB >= Supply Dewpoint	1
Cooling load and O.A. WB < Supply Dewpoint	0.5

Note: Cooling load refers to chilled water plant cooling load, not chiller cooling load. Chilled water plant cooling load equals chiller cooling load if waterside economizer is not specified.

The tower fan demand and energy consumption can, thus, be calculated using Equations 23 and 24, respectively.

$$\text{TOWER FAN KW} = \frac{\text{TOWER FAN BHP} * 0.745 * \text{NO. OF CELLS}}{\text{MOTOR EFF.}} \quad (23)$$

$$\text{TOWER FAN KWH} = \text{TOWER FAN KW} * \text{TOWER RUN-HRS.} \quad (24)$$

Waterside Economizer

If a waterside economizer is specified, the chillers would operate fewer hours while the cooling tower fan would usually operate more hours. The actual number of hours would vary depending on climatic conditions, tower surface area and load. There are no simple equations to determine these hours. However, a simulation program would have already calculated the three variables above and so could determine the chiller and cooling tower fan on/off status on an hour-by-hour basis.

Control Sequences

The concept of a mixed chiller-type plant is to take advantage of the benefits of each chiller type. Electric centrifugal chillers are, in general, more efficient than gas absorption chillers. However, the costs of electricity and gas vary. Control sequences must be developed to operate the electric chillers when electricity is less expensive than gas and to operate gas absorption chillers when gas is less expensive than electricity.

Electricity and gas costs can vary with time of day, month of year, magnitude of energy consumption and/or "flatness" of energy demand profile. For example, the Georgia Power PL-7 rate schedule includes both demand charges and consumption charges. The costs of these two components is based on the ratio of monthly energy consumed to "billing demand". The billing demand is calculated each month. During the months June through September, the PL-7 billing demand is the greater of:

1) The actual peak demand of current month,
2) 95% of the highest actual demand recorded in the previous applicable summer months, or
3) 60% of the highest actual demand recorded in the previous winter months.

During the months of October through May, only conditions 2 and 3 apply.

Assuming that the billing demand is set by the summer peak, the consumer will pay for at least 95% of that peak for the next 11 months even though ensuing monthly peaks may be considerably less. Most commercial buildings in the Atlanta area fall in the brackets where demand charges range from $8.91 to $12.63 per KW. Therefore, it is most advantageous to reduce the summer peak demand by running a gas absorption chiller during the buildings' peak energy demand periods.

Conversely, during non-peak hours most buildings will accrue electricity charges at the low rate of about 2.5 cents per KWH. Assuming a .6 KW/Ton chiller, this equates to 1.5 cents per ton-hour. Gas costs using the Atlanta Gas N-16 rate, interruptible PGA and summer air conditioning credits are about $0.35 per therm. Assuming the use of a direct-fired absorption chiller with a COP of 1.0, results in a ton-hour cost of 4.7 cents. There are other factors to consider. However, it is obvious that electric centrifugal chillers can provide cooling much less expensively than gas absorption chillers during non-peak hours in the Atlanta area.

The chiller control sequence, in this example, must reduce the summer peak electric demand as much as possible by using the gas absorption chiller during peak hours. However, the control sequence must also minimize absorber run-hours during non-peak hours. How much should the peak demand be reduced? How many hours should the absorber run? The answers to these questions depend on the building's demand profile.

For example, a hypothetical building is billed based on the PL-7 rate schedule ratchets of "200 to 400 hours use of demand" and "over 1000 KW billing demand". The incremental rates attributable to these brackets are $8.906 per KW and $0.026 per KWH. Reducing the peak demand of this building by 200 KW could save $20,400 per year in demand charges. However, if this would require operating the absorption chiller 300 hours, this could result in an additional energy cost of $9,800 and a net annual savings of $10,600. (These costs vary with chiller efficiencies and actual run hours. The numbers quoted here are for a typical, but hypothetical building.) Reducing the peak demand of this same building by 300 KW could save $30,600 per year in demand charges. However, if the absorption chiller must run 650 hours, this results in an additional energy cost of $20,700. This results in a net annual savings of $9,900. The optimum run-hours of each chiller will vary depending on the building's demand profile, the efficiency of each chiller and the size of each piece of equipment. Therefore, parametric analyses must be conducted to determine the optimum size of each chiller. The result of the control sequence evaluation should be the mix of centrifugal and absorption chiller sizes which provide the lowest annual energy costs under the applicable rate schedules.

Economic Criteria

Lowest energy cost does not imply optimum mix. Equipment cost and maintenance costs must also be considered. Actual costs should be obtained from manufacturers for the specific chiller sizes and efficiencies selected. Cost per ton values should not be used due to the great variation in this number depending on the size of the chiller. Figure 2 presents chiller costs for four different types of chillers. These values will have changed by the time this paper is printed. However, the relative costs illustrate the difference in first cost between centrifugal and absorption chillers.

There are numerous methods for performing economic evaluations. The method used will depend on the economic criteria dictated by the client. The simplest economic evaluation technique is the "Simple Payback". The simple payback is the number of years it will take to recoup an initial additional investment through annual operating cost savings. Simple payback can be calculated using Equation 25.

$$\text{SIMPLE PAYBACK} = \frac{\text{ADDITIONAL INVESTMENT}}{\text{ANNUAL OPERATING COST SAVINGS}} \quad (25)$$

The additional inital cost is the cost difference between the configuration being studied and a base case configuration. It is best to define the base case as the design with the lowest first cost. In this way, the additional inital cost of all alternatives is a positive number. Negative simple paybacks are not usable.

Other economic evaluation methods include: Discounted Payback, which accounts for the ability to have invested that additional initial cost in some other investment; Internal Rate of Return, which portrays the payback as an annual percentage of the initial investment; and Cash Flow, which provides annual cash outlay or income information. All of these methods consist of iterative calculations best performed by a computer simulation. However, if the analyst uses the methodology outlined above to obtain the annual operating costs and actual equipment costs for various chilled water plant design alternates, he or she can use one of numerous economic tools on the market to present the results in any format desired by the building owner.

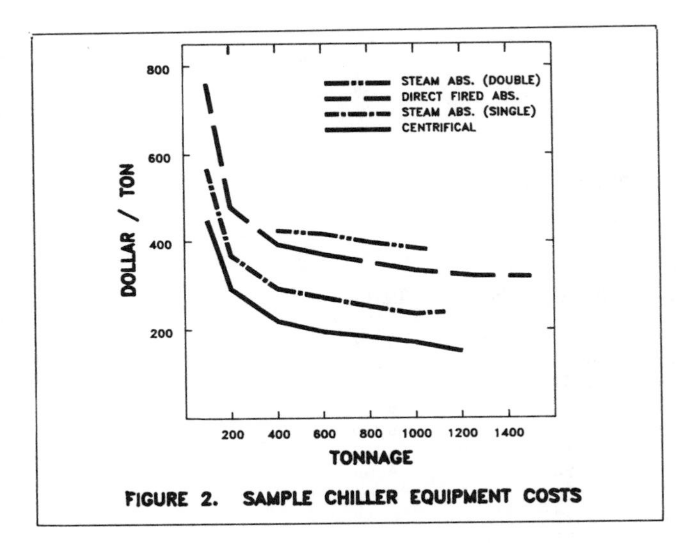

FIGURE 2. SAMPLE CHILLER EQUIPMENT COSTS

SUMMARY

The methodology described in this paper should allow an analyst to properly evaluate the optimum mix in any multiple chiller plant. The methodology stresses hourly simulation and accurate cost figures. Average values such as energy cost per KWH and equivalent cost per ton should not be used. The building, in total, must be simulated to determine appropriate rate schedule ratchets and equipment sizes. To accommodate these time-consuming tasks, Jones, Nall and Davis developed a computer simulation program which includes building load calculations, equipment sizing routines, energy cost calculations and numerous economic calculations. This methodology will permit an accurate assessment of alternate chiller plant design strategies that can be better substantiated than "quick" calculations using typical chiller load profiles and average costs.

TABLE 1

KW/TON = Efficiency of the electric centrifugal chiller

MAX TON = Smaller of the size of the centrifugal chiller and the maximum cooling load of the month.

CENT. CHLR. TON-HRS. = Monthly centrifugal chiller cooling load.

CENT. CHLR. RUN-HRS. = Monthly centrifugal chiller run hours.

LHV = Natural gas lower heating value. Usually 930 BTU/CF.
HHV = Natural gas higher heating value. Usually 1030 BTU/CF.
COP(LHV) = COP of chiller based on lower heating value of natural gas.

ABSORBER TON-HRS. = Monthly absorption chiller cooling load.

ABSORBER RUN-HRS = Monthly absorption chiller run hours.

BTU/# = BTU content in one lb. of steam, program assumes 1,000 BTU/# steam.
BOILER EFF. = Steam boiler efficiency.
DIST. EFF. = Distribution efficiency of the steam.
#/TON-HR. = Steam chiller efficiency.

CHW GPM = Chilled water flow rate in GPM
PEAK COOLING LOAD = Maximum cooling load of the building (BTU/hr.)
CHW dT = Chilled water temperature difference

CHW BHP = Chilled water pump BHP
CHW HEAD = Chilled water flow resistance in feet of head.
PUMP EFF = Pump efficiency.

CHW KW = Chilled water pump KW
Motor Eff = Motor efficiency.

CHW KWH = Monthly chilled water pump KWH
COOLING HRS. = Monthly chilled water plant run hours

HEAT REJECTION = Maximum rate of heat rejection from the chiller (BTU/Hr.)
CENT. SIZE = Centrigual chiller size (ton).

TABLE 1 (Continued)

CW GPM = Centrifugal chiller condenser water flow rate in GPM.

CW dT = Condenser water delta T.

CW BHP = Condenser water pump BHP.

CW HEAD = Condenser water flow resistance in feet of head.

CW KW = Monthly condensed water pump KWH

CW KWH = Monthly condenser water pump KWH

CHILLER COOLING HRS. = Monthly centrifugal chiller run hours ignoring waterside economizer.

ABS. SIZE = Size of the absorption chiller (ton)

BURN EFF. = Burner efficiency in direct fired absorption chiller.

#/TON-HR. = Steam absorption chiller efficiency.

BTU/# = BTU content in a lb. of steam.

ENTERING TEMP = Mixed entering tower water temperature

$(\ )_{cent}$ = For centrifugal chiller

$(\ )_{abs}$ = For absorption chiller

TWR GPM = Total water flow rate to the cooling tower

NO. OF CELLS = Number of cells of the cooling tower.

TOWER RUN-HRS. = Monthly tower fan run-hours under normal operation.

TABLE 2

HITACHI DIRECT FIRED ABSORPTION CHILLER

PARASITIC LOADS (KW)

MODEL NO.	TONS	SOLUTION PUMP	REFRIG. PUMP	PURGE PUMP	BURNER	CONTROL PANELS	TOTAL
10N	100	3.70	0.40	0.40	0.22	1.00	5.72
11N	125	3.70	0.40	0.40	0.37	1.00	5.87
11NL	140	3.70	0.40	0.40	0.37	1.00	5.87
12N	150	3.70	0.40	0.40	0.52	1.00	6.02
12NL	170	3.70	0.40	0.40	0.75	1.00	6.25
13N	200	3.70	0.40	0.40	1.12	1.00	6.62
14G	250	4.50	0.80	0.40	2.24	1.00	8.94
14GL	270	4.50	0.80	0.40	2.24	1.00	8.94
15G	320	4.50	0.80	0.40	3.73	1.00	10.43
15GL	345	4.50	0.80	0.40	3.73	1.00	10.43
16G	400	4.50	0.80	0.40	3.73	1.00	10.43
16GL	430	5.50	1.50	0.40	3.73	1.00	12.13
17G	500	5.50	1.50	0.40	3.73	1.00	12.13
18G	600	7.50	1.50	0.40	3.73	1.00	14.13
19G	700	7.50	1.50	0.40	5.60	1.00	16.00
20G	1000	12.70	1.50	0.40	7.46	1.00	23.06
21G	1250	11.40	1.50	0.40	7.46	1.00	21.76
22G	1500	16.70	1.50	0.40	11.20	1.00	30.80

TABLE 3

STEAM ABSORPTION CHILLER

PARASITIC LOADS (KW)

MANUF.	MODEL NO.	TONS	SOLUTION PUMP	REFRIG. PUMP	PURGE PUMP	CONTROL PANELS	TOTAL
HITACHI	10N	100	3.7	0.4	0.4	1.0	5.5
	13N	200	3.7	0.4	0.4	1.0	5.5
	15G	310	4.5	0.8	0.4	1.0	6.7
	16G	400	4.5	0.8	0.4	1.0	6.7
	17G	500	5.5	1.5	0.4	1.0	8.4
	18G	600	7.5	1.5	0.4	1.0	10.4
	19G	700	7.5	1.5	0.4	1.0	10.4
	20G	1000	12.7	1.5	0.4	1.0	15.6
	21G	1250	11.4	1.5	0.4	1.0	14.3
	22G	1500	16.7	1.5	0.4	1.0	19.6
TRANE	ABTD-03J	385	9.3		0.2	1.0	10.5
(TWO	ABTD-04F	465	10.8		0.2	1.0	12.0
STAGE)	ABTD-05C	527	11.9		0.2	1.0	13.1
	ABTD-05J	590	16.0		0.2	1.0	17.2
	ABTD-06A	656	17.5		0.2	1.0	18.7
	ABTD-07A	750	20.3		0.2	1.0	21.5
	ABTD-08C	852	21.7		0.2	1.0	22.9
	ABTD-09C	935	24.2		0.2	1.0	25.4
	ABTD-10C	1060	27.0		0.2	1.0	28.2
TRANE	ABSC-01A	101	3.8		0.2	1.0	5.0
SINGLE	ABSC-01E	148	4.2		0.2	1.0	5.4
STAGE	ABSC-02A	200	5.4		0.2	1.0	6.6
	ABSC-02F	256	7.0		0.2	1.0	8.2
	ABSC-02J	294	7.2		0.2	1.0	8.4
	ABSC-03F	354	7.5		0.2	1.0	8.7
	ABSC-04B	420	8.3		0.2	1.0	9.5
	ABSC-05C	520	8.8		0.2	1.0	10.0
	ABSC-05J	590	9.5		0.2	1.0	10.7
	ABSC-07C	750	10.4		0.2	1.0	11.6
	ABSC-08C	852	11.0		0.2	1.0	12.2
	ABSC-09D	955	13.9		0.2	1.0	15.1
	ABSC-11A	1125	15.1		0.2	1.0	16.3
	ABSC-12A	1250	17.0		0.2	1.0	18.2

TABLE 4

COOLING TOWER

MODEL NO.	FAN HP	MIN. GPM	MAX. GPM
221-211	15	300	1700
222-511	20	450	2500
222-611	25	550	3000
222-711	40	700	3500
222-721	50	750	3500

Chapter 53

NEW GENERATION NATURAL GAS ENGINE-DRIVEN CHILLER SYSTEMS

D. A. Smith

The energy consumption profile of commercial and industrial buildings has changed dramatically in recent years. Each computer or electronic gadget brought into the workplace brings with it a need for additional air conditioning. Indoor air quality concerns and high-tech manufacturing processes are requiring temperature and humidity control systems unheard of just a few years ago. What is the new technology in gas engine-driven chiller packages and how does it help fill these air conditioning needs?

1. Utility Rate Trends

Escalating electrical energy costs and severe power demand charges have sent many of these air conditioning customers looking for gas driven alternatives. Although utility rates vary across the country, the Georgia Power PL-7 electric rate and the Atlanta Gas Light N-2 rate will be used in this paper as typical for medium size commercial customers.[1] The PL-7 electric rate has a 12 month ratchet clause which can severely penalize the customer who hits that peak load for only one day during the summer.

Disregarding the ratchet clause, customers are billed under PL-7 at a rate of 60% of their actual demand during the winter months (October thru May) and 100% of their actual demand during the remaining summer months. A typical load profile normalized for these factors is shown by the hatched bar areas in figure 1. This shows more than 400Kw difference between the summer and winter billing demands. However, the 12 month ratchet clause actually sets the billing demand at 95% of the highest previous summer month. Instead of enjoying the benefits of reduced demand during the winter months as indicated by the bar areas (around 400 Kw), the customer must actually pay the ratcheted amount as shown by the solid line (around 850 Kw) in figure 1. These demand charges resulting from the ratchet clause can comprise as much as 30% to 80% of the monthly utility bill.

Figure 1 makes it clear that if gas alternatives can be used to reduce electric demand peaks during the summer, then the benefits will extend to the winter months as well. In comparison, the N-2 gas rate has no demand fees or associated ratchet clause that penalizes the customer for heavy summer usage. Assuming that the air conditioning load is the primary contributor to the summer electrical needs in this typical building, then this 400 Kw of demand difference between summer and winter should be looked at closer for possibilities of peak shaving by gas driven air conditioning equipment.

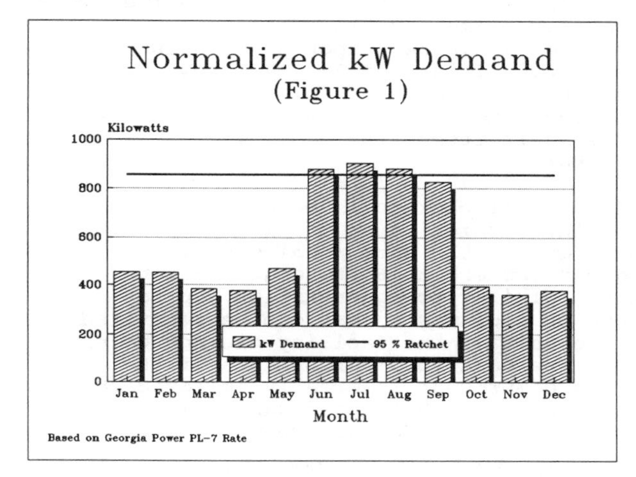

2. Peak Shaving Alternatives

Gas absorption air conditioning has been used quite successfully in this peak shaving mode of operation for many years. However, small to medium size commercial facilities often lack the maintenance staff or know-how to diligently maintain absorption equipment. The majority of HVAC contracting companies specialize in conventional vapor compression refrigeration equipment.

Natural gas engine driven chiller systems conveniently fall into this category. The refrigeration or chiller portion of the system can be exactly the same as with electric driven equipment. The compressor is simply turned by a natural gas internal combustion engine rather than an electric motor.

3. Development of the Engine System

This idea of coupling an engine to a compressor is nothing new. It was done quite extensively in the 1960's. What is new is the engine technology itself and the control systems. The 60's generation equipment was mainly based on expensive industrial grade engines. In additional to high initial cost these engines also required diligent (and often frequent) maintenance by highly skilled technicians. The systems were sometimes custom made for each installation and were not mass produced packages. The new generation systems are based on inexpensive automotive derivative engine sets that are computer microprocessor controlled and require only limited routine maintenance.

Although various engine driven packages are on the market or in final R&D stages, the remainder of this paper will deal with the 150 ton gas engine driven chiller system developed in part as a project of the Gas Research Institue.[2]

4. Engine Chiller Features

This 150 ton chiller package was developed for GRI by Tecogen, Inc. in Waltham, Mass. Tecogen has been in the packaged cogeneration business for a number of years using marine grade automotive derivative engines for prime movers. The system operates on a typical vapor compression cycle using Refrigerant 22. A screw compressor is used to optimize part load performance and most of the components are off-the-shelf items from domestic manufacturers.

The Teco-Chill package is based on a 454-cu-in. automotive engine modified for natural gas -- the same model used in Tecogen's commercial 60 Kw cogeneration module. Similar to cogeneration, up to 750,000 Btuh of waste heat can be recovered from the unit if the facility has hot water needs. The part load performance of the chillers compressor is greatly improved by the engine's variable speed capability. The microprocessor control can modulate the engine speed to track the cooling load over the full range of the equipment (from zero to 100%). No wasteful hot gas by-passing or cycling on and off is necessary to carry small loads. To obtain similar flexibility, an electric chiller would have to be equipped with adjustable speed drive, which would add about $100/ton to its price.[3] An additional advantage of the gas engine is that it can track even higher peak loads (up to around 180 tons) for intermittent periods of time by simply revving up to higher rpms.

5. Part Load Efficiencies

This system lends itself well to computer controlled energy management systems and can even be monitored and controlled via telephone modem.

The variable speed capability of the engine system delivers exceptional performance under part load conditions. The chiller's COP at standard rating conditions is 1.4 at the rated 150 ton capacity. This increases to a maximum of 2.0 as load is reduced to around 30% (see figure 2). The COP's are even higher if credit is given for optional heat recovery.[4] This is nearly twice as efficient as the latest dual stage absorption cooling equipment which reports COP's in the area of 1.0.

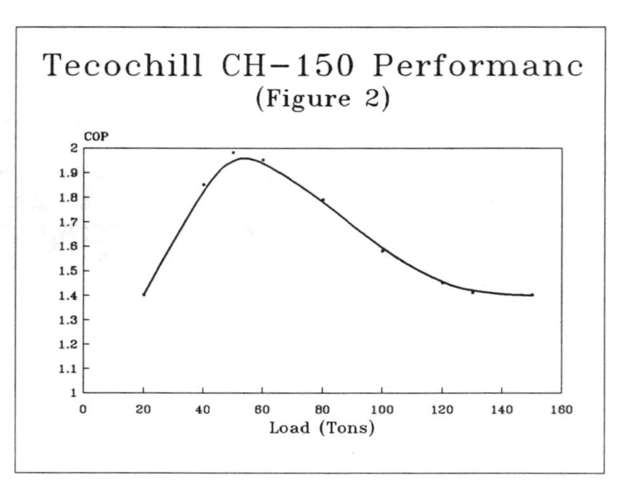

Typical load durations for various types of commercial buildings are shown in figure 3. All of these profiles indicate that the facilities spend most of their time during the cooling season under part load conditions[3] (the same conditions that the engine system operates at most efficiently). Hotels and hospitals are especially good candidates for these systems since they usually have large domestic hot water needs.

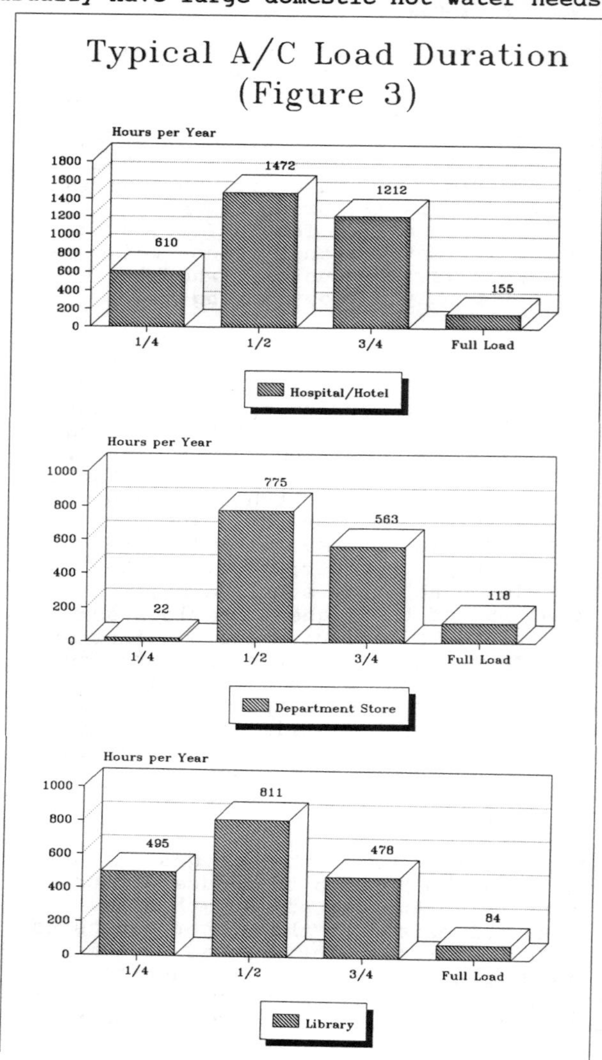

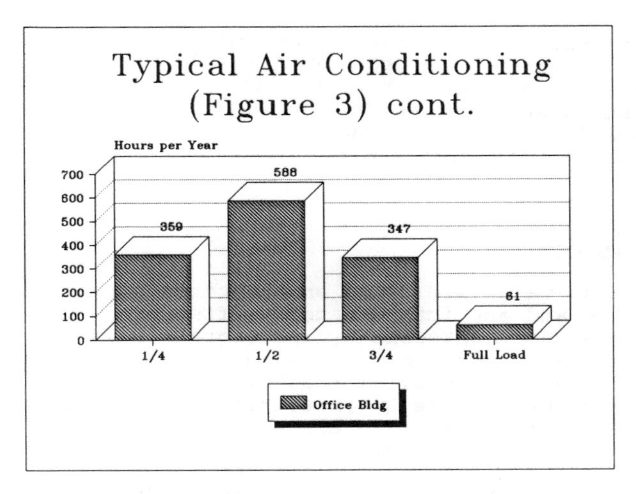

6. System Maintenance

The service and maintenance on the refrigeration portion of these systems is essentially the same as with any conventional chiller. Standard HVAC contractors should recognize all the components and the controls have on-board system diagnostics. Routine engine maintenance consists of changing fluids and filters, spark plugs and wires, and general tune-up checks. This routine maintenance would be scheduled to be done seasonally when the unit would ordinarily be off-line. The engine is very accessible in these packages and the automotive type maintenance can be mastered quickly by the standard contractor.

This same engine in the Tecogen 60kw cogen modules has demonstrated life spans in excess of 20,000 hrs. One of the reasons that Tecogen has so much confidence in this General Motors type engine is that their parent company also owns the Crusader Boat Company -- which has used this same basic engine as their workhorse for many years. In addition, the GRI project included extensive engine research covering the performance and lifespan of this natural gas conversion. Unlike the engine driven systems of the 1960's, these marine grade engines are relatively inexpensive (aprox $3000) and can simply be swapped out rather than overhauled at the end of their useful life. In order to alleviate customer concern with engine reliability Tecogen offers a complete engine warranty package which covers all scheduled and unscheduled maintenance on the engine (including engine replacement in the event of catastrophic failure or when the engine reaches the end of its usable life). The price of this package is $1.75 per equivalent full load operating hour ($ 2625 for a typical 1500 EFLH Atlanta cooling season). The refrigeration system life expectancy is in excess of 20 years.

7. Operating Cost Savings

Figure 4 shows how the Tecochill unit stacks up when compared to various other types of chillers in the 150 ton range. These costs per ton hour are based on the efficiencies and auxiliaries listed in figure 5.

Depending on the cooling application and what step of the electric power rate the customer falls into, the Teco-Chill unit can save from $7000 to $20,000 per year in operating costs over conventional electric systems. Using this chart and including the additional engine maintenance warranty fees provided by Tecogen, the engine chiller demonstrates an annual savings of approximately $ 7725 over comparable electric centrifugal equipment and $ 13,800 over reciprocating equipment. If heat recovery is included the cost savings are even higher.

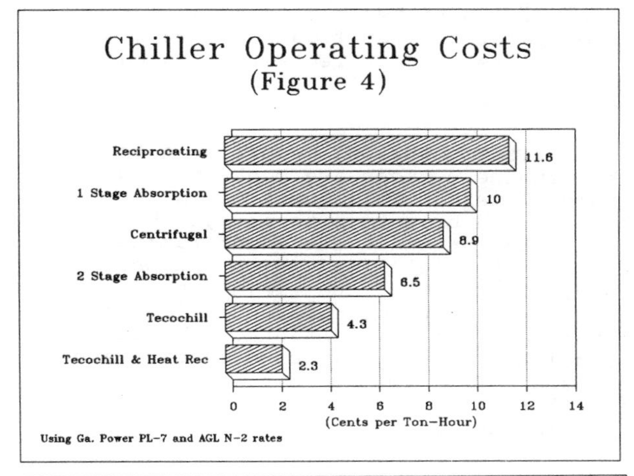

(FIGURE 5) Data Based On :

Georgia Power PL-7 Rate
Atlanta Gas Light N-2 Rate
1500 Equivalent Full Load Hours of Operation

Chiller Specifications:	# Steam /Ton-Hr	Therms /Ton-Hr	kW /Ton	Seasonal COP
Trane Electric Recip			.88	4.0
Trane 1-Stage Steam Abs	18.7	.24		0.5
Trane 3-Stage Centravac			.65	5.4
Hitachi 2-Stage DF Abs		.12		1.0
Tecochill Engine Driven Screw		.08		1.5

Chiller Auxillaries:	Aux Pmp kW	Cond Pmp kW	Twr Fan kW	Total kW
Trane Electric Recip		6.9	4.4	11.3
Trane 1-Stage Steam Abs	4.5	8.5	8.8	21.8
Trane 3-Stage Centravac		6.9	4.4	11.3
Hitachi 2-Stage DF Abs	6.0	13.6	9.5	29.1
Tecochill Eng. Dr. Screw	2.0	7.5	4.6	14.1

8. Field Test Results

This type of savings has been demonstrated by the 7 field test units that GRI has been monitoring this past year. These units were placed in motels, commercial office buildings, hospitals, apartments, and light industrial facilities. The Atlanta test unit was placed at Bleyle of America, a knitware facility in Newnan, Ga. The Bleyle Teco-Chill unit has been on line since May 87 and has demonstrated an excellent on-line availability of 99.5%. The Tecochill replaced multiple 30 ton electric rooftop units. The unit is currently used as a chiller only and does not utilize waste heat. Bleyle data indicates reduction in electrical demand which will amount to over $13,000 per year (nearly 90 Kw load reduction).

9. First Cost Concerns

The problem with gas air conditioning equipment has always been high first costs. With absorption equipment, the facility and the application has to be just right for the payback to be reasonable. Although the Tecochill unit is still in the early production stages, the mature market price is expected to be around $ 400/ton. Electric driven centrifugal equipment in the 150 ton size can be bought for $ 240/ton. Even with this first cost premium the engine driven unit demonstrates a 3 year payback using the operating cost figures shown above (2 years or less if heat recovery is used).

10. Other Questions

One of the first questions people seem to have when engine systems are mentioned is about noise levels. This question is quickly answered by a trip to one of the existing installations. The sound rating is 82 dbA at 20 ft.- full load (with optional enclosure). At most of the field test sites the chiller has been the quietest thing in the mechanical room -- often unnoticeable next to boilers, pumps, air compressors, and exhaust fans. Even in the motel where the unit is relatively close to guest rooms, noise levels have been well within acceptable limits.

What if your needs don't fall in the 150 ton range? Although the 150 ton unit is the only commercially available Tecogen model at this time, other units are currently in the works. A 500 ton model utilizing two of the 454 cid engines and a Carrier based dual centrifugal compressor system is currently being tested at the Tecogen factory. As early as 1989 variations on the original unit are to be introduced which will have nominal ratings in the 120-180 ton range. Small rooftop units are also under development.

11. Conclusions and Future Trends

The Gas Research Institute sees this new gas engine system technology as the trend setter of the 1990's. Power rates around the country are expected to continue to escalate which should spur further interest in gas alternatives to electric driven air conditioning equipment. The state of the art computer control systems coupled to these units are extremely versatile and were built with electronic energy management systems and remote control in mind.

The key to greater savings and rapid payback of any first cost premium lies in utilizing the waste heat recovery. Even facilities without hot water needs can still find innovative ways to use this energy -- to drive desiccant dehumidification units, small absorption air conditioners, reheat systems, or various other equipment.

Major air conditioning equipment manufacturers are currently working with Tecogen on the development of future units. Already Tecochill 'clones' have been spotted on the market that utilize reciprocating compressor units. This technology is expected to grow in popularity and broaden in scope. Anyone with commercial or industrial cooling needs owes this new generation of natural gas engine driven systems a very close look.

References

1. These rates are based on the terms of service and applicable rate tariffs for Atlanta Gas Light Company and Georgia Power Company approved and filed with the Georgia Public Service Commission as of Jan. 1988 (Approx. Utility fees: Electricity, $ 12.62/Kw Demand and 2.6 c/Kwh consumption; Gas, 39.36 c/therm average).

2. The Tecochill 150 ton gas engine driven chiller project; The Gas Research Institute - Chicago, IL.; Bruce B. Lindsay, Project Manager, 1988.

3. The Gas Research Institute Digest, Field Test Highlights; Summer 1988, p.29.

4. Tecogen, Inc. - Waltham, Mass., published performance specifications for Tecochill CH-150 unit.

5. Heating, Piping, and Air Conditioning, How to estimate energy requirements for big building air conditioning; January 1960, p. 157.

6. Tecochill Field Test Managers meeting, Chicago, Illinois; May 1988 - presentation by Mr. Aris J. Papadopoulos, Tecogen, Inc.

SECTION 11
PROCESS ENERGY CONSERVATION

Chapter 54

OPPORTUNITIES FOR WASTE HEAT RECOVERY IN SMALL AND MEDIUM-SIZED MANUFACTURING PLANTS

R. J. Jendrucko, P. S. Miller

INTRODUCTION

The performance of an initial energy audit of an industrial facility, including a complete survey of plant energy using equipment, typically leads to the identification of several distinct opportunities for achieving substantial energy savings. These proposed measures usually include several simply implemented, low- or no-cost actions such as extinguishing unnecessary lighting. In addition, it is common to arrive at recommendations which involve large energy savings but which require substantial capital expenditure and plant equipment modifications. Among the latter are those which identify with the broad area of waste heat recovery. Energy conservation opportunities of this type are relatively common in industry in the United States owing to the wide use of thermal processing.

Although many of the longstanding potential industrial waste heat recovery measures nationwide may have been implemented in recent years or at least considered initially, there are data that suggest that much is yet to be accomplished. Most of this data derives from various government and utility sponsored programs promoting industrial energy conservation. In particular, recent data reflecting the frequency of current opportunities for industrial waste heat recovery have been obtained based on energy audits conducted by the University of Tennessee's Energy Analysis and Diagnostic Center [EADC]. This EADC is one of 13 University-based centers operated under contract to University City Science Center of Philadelphia, which manages the nationwide EADC program for the United States Department of Energy, Office of Industrial Programs. For a period of over 10 years, the EADC program has made available no-cost, on site energy audits of small- to medium-sized manufacturing companies. The selection of audited firms is governed by the following criteria:

- Clients must be small- to medium-sized industrial manufacturing firms within the Standard Industrial Classification (S.I.C.) Code 20-39, who are located within 150 miles of the EADC and who meet at least three of the additional following requirements:
- Have a maximum of $1.5 million/year in energy costs at a particular plant.
- Have a maximum of $50 million/year in gross sales for a particular plant.
- Have a maximum of 500 employees at a particular plant.
- Lack in-house professional expertise in energy use and conservation at the particular plant to be served.

INDUSTRIAL CLASSIFICATION OF AUDITED PLANTS

During a 10 year period from 1977-1987, over 340 energy audits of small- to medium-sized manufacturing plants were conducted by the University of Tennessee's EADC. Since its inception, the Tennessee program has served plants in nearly every category of manufacturing industry reflected in the S.I.C. codes. Among the 100 most recently audited plants, excluding those served during the 1988 contract period, 17 of the 20 general S.I.C. code categories were represented. The one-day plant audits included an analysis of historical energy use patterns and the collection of operational data on energy using equipment.

Among the 100 plants, 28 were found to have waste heat recovery opportunities associated with a total of 31 recommended waste heat energy conservation measures. The distribution of S.I.C. codes for these 28 industrial facilities is shown in Table I.

TABLE I
DISTRIBUTION OF NUMBER AND S.I.C. CATEGORIES OF 100 AUDITED PLANTS IDENTIFIED AS HAVING WASTE HEAT RECOVERY OPPORTUNITIES

Manufacturing Category	SIC Code Number	Number of Plants
Food and Kindred Products	20	10
Textile Mill Products	22	3
Electrical and Electronic Machinery, Equipment, and Supplies	36	3
Chemicals and Allied Products	28	2
Fabricated Metal Products Except Machinery and Transportation Equipment	34	2
Transportation Equipment	37	2
Machinery Except Electrical	35	2
Apparel and Other Finished Products Made From Fabrics and Similar Materials	23	1
Lumber and Wood Products Except Furniture	24	1
Furniture and Fixtures	25	1
Petroleum Refining and Related Industries	29	1
Rubber and Miscellaneous Plastics Products	30	1
Primary Metal Industries	33	1

In all, 13 of 20 possible code categories are represented. Note that Table I appears to list a total of 30 plants. This is due to two of the plants having multiple S.I.C. code listings.

Waste heat recovery opportunities are frequently classified according to the nature of the thermal source streams or the type of recovery equipment required. For the present purpose, the identified opportunities are conveniently grouped according to specific categories of plant equipment since the emphasis of the conducted audits was to analyze energy use efficiency in such equipment.

The 31 waste heat recovery opportunities identified may be conveniently grouped into four general categories including combustion equipment, chillers, air compressors, and a miscellaneous category. The frequency of waste heat opportunities identified for each category are listed in Table II below.

TABLE II
DISTRIBUTION OF WASTE HEAT OPPORTUNITIES AMONG GENERAL CATEGORIES

Equipment Category	Observed Frequency of Opportunities for Waste Heat Recovery
Combustion Equipment	
Boiler	3
Cupola	1
Smelter	2
Furnace	2
Oil Heater	1
Incinerator	1
Oven	3
Chillers	
Process	8
HVAC	1
Air Compressors	7
Miscellaneous	
Flare	1
Hot Castings	1

Waste heat recovery opportunities reflected in Table II data can be further analyzed in terms of the distribution of source temperature. It has become common practice to define three waste heat source temperature ranges as follows: low temperature, below 350°F, medium temperature, 350-1200°F, and high temperature, above 1200°F. For the waste heat recovery opportunities summarized in Table II, the source temperature distribution breakdown is given in Table III.

TABLE III
DISTRIBUTION OF WASTE HEAT SOURCE TEMPERATURES FOR THE 31 OPPORTUNITIES SPECIFIED IN TABLE II

Source Temperature	Number of Opportunities
Low	16
Medium	10
High	5

For the 31 waste heat recovery recommendations cited, the distribution of recommended uses for waste heat is given in Table IV below.

TABLE IV
DISTRIBUTION OF RECOMMENDED USES OF WASTE HEAT

Types of Recommendations	Number of Recommendations
Process liquid heating	13
Plant space heating	9
Combustion air preheating	7
Miscellaneous	2

DISCUSSION

In analyzing the data presented above, it should be noted that the relevant audits were done only for small- to medium-sized manufacturing plants and therefore do not include any indications of waste heat recovery opportunities for plants in the large size range. However, the data presented here were assimilated from a relatively large number of plants over a wide distribution of S.I.C. codes and geographical service area. Thus, the findings discussed here may be taken to be representative for manufacturing plants in general.

The types of manufacturing operations in plants having waste heat recovery opportunities listed in Table I is fairly broad with most categories of plants identifying with one to three opportunities. The Food and Kindred Products (S.I.C. 20) group, however, was observed to have the highest frequency by far. This high frequency may, in part, reflect a somewhat higher distribution of plants of this type in the University of Tennessee's EADC service area. However, it would also appear likely that the high distribution of S.I.C. 20 plants is due to the relatively heavy use of thermal processing of food and kindred products. With further reference to the data in Table I, the omission of seven S.I.C. code categories possibly reflects a relatively low potential for waste heat recovery in plants identified by those codes.

For the equipment classes listed in Table II, the sources of waste heat identified most frequently included combustion equipment stack gas, chiller high pressure refrigerant gas, air compressor cooling air or water, and miscellaneous sources. Among these sources it is clear that two categories (chillers and air compressors) yield primarily low temperature waste heat best suited for process stream heating or facility space heating. Stack gases, on the other hand, span a wider range of available temperatures. For example, most industrial steam boilers produce low temperature waste heat, while industrial furnaces produce waste heat in low, medium, and high temperature ranges. Generally, higher temperature sources are desirable since uses for such heat are expanded to numerous possibilities in many plants. Most sources identified, however, fall in the low source temperature range (Table III). Nevertheless, this observation does not preclude the potential importance of low temperature heat recovery since the total savings resulting from implementation of numerous opportunities in this class may result in the accumulation of a high percentage of total energy savings for a given plant.

Among the many classes of measures for energy conservation which may be recommended for an industrial facility, waste heat recovery is frequently found to be among those yielding the greatest magnitude of energy savings. For this reason waste heat recovery should always be investigated as a possible means for significant energy savings. However, the implementation of waste heat recovery usually involves the purchase of specialized and sometimes custom-made heat recovery devices (e.g. heat exchangers) which can result in a relatively high implementation cost. Thus, in order to evaluate the economic attractiveness of the waste heat recovery measures recommended in our program, simple payback analyses were performed. The resulting payback distributions grouped by source temperature categories are given in Table V.

TABLE V
DISTRIBUTION OF SIMPLE PAYBACK RANGE FOR WASTE HEAT SOURCE TEMPERATURE CATEGORIES

Temperature Source Category			
Payback	**High**	**Medium**	**Low**
Range	0.3-4.6 yrs	1.8-5.5 yrs	0.2-4.4 yrs
Average	2.2 yrs	3.0 yrs	2.0 yrs

The data show a wide distribution of simple, first order paybacks ranging from 0.2 - 5.5 years. Among the 31 measures recommended, 13 were identified with paybacks of two years or under which is usually viewed by industrial clients as relatively attractive. Therefore, although high equipment costs may be associated with implementation of waste heat recovery methods, the amount of energy savings is often such that attractive paybacks can be achieved.

SUMMARY

The data presented above clearly demonstrate that waste heat recovery opportunities are still frequently found in a wide variety of small- to medium-sized manufacturing plants. Thus, it would appear that there remains a substantial need for industries in this size range to seek to identify economically attractive waste heat recovery measures. Finally, since a sizable fraction of the recommendations cited here was associated with long paybacks in excess of two years, there is a need for waste heat recovery equipment vendors to work towards reduction of equipment and installation costs so that future overall recommendation implementation rates can be improved.

Chapter 55

ENERGY CONSERVATION THROUGH CONDENSING HEAT EXCHANGERS

H. Singh, A. K. Mallik, A. Kapur

ABSTRACT

Energy used in the industrial boilers accounts for nearly one-sixth of the total energy consumption in the United States of America. A significant amount of this heat energy is discharged to the atmosphere in the form of flue gases. Efforts should be made to recover this energy to the maximum possible extent so as to enhance the boiler system efficiency and achieve considerable savings. Whereas the non-condensing economizers can capture only a part of the sensible waste heat, the condensing heat exchangers can recover most of the sensible as well as the latent heat from the flue gases. Thus condensing heat exchangers are highly effective in reducing the flue gas temperature and extracting the heat energy therefrom.

In this paper, techno-economic feasibility of condensing heat exchangers is discussed. Data gathered in the above context from various manufacturing facilities in the states of North Carolina and Virginia during industrial visits has been analyzed. The data consists of information about boiler types, combustion efficiencies, flue gas temperatures and other related parameters. A number of case studies are reported to highlight the economic justification for investing in the installation of condensing heat exchanger systems.

INTRODUCTION

In order to increase the efficiency of industrial processes and heating installations like boilers, furnaces, water heaters and fuel-fired heat pumps, it is essential that the exhaust temperature of effluents be lowered below the dew point temperature of products. Not only does this enhance the savings of sensible heat but also brings about additional savings in the form of latent heat recovered from the condensed products [1]. However, the recovery of waste heat from combustion flue gases of the boilers and furnaces has traditionally been limited by the acid dew point of the gases. Below this dew point the condensate formed from the flue gases is highly corrosive and can cause excessive fouling. In recent years, new technologies have emerged [2] which attempt to resolve this problem by the use of special corrosion-proof materials in the manufacture of heat exchangers to enable them to operate in the condensing mode. Some of the materials used in these condensing heat exchangers are ceramics [3], borosilicate glass, Teflon, Viton and fiberglass [4].

In traditional heat exchangers (THE), the inlet temperature of water or air is normally kept above 220°F and flue gas outlet temperature above 325°F to prevent condensation. On the other hand, in condensing heat exchangers (CHE), typically the inlet temperature of water or air could be less than 100°F and exhaust gas exit temperatures could be in the range of 100-130°F [5]. Since the tube heat exchange surface and the exhaust area of the condensing heat exchanger are at a temperature well below the acid dew point of the flue gases and the dew point of water vapor, a continuous "rain" of condensate is produced resulting in latent heat recovery besides the enhanced recovery of the sensible heat. The condensate carries with it substances like sulfuric acid and sulfurous acid besides the particulates which have been scrubbed from the gases and washed from the tubes. Hence, the condensate formation reduces the atmospheric pollution due to the exhaust gases.

HEAT EFFECTIVENESS OF A CONDENSING HEAT EXCHANGER

Total heat effectiveness (e) of a condensing heat exchanger is given by the ratio of actual heat transfer for the given device to the maximum possible heat transfer between the flue gas and the supply fluid systems.

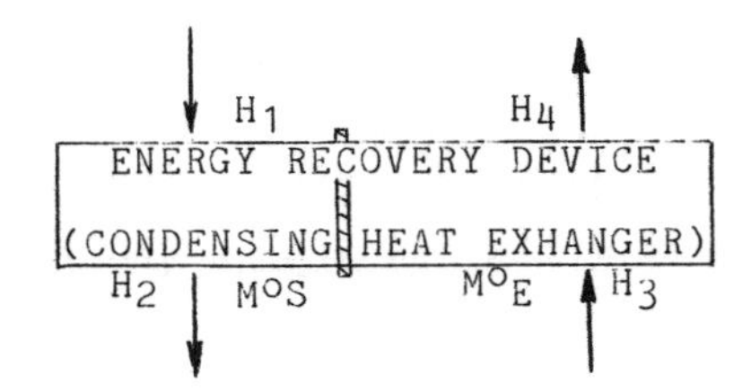

For the sketch given above,
Total heat effectiveness,

$$e = \frac{M^o{}_S\ (H_2-H_1)}{M^o{}_{min}(H_3-H_1)} \quad \text{or} \quad \frac{M^o{}_E\ (H_3-H_4)}{M^o{}_{min}(H_3-H_1)}$$

where
$M^o{}_S$ = Mass flow rate of supply fluid
$M^o{}_E$ = Mass flow rate of flue gas
$M^o{}_{min}$ = Smaller of $M^o{}_S$ and $M^o{}_E$
H_1 = Enthalpy of entering fluid
H_2 = Enthalpy of leaving fluid

H_3 = Enthalpy of entering flue gas
H_4 = Enthalpy of leaving flue gas

Thus, enthalpies of the supply fluid (H_2) and flue gas (H_4), when they are leaving the CHE, are given by

$$H_2 = H_1 + \frac{e \cdot M^o_{min}(H_3 - H_1)}{M^o_S}$$

and

$$H_4 = H_3 - \frac{e \cdot M^o_{min}(H_3 - H_1)}{M^o_E}$$

For the CHE to be more effective, H_2 should be large and H_4 small. This is possible when effectiveness parameter e, is large.

Figures 1.a and 1.b show the characteristic difference between the CHE and THE systems as regards waste heat recovery from the flue gases. Figure 1.a is for the typical THE system. Only sensible heat is recovered. From inlet to outlet at the THE, temperature of flue gas decreases from T_3 to T_4 and that of supply fluid increases from T_1 to T_2. Figure 1.b represents a typical CHE system. The dry bulb temperature of the flue gas can be reduced below the dewpoint, thus latent heat can also be recovered along with enhanced recovery of sensible heat. Figure shows the decrease in temperature of flue gas from T_3 to T_4', and increase in temperature of supply fluid from T_1 to T_2'.

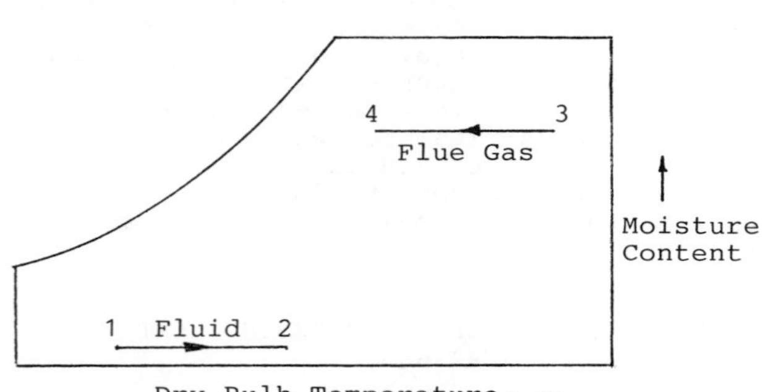

FIGURE 1.a. HEAT EXCHANGE (THE SYSTEM)

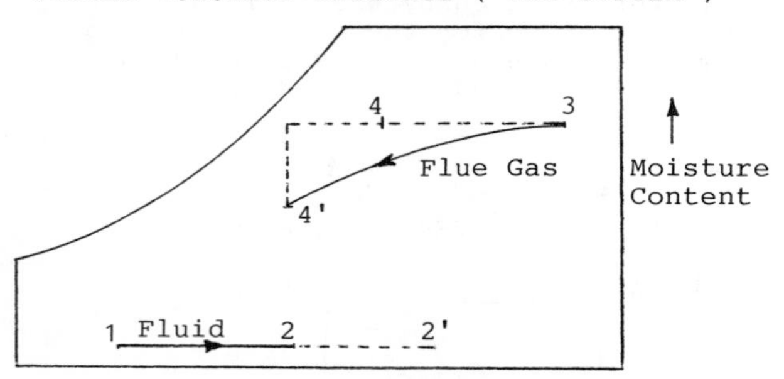

FIGURE 1.b. HEAT EXCHANGE (CHE SYSTEM)

CHE AS A PREHEATER

A schematic sketch (Figure 2) shows the basic circuit for preheating the supply air or water by the combustion flue gases. The traditional heat exchanger, though not an essential component in this system, may be used for recovering heat when the combustion flue gases are at relatively high temperature.

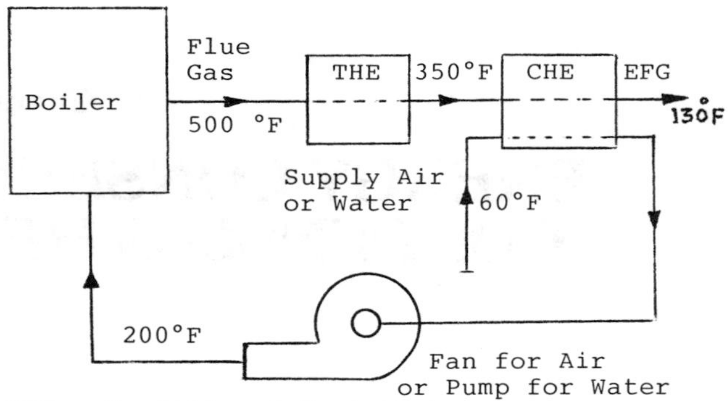

THE - Traditional Heat Exchanger
CHE - Condensing Heat Exchanger
EFG - Exhaust Flue Gas
(temperatures shown are typical only)

FIGURE 2. SCHEMATIC SKETCH OF CHE AS A PREHEATER

CORROSION THRESHOLD TEMPERATURES

As already pointed out, the flue gases comprise of corrosive elements. When the temperature is reduced below a certain limit (Threshold Value) condensation occurs and these elements become a part of the condensate to a large extent. The condensate corrodes the surfaces in contact. Figure 3 gives a typical representation [4] of flue gas corrosion zones versus temperature in a coal or oil fired boiler when approximate sulfur content is 1% to 2%.

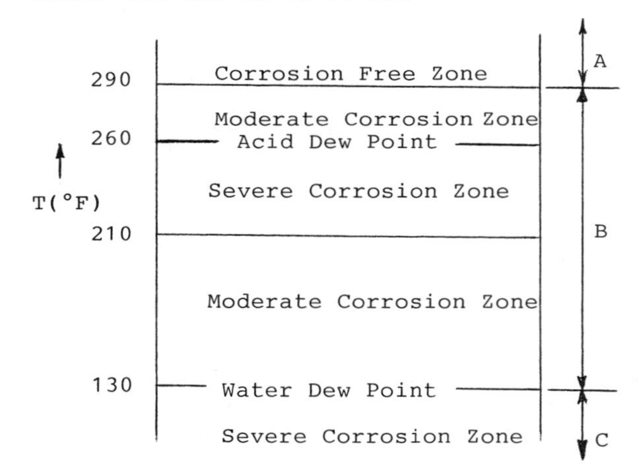

T= Temperature of heat transfer surface
A= Safe zone
B= Sulfuric acid corrosion zone
C= Sulfurous acid corrosion zone

FIGURE 3. FLUE GAS CORROSION ZONES

For the gas fired boilers it is the sulfurous acid zone which is predominant and causes most of the damage.

CORROSION DAMAGE

Severity of the corrosion damage due to flue gas condensate can be expressed in terms of the corrosion attack on the carbon steel as follows:

Oil and coal fired boilers - 300 to 500 mils/year
Gas fired boilers - 50 to 60 mils/year

MATERIALS OF CONSTRUCTION

In view of the extensive corrosion damage due to the flue gas condensate, there has been an intensive search for corrosion-resistant materials all over the world in the past two decades. Metallic materials like special stainless steels and some aluminum alloys have shown promise in certain applications [6,7,8]. However, some of the primary materials of construction [4] being used more often in the condensing heat exchangers are listed below along with their temperature limitations.

(i) Borosilicate glass - up to 700°F
(ii) PTFE (Teflon) - up to 500°F
(iii) Viton - up to 500°F
(iv) Fiberglass - up to 450°F

HEAT TRANSFER CHARACTERISTICS

The materials of construction named above have very low heat transfer coefficients compared to metals and their alloys. However, in a gas to liquid heat exchanger, this is not a serious handicap because the gas side heat transfer coefficient is the dominant factor in the overall heat transfer. This coefficient increases with the increase in the pressure drop of the gas. It is affected significantly by the distance between the tubes and by their pattern [4].

LOW TEMPERATURE HEAT RECOVERY APPLICATIONS

(1) An ideal application of the low temperature heat recovery system i.e. CHE system is to preheat cold boiler makeup water before it enters the deaerator.
(2) To preheat combustion air before it enters the boiler windbox.
(3) A heat sink not connected with the boiler system can also be used with the CHE system if it can be matched with the boiler operating hours and load cycles.
(4) Other less common but practical uses are the operation of a low temperature electric turbine cycle and an absorption chilling cycle.

A TYPICAL BOILER-CUM-CHE SYSTEM

A typical layout of the boiler and CHE system is shown in Figure 4. The flue gases are drawn from the stack through a control damper by an induced draft fan. Typically, the inlet plenum could be made of carbon steel, the exhaust plenum of fiberglass reinforced plastic and the tube could be covered with Teflon.

The boiler-cum-CHE system is fully instrumented and controlled. The exhaust gas flow to the CHE fan is controlled by modulating a damper. This fan has a high temperature construction with shaft seal, heat slinger and high temperature bearings. A cut-out protects the inlet plenum from high temperatures (in excess of nearly 500°F) by shutting down the fan and the damper. A microprocessor based timing system cycles a control valve to intermittently wash down the shell side of the heat exchanger. Typical frequency is 2 minutes every eight hours which ensures peak operating efficiency [5]. For superior long term performance, the entire gas (shell) side of the heat exchanger is made from corrosion-resistant materials like Teflon. The high/low flow cut-out shuts down the system when there is deficiency or excess of water flow.

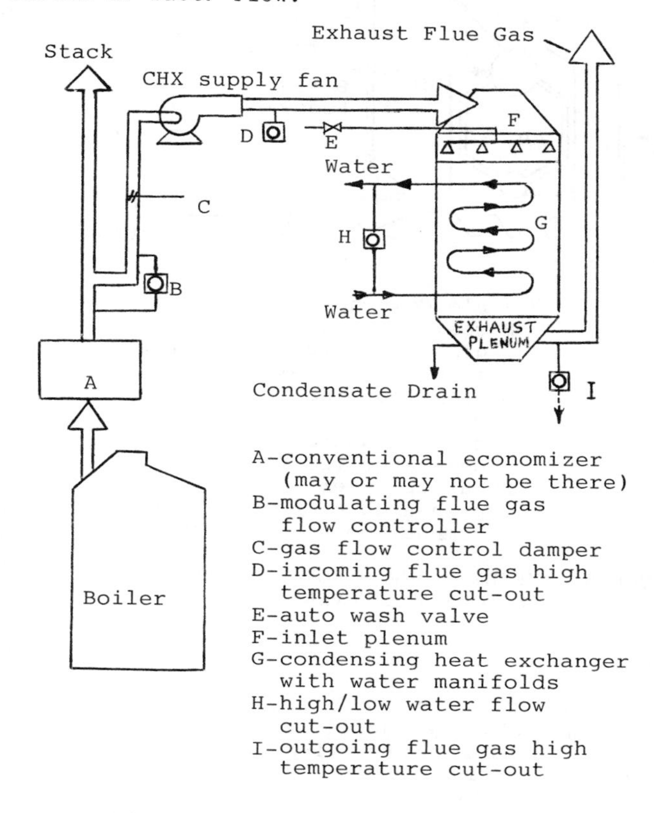

FIGURE 4. A TYPICAL BOILER-CUM-CHE SYSTEM

AN EFFICIENT WASTE HEAT RECOVERY SYSTEM

Figure 5 shows a system diagram for efficient recovery of the waste heat from the boiler. The condensing heat exchanger works in conjunction with a blowdown heat exchanger and an air-preheater. Temperatures shown are only typical. Boiler makeup water passes through the blowdown heat exchanger to extract the waste heat from the blowdown before entering the CHE where it extracts heat from the flue gases. After leaving the CHE the makeup water preheats the combustion air before entering the deaerator.

CASE STUDIES

Some typical case studies are reported below to show the economic benefits of the condensing heat exchangers being incorporated in the manufacturing facilities using boilers.

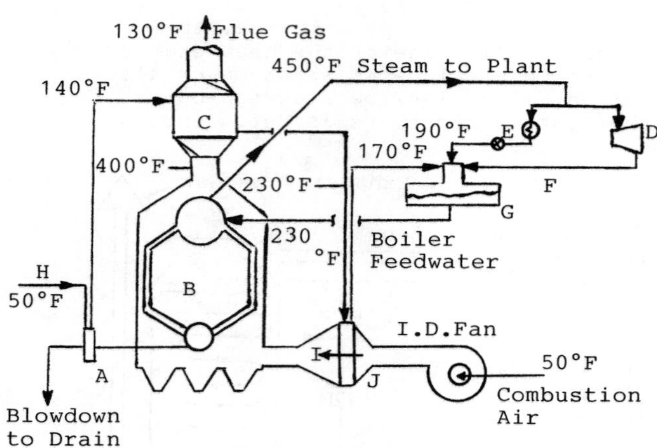

A-blowdown heat exchanger
B-coal fired boiler
C-glass tube low temperature economizer (CHE)
D-steam turbine
E-condensate return
F-back pressure steam
G-deaerator (230°F)
H-makeup water
I-preheated combustion air(173°F)
J-combustion air preheater

FIGURE 5. SYSTEM DIAGRAM FOR EFFICIENT WASTE HEAT RECOVERY [4].

Case Study No. 1

During visits to several small and medium sized manufacturing facilities in the states of North Carolina and Virginia, data was gathered regarding boiler types, combustion efficiencies, flue gas temperatures and other related parameters. Almost all the boilers being used were of fire tube type with the typical average capacity of 250 boiler horsepower, 70 to 80% thermal efficiency and 395°F exhaust flue gas temperature. A representative facility (Poultry Feed Plant) is selected for the assessment of energy savings potential presuming that a CHE system would be incorporated in it. The details of the savings potential calculated with the help of a computerized system design software [5] are as follows.

CHE System Design for a Representative Facility: (Poultry Feed Plant)

Application - To heat boiler makeup water
Total boiler capacity - 350 Boiler HP

Design Parameters:

Average Steamload For Case	12,000	LB/HOUR
Available Flue Gas Mass	13,346	LB/HOUR
Boiler Feedwater Temp.	220.0	DEG. F.
Steam Pressure (Saturated)	25	PSIG
Excess Combustion Air	20.00	PERCENT
Flue Gas Temp At Source	300.0	DEG. F.
Maximum Waterflow to CHE	19	GAL/MIN
Flue Gas Wtr. Vapor Dewpoint	133.4	DEG. F.
Flue Gas Density	0.0526	LB/CU.FT
Specific Heat of Flue Gas	0.2656	BTU/LB DEG. F.
Hours of Operation For Case	8760	HOURS
Fuel Fired	NATURAL GAS	
Fuel Cost	$3.92	$/MBTU
Existing Fuel To Steam Efficiency	81.31	PERCENT
Existing Thermal Efficiency	83.70	PERCENT

Heat Exchanger Performance:

Flue Gas Mass at CHE Inlet.	13,346	LB/HR
Flue Gas Flow at Inlet to CHE	4,228	ACFM
Flue Gas Inlet Temperature	300.0	DEG. F.
Flue Gas Outlet Temperature	134.9	DEG. F.
Waterflow Through CHE	19.0	GAL/MIN
Water Inlet Temperature	60.0	DEG. F.
Water Outlet Temperature	147.8	DEG. F.
Sensible Heat Recovered	585,417	BTU/HR
Latent Heat Recovered	261,986	BTU/HR
Total Heat Recovery	847,403	BTU/HR
Savings For This Case	$35,786	$/YEAR

Engineering Data:

New Boiler Fuel To Steam Eff.	87.17	PERCENT
New Thermal Efficiency	89.55	PERCENT
Efficiency Increase	5.85	POINTS
Fuel Savings	7.20	PERCENT
Waterside Pressure Drop	9.31	PSIG
Theoretical Fan Power	3	HP
Heat Exchanger Flue Gas Pressure Drop	1.69	IN. W.G.
Plenum, Duct And Breeching Loss	0.68	IN. W.G.
Condensate Flow Rate	0.5	GAL/MIN

NATURAL GAS ANALYSIS USED FOR THIS CASE

%C	$\%H_2$	$\%N_2$	$\%O_2$	%S	$\%H_2O$	%Ash
75.20	23.50	1.30	0.00	0.00	0.00	0.00

Payback Period:

Estimated initial cost of CHE system	= $60,000
Installation cost	= $20,000
Total Cost	= $80,000

Savings for this case = $35,786/year

Simple payback = $\frac{\$80,00}{\$35,786}$ = 2.24 years

This shows that the investment on the CHE system can be recovered in a short period of time and ultimately it would pay handsome dividends.

Case Study No. 2

A case study similar to the above was conducted for a facility located in Greensboro where two gas fired boilers (total 1200 BHP) are likely to be retrofitted with a condensing heat exchanger to preheat boiler makeup water from 80°F to 173.9°F, the inlet and outlet temperatures of flue gas being 350°F and 134.1°F respectively. Total heat recovery (sensible and latent) has been estimated to be 1,314,385 BTU/hour with annual savings amounting to $60,950/year against an initial total installed cost of $116,000 for a CHE system. The simple payback period being less than 2 years, the investment is fully justified [5].

Case Study No. 3

At a particular tobacco manufacturing facility, waste steam from the steam turbines is used to heat air. The hot air is passed through a tobacco dryer as shown in Figure 6 and enters the condensing heat exchanger at 270°F where it heats water from 79°F to 112°F.

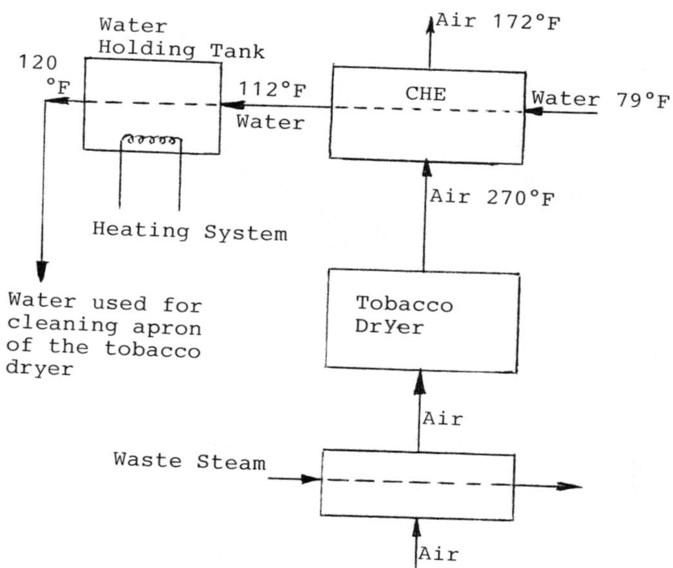

FIGURE 6. WASTE HEAT RECOVERY SYSTEM DIAGRAM FOR A TOBACCO DRYING PLANT

Data and Analysis:

Designed SCFM = 20,000
Measured SCFM = 15,000
Entering Air Conditions = 270°F DB, 142°F WB, 134°F Dew Pt.
Moisture Content = 0.1304 lb per lb of Dry Air
Enthalpy Content = 219 BTU per lb of Air
Leaving Air Conditions = 172°F DB. 127°F WB 123°F Dew Pt.
Moisture Content = 0.0921 lb per lb of Dry Air
Enthalpy Content = 146 BTU per lb of Air
Moisture Condensing Rate= 5.34 GPM

Processed water is heated by using the recovered heat. Water piping material is block steel.

Water Flow Rate = 325 GPM
Entering Water Temperature = 79°F
Leaving Water Temperature = 112°F

Warm exhaust air is flowing vertically in upward direction around the water passage whereas the water is sprayed in the horizontal direction inside the passage. The heated water kept in a holding tank at 120°F is used as shower water to clean the apron of the dryer. The tobacco dryer is operated 24 hours a day, Monday thru Friday. The heat recovery unit is washed every week to remove any fouling material.

Installed cost of the CHE system = $250,000
Steam Cost = $3 per 1000 lbs or $3/MBTU
(Steam is generated from coal fired boilers)

Total heat recovered =

$$\frac{325\ \text{gal}}{\text{min}} \times \frac{8.33\ \text{lb}}{\text{gal}} \times \frac{1440\ \text{min}}{\text{day}} \times \frac{260\ \text{days}}{\text{year}} \times \frac{1\text{BTU}\ (C_p)}{\text{lb}^\circ\text{F}} \times (112 - 79)^\circ\text{F}$$
$$= 3.345 \times 10^{10}\ \text{BTU/year}$$

Annual Savings $= \frac{3.345 \times 10^{10}\ \text{BTU}}{\text{year}} \times \frac{\$3}{10^6\ \text{BTU}}$
$= \$100,350/\text{year}$

Simple payback $= \frac{\$250,000}{\$100,350} = 2.49$ years

Since the payback period is short, the investment for the CHE system is fully justified.

Maintenance Problem:

Since the process water picks up some solid impurities from tobacco, slime was found on the valves. The header is washed once a month to rectify this problem.

Case Study No. 4

For a sample example whose data (in brief) is given below, 'potential' heat recovery was calculated for a wide range in the values of flow rate and inlet temperature of the boiler makeup water. The potential heat recovery has been taken as the more restrictive of the following two factors.

(1) Heating the boiler makeup water available to a maximum of 200°F(2) No greater than a 3-year simple payback on the incremental cost for the condensing heat exchanger.

CHE System Design Data:

Boiler Name Plate	- 40,000 lb/hr
Fuel Fired	- natural gas
Average Steam Load	- 20,000 lb/hr
Available Flue Gas Mass	- 23,106 lb/hr
Flue Gas Temp. at Source	- 380°F
Flue Gas Water Vapor Dew Pt.	- 133.67°F
Flue Gas Density	- 0.0476 lb/ft^3
Specific Heat of Flue Gas	- 0.2657 BTU/lb°F
Hours of Operation	- 8760 hrs/yr
Fuel Cost	- $2.6/MBTU
Existing Fuel to Steam Eff.	- 79.18%

The calculations for this example were done with the help of a computerized system design software [5] and the results obtained are given in the tables as follows.

Table of Results

Q° gpm	WIT °F	WOT °F	GOT °F	LHR BTU/hr	THR BTU/hr	LTR/THR %
20	40	200	168	297,930	1,599,577	18.63
30	40	200	130.8	843,283	2,373,518	35.53
40	40	185.2	125.9	1,286,069	2,846,434	45.18
20	50	200	176.6	250,542	1,499,434	16.71
30	50	200	132.1	710,869	2,233,235	31.83
40	50	187.5	127.4	1,144,262	2,695,511	42.45
20	60	200	186	207,989	1,399,548	14.86
30	60	200	133.3	580,133	2,095,111	27.69
40	60	179.9	131	845,772	2,374,952	35.61

20	70	200	196.1	170,140	1,299,657	13.09
30	70	200	139.1	470,334	1,949,500	24.13
40	70	177.6	133.1	630,988	2,146,973	29.39
20	80	200	203.5	141,110	1,199,608	11.76
30	80	200	148.2	375,898	1,799,434	20.89
40	80	182.8	133.7	543,080	2,055,371	26.42

Q^o = Makeup water flow rate
WIT = Water temperature at inlet to CHE
WOT = Water temperature at outlet from CHE
GOT = Flue gas temperature at outlet from CHE
LHR = Latent heat recovery
THR = Total heat recovery (latent and sensible)

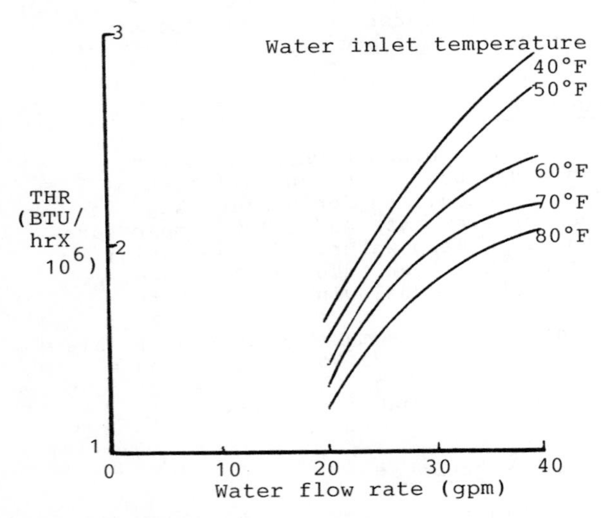

FIGURE 7. TOTAL HEAT RECOVERY (THR) VS FLOW RATE FOR DIFFERENT INLET TEMPERATURES

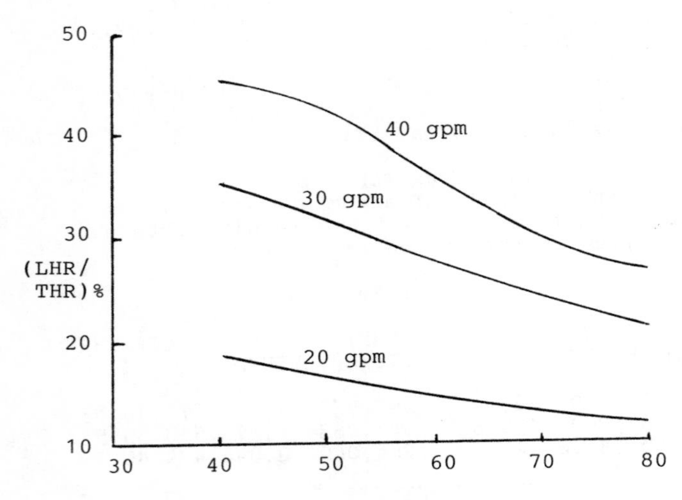

FIGURE 8. CONTRIBUTION OF LATENT HEAT TO TOTAL HEAT RECOVERY

From the above table of results, curves are plotted (Figure 7) to show the effect of water inlet temperature and flow rate on the total heat recovery by the condensing heat exchanger. It is seen that when the heat sink is large (low inlet temperature and high flow rate of makeup water), the heat recovery is also large. Figure 8 shows the contribution of latent heat to the total heat recovery. It is observed that LHR becomes more significant as the heat sink becomes larger.

CONCLUSION

From the case studies reported it is seen that the condensing heat exchanger is a useful device for energy conservation through the process of low temperature waste heat recovery. Thermal efficiency of the boiler can be increased considerably by the incorporation of the CHE system. The payback periods are short. Thus, the investment for CHE system would in general,be attractive and justifiable.

References

1. Grehier, A. and Rojey, A., "Condensing Heat Exchanger for Heat Recovery From Flue Gases", International Symposium on Condensing Heat Exchangers, 1987. *

2. Preto, F., Lee, S.W., and Hayden, A.C.S., "Performance of Domestic Oil-fired Condensing Systems," International Symposium on Condensing Heat Exchangers, 1987.*

3. Heinrich, J., Huber, J., and Schelter, H., "Compact Ceramic Heat Exchangers: Design, Fabrication, and Testing," International Symposium on Condensing Heat Exchangers, 1987.*

4. Shook, J.R. and Luttenberger, D.B., "Heat Recovery: Low Temperature Heat Recovery for Boiler Systems," District Heating and Cooling, Vol. 71, No. 2, 4th Quarter, 1985.

5. "Heat Recovery System," Condensing Heat Exchanger Corporation, Warnerville, New York (Personal Communication).

6. Kobussen, A.G.C., Oonk, A. and Hermkens, R.J.M., "Corrosion of Condensing Heat Exchangers and Influence of the Environment", International Symposium on Condensing Heat Exchangers, 1987.*

7. Joseph, A.L., Schaus, O.O. and Overall, J., "Evaluation of Stainless Steel Materials for Resisting Chloride-Induced Corrosion in Condensing and Partially Condensing Gas-fired Appliances," International Symposium on Condensing Heat Exchangers, 1987.*

8. Hindin, B. and Agarwal, A., "Corrosion Behavior of Some Candidate Alloys in Solutions of Concentrated Flue-Gas Condensate," International Symposium on Condensing Heat Exchangers, 1987.*

* Proceedings available as D.O.E. Report No. BNL-52068 from N.T.I.S.

Chapter 56

THE NEW CONCEPT AND NEW WAY FOR ENERGY USE

Z. Yunming

ABSTRACT

On the basis of heat efficiency and exergy efficiency, a new concept named INTEGRATED BENIFIT AND EFFICIENCY is suggested in this article for trying to improve the evaluation of energy use. From the viewpoint of systems engineering, a new way for energy use is also suggested.

INTRODUCTION

We live in such a world which is not only in shortage of energy supply but also is in great waste in energy use. For seeking the way to decrease the waste of energy, the balance of energy which based on the first law of thermodynamics is calculated, and the heat efficiency is suggested for evaluate the perfection of energy use. Later, the principle is recognized, that is that the energy is not only has the difference in quantity, but also has the difference in quality. Then the concept of exergy based on the second law of thermodynamics is suggested for calculating the "work lost" and use the exergy efficiency to evaluate the perfection of energy use, as a supplement. But both of them sometimes won't be able to reflect the difference of economic benifits sufficiently in the processes of energy converting and are often calculated in the confines of a certain process. Therefore, if the "work lost" is calculated from the angle of systems engineering, not limited in the confines of a process, then some new ways for improving the energy use may be found out and more benifits will be obtained. Here some examples are put forward to explain its principle. This kind of work might be done before unconsciously but I give a formal name, INTEGRATED BENIFIT AND EFFICIENCY for arousing the attention from people.

COMBINED CYCLE COGENERATION

Combined cycle cogeneration is a way for improving the perfection of energy use. For example, if the fossil fuel is burned in a boiler for generating electricity, the efficiency from heat to work is about 40% in the advanced supercritical turbogenerator but in combined cycle cogeneration, the efficiency reaches to 50%. From the viewpoint of exergy, the reason is that the high grade energy(high temperature gas) is used for producing electricity in combustion turbine generator first, then the low grade energy(low temperature waste gas) comes from the outlet of combustion turbine, is used for producing electricity again in a turbogenerator. Thus the heat lost to environment is decreased. It means that the irreversible lost from the viewpoint of thermodynamics is decreased. So the exergy efficiency is increased from about 25% to about 35%.
But this improvement mentioned above, just reflects the rationality of a process, not reflect the difference of economic benifits caused by different fuels(oil, coal, natural gas).
In general speaking, the economic benifit is much higher as the oil and natural gas are used for producing chemicals than to produce electricity. But for coal is not so apparently. Therefore if the oil or natural gas is burned in boilers as done in many cuntries or oilfields, and the coal is used for producing chemicals, that is a great waste for a society. But this lost won't be able to reflect in the calculations of heat efficiency and exergy efficiency. It needs to integrate both benifit and efficiency to take into account.

For the utilization of coal, it is a great progress as the coal is gasified and burned to drive the combustion turbine - steam turbine in a combined cycle power plant(IGCC Fig. 1). Moreover it is a way to avoid SO_x and NO_x emission in air.

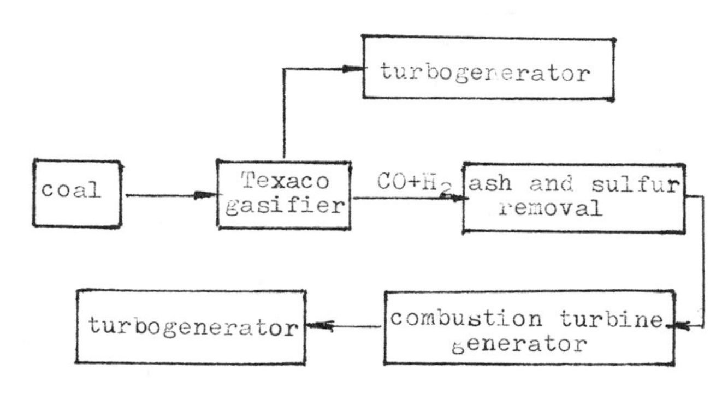

FIGURE 1 IGCC PROCESS

But because of the special demands of a combustion turbine, at the present time, the production cost of electricity in IGCC is higher than in usual[1,2] However, if the power plant produce both electricity and methanol with the gas(CO + H_2) generated from gasifier, then the cost will be decreased apparently. Because of thr gas(CO + H_2) goes through a reactor the yield of methanol is about 5% (100 atm and 230-270 C), it needs a recycle compressor in a methanol plant[3] (Fig. 2).

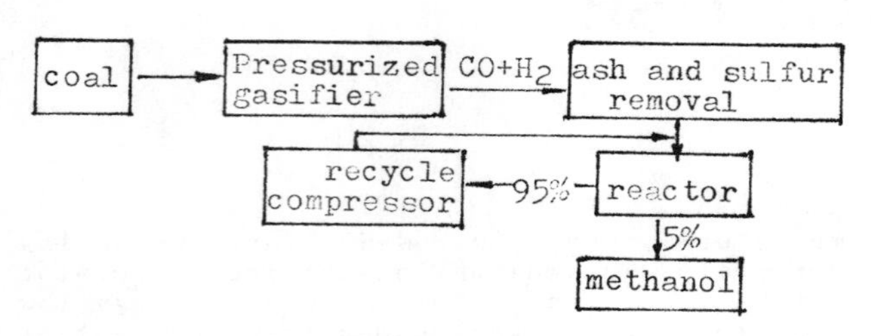

FIGURE 2 METHANOL PRODUCTION

But for a combined power plant, the unconverted gas is taken to burn for electricity, the recycle compressor is saved. That means the investment and producing cost in an integrated complex is cheaper than in two plants which produce the same amounts of electricity and methanol respectively with using same coal(Fig. 3).

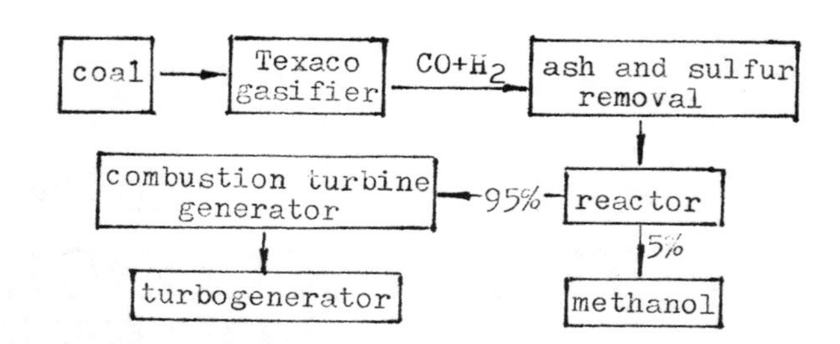

FIGURE 3 COMBINED POWER PLANT

This point won't be shown in the calculations of energy or exergy balance in the confines of a process, but shown from the angle of systems engineering.

In the light of "town gas + electricity" and "chemicals + electricity", Mr.Yang and Mr.Luo suggested a combined complex using coal to produce methanol 3 x 10^5 tons/year, power 5 x 10^4 kw, town gas 3 x 10^6 HM^3 / day (15.9 x 10^6 J/ HM^3), steam 100 t/h (7.8 x 10^5 Pa). They found that the investment of the integrated complex will save 60% and the producing cost will save 20% than individually to build several plants producing the same amounts of methanol, electricity, and steam. Moreover the problem of waste water pollution is also resolved[4] (Fig. 4).

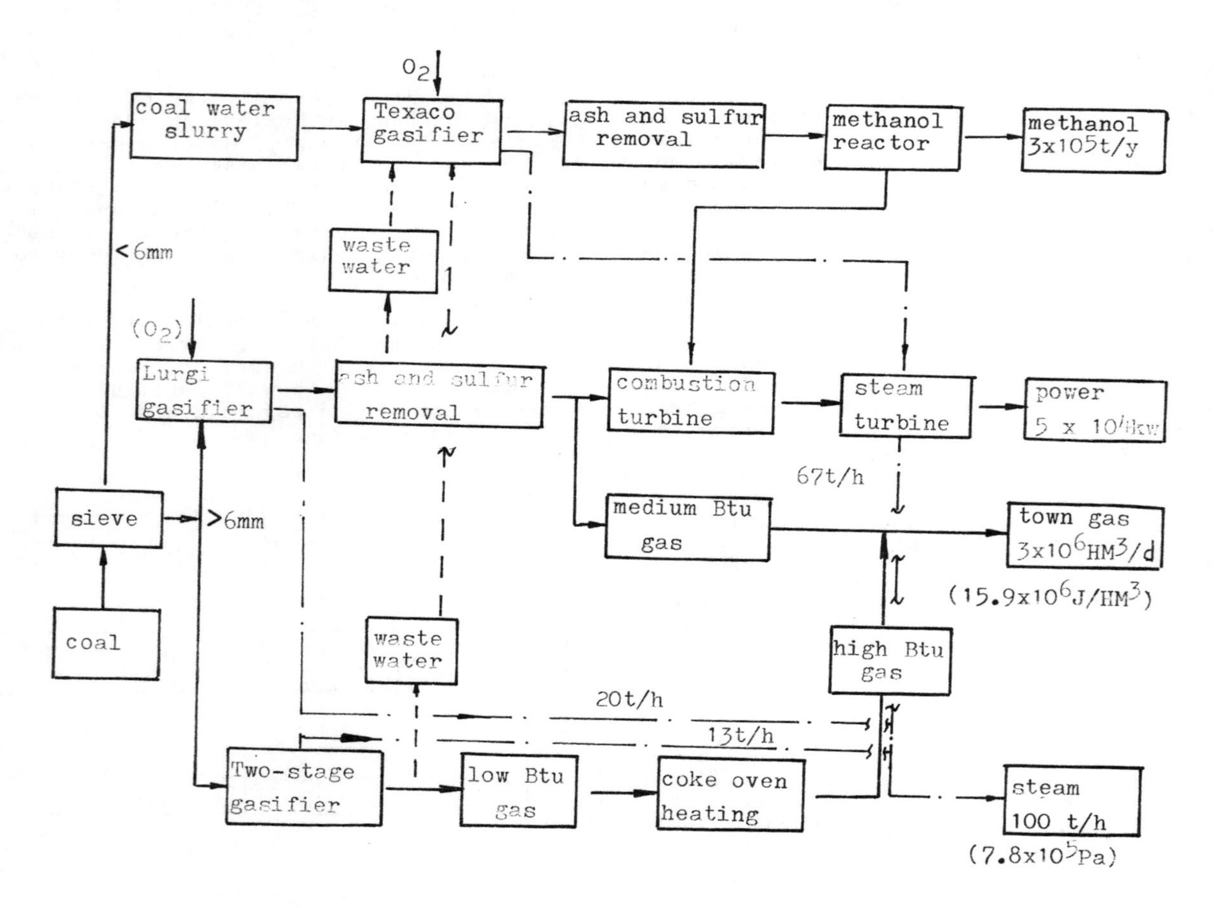

FIGURE 4 A PROJECT OF AN INTEGRATED COMPLEX

The reason is that all the various grade energies are put in the suitable positions for reasonable utilizations. This is a good sample showing how to integrate the benifit and efficiency to take into account from the angle of systems engineering.

UTILIZATION OF RENEWABLE ENERGY

The energy comsumption is about 2.5×10^{20} joule in 1984 and 95% of that comes from fossil fuels.
Solar energy put on the earth is about 2×10^{24} j/y and converted by plants is estimated about 3×10^{21} j/y. It is ten times of the world energy comsumption.
Over the past 100 years, the fossil fuels are extracted with mankinds's utmost but for the utilization of renewable energy, the progress is made slowly (Fig.5).

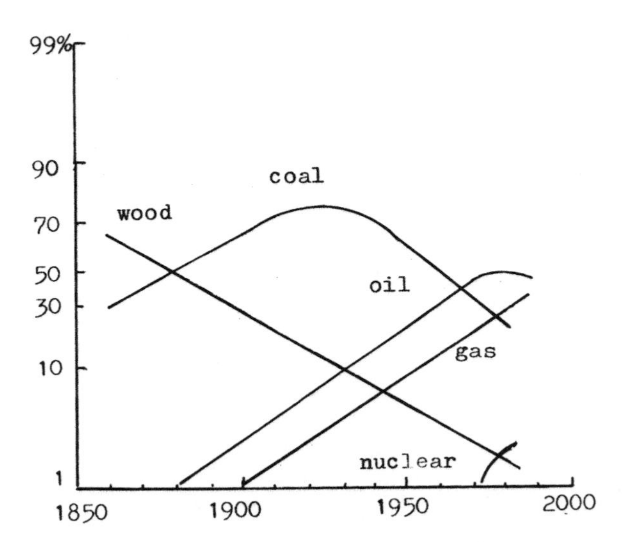

FIGURE 5 WORLD ENERGY CONSTITUTION

However, the fossil fuels will be exhausted someday in the future, the utilization of renewable energy has to be paid great attention. It is not only for energy supply but for air pollution and chemicals supply as well.
Forest has a higher eficiency on the utilization of solar energy than general crops. So a planned energy forest plantation and development program ought to be thought highly of. On the other hand, several plants which offer more oil, ethanol, and Btu than others, such as energy cane, Euphorbia lathyria are reviewed [5].

Besides, the reasonal utilization of traditional forest and agriculture waste is also an important problem.
It is estimated that the Btu of traditional forest and agriculture waste(branches, barks, wood scraps, corn cobs, bagasse, sun flower husks, etc.) is equal to 1/5 of the world energy comsuption. The heat efficiency is about 10% as the waste taken to direct burning for living; nevertheless the heat efficiency reaches to 60-80% as the waste burned in boilers to generate electricity and heat. For instance, the heat efficiency is 85% in a mordern sugar plant boiler (Fig. 6). However the exergy efficiency is just 26.8%.

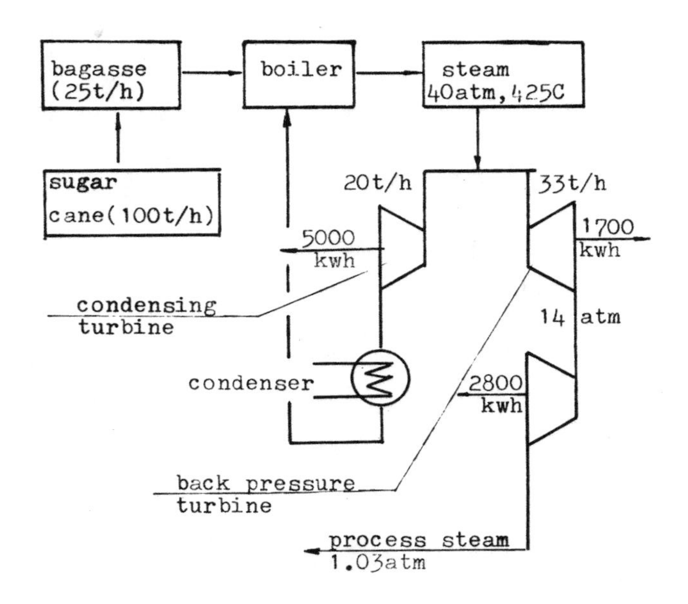

FIGURE 6 A MODERN SUGAR PLANT POWER/ HEAT SYSTEM

The problem is that the high temperature flue gas with more exergy just produce 425 C steam with less exergy. From the viewpoint of exergy, there is a great irreversible lost in it. For improving, I suggest to use combined cycle cogeneration in sugar plant. Because of bagasse only has a little ash and sulfur which is similar as the most forest and agriculture waste, it seems possible to drive turbine with flue gas at high temperature. After bagasse burning in a pressurized combustor, the ash particles may be removed with cyclones and woven or felted ceramic units, as was done at Westinghouse's Waltz Mill facility. In initial tests of that ceramic unit at 11 atm and 843 C, 99.7% of coal ash particals above lum were removed and the bags were successfully pulse-cleaned on line [6]. Although preliminary, these results are encouraging for sugar plant combined cycle cogeneration. After filteration, high temperature flue gas goes into combustion turbine generator, then enters into boiler as shown in figure 7. The exergy efficiency will reach to 30% or more. Deducting the power of compressor used for pressurized combustor, the electricity will increase 1860 kwh/25 tons bagasse(74.4 kwh/t). The same concept may be suitable for similar materials.

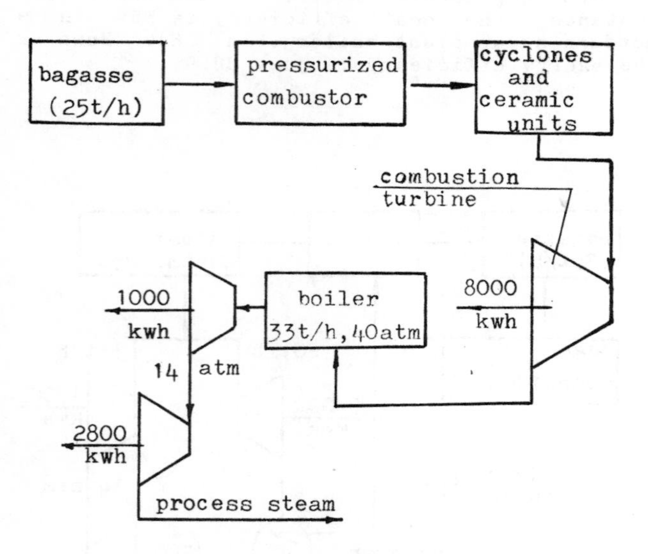

FIGURE 7 AN IMPROVED DESIGN FOR SUGAR PLANT POWER / HEAT SYSTEM

However, as well known, a wide range of chemicals can be produced from the basic components(pentose, cellulose, lignin, tanin, rosin, terpenoids, etc.) of forest and agriculture waste [7]. So from the angle of systems enginering, if first we extract useful chemicals (oil, methanol, ethanol, glucose, furfural, acetic acid, etc.), then burn the residue in an effective way for producing power and heat, the economic benifit will be much more than burn it all. The integrated benifit and efficiency will be higher in an integrated complex than individually producing the same chemicals and energy in several plants. Therefore this model may become the one of the main patterns for supplying energy , chemicals, and their derivativs, after the fossil fuels exhausted someday in the future.

REFERENCES

1. B. M. Louks, "Costs of Coal Gasification Combined Cycle Power Plant" EPRI Journal March, 1983, p34-35

2. John Douglas, "Cool Water, Milestone for Clean Coal Technology" EPRI Journal, Dec., 1984, p.17-25

3. D. D. Metha, Hydrocarbon Processing, 55, No.5, 165(1976)

4. Yang Keshan, Luo Guangliang, " 'Three Combined supply' and Integrated Utilization of Coal" Technology of Energy Resources 1, 39-42(1987)

5. Melvin Calvin, "Renewable Fuels for the Future" Energy, Resources and Environment ------Proceeding of the First U.S.-China Conference, 1982, p.11-21

6. William Slaughter, "Pressurized Fluidized Bed Combustion" EPRI Journal July/Aug., 1984, p.31

7. Hepner, L., Eurochem Conf. IChemE, June 1977, "The feasibility of basic chemicals for fermentation processes"

SECTION 12
COGENERATION RETROFIT

Chapter 57

EVALUATING COGENERATION SYSTEM RETROFIT ALTERNATIVES

T. Bantz

ABSTRACT

For a given retrofit opportunity, a large variety of cogeneration system alternatives may exist. The evaluation and ranking of alternatives can be performed using conventional microcomputer software. Specific elements of the analysis include: identification of critical data requirements; spreadsheet development and formatting techniques; shortcuts for modelling system thermodynamics; sensitivity analysis techniques. The analysis draws upon Stanford's experience in developing project design criteria for its combined-cycle cogeneration system now in operation.

Cogeneration System Options

Cogeneration system options for a given site can be defined in terms of size and operating regime. With respect to size, the following alternatives can be considered in the initial feasibility assessment.

A plant can be sized to satisfy for a given site:

(s1) maximum (peak) electric demand (i.e. kilowatt demand)

(s2) minimum electric demand

(s3) peak thermal demand (i.e. steam demand = lb/hour)

(s4) minimum thermal demand.

With respect to operating regime, the following alternatives can be considered in the initial feasibility assessment:

(r1) plant operates at flat electric and thermal output

(r2e or r2t) plant operates in electric or thermal load following mode.

These alternatives once identified, serve as a guide in the selection of specific equipment options (e.g. prime movers) for further evaluation.

Evaluation of Alternatives

By combining the operating and sizing options described above, 10 plant size/operational alternatives can be defined. These are proposed only as a starting point in what is essentially an iterative process.

This process may ultimately lead to the identification of additional options (e.g. plants of intermediate size or operating regime). Alternatively, it may be possible to eliminate the need for further analysis on several of the 10 options, based on the known characteristics of a specific site.

Salient features of several of the 10 options are discussed below.

(1) s1r1: plant sized for peak electric demand, operates at full output continuous

Under this option the plant exports power to the local utility grid. The economic feasibility of this option directly depends on utility export prices and policies.

Frequently export prices are tied to the utility's short-run marginal fuel. If the utility's marginal fuel is the same as the cogenerator's (e.g. natural gas), feasibility would then depend on cogen vs. utility comparative heat rates (HR). To the extent that cogen exceeds utility HR, economic feasibility is adversely affected (since export revenues are not sufficient to offset fuel costs).

(2) s1r2e: plant sized for peak electric demand, operates in electric load following mode

This option eliminates the problem associated with low export rates. However, cogen heat rate remains at issue in this option, i.e. this choice is viable to the extent that cogen thermal output does not exceed site thermal requirements. If site thermal requirements are comparatively low, other options are likely to be competitive.

(3) s1r2t: plant sized for peak electric demand, operates in thermal load following mode

This option may or may not eliminate the problem associated with low export rates. Usually site thermal and electric demands are not precisely coincident. A situation may therefore arise where the plant is satisfying peak thermal load while exporting power. In this state of operation, the plant operates at optimal heat rate. The export rate problem may not therefore be significant unless export prices are low during these electric export periods.

(4) s2r1: plant sized for minimum electric demand, operates at full output continuous

This option (logically the same as s2r1e) eliminates any consideration of export rates and policies since there is no export. This is a competitive option to the extent that cogen thermal output does not exceed site requirements, i.e. to the extent that plant HR is low. (As further discussed below, combined cycle systems are one way to utilize excess thermal, although at some additional equipment investment.)

This option (s2r1) is less likely than (3) to benefit from scale economies given that unit capacity cost ($/kw of installed capacity) typically declines as plant size increases.

(5) s2r2t: plant sized for minimum electric demand, operates in thermal load following mode

Like the previous option, this one eliminates the export issue, and is also less likely to capture economies of scale. It is still possible that site thermal load could at times be exceeded in this option. Feasibility depends on the extent to which this occurs.

For a given site, this option will probably have a better HR than option (3) and is a comparatively better option given that capital costs will be less than or equal to (3).

(6) s3r1: plant sized for peak thermal demand, operates at full output continuous

This facility will operate at times with excess thermal output. This excess can either be vented (at some economic loss depending on the magnitude of the excess), or converted to electric output using a combined cycle cogen system (e.g. steam turbine add-on).

If combined cycle is selected, economics will be sensitive to the value of electric output, and the incremental investment in the steam turbine. To the extent that cogen electric output exceeds site demand, economics are sensitive to export prices and policies.

For a site where s3r1 is a viable alternative there are likely to be other competitive alternatives (e.g. (8), (9)) as will be discussed below.

(7) s3r2e: plant sized for peak thermal demand, operates in electric load following mode (no comment)

(8) s3r2t: plant sized for peak thermal demand, operates in thermal load following mode

In this option there is no excess thermal output so plant HR will be comparatively high. (This alternative is not combined cycle by definition.) This option benefits from scale economies, and is affected by export prices if cogen electric output exceeds site demand.

(9) s4r1: plant sized for minimum thermal demand, plant operates at full output continuous

This option (logically the same as s4r2t) produces no excess thermal output. This option does not benefit from scale economies (relative to (8)), and is affected by export prices if cogen electric output exceeds site demand. Where export prices are insufficient to offset marginal fuel costs, this is a better option than (8).

(10) s4r2e: plant sized for minimum thermal demand, plant operates in electric load following mode (no comment)

Data Requirements

Data required for quantitative analysis of alternatives includes (1) thermal and electric site demand data, (2) cogen system performance specifications, and (3) energy pricing data.

With regard to (1), the size of the required data set varies depending on specifics of the site. For example, a site with flat and continuous loads is easily characterized, for purposes of this analysis, by two points, thermal load (e.g. 50,000lb/hour) and electric load (15,000 kw).

For a site where loads vary significantly, and where time-of-use utility rates apply, hourly load data is required in the analysis. The data set in this case would require 2 x 8760 points to calculate annual cash flows. Sites with multiple (n), variable loads (e.g. electric system plus multiple steam systems at different pressures) require (n) x 8760 points to complete the analysis.

Several practical issues arise in the acquisition and storage of site load data. For example, thermal load data may not be consistently tabulated on an hourly basis. This is problematic if loads are highly variable. Anecdotal information on plant operations can often be used to extrapolate from limited data. Frequently there are well established cycles of demand which can serve as a basis for extrapolating from limited data.

A simple extrapolation technique involves first transferring what limited data exists to a spreadsheet program. The first column (column A) of the spreadsheet is labeled "hour". The first row entry (under the title row) in this column is the first hour of the year. A 2 is entered in the second row corresponding to the second hour, 3 in the third row, etc. This process is automated, and virtually instantaneous in most spreadsheets.

In the next columns to the right of column A, thermal and electric load data is entered, corresponding to each hour in which the load occurred. For example, column B might list steam loads, e.g. hour 1 - 50,000, hour 2 - 48,000, etc. Column C would list electric loads, e.g. hour 1 - 15,000, hour 2 - 14,900, etc.

It is possible to avoid manual data entry if the data is already stored in a computer. Many district steam systems monitor loads using computerized data tracking systems. Usually there is a standard procedure for transferring data from the data tracking computer to the spreadsheet.

It is possible that the spreadsheet will not have sufficient memory to store a complete year of hourly data. At this writing most spreadsheets are limited to about 400K bytes. This limit can to some extent be surmounted using memory expansion and extension hardware, or the load data can be stored in 1 month (per spreadsheet file) segments as I do. This obstacle is soon to be removed with the release of new software which increases spreadsheet size by several orders of magnitude.

Once the data is arrayed in the spreadsheet, any gaps will be easily identified. (Breaking up the data into monthly segments will facilitate this review.) It is likely that rules for extrapolating missing data will become evident at this point also.

Engineering-Economic Analysis

The first step in the analysis of alternatives is to identify maximum (i.e. peak) and minimum (min) loads. Most spreadsheet programs have a built-ins procedure for automatically determining peak and min loads (e.g. in Lotus 1-2-3 the "@min" and "@max" functions are used in this regard).

In the simple case where there are only two loads, one thermal and one electric, 4 numbers are generated with the min/max analysis, i.e. (1) max and (2) min thermal loads, and (3) max and (4) min electric loads.

Based on these 4 numbers, 4 plant configurations are defined, s1 through s4, as described above. The next step is to obtain cogen plant performance data (usually from vendors) for the 4 equipment options, which more-or-less matches each of these load requirements (keeping in mind that the systems selected for the analysis need not precisely match electric and thermal loads).

For example, a site with max/min electric of 20/10 MW (megawatts) and max/min thermal of 160,000/30,000 lb/hr, would use gas turbine based cogen systems of 20 and 10 MW, corresponding to the electric load max/mins. Corresponding to the thermal max/mins would be gas turbines rated at 40 and 8 megawatts respectively, i.e. 40 and 8 MW systems will supply sufficient thermal output to generate 160,000 and 30,000 lb/hr steam.

System performance specifications come in a variety of formats. The most useful format for this analysis is one which uses two linear equations of the form $f = me + b$, where for equation (1):

- f=fuel input (e.g. kw, btu/hr, therms/hr)
- e = electric output (kw)
- m and b = empirically derived constants.

For the second equation (2) fuel input is stated as a function of thermal output ($f = mt + b$):

- f = fuel input
- t = thermal output (e.g. kw, lb/hr, btu/yr)
- m and b = empirically derived constants.

Vendors may not necessarily supply performance equations as described above. However, they will typically supply a series of f, t and e values which can be then be transformed into the referenced linear equations. It is important that the values supplied reflect plant performance assuming whatever emission control system is likely to be required.

Also, performance data, particularly for equation 2, may deviate slightly from strict linearity. In that case it will probably be sufficient to use an estimated linear relationship. Most spreadsheet programs will identify regression lines for a given data set in this regard.

Once the performance equations have been identified for equipment options s1 to s4, plant operating parameters can then be established for the different operating regimes described above, i.e. r1, r2t, and r2e. Plant parameters include electric and thermal output and fuel input. The procedure for calculating plant parameters for a given plant operating regime is described below.

For example, the s1r1 option (or for any r1 option), plant operating parameters are constant for every hour. The e value of equation (1) is simply peak electric demand. The calculation of fuel input as a function of peak electric demand would be performed in column D of the spreadsheet.

Once fuel input is calculated for s1r1 then, using equation (2), thermal output can be calculated (in column E). In this case the terms of equation (2) are rearranged to $t = (f-b)/m$, where t is thermal output. This equation is then written in column E of the spreadsheet, referencing the value f (i.e. fuel input) as calculated in column D.

In the case of an r1 option, f, fuel input for equation (1), is constant for every hour of plant operation. So t, thermal output per equation (2), will be constant for every hour of plant operation.

The top left hand corner (typically referred to as the "home" position of the spreadsheet) would have the following contents, pursuant to the foregoing instructions.

```
column --->    A          B            C
row|                   "site loads"
  V                   STEAM        ELECTRIC
              HOUR    DEMAND(LB)   DMND (MW)
  1            1       160,000       40
  2            2       147,000       38


column --->    A....... D             E
row |                  "plant parameters"
  V           HOUR  FUEL(THERM)   THERMAL OUT
                     f = me+b     t = (f-b)/m
  1            1     4800            160,000
  2            2     4800            160,000
```

The plant electric output parameter would appear in column F and could be entered as a constant (e.g. 40 MW) as specified per s1r1. It could also be entered as a function of f, i.e. $e = (f-b)/m$.

The s1r1 option has no variable plant parameters. The values appearing in columns D, E and F may therefore be entered as constants rather than as functions of site load parameters. (It does in fact save spreadsheet memory to do so.) In the case of an r2 option however, plant parameters vary as a function of site loads. So it is necessary to include the prescribed functional relationships in the appropriate spreadsheet cells.

For example, in the case of an s1r2e option, f in column D is a function of the electric demand variable listed in column C. Because values in column C vary by hour, column D values will likewise vary. Likewise column E and F values will vary according to the functional relationships prescribed above. (Since the plant is simply following electric demand in the s1r2e option, the values in column F will be the same as appearing in column C.)

There is a third data set required for the analysis (in addition to site load data and plant specifications), namely energy pricing data. This data is arrayed to the right of the "home" position in columns G, H, and beyond. Energy pricing data includes electric retail rates (i.e. rates payed by site user for on-site power), electric purchase rates (i.e. rates payed for export power), and fuel costs.

Export power rates often incorporate two components, typically referred to as energy and demand. The rate may vary over time, moving from a lower price to a higher priced period on a daily, weekly or seasonal cycle.

The export power rate is entered into column G (and additional columns if more than one component) of the spreadsheet. It is not necessary to manually enter each hourly value given that the rate cyclically varies. It is necessary only to manually enter a full cycle and then use the spreadsheet "copy" command to copy the cycle to the remaining hours of the spreadsheet. The copy process, if optimized, is virtually instantaneous.

The retail power rate, entered into column H, is modelled in the spreadsheet in essentially the same way as the export rate. The rate may vary over time, moving from a lower price to a higher priced period on a daily, weekly or seasonal cycle.

Fuel cost, entered into column I, is also modelled in the same way as power rates, although variation is usually not on a daily cycle. In some cases, there is a pricing distinction between fuel burned in a cogeneration system versus that burned in a boiler. It may therefore be necessary to make this distinction in the spreadsheet.

Based on the foregoing, the spreadsheet would incorporate the energy pricing data as follows (to the right of plant parameters):

```
column --->  A ......G                 H
row |                "energy pricing data"
    V                EXPORT PWR     RETAIL
           HOUR      ($/KWH)        POWER
    1        1       0.02450        0.0445
    2        2       0.03455        0.0566
```

```
column --->   A .......I
row |
    V                      FUEL COST
              HOUR         ($/THERM)
    1           1          0.2965
    2           2          0.2965
```

Ranking Alternatives

The data arrayed in columns A to I is used to calculate the net cost or savings of each alternative, i.e. the "net" of expenses and income. The "net" is then used as a basis for ranking alternatives. The cost accounting approach used in the analysis of alternatives may vary according to the specific requirements of the analyst. For purposes of this discussion, expense and income categories are defined as follows.

Income is: (1) revenues from export power sales; and (2) the economic value of any cogen plant output (i.e. electric and thermal) used on site. Expense is limited to fuel cost. (There are of course other operating costs, e.g. maintenance, debt service, etc., which are ignored in this analysis as discussed below.) The procedure for calculating income and expense for a given option is described as follows.

The income item, export sales revenue, is calculated in column J. The calculation is based on data in column G (export power price), column C (site electric demand), and column F (plant electric output). Using the column letter designations to represent cell content, the following formula is used to calculated export power sales revenue = J for a each operating hour (1 to 8760 hr/yr):

$$J = G \times (F - C), \text{ where } J = 0 \text{ if } F < = C.$$

Note that the above equation and logic condition can be conveniently written (and then copied to other cells) in a standard spreadsheet cell formula as (syntax varies according to software):

```
@if(F<=C,J=0,J=Gx(F-C).
```

Cogen electric output (value) for on-site use, is also treated as an income item and is calculated as follows in column K:

```
@if(F<=C,K=FxH,K=HxC)).
```

Cogen thermal output (value) for on-site use, is also treated as an income item and is calculated as follows in column L:

```
@if(B<=E,L=IxkxB,L=IxkxE).
```

(The factor k above is the conversion factor used to convert from steam lb to therms. It is a function of site conditions, i.e. boiler efficiency, and saturated steam pressure.)

The single expense item, fuel cost, is calculated in column M as follows: $M = I \times D$.

Based on the foregoing, net of income and expense is calculated in column N as: $N = J + K + L - M$. Annual "net" is then the sum of all N (for each hour 1 to 8760 hr/yr).

This then is the basic framework in which each cogen plant alternative is evaluated. To move from the evaluation of one alternative to the next, the cogen plant parameters formulated in columns D, E, and F are simply altered to reflect the performance specification and operating regime of the next option to be evaluated. The formulas and data in the other columns remains unchanged.

Once the annual "net" has been identified for each option, plant alternatives can be ranked according to "net". It is useful for comparative purposes to array values for net and other relevant data as follows.

Cogen Option	Electric Capacity (MW)	Net mm$/yr
s1r1	40	3
s2r1	10	4
s3r1	60	-1
s4r1	34	3
s1r2e	40	4

Based on the foregoing data, several alternatives can be eliminated from further analysis. For example s3r1 is eliminated based on a negative net. S2r1 is likely to be the optimal alternative given that it has the highest net, and most likely the lowest debt service requirement (given that it is the smallest plant of the options evaluated.)

Note that the process for ranking alternatives is performed without reference to other operating expense items (maintenance, staff, debt service, etc.) beyond those defined above. Whether this process is definitive in this regard depends on site specific data, i.e. it may be necessary to consider other operating costs in the event that alternatives are more nearly equivalent. It may also be useful to consider the impact of alternate energy pricing scenarios on comparative economics. (This is conveniently accomplished in the spreadsheet by changing the energy pricing data in columns G to I).

As noted above, this exercise is intended only as a starting point in what is essentially an iterative process. It offers a convenient and low-cost method for examining and comparing (at the beginning of the planning process when planning resources are limited!) the wide range of possible plant options. The process is greatly facilitated by, if not impossible without, the use of the personal computer and commonly available spreadsheet software.

Chapter 58

THE MIDLAND COGENERATION VENTURE: REGENERATION THROUGH COGENERATION

R. C. Lincoln

An old truism in politics states that "where you stand depends upon where you sit." My Company began the project development phase of what became the Midland Cogeneration Venture flat on its back. Since then, Consumers Power -- America's fourth largest combination gas & electric utility -- has experienced a dramatic turnaround, -- climbing from the brink of bankruptcy about three years ago to being named the most improved U.S. utility of 1986 by Forbes and "Utility of the Year" in 1987 by Electric Light & Power. Our stock rose to a high of about $20 during this period after bottoming out at about $4 in 1984.

A key in this turnaround has been our Midland facility, crucified as a nuclear plant, then resurrected as a gas-fueled, combined-cycle cogeneration plant.

It's appropriate that Midland will be an important part of our salvation, since it caused our financial crisis. It was started twenty years ago as a nuclear cogeneration plant. For a combination of reasons, the nuclear plant's cost then ballooned to over $4 billion and the construction schedule stretched across two decades. After about twenty years' effort, it had generated nothing but controversy.

Our customers, alarmed by the potential rate consequences of a $4 billion plant, were up in arms. That, in turn, worried politicians, who relied on the votes of the six million people we serve to stay in office. The few politicians who stood by us found their support becoming a campaign issue against them. Our largest industrial customers joined with Michigan's most liberal politicians in a coalition against completing the plant. One of those major customers, Dow Chemical, was squared off against us in a major lawsuit over Midland. And if we felt we were the Christians in that fight, everyone in the coliseum was rooting for the lions.

As you might imagine, the media seemed united against us, fueled by a growing anti-nuclear sentiment following Three Mile Island and what they saw as a serious lack of honesty on our part regarding Midland. Eventually, in 1984, when we could no longer raise needed funds in capital markets, we were forced to shut down the project.

Our employees were severely demoralized by this seemingly unanimous public hostility, and by the years of budget restrictions from diverting needed operating dollars to what looked like a fruitless effort to complete Midland and shareholders were outraged when the Company was forced to suspend common stock dividends to stave off bankruptcy.

With that backdrop, it is easy to understand how our board reached out for new management in November 1985. Since the over $4 billion the Company had spent on Midland represented some 40 percent of our total assets, it didn't take us long to figure out that the key to our success was putting that investment to work and putting the controversy to rest. This would require overcoming both engineering and communications challenges.

The entire Midland issue had been so ferociously politicized that our new CEO was heavily criticized for even proposing to study what, if anything, could be done with the plant. The new management set the tone for what was to come by confronting the controversy squarely and backing down those opponents who were trying to keep us from even studying the available options for Midland.

The engineering challenges were addressed in a comprehensive options study to identify how best to use Midland to stave off the Company's impending shortage of generating capacity -- the whole reason Midland was built in the first place. Our reserve margin is now about 13% -- about half the level even our regulators endorse.

We looked at either abandoning the facility altogether and purchasing needed electricity from other utilities (if available), or putting it to work by either completing it as a nuclear plant or converting it to either coal or gas.

As you may know, we ultimately concluded that natural gas had a number of advantages. It required the least capital investment and had the shortest construction time. It provided the potential for high plant availability and efficiency. It was proven technology. Gas-market conditions were favorable. And natural gas had the least environmental impact.

Subsequent agreements with Dow Chemical -- with whom we'd been in court for two years -- have enabled us to bring the additional benefits of cogeneration to Midland. This, by the way, was the reason the nuclear plant had been sited in Midland in the first place.

The communications challenge was as formidable as the engineering challenge, and we had to address both at the same time.

A big part of our communications challenge stemmed from the fact that the Company had never really made the case that Michigan needed the power Midland could provide. We then had to show the benefits of making this power in Michigan at Midland, versus trying to either buy it outside the state or starting all over to build a new plant somewhere else in Michigan.

We had to remind the state that its future economic growth depended on the ability of Consumers Power --

Michigan's largest utility serving 6 of the state's 9 million residents in 67 of 68 Lower Peninsula counties -- to do its job of powering the state's progress.

As 1986 began, our new Communications Department developed the Company's first-ever strategic communications plan. The immediate goal was to build a base of public understanding and support for the Company's proposal to put Midland to work that would be announced later in the year.

The assessment of our resources was not what most political strategists would call ideal. The hour was late. There was neither the time nor money to enlist and educate sophisticated outside communications experts. It would also have been politically inappropriate -- not to mention financially out of reach -- to run a slick advertising campaign. In the end, we had to develop a home-grown, grass-roots campaign that utilized the only real resources we had available to use -- approximately 10,000 employees and more than 100,000 Michigan shareholders.

Strategically, these resources were well-positioned, since they lived and worked throughout Michigan in virtually every Lower Peninsula legislative district. We also found that our shareholders were the very people we needed to reach -- business and civic leaders and others in influential positions. Many of our employees had their own strong, individual ties to key politicians and community leaders. They were a largely untapped resource, since we never before had made such a concentrated effort to reach, inform, motivate and support them to help the Company solve its communications problems.

The new communications initiative was kicked off with a videotape which was shown to all employees. It was a pep talk to motivate them into action. It also crystallized the messages we wanted them to carry and demonstrated tools we had produced for them to use.

We better focused our employee communications program by consolidating seven different publications into two - a monthly magazine called Progress and a weekly newspaper called CPWeekly. Progress is mailed to employees at home so we can reach their families; we also send it to all retirees and certain VIPs in the state. CPWeekly is distributed at work.

The campaign itself was led by the CEO and other senior officers who briefed state government officials, beginning with the governor and including every Michigan legislator in both Lansing and Washington. We briefed our major customers and labor leaders. We also briefed every major media outlet in Michigan at least once and had follow-up sessions with most of them.

We actively solicited supportive editorials from the media for the general proposition of putting Midland to work rather than simply walking away from it. We encouraged business, labor and civic leaders to write supportive letters to the state's political leaders on this same point. Management led the charge at the national and state levels, while employees took the lead at the regional and local levels. Again, the basic idea was to lay the groundwork for whatever specific Midland proposal we'd make in the spring following the options study.

As a constant reminder to employees and the public of the key role the Company plays in our state's economic growth, the phrase "Powering Michigan's Progress" was added to our logo. The slogan is our job description and underscores the partnership we have with the state as its largest utility. It was incorporated on all new and existing signage throughout the Company.

Simultaneously, we greatly expanded the use of press releases stressing Michigan's need for the power that Midland could provide. The Company issued a press release the equivalent of every business day in 1986 and handled about 4,000 telephone inquiries from the media.

The results of our activities were immediate and rewarding. News stories explaining our side of the issue began appearing. Our key audiences began to respond to arguments from us which they'd not heard before. Throughout Michigan, there was genuine appreciation for our efforts to open a dialogue on how best to use Midland to make the power Michigan required.

A major milestone in the Midland story developed late in 1986, when Consumers Power and Dow Chemical announced an agreement to form a partnership that would complete, own and operate Midland as a cogeneration plant. Dow and CPCo had originally been partners in Midland's nuclear construction phase, but, as I've said, that original relationship had dissolved into bitter lawsuits in the last several years.

Dow's support represented another dramatic "turnaround" example of a former adversary's conversion for Midland's own conversion.

A firm consensus of support now exists for putting Midland to work as we've proposed. It is evidenced by letters of support signed by the bipartisan leadership of Michigan's House and Senate, both U.S. senators and 17 of Michigan's 18 congressmen.

All in all, over 500 resolutions and letters of support were forthcoming from key governmental, business, labor and civic leaders throughout Michigan.

Our work has generated more than 150 supportive editorials -- including several in every important newspaper in Michigan -- supporting the Company's conversion effort.

This very visible communications effort has contributed a great deal to our success. But it obviously hasn't been the only factor. A great deal of behind-the-scenes work has been needed to make the Midland Cogeneration Venture a reality.

First and foremost was structuring the project in a way that made it affordable for Consumers Power, yet still attractive to other investors. Since we were in such poor shape financially, we needed to develop a project structure that required a minimum of cash from the Company. Our strategy was to attract equity partners who would have a vested interest in Midland. We felt that their dual interests in construction obligations and potential return on their investment in a successful project would make for a very strong consortium. In short, our basic strategy has been to require those who hoped to sell products or services to the project to invest in it.

In every phase of our project, we had people bid against each other for different parts. We had Dow and Fluor bid against each other on contracting and constructing the project. Brown Boveri, GE and Westinghouse were competing for the turbines. Combustion Engineering was competing with four other manufacturers for the waste steam recovery system. Panhandle and Coastal, the two major pipelines that supply the State of Michigan, were competing against each other for the new business in this plant, which will utilize 80 billion cubic feet of gas a year.

For every partner we've gotten into this project, we have negotiated with anywhere from two to four parties simultaneously. Each of the parties knew that the terms and conditions, as well as the size of the equity investment, would affect our decision on the successful bidders. As a result, the partnership has seven multibillion-dollar corporations. The partners are subsidiaries of Brown Boveri, CMS Energy, The Coastal Corporation, Combustion Engineering, The Dow Chemical Company, Fluor Corporation and Panhandle Eastern Corporation.

The capital structure of the project is also unique, since you're starting with Consumers having $1.5 billion of usable assets already on the site. We're contributing between $300 to $400 of those assets to the partnership for our 49% equity ownership in the partnership. The remaining $1.2 billion of assets have been sold to the partnership in exchange for market-type notes that Consumers Power Company will own. The other investors are contributing a portion of their cash equity, with the balance due upon operation.

We also needed a supply of gas with a predictable price - preferably long-term contracts with some kind of controlled escalation tied to coal prices. Ultimately, the project needs 200 million cubic feet of gas per day, or, as I said, about 80 bcf per year. The gas supplies we're looking at are in both Canada and the U.S. We're very fortunate that with Panhandle, ANR, and ANR's subsidiary, Great Lakes Gas Transmission Company, we have access to all the major producing areas in the U.S. and Canada. So far we've signed up all of our total requirements in 1990 and more than 75 percent of our ultimate requirements in 1994. The MCV initially will use about 150 million cubic feet of gas per day, climbing to 200 million cubic feet per day in 1994.

The cogeneration rates that will be charged to Consumers Power Company are based on avoided costs, as most of you know. So far, the Michigan Public Service Commission has approved two cogeneration contracts - one with a capacity cost of about 4.9 cents per kilowatt-hour and the other one at about 4 cents. The rate that we're seeking in the purchased power contract between the MCV and Consumers Power is slightly less than the lowest end of that range when it is escalated to 1990. The impact on Consumers' rates are minimal. The MCV will have a rate impact in 1990 of about 14% higher than today's rates, which are currently lower in nominal terms than our electric rates were in 1984. In today's marketplace, our electric rates are about 18% lower than Michigan's other major electric utility. We believe that gap will widen because of new power plants coming on line in Michigan. When you look at the change in electric rates between now and 1991, we're really looking at an annual compound growth of 3%, which is less than the rate of inflation. The MCV will be competitive ($1,500/kw versus Michigan and U.S. averages for new plants of $2,600 and $1,900 per kw, respectively).

Our strategy here is not to simply go after the rates that we think we're entitled to. We're trying to find a formula that's acceptable in the marketplace and which can create a win-win scenario between ratepayers, cogenerators and Consumers Power Company.

From a technical standpoint, we really have upside potential in this project. We're only using the smaller of the two steam turbine generators remaining from the old nuclear plant. We have the opportunity to convert the second nuclear turbine, which is about 880 megawatts, with the addition of about 16 more gas turbines, to generate an additional 2,000 megawatts of power. That's available when and if the market needs it.

Also, all of our gas turbine units can be retrofitted for coal gasification if the gas market isn't there - either at the end of our gas contracts or when the opportunity for the second unit exists. That's kind of our insurance policy in this project. If gas prices get too high, there would be the opportunity to switch to low-Btu coal gasification, and Dow Chemical happens to possess some of the best technology in that area today.

We believe the power production industry is changing dramatically. Real competition to utilities is coming from independent bulk power producers. Both utility and independent power producers will have to fight for industrial customers who can build their own cogenerating facilities.

We think there are tremendous future opportunities for companies that are innovative and decisive. Cogeneration represents one of those opportunities. As the Midland Cogeneration Venture demonstrates, this can be an exciting time for all of us.

Chapter 59

OPTIMIZING HEATING AND COOLING WITH AN EXISTING COGENERATION FACILITY

W. Fierce, J. Behrendt

ABSTRACT

A low temperature hot water system utilizing extraction steam and condenser water from steam turbine generators is used to heat a 490,000 square feet manufacturing building previously heated by natural gas. The $1,210,000 project reduced building energy costs for heating by $510,000/year and reduced maintenance costs by $119,000/year.

Return water from the heating system provides "free" cooling to a new E.coat paint system during the heating season. Two absorption chillers operating on extraction steam from the turbine generators were added to provide cooling during warm weather. The $830,000 cooling addition to the original project saves $239,300/year through reduced electric purchases and eliminated the $400,000 electric chiller installation originally scoped with the paint system.

An additional 130,000 million BTU's of thermal energy is extracted from an existing cogeneration facility to provide building space heat, process cooling, and office air conditioning.

PROJECT DESCRIPTION

The John Deere Harvester Works in East Moline, Illinois is a 5,082,000 square foot facility that produces John Deere harvesters (combines) for the North America market. The facility has a central powerhouse that provides steam, compressed air, process cooling water and approximately 50% of the electricity required by the facility. The powerhouse burns inexpensive Illinois coal ($1.30/MMBTU) to produce 175 psig and 600 psig steam which powers steam driven air compressors and steam turbine electric generators. The powerhouse is shown schematically in Figure 1.

POWERHOUSE SCHEMATIC
JOHN DEERE HARVESTER
FIGURE 1

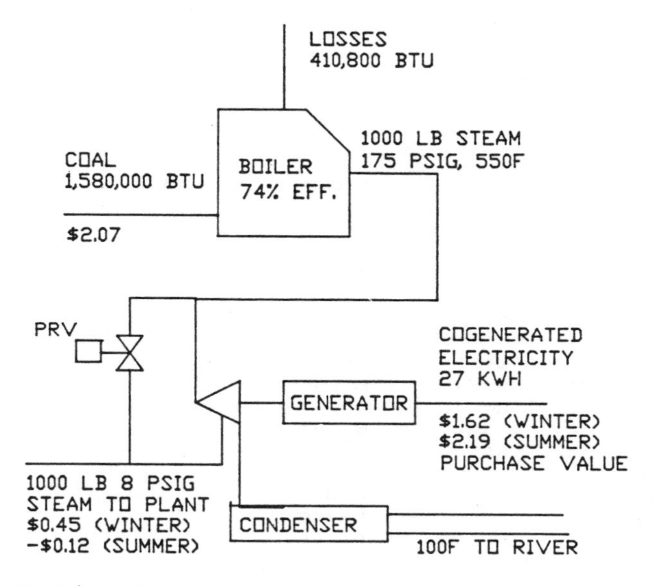

Heating System

In 1986 the 490,000 square foot combine assembly building at John Deere Harvester Works was converted from direct fired natural gas heat to low temperature hot water. The project, which included installation of a plant-wide central control and monitoring system, will save $629,000/year. The system is shown in Figure 2.

Condenser water and 8 psig extraction steam from the powerhouse turbine generators are used as a heat source. A hot 60% water/40% glycol solution at temperatures up to 170°F is supplied to the combine assembly building from the powerhouse. The water is piped to 24 rooftop 100% outside air make-up units retrofitted with hot water heating coils. The air is heated as it passes over the coil and the cooled water/glycol solution is returned to the powerhouse at 60°F or colder. The return solution is heated to approximately 95°F in a plate and frame heat exchanger by 100°F water being returned from the turbine generator condensers to the river. The water is heated further to as much as 170°F by 8 psig steam. The heated water is then pumped back to the combine assembly building and the cycle is repeated.

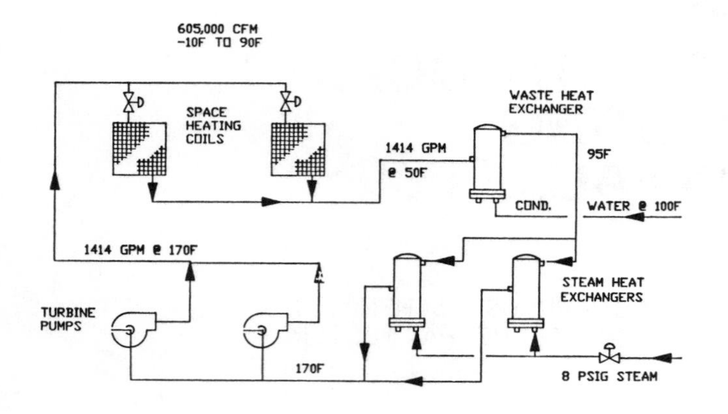

HEATING SYSTEM
FIGURE 2

Cooling

A new paint system was added to the main combine assembly building in 1986. The paint system requires 326 tons of cooling to maintain 80,000 gallons of paint at $75^{\circ}F \pm 1^{\circ}F$. The return water/glycol solution from the low temperature hot water system is diverted to the paint system to provide cooling while recovering heat. Two 450 ton absorption chillers, operating on 8 psig steam, were added in the powerhouse to provide cooling during the summer months for the paint system. These same chillers also provide 130 tons of cooling water for factory spotwelders and 286 tons of office and computer room air conditioning. The system also provides cooling by direct exchange with river water when river water temperatures are below $65^{\circ}F$. The cooling system is shown schematically in Figure 3. The combined system is shown in Figure 4. The cooling system saves $239,000/yr. and eliminated a $400,000 capital expenditure.

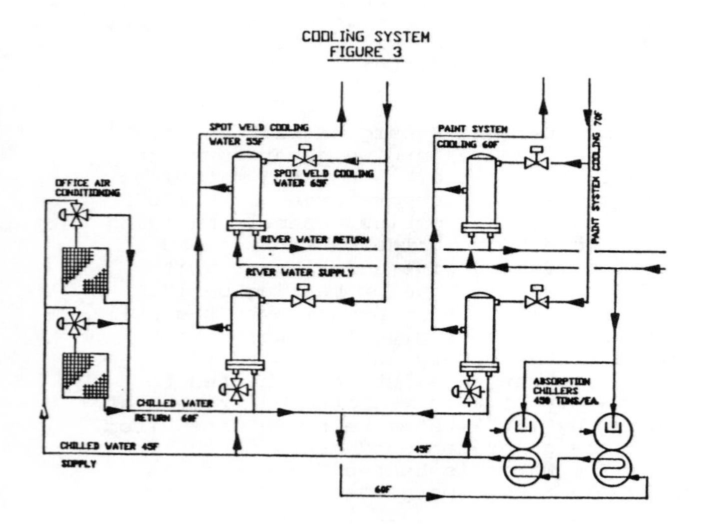

COOLING SYSTEM
FIGURE 3

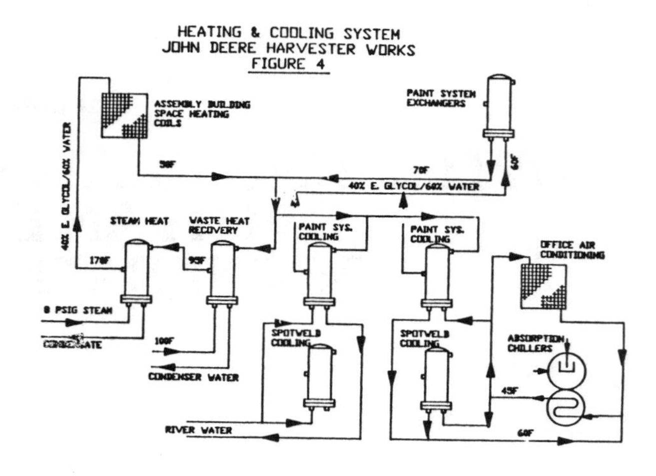

HEATING & COOLING SYSTEM
JOHN DEERE HARVESTER WORKS
FIGURE 4

PROJECT OBJECTIVE

The objective of the project was to utilize the existing powerhouse to convert low priced coal into steam to replace higher cost natural gas heat and electric refrigeration. The respective objectives of the heating system and cooling system are stated below.

Heating System

The combine assembly building was constructed with direct-fired natural gas air make-up units in the late 1960's when natural gas was relatively inexpensive. Since 1973, the cost of natural gas has increased from $0.68/million BTU's to $4.63/million BTU's in 1986. In 1985, the $596,000 in purchased natural gas to heat the combine building was 17% of the total facility energy costs.

The objective of the heating portion of the project was to minimize costs subject to the following constraints:

1. Limit increases in peak powerhouse boiler steam output to 40,000 lb./hr.
2. Minimize effect on low pressure steam distribution system.
3. Maximize electric generation.
4. Maintain or improve comfort levels in the combine assembly building.

Cooling System

Since 1982 the cost of purchased electricity has increased 90% to a summer rate of $0.081/KWH. The objective of the cooling portion of the project was to reduce costs by replacing electric refrigeration with steam absorption refrigeration.

The cooling system was required to meet the following criteria:

1. Provide 326 tons of reliable cooling to maintain 80,000 gallons of paint at 75°F ± 1°F. Variations in paint temperature cause poor paint quality and higher paint use.

2. Provide 130 tons of office air conditioning.

3. Provide 286 tons of office air conditioning.

4. Maximize electric generation.

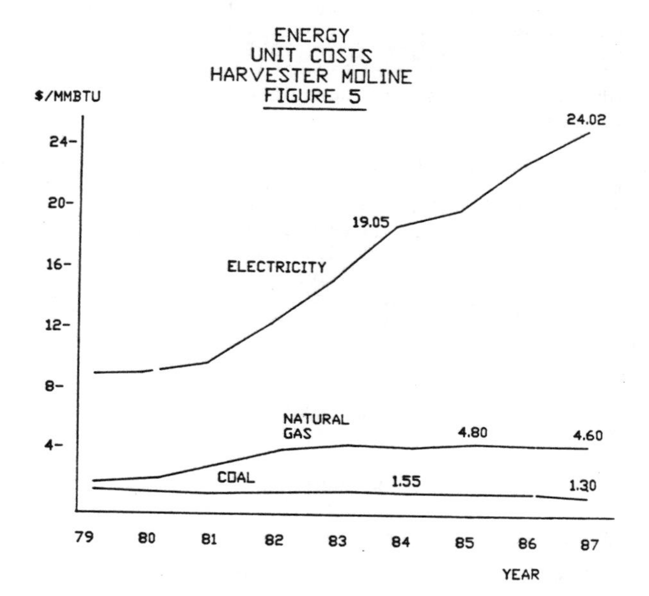

ENERGY COSTS
$/MILLION BTU

TABLE 1

	COAL	GAS	ELECTRICITY
1987	1.30	4.60	24.02
1986	1.47	4.63	23.09
1985	1.52	4.80	19.89
1984	1.55	4.62	19.05
1983	1.65	4.67	15.30
1982	1.72	4.07	12.64
1978	1.51	1.67	8.38
1973	0.46	0.68	4.28
1972	0.46	0.61	4.27

DESIGN

Heating System

The peak heating load for the main combine assembly building is 64 million BTUH or approximately 65,000 lb./hr. of 8 psig steam. Adding this heating load would increase boiler loading to a point where reliability of overall plant operation would be threatened or a new boiler would need to be added. Adding a new coal fired boiler was immediately rejected since the cost would be in excess of $5,000,000 and could not be justified by anticipated savings nor future expansion plans. It was determined, after evaluating powerhouse and plant operations that any peak steam load addition must be less than 40,000 lb./hr.

To keep the peak steam load below 40,000 lb./hr. it was necessary to either reduce the heating load by some form of conservation within the combine assembly building, find and utilize some type of waste heat or retain some gas fired units. The combine assembly building is a concrete panel building with no windows and minimal doors. The primary energy loss is from exhausting air and the subsequent air make-up. The exhaust is 698,000 CFM or 93,000 CFM more than the 605,000 CFM make up. This 93,000 CFM shortfall is made up by air infiltration around doors and through cracks between the concrete wall panels. This infiltrating air causes drafts and cold spots in the building. Any attempt to conserve energy in the building by converting the 100% outside air make-up units to recirculating air units would make the building more negative and aggravate the already existing draft and cold spot problems. Adding insulation to the building was also considered but did not offer sufficient energy reduction potential to bring the peak load below 40,000 lb./hr.

The only single source of waste heat of sufficient size to offer potential for reducing the peak steam load below 40,000 lb./hr. was the condenser water being returned from the powerhouse to the river at $100^{o}F \pm 5^{o}F$. This is a low grade source since its temperature is below the normal supply temperature of a typical hot water system supply temperature, but it does have the advantage of being a relatively steady source (2,000 GPM, minimum) at a relatively constant temperature.

To utilize the condenser return water as a heat source, it was necessary to design a hot water heating system with a low enough return water temperature to recover heat from the $100^{o}F$ condenser return water. Review of water heating coil design data showed that water could be returned at $50^{o}F$ at outside air temperatures below $20^{o}F$ on a practical basis. Utilizing a system of this type would require relatively low supply temperatures and large flow rates resulting in larger pipe sizes, pumps and heat exchangers relative to a typical hot water heating system. However, the ability to utilize the condenser return water as a heat source reduced peak steam loads from 65,000 lb./hr. to 34,000 lb./hr. and made it possible to eliminate gas from all the combine assembly building air make-up units.

The only practical location for recovering the condenser water heat is at the powerhouse, prior to the condenser water entering the buried river water return line. While this required installation of approximately 1000 feet of buried pipe, it allowed for placement of the steam heat exchangers and circulating pumps in the powerhouse. Placing the steam heat exchangers in the powerhouse allows 8 psig extraction steam to be used as a heat source without adding additional load to the plant low pressure steam distribution system. If the steam exchangers were placed in the combine assembly building, it would be necessary to increase the low pressure steam main capacity or use more expensive high pressure steam.

Locating the circulating pumps in the powerhouses made steam turbine driven pumps the obvious choice. A steam turbine driven pump does have a higher initial cost than a single speed electric driven pump. However, with the volume of water being pumped and wide variations of water required, a variable speed pump is the proper choice. A turbine driven pump can be speed controlled by specifying a relatively inexpensive electronic governor. A 60 BHP variable speed electric driven pump is approximately the same cost as a steam turbine driven pump, but the steam turbine pump costs approximately $0.006/HP/hour to operate versus $0.046/HP/hour for an electric pump.

Finally, by locating the steam heat exchangers in the powerhouse, the low pressure steam distribution system is not affected and turbine extraction pressure does not have to be increased. This configuration allows for maximum steam extraction at a given steam load and, therefore, maximizes electric generation.

Cooling System

Absorption chillers were installed to replace seven electric office air conditioners (286 tons), a 130 ton steam jet refrigeration system and to avoid the installation of two 300 ton electric chillers included with a new paint system.

During warm weather, low pressure steam load is low and purchase electric rates are high. Installation of absorption chillers using 8 psig steam as a heat source reduces cooling costs from $0.98/ton to -$0.001/ton. The negative operating cost of the absorption chiller is due to the summer electric rate of $0.081/KWH and the low coal costs of $1.30/MMBTU. The electricity generated is worth more than the incremental cost of the steam. At these fuel prices, the cost of 8 psig steam, after subtracting the value of the electricity is -$0.12/Mlb.

The absorption chillers provide 1,231 GPM of 45°F chilled water. Up to 456 GPM of chilled water is pumped through chilled water coils retrofitted in the existing office air handling units to provide air conditioning. At design conditions the 456 GPM of water is returned to the chillers at 60°F.

In addition to the air conditioning provided above, up to 755 GPM of chilled water is used to provide spot welder cooling and paint system cooling. For the paint system, a plate and frame heat exchanger is used to cool 1,000 GPM of a 40% E. glycol/60% water solution from 70°F to 60°F using 472 gpm of 45°F chilled water at design conditions. A similar system is used to cool 350 GPM of spotweld cooling water from 65°F to 55°F using 303 GPM of 45°F chilled water. The absorption chillers are operated from May through September.

During the winter, return water from the assembly building is used to cool the paint system through a plate and frame heat exchanger. This heat removed from the paint system, 3.9 MMBTUH at design painting rates, is used to heat the building. During spring and fall, when building heat is not required and the chillers are not operating, paint system cooling is provided by exchanging river water with the paint system cooling solution. River water is used to cool the spotwelders at all times, except summer.

OPERATION

Heating System

The maximum amount of waste heat is recovered when the return glycol/water solution is coldest. Optimum operation is when the supply temperature is just high enough to meet space temperature demands and pump flow is just adequate to satisfy heat load. With the computing capabilities available in the central control and monitoring system, virtually any type of control scheme can be devised and implemented. The control scheme that has worked best is to reset supply temperature based on outside air temperature and maintain a constant supply head.

Building heat is not required above 55°F outside air temperature since internal heat sources provide sufficient heat to keep the space temperature above 65°F when the building is occupied. When the building is unoccupied, the space temperature is maintained at 55°F. The supply temperature varies from 85°F at 55°F outside air temperature to 170°F at -15°F outside air temperature. Flow-through the system varies from 650 gpm to 1,414 GPM. This control scheme keeps the return temperature between 50°F and 60°F at out-side temperatures below 45°F allowing waste heat to provide 39% of building heat.

Cooling System

The chillers provide a constant flow chilled water loop that is reset based on return chilled water temperature and air conditioned space temperatures. The temperature of each process cooling stream is controlled by a bypass loop at each of the exchangers to maintain a preset supply temperature.

PROJECT ECONOMICS

The project uses waste heat and coal-fired steam extracted from turbine generators and direct coal-fired steam in place of higher priced natural gas to heat the main combine assembly building. Extraction steam is also used to provide process cooling and office air conditioning that would otherwise be provided by electric chillers. Unit costs for coal, gas and electricity are shown in Table 1 and Figure 5.

ENERGY COST
COGENERATION
(EXTRACTION STEAM)

TABLE 2

Coal Price	=	$1.30/MMBTU
Boiler Efficiency	=	74%
Turbine Steam Rate	=	37 lb./KW/Hr.
Cost of 175 psig, 550°F Steam	=	$2.07/Mlb.
Winter		
Value of Electricity @ $0.06/KWH = 1,000 lb/37 lb./KW/Hr. X $0.06	=	$1.62
Cost of 8 psig, 250°F Steam	=	$0.45/Mlb.
Latent Heat at 8 psig	=	955 BTU/lb.
Energy Cost = $0.45/955 BTU/lb. X 1,000 lb.	=	$0.47/MMBTU
Summer		
Value of Electricity @ $0.081/KWH = 1000 lb./37/KW/Hr. X $0.081	=	$2.19
Cost of 8 psig, 250°F Steam	=	-$0.12/Mlb.
Latent Heat at 8 psig	=	955 BTU/lb.
Energy Cost = -$0.12/955 BTU/lb. X 1000 lb.	=	-$0.13/MMBTU

The effective incremental cost of steam extracted from the turbine generators is a function of the price of coal adjusted for boiler efficiency and radiation losses and the cost of electricity. The cost of extraction steam is $0.45/M lbs. during winter electric rate periods and -$0.12/M lbs. during summer electric rates as shown in Table 2. Direct coal-fired steam cost is $1.83/M lbs. as shown in Table 3.

Above 12°F extraction steam at $0.45/M lbs. is used to heat the building. During production hours, when outside air temperatures are below 12°F, additional steam is provided through a pressure-reducing valve. The unit cost of incremental steam supplied to the heating system increases from $0.45/M lbs. to $1.83/M lbs. as the temperature drops below 12°F. For a normal heating season, the fuel cost for the converted system is $117,400/year versus $566,400 in natural gas for the original direct-fired heating units.

Installation of a central control and monitoring system has reduced fan operating hours by shutting off air make-up units during non-production hours and cycling specific units during production hours. The air make-up units are two speed units with summer make-up being approximately double winter make-up. A survey prior to installation of the new central control and monitoring system showed that several units were running at high speed during the winter. The survey also showed that during the summer several fans could be operated on low speed and still furnish adequate make-up air. The building operates three shifts per day. However, the first shift has by far the largest summer ventilation requirement. The survey showed that air make-up on second shift was essentially the same as it was on first shift. Supervisory personnel in the building were instructed to shut fans off when ventilation was not required and to operate the air make-up units at low speed

ENERGY COST
DIRECT FIRED STEAM
TABLE 3

Coal Price	=	$1.30/MMBTU
Boiler Efficiency	=	74%
Cost of 175 psig, 550°F. Steam	=	$2.07/MMBTU
1000 lb. 175 psig, 550°F Steam Plus 132 lb. 160°F Condensate Produces 1,132 lb. of 8 psig steam		
Cost of 8 psig, 250° Steam $2.07/1,132 lb.	=	$1.83/Mlb.
Latent Heat at 8 psi	=	955 BTU/lb.
Energy Cost = $1.83/955 BTU/lb. X 1,999 lb.	=	$1.92/MMBTU

during the winter. While improvements were made, the difficulty of reprogramming the 19 year old Honeywell System 6 building control system made it impractical to shut the units on and off when appropriate. After installation of the new Johnson Controls JC8540 central control and monitoring system it is considerably easier to shut down units when make-up air is not required. The electrical savings from shutting down fans with the new central control and monitoring system is $61,000/yr.

The new central control and monitoring system replaced a Honeywell System 6, a Honeywell Delta 1000 and a Johnson Controls JC80 system. These three systems had annual full maintenance contracts of $90,000/yr. The new central control and monitoring system has a full maintenance contract cost of $35,000/yr. This $35,000 maintenance contract cost is fixed for the next five years and results in a $55,000/yr. savings in contract maintenance costs.

The natural gas burner train with its safety interlocks flame sensors, ignition rod and firing controls was the major maintenance expense on the air make-up units. Installation of hot water coils with a simple control valve eliminated most of the maintenance on the air make-up units saving $64,000/yr.

Total annual savings for the heating portion of the project is $629,000/yr. as shown in Table 4.

Two 450 ton absorption chillers replaced 612 tons of electric refrigeration. The cost of electric refrigeration is typically $0.08/ton (1KWH/ton/hour) versus -$0.001/ton for the absorption unit. The absorption chillers also replaced 130 tons of steam jet refrigeration. The steam jet refrigeration used a turbine condenser in its operation, reducing generating capacity by an average 280 KW for the four summer months. The savings for the chiller project are $239,300/year as shown in Table 5. Installation of the two absorption units avoided a $400,000 capital expenditure by eliminating the purchase of two 300-ton chillers with cooling towers included in the new paint system.

Total project cost was $2,040,000. Savings are $868,300/yr. based on current energy prices.

HEATING SYSTEM SAVINGS

TABLE 4

Cost Reduction		
Natural Gas (123,140 x 10^6 BTU x $4.60/106 BTU)		$566,400/Yr.
Electricity (Fan HP)		61,000/Yr.
HV Maintenance		78,000/Yr.
Contract Maintenance		
Delta 1000	$67,000	
System 6	8,000	
J.C. 80	15,000	90,000/Yr.
		$825,000/Yr.
Cost Increases		
Coal (Steam)		$117,400/Yr.
HV & Equipment Maintenance		14,000/Yr.
Contract Maintenance New CC&M		35,000/Yr.
		$196,000/Yr.
Net Savings Before Tax		$629,000/Yr.

COOLING SYSTEM SAVINGS

TABLE 5

Savings:	
Electric cost 1,800,000 tons/yr @ $0.08/ton	$144,000/Yr
Contract Maintenance on Existing Units	82,400/Yr
Additional Electrical Generation due to Steam Jet Refrigeration Removal	17,500/Yr
	$243,900/Yr
Less:	
Steam for Absorption Chillers 1,800,000 tons/yr @ -$0.001/ton	- 1,800/Yr
Maintenance	2,800/Yr
	1,000/Yr
Net Savings Before Tax	$242,900/Yr

Chapter 60

NEW YORK STATE ENERGY OFFICE'S EXPERIENCE WITH RETROFIT COGENERATION UNDER THE INSTITUTIONAL CONSERVATION PROGRAM

R. P. Stewart, N. Ghiya

INTRODUCTION

The Institutional Conservation Program (ICP) is the largest grant program administered by the New York State Energy Office (SEO). Established under the Federal Institutional Buildings Grant, the ICP awards 50 percent matching grants for technical assistance studies and for energy conservation measures (ECM) in schools and hospitals.

The ICP receives Technical Assistance Study and ECM applications each year. The applications are evaluated for their technical accuracy and completeness, and once approved, are ranked based on simple payback, energy savings, climate and other criteria. Grant awards start with the highest ranked application and proceed down the list until either the applications or the funds are exhausted. Since 1979, approximately $92 million have been awarded in ten grant cycles, of which $62 million were federal funds and $30 million from petroleum overcharge settlements.

Awards are limited to a set amount known as the grant or institutional cap. The grant cap is determined by the amount of funds available in any one grant cycle. In cycles one through eight the grant cap was $300,000 or less, in cycles nine (1986-87) and ten (1987-88), with the availability of petroleum overcharge funds, the grant cap increased to $1.275 million and $850,000 respectively. The cap in cycle 11 is $765,000.

During the first eight cycles, cogeneration received less than $1 million out of a total of $60 million. In cycles nine and ten, cogeneration awards increased to $15 million or 50 percent of the total grant awards. In these two cycles, hospitals accounted for 68 percent of the total cogeneration awards and the remaining 32 percent went to universities and school districts. The increase in cogeneration applications can be attributed in large part to the increase in the grant cap. The maximum grant award per institution prior to cycle nine was a small fraction of the capital required for cogeneration. In addition, implementation of cogeneration requires integration of complicated building energy systems and components. Cogeneration was neglected in favor of a large number of conventional, less capital-intensive measures such as building envelope improvements, HVAC systems improvements and energy efficient lighting. Heat-recovery from hospital-waste incineration was consistently the most expensive ECM funded prior to cycle nine.

Other factors that may have limited the number of cogeneration applications prior to cycle nine arise from the relative immaturity of the cogeneration industry itself. The facility and consulting engineering community did not possess the operating or design experience required for successful implementation of cogeneration systems. They lacked knowledge of available cogeneration equipment. In addition, a lack of good manufacturers' catalog materials, published data on cogeneration equipment and experienced sales personnel, may also have played a role in inhibiting cogeneration development in earlier years. As demonstrated by the large number of cycle nine and ten cogeneration applications, as the industry matured, engineering analysts began to assess the viability of cogeneration systems. Even so, applications for cogeneration still came from a small number of consulting engineers and facilities.

Institutional cogeneration is attractive in New York State for the reasons that prevail throughout the rest of the nation, i.e. the large differential between electricity prices and engine generator fuel prices. New York State has some of the nation's highest electricity costs, which makes cogeneration especially attractive as a way to increase an institution's energy efficiency and reduce energy costs. Hospitals make excellent cogenerators because they are continuous energy users. Not surprisingly, most of the cogeneration applications received through the ICP are from hospitals.

The remainder of this report is divided into three sections. The first is a summary of the types of hospitals that have received either a cycle nine or cycle ten grant to implement cogeneration projects. It is intended to provide comparative information to assist engineers during the preliminary stage of assessing a hospital's cogeneration potential. The second section discusses specific issues associated with hospital cogeneration analysis and implementation. The third and final section evaluates the state's remaining hospital cogeneration potential.

COGENERATION UNDER THE ICP

Based on a satisfactory technical review and ranking, the SEO, in cycles nine and ten, approved 16 hospital cogeneration applications. Data were extracted and analyzed from these applications only. Hospital characteristics are listed in Table 1 in ascending order based on the number of beds. Hospital size ranged between 75,000 and 880,000 gross square feet and between 120 to 670 beds. The total annual fuel consumption ranged from 26 x 10^9 Btu to 340 x 10^9 Btu expressed in terms of source Btu (source Btu for electrical energy is calculated using the conversion factor of 11600 Btu per Kwh). The average electric demand, calculated by dividing the annual electric consumption by 8760, varied from 175 Kw to 2000 Kw. The average ratio of peak demand to average demand for the sample is 1.7.

The amount of floor space per bed, indicated by the area to bed ratio, ranged from a high of 1600 to a low of 462. The area to bed ratio can be a useful barometer in a hospital cogeneration study. The variation in the ratio is caused by differences in the type of medical treatment or hospital services provided. A hospital can be a teaching or research hospital, or it may provide laboratory, diagnostic, or mental health services in addition to offering acute medical care. These differences in hospital function, expressed by the area to bed ratio, effect both energy consumption and demand for energy services. For instance, seven of the nine hospitals with area to bed ratios greater than 900, exhibited the highest peak and average electric demands as well as the greatest annual energy consumption. A high electric demand means more electric capacity can be designed into the cogeneration system provided the thermal base can absorb the additional heat of a larger unit.

All hospital cogeneration projects approved in cycles nine and ten were engine-driven systems (Table 2). Approximately 50 percent of the systems utilized dual-fuel engines with natural gas as the primary fuel to take advantage of interruptible gas rates. Only one application was for diesel fuel. System sizes ranged from 180 Kw to 2520 Kw with a total capacity of 17,000 Kw. Excluding hospital No.1, where the estimated cost was for equipment purchase only, the average cost per installed Kw was almost $1,400. The total energy savings from these projects amount to 631,000 x 10^6 Btu or equivalent to 20 percent of the group's present annual energy consumption. Most of the selected sizes were base-loaded to optimize utilization rather than to satisfy the peak load requirements of the facility (Figure 1).

Data were extracted in order to formulate a mathematical model to estimate electrical and thermal profiles based on hospital size-related variables such as number of beds and area. At first, an attempt was made to establish a relationship between building area and annual energy consumption (Figure 2). After discarding the hospital with extreme consumption of more than 700,000 btu per square foot, the sample mean was 453,500 Btu per square foot with a standard deviation of 64,995. An analysis of the buildings' peak and average electric demand with respect to size-related variables was not satisfactory because of the effect of climate and location. For example, a downstate hospital with about the same floor area as an upstate hospital had 30 percent greater peak electric demand. After considering climatic effects on hospital energy consumption, relating electric demand to hospital size remained elusive due to facility-specific, system-related variables, such as the effects of mechanical versus absorption cooling. For example, hospital No.6 has twice the summer peak electric demand of hospital No.5 even though they are in the same geographic region and have comparable area to bed ratios.

A regression analysis was performed to relate the number of beds and area to the annual fuel consumption (Figures 3 and 4). The relationship obtained was more accurate in terms of area than number of beds because a hospital with a high area to bed ratio tends to consume more energy than one with a smaller number of beds. It is difficult to formulate a reliable model due to the limited sample size. However, the results obtained from the analysis of Btu per square foot approximate the average energy use of 487,000 Btu per square foot established in a previous SEO review of applications in grant cycles one through four.

SPECIFIC ISSUES IN HOSPITAL COGENERATION

Engineers must be prepared to address many issues when analyzing hospital cogeneration. A few important ones are: hospital codes; cogeneration compatibility with existing plant; financing options; and the role of a hospital as a "qualifying facility".

Hospital Codes: "Essential services" are the hospital services that are to be maintained in the event of an emergency, such as when the primary source of electric power is disrupted. In emergencies, the essential services load is usually met by emergency generators. The relationship between the emergency generators and the proposed cogeneration system warrants careful consideration because it may be possible to use the emergency generators as standby to the cogeneration equipment, or as "peakers" where demand charges are large. Most emergency generators are not designed for prime-power service but could supplement a base-loaded system with peak-shaving and still provide the cogeneration system with maintenance power. Hospital codes prohibit the use of emergency generators as peak-shavers where the normal supply is from the grid. However, the status of the emergency equipment is changed with cogeneration. According to the New York State Hospital Code, for instance, the source of emergency electricity supply, for hospitals where power is produced on-site, may be either an emergency generator set or the central station transmission line. With the utility providing emergency electric service, the emergency generators are usually available to supply power during scheduled maintenance of the cogeneration equipment or to supplement cogeneration power production. Investment in the primary plant is reduced and peak demand charges for stand-by or emergency power from the grid are minimized.

Compatibility with Existing Plant: A cogeneration evaluation is not complete without an evaluation of the relationship, in quantity and in time, between the facility's electric and thermal loads. There is a fine line between an optimally designed plant and one that leaves energy and energy cost savings on the drawing board. Attention should be given to the temperature and pressure of existing thermal distribution systems. The locations of high and low grade heat sinks also need to be integrated into the design of the cogeneration system.

Financing Options: There are ways to use savings-based agreements to finance a hospital's matching contribution for ICP grants. One such arrangement, approved by the United States Department of Energy and used in New York, involves either equipment vendors or third party investors providing an ICP grantee with its matching contribution, in exchange for a commitment from the grantee to pay for the costs of the project based on future savings. Since savings-based agreements traditionally provide all financing for a project, it is likely that more favorable terms can be contracted for when the grantee is paying for part of the cost of the project with ICP grant funds. In the case of leased equipment, title to equipment specified in the ICP grant award is to pass to the grantee before the close out date of the grant; usually no more than two to three years.

Hospitals as "Qualifying Facilities": The increase in system efficiency, between hospital cogeneration and the separate purchase of electric energy and the production of thermal energy, entitles hospitals to qualifying status under federal rules implementing the Public Utilities Regulatory Policies Act-1978. At a qualifying facility, electric utilities must interconnect with the hospital to permit electric energy purchases from or sales to the hospital. Selling power back to the grid is not allowed under the ICP but interconnection for the purpose of purchasing power from the grid is. The cost of interconnection and the price for backup electric service are costs over which energy engineers have little control. An investigation of the interconnection costs and backup service rates of the electric service utility should be a part of hospital cogeneration analyses.

HOSPITAL COGENERATION POTENTIAL

Department of Health statistics show 300 public and non-profit hospitals in New York State. A study conducted by the SEO of 234 of the 300 hospitals reveals an average area of 270,000 square feet and an average number of beds of 311. Based on the sample mean of 453,500 Btu per square foot, total hospital energy consumption in New York is approximately 3.65×10^{13} Btu. With a calculated average energy savings of 20 percent, there is a potential to save 7,300,000 mmBtu of source energy through cogeneration. The kilowatt capacity potential would depend on whether the systems were selected to satisfy a base or peak electric demand.

Figure 5 presents the mathematical model for average electric demand based on annual energy consumption. Multiplying the total hospital energy consumption (3.65×10^{13} Btu) by the constant yields 182,500 Kw of average electric demand.

With a ratio between peak demand and average demand of 1.7, if cogeneration systems were installed to meet peak electric requirements, hospital cogeneration capacity increases to approximately 310,000 Kw. At an average of $1,400 per Kw, the market for hospital cogeneration in New York State is between $255.5 million and $434 million.

In terms of ICP funding for cogeneration, the level of activity under future grant cycles is uncertain: petroleum overcharge funds dedicated to the ICP are almost exhausted and the grant cap in cycle eleven has decreased slightly. Though institutional cogeneration is an excellent energy conservation measure, it is uncertain that penetration in the New York institutional market will be maintained at the levels of the past two years. Even with the currently available funds and substantial matching grants, the number of cogeneration projects has been much less than the economic potential. Notably, outstanding opportunities exist for cogeneration in New York's largest hospitals, where, though some effort has been made by systems designers, very little actual capacity has been installed. With regard to the ICP, it should be noted that for these larger institutions, some of which might appoach $20 million in project cost, the project cap, while not 'peanuts', will be much less significant as a motivator. For these large projects especially, the building program, the evolution of clinical technologies that require large amounts of electricity and the interaction of the cogeneration system with building HVAC systems, all tend, in combination with the large amount of capital involved, to inhibit positive decisions, even though the economics, as of today, are very favorable.

Table 1. Summary of Building Energy Consumption

No.	County	# of Beds	Area Sq. Ft.	Area per Bed	Peak Demand Kw	Avg. Demand Kw	Annual mmBtu Consmp.	Btu Per Sq Ft
1	Dutchess	120	74778	623	378	178	2.6E+10	347695.8
2	Ontario	132	157000	1189	500	395	6.8E+10	433121.0
3	Chautauqu	176	126912	721	500	293	5.5E+10	433371.1
4	Suffolk	194	216200	1114	900	483	1.0E+11	484273.8
5	Suffolk	271	262135	967	676	476	1.2E+11	457779.3
6	Nassau	278	219434	789	1228	743	1.2E+11	546861.4
7	Queens	282	132280	469	648	296	6.0E+10	453583.3
8	Broome	310	450000	1452	1776	1119	2.2E+11	488888.8
9	Kings	343	396723	1157	1008	540	1.8E+11	453717.0
10	Suffolk	350	226762	648	1045	680	1.3E+11	573288.2
11	Suffolk	398	355000	892	1955	1239	1.2E+11	338028.1
12	Kings	457	749370	1640	3524	1913	3.4E+11	453714.4
13	Kings	532	505800	951	2550	1885	2.7E+11	533807.8
14	Nassau	533	495560	930	2534	1612	3.6E+11	726450.8
15	Kings	567	454994	802	1400	885	2.1E+11	461544.5
16	Kings	670	881834	1316	2582	1493	3.3E+11	374220.0

Table 2. Characteristics of Cogeneration Projects

NO.	Peak Demand Kw	Average Demand Kw	Total Cogen Install	Fuel for Cogen	Cost of Cogen Install	Cost Per kw	mmBtu Savings
1	378	178	750	#2 Oil	306130	$408	1106
2	500	395	375	N.Gas	501473	$1,337	9904
3	500	293	180	N.Gas	260000	$1,444	12000
4	900	483	511	DF-N.Gas	588000	$1,151	23689
5	676	476	560	DF-N.Gas	741500	$1,324	23413
6	1228	743	1232	DF-N.Gas	1422500	$1,154	34032
7	648	296	800	DF-N.Gas	966000	$1,208	22338
8	1776	1119	1300	DF-N.Gas	1367400	$1,052	50596
9	1008	540	1000	DF-N.Gas	1320000	$1,320	36871
10	1045	680	850	DF-N.Gas	2002600	$2,356	26400
11	1955	1239	1260	DF-N.Gas	1488500	$1,181	71178
12	3524	1913	2100	N.Gas	4050000	$1,929	75900
13	2550	1885	1330	DF-N.Gas	1595000	$1,199	45823
14	2534	1612	2520	N.Gas	3904000	$1,549	97424
15	1400	885	1000	DF-N.Gas	1250000	$1,250	38920
16	2582	1493	1600	N.Gas	1763000	$1,102	61399

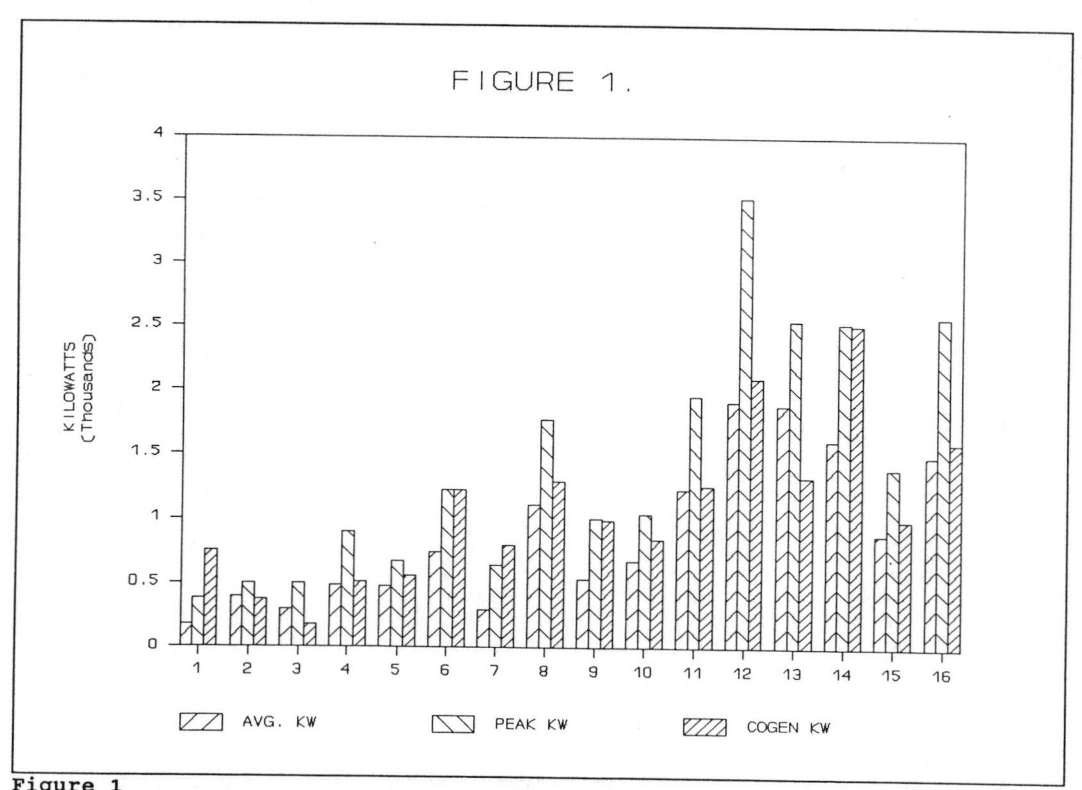

Figure 1
Electric Demand and Cogeneration Capacity at 16 New York State Hospitals

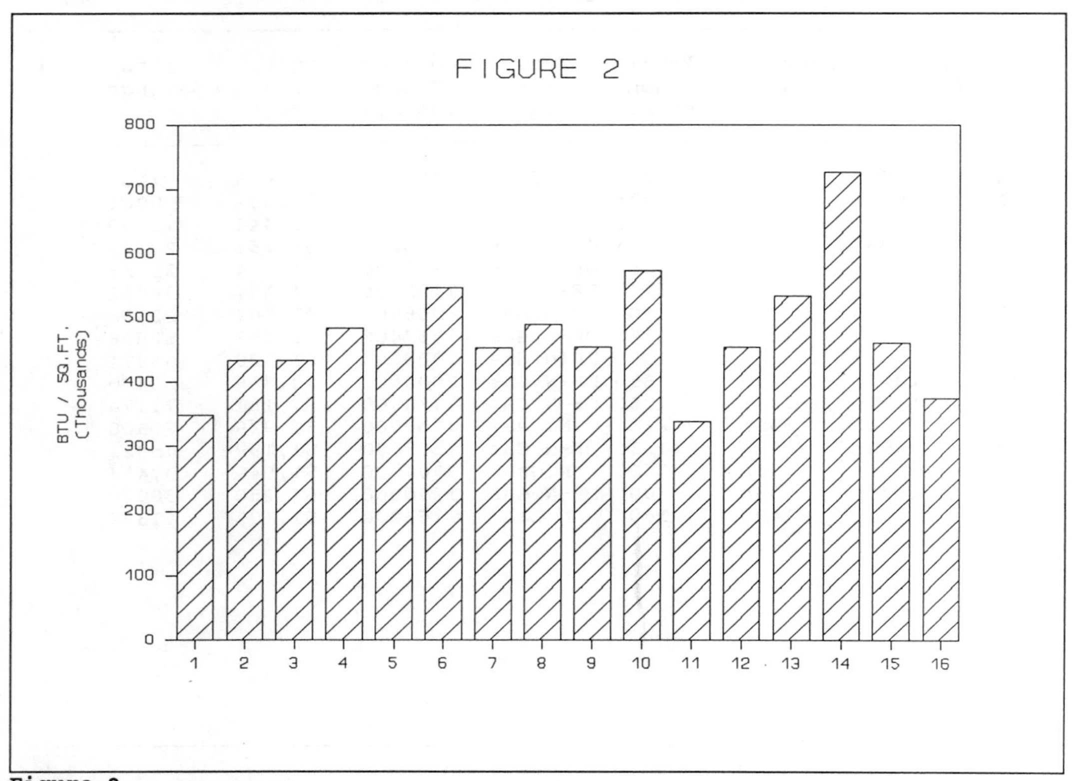

Figure 2
Energy Usage at 16 New York State Hospitals

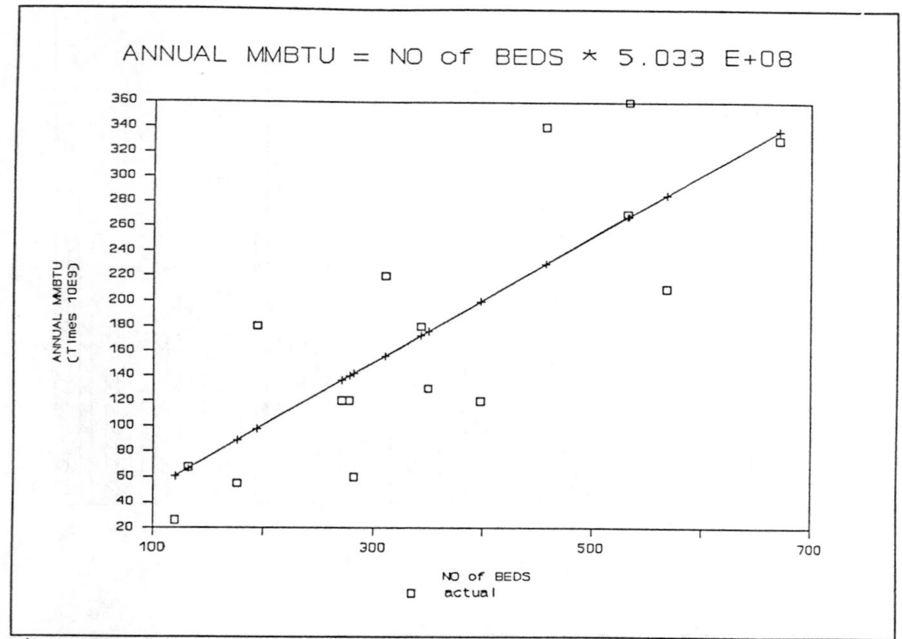

Figure 3 Linear regression for determining annual energy consumption based on number of hospital beds.

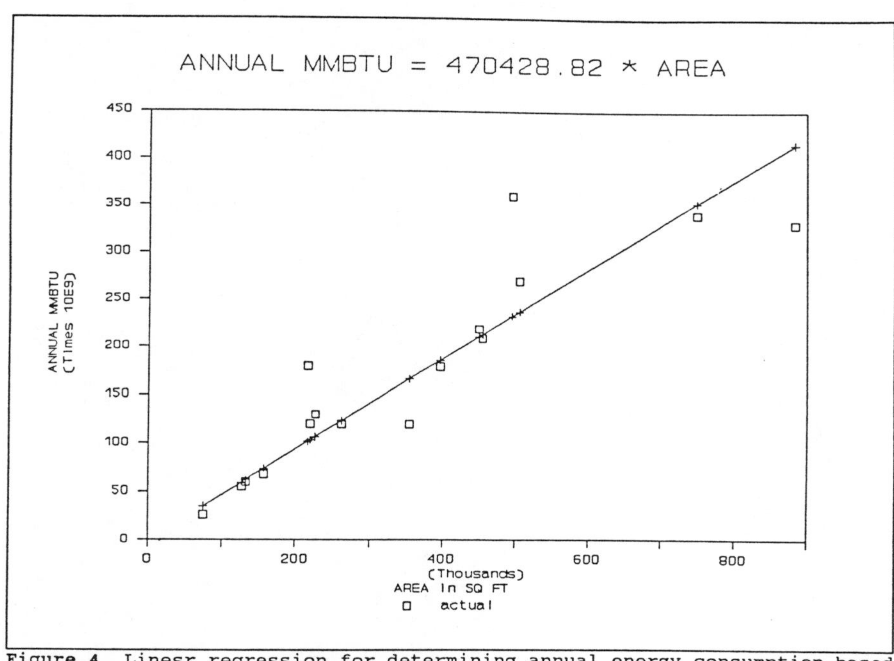

Figure 4 Linesr regression for determining annual energy consumption based on hospital gross square footage.

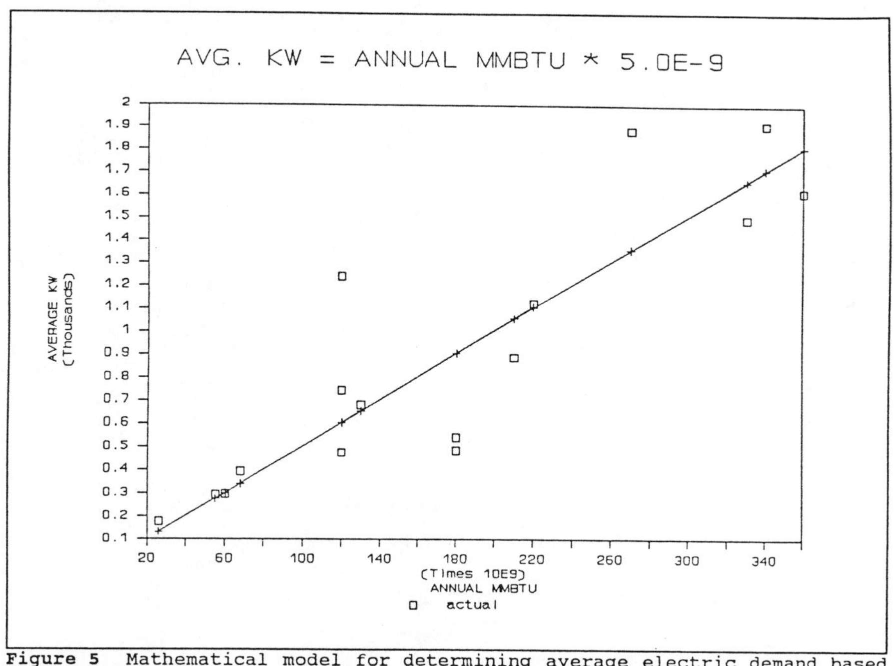

Figure 5 Mathematical model for determining average electric demand based on annual energy consumption.

SECTION 13
COGENERATION PROJECT DEVELOPMENT

Chapter 61

ADVANCED COGENERATION CONTROL, OPTIMIZATION, AND MANAGEMENT

F. Hinson, D. Curtin

ABSTRACT

The performance of cogeneration power plants can now be assessed on-line and in real time using a distributed microprocessor-based data acquisition and control system. A representative implementation is described for cogeneration power in a food processing plant.

The COPA (COgeneration Performance Assessment) package comprises separate, distributed control modules for data input, performance analysis for each plant device, overall plant performance summary, and operator displays. Performance of each of the respective cogeneration devices is assessed relative to a performance model of the device, thus an accurate assessment of performance is provided under all load conditions. Operator displays provide real time depiction of the performance of each device and the overall plant performance.

Deterioration of performance of a device is quantified in terms of the cost of additional fuel requirements and/or the value of power not produced, and this cost is accumulated.

ADVANCED COGENERATION CONTROL, OPTIMIZATION, AND MANAGEMENT

Accurate and up-to-date knowledge of performance and condition of plant components is vital to safe, reliable, and economical operation of any size of a cogeneration plant. Early detection of mechanical problems is important for averting the development of more serious damages. Preventive maintenance based on performance degradation reduces unscheduled outages and is the most cost-effective way to maintain maximum efficiency of the plant. Accurate information on component performance characteristics is also essential for overall optimization of the plant. Therefore, in order to make timely and informed decisions, real time monitoring and assessment of pertinent operating data are necessary.

Bailey Control Company's cogeneration modeling and performance assessment tool takes advantage of some unique features of their microprocessor-based distributed control system network to monitor operating parameters and to concentrate the collected data into relevant performance measures. The system compares the calculated performance indicators to values expected under similar operating conditions to reflect the actual deterioration rather than simply changes in operating conditions. The ultimate objective is to provide early indication of internal damage, so as to provide quantitative information on the effect of deterioration on the overall performance of the plant, and to generate models for plant optimization.

A cogeneration plant where the performance assessment system is installed is a typical industrial application consisting of the following main components:

- Gas Turbine
- Heat Recovery Steam Generator
- Extraction-condensing Steam Turbine
- Surface condenser
- Cooling tower

The gas turbine is a simple cycle, single-shaft unit consisting of a multistage axial flow compressor, 3-stage turbine, dual fuel system (natural gas/ no. 2 fuel oil), hydrogen-cooled generator package, inlet air system with evaporative cooler, steam injection for NOx control and accessories. The design power output of the unit is 89 MW.

The heat recovery steam generator produces steam at two pressure levels. The high pressure steam system pressure fluctuates with load, while the low pressure steam pressure is constant at all loads. The nominal steam production of the HRSG is 286 Klbs/hr of high pressure steam and 97 Klbs/hr of low pressure steam.

The steam produced in the HRSG is received by the double-admission extraction-condensing steam turbine. The unit is equipped with one non-automatic extraction to supply gas turbine injection steam, and with one automatic extraction to provide process steam. The design output of the steam turbine is 38.5 MW.

The surface condenser cooling water is supplied by a cooling water pump, circulating the water in the condenser-cooling tower cycle. In addition, the condenser receives return condensate from the process, as well as demineralized water to make up the steam cycle material losses.

The cooling tower is a counter-current, induced draft unit consisting of three cells. The fans are operated by constant speed motors. In addition to the hot circulating water, the cooling tower receives make-up water to compensate for the spray loss, evaporation, and blowdown loss.

Because of the modular design of the performance assessment package, it is possible to apply the system to analyze cogeneration plants of virtually any complexity and size. Accordingly, the software package delivered to any application provides for possible future expansions. Depending on the type and number of the new components to be analyzed, additional hardware may or may not be needed.

SYSTEM ARCHITECTURE

The Bailey performance assessment system for cogeneration plants is composed of multiple, multi-function controller modules. These modules provide an interface to the plant network for on-line data acquisition, an interface to a CRT for displaying computed results and accommodating operator interaction with the system, and the requisite environment for performing the computational analysis of the respective devices in the cogeneration plant. In this architecture, the modules are labeled to indicate their primary functions, which are briefly described as follows:

1. The INS is the input supervisor. This module inputs data from the network, performs unit conversions, validates data integrity, and formats the data into files for transmittal to the DAM's.

2. The DAM's are device analysis modules. These modules do the performance analysis for the respective devices, gas turbine, heat recovery steam generator, steam turbine, condenser, and cooling tower, then format the computed results into files for transmittal to the PAP.

3. The PAP is the plant performance assessment program. This module combines the results from the respective devices to provide an assessment of the overall plant performance and provides appropriate displays to the CRT.

The INS and the PAP modules are customizable, that is, portions of the programs in these modules are tailored to accommodate the specific requirements of each application. For example, addresses for data input from the network are application dependent, and CRT displays are tailored to the application and user requirements.

The DAMs do the performance analysis computations for each of the respective devices in the plant. One DAM does the calculations for gas turbines, a second DAM does the calculations for HRSG's, a third and fourth DAM do the calculations for steam turbines and condensers, and a fifth DAM does the calculations for cooling towers. The computational procedures in the DAMs are so designed that they can be universally applied, without modification, to most cogeneration plant configurations.

THE INPUT SUPERVISOR

The primary task of the input supervisor (INS) is to input process variable data from the network. Input data are subjected to unit conversion calculations as required. Integrity checks are made to ascertain good or bad point quality, data out of normal operating ranges, values manually substituted or values estimated. Subsidiary tasks include estimation of variables not measured, averaging multiple measurements of variables, and computation of thermodynamic properties of selected air, steam, and water variables.

After conditioning the data obtained from the network, the INS then formats the data, comprising the values of the process variables and the quality parameters, into files for transmittal to the data analysis modules (DAM). In order to permit mapping field data from random source addresses to formatted files, application data files are used to specify the mapping. These files are constructed off-line at a work station, and are loaded into the multi-function modules at system configuration time. One type of file lists the tag name, source address, units of measurement, and range limits for all the inputs. Other application files specify computations to be performed. The use of application data files permits the design of computational procedures which are relatively independent of plant configuration.

THE DATA ANALYSIS MODULES, MODELING TECHNIQUES

Thermodynamic performance assessment of individual power plant components is traditionally carried out according to detailed guidelines set forth by the respective ASME performance test codes. The resulting performance measures are normally reliable and repeatable since the tests are carried out under well controlled conditions. The ASME test codes pay much attention to minimizing measurement uncertainty by giving guidelines for sufficient number of test runs and durations, constancy of test conditions, and calibration of instruments. Actual operating conditions are often different from test conditions. The ASME test codes were originally developed for acceptance tests for which the units are operated at specified loads, whereas actual loads may vary continuously. During normal operation the other operational variables may be considerably different from test conditions. Such variables include such as, steam, water and ambient air conditions, and fuel characteristics. In order to be useful for on-line performance evaluation, the calculation methods must be modified.

The ASME test codes require corrections for off-design operating conditions and provide some guidance how to perform such corrections. However, the test codes rely predominantly on correction curves provided by the manufacturer. This approach has the following disadvantages:

- Complete manufacturer's curves may not be available.

- o Equipment manufacturers often incorporate ample safety margins into the performance curves. Therefore, such curves do not depict the true expected performance.

- o After an extended period of operation, it is seldom possible to restore the unit to the original, new, clean condition. Therefore, the original curves issued by the manufacturer may no longer provide realistic values for expected performance measures.

In order to overcome these problems, Bailey's performance assessment system is based on extensive modeling techniques with minimal requirements for manufacturer's data.

Basically, the system first determines the actual performance at current operating conditions. Thereafter, the program uses mathematical models of the plant components to calculate performance measures that could be expected from clean units under similar operating conditions. The actual performance measures are then compared to expected values to reveal deterioration in actual mechanical conditions.

After a maintenance and cleaning shutdown, the component performance is identified. To accomplish this, the user sets an "IDENTIFY" flag at the operator's console. This starts the model identification routine, and new performance data are collected into data files. Then by regression analysis, the program calculates the model parameters that provide the best fit to the collected data. After a sufficient number of observations have been collected, the user resets the "IDENTIFY" flag. This establishes a new reference model which will be used in subsequent calculations. As the model identification is done after each maintenance or cleaning, the resulting reference model provides the most realistic values for expected performance measures.

Types of models used in the performance evaluation programs include models for heat rate, isentropic efficiency, thermal transmittance, number of transfer units, and flow characteristics. From these basic models, the device analysis programs derive additional redundant performance measures to further quantify deterioration.

Because of the strong correlation between the mechanical condition and the thermodynamic behavior of the plant components, it is possible to evaluate the condition of the units from thermodynamic performance measures. The process inputs for such calculations include pressures, temperatures, flows, and other quantities that do not require special instruments. Since the purpose of performance assessment is to reveal changes in mechanical health of the components rather than absolute performance level, normal plant instrumentation is sufficient for on-line monitoring.

Mass and energy balance equations around individual components are used to determine unknown quantities that are difficult or impractical to measure directly. Another equally important purpose of the balance calculations is to verify the integrity of the input data, including the primary sensors which provide input data.

THE GAS TURBINE DAM

Gas turbine performance calculations are performed according to the ASME Power Test Code PTC 22, Test Code for Gas Turbine Power Plants. The primary objective, as specified by the code, is to determine:

- o Capacity under specified operating conditions

- o Thermal efficiency under specified operating conditions

Additional performance measures include:

- o Compressor isentropic efficiency and capacity

- o Turbine isentropic efficiency and capacity

- o Ratio of combustor pressure drop to combustor inlet pressure

The purpose of the gas turbine models is to analyze the unit operation at off-design conditions. The models are calibrated automatically, using on-line process measurements and regression analysis as described previously.

The overall model permits calculation of mechanical capability, generator capability, and thermal efficiency. The maximum mechanical output of the compressor/turbine combination depends on the mechanical condition of the unit and the ambient air conditions. Decreasing inlet air temperature increases density of the air, which increases compressor throughput. At the same time, the relative compressor work decreases, improving overall heat rate of the unit. Therefore, the maximum capacity of the unit increases rapidly with decreasing inlet air temperature. Under certain conditions the maximum mechanical output of the gas turbine may exceed the generator capability. To avoid generator overheating, the unit load has to be kept within certain limits. The overheating limit depends on the power factor and the coolant temperature of the generator hydrogen coolant.

The maximum generator output is calculated and compared to the mechanical capability. The smaller of the values is reported as the overall capability of the unit. Thermal efficiency of the gas turbine depends on the mechanical condition, unit load, and ambient air temperature. Efficiency of the gas turbine decreases and heat rate increases with decreasing load.

Inlet vane control on the gas turbine maintains compressor volumetric flow, pressure ratio, and isentropic efficiency practically constant over the normal operating range. The fundamental measures of turbine performance are the nozzle area index number (NAIN) and turbine isentropic efficiency. Nozzle area index number remains essentially constant at all loads. Therefore, the actual NAIN is compared to the expected NAIN. The expected value is determined as an average of the values calculated during model identification.

Actual output of the gas turbine is modeled as a function of isentropic output. The coefficients for the model are determined (by on-line linear regression) during "identify mode". The model is then used to calculate expected values for isentropic efficiency.

THE HRSG MODULE

In order to comply with the ASME Power Test Code PTC 4.4, Code on Gas Turbine Heat Recovery Steam Generators, all flows, temperatures and enthalpies of the gas and working fluid streams must be determined, as well as the heat losses from the HRSG surface. The test code recognizes the difficulties in obtaining reliable measurements for the gas flows and temperatures. Therefore the code suggests energy balance methods to solve for the missing quantities. Gas flow is computed from the gas turbine overall energy balance as a part of gas turbine evaluation. HRSG gas temperatures are determined from the energy balances around the respective heat transfer sections. These computations require gas composition as input data, which are also obtained from gas turbine computations.

The HRSG performance calculation software permits inclusion (or future addition) of duct burner or exhaust gas bypass systems. For such applications, the HRSG mass and energy balance calculations include duct burner combustion balances and overall energy balance of the HRSG to obtain augmenting air flow, temperature after duct burner, and bypass gas flow.

Power Test Code 4.4-1981 recommends steam flows should be calculated from water flow measurements where possible. Therefore, steam production is computed from the feedwater, blowdown, and desuperheating water flows. Computed gas temperatures and steam flows are compared to the actual measurements and the relative error between the two is displayed.

The HRSG is composed of the following heat transfer systems:

- HP Superheater
- HP Evaporator
- HP Economizer #1
- LP Evaporator
- HP Economizer #2
- LP Economizer

This design is representative of units of average complexity.

Superheated steam temperature is controlled in the desuperheater. HP steam system pressure is not controlled, and depends on the HP steam production rate. The pressure of the saturated LP steam is controlled by the steam turbine LP admission valves.

A most simple HRSG configuration could be a single pressure level unit consisting of two sections, an economizer and an evaporator. An example of a most complex configuration would be a three pressure level unit comprising eleven sections. In addition to the working fluid heat transfer sections, the HRSG may include external heat recovery devices. The purpose of such heat transfer sections is to utilize part of the remaining energy for low level evaporation or feedwater preheating. Industrial heat recovery steam generators are often equipped with bypass systems and auxiliary duct burners. The bypass damper and stack allow independent operation of the gas turbine during periods of low steam demand. The duct burner provides additional steam generating capacity.

The fundamental measures of HRSG overall performance are defined by the ASME power test code as the following:

- Efficiency, input-output method
- Efficiency, thermal-loss method
- Overall effectiveness

Additional performance measures include the following:

- Pinch point
- HP steam production
- LP steam production
- Effectiveness of each section

Input-output efficiency is obtained by dividing the thermal output by the input. The ASME test code defines thermal output as the heat absorbed by the working fluid. Thermal input is defined as the sensible heat in the exhaust gas supplied to the HRSG plus the heat of combustion of the supplemental fuel (if applicable). Thermal-loss efficiency is an indirect way of calculating the overall efficiency of the unit. Various losses are calculated as a percentage of the thermal input. The power test code defines heat losses as follows:

- Heat loss in moist exhaust gas
- Heat loss due to surface radiation and convection

Effectiveness is defined as the ratio of the actual heat transfer rate in a heat exchanger to the thermodynamically-limited maximum possible heat transfer rate which could be realized only in a countercurrent heat exchanger of infinite heat transfer area. In the HRSG effectiveness is determined separately for the individual heat transfer sections and for the entire unit. Actual effectiveness is computed directly from the gas and working fluid temperatures. Expected effectiveness is calculated from theoretical heat transfer equations calibrated during model identification mode. Maximum heat transfer rate in an individual section occurs when the gas outlet temperature reaches the working fluid inlet temperature, or when the working fluid outlet temperature reaches the gas inlet temperature. The expected effectiveness of each section is calculated from the conductance model for the section. Conductance models are derived as a function of hot gas flow through the section, and are calibrated during model identification. The calibration is based on observed heat transfer rates.

Overall effectiveness of the HRSG is defined as the ratio of the enthalpy drop of the gas to the maximum enthalpy drop of the gas which is theoretically possible. The maximum enthalpy drop occurs when the gas and water temperatures coincide at one or more points (pinch points). The definition of overall effectiveness is similar to that for individual sections. The problem with overall effectiveness is that the location of the pinch point is generally not known in advance. Maximum theoretically possible temperature and enthalpy drop is determined by comparing gas temperature drop lines drawn to all possible pinch locations. The point that allows the least drop in gas temperature is the critical point defining the actual output of the unit. This is the pinch point location that is used to determine overall effectiveness.

Expected overall effectiveness is computed as the ratio of the expected gas temperature drop to the theoretically maximum possible temperature drop. The magnitude of the pinch is defined as the minimum temperature difference between the gas and working fluid streams. The location of the pinch is known from previous calculations; consequently the magnitude can be determined.

In order to quantify HRSG performance deterioration in terms of fuel consumption, the program computes the steam production rates that could be expected from a clean unit. It is assumed that any improvement in overall effectiveness (from a clean unit) has the same relative impact at all pressure levels. Therefore, the expected steam flow at any pressure level can be calculated from the corresponding actual flow, corrected for the change in effectiveness.

Performance information developed for HRSG assessment is used to derive a thermal output model which can subsequently be used for overall optimization of the plant.

THE STEAM TURBINE DAM

The ASME Power Test Code PTC 6 - Steam Turbines, requires accurate determination of all mass and energy flows within the turbine envelope boundary. Because of the difficulties in installation and calibration of accurate steam flow measuring devices, PTC 6 recommends the use of water flows for determining primary flows to the turbine. Since steam flows were computed in the HRSG DAM from the corresponding feedwater, blowdown, and desuperheating water flows, these flows are used for steam turbine calculations.

Based on the test code PTC 6, extraction flows can be determined by heat balance calculations. NOx bleed flow is computed from the desuperheater mass balance, and the process extraction flow is a direct measurement. As the flows are already available, the desuperheater energy balance is used to verify the consistency of input data.

ASME Test Code PTC 6S, Simplified Procedures for Routine Performance Tests of Steam Turbines, recognizes that it is often impractical to measure the secondary mass and energy flows, and allows the use of manufacturer supplied curves. Manufacturer's data are therefore used to obtain generator and mechanical losses and steam leak-offs as a function of load.

Test code PTC 6S also includes recommendations for procedures to monitor changes in overall turbine cycle performance. Performance measures to be monitored include:

- Pressure/Flow relationships
- Enthalpy drop efficiency characteristics
- Heat rate at specified loads

Test code PTC 6S emphasizes the importance of corrections for operating conditions. The performance evaluation program uses curve fitting routines to model turbine performance at different steam loads. Thermodynamic methods derived from DIN-1943 Test Code for Steam Turbines are used to correct for different steam conditions. Such adjustments are used to develop expected performance values for the current operating situation. Expected values are compared to the actual values to evaluate the potential increase in steam turbine power output at the current operating conditions. In order to reveal the rate of deterioration, the values of isentropic efficiency and turbine capability are corrected to given reference conditions.

Mass and energy balances include balances of the desuperheaters, estimation of leakage steam, determination of steam to the condenser, estimation of mechanical losses in the turbine, and losses in the generator. In the desuperheater, it is necessary to know enthalpy and flow of the extraction steam before the desuperheater in order to evaluate turbine performance. The desuperheater balance calculates flow and enthalpy where process measurements are not available, or verifies the accuracy if the measurements are available. As the flow and enthalpy of condenser steam cannot be determined by direct measurements, the unit overall mass and energy balances are used to determine these quantities; the same principle is also used for backpressure turbines. Steam flow to the condenser is the sum of admission flows minus the sum of extraction flows. Enthalpy of the condenser steam is the sum of admission enthalpy minus the sum of extraction enthalpy minus generator power output minus generator losses minus mechanical losses in the turbine.

Internal performance of a steam turbine is subject to deterioration due to any of the following factors:

- Blading deposits
- Increased external leakage

- o Increased internal leakage
- o Blade erosion
- o Overload-valve leakage

The evaluation of turbine performance is based on mathematical models of the unit. The purpose of the models is to analyze the unit operations at off-design conditions. The models are calibrated automatically using on-line process measurements and automated curve fitting procedures. The program includes the following types of models:

- o Pressure drop models
- o Efficiency models
- o Mechanical capability models
- o Generator capability models

Pressure drop models are used to determine stage pressures as a function of the steam flows. In a given flow situation, higher than expected control stage pressure is indicative of blading deposits, while lower than expected values may indicate increased leakage or blade erosion. Efficiency models are used both to detect deterioration and to quantify it in terms of lost power output or increased steam consumption. Efficiency models also provide a basis for plantwide energy optimization. Mechanical capability models combine the flow and efficiency models and predict the maximum achievable mechanical output of the turbine. Generator capability models are used to determine the maximum possible generator output with the present power factor and coolant temperature.

Steam flow through a multistage turbine section can be expressed as a non-linear function of inlet and outlet pressures. In some instances some of these variables may not be known, especially with non-automatic extractions where the extraction stage pressure depends on the flow. In order to analyze turbine sections with missing inlet or outlet pressure measurements, adjacent sections are lumped together to form models of larger sections with known inlet and outlet conditions. Two such models are required for this application; a model for the HP part and a model for the LP part of the turbine. During model identification mode, the models are saved to be used as reference models in subsequent calculations. The reference models can be used to determine the flow _vs_ pressure characteristics that could be expected from a clean, healthy unit. These procedures are used only where actual measurements are not available. The use of actual pressure measurements is a recommended practice because of better accuracy.

The most important energy losses related to turbine operation are the following:

- o Mechanical losses
- o Generator losses
- o Condenser energy losses
- o Gland steam losses

Knowledge of external losses at various loads is not sufficient to describe turbine performance characteristics. It is equally important to know how the internal losses change the amount of energy the steam has available for conversion into work at different loads and operating conditions. All internal and external losses can be described with a constant loss term and a term proportional to steam flow.

Variations in inlet pressure and temperature and outlet pressure change the amount of energy that a turbine has available for energy conversion. The maximum available kinetic energy equals the isentropic enthalpy drop, and therefore the changes in power output are proportional to the changes in isentropic enthalpy drop. Because of the volumetric flow changes, the internal efficiency varies slightly with different isentropic enthalpy drops.

From the models, turbine heat rate and isentropic overall efficiency can be calculated for any given operating situation.

To assess a wide variety of industrial turbines, a generic extraction/ admission turbine model for mechanical capability was developed. The generic model describes units with up to three "parts" separated by admission valves. Each part (HP, IP, or LP part) includes up to two intermediate admission/ extraction points. A "section" is the portion of a "part" between a flow inlet and a flow outlet. Mechanical capability is defined here as the maximum power output that can be obtained with given extraction rates, assuming unlimited supply of admission steam. This is a constrained optimization problem which can be stated in the following simplified form:

- o Find the combination of admission flows that provides the maximum steam flow to the condenser, with given extraction flows, without exceeding the pressure _vs_ flow limitations of the turbine.

The combination of admission and throttle flows that produce the maximum mechanical output is determined. Using these flows and the given extraction flows, the actual mechanical output can be computed from the overall input/ output model. In order to determine the mechanical capability that could be expected from a clean, healthy unit, the same procedure is repeated, using reference model coefficients with the flow model and input/ output model.

The generator capability is the maximum MW output at the current operating conditions. The steam turbine DAM uses the same generator capability model as is used for the gas turbine.

Models for overall plant optimization are derived from models and data calculated during the turbine performance assessment analysis. The information required is obtained from the input/ output model, from mass balance computations, and physical constraint computations.

In order to quantify the deterioration of various power plant components, it is important to know the overall effect of any deviations in performance. To simplify the task, the respective component DAM's determine the sensitivity of the key variables to variations in external conditions. Sensitivity of power output is determined for variations in the following quantities:

- Condenser pressure
- HP throttle flow
- Interstage admission flows

Sensitivity of power output to condenser pressure variations is necessary to quantify condenser or cooling tower deterioration in terms of lost power generation. Sensitivity of power output to HP throttle flow is needed to quantify HRSG deterioration. As the extraction flows are fixed, any deviations in HRSG steam output can be seen in condenser steam flow.

THE CONDENSER MODULE

Assessment of condenser performance is based on theoretical heat transfer calculations. The ASME Test Code on Steam Condensing Apparatus, PTC 12.2-1983, recognizes the difficulties in obtaining values for the cooling water flow and steam enthalpy. The code suggests energy balance methods to solve for the missing quantities. Steam flow and steam enthalpy are computed in the steam turbine DAM, and are used in conjunction with measured variables to determine cooling water flow in the condenser energy balance.

The primary measures of condenser performance defined by Power Test Code PTC 12.2-1983 are the following:

- Condenser pressure
- Overall thermal transmittance
- Condensate subcooling

Other redundant performance measures include:

- Cleanliness factor
- Fouling resistance

Condenser pressure is the most important performance measure as it is closely related to the useful expansion of steam in the turbine. Measured condenser vacuum is compared to expected vacuum to evaluate the potential increase in steam expansion and power generation. Expected condenser vacuum is calculated from theoretical heat transfer equations. Overall thermal transmittance is calculated directly from current operating parameters and is compared to the expected value obtained from heat transfer models.

Cleanliness factor is the ratio of thermal transmittance of used tubes to transmittance of new tubes at the same operating conditions. Cleanliness factor is computed both for actual and for expected performance using the actual, expected, and theoretical values of the overall thermal transmittance. Fouling resistance (resistance attributed to fouling) is the difference between actual and theoretical (theoretical implies new, clean tubes) overall resistances. Overall resistance is the reciprocal of overall transmittance. Condensate subcooling is the difference between the saturation temperature corresponding to the condenser pressure and the temperature of the condensate in the hotwell. It is important to minimize condensate subcooling, therefore the expected value for subcooling is zero. Any deviation from this is an indication of performance deterioration.

The condenser performance models consist of theoretical heat transfer equations. Unknown parameters in the equations are identified from process measurements, using automated curve-fitting procedures. The condenser reference model is developed from the data obtained during test runs after condenser cleaning. This model is used to estimate expected performance values for the current operating situation. Another model is derived from current operating data to predict the performance under reference conditions.

Overall thermal transmittance is the most important performance criterion as all other measures are derived from it. Overall transmittance is computed directly from the process variables. Expected transmittance is obtained by correcting the reference transmittance from the reference conditions to the actual conditions. Cleanliness factors and fouling resistances are derived from actual and expected overall transmittances. Condenser pressure is calculated from mass and energy balances and heat transfer equations.

Sensitivity of condenser saturation temperature to a change in cooling water temperature is essentially unity. This result is useful in quantifying cooling tower performance deterioration. Sensitivity of condenser pressure to steam flow variation is useful for modeling condensing turbines; this factor is determined.

THE COOLING TOWER MODULE

In order to correctly assess cooling tower performance, air and water flows and temperatures must be known. Cooling tower air flows are seldom measured, and circulating water flow usually is not measured. Cold water temperature at the cooling tower outlet cannot be measured directly, and water basin temperature must be used instead, which can introduce error due to make-up water input.

Since on-line measurement of cooling tower air flow is not available, the performance assessment program prepares air flow estimates by correcting the design air flows for the number of cells in operation and for the actual *vs* design fan speed. Circulating water flow is obtained from the condenser performance calculations. The ASME Power Test Code PTC-23 recognizes the difficulties in obtaining the actual cold water temperature, and requires appropriate corrections. The program determines such corrections from the dynamic energy balance of the cooling tower water basin.

Cooling tower performance calculations are performed according to the CTI Test Code, "Acceptance Test Procedure for Industrial Cooling Towers," Bulletin ATP-105, and comply with the ASME Test Code PTC 32, Test Code for Atmospheric Water Cooling Equipment. The fundamental measure of thermal performance of the cooling tower is:

- o Approach

Other redundant performance measures include:

- o Cold water temperature
- o Range

Approach is the difference between the cold water temperature and the ambient wet-bulb temperature. The actual approach is obtained from the measured process variables, while the expected value is computed using theoretical heat transfer models. Cold water temperature is the most important performance measure as it is closely related to the condenser pressure and turbine power generation. The expected cold water temperature is calculated from the same heat transfer equations as the expected approach. Range is the difference between hot and cold water temperatures. As the circulating water flow is frequently constant, range is a direct measure of the cooling duty. In order to reveal the rate of deterioration, the value of approach is corrected to given reference conditions.

The cooling tower performance evaluation is based on mathematical models of the unit. The purpose of the models is to obtain values for unmeasured variables and to analyze the unit operations at off-design conditions. The system includes the following models:

- o Fan models
- o Dynamic energy balance of water basin
- o Cooling tower heat transfer models

The fan model is used to estimate the air flow as a function of fan speed (because air flow is not measured), and to estimate power consumption. A dynamic energy balance is computed in order to estimate the cold water temperature at the cooling tower outlet.

The heat transfer models are used to analyze the cooling tower performance at off-design conditions. The models consist of theoretical heat transfer equations. The unknown parameters in the equations are identified from process measurements using curve fitting procedures. The reference model is developed from data obtained during initial operation of the unit or from test runs after maintenance. This model is used to estimate expected performance values for the current operating situation. Another model is determined from current operating data to predict performance under reference conditions. The heat transfer models are built upon the notion that cooling tower performance can be characterized in terms of number of transfer units. During model identification, the number of transfer units is determined at the reference conditions. As air and water flows and ambient conditions change, the heat transfer characteristics of the cooling tower change. These variations are accounted for by correcting the reference transfer units into the actual operating conditions. The corrected value becomes the expected number of transfer units available. The expected cold water temperature is determined so that the number of transfer units required equals the number of transfer units available. The expected range and approach are then obtained directly.

One of the most important considerations in cooling tower operation is the optimum number of cells to use. Another similar factor is the optimum fan speed, applicable to units equipped with variable speed drives. When increased power generation in the condensing turbines (due to lower condenser temperature and pressure) exceeds the additional fan power consumption, a new cell should be activated or fan speed should be increased, or conversely, one of the active cells should be shut down or fan speed decreased. The cooling tower performance computations include calculations for the requisite information needed for cooling tower optimization.

THE PLANT PERFORMANCE ASSESSMENT MODULE

As discussed previously, the plant performance assessment module (PAP) is customized for each application, that is, displays may be customized, and data addresses for preparation of the displays may be unique for an application. However, the following functionalities are always included in the module:

- o Provide overall performance analyses. Derived from the results calculated by the respective DAM's, the PAP calculates overall performance analyses for the entire plant.
- o Quantify performance deterioration. The PAP analyzes the deteriorated performance of each device and then quantifies the losses in terms of lost power output or increased fuel consumption in both engineering and economic units.
- o Outputs all computed data to the display device. This includes the computed results from all the DAM's plus the results from the PAP.

The PAP combines the calculated results from the respective devices to determine the overall performance for the entire plant. Cogeneration power is defined as the power generation that can be truly credited to process steam use. Supplemental power is defined as the power generation resulting from the use of condensers. Plant heat rate is the ratio of total fuel consumed divided by total power output.

The PAP analyzes the performance of each device to quantify performance degradation in terms of megawatts and Btu's. These losses are also accumulated for each device since the time of its last maintenance and model identification, and are displayed on the CRT to the operator. The economic summary displays the cost of performance degradation. In order to quantify performance deterioration, the PAP first determines the operating mode of the plant. In maximum power output mode, the objective is to produce maximum possible power. In load following mode, the objective is to produce the specified output, either electrical or thermal, which is less than maximum capability. In maximum output mode, performance degradation is quantified as the value of lost power; in load following mode, performance degradation is quantified as the cost of increased fuel consumption.

The Bailey COgeneration Performance Assessment (COPA) package provides new capability for examining and quantifying the performance of cogeneration plants. The package also provides an integrated route for optimization of the plant. Accurate determination of component performance is a valuable aid to operating engineers, maintenance engineers, and plant management for ensuring reliable and efficient plant performance. Some of the benefits of COPA may be summarized as follows:

- Early detection of mechanical problems, preventing development of more serious damage.
- Optimal maintenance scheduling.
- Prevention of unscheduled device shutdowns, eliminating additional energy costs or revenue losses.
- Accurate and up-to-date analysis information available for overall optimization of the cogeneration plant.
- Accurate and up-to-date cost information of performance degradation available for evaluating the performance of each device.

Chapter 62

THE PERFORMANCE AND ECONOMIC FEASIBILITY OF PACKAGED COGENERATION SYSTEMS

P. A. McBride, B. D. Wood, R. L. Sears

Most of today's cogeneration installations are in large industrial facilities. These systems were custom designed for particular sites because of the high degree of variability in energy loads and utility interconnection requirements. Recently there has been a great deal of interest in smaller systems (particularly those with less then 1500 kW electrical output) that can be mass produced for standard applications. These so-called packaged units usually consist of a reciprocating or gas turbine engine (the prime mover), a heat recovery system, an electric generator, a control system, and electrical and thermal interface equipment. It is hoped that the mass production economies can be as significant as the economies of scale associated with the larger industrial systems. The high costs (both in money and in time) of project engineering can be greatly reduced.

The applications for small systems are largely in the commercial and institutional markets: hotels, health clubs, hospitals, restaurants, nursing homes, industrial facilities, shopping centers, university complexes, office buildings, and apartment buildings. Many of these facilities are operated in chains, each building having similar and fairly predictable energy demand patterns. This simplifies the process of designing standardized systems.

Potential customers of small systems often are not aware of the benefits of cogeneration. Energy is typically viewed as having a fixed overhead cost and there is an unwillingness to take a more active role in the energy production process. Clearly there is a need for a fast, simple, and inexpensive method to estimate the economic feasibility of a small packaged system. The information from such a method could provide the incentive to proceed with a more detailed analysis.

Software has now been developed so that managers of commercial and institutional facilities can determine the economic feasibility of a packaged cogeneration system sized for their particular energy requirements. In addition, it provides a way to test the sensitivity of the economic performance to escalating fuel and electricity costs.

A variety of technical and economic information about currently available cogeneration systems was gathered to provide data for the software. This information by itself is a new and valuable resource to engineers responsible for choosing an particular packaged unit.

SYSTEM PERFORMANCE MODELS

Accurate information on the cost and performance of cogeneration systems is essential for any level of economic analysis. Calculation of the economic savings associated with using a cogeneration system requires knowledge of the relative amounts of electricity and thermal energy which can be produced. The cost of operating a system is directly related to its fuel consumption. A detailed economic analysis requires that the specific performance characteristics of a particular cogeneration system be used. The process of choosing one system over another for this purpose can be difficult and time consuming.

Cost and performance information about packaged cogeneration systems that use natural gas reciprocating engines has been collected. By itself, this data can be used to assist in the process of choosing one particular packaged system over others. The data has also been analyzed to determine the typical performance levels to be expected from systems within the 6 to 1500 kW size range. During the period of February through April, 1987, cogeneration system manufacturers were contacted and requested to provide the information shown in Table 1 for each system that they sell.

A substantial effort was made to contact every current manufacturer of packaged cogeneration systems. Eleven companies sent data sufficiently detailed to be included in the survey, which contains data from 55 hot water units and 9 steam units. Eight manufacturers either did not respond, did not have readily available information, or did not sell systems of the type considered in this survey.

Information about system costs deserves special note. Cogeneration systems are built from many components. Manufacturers often assume that because of the large variety of applications, some components will not be needed on all of the systems they sell; hence they are not included in the regular purchase price. Examples of such "optional" components include weatherproof enclosures, special metering, custom designed control systems, batteries, starting motors, and special safety and protection equipment. There is little uniformity among manufacturers concerning standard equipment and quoted prices reflect this fact. Table 1 includes two items that can make a significant difference in total system cost: the type of generator included and whether provisions are included for disposal of waste heat

Table 2 is one example of the twenty lists developed from the system data. Each is arranged according to one type of performance or cost parameter. For those instances where the inclusion of multiple systems has an effect on the listing order, separate listings were compiled. For the purpose of the correlations developed here, installations with up to four identical units are considered.

The assessment of the economic feasibility of packaged cogeneration systems requires performance and cost information. For this purpose it was necessary to establish relationships between fuel input, thermal output, electric output, and system cost. The first step in this process involved viewing the data graphically to establish the form of the functional relationships between the variables (Figs. 1 - 9). Because of the large concentration of systems near the lower end of the range, Figs. 2, 4, and 6 were plotted for systems 200 kW and below. All plots have electric output on the abscissa since this is the most common reference for system size. Figures 1 through 4 are observed to be nearly linear, with more scatter visible in Figs. 3 and 4. One would expect that variations in heat recovery sources and equipment could lead to this scatter. In contrast, since the fuel consumption of engines and the efficiency of generators are relatively standard, this would explain the low scatter in Figs. 1 and 2. The linear appearance of the relationship has been enhanced by the consideration of multiple units (e.g., the fuel consumption of two units is essentially equal to twice the fuel consumption of one unit). The cost graphs, Figs. 5, 6, and 9, show a great deal of scatter but a somewhat linear relationship, particularly for the smaller hot water systems. The scatter results from the previously mentioned problems associated with quoted system cost figures.

The method of least-squares was used to fit straight lines to the data. Table 3 shows the resulting least-squares equations and the lines are shown on Figs. 1 - 9. A separate equation was found for the hot water system cost data for systems of 30 kW or less.

The maintenance cost, on average, was 1.36 cents per kWh. The average design inlet and outlet temperatures were found to be 169 and 194^0 F, respectively, which results in an average temperature difference of 25^0 F.

OPERATING STRAGEGIES

Any feasibility analysis must have an assumption pertaining to the way in which the cogeneration system will be operated. Options include thermal or electric load following, base load operation or peak following, and the sale of electricity to the electric grid. The particular options chosen can make a substantial difference in the economic feasibility of a particular system.

Every facility that is a potential application for a cogeneration system has unique energy load profiles. In general, both the thermal and electrical loads vary throughout the day and season. Usually there is a pattern to both loads that is fairly predictable; these patterns are important criteria for the selection of an operating mode. Ideally there should be a match between the electrical and thermal loads, i.e., as one load increases, so does the other. This will insure the utilization of all the energy output from the cogeneration system and therefore result in the most efficient operation.

The cogeneration system can be sized to follow either the thermal or electric load (called, respectively, thermal or electric tracking), following either the base load or the peak load. Thermal tracking has two principal advantages: 1) auxiliary thermal sources and heat disposal systems are not needed, and 2) system operation at peak efficiency is insured since all of the thermal energy is used. This assumes, of course, that all of the generated electricity can be utilized, but this is usually the case. If not, the possibility exists for selling the excess power to the electric utility. Electric tracking would not involve any power sales to the utility, but any excess heat produced would have to be stored or rejected. Isolated cogeneration systems, i.e., those with no connection to the utility grid, must operate in this mode.

Packaged cogeneration systems based on reciprocating engines produce more thermal energy then electrical. This does not match well with many commercial applications that have a greater need for electricity. It has been found that in many circumstances the most economical operation mode involves tracking the thermal base load (GRI, 1985 and OTA, 1983). Thermal peaking can be provided by conventional heating units. Normally only part of the electric power requirement would be met with the cogeneration system-- the remaining electric power would be provided by the utility. Running the reciprocating engine at a constant level will insure maximum fuel efficiency.

ECONOMIC MODEL

The simplest method for assessing the economic attractiveness of a cogeneration system is the simple payback method. This method yields a number which represents the number of years required for the invested capital to be offset by the resulting benefits. It is calculated by dividing the incremental initial cost by the net annual cash inflow. The incremental initial cost refers to the installed cost of the cogeneration system, subtracting the cost of similar equipment designed to achieve the same results without cogeneration, if there was already a need to purchase such a unit. The annual cash inflow is the energy cost savings minus operating and maintenance costs.

IMPLEMENTATION OF THE ECONOMIC MODEL IN A COMPUTER PROGRAM

A computer program has been developed based on the foregoing information (and additional assumptions presented in this section). Information about fuel and electricity costs, operating mode, and type of thermal output required constitute the data input to the program. The program will calculate the typical cost, simple payback period, and several other economic parameters. Thermodynamic and cost relationships are taken from those presented in Table 3 and hence reflect actual system performance.

In order to calculate the simple payback period, the energy cost savings is required. Referring to Fig. 10, two types of energy savings are apparent: 1) the cost savings associated with not purchasing electric energy from the utility in the amount generated annually by the cogeneration system (purchased power savings), and 2) the savings associated with not purchasing the fuel required for thermal energy production in the amount produced by the cogeneration system.

The net annual cash inflow is the sum of the net savings figures minus operating and maintenance costs. The operating cost is the annual cost for the fuel required by the cogeneration system. This calculation involves the use of the fuel consumption relationships developed previously. The maintenance cost is assumed to be 1.36 cents per kWh.

System cost figures were adjusted for installation by adding 30% to the packaged system cost figures. Labor and material costs for installation are very site specific and can add anywhere from 20 to 50% to the system cost (A.D. Little, 1985). The relationships previously developed were used for the cost estimates. Because of the scatter noted in the determination of the cost relationships, it was decided to have the program provide three cost figures: maximum, minimum, and typical.

Several assumptions inherent in the program should be emphasized. All the thermodynamic and cost relationships represent average values, i.e., the performance indicated is representative of current packaged cogeneration systems but is not necessarily achievable in particular circumstances. It is assumed that cogeneration systems of exactly the requested size are available; in reality systems come in discrete sizes. With the expanding packaged cogeneration system market, it should be possible to match closely the desired size with an actual unit. The program will not consider systems outside of the 6 to 1500 kW range. As discussed in the previous section, constant base load operation is assumed with no electricity sales to the utility.

PROGRAM USE

The program is written in Microsoft BASIC for the IBM personal computer. Since the program is not iterative, the logic is quite simple and amenable to modification. Wherever possible, multiplying factors are not combined so that the underlying components are clearly visible in the program listing. The program is well documented. English units are used, reflecting the current practice.

The program is user-friendly and can be used by someone with limited computer experience. Questions that require a non-numeric response have acceptable responses shown in parenthesis (upper case letters must be used). Where a numeric response is called for, the proper units are shown in parenthesis. The program asks for a usage factor, which is the estimated fraction of the year that the cogeneration system will be in operation. The calculated data is displayed on the screen immediately after everything is entered, with descriptive labels and units shown.

The program can be quickly run again for different input data. Sensitivity to energy cost escalation can be assessed by varying each cost figure separately over an estimated range. The effect on the simple payback period can then be noted. The effect of the operating mode (thermal or electric tracking) can also be determined.

As an example, consider the following scenario: a small

industrial facility wishes to establish the economic feasibility of installing a packaged cogeneration system. The facility has a 500 kW base electric load. A natural gas fired hot water heater presently provides 3 million Btu per hour of hot water. Electricity rates are presently 7.3 cents per kWh and natural gas costs $4.65 per million Btu. It is estimated that the cogeneration system could operate 50 weeks each year, yielding a usage factor of 0.96. Table 4 shows the program output. The thermal output of 2.9 million Btu per hour is quite close to the requirement of 3 million Btu per hour (it is possible that an actual packaged system might be offered with the required output). With a net savings of about $156,000 per year, the simple payback period is estimated to be between 1.7 and 5.3 years, with 3.4 years being typical. The program was also run based on the same scenario but with slight variations. If thermal tracking is used, the output is only slightly different since the ratio of the electric and thermal loads are compatible with current packaged systems. If the cost of electricity goes up to 12 cents per kWh, the typical simple payback period drops to 1.5 years, reflecting the additional value of the cogenerated electricity. If the price of natural gas goes up to $8 per million Btu, the payback period climbs to 6 years, which is probably unacceptable to most small industries.

Complete information on the topics presented herein, including a copy of the computer program, is available from the Center for Energy Systems Research for a nominal fee.

REFERENCES

A. D. Little, Inc., 1985, "System and Technology Concept Evaluation for Small Commercial/Residential Cogeneration Applications," draft of final report to Gas Research Institute, Chicago, IL, pp. 4-6, 5-18 and 6-4.

Gas Research Institute (GRI), 1985, "The Quest for Smaller Units," ASHRAE Journal, Vol. 7, No. 7, pp. 18-23.

Office of Technology Assessment (OTA), 1983, Industrial and Commercial Cogeneration, Congress of the United States, Washington, DC, Chapter 2, p. 35.

TABLE 1 INFORMATION REQUESTED FROM SYSTEM MANUFACTURERS.

model number	-
fuel input	Btu/hr
electric output	kW
thermal output	Btu/hr
type of thermal output	-
generator type	-
provisions included for waste heat disposal	-
cost	1987 U.S. $
estimated maintenance cost	cents/kWh
design inlet and outlet temperatures	^{0}F
final exhaust temperature	^{0}F
configuration of steam systems	-

Table 2 Arranged listing example -- hot water systems arranged by thermal output.

model	HW out MB tu/hr (M=1000)	model	HW out MB tu/hr (M=1000)
VW1	63	WA6	986
VW2	97	JB1	1,004
VW3	106	MN4	1,099
VW4	118	VW11	1,156
VW5	138	TC2	1,182
MI1	170	JB2	1,339
TE1	187	MN5	1,345
WA1	193	WA7	1,430
VW6	198	VW12	1,455
EM1	256	EM4	1,464
VW7	274	MR1	1,636
MN1	283	MN6	1,638
WA2	303	MR2	1,701
VW8	311	MR4	1,928
WA3	360	CO3	2,000
EM2	393	JB3	2,033
TC1	440	MR3	2,039
WA4	443	WA8	2,112
EN1	540	MR5	2,580
MN2	580	MR6	2,667
VW9	624	MR7	2,690
CO1	630	JB4	2,711
EM3	635	MR8	2,900
MN3	666	CO4	3,000
WA5	709	MR9	3,430
CO2	830	MR10	3,883
EM4	922	CO5	4,000
VW10	929		

TABLE 3 LEAST SQUARES EQUATIONS.

Hot Water Systems:

electric output (x, in kW) vs. fuel input (y, Btu/hr)
$y = 11500\,x + 52400$

electric output (x, in kW) vs. thermal output (y, Btu/hr)
$y = 5610\,x + 99500$

electric output (x, in kW) vs. cost (y, in $)
$y = 801\,x + 7990$

electric output (x, in kW) vs. cost (y, in $) for 30 kW or less
$y = 1100\,x + 23.9$

Steam Systems:

electric output (x, in kW) vs. fuel input (y, Btu/hr)
$y = 11900\,x - 47200$

electric output (x, in kW) vs. thermal output (y, Btu/hr)
$y = 5190\,x - 34000$

electric output (x, in kW) vs. cost (y, in $)
$y = 888\,x + 39700$

TABLE 4 SAMPLE PROGRAM OUTPUT.

```
PRELIMINARY FEASIBILITY ANALYSIS FOR 500 KW COGENERATION SYSTEM

THERMAL ENERGY FORM: HOT WATER
THERMAL OUTPUT                          2,903,907. BTU/HOUR
FUEL INPUT                              5,807,729. BTU/HOUR
ENERGY COSTS:
    ELECTRICITY COST                        7.3 CENTS/KWH
    PRESENT FUEL COST                       $4.65/MILLION BTU
    COGENERATION FUEL COST                  $4.65/MILLION BTU
USAGE FACTOR                                0.96
TYPICAL SYSTEM COST, UNINSTALLED        $408,585.
TYPICAL SYSTEM COST, INSTALLED          $531,160.
HIGH COST, INSTALLED                    $821,795.
LOW COST, INSTALLED                     $265,070.
OPERATION AND MAINTENANCE COST           $57,185./YEAR
PURCHASED POWER SAVINGS                 $306,950./YEAR
ANNUAL COGENERATION FUEL COST           $277,109./YEAR
SAVINGS FROM PRESENT BOILERS            $133,596./YEAR
NET SAVINGS                             $156,252./YEAR
TYPICAL SIMPLE PAYBACK                      3.40 YEARS
HIGH SIMPLE PAYBACK                         5.26 YEARS
LOW SIMPLE PAYBACK                          1.70 YEARS

DO YOU WANT TO RUN THE PROGRAM AGAIN? (Y OR N):
```

PERFORMANCE AND COST RELATIONSHIPS
FOR HOT WATER SYSTEMS

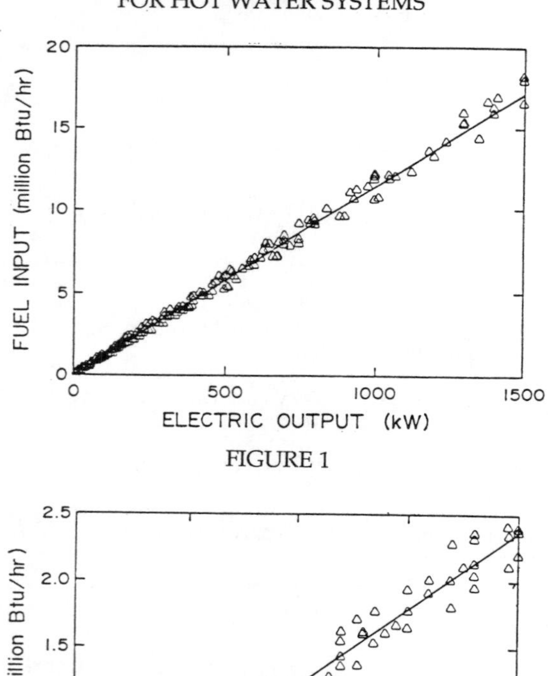

FIGURE 1

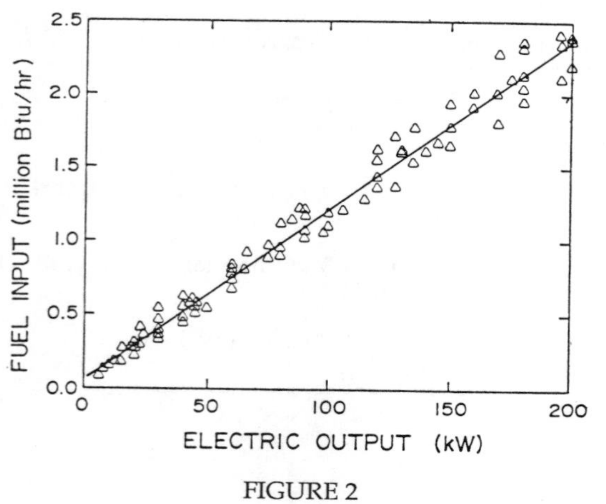

FIGURE 2

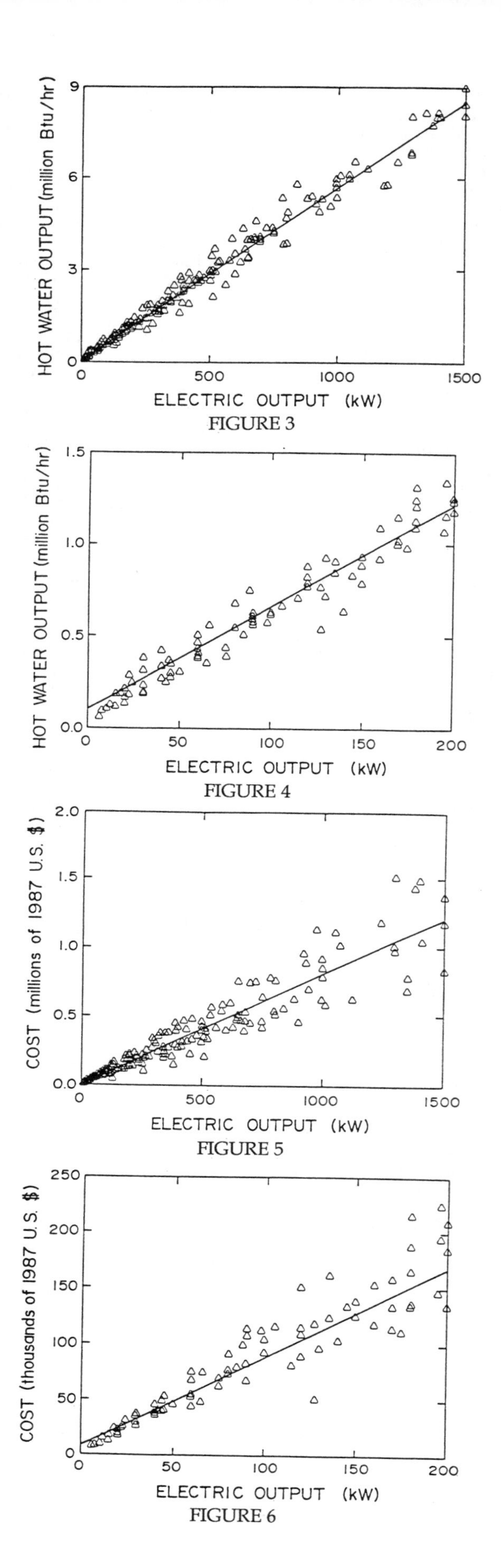

PERFORMANCE AND COST RELATIONSHIPS FOR STEAM SYSTEMS

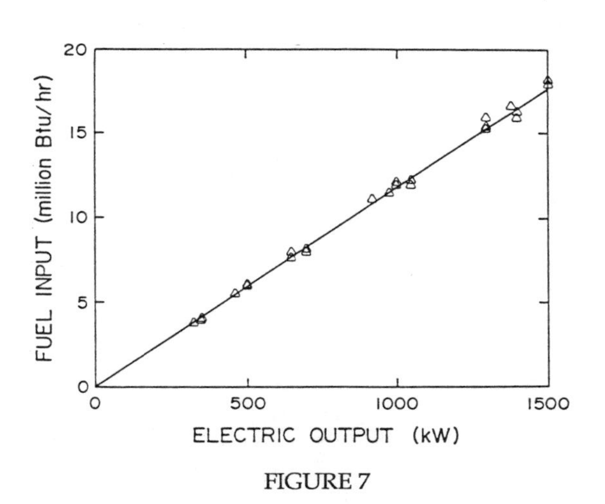

FIGURE 7

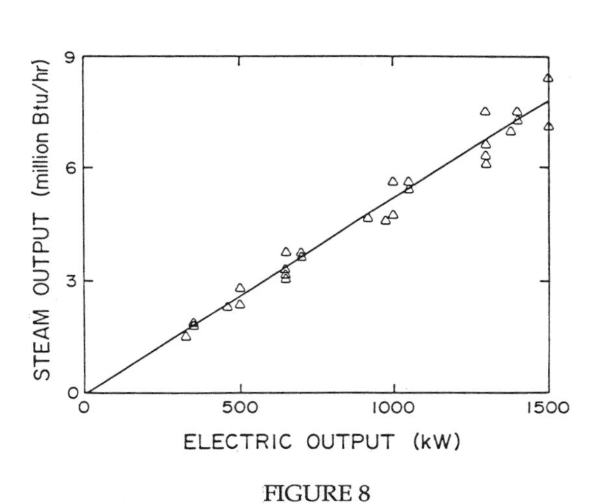

FIGURE 8

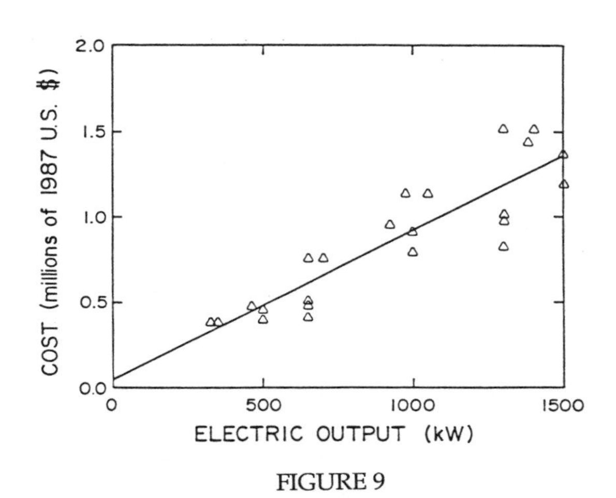

FIGURE 9

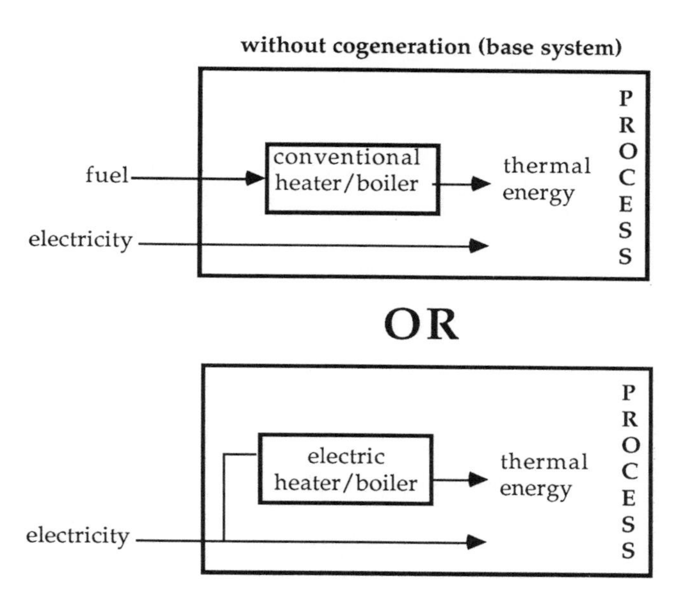

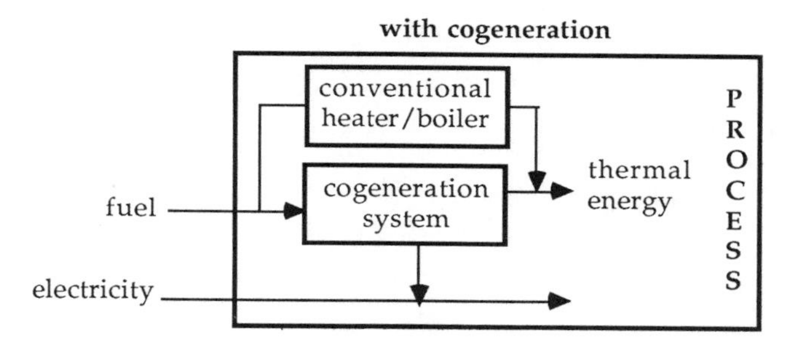

FIG. 10 SYSTEM MODELS

Chapter 63

SMALL COGENERATION THERMAL DELIVERY

R. W. Persons

INTRODUCTION

Reciprocating engines in the 15 to 150 HP range are being packaged with generators and chillers by several manufacturers. The packaging simplifies the design of small cogeneration systems by pre-engineering solutions to the integration of prime mover, generator, input and output components, and controls. The installing engineer's job is reduced to designing for five site-dependent input and output subsystems:

Fuel input
Electrical or chilled water output
Useful Heat output
Engine exhaust output
Generator waste heat output

For the ideal site, all of these subsystems can be very simple. Usually, though, the heat output subsystem is relatively complex. It may include several heat exchangers, pumps, thermostats, and control valves together with their associated controls.

Proper design of a heat output subsystem begins with an understanding of the idiosyncracies of the engine which distinguish it from a hydronic heater ("boiler"). These idiosyncracies include the warmup period, the limited modulating ability in the high-efficiency package design, the associated harmfulness of too many short run cycles, the cool-down period, and the pressure-temperature-flowrate input window. They are discussed in Part I.

The application of the engine's heat output to typical thermal loads is discussed in Part II. The host sites for packaged equipment are typically commercial or institutional buildings with two or three thermal loads which must be addressed to make cogeneration economically favorable. Matching engine output with multiple loads requires an understanding of the operation of building systems as well as the engine. A dump or "heat balance" radiator can be used to sidestep this understanding, usually with a premium in first cost and a reduced rate of return. However, a control strategy is presented which has been used to design many successful systems which operate without radiators.

PART I
THE PACKAGED ENGINE AS A HEAT SOURCE

An engine is different from other hydronic heaters in several fundamental ways:

> When started cold, the engine cannot deliver any heat to load(s) until it has reached its design operating temperature, and it must be maintained at that temperature throughout the run cycle.

> The engine must deliver <u>all</u> of its heat output (at a given shaft output) to thermal load(s).

> The engine should be operated as continuously as possible, and should never be run through repetitive "on" cycles of less than 20 minutes.

> At the end of each operating cycle, the engine must be allowed to idle while continuing to transfer heat to a load as it cools down.

> The pressure, temperature, and flowrate specifications for the selected package must be accommodated by the design.

These idiosyncracies derive mostly from the need to keep the engine temperature between the minimum and maximum design points in order to prolong engine life. Attention paid to them during system design translates into financial success for the project.

Warmup

With respect to warmup, the difference between an engine and a boiler lies in the location of the temperature control point and in their relative internal thermal storage. A boiler is controlled on and off to maintain a fixed temperature within itself. An engine is controlled on and off to maintain a fixed temperature somewhere outside itself - in a tank or a swimming pool, for instance. A boiler has a much greater internal thermal storage than an engine. Unlike a boiler, an engine which spends part of a day off cools nearly to ambient temperature in a few hours.

Prompt engine warm-up is important because lubrication and combustion are meant to work within a specific temperature range. If an engine runs too cool, then the metal parts (piston rings and cylinder walls, for instance) do not slide against each other as easily as they should. The results are excessive wear on all of the engine's bearing surfaces, shortened engine life, and reduced efficiency.

Proper engine temperature can be obtained quickly at startup and maintained during operation through the use of a simple, automotive-type, self-contained, three-way thermostatic mixing valve, as illustrated in Figure 1. Figure 1a illustrates the water circuit in a popular packaged cogeneration unit. Figures 1b and 1c show operation with and without a mixing valve.

Without a mix valve, the water returning to the engine from the load may not reach a temperature suitable for the engine until the load (hot water tank, building space, pool, etc.) has reached design temperature. Instead, the mix valve recycles some or all of the supply water back to the return.

The mix valve must sense the return temperature, which is why it is used in a mixing rather than a diverting configuration - as a diverter valve, it would be placed on the supply line and would require a remote bulb to sense return temperature. The pump may be

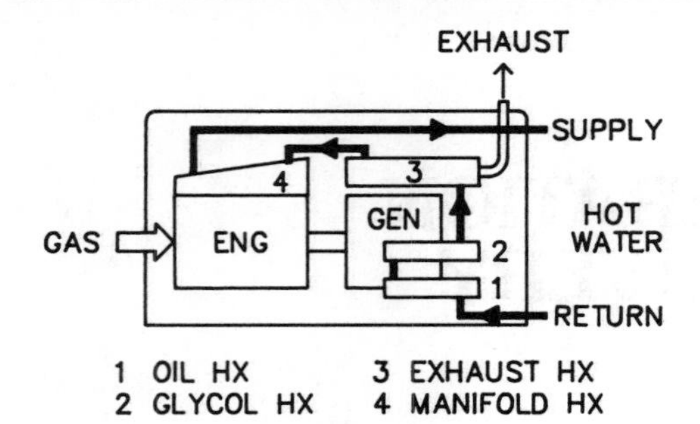

FIGURE 1a — COGENERATOR WATER CIRCUIT

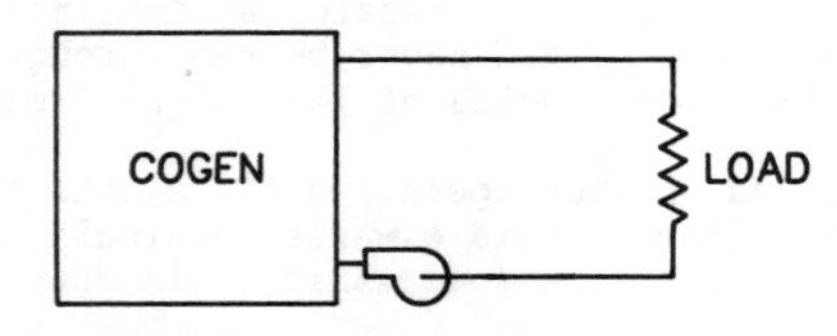

FIGURE 1b — WITHOUT MIX VALVE

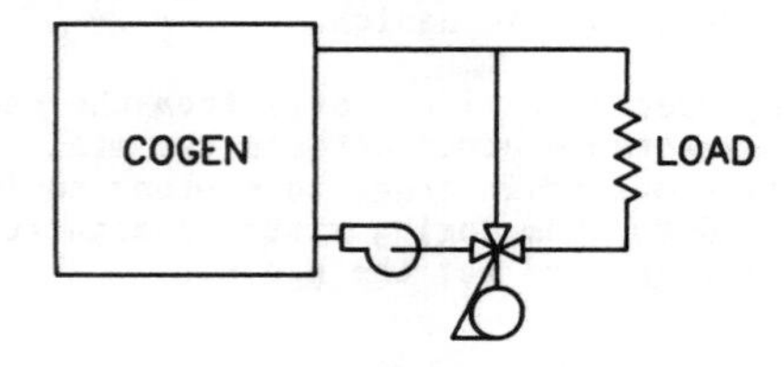

FIGURE 1c — WITH MIX VALVE

placed at either the inlet or the outlet of the cogenerator - the cogenerator inlet and outlet pressures will be about the same in either case. It is usually placed at the inlet, since the lower temperature there allows the pump seal to last longer.

In addition to causing prompt engine warmup, the mixing valve intervenes if the return temperature from the load falls below the setpoint of the valve. When this happens, the flowrate through the load is reduced. The flowrate through the cogenerator remains constant. The difference between them is the flowrate through the bypass (side) port of the mixing valve. The full heat output of the cogenerator is still being delivered to the load - heat delivery to the load takes place at a lower flowrate, but a higher temperature difference, than if there were no bypass.

Heat Balance

The difference in thermal storage between an engine and a boiler account for another idiosyncracy of the engine. It's very simple: Any heat which is produced within an engine and not removed from it results in a rise in the temperature of the engine. If the temperature of the engine gets too high, then undesirable things happen: steam may be produced in passages intended to hold only water, resulting in reduced heat transfer rate, further temperature rise and elevated pressures; lubrication breakdown; etc.

The need, then, is to remove heat from the engine at the rate at which it is produced at specified operating conditions. That is done by providing for suitable temperature and flowrate conditions in the Thermal Delivery System (TDS). Part of a typical TDS is shown in a simplified schematic in Figure 2. The system includes, in addition to the cogenerator, a heat exchanger, a load, and piping loops to carry heat from

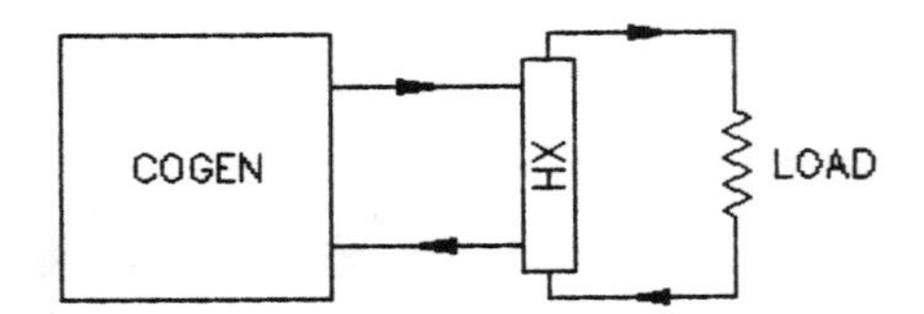

FIGURE 2 — TYPICAL THERMAL DELIVERY SYSTEM

the cogenerator to the heat exchanger and from the heat exchanger to the load. The load which is represented conceptually in Figure 2 could consist of a tank, a pool, a hydronic space heat system, a radiator, etc.

Thermal specifications for the system of Figure 2 might be as shown in Table 1. The engine-side specifications come from the cogeneration unit manufacturer's recommendations for supply and return temperatures and water flowrate. These specifications are for a unit having a thermal output of 440,000 Btu/hr.

TABLE 1
THERMAL SPECIFICATIONS: SYSTEM OF FIGURE 2

	HX Engine Side	HX Load Side
Flowrate, gpm	20	44
Temperature In, F	210	140
Temperature Out, F	166	160
delta T, F	44	20
gpm x delta T	880	880
Heat In or Out, MBh	440	440

Note that the load side performance of the heat exchanger has the same product of flowrate by temperature change as the engine side, 880 gal-F/min. If the load-side temperatures shown have been chosen suitable to operate at these conditions, and the specified flowrates are maintained, then the system will operate as intended. A good cogeneration system cannot be designed without recognition of the importance of designing for heat balance.

Occasionally, it occurs to someone that the concern over heat balance could be eliminated either by designing the cogeneration unit to modulate its output or by using a "heat balance" radiator to reject any heat not taken by the load. Either of these measures conflicts with the interest of the system investor. The disadvantages of using a radiator are discussed in Part II.

Modulation is undesirable in small cogeneration units because it would add complexity and cost while decreasing the value of the owner's investment. The units are generally designed for high efficiency at full output. As a result, their exhaust temperatures are quit low, and precautions must be taken to avoid problems with exhaust condensation. Any decrease in the cogenerator output without control of exhaust temperature would cause unacceptable problems in the reliability and durability of the unit. Control of exhaust temperature would have to be adjustment of heat transfer area or water-side flowrate, neither of which can be done simply and reliably.

Duty Cycle

Boilers have very few moving parts: one or two gas valves or an oil pump; perhaps a fan and a damper. Engines have many moving parts: pistons, wrist pins, connecting rods, valves, camshafts, rocker arms, valve lifters, oil and coolant pumps, etc. Short cycling a boiler reduces its seasonal efficiency, but has a much less significant effect on its life. Short cycling is very harmful, however, to the engine driven cogenerator. It is accompanied by temperature cycling of all parts, wear on the starter and ring gear, condensate accumulation in the exhaust system, etc.

The mechanically (and financially) ideal duty cycle for an engine is round-the-clock operation, shutting down only for maintenance. This ideal can be achieved only if

- o Thermal load always exceeds cogenerator output, or
- o Enough energy storage is available to keep the engine running constantly in spite of load fluctuations.

Of course, a "dump" radiator may be used to ensure the first of these conditions, but the use of such a radiator has undesirable consequences which are discussed in Part II.

The majority of practical small cogeneration applications entail some cycling of the engine. This brings up the questions:

- o How short can a run cycle be without seriously affecting engine life?
- o How often can this shortest run cycle be tolerated?

Considerable experience has shown that minimum acceptable run cycle length is 15 minutes. A cogeneration unit which is running for such short cycles during a given time of day should do so not more than once per hour. Run cycles of 15 minutes separated by 3-minute off cycles, for instance, should be a rare occurrence.

A short cut for applying this rule of thumb is to express the output of the cogeneration unit in the peculiar, but convenient unit of gal-F/min. This is done simply by dividing the output, in Btu/hr, by five hundred:

$$\text{Gal-F/min} = (\text{Btu/hr})/500$$

The result is fairly accurate if the fluid is water. For instance, an output of 440,000 Btu/hr would be restated as 880 gal-F/min. A cogeneration unit of this output could be said to:

- o Heat a flowrate of 880 gpm by 1 F, or
- o Heat a flowrate of 10 gpm by 88 F, or
- o Heat an 880 gallon tank of water by 1 F per minute.

This short cut can be used to illustrate the importance of energy storage, or thermal capacitance, in a system which does not have a constant load. Suppose the 880 gal-F/min cogeneration unit is used to heat a 600 gallon spa (or Roman bath, whirlpool, Jacuzzi, etc.). Suppose that the spa is equipped with a thermostat having a setpoint of 105 F and a differential (or deadband, or hysteresis) of 1.0 F. Due to standby loss, the spa has cooled below the thermostat setpoint by the amount of the thermostat's differential, i.e. to 104 F. The cogeneration unit run time taken to restore the spa temperature to 105 F will be

$$(500 \times 1.0)/880, \text{ or } 0.57 \text{ minutes.}$$

Since the calculated run time violates the aforementioned rule of thumb, the designer should not use the spa as a load, or at least should not plan to run the cogenerator at any time when the spa is the only load.

This discussion of duty cycle points out the importance of adequate storage for the cogeneration system which is not truly baseloaded. Storage is an expensive feature to add, however. This often results in a conflict between the assessment of a site's cogeneration potential, economic constraints, and technical adequacy. It should be understood that a sacrifice in technical adequacy will likely lead to dissatisfaction with economic performance.

It should be noted that small cogeneration systems operating on thermal demand will have randomly distributed "on" periods during times of low thermal load. A large enough population of such systems will thus act electrically like a single, modulating central plant. This point is too often missed by utility regulators when they investigate the availability of small cogeneration systems.

Residual Heat

A running engine has a higher temperature at its cylinder walls than in its coolant passages. An engine which is shut down abruptly experiences what is known as "heat soak" - coolant flow stops, and the temperature gradient causes a rise in temperature of the coolant. The result can be boiling and/or deterioration of the coolant. This problem can be avoided by shutting the engine down in stages. After the load has received all of the heat it needs, the engine's electrical output is ramped down to zero. But the engine continues to idle, pumping coolant and removing heat from the block faster than the heat is generated. The engine is finally shut down only after block temperature has dropped below the point where damaging heat soak can occur.

The load must have enough thermal storage to accept the residual heat. Otherwise, the load temperature would rise excessively beyond the setpoint temperature. The spa discussed above, for instance, would obviously overheat during residual heat rejection if only 0.57 minutes of full output were required to restore it to the setpoint temperature.

Small boilers can be shut down abruptly without damage, although parts of the combustion chamber may be much hotter than their surroundings. Larger boilers are sometimes shut down in stages, although this may be done for purging as much as for combustion system cooling.

Package Specifications

Engines are generally less flexible than boilers as a heat source. For the engine, as for the boiler, the manufacturer typically recommends a Pressure-Temperature-Flowrate envelope governing input/output conditions. Adherence to those recommendations during design is vital to trouble-free operation of the system.

The recommended minimum static pressure (cold system) for an engine is typically 20 psi for a system as shown in Figure 1a, which employs manifold exhaust heat recovery [1]. That is because the irregular geometry and high heat flux of the manifold results in localized water temperatures near 260 F. The 20 psi static pressure is necessary to avoid vaporization at that temperature. A boiler, by contrast, typically has lower and more uniform heat flux. Most small boilers

[1] Manifold exhaust heat recovery has the very desireable feature of enabling supply temperatures in the 225-235 F range. Without manifold exhaust heat recovery, supply temperature is limited to the 195-205 F range.

are set for a minimum static pressure of 12 psi. Note that a direct (without an intermediate heat exchanger) connection of an engine with a boiler system may thus entail raising the minimum static pressure of the boiler system. In an engine system charged with water, the minimum static pressure is maintained by a pressure reducing valve, or regulator, the same as in a boiler system.

The recommended maximum pressure (hot system, pumps running) for an engine is typically 60 psi for a system which employs manifold exhaust heat recovery or through-the-block jacket heat recovery. This pressure is higher than for cast iron boilers and lower than for fin tube boilers. As for a boiler system, the maximum pressure is regulated by the design provision of adequate expansion tank capacity. Overpressure damage is likewise prevented by a pressure relief valve.

The recommended return temperature entering a packaged engine is in the 150-170 F range for most systems. This requirement is related to maintaining proper engine operating temperature, as discussed earlier. The return temperature to boilers is generally more flexible, in the 70-190 F range.

The difference between supply and return temperatures for an engine is generally greater than for a boiler. A 20 F rise is common for boilers. For engines, a 40-50 F rise is common. A low rise is desireable in the space heating duty for which boilers are most often used. The minimal variation in temperature of the heating medium throughout the building makes terminal unit design and control manageable. Heat transfer from the engine, however, was historically a worthless necessity best done at maximum economy. The high temperature rise of the engine system minimizes the flowrates and surface area required for heat transfer. These effects are usually beneficial in system design. Integration with low temperature rise space heat systems is not a problem, since the space heat flowrate is usually several times the cogenerator flowrate.

Specification of temperature rise, of course, specifies the water flowrate for a given engine heat output. Flowrate, however, is also related to package inlet pressure. Due to the compactness of its heat transfer surface, the cogenerator is likely to be the major pressure drop in the system. The pressure drop through the cogenerator increases as the square of the flowrate, as for any other static device. Therefore, a recommended minimum static pressure and maximum operating pressure imply a maximum flowrate. Perhaps a typical design dilemma will help to illustrate the relationship. A designer is faced with a high-temperature load which he is thinking of supplying from a cogeneration source. He plans the system to operate at the highest supply temperature allowed by the manufacturer of the cogenerator. He then plans to maximize the return temperature by maximizing the flowrate. Of course, he must be certain that the return temperature also does not exceed the manufacturer's recommendation. What is less obvious is that the pump discharge pressure required to yield the desired flowrate must also not exceed the recommended maximum inlet pressure to the cogenerator.

The cogenerator manufacturer's recommendations regarding flowrate and inlet and outlet pressure and temperature should be sought and observed in system design. Of course, the most successful design will be the conservative one which stays well within these guidelines and does not needlessly push the limits.

PART II
USING THE ENGINE'S HEAT OUTPUT

Several features of cogeneration retrofits combine to demand fairly specific and complex thermal system design strategies. The engine as a heat source is less forgiving than a boiler, as discussed in Part I. It is often connected to multiple thermal loads (such as domestic water and space heat) to achieve economically viable annual operating hours. The control strategy of each load must be accommodated while allowing the cogeneration system to baseload and the primary system to provide peaking, all without any discomfort to the building occupant. Installed cost of the system must be as low as possible, or the cost per kW of these small systems can become excessive.

System design could be simplified by using a radiator as the last load in the loop. Excess heat dumping must be avoided, since that would violate the interest of the system investor and the intent of PURPA. These factors, combined with the cost of a radiator subsystem in freeze-prone climates, argue in favor of a "determinate" control strategy -- the engine is run whenever its full heat output can be used by one of the loads.

A design philosophy which meets the needs of small systems without requiring a radiator is presented, along with examples of its application.

Multiple Thermal Loads

The return on investment in cogeneration is greatest for a system which operates on thermal demand and which usually has a demand greater than the cogeneration heat output. A result is that most sites suitable for packaged cogeneration systems in the 15 to 150 HP range could be classified as medium to large commercial or institutional facilities: hospitals, nursing homes, dormitories, YMCA buildings, municipal pools, and athletic clubs are some examples. The great heat demands of such facilities often stem from several sources. Space heating, domestic water heating (including laundry and kitchen), and swimming pool heating are the most common.

This multiplicity of loads is an engineering challenge. The cogeneration system design must take account of the unique characteristics of each type of load: flowrate, temperature setpoint and differential, temperature rise, thermal storage, safety issues, etc. The design must also provide the simplest possible solution to the complex multiple load system, or the cost of the system will be too great to justify its realization.

The delivery of heat to multiple hydronic loads, sometimes from multiple cogeneration units, often calls into play a concept known as primary/secondary pumping. Figure 3 is a schematic illustration of a primary/secondary system. Primary pump PP1 typically runs continuously. Secondary pumps SPA and SPB run whenever their associated cogeneration units run. Secondary pumps SP1 and SP2 run whenever their associated loads are in operation. Each connection between a secondary loop and the primary loop consists of two tees separated by a short section of "common piping", i.e. a section of piping which is common to the primary and secondary loops. The pumps are selected so that each secondary pump will move the same flowrate as the primary pump - when the secondary pump operates, the flowrate in the common piping becomes zero. The advantage of this type of system is that each pump's performance is entirely independent of the on or off status of the other pumps. Note that the two cogeneration units are piped in parallel - the primary flow is split equally between two "crossover" pipes, which include the common piping sections. The primary/secondary pumping strategy is fully explained in the Primary Secondary Pumping Application Manual, Bell & Gossett Bulletin No. TEH-775, ITT Fluid Handling Division, 1968.

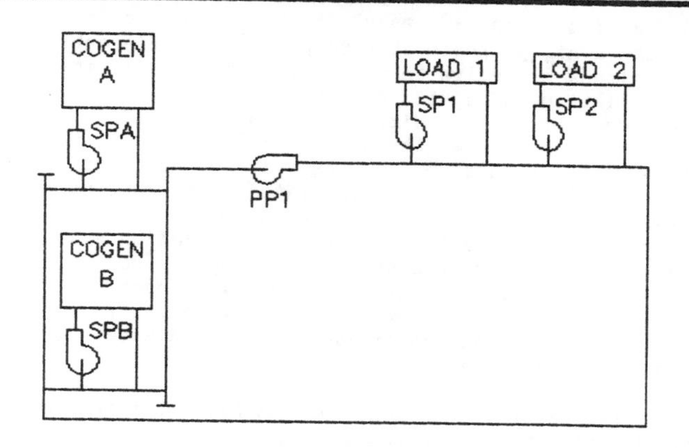

FIGURE 3 – PRIMARY/SECONDARY SYSTEM

Load Characteristics and Cogeneration Tie-In

Successful integration of the cogeneration system into the existing building heating systems absolutely demands that the engineer become familiar with the operation of those systems. While it may seem like no two building heating systems are the same, they really are composed of variations on several major themes. The common elements are the heat source, a pumped loop, and some form of thermal storage through which heat passes on its way to the end use.

Swimming Pools: Swimming pools are the simplest load that can be served by cogeneration. There is not much variation among pool heating systems, compared to domestic water and space heating systems. The vast thermal storage and low temperature of the pool simplify engineering of the cogeneration tie-in. Pool heating will be examined in some detail to show how even such a simple load can entail many design questions. Figure 4 is a simplified schematic of a pool heating/filtration system. The "original" components are shown in dashed lines, and the "new" components added for three types of cogeneration tie-in are shown in

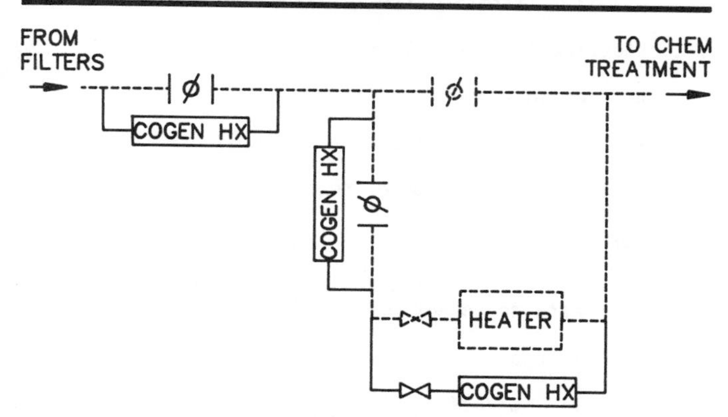

FIGURE 4 – ALTERNATE POOL HEATING TIE-INS

solid lines. The original heat source may be gas, electric, oil, steam, or hot water. The pumped loop is usually designed primarily for filtration, providing much more flow than is necessary for heating. The thermal storage is, of course, the very large volume of water in the pool. The end use is maintenance of the pool at a comfortable temperature for swimming - about 80 F.

The heater is controlled by a thermostat which is usually located upstream of the heat input, either internal or external to the heater. The exact location of this sensor matters in choosing a means of cogeneration tie-in.

Since the filtration flow is much greater than that needed for heating, the heater is placed in a "shunt" off the filtration loop. The butterfly valve in the main loop is throttled only enough to force an adequate flow through the heater.

The cogeneration tie-in for pool heating may be made in any of several ways. It can be arranged upstream of and in series with the existing heater as a "sub-shunt" from the original shunt, or as a separate shunt off the main filtration line, or in parallel with the heater on the original shunt. All of these approaches entail a change in the relationship of pressure drop vs. flowrate for the filter system. For this reason, it is wise to select pool cogeneration heat exchangers for minimum pool-side pressure drop [2]. The choice of tie-in type depends on the filtration flowrate, pipe sizes, and the relative strategic importance of rapid recovery at times of heavy water makeup (e.g. filter backwashing or pool wall cleaning).

It is important to note that no matter which method of tie-in is used, the thermostat which calls for heat from the cogenerator is located on the load side of the heat exchanger. This is true for any load except, of course, one which is connected to the cogenerator without a heat exchanger. In the case of the pool, for instance, the cogenerator should deliver heat to the pool whenever the pool temperature gets too low. It is a common error to think that the thermostat should be placed on the cogenerator return, and the cogenerator should run whenever the return cools. It is true that the cogenerator return should not exceed a certain maximum temperature, but that should be prevented by selecting the proper heat exchanger in relation to the load-side thermostat setpoint.

The sub-shunt and separate shunt are conceptually similar. The separate shunt is best used when there is doubt as to whether the total pool filtration flow is enough to accept the full output of the cogeneration plant at a reasonable temperature rise, e.g. 30 F. The sub-shunt should be used when the flowrate is more than adequate for cogeneration tie-in, and there would be a significant savings realized by installing the butterfly shunt valve in the heater shunt piping rather than the larger main filter piping. When the sub-shunt or separate shunt arrangements are used, the thermostat for the original heater must be moved upstream of the cogeneration heat exchanger if the original heater is to assist with recovery during heavy makeup periods. It must also be verified that the pool-side outlet temperature of the cogeneration heat exchanger is acceptable as an inlet temperature for the original heater.

The parallel shunt favors systems in which the original heater must operate in concert with the cogen-

[2] If there is any doubt about the reserve capacity of the filter pump, it is best to measure the pre-cogeneration operating point of the filter pump, establish the system curve, and estimate the operating point with cogeneration. If the change in flowrate threatens to adversely affect pool filtration or heating, then a secondary loop should be used for pool heating rather than a shunt loop.

erator to provide rapid recovery, and the thermostat of the original heater cannot be conveniently moved upstream of the cogeneration tie-in. Care must be taken to avoid overloading the filter pump by the addition of the parallel path. As for the other types of shunts, the several valves must be balanced so that the necessary flow through the original heater is maintained and the design flow through the cogeneration heat exchanger is obtained.

Domestic Hot Water: A domestic hot water (DHW) system may take one of the forms shown in Figure 5. For all of them, the end use is a one-way (except for recirculation, which is discussed briefly later) transfer of hot water to the fixtures. Figure 5a shows a sidearm heater arrangement. This form typically is used with a gas or oil heat source. There is a pumped loop between the heater and the tank. The tank provides thermal storage. This type of system can be equipped with cogeneration easily. The cogeneration heat exchanger can be placed in series or in parallel with the original heater, raising concerns similar to those cited for pools earlier.

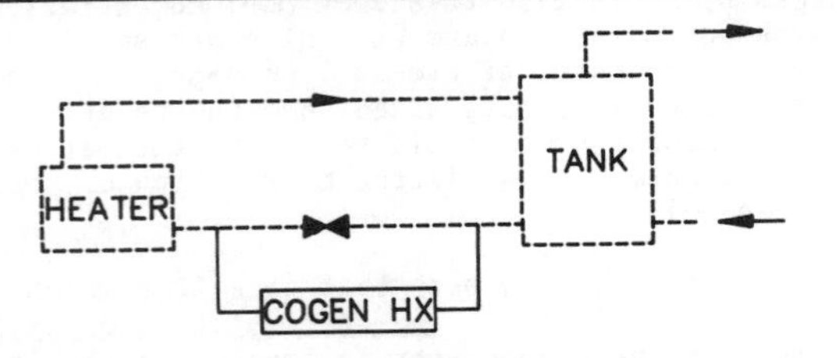

FIGURE 5a – SIDEARM DOMESTIC WATER HEATER

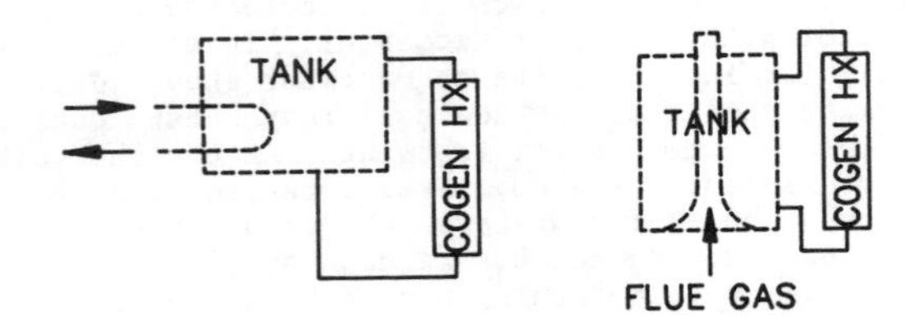

FIGURE 5b – STORAGE DOMESTIC WATER HEATERS

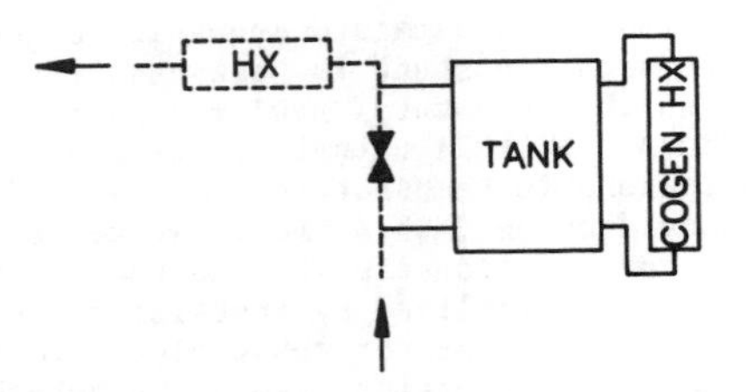

FIGURE 5c – INSTANTANEOUS DOMESTIC WATER HEATER

Figure 5b shows two types of storage heater. The heat source may be a coil circulating steam, hot water, or electricity, or it may be flue passages carrying gas or oil combustion products. There is no pumped loop. Again the tank provides some thermal storage. This type of heater can also be simply connected to cogeneration. A pumped loop is created, arranging the cogeneration heat exchanger as a sidearm heater. The main problem is often finding tank penetrations which can be used for the loop connections.

Figure 5c shows an instantaneous heater. Its heat source is usually steam. It has no pumped loop or thermal storage. Delivery temperature is regulated by a steam flow control valve or a water temperature control valve. The usual means of adding cogeneration to this type of system involves the addition of storage which is preheated by the cogenerator and feeds the instantaneous heater. The instantaneous heater operates only during times of peak load and cogenerator service.

For any type of DHW system, the designer may consider making a limited amount of storage more effective by using a higher thermostat setpoint and differential, and adding a mix valve to the tank. This modification to an existing DHW system can become a liability if it is not done properly. Situations must be avoided in which the pressures at the hot and cold inlets to the mix valve are unbalanced, or the valve cannot control its outlet temperature. Recirculated water must be allowed access to the mix valve cold port, as well as the hot port via the tank, or the recirculation system will become starved or overheated during periods of low draw. Other problems that can occur with added tempering valves are deterioration of recirculation flowrate and lack of responsiveness as hard water fouls the temperature sensing element.

Space Heat: Space heat systems have heat sources, pumped loops, thermal storage, and end uses as do pool and DHW systems. They are more varied and complex, however. The variations are presented in the outline below. Following the outline, a few typical systems and their adaptation to cogeneration are described.

I. Heat Sources for Space Heat

A. Steam boiler, possibly with steam-to-water converter or coil in duct, control always fixed pressure type

B. Hydronic heater ("boiler")

1. Fixed temperature
2. Reset temperature

II. Distribution Media

A. Water

B. Air

C. Steam

D. Combined

1. Steam-to-water
2. Steam-to-air
3. Water-to-air

III. Distribution Control Method

A. Fixed temperature

B. Reset temperature

C. Variable volume or reheat

IV. Terminal Units

A. Convection - naturally modulated - energy rate dependent on distribution temperature

B. Forced

1. On/off
2. Modulated via discharge temperature, volume, or reheat

The simplest type of space heat system to which to connect a packaged cogeneration system is one which employs a hydronic heater, fixed temperature water distribution, and forced terminal units in a one- to five-story building. In such a case, it may be possible to connect the cogenerator directly (without a heat exchanger) in series as a secondary loop from the main return [3]. A system like this is shown in Figure 6. If there is an outdoor air temperature thermostat which shuts down the main space heating pumps on warm days, then another thermostat should be added to disable the cogeneration space heat input at a slightly lower temperature. If the boiler thermostat T1 is of the outdoor reset type, then the cogeneration space heat thermostat T2 should be of the same type, with the same reset ratio and a slightly higher setpoint.

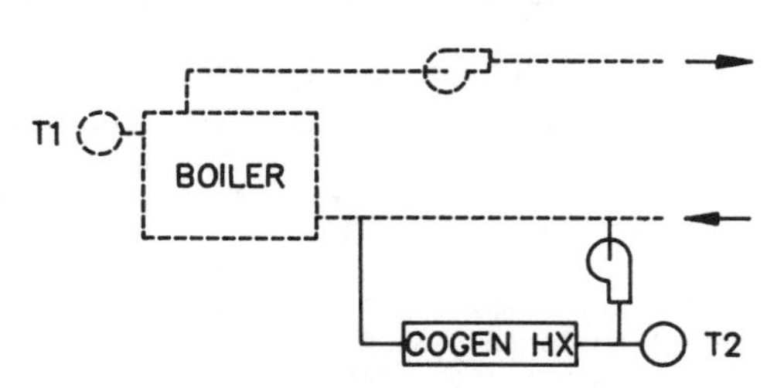

FIGURE 6 – SIMPLEST SPACE HEAT TIE-IN

Another common type of space heat system uses a fixed temperature hydronic heater, outdoor reset water distribution, and convection terminal units. This type of system is shown in Figure 7. The mixing valve bypasses some of the return water around the boiler to keep the supply temperature only as high as required by the outdoor temperature. Therefore, the flowrate through the heater could be more or less than the recommended flowrate through the cogenerator. In this case, it is usually best to connect the cogenerator in parallel with the heater. When the space heat return flowrate to the heater exceeds the cogeneration flowrate, then load is transferred from the heater to the cogenerator as the latter shunts off and heats some of the flow to the mixing valve. When the space heat return flowrate to the heater is less than the cogenerator flowrate (i.e., warm days), then the cogenerator heats all of the flow to the mixing valve and there is also backwards flow through the heater. This allows the storage in the heater to prevent short-cycling of the cogenerator in low-load situations. Note that the location of cogeneration thermostat T3 is crucial for correct operation, due to the bidirectional flow which can occur in the heater.

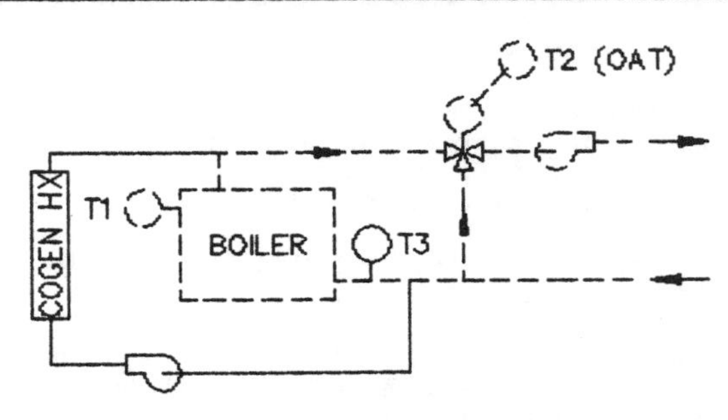

FIGURE 7 – RESET SPACE HEAT TIE-IN

Sometimes, there are reasons for which the cogenerator cannot be connected in parallel with the boiler as shown in Figure 8. Examples are systems in which the boiler temperature must be maintained higher than the maximum allowable cogenerator return temperature, and systems with hydronic heaters with too little thermal storage. In these cases, the cogenerator can be connected as a secondary loop from the space heat return, upstream of the tee to the mixing valve. This approach uses the thermal storage in the distribution piping to prevent short cycling of the cogenerator, so it only works if that piping contains enough volume. It also entails control of the cogenerator input with a redundant outdoor reset thermostat, as for the reset-controlled boiler.

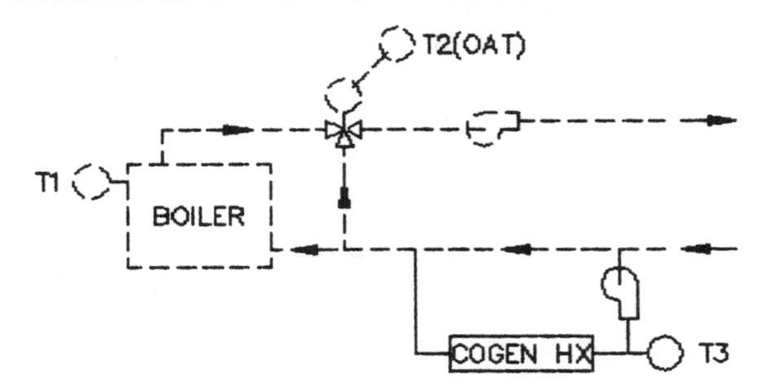

FIGURE 8 – BUILDING-SIDE SPACE HEAT TIE-IN

Some space heat systems cannot be connected directly to the cogenerator. An example is systems in tall buildings (more than 5 stories) in which the pressure at the boilers, added to the cogenerator pump head, can exceed the maximum allowable cogenerator inlet pressure. The addition of a heat exchanger and pump provides pressure isolation.

Some space heat systems must be equipped with added storage in order to be heated by cogenerators. An example is a system with a steam heat source, multiple zone controlled water distribution, and convection terminal units. Since very little water mass is involved even if all of the zones could be consolidated, the added storage prevents short cycling of the cogenerator.

The most difficult type of space heat system to which to connect a cogenerator is one having a steam or electric heat source and air distribution. The cogeneration heat input to such a system is via a hydronic coil added in a duct. Although coils are often less expensive than shell-and-tube heat exchangers, the associated engineering and installation costs are much greater than for any other type of load. The reasons are that the design must avoid disturbance of original flow conditions and installation involves extensive custom sheet metal fabrication in tight spots. Strictly speaking, the design should include: measurement of original air flow rate by an accurate means such as a hot-wire anemometer; measurement of original pressure rise of fan; establishment of system characteristic and operating point on fan curve; coil selection based on cogenerator heat output rate and temperatures, water and air flowrates, entering air temperature and flowrate, cross-sectional area of duct section having space available for coil, and smallest reasonable air-side pressure drop; provision for maintenance of original air flowrate by changing fan pulleys or motor, or by adding a booster fan; and provision for freeze prevention where applicable.

Peaking, Backup, and User Comfort

Proper control of the original system heater is necessary to assure that as much of the load as possible is transferred to the cogenerator. With proper adjustment of the thermostats, the original heater acts to meet thermal load peaks which cannot be met by the cogenerator, and acts as a backup to the cogenerator in the event the unit is being serviced. Basically, all that has to be done is to set the original thermostat lower than the cogeneration

[3] It can be assumed that the space heat flowrate is several times the cogeneration flowrate if the cogenerator's output is matched to the space heat load.

thermostat for each load. The allowable difference between the setpoints and the differential associated with each setpoint depend on the requirements of the process. The relationships are shown graphically in Figure 9. If the thermostats are set less than 5 F apart, it should be verified that their dial readings agree with their true setpoints. The differential on the cogeneration thermostat should be as great as possible (in keeping with user comfort), to provide maximum thermal storage and minimum short cycling. The differential on the original thermostat should be as little as possible, to minimize user discomfort or inconvenience.

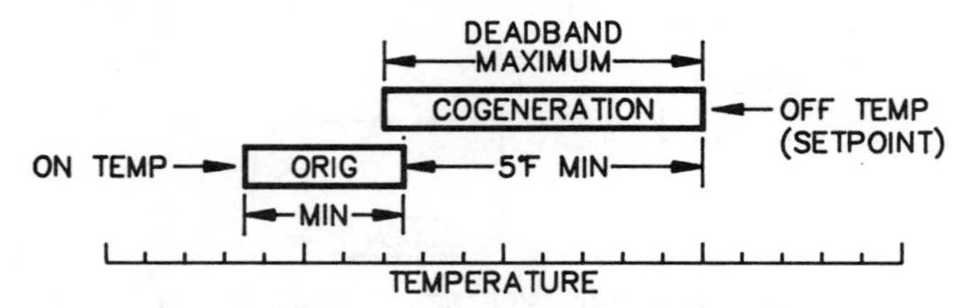

FIGURE 9 – SETTING OF ORIGINAL & COGEN THERMOSTATS

Installed Cost

The preceding discussion has shown that the types of thermal systems to which small, packaged cogenerators are connected are quite complex. Together with the fact that economies of scale work against these small systems, this makes it difficult to install them at a reasonable cost (e.g., 900-1300 $/kW including cogenerator). This difficulty can be mitigated by the formulation and packaging of a strategy to standardize the treatment of various thermal loads. The following discussion addresses candidate strategies and the packaging of the preferred strategy.

The Indeterminate Strategy

The indeterminate strategy is one means of reconciling variable heating loads with the constant output of the cogenerator. It is illustrated in Figure 10. When the cogeneration module runs, its heat delivery to the load(s) is governed by a modulating temperature control valve which controls the load-side supply temperature. Any heat produced by the cogenerator and not used by the load is dumped to the atmosphere by a radiator. The amount of heat dumped is governed by a second temperature control valve which controls the cogen-side return temperature. The load(s) can receive any fraction of the cogenerator's full output at any time when the cogenerator is running. The amount of heat actually delivered to the load can be known only by placing an accurate Btu meter across the load. The heat exchanger may be selected on the basis of a once-through inlet temperature, such as 60 F groundwater for a DHW system. The heat transfer capacity need not have any exact relationship to the output capacity of the cogenerator, although it would be pointless to make it any larger.

The indeterminate strategy makes sense only under rare circumstances: The cogenerator is truly baseloaded (in which case no radiator is necessary); the controls are capable of detecting that the sum of the available, variable loads exceeds the cogenerator output; the cogenerator is capable of modulating continuously between 0-100% of its capacity; or system economics favor use of only some of the heat output.

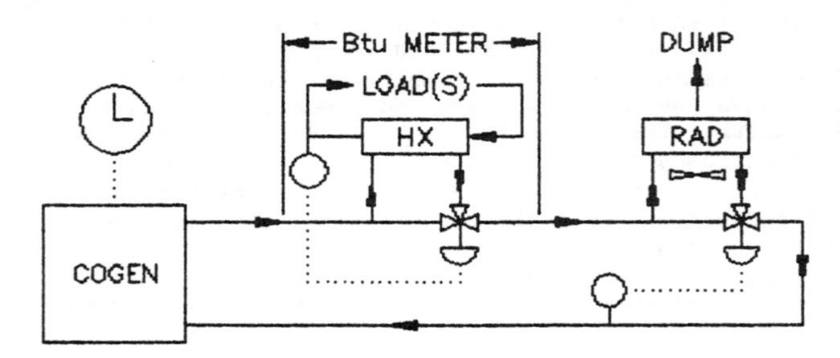

FIGURE 10 – INDETERMINATE STRATEGY

A truly baseloaded cogenerator will function well under an indeterminate or determinate design strategy. Controls which can add the demands of several thermal loads and run the cogenerator when the sum exceeds its output would be very complex. They could also allow heat to be supplied to each load by a non-cogeneration source when the sum was less than the cogenerator output, displacing potential cogenerator run time.

As discussed earlier, small, packaged cogenerators are designed to run at maximum efficiency at the full output design point. Such high efficiency is accompanied by nearly condensing exhaust temperature. Modulation to less than full output, without reduction of heat transfer area or temperature difference, could result in prolonged exhaust condensation, exhaust system corrosion, condensate discharge, engine flooding, and other problems.

Occasionally, system economics DO favor the production of electricity alone by the cogenerator (e.g., areas of high electric/gas price ratio). Even in these cases, the indeterminate strategy makes it difficult for the system owner to tell how much heat energy he is saving. The indeterminate strategy also makes it difficult to assure that the cogeneration system is a Qualifying Facility under the PURPA regulations (i.e., that its annual electrical efficiency plus half of its annual thermal efficiency, both based on low fuel heating value, add to greater than 42.5%). At this early stage in the introduction of packaged cogeneration technology, it would be prudent to employ practices which build the image of the technology in the public eye.

The Determinate Strategy

The determinate strategy is the preferred means of reconciling variable heating loads with the constant output of the cogenerator. It is illustrated in Figure 11. It entails the following criteria:

- The cogenerator operates at full output whenever heat is required.

- Each thermal load is designed to demand heat input only when it can accept the full thermal output of the cogenerator for a period long enough to prevent short-cycling of the cogenerator.

- Loads are served one at a time to maximize annual cogenerator running time by avoiding "heat robbing" of one load by another.

- Loads are served in an order of priority which maximizes annual cogenerator running time by capturing peak demand periods of erratic loads while the more steady loads "coast" on their thermal storage.

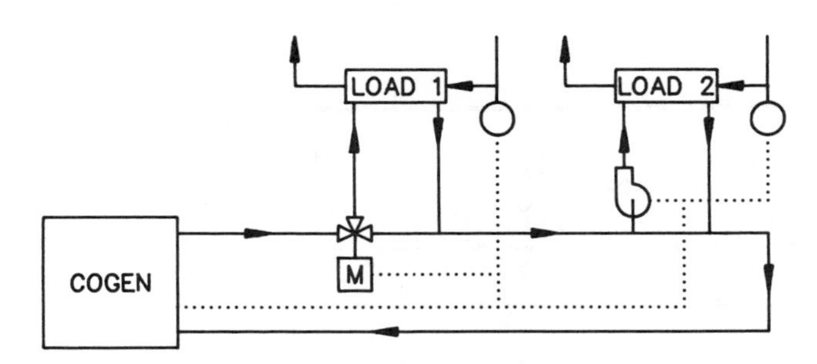

FIGURE 11 — DETERMINATE STRATEGY

With respect to the system of Figure 11, the determinate strategy is implemented as follows. Each load is monitored by a thermostat which is somehow in contact with the load's thermal storage. The thermostat may be placed in a well in a tank, or in a continuously pumped loop from the tank, for instance. When either load's thermostat signals a "call for heat", the cogenerator starts. The cogenerator's heat output is directed to the active load by a diverter valve or by a secondary pump, as shown in the left and right loads of Figure 11, respectively. As discussed in Part I, the thermostat differential and storage volume for each load are arranged so that the cogenerator cannot be caused to short cycle by responding to the load, even if it is only making up standby losses from the load.

There is a precise rationale to heat exchanger selection under the determinate strategy: **EACH LOAD'S HEAT EXCHANGER IS SELECTED SO THAT IT WILL TRANSFER THE FULL THERMAL OUTPUT OF THE COGENERATOR TO THE LOAD WHEN ITS LOAD-SIDE INLET TEMPERATURE EQUALS THE THERMOSTAT SETPOINT.** This selection is not difficult, given the automated design methods available to manufacturer's representatives. The only rationale that could be simpler would be to select a heat exchanger at random.

The implications of designing each load to accept the full output of the cogenerator can be explained in terms of Figure 12 and Table 2. Suppose that the load is a well mixed tank of domestic hot water [4]. The heat exchanger is selected to maintain the tank at a temperature of 140 F. One might think that this heat exchanger, with a design outlet temperature of 160 F, could actually maintain a tank temperature of 160 F. The tank temperature is also the heat exchanger inlet temperature, however, and this heat exchange clearly would not transfer the full output of the cogenerator with an inlet temperature 20 F higher than the design point.

TABLE 2
THERMAL SPECIFICATIONS: SYSTEM OF FIGURE 12

	HX Engine Side	HX Load Side
Flowrate, gpm	20	44
Temperature In, F	210	140
Temperature Out, F	166	160
delta T, F	44	20
Heat In or Out, MBh	440	440

[4] It might seem tempting to use thermal stratification as a means of extending cogenerator duty cycle and maximizing captured load. However, achieving stratification entails placing thermostats at the top and bottom of the tank (it's often impossible to find even one spare opening on a tank), the necessity of a tempering valve, and demand control of the DHW pump (vs. continuous operation, which allows placement of the thermostat outside of the tank).

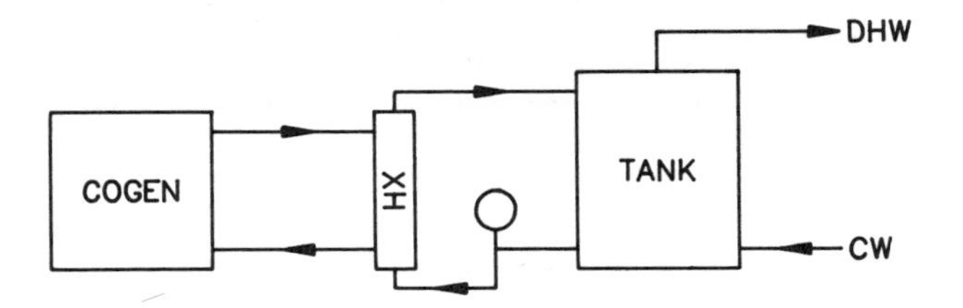

FIGURE 12 — TYPICAL DHW SYSTEM

Note that the load-side flowrate shown in Table 2 is about twice the engine-side flowrate. As a result, the load-side temperature rise is about half the engine-side temperature drop. This strategy, together with the use of single-pass, counterflow heat exchangers, maximizes the temperature of the water which can be stored in the tank. It cannot be carried too far, however, or the heat exchanger selection would be dominated by the need to keep load-side pressure drop reasonable.

Logical analysis will generally reveal that it does not make sense to try to serve more than one load at a time from the cogenerator. Suppose that the left-hand load in Figure 11 is a DHW load, and the right-hand load is a pool load. Suppose that the DHW tank, which happens to be a preheat tank for an instantaneous heater, has cooled to 50 F while the cogenerator was being serviced. When the system is restarted, the cogeneration loop water emerges from the DHW heat exchanger so cold that it actually removes heat from the pool, rather than adding it! Even if the cogenerator did heat the DHW and the pool at once, the pool heat exchanger exit temperature would be so low that the engine mixing valve would reduce the cogeneration loop flowrate -- the result might be that both the pool and DHW original heaters would both activate to provide peaking heat input.

Instead, the cogenerator should serve only one of the loads at a time. If it is heating the pool, and the DHW load calls for heat, it should satisfy the DHW load and then return to heating the pool. The reason is that the pool, with its great thermal storage, is less likely to call upon its original heat source for backup while the cogenerator is busy heating DHW. If the pool were heated preferentially, then the DHW load, with its lesser storage, could cycle its original heater several times while waiting for the cogenerator to finish with the pool.

Extending the Module Principle

The packaging of cogeneration engine/generator and engine/chiller sets has brought cogeneration technology within economic reach of many commercial establishments. It has done so by standardizing the sophisticated engineering involved in combining the components and controls. The packaging concept can be carried a step further to include standardization of the components which link the cogenerator to the thermal loads. In order to include the thermal system controls in the package and accommodate the special requirements of the engine heat source, this further modularization is best guided by the determinate strategy.

A modular thermal delivery system has been made available by at least one manufacturer of packaged cogeneration equipment. A "load module" for a one-cogenerator, two-load system is shown as a piping schematic in Figure 13. In its first few installations, this concept has shown its ability to reduce the costs of system design, installation, startup, and service. Continued standardization and packaging is recommended as a potentially powerful stimulus to growing acceptance of packaged cogeneration.

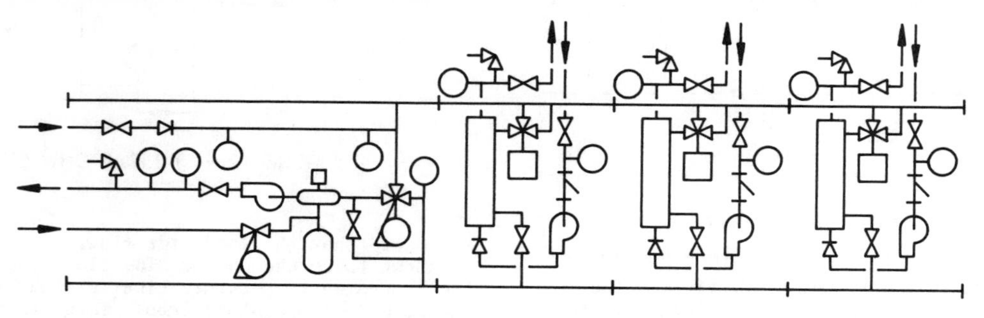

FIGURE 13 — PACKAGED THERMAL DELIVERY SYSTEM

Chapter 64

PACKAGED COGENERATION INSTALLATION COST EXPERIENCE

J. A. Mulloney, Jr., M. T. Panich, S. M. Knable

Authors Note: This presentation is a report of a study recently completed by EA-Mueller, Inc. (formerly Mueller Associates, Inc.) for the Gas Research Institute (GRI). The full study, "Cost Reduction in the Manufacture and Installation of Gas-Fired Cogeneration Systems" aims to reduce manufacturing and installation costs of these systems from the present \$1,200/kWe±\$200 kWe for simple systems to about \$600 to \$700/kWe, at industry maturity. Readers who are interested in obtaining the GRI report should contact the National Technical Information Service (NTIS) at (703) 487-4650 and ask for publication PB88183033. The cost for a printed report is \$29.95 plus \$3.00 for shipping and handling.

First cost is a significant economic factor in cogeneration choice in many of the target markets for packaged cogeneration systems. This is especially true in several of the commercial submarkets, such as multi-family residential and "spec" (speculative) buildings, where low first cost is deemed more important than life-cycle costs which can be passed through to the tenants. Early indications from the marketplace were that prepackaged systems, as installed, are perhaps 50% higher than the generally recognized, industry-wide installed cost target for widespread adoption of cogeneration (of about \$1,000/kWe) in this size range. However, these early indications were not publicly well documented and sometimes represented biased or conflicting claims.

The Engine-Driven Cogeneration Systems area at GRI has as its objective the development of efficient, cost-effective, gas-fueled prime mover cogeneration equipment for commercial, institutional, industrial, and multi-family residential applications. The program incorporates efforts aimed at a better understanding of the characteristics of potential cogeneration applications, the development of standardized prepackaged cogeneration units for major application groupings with common characteristics, and development of advanced system components to improve efficiency, reliability, and reduce first cost. In the course of this program, the costs of fabricating and installing packaged cogeneration units have been identified as possible barriers to cost-effective implementation of natural gas cogeneration.

Given the above concern and the long term implications of high first cost on market acceptance of prepackaged cogeneration systems, GRI established the study project reported herein on cost reduction approaches in the manufacture and installation of prepackaged cogeneration systems. The project objectives were as follows:

- Develop a cost data base of existing prepackaged cogeneration sites.
- Identify and evaluate cost reduction concepts/approaches.
- Estimate future total installed costs.

A project team was assembled as follows to carry out the project:

CONTRACTOR: Mueller Associates, Inc. (MAI) - Consulting engineers with broad experience in traditional HVAC design and emerging alternative energy systems.

SUBCONTRACTORS: Cogenic Energy Systems, Inc. (Cogenic) - A leading manufacturer and marketer of prepackaged cogeneration systems.

Pioneer Engineering and Manufacturing Company - Experienced in automotive and other manufacturing technologies--automation, robotics, just-in-time systems, etc.

CONSULTANTS: Leroy H. Lindgren - A manufacturing cost estimating consultant.

Leonard A. Hall - Consultant with broad based HVAC (heating, ventilating and air conditioning) industry experience with Carrier Corporation.

A variety of data collection approaches were utilized including:

- A mail survey of sites that provided an initial cut at three major cost categories of interest:
  - o Unit costs, F.O.B. factory
  - o Installation (site) cost
  - o Integration (soft) costs.
- Site visits to provide dimension and understanding/insight about the costs (40 site visits).
- Industrial consultants (Lindgren, Pioneer and Hall) used to scope out target manufacturing costs to produce and distribute units.

- Installation costs and insights from Cogenic's experience.
- Installation cost baseline characterization via MAI design studies.
- A workshop of key participants in the industry to provide focus and commentary on the industry's views on important cost considerations.
- Various personal contacts with manufacturers, designers, developers, GRI staff, owners, installation operators, maintenance servicemen, etc. also provided valuable information and insight on prepackaged cogeneration.

Existing systems costs were developed as a data base to provide a baseline for subsequent reference. The data base contains three cost categories that sum to total installed costs, namely:

- Cost of the prepackaged cogeneration unit--F.O.B. manufacturer's dock.
- Cost of installation--applicable freight and the physical site effort to place the unit into service.
- Integration costs--the soft costs including such things as utility negotiations, determination of thermal interface and electrical interconnect requirements, permitting, engineering design, and equipment start up.

While the integration costs cited above are normally considered part of installation costs, it was decided to place them in a separate cost category for reporting purposes in order to highlight their importance. This highlighting was felt from experience to be appropriate because the impact of these integration costs tend to be under-appreciated and not properly accounted for or budgeted in many projects.

The technical criteria for site selection were as follows:

- Prepackaged system intent (i.e., not a custom system)
- 50 to 500 kWe unit or multiples thereof
- Natural gas-fired (i.e., not diesel or biogas)
- Not a prototype, trail, or subsidized unit
- Installed and operating (i.e., a working system).

Well over 200 sites were screened. Fifty-three (53) sites met the technical requirements, were cooperative, and provided sufficient site information for inclusion in the data base. This was roughly 40% of all sites meeting the technical criteria at the time (Fall 1986 - Winter 1987). Many other cogeneration sites exist as non-criteria custom, biogas, etc. sites that do not fit the concept of prepackaged and were, accordingly, excluded from the study.

The characteristics of the 53 sites in the data base are as follows:

- Mostly 50 kWe units (63% of units, 57% of sites).
- 1.5 units/site on average with as many as five units at a site.
- 1.8 heat applications/site on average (e.g., domestic hot water plus pool plus spa would be three applications).
- 89% retrofit situations.
- Mostly Southern California and New England locations.

Table 1 presents the average and range of the data base site costs by cost category. The distribution of the total installed cost is presented as a low-to-high display on Figure 1 which also includes a breakout of component costs. Note the plateau of unit costs.

COST CATEGORY	AVERAGE	RANGE
F.O.B. UNIT COST	729	500 - 1,100
INSTALLATION (SITE) COSTS	541	100 - 1,900
INTEGRATION (SOFT) COSTS	94	0 - 500
TOTAL INSTALLED COST	1,344[1]	600 - 2,800

NOTE 1. The first three cost categories are not additive to the total because some sites did not report each category but all sites reported total installed cost.

TABLE 1. AVERAGE DATA BASE SITE COSTS, $/kWe

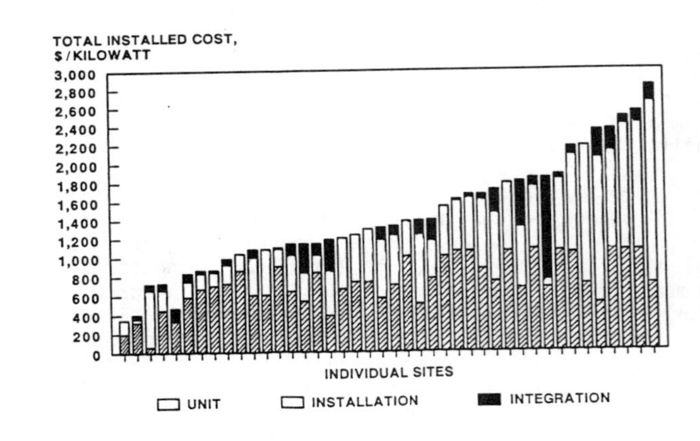

FIGURE 1. LOW-TO-HIGH DISPLAY OF TOTAL INSTALLED COST

As indicated by Figure 2, the primary reason for the large variation in total installed cost is the large variation of installation (site) cost. In order to develop a better understanding of some of the parameters involved in installation (site) cost variability, MAI performed an analysis of the effect on cost of system complexity that resulted in the reference cost data points for installations considered simple (one heat application and no building modification), moderately complex (three heat applications with building modifications), and complex (moderately complex plus absorption chilling) as shown on Figure 3.

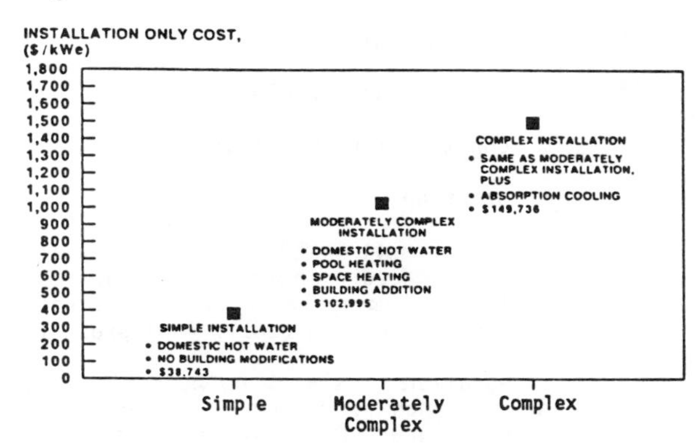

LEVEL OF COMPLEXITY (100 kWe UNIT)

FIGURE 2. LOW-TO-HIGH DISPLAY OF DATA BASE INSTALLATION (SITE) COSTS

With respect to manufacturing costs of the prepackaged cogeneration units and the manufacturer/distribution system mark-ups necessary for satisfactory on-going manufacturing operations (as opposed to low volume manufacturing startup), several approaches were used including a purchased component/shop fabrication elemental cost buildup, the application of learning curve concepts, and reference to both automobile and HVAC industries for overhead factors. It was found that prepackaged cogeneration unit manufacturing is amenable to mass production techniques where learning curve cost improvements of 85% to 90% (i.e., decreases of 10% to 15% in cost with each doubling of output) are possible.

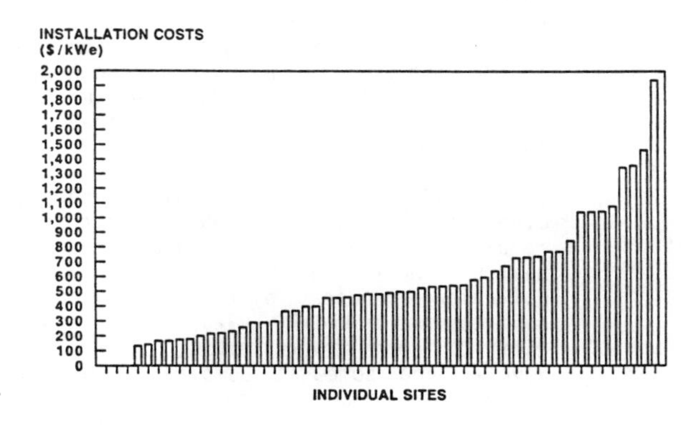

FIGURE 3. THE COST OF SYSTEM COMPLEXITY

One unanticipated finding was there there appears to be come diseconomy of scale with respect to prime movers. The particular prime movers chosen for the size levels examined and the reasons for their choice in the manufacturing cost study are as follows:

- 60 kWe--This is a high production level, mass produced Chevrolet 454 engine from the automotive industry. It is modified and de-rated to operate on natural gas. This particular engine was chosen because of its current use in the cogeneration industry. The cost of this prime mover is $3,500 or $58/kWe.

- 100 kWe--This Caterpillar 3306 engine is also currently used by the cogeneration industry. It is produced by the manufacturer in a natural gas fired configuration. The engine is produced in volume for truck, generator set, construction, marine and other applications. It appears to be oversized for the application since it is also used to power 150 kWe generator sets. The comparable price is $135/kWe.

- 500 kWe--This selection was from the Caterpillar 398 series of low speed (1,200 RPM, V-12) engines used for primarily stationary and marine applications. Typically, production levels on this general class of engines are low relative to the others and possibly with dated equipment and techniques. The comparable price is $195/kWe.

While it is easy to generalize on the above prime mover types and conclude that high volume engine production is one key to lower cogeneration costs, there are a host of other considerations and trade-offs that must be taken into account by cogeneration system manufacturers. Among these considerations are efficiencies, pollution controls, specific engine features, engine controls, durability (lifetime) and RAM (reliability, availability, and maintainability). These considerations and trade-offs notwithstanding, the primary capital cost driver in the development of prepackaged cogeneration units is the cost of the prime mover. Figure 4 makes this point graphically by way of a key component percent cost breakout for three representative prepackaged unit sizes.

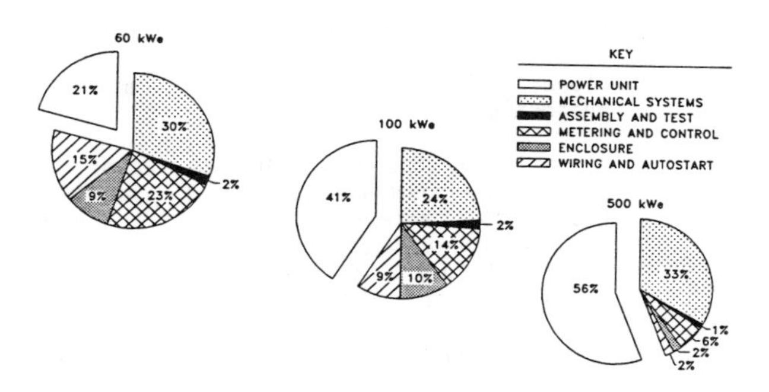

Note: The power unit consists of the engine and the generator.

FIGURE 4. KEY COMPONENT CAPITAL COST BREAKOUT BY PREPACKAGED UNIT SIZE

Given the industry prime mover choices to date (i.e., as systems get larger the trend is toward heavy duty industrial engines), the implication of this diseconomy of scale with respect to prime mover size is that the total prepackaged unit capital cost (the sum of the prime mover cost and the balance of unit costs) may not have an economy of scale with respect to size. Indeed, the manufacturing cost estimates indicated a somewhat flat size scale factor with diseconomies of power unit size scale balancing economies of balance of unit size scale. These estimates also showed that individual highs and lows of cost ($/kWe), depending on specific prime movers, had greater variation in total prepackaged unit capital cost than any trend over the 50 to 500 kWe size range of interest.

Other key cost drivers in addition to the installation complexity, manufacturing learning curve, and the prime mover diseconomy of scale were found to be as follows:

- Affordability - Affordability is a non-technical term coined in this report to convey the situation where a project's final criterion (payback, ROI, etc.) is such that a very high first cost is affordable. For example, there are a number of areas of the country where high electricity rates have encouraged the introduction of prepackaged cogeneration systems. The early-entry prepackaged system cogenerators in these situations were from the entrepreneurially oriented promoters, developers, and equipment manufacturers offering third party shared savings or similarly attractive projects. The magnitude of affordability as a cost driver is demonstrated when the site data base total installed cost figures, $/kWe, are separated into two groups as follows:

    o Sites known to be developer or wrap-around (total service provider) projects, and

    o All other sites (which may also include some unrecognized developer and wraparound projects).

    It is clear that the developer/wraparound group (19 projects; $1,810/kWe average) represent a distinctly different cost population than the group of all the other sites (34 projects, $1,113/kWe average). There are a number of reasons for this total installed cost difference, developed in the report in some detail, including a tendency on the part of developers to "load in" capital costs to maximize project profit.

    Accordingly, the affordability factor has a strong influence on total installed cost. MAI believes that this has caused much of the GRI concern about high total installed cost. Figures reported in the media only as summary statistics do not tell the complete story. Cost reduction efforts must focus on reducing basic packaging and installation costs through technology and other means without consideration of affordability situations.

- Integration Costs - Integration costs include costs associated with the following:

    o feasibility study

    o utility interface negotiations

    o permitting

    o legal/financial/insurance technical support

    o design engineering

    o bid and award cycle

    o construction oversight

    o start up

    o other soft costs such as raising prepackaged cogeneration awareness levels among parties participating directly or indirectly in a project.

    Integration costs can become a cost driver when they begin to exceed the expected norm for whatever reason. The costs accrued as integration costs are primarily labor, fees for permits, and construction financing interest. These costs arise both as active costs in efforts to complete the project or as holding costs until the project is complete and project revenues begin. Either type of cost can become a cost driver especially on the smaller end of the size range.

    Each of these cost elements are described in some detail in the report. Key integration activities include accurate assessment of loads via site reviews and loads monitoring as well as site reviews for "fatal flaw" obstacles such as excessive amounts of existing electrical switchgear and circuitry that is not up to current code and must be either grandfathered or replaced as part of the installation. Negotiations with the local electric utility can be a source of great frustration and cost when the utility resists the prepackaged cogeneration project. Tales abound of delays, footdragging, excessive justification, midstream changes in requirements, incorrectly ordered interconnection equipment, etc. which run up the integration and construction financing cost of the installation.

    In addition to local building codes, other permits can include meeting the requirements of architectural boards, health care facility boards, air pollution agencies, planning and zoning commissions, tidal plain and port district commissions, etc. Given the novelty and somewhat complex nature of prepackaged cogeneration installations, codes inspectors have caused significant time delays and cost, for example, by requiring testing laboratory certification of a non-rated installation component. In other instances, inspectors may choose not to sign off on the system because they do not understand it or are afraid to make a decision. In addition to dealing with the above labor and running costs, obtaining permits cost money for the permit itself, the amount of which is dependent on the local jurisdiction and is usually quite minor. The time and effort involved in satisfying these jurisdictions may not be minor, however.

- Value Added Features - Prepackaged cogeneration units can be operated in conjunction with other equipment such as absorption chillers, thermal storage systems, desiccant coolers, etc. to supply added value to the project but only at the expense of higher initial costs. These may be valid additions with respect to life-cycle costs but are not within the scope of this capital cost reduction study. In practice, it is usually difficult to sort out project costs for the value added feature which often results in the cogeneration project segment being overcharged.

- New Versus Retrofit - Substantial savings may accrue by designing-in the prepackaged cogeneration system when a commercial property is initially designed compared to a subsequent retrofit with the same system. While this comparison is highly site-specific, there is one actual installation comparison that is illustrative. The InterContinental Hotel in San Diego, California has two residence towers, one of which was retrofit with an 800 kWe cogeneration system and the other equipped with a similar 800 kWe system that was designed-in to integrate most efficiently with the mechanical equipment. The cost saving was approximately 40%.

A major strategy followed throughout the subject study was to elicit cost reduction comments from as many cogeneration industry participants as possible during and subsequent to the study investigations. In this way, a clear focus on industry concerns could be achieved. As one means to implement this strategy, a workshop was conducted in order to directly obtain industry feedback in a structured format. It was correctly anticipated that the workshop forum would provide industry an opportunity to discuss mutual cost problems and identify potential solutions.

Another logical approach to assessing cost reduction is to first determine a common technological baseline for cogeneration packaging and installation technology. However, a close examination and comparison of the systems available today in the marketplace illustrates fundamental differences between manufacturers with respect to design philosophies. For example, some manufacturers design systems with durability the principal design criteria. Others try to minimize the first cost of the package (at the expense of durability perhaps). Still others attempt to minimize the overall life cycle costs of operation and maintenance. Accordingly, it is difficult to determine a common technological baseline for competing cogeneration systems. In particular, it becomes clear that the reality of this range of design (and marketing) philosophies will complicate any analysis of cost reduction potential. That is, it becomes very difficult to assume that specific cost reduction ideas will have generalized industry applicability, or to what degree.

Having qualified some of the difficulties to be found in looking at specific cost reduction ideas, the project management undertook to do just that, i.e., to list, describe, and assess cost reduction approaches, ideas, and specific suggestions. A number of potential cost reduction concepts were identified and evaluated in some detail. These concepts originated/evolved from multiple sources as previously described plus Gas Research Institute staff professionals and study team members' respective in-house expertise.

In order to provide some insight into the nature and economic potential of these concepts and approaches, a structured process for evaluating each concept relative to the others was devised. This structured process is schematically represented in Figure 5. Several hundred potential cost reduction concepts were first screened to eliminate concepts that were redundant or overlapping as well as those that were minor in impact or seemed to be impractical. The sixty-seven (67) that survived the screening were given more detailed consideration and analysis during which each concept was defined/described and its impact described in text form in sufficient detail to convey the nature of the concept to both the technical and non-technical reviewer.

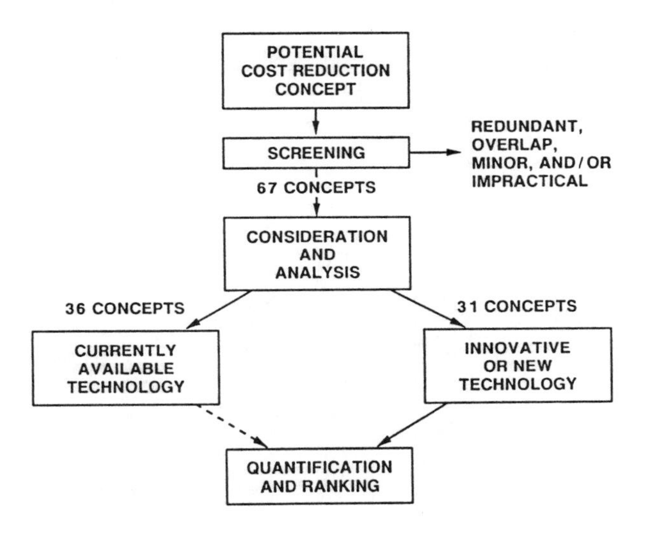

FIGURE 5. PROCESS FOR EVALUATING COST REDUCTION CONCEPTS

A structured cost reduction assessment was designed and implemented in order to arrive at the relative economic importance of this highly diverse group of cost reduction concepts. Table 2 presents a number of the better cost reduction concepts along with an estimate of their relative potential worth in reducing total installed cost. The concepts listed in Table 2 are more fully described in the report.

Assuming that anticipated learning curve, technology transfer and RD&D activities take place, Figure 6 provides some insight into the potential for future installed costs (at Fall 1987 dollar values) as the industry matures. The first bar, on the left side, represents the data base finding of $1,344/kWe total installed cost broken down into F.O.B. unit cost ($729/kWe) and installation cost ($615/kWe).

COST REDUCTION CONCEPT	RELATIVE SAVINGS POTENTIAL, $/kWe
Appropriate Facility	>73
Configuration Management	22 to 73
Product Design/Update	22 to 73
Subassembly Manufacture	7 to 22
Fuel Charge Density Improvement	>33
Increase Engine-Generator Speed	>33
Cogeneration Designed Engine	>33
Direct Fuel Injection and Diesel Cycle Engine	>33
Multi-Engine Package vs. Multi-Systems	>18
Experienced Construction Installation Teams	>25

TABLE 2. MAJOR COST REDUCTION CONCEPTS

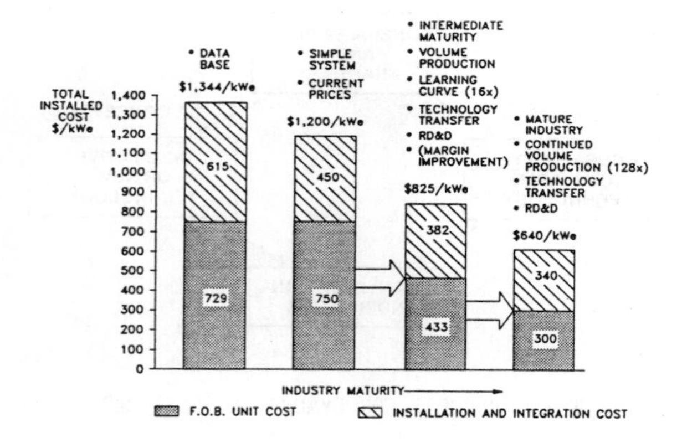

FIGURE 6. COST REDUCTION POTENTIAL OF PREPACKAGED COGENERATION SYSTEMS

The second bar shows that for simple systems at the lower end of the size range and current prices as discussed earlier in this section, a cost range of $1,200 ± $200/kWe could be expected. The cost range of ± $200/kWe is important in that is indicates the variability to be expected even with simple systems.

The third bar indicates that, with assembly line techniques and volume production at 16 times current level and assumptions of an 85% to 90% learning curve for manufacturing and 96% for installation and integration costs, an average total installed cost of $825/kWe can be achieved. Research, development, and demonstration and technology transfer will speed up the cost decline but necessary margin improvement for a viable industry will counterweigh some of the cost improvements before the F.O.B. unit prices come down. This somewhat arbitrary intermediate maturity level also assumes simple systems and current prices.

The last bar on the right is indicative of a mature industry with continued volume production, technology transfer, and R&D improvements. This level of $640/kWe average may not be attainable without sufficient RD&D to support it. Note that the F.O.B. unit cost ($300/kWe) is less than the installation cost ($340/kWe). Indeed, some of the costs of installation may be moved on board the unit and simplified. As a test of reasonableness given the substantial reductions presented by the progressive industry maturity bars of the Figure, the F.O.B. unit price of $300/kWe includes less than $60/kWe for its automotive derivative prime mover.

Other key findings of the project and some of their implications are as follows:

- Overview of Prepackaged Cogeneration Awareness/Understanding/Knowledge as Relates to Cost Reduction

  The level of understanding of prepackaged cogeneration by the interfacing publics (owners, property managers and maintenance personnel, architect and engineer design professionals, codes officials and inspectors, etc.) remains a hurdle to lowering total installed costs. Despite all that has been written about the virtues of cogeneration, there remains a significant knowledge gap at the working level. This knowledge gap was commented upon in virtually every activity undertaken during the study. The workshop participants expressed it in the recommendation for a series of guidebooks. In site visits, the criticism of system mistakes was generalized into a knowledge problem. Industry professionals stressed the frustrations and delays encountered as a result of lack of understanding on the part of the interfacing publics. Project developers lamented the time and effort needed to educate not only prospective clients but everyone else connected with a project. The data base indicated few traditional architect/engineer design installations which appears to be due in great measure to the lack of understanding of the potential and design parameters of prepackaged cogeneration systems on the part of traditional architect/engineer designers.

  While lack of understanding is a problem with all new technologies and there are certain market forces, such as competition and the example of others, that work to increase understanding, the critical dimension is time. Technology transfer and educational efforts can accelerate the learning curves with respect to cost reductions in installation (site) and integration (soft) costs while, at the same time, improving the acceptance of these systems as commercial equipment.

- Current Competitive Market Pricing Implications

  Manufacturer product and price list announcements in the Summer and Fall of 1987 have narrowed the competitive price range for prepackaged systems. While there are manufacturers with products priced outside this competitive range, sales volumes are low, at least in one case because the product is priced too high. This current competitive pricing range, F.O.B. factory, for prepackaged cogeneration systems is $700 to $800/kWe at the low end of the 50 to 500 kWe range of interest and $600 to $700/kWe at the high end.

  Given the relatively low sales of units at this point in time of this emerging industry, the above current pricing range should be viewed in terms of current, low-volume production (with one notable single product exception). The target learning curves of the manufacturing cost study indicate target F.O.B. factory prices for cumulative production in the range of 20 to 30 units to have approximately this price level.

  The implications of this competitive pricing range are that, with the one notable exception, manufacturers cannot yet be covering their overheads even if they have appropriate, cost-favorable manufacturing facilities.

- Emerging Industry Infrastructure

  The manufacturer to end user support infrastructure is in a state of flux primarily because the industry is so new and the above-mentioned knowledge gap occurs in many diverse areas. In particular, the traditional architect/engineering specifying firms and individuals that spec out new buildings and renovations are not actively considering packaged cogeneration as yet.

- Design Philosophy/Reliability, Availability, Maintainability (RAM) Tradeoffs

  As indicated in various places above, the first cost versus operating cost issue is alive and well with respect to these units.

- Inter-tie Problem

  The main problem area unique to prepackaged cogeneration involves electrical utility resistance in the form of prolonged negotiations, lengthy approval schedules, excessive intertie requirements and costs, and various other activities.

In summary, to achieve the project objectives with respect to cost understanding and cost reduction, it was necessary to take a "snapshot" of the industry to use as a point of departure. Various approaches were then employed to address not only current but future costs. A number of specific cost reduction concepts were described and ranked for consideration by installers, developers, manufacturers and others. The major finding is that the total installed cost for these systems are amenable to cost improvements and can be significantly reduced by R&D, technology transfer and learning curve activities, especially at the lower end of the 50 to 500 kWe size range.

ABOUT THE AUTHORS

Joseph A. Mulloney, Jr., P.E., CCP, with over 20 years of broad industrial experience, has been involved in several technology-oriented small businesses. He is a member of the Association of Energy Engineers, the American Institute of Chemical Engineers, and a director of the MIT Enterprise Forum of Washington-Baltimore. He is a graduate of MIT, with a Masters of Engineering Administration from Washington University.

Michael T. Panich, P.E., directs the Energy Systems Division of EA-Mueller, Inc. He has over 18 years experience in engineering and the management of engineering projects including private sector and government research efforts. Mr. Panich is a member of ASME and several honorary engineering societies. He is a graduate of Ohio State University and holds a Masters from the University of Maryland, both in Mechanical Engineering.

EA-Mueller, Inc. (formerly Mueller Associates, Inc.), Baltimore, Maryland, is a consulting engineering firm committed to the concept of blending traditional and emerging technologies to best meet a clients' state-of-the-art requirements. EA-Mueller, Inc. is a subsidiary of EA Engineering, Science and Technology, Inc.

Steven M. Knable is Project Manager, Cogeneration Applications for the Gas Research Institute (GRI). Since 1982, Mr. Knable's activities have focused on the markets and business assessments for on-site fuel cell systems. In addition, he has coordinated the market assessment activities for the 40 kWe fuel cell field test program with the participating utilities. Most recently, Mr. Knable's activities have focused on engine-driven cogeneration systems and the coordination of GRI's field test activities. He is a graduate of Southern Illinois University, Carbondale and the University of Illinois, Chicago, with a Masters Degree in Business Administration-Marketing Concentration.

Chapter 65

DIMENSIONING OF COGENERATION IN COMPLEX COMPETITIVE SYSTEM SURROUNDINGS

L. Backlund

Small scale cogeneration in Avesta

The National Energy Administration has by direction from the Swedish Government made a study of the conditions for future small scale cogeneration of electricity and heat in Sweden.
Small scale cogeneration is there defined as cogeneration plants not larger than 25 MW_e. [1]
The analysis in this section deals with an energy system where small scale cogeneration is one of several ways to provide electricity and heat in an industrial area in Avesta, Sweden.
The cogeneration heat has to compete with the waste heat from industrial processes and low temperature heat from the river and ambient air. There is also the municipal district heating system to take into consideration for the supply of heat to the area. In this competitive situation it is motivated to make an analysis of the system in order to find out how to use the different possibilities of energy supply. The time period covered by the analysis is 10 years from 1988 to 1997.

The energi supply situation in the northern industrial area in Avesta

Representatives for Avesta Industristad AB and Avesta AB are making a study of the alternative possibilities of supplying heat to the northern industrial area and the southern industrial area in Avesta. One alternative is to be connected to the district heating system in the city of Avesta. Another alternative is to install a diesel engine based cogeneration plant in the northern industrial area. Combined with the cogeneration production the industrial waste heat may be utilized for heating purpose. By connecting the northern area to the southern area by a heat culvert the waste heat of the southern area can be used in both areas for space and tap water heating.

The structure of the energy system

In figure 1 the energy system has been mapped into a formal structure. One of the energy supply options is the river Dalälven. The temperature of the water in the river is very low +0.03 °C in wintertime. Therefore it is not possible to simply circulate the water through the condensor of the heat pump and thereby extract heat from the water. Yet it is possible to extract heat out of the river by applying a pipe system on the bottom of the river. By circulating a brine through the pipe system connected to the condenser it is possible indirectly to extract low temperature heat from the bottom and the water of the river. Similar heat extraction systems have been built in lakes and rivers in the north of Sweden. The system analysis will show if this kind of heat extractor is able to compete with other options in this case.

The competition between heat production of cogeneration and industrial waste heat

The cogeneration is exposed to competition from industrial waste heat if the heat culvert is built between the southern industrial area and the northern industrial area where the cogeneration plant may be situated. On the other hand the heat culvert opens the possibilities of transferring heat from cogeneration to the southern area if heat from cogeneration can compete with the industrial waste heat. The time functions of waste heat and heat load combined with the electric load and the time function of the price of electricity establish a complex system situation. On top of that the different energy equipment have their own time functions of efficiency and availability. The long term changes in the ratio between the prices of electricity and fuels will further complicate the picture. Yet the degree of complexity is irrelevant for the possibilities that the system as such may contain valuable potential of unused energy and energy quality. However the complexity of the system is relevant when deciding on ways of discovering the hidden potentials of energy and energy quality. [2]

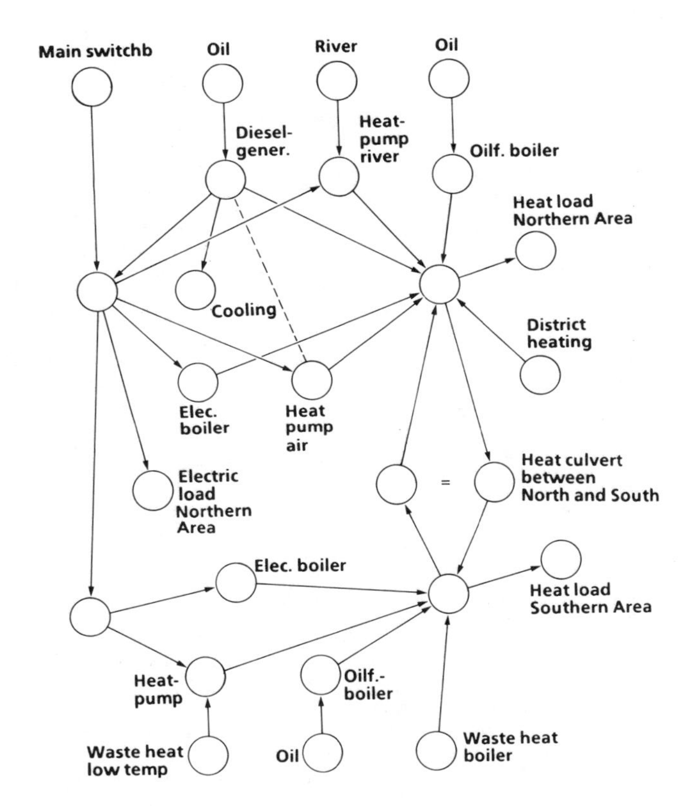

Fig. 1. System structure

System solution

The result of the analysis is presented in tables. The system model Modest has been used. A ranging operation has been carried out as an aid when deciding how robust the solution is with regard to the boundary conditions used in the analysis.
These conditions should be observed when judging the result of the analysis. Certain values have been given to parameters of price changes, inflation, interest rates, price of equipment etc.

Small scale cogeneration

The solution shows that small scale cogeneration is competitive in this system environment. This is not a priori selfevident because of the competition from industrial waste heat.

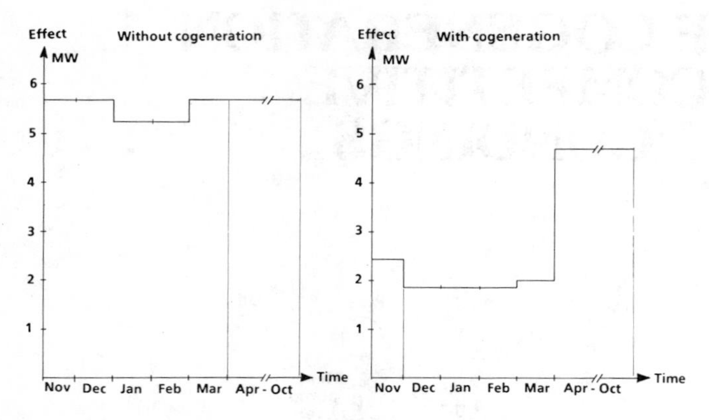

FIGURE 2. UTILIZED OUTPUT FROM THE ELECTRIC NET

An important factor in favour of the cogeneration is the time function of output cost from the electric net. The time functions of waste heat and low temperature heat from ambient air work in the same direction. Characteristic of the time functions of the latter is that they have time periods of low output or no output at all mixed with time periods of full output. This is to a certain degree a disadvantage for these heat sources. The system model takes this into consideration together with other facts when deciding the dimensioning of the energy equipment in the system.

Heat culvert connection

The optimal solution contains the construction of a heat culvert. This culvert connects the southern industrial area to the northern area. The heat culvert is used to transfer heat from the cogeneration plant in the northern area to the southern area during time periods when the cogeneration is in operation.
During the time period May, June, August and September the heat culvert is used to transfer heat from south to north. This means that in this period industrial waste heat is used to heat the northern area as well as the southern area.
During the summer vacation when no industrial waste heat is produced the heat pump based on ambient low temperature heat in the northern industrial area provides heat to the southern area by the heat culvert.
It should be noticed that the obtained dimensioning of the heat culvert is low, only 383 kW. In reality the given specific cost of the heat culvert 2500 SEK/kW_{th} is not valid for the construction of such a small heat culvert. By ranging it could be seen in table 1 that the cost may be raised to 2512 SEK/kW_{th}. The difference is so small that it does not really matter in this context. The heat culvert between the northern industrial area and the southern industrial area will be omitted in the optimal solution if the cost of the cogeneration plant goes up above the upper level of the ranging value. This highlights the fact that the results of the system analysis must be scrutinized for error detection and plausibility check. As soon as there is an analysis result a check has to be made whether the used boundary conditions and linear functions are relevant for the obtained result. If this is not the case a new analysis has to be made with the new relevant boundary conditions and linear functions. In this case the reason for the difference between the used specific cost of the heat culvert and the real specific cost of the heat culvert is that the analysis result gave a small size culvert. The linear function for specific heat culvert cost assumed that a large size culvert might be used, if any, in the solution.
The remedy to this is to use a linear function that maps the real specific cost function in a wider range of heat culvert size. One way of doing this is to use two or more separate linear functions for different ranges of heat culvert sizes. The linear functions are then combined with conditions that the optimal solution can comprise at

Table 1 Result of analysis
The energy equipment. The optimal solution.
Ranging

Equipment	Size	Investm cost used value	Ranging Upper limit	Lower limit
Cogenerator	3144 kWe	1977 SEK/kWe	2000 SEK/kWe	1780 SEK/kWe
Electric boiler North	0	650 SEK/kWe	8 SEK/kWe	
Heat culvert	383 kW_{th}	2500 SEK/kW_{th}	2512 SEK/kW_{th}	2395 SEK/kW_{th}
HP river	0	4500 SEK/kW_{th}	4148 SEK/kW_{th}	–
Oilf boiler North	7647 kW_{th}	0 SEK/kW_{th}	–	–
*) District heat	0	240 SEK/kW_{th}, year	6.34 SEK/kWth, year	–
HP ambient air	1440 kWth	4200 SEK/kW_{th}	4337 SEK/kW_{th}	4185 SEK/kW_{th}
Electric boiler South	3417 kWe	0 SEK/kWe	–	–
Waste heat <70	3433 kW_{th}	0 SEK/kW_{th}	–	–
Waste heat boiler	2000 kW_{th}	0 SEK/kW_{th}	–	–
Oilf boiler South	16244 kW_{th}	0 SEK/kW_{th}	–	–
HP Waste heat	4806 kW_{th}	4000 SEK/kW_{th}	4134 SEK/kW_{th}	33985 SEK&kW_{th}

*) Subscription fee

the most one range out of the given ranges of heat culvert sizes. To solve this problem mixed integer linear programming has to be used. Although this can be done it is easier to start with an initial analysis using simpler linear functions. The analysis result will then indicate which relations have to be mapped more thoroughly. The new analysis will then show if the improved mapping has given a more realistic result. By doing this in an iterative way the result will be an optimal solution with an appropriate accuracy. [2]

The district heating system

The optimal solution did not contain a connection of the analysed energy system to the district heating system of the city of Avesta. The ranging operation gave the upper price levels of output and energy to make the utilization of the district heating system possible. (See table 3).
The given energy rate 0.19 SEK/kWh is low enough to consider utilizing district heat at peak load time (see table 3). The reason why this is not the case in the optimal solution of the system is that one more condition has to be met by the district heating system. The cost of output from the district heating system has to be decreased from 240 SEK/(kW, year) to just above 6 SEK/(kW, year) (see table 1).
If this is done the optimal solution contains the connection of the analysed energy system to the district heating system.
Energy from the district heating system is then utilized during the peak load periods. In addition to decreasing the output cost the energy prices have to be lowered below the limits in table 3 to make district heating worth while during off peak load periods.
The connection of district heating to the analysed system does not force the cogeneration out of the optimal solution. The result will be a smaller cogeneration plant which will be used in much the same way as before. A new analysis with the new output and energy costs of district heating will give the optimal solution with all relevant facts for the new situation.

The system cost for the present system in Avesta

To be able to compare the total discounted system cost of an optimal solution with the total discounted cost of the present system an analysis is made of the optimal use of the present system. (See table 4).
An analysis has also been carried out of an alternative system situation where the cogeneration possibilities have been blocked. (See table 4).

Table 2. Result of analysis
Utilized electric output from the main electric network.

Month	The maximum output (kW) from the net at optimization Without diesel electric generator	With diesel electric generator
November	5672	2358
December	5672	1858
January	5249	1858
February	5249	1858
March	5672	1957
Apr - Oct	5672	4649
Subscribed effect	5672	3503
High load effect	5566	2007

Table 3. Result of analysis
Municipal district heating. Ranging

Time element			Used energy price SEK/kWh	Ranging Upper limit SEK/kWH
Nov	Mon-Fri	22-06	0.190	0.141
		06-22	0.190	0.047
	Sat, Sun		0.190	0.180
	Peak load		0.190	0.197
Dec	Mon-Fri	22-06	0.190	0.149
		06-22	0.190	0.149
	Sat, Sun		0.190	0.180
	Peak load		0.190	0.197
Jan	Mon-Fri	22-06	0.190	0.197
		06-22	0.190	0.197
	Sat, Sun		0.190	0.187
	Peak load		0.190	0.197
Feb	Mon-Fri	22-06	0.190	0.187
		06-22	0.190	0.187
	Sat, Sun		0.190	0.180
	Peak load		0.190	0.197
Mar	Mon-Fri	22-06	0.190	0.149
		06-22	0.190	0.086
	Sat, Sun		0.190	0.180
	Peak load		0.190	0.197
Apr, Oct	Mon-Fri	22-06	0.190	0.180
		06-22	0.190	< 0
	Sat, Sun		0.190	0.180
	Peak load		0.190	0.197
Maj, Jun, Aug, Sep				
	Mon-Fri	22-06	0.190	0.111
		06-22	0.190	0.081
	Sat, Sun		0.190	0.191
	Peak load		0.190	0.194
July	Mon-Fri	22-06	0.190	0.047
		06-22	0.190	0.047
	Sat-Sun		0.190	0.047
	Peak load		0.190	0.219

Table 4. Result of analysis.
The total discounted system cost

Analysis	The total discounted system cost	
The present system optimally run	218	MSEK
The optimal solution when the possibility of installing and using cogen, is blocked	191	MSEK
The optimal solution where the cogen is used	183	MSEK

Result. Conclusions

- Cogeneration is competitive in this system environment.
- The optimal solution did not contain a connection of the analysed energy system to the district heating system of the city of Avesta.
- An important factor in favour of the cogeneration is the time function of output cost from the electric net.
  The time functions of waste heat and low temperature heat from ambient air also favour cogeneration.
- The optimal solution contains the construction of a heat culvert which connects the southern industrial area to the northern area.

References:

1. Small-Scale Heating and Power Production.

   National Energy Administration,
   Sweden 1986:1 (In swedish).

2. Optimization of Dynamic Energy Systems with Time Dependent Components and Boundary Conditions

   Backlund E. L.
   LITH 1988

Chapter 66

COGENERATION VS. CENTRAL POWER PLANTS: A DIFFERENT PERSPECTIVE

C. F. Carver

In the last decade, cogeneration has been rediscovered. From its embryonic beginning, more than 70 years ago, cogeneration has come back. The cogeneration concept is challenging large central power plants. With improved diesel and combustion turbines, cogeneration has improved its reliability, and it has provided the hardware to utilize renewable energy resources. There is a place in our future for cogeneration.

The most striking feature of cogeneration is its near total energy utilization. If you consider the energy utilized for shaft work, plus the energy recovered for thermal use, you would believe that a cogeneration system is about 70% efficient. Since the standard power plant is only about 35% efficient, and distribution/transmission loss another 5%, it is very easy to come down on the cogeneration side.

This is what our politicians, federal and state agencies have done. And why not? Surely it is better to operate a 70% efficient system than a 30% efficient central power system. After all, if the nation would burn 40% less imported oil, the economy would improve; our national defense would be better; and our environment would be cleaner. This must be true, as the Europeans and Japanese have been successfully cogenerating for years.

However, there is a problem. In many cogeneration operations, the 30%-70% efficiency split "aint" true. In some cases there are only 17 percentage points difference.

In the future, the most probable common cogeneration application will be a diesel or combustion turbine, which is base loaded (providing constant, steady power), and providing steam/hot water for heating, air conditioning and domestic hot water. These units are already on the market. They may be fairly easy to install, and can be skid mounted or totally packaged.

The problem is that they will burn nearly the same amount of natural gas or oil as a central power plant. The reason is that not all BTU's have the same capacity. I know I just contradicted the laws of thermodynamics, or did I? Actually, the laws of thermodynamics state that all sources of energy have the same capacity to perform equal amounts of work. A machine is a device that transforms energy into work.

The key is that some machines are more efficient than others. Because of the machinery involved, you can generally obtain more work out of electrical BTU than a thermal BTU....What a concept!!!

In my cogeneration scenario, the culprit is the absorption air conditioner. An absorption chiller is not as efficient as a centrifugal chiller. Even disregarding parasitic losses, the best absorption unit (COP = 1.3) cannot compare with a good centrifugal chiller (.6 kW/ton).

Capturing waste energy and obtaining work from the waste energy, is cogeneration's main attraction. And yet, it is this feature than cannot compete with purchased power.

There are many good appliications for cogeneration. A cogeneration package, utilizing absorption units, can be economically justified when electric costs (¢kWH and $kW) are high and gas/oil prices are low, or when a waste energy source is available. It is harder to justify cogeneration-absorption units when minimizing total energy consumption is your main concern. Politicians, federal and state agencies should take note.

Please refer to figures 1 and 2.

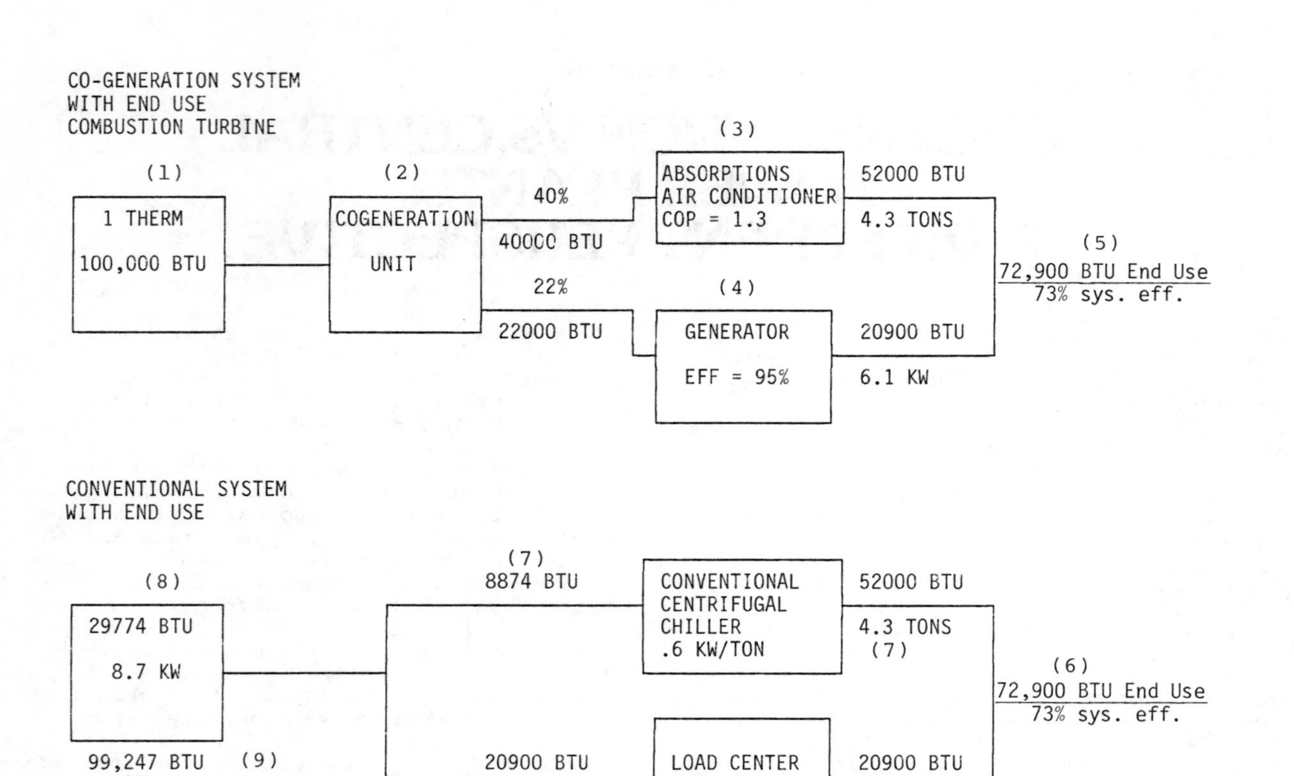

FIGURE 1.

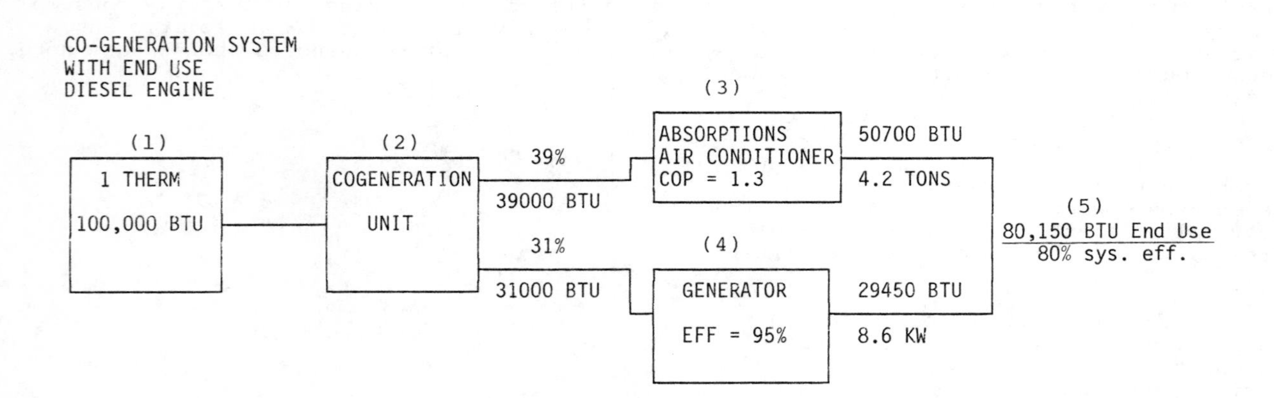

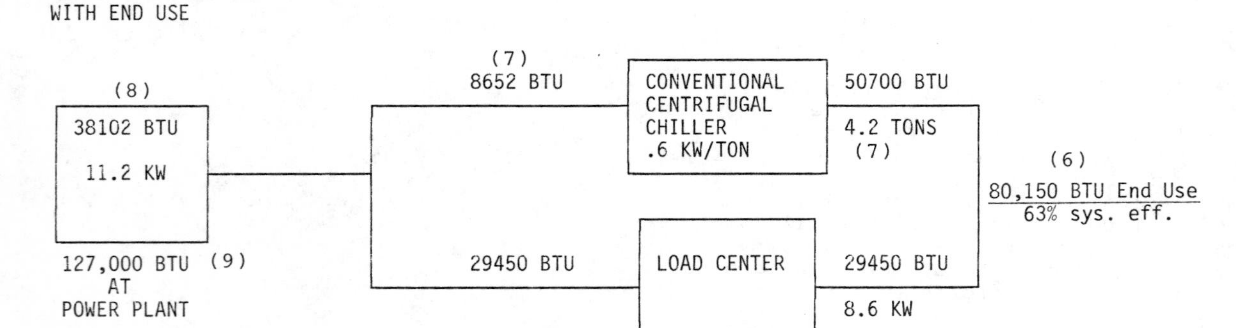

FIGURE 2.

QUESTION - How many BTU's are required, at a central power plant, to perform the same amount of work as one Therm used in a cogeneration System?

ANSWER - Combustion turbine cogeneration unit - 99,247 BTU
- Diesel cogeneration unit - 127,000 BTU

1. One Therm (100,000 BTU) is burned in the prime mover of a cogeneration unit.

2. Combustion turbine - 40% recovered as hot water/steam
22% recovered as shaft work

Diesel - 39% recovered as hot water/steam
31% recovered as shaft work

3. Hot water/steam is utilized by an absorption air conditioner to make chilled water:

Combustion turbine - 4.3 tons (52,000 BTU)

Diesel - 4.2 tons (50,700 BTU)

(parasitic electrical loads not included)

4. Shaft work turns a generator:

Combustion turbine - 6.1 kW (20,900 BTU)

Diesel - 8.6 kW (29,450 BTU)

5. Net effect (work accomplished):

Combustion turbine - 72,900 BTU (73% system efficiency)

Diesel - 80,150 BTU (80% system efficiency)

6. Same amount of works is accomplished by a conventional power plant.

7. As compared to:

Combustion turbine - 8,874 BTU's are required for 4.3 tons

Diesel - 8,652 BTU's are required for 4.2 tons

Conventional contrifugal chiller = .6kW/ton = COP = 5.9 - parasitic electrical loads not included.

8. Power required at the meter as compared to:

Combustion turbine - 8.7 kW (29,774 BTU)

Diesel - 11.2 kW (38,102 BTU)

9. Energy required at central electric power plant (as compared to):

Combustion turbine - 99,247 BTU

Diesel - 127,000 BTU

Central power plant efficiency = 35%
Transmission/distribution efficiency = 95%

Chapter 67

COGENERATION OPPORTUNITIES IN DEVELOPING COUNTRIES

J. A. Rodriguez

INTRODUCTION

A sizable portion of Electric Utilities in developing countries were owned and operated by American and European holding companies using rates and procedures essentially based on the Edison Electric Institute guides until the 1960's.

After the $320 million uncompensated expropriation of the Cuban Electric Company in 1960 the electric utilities were gradually turned over completely to government agencies

These agencies were not able to resist the political pressures to set rates on a populistic rather than economic basis. Proper project evaluation has not been practiced and sound administrative procedures have been left aside. As result, and time has passed, service has deteriorated.

(1) Approximately 35% of the total debt of developing countries is related to the electrical sector of the economy.

DESCRIPTION OF THE OPPORTUNITIES

In most developing countries there is no equivalent of PURPA, so for now and the near future Cogeneration will be limited to in house power and energy requirements. As we may see in IEEE standard 739-1984 "Approximations for Determining Cogeneration Feasibility" shown on Fig. 1:

When gas is less than $3.oo/MBTU and electric cost is more than 25MIL/KWH the two major factors are in the favorable range. It is presumable that most process industries will be viable for Cogeneration.

As authors of this standard indicate "if the cost of electricity is 30MILS per KWH and fuel at the industrial plant is $2.oo/MBTU then economic feasibility for Cogeneration is likely".

Numerous process industries that use steam or heat, not using Cogeneration, are paying for power between $.05 to $.08 per KWH, or even more, and a very large portion have natural gas available at less than $1.oo per MBTU. Many operate over 7000 hours per year. Their load generally ranges from 0.5MW to 10MW.

In spite of these conditions there is practically no Cogeneration at present in Latin America, as most industries and institutions still depend on central power. When the energy crisis of the 70's became a reality, over protected industries in Latin American were not able to adjust and they were spared from the crisis only by government subsidies and/or by protective tariffs.

Industry and other private and public institutions are starting to recognize that it is essential for them to have more reliable power. We must make them aware of the fact that Cogeneration will provide them with good quality power with appreciable savings in their energy expenses.

Exceptional opportunities exist in this area for Cogeneration packagers and for firms able to provide turn-key solutions for specific projects. Integration with existing components and/or local suppliers may be necessary.

SUMMARY OF MAIN FACTORS

1) Most rates applied by electric utilities do not recognize high load factors as a basis for incentive discounts.
2) Most rates do not allow a discount for energy use on Sundays, Holidays or nights.
3) Natural gas rates are generally low and in many locations gas is available at less than $1.oo/MBTU.
4) Process industries pay high rates per KWH and this power could be cogenerated at a much lower cost.
5) Numerous process industries using steam or heat in some form are now buying electric power at rates between $0.05 and $0.09 per KWH.

REFERENCES

1) IEEE Recommended Practice for Energy Conservation and Cost-Effective Planning in Industrial Facilities. IEEE Standard 739-1984.
2) N.E. Hay Editor. "Guide to New Natural Gas Utilization Technologies". American Gas Association, Second Printing April 1986.
3) Jay Althoff-The Trane Co. "Marketing and Productivity Opportunities of Computer Aided System Design.
4) "Integration of Efficient Design Technologies". Proceedings of the 10th World Energy Engineering Congress 1987 - Section 2.
5) P. R. O'Connor. "Competition and Functional Segmentation in the Electric Utility Industry" IEEE Power Engineering Review - December 1987 - Pages 3, 4 and 5.
6) Servir la Deuda y Servir a la Economía Social" (Serving the Debt and Serving the "Social Economy") - "Perfil de coyuntu ra económica". September 1987 Nº. Univesidad de Antioquia A A 1226 Medellín, Colombia pages 35 to 38.
7) Directory of Utilities "1987 IPS-International Power System" By the Editors of "Electrical World" and "Power" pages 121 to 125
8) G.R.Guinn Ph. D.,"ANALISYS OF COGENERATION SYSTEMS USING A PUBLIC DOMAIN SIMULATION". Director of the Alabama solar Emergy Center the University of Alabama in Huntsville
9) G.R.Guinn Ph.D "Preliminary Review of Cogeneration Feseability Studies for Gulf States Steel" University of Alabama Huntsville, Sept. 11,1987.
10) G.R. Guinn Ph.D."A study of the feasibility of Cogeneration in Courtaulds North America" March 31, 1988 Alabama Dept. of Economic and Community Affairs Science, Technology And Energy Division.

Fig. 1
Approximations for Determining Cogeneration Feasibility

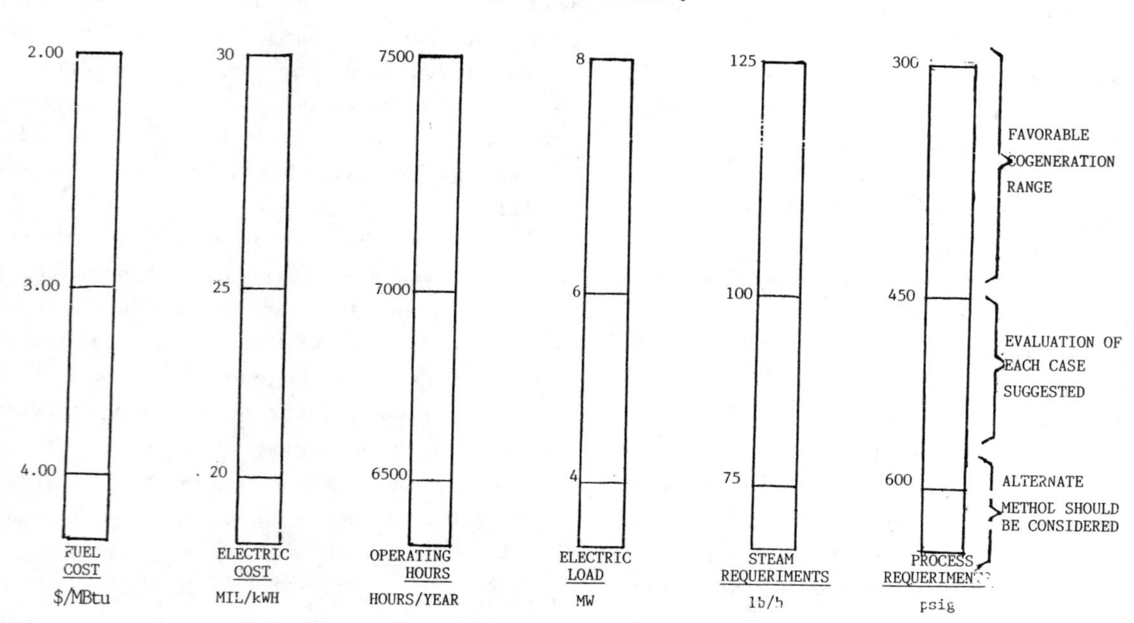

11) G.R.Guinn Alabama Cogeneration manual.
12) "INTERCONEXION ELECTRICA S.A. INFORME DE OPERACION" 1986 Medellín, Colombia 1987.
13) "Situacion Energética de la Costa Atlantica-El Sector Electrico" Corelca-Abril 1987. Barranquilla, Colombia-1987.
14) "Desarrollo Futuro del Sector Electrico by Oscar E. Mazuera Gonzalez Director Ejecutivo Corporation Autonoma Regional del Cauca -CVC". Plan de Economía Social Direccion Nacional de Planeación Bogotá, Colombia August 1987.

SECTION 14

ENVIRONMENTAL IMPACTS OF COGENERATION

Chapter 68

ENVIRONMENTAL IMPACTS OF COGENERATION AND SOME BAD ALTERNATIVES

J. R. Ross

ABSTRACT

This paper identifies various cogeneration types, and places in perspective the environmental impacts of cogeneration as a preferred alternative to other forms of heat and power generation with emphasis on waste heat and fuel as opposed to fossil fuel use. A methodology for deciding about cogeneration and the avoidance of environmental problems is offered.

Types of Cogeneration

There are many possible definitions for cogeneration, however the most generic categories are topping and bottoming systems. A topping system primarily generates electricity, and an alternate use is found for the exhaust steam through back pressure or extraction turbines or for exhaust gases to produce steam or other usable heat. A bottoming system produces heat to facilitate a process, and the excess is captured for electric power generation or other uses as a by product.

In more specific terms, some typical cogeneration methods include use of waste process gases to drive gas turbines for electric generation, recapture of low pressure steam for electric generation or driving machinery and use of waste steam or heat for electric generation and peak shaving. Incineration of waste materials as fuel for steam generation, process heat, district heating or air conditioning or electric generation are cogeneration options which are becoming increasingly popular.

Heat Recovery From Industrial Processes

A common form of cogeneration is recovery of heat which has been used for an industrial process. The fuel has already been added to the process by the time heat is to be recovered, and it is simply good business to make recovery efforts. Normally, no additional fuel must be added to create the cogeneration effect. The environmental impact of this cogeneration process is essentially zero beyond that of the original process. If the basic process has been designed to avoid negative environmental impacts, the recovery of heat will have little further environmental impact.

Heat Recovery From Diesel Engines or Turbines Driving Electric Generators

Recovered heat from diesel engines or turbines used for electric generation is a form of cogeneration which is being used successfully for heating buildings and generating process heat. While this concept is a recognized part of cogeneration, its main thrust is to generate electricity at about half the cost of commercial power. Diesel engines still use fossil fuel, and their use for electric power generation may not be as efficient from a thermal efficiency perspective. This condition could contribute further to the decline of oil reserves.

There is also a concern about the impact on the environment from the exhaust gases of the engines or turbines. Notable among the gases in the exhaust stream is Nitrous Oxide (NOx). A subsequent paper will address this pollutant. Sulphur dioxide may also be present in the exhaust gas stream. Unburned hydrocarbons are contained in the exhaust stream to the extent that the engine or turbine is inefficient or that waste gases are not recycled into the intake of the unit or neutralized to prevent air pollution.

The use of diesel engines in cogeneration facilities also produces waste oil which must be disposed of as a hazardous waste because of its heavy metal content.

Organizations designing units of this type should be aware of the need to reduce gas emissions and to the extent possible to minimize the impact on the environment.

Refuse Burning

A major thrust currently underway in many countries is the burning of municipal waste as a means of producing energy in the form of steam, electricity or both. Energy production is a secondary purpose of these facilities commonly referred to as resource recovery plants. Some major success stories and some terror tales have developed from the increased use of this form of cogeneration.

Nashville, TN for example heats and cools the entire downtown area by burning municipal waste in a thermal transfer plant. Although environmental problems, primarily, odors beset this plant in the early stages of operation, it is essentially a trouble free operation at this time.

A smaller city, Gallatin, TN performs a resource recovery process for metal and glass removal, and burns the combustibles for process steam at adjacent manufacturing plants with the excess being used to generate electricity. This plant too had initial air pollution has been resolved.

Some resource recovery plants have failed due to technical problems such as improper design, and others due to mismanagement or unrealistic expectations. Other have failed due to public outcry, as nobody wants a garbage burning plant next door to one's home. Once the perception of such facilities becomes negative, technical excellence and environmental acceptability become inconsequential.

Landfill vs Waste Burning

Landfills have become popular with self styled environmentalists because there is no air pollution involved. The battle cry of environmental extremists in the seventies was to stop burning.

To them, any air pollution was intolerable, and the only alternative offered was to put all waste into landfills. We are only now beginning to see the folly of this bad alternative.

First, landfills are becoming filled to capacity in many locations, sites for landfills are becoming increasingly difficult to find as land with suitable characteristics does not exist in many locales. Florida, for example has few hollows, and is forced to build mounds of garbage on flat sandy soil where the water table is about four feet below the bottom of the landfill. Most of the waste in these landfills could be burned with little impact on air quality . Other landfills are located within a mile of fresh water lakes from which drinking water is pumped. The impact on water quality in the next century may be devastating due to the unknown long term effects of leaching of plastics and other chemicals which may be greater in the long run than a temporary air quality problem.

Technology has kept pace for the solution of air quality problems, even though the solutions are too expensive to implement in many cases. Advocates of landfills contend that technology will keep up, and that by the time this gets to be a real problem, there will be a technological solution. This same argument when used in defense of nuclear power plants is loudly rejected by the anti-nuclear activists, many of whom are also landfill advocates.

How to Dispose of Waste

Waste is something that cost money. For example, packaging materials which end up in the trash, have been paid for as part of raw material cost. The beginning of the waste stream is typically far away from the landfill or trash dock, and is really the result of a decision process which begins in product design, purchasing, engineering and even on the top manager's desk of manufacturing firms which are both customers and sellers of packaged goods. It should never be assumed that any waste item cannot be eliminated, utilized more effectively or recovered. If the organization recognizes this concept, much of the waste can be eliminated at the source, and recovery efforts will be unnecessary. If the organization has a consistent value system and strategy aimed at waste reduction, much can be accomplished.

Paper and wood which represent almost half of most municipal and industrial waste streams can be burned with essentially no negative environmental impacts. Wood and paper smoke contains no sulphur, dioxins or NOx, and while visible are essentially harmless. Some plastics may be burned with little impact while others produce noxious gases which must not be emitted into the atmosphere. The best use of plastics is to sort and recycle them for conversion into specific recycled plastic or into a usable generic plastic rather than risking their decomposition in a landfill or burning them with adverse effects on air quality.

Metals should be segregated from the waste stream and not subjected to landfilling . Some landfills actually bury old washing machines, refrigerators, conveyors or other heavy metal objects which could be sold to a metal recycler. Generally, the landfill operator can profit from eliminating metals from the landfill and selling them to a recycler. The technology for segregating aluminum from steel is now developed, and if done, burning of trash can be made more efficient and profitable.

Communities which use curbside segregation of trash (metals, glass and plastic)not only get the value of the recycled materials, but also reduce the volume of materials which are passed through the incineration process, eliminate these materials from landfills and avoid negative environmental impacts of mixed burning or landfilling of unsegregated trash. Some communities which do not use curbside trash segregation use prison or welfare recipient labor for trash segregation to reduce the cost of the operation.

If waste burning includes plastics, tires or other substances which produce toxic fumes which may damage the ozone layer or create health problems for people, recycling is a preferred alternative. In any event, adequate safeguards must be taken to avoid damage to air quality.

While air quality is only slightly involved, further segregation of municipal waste should remove food waste, grass clippings and similar organic material which does not burn well, but which can be composted successfully, and the compost used for organic fertilizer.

Hazardous Waste Burning

In some cases, volatile hazardous material can be burned in an incinerator on site and the heat so generated can be reclaimed thereby qualifying the facility as a cogeneration installation. Hazardous volatile wastes are often blended in with other fuels, and can be burned with minimal environmental impact.

There is a whole continuum of options relating to the disposal of hazardous waste. Burning of hazardous waste is one of the less desirable options on the table as identified by Turner etal in Reference # 22.

If a hazardous waste problem exists, the options of eliminating it at the source or recycling on site are much more desirable both from an economic and environmental perspective. On balance, however, it is often less costly to burn hazardous volatile wastes on site than it is to pay a disposal service to haul it off and dispose of it, depending on volume and type of waste. The technology for hazardous waste burning is rapidly catching up to the demands of industry.

Electric Generation Using Fossil Fuels

Commercial electric utilities chiefly use coal, oil or natural gas as fuel for electric generation. Anyone who listens to television news or reads a newspaper will know that burning of coal is a problem due to acid rain which is produced by sulphur dioxide and nitrous oxides mixing with rain water. Coal became a popular substitute for oil as fuel for electric generation in the mid-seventies due to the oil shortages perceived in that era, even though many electric utilities had only a few years before been forced by the E.P.A. to convert to oil to comply with air pollution regulations. This capital drain caused by switching fuels in quick succession has precluded implementation of many more attractive cogeneration alternatives which could have saved energy and money. Electric utilities could augment coal with waste fuels, and reduce the percentage of sulphur dioxide in the emissions to more acceptable levels.

A very viable solution is to mix wood waste or sawdust with coal. Another solution would be for the utilities to become contract waste burning facilities.

Nuclear Power Generation

The ultimate environmental impacts of nuclear power are yet to be determined. Nuclear plants are built to the state of the art and limits of knowledge which exist at the time of their construction, but this is continually changing. At the moment of generation, there is essentially no environmental impact from nuclear power generation, however the impact on the environment may occur many years in the future if a solution to the nuclear waste problem is not found. The issue of nuclear waste is continually with us, and ultimate disposal concepts are still years away.

In building a nuclear plant, most companies must spend far more money than planned to comply with ever more stringent regulations and enforcement procedures. Little thought has therefore been given to opportunities for cogeneration at nuclear plants. Hopefully, as this industry matures, further effort to take advantage of opportunities for cogeneration in nuclear power production will yield positive results.

Wood Waste Cogeneration

One of the bright spots in the cogeneration vs. environment picture is the wood products industry. Wood contains no sulphur or other injurious fractions. It even smells good when burned, and the smoke can be burned without trace if temperatures are sufficiently high.
A very successful cogeneration installation was made in 1980-81 by a small wood products company. The manager was looking for ways to utilize more of the wood waste, much of which was being sent to a landfill or sold at a loss for papermill fuel. He began investigating cogeneration, and since a boiler was already installed, the step to cogeneration was relatively easy.

Regulatory hurdles and contract negotiations with the utility had to be surmounted. As this was a new concept to the utility, a cogeneration policy had to be developed.

Using funds obtained from a DOE Appropriate Technology Grant, the client purchased a used steam engine and generator set, and proceeded to install the unit. New interface equipment was also installed. While some downtime to get parts for the steam engine has been experienced, the client has received about $50,000 per year in revenues for electricity sold to the utility. This is just one of many cases where cogeneration from wood waste has been made profitable without damage to the environment.

A larger wood products plant installed large steam boilers, turbines, electric generators and pollution control equipment, and effectively isolated itself from the outside utility with the result that wood waste was being disposed of productively and electric costs were reduced significantly. A major chemical company installed a wood fueled power station at its major facility, and obtains waste wood from a 100 mile radius of the chemical plant. Adequate pollution controls are incorporated into the power station.

Utilities which burn coal as fuel for electric production have experimented with blending wood shavings with pulverized coal with the result that the sulphur content of emissions was reduced.
In some cases the blending of wood with coal brought the total emissions within acceptable tolerances not being achieved with coal alone.

Some notable failures occurred in the late seventies in attempts to utilize wood waste through pyrolysis, however these failures increased the body of knowledge, and will ultimately lead to future successes.

Other Cogeneration Options

As research continues, other types of cogeneration will be discovered, however there will be environmental objections to each both valid and fictional. Engineers must continue to search for new solutions to energy and environmental problems with the objective of extending the life of this planet and stretching its resources.

Bad Alternatives to Cogeneration

We have already discussed the negative aspects of burning coal which produces acid rain, and the pollutants present in diesel exhaust emissions.

We have also mentioned the long term negatives of landfills as a means of disposal of solid waste, but have also discussed the options of cogeneration, recycling and resource recovery which can make solid waste disposal environmentally acceptable.

Since 1981, oil has been plentiful, even though sources of supply have been in jeopardy from a military standpoint. Oil prices have moderated, and the higher costs have been incorporated into corporate and individual budgets. The goals of energy independence so loudly espoused in the late seventies have been dropped due to this fortunate turn of events, however the world's oil supply is limited, subject to political manipulation and undependable. Cogeneration can reduce dependence on oil, and can do so with less impact on the environment than continued oil consumption.

Cogeneration is a viable option to the above bad alternatives, and it behooves industry and government to act decisively to promote cogeneration as an environmentally preferable alternative to wasteful or polluting practices.

Considerations About Cogeneration

Organizations considering cogeneration should go through a rigorous decision process which would address corporate strategy, capital availability, costs vs. benefits, dependability of waste fuels or energy forms, environmental impact, public perception of the project, commitment by management and availability of management people to keep the system going long term. (See Ref. #s 17,18)

The following additional concerns should impact the decision about cogeneration:

A. Determine first if all or part of the waste to be recovered be avoided. If the process is designed properly, and the proper decisions are made about waste avoidance, perhaps the entire cogeneration project can be avoided. You don't have to cogenerate if you don't have the waste in the first place.

B. Isolate any part of the waste stream, either material or gases which can be recycled in other ways to reduce the waste stream. If the waste stream can be reduced, the size of the cogeneration facility can also be reduced to handle what's left.

C. Get the maximum benefit from the cogeneration and recovery effort by finding every applicable waste, and adding more input to the system as opportunities become available.

D. Avoid damage to the environment by following steps A and B, and by installing environmental safeguards at every level of the recovery process.

CONCLUSION

As with any innovative concept, cogeneration has some problems which must be resolved, including those associated with environment. Too often critics of any new idea will see only the negatives, and will ignore the positives which far outweigh the a problems. Where environment is concerned, people often with hidden agendas, may disrupt worthwhile projects because they ignore the positive benefits.

Despite some negative environmental impacts, cogeneration is an attractive alternative to other forms of heat and power generation. By using the same energy more than once, cogenerators actually reduce environmental impact as well as save energy, and the environment and the energy supply are both positively impacted. The net impact of cogeneration on the environment is significantly less than other forms of energy consumption and production.

Organizations wishing to cogenerate should consider the environment, and take all necessary safeguards. In some cases, an objective of cogeneration may be to improve the environment as well as save energy or reduce energy costs. Future cogenerators should accept environmental problems as opportunities, and should proceed to obtain the significant benefits which cogeneration can yield.

BIBLIOGRAPHY

1. Baker, Andrew J., and Clark, Edward H., "Wood Residue as an Energy Source--Potential and Problems," Forest Products Laboratory, USDA Forest Service, Madison, WI

2. Brunner, Calvin R., P.E., "Small Scale Resource Recovery: An Overview", Waste Age, November 1985

3. Cheremisinoff, Paul N. and Morresi, Angelo C., Energy from Solid Wastes, M. Dekker, NY (1976)

4. Dorgan, Charles E. and Baker, William R., "Cost Analysis of Energy Proposals," Proceedings of "Energy Management Workshop", Virginia Polytechnic Institute and State University, Blacksburg, VA (1977)

5. Ellis, Thomas H., "Should Wood Be a Source of Commercial Power", Forest Products Journal, (Oct 1975)

6. Forest Products Research Society, Wood Residue as an Energy Source, U.S. Department of Agriculture, Madison, WI

7. Halzhauer, Ron, "Selecting Waste Heat Recovery Equipment", Plant Engineering, (Aug 1978)

8. Hill, Christopher T. and Overby, Charles M., "Improving Energy Productivity through Recovery and Reuse of Wastes", Energy Conservation and Public Policy, Prentice Hall, (1983)

9. Jackson, Frederick R., Energy From Solid Waste, Noyes Data Corp., Park Ridge, NJ (1974)

10. Kulvicki, Ray, "Cogeneration: One Option in the Search for Energy", Timber Processing Industry, (April 1978)

11. Lawn, John, "Cogenerators See New Promise, Face Old Problems, Energy Management, (Dec/Jan 1981)

12. Lohius, D.J., "Energy from Biomass Wastes", AEE World Energy Engineering Congress, (1978)

13. Nemerow, W.I. and Nelson, Leonard, Industrial Solid Wastes, Ballinger Publishing Company, (1984)

14. Rimberg, David, Municipal Solid Waste Management, Noyes Data Corporation, New Jersey and England (1975)

15. "Resource Recovery in Sumner County, Tennessee", Sanders & Thomas, A STV Engineers Professional Firm, Nashville, Tennessee (1977)

16. Resource Recovery Through Incineration, American Society of Mechanical Engineers, Incineration Conference Proceedings, (1974)

17. Ross, James R., "Cogeneration - A Concept for Today", American Institute of Industrial Engineers Spring Annual Conference and World Productivity Congress, 1981.

18. Ross, James R., "Strategic Planning for Solid and Hazardous Waste Management and Resource Recovery", Institute of Industrial Engineers Integrated Systems Conference, 1987

19. Streb, Alan J., "Cogeneration-What's Ahead", Proceedings AEE World Energy Engineering Congress, (1978)

20. Turner, Wayne C. and Blackshare, Derek; Webb, Richard E. and Shirley, James M., "An Industrial Engineering Approach to Hazardous Materials/Waste Management", Institute of Industrial Engineers World Productivity Forum & 1987 International Industrial Engineering Conference Proceedings,

21. Turner, Wayne C. and Blackshare, Derek; Webb, Richard E. and Shirley, James M., "The True Cost of Hazardous Materials Chemicals and Wastes", Institute of Industrial Engineers World Productivity Forum & 1987 International Industrial Engineering Conference Proceedings,

22. Turner, Wayne C., Webb, Richard E. and Shirley, James M., "Management of Hazardous Materials, Chemicals, and Wastes. Case Studies Involving Large Savings", Institute of Industrial Engineers 1986 Fall Industrial Engineering Conference Proceedings

23. Wilson, David C., Waste Management: Planning, Evaluation, Technologies, Clarendon Press Oxford (1981)

BIOGRAPHICAL SKETCH

James R. Ross is a Management Consultant with The University of Tennessee, Center for Industrial Services consulting in IE, productivity, energy and CIM areas. He holds a B.S. in Industrial Engineering from Geneva College, and an M.B.A. degree from Tennessee Technological University. He was previously Chief Engineer with Oxco Brush Division, Vistron Corporation and has held a variety of industrial engineering and consulting positions.

Ross is a senior member and long time director of the Middle Tennessee Chapter and past Nashville Chapter president of I.I.E., former chapter vice president of A.P.I.C.S. and is a registered Professional Engineer in Tennessee. He is 1988-89 Director of the I.I.E. Energy Management Division.

Chapter 69

TURNKEY SYSTEM FOR NOx REDUCTION

M. Grove, W. Sturm

ABSTRACT

The CER-NOx process[1] (see last page: Endnotes) for the selective catalytic reduction (SCR) of NOx uses ceramic molecular sieve catalyst modules. The CER-NOx process overcomes many of the field problems[2] such as poisoning, plugging and masking encountered with heavy/noble metal type (SCR) catalysts, introduced approximately ten (10) years ago in Japan and the USA. The CER-NOx (SCR) NOx abatement process can be combined with volatile organic compound (VOC) and sulfur oxide (SOx) abatement processes. This paper presents:

- The urgent need for NOx reduction based on European research on Cause and Effect of dying forests, lakes, and coastal areas, and its impact on wild and marine life.

- The CER-NOx (SCR) NOx abatement process, including abatement cost data and typical applications.

I. INTRODUCTION:

In the past, energy engineers have primarily been concerned with the cost effective generation and distribution of energy. Like chemical engineers, who had to face, together with the public, the effects of many years of soil, air and water pollution (super fund legislation, etc.), also, the energy engineers are now facing the effect and the cost of polluting the environment.

Also, in Europe, the air and water pollution became the top political issue in recent years, and even more so in recent months.[3]

Scandinavian countries are suing Great Britain, probably the biggest European polluter at the International Court in Holland.

- The ministers for environmental control of the twelve countries of the European Community (EC) met in emergency sessions in June 1988 "to save the North Sea."

West Germany will complete it's ambitious multi billion dollar program for SO_2, NOx, etc. abatement systems by 1991 to increase its pollution control leverage against polluting neighboring countries.

Flue and waste gases from any combustion system and many chemical and manufacturing processes contain nitrogen oxides, which together with VOC form ozone at low altitude through photosynthesis. (See Figure 1: What One Should Know About Air Pollutants). Ozone (O_3) is the most potent oxidant attacking and destroying organics and enhancing the effect of acid rain. NOx also forms nitrates, a fertilizer feeding killer algae, killing other marine life.

NOx and SO_2 form nitric acid and sulfuric acid rain, etc., dissolving heavy metals which poisons plants, animals and human beings at an alarming rate.

The most impact of this air pollution, however, is yet to come because of the slow chemical reaction with Half Life of five to twenty years. That is, the pollutants of today will have their full impact in five to twenty years.[4]

Europe and Japan have enacted regulations after thorough research on causes and effects of acid rains and ozone formation at low altitudes and dying forests, lakes and coastal waters. Both countries will have completed the implementation of air pollution abatement systems, including cogeneration facilities burning gas, oil and waste and coal burning power plants in the early 1990's.

Canada is suffering severely under the pollutants generated by the power and other industries of the mid western states of the USA. Here in the US, industry lobbies are still fighting to avoid the added "unnecessary" cost to their "bottom line" at everyone's and their children's long term expense.

WHAT ONE SHOULD KNOW ABOUT AIR POLLUTANTS

Air Quality in low altitudes is primarily influenced by:

- Sulfur dioxide (SO_2)
- Nitrogen oxides (NO_x)
- Volatile organic compounds (VOC) such as hydrocarbons, and carbon monoxide, ammonia, etc.

resulting in periodical high concentration of:

- acid rain (pH 3.1 through 5.6 have been measured) is formed with SO_2/SO_3 (sulfuric acid) and with NO_2 (nitric acid).
- ozone (O_3) the most potent oxidant, is formed through complex photo chemical reaction, involving (VOC) and (NO_x) at low altitudes.[1]

- The balance in the atmosphere

  - o NO_2 and O_2
  - o NO and O_3

  through photosynthesis.

- is changed in the presence of:

  - o hydrocarbons
  - o hydroxyl

Approximately 80 million people in the USA still live in over 50 metropolitan areas that exceed the federal ozone limit of 0.132 ppm set for December 31, 1987.

The increase of O_3 at low altitude also influences the O_3 in the stratosphere (ozone hole), causing increased ultraviolet (UV) - radiation, in turn increasing the photosynthesis reaction at low altitudes.[2] The reactions are as follows:

$$RH + OH + O_2 \longrightarrow RO_2 + H_2O$$

$$RO_2 + NO \longrightarrow RO + NO_2$$

$$RO + O_2 \longrightarrow \text{Aldehyde} + HO_2$$

$$HO_2 + NO \longrightarrow OH + NO_2$$

The reaction of hydrocarbon and hydroxyl causes the formation of NO_2 which in turn shifts the equilibrium towards the ozone formation.

$$NO_2 + h\nu\ (UV) \longrightarrow NO + O$$

$$O + O_2 \longrightarrow O_3$$

$$O_3 + NO \longrightarrow O_2 + NO_2$$

$$NO_2 + O_2 \longrightarrow NO + O_3$$

In the photostationary stage there is an equilibrium at low altitude in the air.[3]

Figure 1

WHAT ONE SHOULD KNOW ABOUT AIR POLLUTANTS

---

[1] MacKerren, Conrad B., Process Industries News Network, Chemical Engineering, May 9, 1988.

[2] According to Professor P. Builtjes, The Netherlands and R. Stein, West Germany, European Federation of Energy Management Association (EFEM) 1988, and Business Week, June 27, 1988: Proven Chlorine Is The Culprit That's Eating The Ozone.

[3] Pries, Meike, Chemische Rundschau, West Germany, June 3, 1988 "Das Gleichgewicht Wird Gestoert." (The equilibrium is changed).

II. NOx REDUCTION:

Alternate Measures in the Reduction of NOx in Engines and Turbines

What is needed today are primary and secondary NOx pollution control measures.

- Primary: Preventing formation of NOx through boiler gas turbine and engine design changes, recycling flue gas, water injection, etc.. However, these measures will in many cases not meet the more stringent regulatory requirements of today and/or tomorrow.

- Secondary: The catalytic reduction of NOx, for which adequate technology is available today to reduce up to 95% of NOx emissions, using metal catalysts (Vanadium, Iron, Tungsten and other metal Oxides)[5] and ceramic catalysts.

The metal type catalysts, however, have major disadvantages. They:

- Poison (As, Cd, Hg, Pb, alkalines, etc.).

- Mask due to SO_2/SO_3 conversion, phosphorous and zinc compounds, which in some cases is even promoted by the catalyst material.[6]

CER-NOx, (SCR) NOx Abatement Process

The CER-NOx process is using molecular sieve catalysts for the selective catalytic reduction (SCR) of NOx. This ceramic catalyst development was started in the 1950's in West Germany at Steuler Industriewerke GmbH with the objective to reduce nitric acid vapors of stainless steel pickling lines. Steuler introduced the first NOx (SCR) abatement system to the European market in 1982. Since then, approximately 100 NOx (SCR) reactors were installed in Europe in various industries, including approximately 50% in the engine (back-up power, locomotive, etc.) and co-generation industry.

Because the selective catalytic reaction of NOx and ammonia takes place in the micro pore structure of the CER-NOx ceramic molecular sieve catalyst honeycomb module it is, unlike metal type SCR catalysts, virtually immune to poisoning, plugging and masking. (See Figure 2: What is a Molecular Sieve Catalyst).

The high electrostatic forces in the pores reduce the reaction potential of nitrogen oxide with ammonia, allowing a reaction temperature as low as 300C/572F.

The exothermic reaction of NOx with NH_3 in the small pore space generates nitrogen gas and water/steam, expelled from the pore structure back into the gas stream, "self-cleaning" the catalyst surface.[7]

The operating temperature of this CER-NOx (SCR) abatement system is:

-Lower limit: 300C/572F due to the formation of ammonia bisulfate.

-Upper limit: 480C/896F due to the thermal decomposition of ammonia which could cause a "run away reaction" of the ammonia above 520C/968F, eventually burning up all injected ammonia.

III. CER-NOx APPLICATIONS:

Figure 3 pictures the functional diagram of the CER-NOx (SCR) equipment of a co-generation facility. The wide operating temperature window allows the combination of NOx and VOC catalyst in one reactor housing. (See Figure 3: Functional Diagram of a Co-Generation Plant). The CER-NOx process is described below, using a co-generation facility as an example.

CER-NOx Process Description

At first the nitrogen oxides (NOx) and the injected ammonia (NH_3), contained in the flue gas will be removed into the micro pore structure of the catalyst through adsorption based on the concentration gradient.

The electrostatic forces generated in the inside of the micro pores, the activation energy for the reaction of nitrogen oxide to nitrogen and water vapor using ammonia is decreased so that the reaction can safely take place at temperatures as low as 300 to 480C/572 to 896F.

Thereafter, the reaction products, nitrogen and water vapor, will be ejected out of the pores and dissipate back into the flue gas stream.

The metering of the injected ammonia quantity into the flue gas stream in front of the CER-NOx (SCR) reactor is determined by the residual NOx concentration, measured behind the reactor. The NOx measurement quantifies photometrically the residual NOx concentration of the gas phase in the UV area of the spectrum. For this purpose, a small flue gas quantity is pumped out of the gas stream continuously, processed and fed into the measuring cell periodically where the residual NOx concentration is measured. The NOx measurement unit controls a motor driven valve of the ammonia supply system. Then again, the right ammonia quantity, injected into the flue gas stream in front of the CER-NOx reactor is adjusted to meet the process guaranteed emission level.

There is only one NOx measurement control needed to supervise the residual NOx output concentration of up to four (4) CER-NOx reactors. For this purpose the individual flue gas streams are called off separately every thirty (30) seconds maximum and fed into the NOx measurement and control system. The lapse time is bridged by the dampening effect of the large micro pore structure of the molecular sieve catalyst.[8]

As stated earlier the NOx measurement of the reactor also activates the ammonia supply system, and together with the temperature control systems assures that no ammonia is injected into the reactor at below 300C/572F or above 520C/968F, avoiding aforementioned problems.

The ammonia can be supplied both as aqueous ammonia (25 to 29% NH_3 and H_2O) or as ammonia gas (100%).

WHAT IS A MOLECULAR SIEVE CATALYST?

- The molecular sieve is made of natural or synthetic/manufactured zeolite. Zeolites are crystalline micro porous solids containing cavities and channels of molecular dimensions - 3 to 10 Angstrom (A) - synonymously called molecular sieves with:

  - o uniform pores with one or more discrete sizes
  - o very high internal surface area (greater 600 sq. meter/gram or 6,600 sq. ft./gram).
  - o able to absorb and concentrate hydrocarbons and other organics, small enough in size to enter into the pore structure.

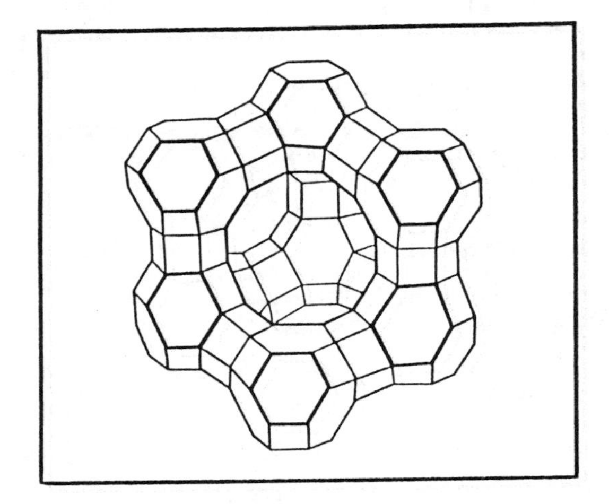

- Zeolite compositions include about 13 elements; Si, Al, Ga, Ge, and Fe, in addition Li, Be, B, Mg, Co, Mn, Zn, P, As, and Ti.

- There are about 60 different structures known today with tens of thousands of theoretical structures. Approximately 1,000 different zeolite patents are issued each year (about 1/3 to Mobil Research & Development).

- Most structures have high thermal and chemical stability as other ceramic materials.

WHERE CAN IT BE USED?

- Zeolite has been used extensively as catalysts in the organic chemistry since the development of the synthetic zeolite chemistry in the 1950's. The petroleum (catalytic cracking), petro chemical and pharmaceutical industries are main users today.

- Zeolite facilitates unprecedented molecular control of chemical reaction and the ultimate in "molecular engineering" of molecules less than 10 Angstrom in size[4]

- The CER-NOx (SCR) Process:

  - o Research and development groups in many parts of the world have been working for years also on applications such as the selective catalytic reduction (SCR) of NOx in flue gases, chemical process waste gases and nitric acid pickling vapors. The CER-NOx Abatement System has been the first commercially available turnkey NOx abatement system using molecular sieve catalysts since 1982.

  - o The CER-NOx process uses a combination of zeolites, including a proprietary Mobil zeolite, meeting all long term reactivity and stability requirements for a NOx (SCR) abatement system.

  - o NH_3 and NOx are adsorbed into the micro pore structure of the molecular sieve and reacts, following reactions such as below[5] to N_2 and H_2O vapor. The reduction is highly exothermic (Δ H = 665 kJ/mol).

  - o The zeolite based Mobil-Steuler CER-NOx process is using molecular sieve catalyst honeycomb modules which are more efficient and less sensitive to contaminants then metal type catalyst, using vanadium, iron, tungsten and other noble metal oxides, because the micro pores are "self-cleaning."

Figure 2

WHAT IS A MOLECULAR SIEVE CATALYST?

---

[4] D. E. W. Vaughn, President of the International Zeolite Association, Exxon Research and Engineering Company.

[5] $4\ NO + 4\ NH_3 + O_2 \rightarrow 4\ N_2 + 6\ H_2O$
$2\ NO_2 + 4\ NH_3 + O_2 \rightarrow 3\ N_2 + 6\ H_2O$
$2\ CO + 2\ NO \rightarrow 2\ CO_2 + N_2$
$NO + NO_2 + 2\ NH_3 \rightarrow 2\ N_2 + 3\ H_2O$

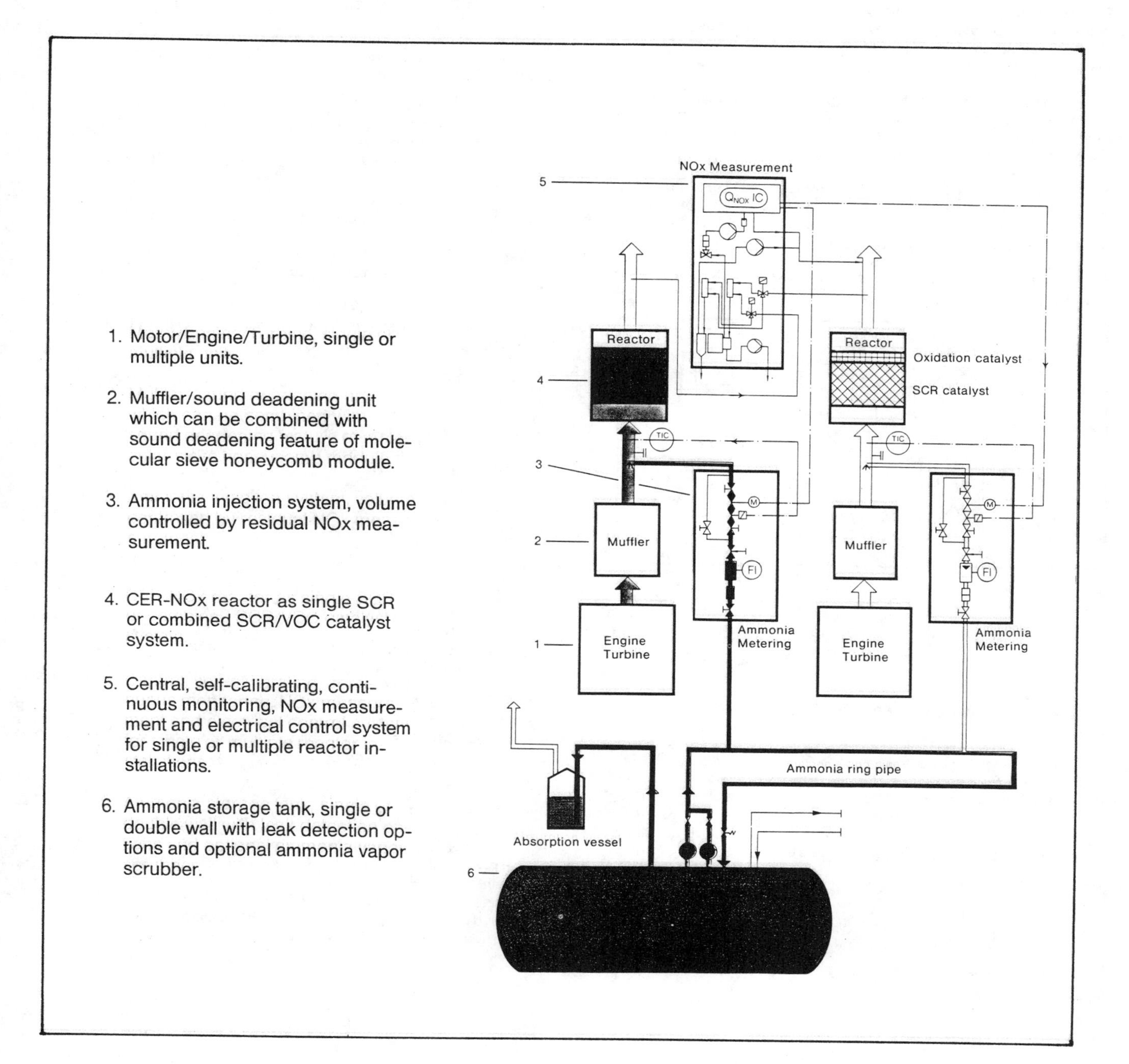
1. Motor/Engine/Turbine, single or multiple units.
2. Muffler/sound deadening unit which can be combined with sound deadening feature of molecular sieve honeycomb module.
3. Ammonia injection system, volume controlled by residual NOx measurement.
4. CER-NOx reactor as single SCR or combined SCR/VOC catalyst system.
5. Central, self-calibrating, continuous monitoring, NOx measurement and electrical control system for single or multiple reactor installations.
6. Ammonia storage tank, single or double wall with leak detection options and optional ammonia vapor scrubber.
NOx Measurement
5
4
3
2
1
6
Reactor
Reactor
Oxidation catalyst
SCR catalyst
TIC
TIC
M
M
FI
FI
Muffler
Muffler
Engine Turbine
Engine Turbine
Ammonia Metering
Ammonia Metering
Ammonia ring pipe
Absorption vessel

Figure 3

FUNCTIONAL DIAGRAM OF A CO-GENERATION PLANT

The aqueous ammonia is preferred by European co-generation operators for safety reasons. The storage tank is usually designed as a double-wall tank with a leak detection and control system.

To prevent the highly odorous ammonia gas from escaping during operation (filling operation and breathing during temperature cycling), an optional absorption vessel can be installed.

Pilot plant tests at coal burning boilers for both antracite and lignite coal showed that the CER-NOx (SCR) NOx abatement system can be located at either the high dust or the low dust side (See Figure 4: Flow Diagram of Steam Generator System With CER-NOx). The test also showed an extended service life versus metal type catalysts.

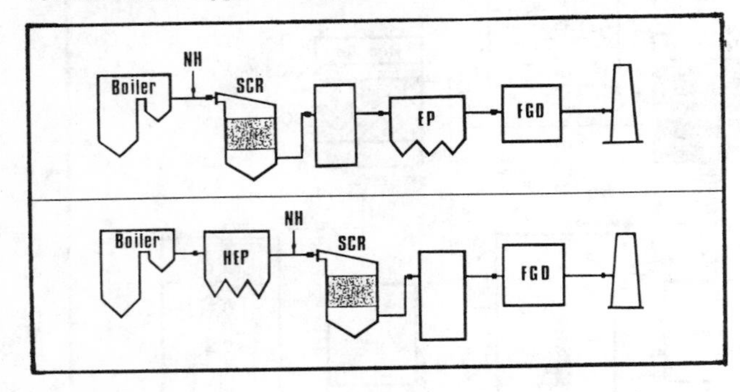

Figure 4

FLOW DIAGRAM OF STEAM GENERATOR SYSTEM WITH CER-NOx

Cost Effectiveness of The CER-NOx (SCR) Abatement System

In a recent cost analysis of the CER-NOx (SCR) NOx abatement system it was determined that the cost of the CER-NOx abatement system can be as low as 5% of the total co-generation project cost. See Figure 5, (CER-NOx (SCR) equipment cost as a percent of the total co-generation equipment cost).

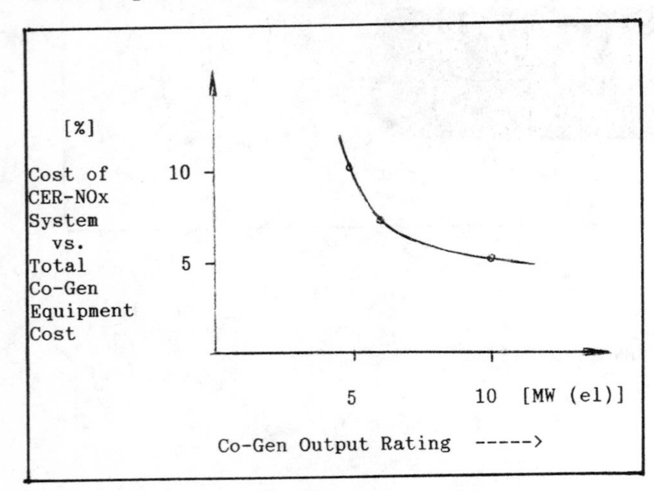

Figure 5

CER-NOx (SCR) EQUIPMENT COST AS A PERCENT OF THE TOTAL CO-GENERATION EQUIPMENT COST

Typical CER-NOx Abatement Projects

Approximately 100 ceramic molecular sieve CER-NOx (SCR) NOx abatement systems have been installed to date in:

- Fossil burning steam generators.
- Waste burning and chemical process facilities.
- Engine plants and co-generation facilities.

In the following, a few projects visited by US engineers and journalists in Europe in 1988:

- Two (2) co-generation projects.
- One (1) glass melting furnace.

1. LZB Hannover: Duel fuel/gas diesel engine, four cycle, (1) CER-NOx (SCR) reactor with (1) in line VOC/CO-oxidization catalyst:

	Diesel/ Nat.Gas	Diesel
- MWM - Co-Gen Project		
- Kilowatt (el) rating:	630 kw	
- Flue gas volume (Nm^3/h):	2,650	3,410
- Flue gas temperature (C):	442	380
- NOx-Inlet (ppm) max:	840	1,380
Calculated at NO_2 at 5 Vol % O_2 (mg/Nm^3):	2,399	4,476
- CO-Inlet at 5 Vol % O_2 (mg/Nm^3):	720	235
- NOx-Outlet Calculated as NO_2 at 5 Vol % O_2 (mg/Nm^3):	310	420
- CO-Outlet at 5 Vol % O_2 (mg/Nm^3):	340	192
- NOx Reduction (%):	> 87	> 90
- NH_3-Slip[9] (mg/Nm^3):	8	10
- Mol Ratio NH_3/NOx:	0.9	0.93
- Reactor start-up:	Nov. 1986	
- (SCR) Catalyst:	CER-NOx Honeycomb Modules	
- Operating Time (hours):	2,400	600

Operating Experience:

- Minimal maintenance requirements.
- High readiness/no down time due to NOx abatement system.
- Small NH_3 consumption/minimal NH_3 slip.
- No diminishing NOx and CO (SCR) reduction rate since 1986.

2. MHKW - Peissenberg
Duel fuel, gas/diesel, 4-cycle engine with (1) CER-NOx (SCR) NOx reactor

		Duel Fuel	Diesel
-	KRUPP-MAK Co-Gen Project		
-	Kilowatt (el) rating: 6,127 kw		
-	Waste gas volume maximum, damp (Nm^3/h):	32,300	37,850
-	Gas temperature: 330 - 520C		
-	NOx inlet, maximum (ppm) at 5% O_2:	2,400	5,030
-	NOx outlet: 140 ppm		
-	NOx reduction (process guarantee):	> 82%	> 90%
-	NH_3 slip: smaller than 10 mg/Nm^3		
-	CER-NOx pressure drop (mb):	8	10
-	Start-Up: December 1987		
-	Catalyst: Honeycomb CER-NOx Modules		
-	Operating time:	2,250	350

Operating Experience: Same as (1).

3. Glass Melting Furnace with (1) CER-NOx (SCR) NOx Reactor.

- Bavaria Glass Project

-	Flue gas volume maximum, damp (Nm^3/h):	50,000
-	Flue gas temperature (C):	340-360
-	NOx-Inlet Calculated as NO_2 (mg/Nm^3):	1,900
-	NOx-Outlet, maximum Calculated as NO_2 (mg/Nm^3):	400
-	NOx Reduction, approx. (%):	80
-	NH_3 Consumption, max, (kg/h):	30 @ 100% NH_3 125 @ 25% NH_3/H_2O
-	CO_2 approx. Vol %:	8
-	O_2 approx. Vol %:	6.7
-	N_2 approx. Vol %:	74
-	H_2O approx. Vol %:	11
-	SO_2 concentration (mg/Nm^3):	700
-	SO_3 concentration max. (mg/Nm^3):	10
-	Particulate concentration max. (mg/Nm^3):	20
-	Reactor start-up:	1987
-	(SCR) Catalyst: CER-NOx Honeycomb Modules	
-	Operating time (hours): approximately	4000

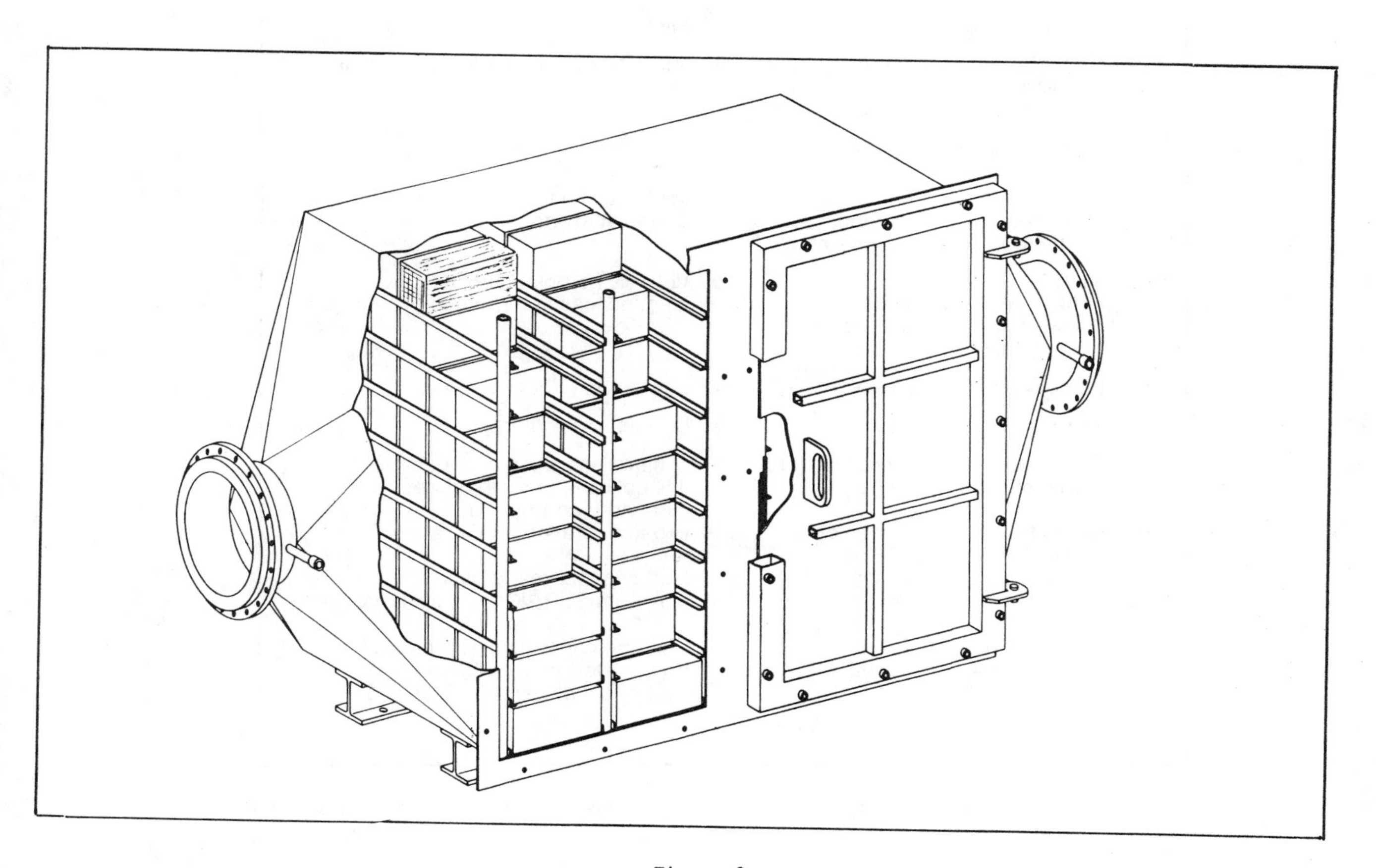

Figure 6

CER-NOx (SCR) REACTOR WITH
CERAMIC MOLECULAR SIEVE CATALYST

IV. CONCLUSION

The CER-NOx (SCR) NOx abatement system has a successful track record of five (5) years in Europe. Currently several companies are working on competitive systems in Europe, the US, and Japan, following this over ten (10) year development work and five (5) year field experience with molecular sieve catalyst for the (SCR) NOx abatement process.

The features and benefits of the CER-NOx abatement system are shown by Figure 7: Features and Benefits of the CER-NOx Process.

CER-NOx FEATURES	YOUR BENEFITS
1. – High reduction efficiency	▸ Minimal residual NOx emission
2. – No surface reaction, but reaction in micro-pore structure	▸ Minimal wear, extended service life
3. – High space velocity values	▸ Minimal space requirement, compact equipment design
4. – Minimal ammonia slip, computer controlled operation, continuous NOx measurement	▸ Minimal influence on electro filter and heat exchanger
5. – No SO2 to SO3 conversion, soot, heavy metal or other masking compound deposits (self-cleaning)	▸ Virtually no poisoning, no plugging, no masking problems
6. – Low specific catalyst weight	▸ Low equipment weight
7. – Minimal pressure drop	▸ Low energy requirements
8. – Reactor may be located on high dust side	▸ Minimal operating cost because it is not necessary to reheat cleaned (scrubbed) gases
9. – Broad temperature range of 300 to 520C/572 to 968F.	▸ Broad operating temperature range ▸ Minimum operating cost
– SCR & VOC catalyst may be located in one reactor	▸ Minimum SCR/VOC catalyst reactor cost
10. – Ceramic body molecular sieve SCR catalyst without heavy metals	▸ Used up catalysts are nonhazardous, safely disposable or recyclable as ceramic filler

Figure 7

FEATURES & BENEFITS OF THE CER-NOx PROCESS

REFERENCES

1. ASME Publications:

Dupaski, R.B., and Padar, F., 1981, "NOx Reduction on Large Bore SI Engines Using Catalytic Converters," Cooper Energy Services, Springfield, Ohio.

Cheesman, Randolph M., 1987, "Non-Selective Catalytic System Application to a 650 KW Reciprocating Engine," Johnson Matthey Catalytic Systems Division, Wayne, Pennsylvania.

Burns, Kenneth R., and Collins, Martin F., 1983, Engelhard Industries Division, Engelhard Corporation, Union, New Jersey, and Heck, Ronald M., Engelhard Industries Division, Engelhard Corporation, Menlo Park, New Jersey, "Catalytic Control of NOx Emissions From Stationary Rich-Burning Natural Gas Engines."

Falletta, C.E., D'Alessandro, A.F., and Thomas, L.R., 1983, "Automated Catalytic NOx Reduction Systems For Gas-Fueled Stationary Engines," Johnson Matthey, Inc., West Chester, Pennsylvania.

Eckhard, David W., and Serve' J.V., 1987, "Maintaining Low Exhaust Emissions with Turbocharged Gas Engines Using a Feedback Air-Fuel Ratio Control System," Cooper Industries-Superior, Springfield, Ohio

2. Other Publications:

Wasser, John H. and Perry, Richard B., "Diesel Engine NOx Control," Air and Energy Engineering Research Laboratory, U.S. Environmental Protection Agency, Research Triangle Park, North Carolina.

ENDNOTES

1. TM Steuler: The CER-NOx Process is a turnkey NOx (SCR) abatement process using ceramic molecular sieve catalyst modules.

2. CER-NOx operating experience with heavy oil operating co-generation engines, including 3.5% sulfur and heavy metals (15 mg/Nm^3 in flue gas).

3. Der Speigel, West Germany, June 1988: Not just fish and trees, but hundreds of seals recently died in the North Sea, poisoned by heavy metals, dissolved by acid rain. Shabecoff, Philip, 1988, "Acid Rain Called Peril To Sea Life On Atlantic Coast," The New York Times, New York, NY.

4. Measurements of Pollution in Japan and West Germany showed that it took five (5) years after introducing abatement systems to see their positive effects.

5. See Japanese Patent Filings 4051966/1975 and 52122293/1977, etc. and South Coast Air Quality Management District, (California).

6. Vanadium also acts as a catalyst in the conversion of SO_2 to SO_3.

7. The exothermic reaction increases the temperature and with it the pressure, driving the gas out of the pores.

8. Sponge effect of 600 m^3/gr or 6,600 sq. ft./gr of zeolite catalyst material.

9. Ammonia slip: The excess/unreacted NH_3 passing through the SCR reactor to be minimized by the CER-NOx process.

Chapter 70

ENVIRONMENTAL EMISSIONS FROM THE RESOURCE RECOVERY PLANT IN THE SUMNER COUNTY COGENERATION FACILITY

T. M. Cathcart

Perspective

Environmental impacts of any particular cogeneration facility should be compared to other available options. Energy from waste facilities have been attacked as polluters, but a broader understanding of the situation should be applied. Studies on municipal solid waste energy recovery facilities have shown potential problems with particulates, acid gases, dioxin, and heavy metal concentration.

However, these concerns are being addressed and newer facilities have included methods and equipment at great expense which are designed specifically to control emissions.

Meanwhile, there seems to be a lack of attention to environmental emissions resulting from other municipal waste disposal or conversion methods. Landfilling of waste results in loss of process control since the operator cannot control temperature, moisture or mixing. Most existing landfills have no method of monitoring or abating environmental emissions. Composting is another method that results in reduced process control and monitoring. Source separation and recycling of usable material is welcomed when a critical analysis of the tradeoffs between environmental benefits and costs is completed, including any special problems with disposal of the remaining unusable material.

This paper presents a case study of the environmental emissions from a municipal solid-waste energy recovery facility. The nature of an environmental emissions study is to document potential problems, but the reader should retain the perspective that other power generating facilities and other waste disposal methods also result in potential environmental problems.

PROJECT BACKGROUND

The Resource Authority in Sumner County (representing the county and the cities of Gallatin and Hendersonville) was established in May 1979 to coordinate and operate disposal of all municipal solid waste (MSW) collected in Sumner County, Tennessee. The Resource Authority constructed, owns, and operates a resource recovery facility which burns MSW to produce process steam to sell to nearby industry, and to sell electricity to the Tennessee Valley Authority (TVA). TVA has provided technical assistance to the Resource Authority and monitored the project throughout the development, planning, design, construction, and operation of the facility.

Major project funding was from the following sources:

Resource Authority in Sumner County Bonds	$12,000,000
Tennessee Valley Authority Loan	$ 2,000,000
U.S. Economic Development Agency Grant	$ 1,500,000
U.S. Department of Energy Grant	$ 250,000

Construction began in June 1980 and the first unit began operation in December 1981. Construction was completed in April 1982 and the plant dedication was held in June 1982.

FACILITY DESCRIPTION

The resource recovery plant is located on a 5-acre site in the Gallatin Industrial Park, within the corporate limits of the city of Gallatin, Tennessee. The plant building houses general offices, a conference room, employee facilities, maintenance shop, incinerator-boiler plant, turbine generator plant, water treatment facilities, central control room, a waste receiving pit, fuel handling cranes, air pollution control equipment, and ash handling and residue removal systems. Other onsite facilities include receiving scales, roadways, parking pads, parking areas, and landscaping.

The waste processing facilities include two parallel incinerator-boiler units. Each unit is capable of processing 100 tons per day of municipal solid waste, producing a maximum of 27,500 pounds per hour of steam. A 550-kW back-pressure turbine generator is supplied steam from a common steam header and reduces steam pressure from 450 psig to a range of 175 to 200 psig for delivery to the process steam customers and for driving plant auxiliary equipment. The incinerator boiler facilities are designed to process as-delivered waste without prior treatment and is classified as a "Mass Burn" facility.

A privately owned satellite proof-of-concept project, located adjacent to the refuse receiving pit, is in operation to demonstrate the benefits of front-end processing of MSW before incineration. The process includes metals and glass removal. The front-end processing system was developed by National Recovery Technologies and is independently operated through arrangement with the Resource Authority.

The incinerator is an O'Connor water-cooled rotary combustor now marketed by Westinghouse Corporation. The air pollution control equipment originally included an electrostatically assisted baghouse, designed and constructed by the Apitron Division of American Precision Industries. The baghouses were eventually removed from service and electrostatic precipitators were installed.

Bottom ash passes from the rotary combustor into the quench tank. Fly ash also is dumped into the quench tank from the baghouse hopper; wetting the fly ash reduces fugitive dust emissions. The mixed ash is conveyed from the quench tank to large metal receptacles. The receptacles are trucked to a sanitary landfill and the ash is buried. The mixed fly and bottom ash is the only solid waste which leaves the site.

PERFORMANCE AND EMISSIONS TESTS

Field performance and emissions tests were conducted by TVA in 1982 and 1983. Preliminary testing was done prior to July 1982 to determine approximate airflow rates and metal concentrations in the ash. The first phase of performance testing was conducted during July 27 and 30, 1982 and the second phase during June 14 to 17, 1983.

Performance testing included waste stream characterization and weight reduction, particulate emission, and boiler efficiency measurements. Performance test results are summarized in Table 1.

The combustor/boiler demonstrated its ability to process municipal waste at its maximum rated capacity of 100 tons/day with various moisture contents and heating values of the waste. The lower feed rate during the 1983 tests were primarily due to the high pressure drop through the baghouse. Volume reduction of municipal waste to dry mixed ash was near the expected value of 90 percent. The 1982 tests did not include particulate emissions measurements because of burn holes in the filter bags.

Fly and Mixed Ash Extraction Tests

The fly ash and mixed ash were subjected to the Environment Protection Agency (EPA) extraction procedure. Extracts were analyzed for eight elements: arsenic, barium, cadmium, chromium, lead, silver, and selenium. If concentrations of any of these elements exceed 100 times the National Drinking Water Standards, the waste stream is characterized as having EP toxicity. Because discharges from energy from waste facilities were specifically excluded by EPA from being regulated as hazardous wastes, the EP toxicity had no direct regulatory effect on facility operation. However, if high concentrations of any of these elements should be extracted, it is likely that these elements are in a form which can leach into surface and groundwater.

ASSESSMENT RESULTS

Mixed Ash

High levels of extractable lead in some samples (Table 2) gives the mixed ash the characteristic of EP toxicity. The baghouse hopper fly ash also has EP toxicity, due to higher than allowable levels of cadmium and lead. This implies that the cadmium and lead content in the ash could pose a hazard to human health and the environment if these elements are allowed to leach from landfill areas into the groundwater.

Liquid Effluent

Both pit and quench water liquids wastes are discharged to the municipal sewage system without pretreatment. After dilution by the mainstream flow of the sewage treatment plant influent and absorption in sewage sludge, the quench and pit water should be rendered relatively innocuous. However, because the levels of cadmium, copper, lead, nickel, zinc, iron, selenium, manganese, and silver are present in concentrations of potential environmental and health concern (Table 3), and because of the extremely high dissolved solids (Table 4), some pretreatment of liquid wastes may be required in the future.

Gaseous Emissions

In May 1982, TVA sampled the stack emissions and recorded the range of concentration for various gases found (Table 5). Additional testing by Cooper Engineers showed average concentrations that were generally near the mid-point of the ranges shown. In addition, they found HCl average concentration to be 509 ppm on a volume basis, corrected to 7 percent O_2.

DISCUSSION

Municipal wastes have a chemical composition which is uniquely variable. Chemical analyses vary greatly from hour-to-hour, day-to-day, and season-to-season as the sources of the waste change. Short-term monitoring of the chemistry of the thermally processed effluents only serves to emphasize this variability. The standard deviation has exceeded the mean in several of our analyses. This inherent variation must be taken into account when developing pollution control systems and when determining the risks associated with unanticipated releases to the environment.

TABLE 1

PERFORMANCE TEST SUMMARY

MSW Data	1982 Tests	1983 Tests	Predicted
Feed Rate (tons/day)	99.80	83.70	85
Higher Heating Value (Btu/lb)	4482.00	6107.00	--
Ash Content (%)	25.60	26.80	27
Moisture (%)	31.70	13.10	20
Performance Data			
Weight Reduction --Wet Ash (%)	52.70	52.40	--
Volume Reduction --Wet Ash (%)	84.20	84.10	--
Weight Reduction --Dry Ash (%)	82.80	67.30	--
Volume Reduction --Dry Ash (%)	94.30	89.10	90
Particulate Emission (lb/hr)	--	3.35	--
Ash Carbon Content (%)	17.30	11.80	--
Excess Air (%)	69.00	74.00	50
Temperature at Baghouse Inlet (^{o}F)	424.00	435.00	400

TABLE 2

COMPARISON OF EP TOXICITY TEST RESULTS OF MIXED ASH AND FLY ASH SOLID EFFLUENT STREAMS

Metal	100 X National Drinking Water Standards (mg/L)	Mixed Ash Extraction Mean Concentration (mg/L)	Fly Ash Extraction Mean Concentration (mg/L)
As	5.0	0.06	0.008
Ba	100.0	0.11	0.13
Cd	1.0	0.79	32.0*
Cr	5.0	0.03	0.02
Pb	5.0	6.98*	14.5*
Hg	0.2	0.003	0.002
Se	1.0	0.001	0.02
Ag	5.0	0.02	0.12

*Indicates the metal concentration exceeds 100X the National Drinking Water Standard concentration.

TABLE 3

COMPARISONS OF METAL CONCENTRATIONS IN PIT WATER AND QUENCH WATER WITH NATIONAL DRINKING WATER STANDARDS

	Concentration (mg/L)		
Metal	Pit Water	Quench Water	National Drinking Water Standards
As	0.0088	0.0031	0.05
Cd	0.0032	0.067	0.010
Cr	0.0455	0.0138	0.05
Pb	0.103	0.688	0.05
Hg	0.0006	0.0089	0.002
Se	0.001	0.0362	0.01
Ag	<0.010	0.015	0.05

TABLE 4

WATER QUALITY ANALYSES OF PIT WATER AND QUENCH WATER

Parameter	Pit Water Mean Concentration (mg/L)	Quench Water Mean Concentration (mg/L)	Water Quality Criteria (mg/L)
BOD_5	557	154	30
COD	933	254	N/C*
Oil & Grease	4.04	9.88	30
TDS	3870	9250	500
TSS	192	295	40
Phenol	0.03	0.03	1.0

*N/C denotes that no criteria or standard was given.

TABLE 5

GENERAL RANGE OF GAS CONCENTRATIONS

O_2 = 7-11%
SO_2 = 100-250 ppm
CO = 250-900 ppm
NO_x = 80-150 ppm
HC = 50-75 ppm

REFERENCES

1. F. G. Parker, T. M. Cathcart, and R. A. Montano, "Sumner County Solid-Waste Energy Recovery Facility, Volume 1: Feasibility Studies, Design, and Construction," EPRI CS-4164, Vol. 1, August 1985.

2. F. G. Parker, J. C. Duggan, and T. M. Cathcart, "Sumner County Solid-Waste Energy Recovery Facility, Volume 2: Performance and Environmental Evaluation," EPRI CS-4164, Vol. 2, September 1985.

3. T. M. Cathcart, M. N. Jarrett, "Sumner County Solid-Waste Energy Recovery Facility," 13th Energy Technology Conference, March 1986.

4. J. L. Hahn, D. J. Kmetovic, Cooper Engineers, "Air Emissions Tests of Solid-Waste Combustion in a Rotary Combustor/Boiler System at Gallatin, Tennessee," West County Agency of Contra Costa County, December 1983.

SECTION 15

COGENERATION AND ENERGY SERVICES FINANCING

Chapter 71

THE IMPACT OF THE TAX REFORM ACT OF 1986 ON THE DEVELOPMENT OF COGENERATION AND ALTERNATIVE ENERGY PROJECTS

L. M. Goodwin

The Tax Reform Act of 1986 (the "Act") made major changes in the Federal tax treatment of cogeneration and alternative energy projects. As a result, many of the significant tax advantages previously enjoyed by the cogeneration and alternative energy industries were lost under the Act. This paper summarizes the key provisions of the Act with regard to cogeneration and alternative energy projects, and analyzes the impact the Act will have upon the development of cogeneration and alternative energy projects.

TAX RATES

The Act significantly reduced tax rates on both the individual and corporate level. However, as with many provisions of the Act, the reduced rates are phased in over a period of time. Under prior law, the top individual tax rate was 50 percent, while the top corporate tax rate was 46 percent. The Act reduced the top individual rate to 38.5 percent in 1987 and 28 percent in 1988. The top corporate rate was reduced to 34 percent for taxable years beginning on or after July 1, 1987, with blended rates being used for taxable years which begin before or end after July 1, 1987. The blended rates are computed based on the proportion of the number of days in the taxable year before July 1, 1987 to the total number of days in the taxable year. For example, a corporation with a calendar taxable year would have a top rate of approximately 40 percent for 1987.1/

DEPRECIATION AND TAX CREDITS

Depreciation

Under prior law, most cogeneration and alternative energy projects which were Qualifying Facilities under the Public Utility Regulatory Policies Act of 1978 ("PURPA") were included in the 5-year Accelerated Cost Recovery System ("ACRS") category. Utility owned generating projects were included in the 10- or 15-year categories. Components of cogeneration and alternative energy projects which were classified as "building components" were included in the 19-year real property category.2/ Depreciation rates for property in the 5-, 10-, and 15-year categories were computed using the 150 percent declining balance depreciation method, and rates for property in the 19-year real property category were computed using the 175 percent declining balance method.

The Act modified ACRS for property which is placed in service after December 31, 1986. Under the Act, eight property categories are provided. Property classified in the 3-, 5-, 7- and 10-year property categories is depreciated using the 200 percent declining balance method. Property classified in the 15- and 20-year property categories is depreciated using the 150 percent declining balance method. Property classified as residential rental and non-residential real property is depreciated using the straight-line method over recovery periods of 27.5-years and 31.5-years, respectively. Generally, the category in which property will be classified is based on the property's class life under the Asset Depreciation Range ("ADR") depreciation system.3/ Property that does not have an ADR class life is classified as 7-year property under the modified ACRS depreciation system.

Under the Act, depreciation for most cogeneration and alternative energy projects is substantially less favorable than under prior law.

1/ Under the Act, the benefit of graduate rates is phased-out for corporations which have taxable income in excess of $100,000. See Tax Reform Act of 1986 § 601(a).

2/ For Federal income tax purposes, the term "building generally includes any permanent structure, or component thereof, unless the structure is an integral part of a qualified activity and can be expected to be replaced when the property it initially houses is replaced. Qualified activities include manufacturing, production or extraction, or the furnishing of transportation, communications, electrical energy, gas, water or sewage disposal services. See Treas. Reg. § 1.48-1-(e)(1). An example of a "building" that is typically associated with cogeneration and alternate energy facilities is a control building.

3/ The ADR depreciation system is the depreciation system which was used prior to the adoption of the ACRS depreciation system as part of the Economic Recovery Tax Act of 1981. Under the ADR depreciation system, groups of assets were categorized into classes according to the function for which the assets were primarily used. Each group of assets was assigned a guideline life over which the assets were to be depreciated.

Many cogeneration projects, including coal-fired cogeneration projects and gas turbine projects operated in a combined cycle configuration, are included in the 20-year depreciation category, with depreciation rates based on the 150 percent declining balance method. Hydroelectric projects and utility-owned biomass and waste-to-energy projects are also included in the 20-year category. Diesel cogeneration projects and single cycle gas turbine cogeneration projects are included in the 15-year category, with depreciation also based on the 150 percent declining balance method. Large scale (greater than 500 Kw) industrial cogeneration and alternative energy projects whose output is for the owner's internal consumption, and not primarily for sale to third parties, are also included in the 15-year category. Smaller industrial projects are included in the depreciation category which applies to the productive assets of the particular industry involved. For cogeneration and alternative energy projects which are classified as building components, depreciation is based on the straight-line method, over a 27.5 year life in the case of residential rental property, and a 31.5 year life in the case of other real property. These new depreciation rules generally apply equally to utility and non-utility projects.

Special rules are provided for non-utility solar, wind, geothermal and ocean thermal projects, and for portions of biomass projects which are qualifying small power production facilities under PURPA. These projects are included in the 5-year category, with depreciation rates based on the 200 percent declining balance method. For these types of assets, the Act actually provides more favorable depreciation than was available under prior law. The following summaries indicate the depreciation treatment for various types of cogeneration and alternative energy projects under the Act.

Biomass Facilities: In order to qualify for 5-year depreciation, biomass property must satisfy two requirements. First it must be part of a small power production facility under PURPA. Biomass cogeneration projects which also satisfy the requirements for qualification as small power producers under PURPA may be eligible in part for 5-year depreciation under this rule. However, it is important to bear in mind that not all biomass cogenerators qualify as small power production facilities under PURPA. Under PURPA, the term "small power production facility" means a facility: (1) which produces electric energy solely by the use, as a primary energy source, of biomass, waste, renewable resources, geothermal resources, or any combination thereof; (ii) which has a power production capacity, together with any other facilities located at the same site, of not greater than 80 megawatts; and (iii) which is less than 50 percent utility owned.

The second requirement is that the property must be included in the "biomass property" category under the Internal Revenue Code. Biomass property includes boilers, burners, fuel loading and handling equipment, and related pollution control equipment. However, it does not include electric generation or distribution equipment or thermal energy distribution equipment. Thus, while a portion of biomass projects may be elible for 5-year depreciation under this rule, other equipment, such as turbines, generators, transformers and steam or hot water pipes will not be eligible for special treatment as "biomass property."

"Biomass" generally means any organic substance other than: (i) oil, natural gas or coal; (ii) a product of oil, natural gas or coal; or (iii) synthetic fuels or other products that are produced from biomass. The definition of biomass includes waste, sewage, sludge, grain, wood, oceanic and terrestrial crops and crop residues and includes waste products which have a market value. Moreover, the definition of biomass does not exclude waste materials, such as municipal and industrial waste, which include processed products of oil, natural gas or coal, such as used plastic containers and asphalt shingles.

Waste Reduction And Resource Recovery Facility: The "Waste Reduction and Resource Recovery Plants" ADR class had a guideline life of 10 years under the ADR depreciation system. Accordingly, assets included in this class are included in the 7-year category under the Act. This category includes assets used in the conversion of refuse or other solid waste or biomass to heat or to a solid, liquid or gaseous fuel. This ADR class also includes all process plant equipment and structures at the site used to receive, handle, collect, and process refuse or other solid waste or biomass. However, this class does not include any package boilers, or electric generators and related assets such as electricity, hot water, steam and manufactured gas production plants classified in the Electric Utility Steam Production Plant ADR class, the Industrial Steam and Electric Generation and/or Distribution Systems ADR class, and the Central Steam Utility Production and Distribution ADR class. These categories are described in detail below. This class also does not include biomass facilities which are qualified small power production facilities under PURPA, and are included in the 5-year property category.

Gas Turbine/Diesel Cogeneration Facilities: The "Electric Utility Combustion Turbine Production Plant" ADR class had a guideline life of 20 years under the ADR depreciation system. Accordingly, assets included in this class are included in the 15-year category under the Act. This category includes assets used in the production of electricity for sale by the use of such prime movers as jet engines, combustion turbines, diesel engines, gasoline engines, and other internal combustion engines, their associated power turbines and/or generators, and related land improvements. However, this class does not include combustion turbines operated in a combined cycle with a conventional steam unit. The meaning of the term "conventional steam unit" is unclear. However, it is reasonable to interpret this term to mean a typical combined cycle unit in which waste thermal energy is used in a waste heat boiler to generate additional electricity.

Cogeneration Facilities Which Produce Steam And Electricity For the Taxpayer's Own Use and Not Primarily For Resale: The "Industrial Steam and Electric Generation and/or Distribution System" ADR class had a guideline life of 22 years under the ADR depreciation system. Accordingly, assets included in this class are included in the 15-year category under the Act. This category includes assets used in the production and/or distribution of electricity with a rated total capacity in excess of 500 kilowatts, and/or assets used in the production and/or distribution of steam with a rated total capacity in excess of 12,500 pounds per hour, for use by the taxpayer in its industrial manufacturing process or plant activity and not ordinarily available for sale to others. However, this class does not include buildings and structural components, or assets used to generate and/or distribute electricity or steam of the type described above, but of a lesser rated capacity. Instead, such assets are included in the appropriate ADR class for the specific industry in which they are used. This class does include electric generating and steam distribution assets, which may utilize steam produced by a waste reduction and resource recovery plant, used by the taxpayer in its industrial manufacturing process or plant activity. The limitations on this category exclude third party financed industrial cogeneration projects.

Combined Cycle And Other Steam Electric Cogeneration Facilities: The "Electric Utility Steam Production" ADR class had a guideline life of 28 years under the ADR depreciation system. Accordingly, assets in this class are included in the 20-year category. This category includes assets used in the steam production of electricity for sale, combustion turbines operated in a combined cycle with a conventional steam unit, and related land improvements. This category also includes package boilers,4/ electric generators and related assets such as electricity and steam distribution systems as used by a waste reduction and resource recovery plant, if the electricity is normally for sale to others. This category may also include certain components of biomass projects used for the production of electricity and that are not included in the 5-year category. This category will include most solid fuel third party financed cogeneration projects.

Central Steam Utility Production And Distribution Facilities: The "Central Steam Utility Production and Distribution" ADR class had a guideline life of 28 years under the ADR depreciation system. Accordingly, assets included in this class are also included in the 20-year category. This category includes assets used in the production and distribution of steam for sale. This class may also include certain components of biomass cogeneration facilities involved in the production of steam. However, this class does not included assets that are included in the Waste Reduction and Resource Recovery Plants ADR class.

Cogeneration Facilities Which Are Building Components Of Non-residential Real Property: Cogeneration facilities which are building components of non-residential real property, are included in the 31.5-year real property category. This category includes Section 1250 property which is not residential real property, and which had an ADR class life of 27.5 years or more. The term "building" generally includes any permanent structure, or component thereof, unless the structure is an integral part of a qualified activity and can be expected to be replaced when the property it initially houses is replaced. Qualified activities include manufacturing, production or extraction, or the furnishing of transportation, communications, electrical energy, gas, water or sewage disposal services. Generally, office buildings and general purpose structures built in conjunction with a cogeneration project will be classified as "buildings," while project specific coverings and housings for specific items of equipment may escape such classification. Structural components are those structural elements that relate to the ordinary functions of the building. Structural components include central heating and air conditioning systems, regardless of whether they are located in, on, or adjacent to the building. Under this rule, certain cogeneration systems for commercial or residential structures may be classified as structural components, and may have to be depreciated using the recovery period prescribed for real property improvements. The determination of whether a cogeneration system is a structural component is difficult, and will vary from case to case. A cogeneration system which generates steam to run a building's heating and hot water system, and which also produces electricity for use in running the building's air condition system could be classified as a structural component under this rule. However, a cogeneration system whose primary purpose is to generate electricity for sale may escape classification as a structural component even though the waste heat from the system is used as a source of energy for a building's heating and cooling system.

Cogeneration Facilities Which Are Building Components of Residential Rental Property: Cogeneration facilities which are building components of residential rental property are included in the 27.5-year real property category. The residential real property category includes buildings, structures and structural components, if 80 percent or more of the gross rental income from such buildings or structures for the taxable year is rental income from dwelling units. If any portion of such a building or structure is occupied by the taxpayer, the gross rental income from such building or structure includes the rental value of that portion as well.

Investment Tax Credit

Under prior law, a regular investment tax credit of 10 percent was allowed for investments in many items of business property, including most cogeneration and alternative energy projects. The Act repealed the regular investment tax credit for most property placed in service after December 31, 1985, although transition rules are provided, and carry-forwards from prior years may still be used after 1985. However, for tax years beginning after June 30, 1986, the Act reduces the regular investment tax credit for both unused credits which are carried forward from prior years and credits allowed for transition period property. The regular investment tax credit allowable for a taxable year beginning on or after July 1, 1987 is

4/ The term "package boilers" has not been defined in the Internal Revenue Code of 1986, Treasury Regulations or any public or private revenue rulings.

reduced by 35 percent, and a prorated reduction applies to credits allowable for a taxable year which begins before and ends after July 1, 1987. The amount by which the credit is reduced cannot be used as a credit for any other taxable year. In addition, for property placed in service after December 31, 1985, the property's depreciable basis must be reduced by the full amount of the investment tax credit allowed for the property (after application of the credit reduction described above).

The Act extends the biomass energy tax credit for two years and the solar, geothermal and ocean thermal energy tax credits for three years, all at declining rates. Under prior law, all of these credits expired at the end of 1985. Although other energy tax credits were not extended, the existing affirmative commitment rules for certain other types of energy property, including hydroelectric projects, were retained. However, energy tax credits claimed pursuant to these affirmative commitment rules are subject to the percentage reduction and basis adjustment rules described above with respect to the regular investment tax credit.

Effective Dates

The depreciation provisions of the Act generally apply to property placed in service after December 31, 1986, and the repeal of the investment tax credit is generally effective for property place in service after December 31, 1985. However, the Act provides transition relief for a variety of items, including:

- Property acquired, constructed or reconstructed pursuant to a written contract that was binding as of March 1, 1986 (December 31, 1985 for purposes of the investment tax credit);
- Property constructed or reconstructed by the taxpayer if the lesser of $1,000,000 or 5 percent of the cost of such property has been incurred or committed by March 1, 1986 (December 31, 1985 for purposes of the investment tax credit), and if construction or reconstruction of the property has begun by that date;
- Property which is part of a project certified by FERC as a qualifying facility on or before March 1, 1986. The legislative history expressly states that a project which has sought self-certification -- i.e. a project which "a developer has simply put FERC on notice is a qualifying facility -- is "not certified as a qualifying facility" for the purposes of this rule.
- Property which is part of a project which was granted a hydro-electric license by FERC on or before March 1, 1986;
- Property which is part of a hydroelectric project of less than 80 MW, for which an application for a preliminary permit, a license or an exemption from licensing was filed with FERC on or before March 1, 1986; or
- Property necessary to satisfy a service contract or lease which was binding on March 1, 1986 (December 31, 1985 for purposes of the investment credit). While electric power sale contracts qualify as service contracts, it is not clear whether this exception will apply to typical PURPA-based power sale contracts, since such contracts generally do not impose a penalty on the developer if no power is actually delivered.5/

The transition rules for depreciation do not apply to any property unless the property has an ADR midpoint life of 7 years or more and is placed in service before January 1, 1989 in the case of property with an ADR midpoint life of less than 20 years, and January 1, 1991 in the case of property with an ADR midpoint life of 20 years or more. Most cogeneration and alternative energy projects which had an ADR life had a life of 20 years or more. For purposes of the investment tax credit, the applicable placed-in-service dates are:

- For property with an ADR midpoint life of less than 5 years, July 1, 1986;
- For property with an ADR midpoint life of at least 5 but less than 7 years, January 1, 1987;
- For property with an ADR midpoint life of a least 7 but less than 20 years, January 1, 1989; and
- For property with an ADR midpoint life of 20 years or more, residential rental property, and nonresidential property, January 1, 1991.

Accordingly, to qualify for transition relief, a facility, which otherwise qualifies for transition relief, that is included in the Waste Reduction and Resource Recovery Plants ADR class must be place in service before January 1, 1989. Whereas, a facility that is included in the Electric Utility Steam Production Plant ADR class, the Industrial Steam and Electric Generation and/or Distribution System ADR class, the Electric Utility Combustion Turbine Production Plant ADR class or the Central Steam Utility Production and Distribution ADR class must be place in service before January 1, 1991 -- two years after the placed in service date for facilities included in the Waste Reduction and Resource Recovery Plants ADR class.

For transition period projects, prior law ACRS depreciation deductions and the reduced investment tax credit would remain available. Property which qualifies for transition relief may generally be transferred to a third party before it is placed in service, or may be sold and leased back within 90 days of being placed in service, without losing the benefit of transition period status.

5/ The Internal Revenue Service recently issued a private letter ruling that held an electric generating facility which was readily identifiable with and necessary to carry out an executed power supply agreement qualified for transition relief.

ALTERNATIVE MINIMUM TAX

Under the Act, an alternative minimum tax of 20 percent is imposed on corporations. The alternative minimum tax base is equal to regular taxable income, plus tax preferences, less certain deductions. The amount by which depreciation calculated under the ACRS depreciation system exceeds depreciation calculated under the Alternative Depreciation System is a tax preference.6/ Corporations are allowed an exemption of $40,000 less 25 cents for each dollar that alternative minimum taxable income exceeds $150,000.

The Act also imposed an alternative minimum tax of 21 percent on individuals. The alternative minimum tax base for individuals is equal to regular taxable income, plus tax preferences, less certain deductions. The individual exemption amount is $40,000 for joint returns, $20,000 for trusts and married taxpayers filing separate, and $30,000 for single taxpayers. The exemption is reduced by 25 cents for each dollar that alternative minimum taxable income exceeds $150,000 for joint returns, $75,000 for trusts and married taxpayers filing separate, and $112,500 for single taxpayers.

PASSIVE LOSS RULE

The Act imposes a limitation on the deduction of passive losses by individual taxpayers which could have a significant impact on cogeneration and alternate energy project development. The new rule applies to losses by pass-through entities, such as Subchapter S corporations and partnerships, as well as to sole-proprietorships. Under this provision, passive losses, including losses of partners in limited partnerships which invest in cogeneration and alternative energy projects, can only be deducted against income from passive investments, and cannot be deducted against income from other sources. Similarly, tax credits (including the investment and energy tax credits) can only be claimed against the tax on passive income.

Losses and credits which cannot be used in the year in which they are incurred must be carried forward, and can only be used to offset income from, or tax on, the investment itself or other passive investments. This provision applies to losses and credits from any source, including losses and credits from transition period projects and other activities commenced before the consideration of the Act. However, for investments made before the date of enactment, the provision is phased in, so that it only applies to 35 percent of losses in excess of passive income for tax years beginning in 1987, 60 percent in 1988, 80 percent in 1989, and 90 percent in 1990. Beginning in 1991, all losses are covered. This provision could significantly alter the pattern of capital formation for all types of capital investments, including cogeneration and alternate energy projects, and could diminish the benefit of many of the transition rules.

IMPACT OF TAX REFORM ON COGENERATION AND ALTERNATIVE ENERGY PROJECT DEVELOPMENT

At this point, although it is obvious that the Act deprives cogeneration and alternative energy projects of many important tax advantages that they enjoyed under prior law, it is too early to tell the ultimate impact of tax reform on cogeneration and alternative energy project development. Nevertheless, there are several general observations which can be made:

- The impact of tax reform varies depending on the type of technology involved: The Act has the greatest impact on capital-intensive cogeneration and alternative energy projects, such as hydroelectric and coal-fired projects, for which capital costs are a major portion of overall project operating costs. By contrast, for projects for which capital costs are low relative to operating costs, such as diesel or gas turbine cogeneration projects, the changes made by the Act may be much less significant. In fact, depending on the relationship between effective income tax rates and capital cost recovery allowances, certain cogeneration and alternative energy projects could actually be more attractive now than they were before tax reform. The impact of tax reform on particular technologies is also affected by the extension of the energy tax credit and other incentives for certain alternative energy technologies, such as biomass and renewable energy facilities.

- Cogeneration and alternative energy projects are generally worse off under tax reform than comparable capital investments: One of the novel features of ACRS as it was originally enacted, when compared with prior depreciation systems such as the ADR depreciation system, was that it provided a uniform 5-year depreciation period for all capital investments, regardless of their actual economic useful life. Thus, under ACRS as it was originally enacted, most types of capital investments did not enjoy a tax advantage relative to other competing types of capital investments. However, the Act returned to the concept of useful lives as a measure of allowable depreciation. Accordingly, cogeneration and alternative energy projects, which have traditionally been thought to have longer useful lives than comparable investments, now have

6/ Under the Alternate Depreciation System property is depreciated using the straight-line method and an extended recovery period. The depreciation is computed without regard to salvage value and using either a half-year or mid-quarter convention. The recovery period is equal to the property's ADR class life, or 12 years if the property did not have an ADR class life. Real property is depreciated over a 40 year period. Property included in the Waste Reduction and Resource Recovery Plants ADR class will be depreciated over 10 years, property included in the Electric Utility Steam Production Plant ADR class will be depreciated over 28 years, property included in the Industrial Steam and Electric Generation and/or Distribution system ADR class will be depreciated over 22 years, property included in the Electric Utility Combustion Turbine Production Plant ADR class will be depreciated over 20 years, and property included in the Central Steam Utility Production and Distribution ADR class will be depreciated over 28 years.

longer depreciation periods than other unregulated capital investments. As a result, the Act made these cogeneration and alternative energy projects relatively less attractive, compared with other capital-intensive investments.

- Lower rates and a stiff minimum tax diminish the overall significance of all tax benefits: Much concern has been expressed regarding the impact of extended depreciation periods on cogeneration and alternative energy projects. However, this concern should be placed in context with the impact of reduced tax rates. Under prior law, each dollar of tax deduction was worth 46 cents of tax savings to a corporate taxpayer, and 50 cents of tax savings to an individual taxpayer. The Act reduces this to 34 cents, and 28 cents respectively. A stiff minimum tax could further reduce this savings. Because the reduced tax rates reduce the benefit from all tax incentives, all projects, including those which have retained more favorable depreciation rates, now have to rely more heavily on actual economic return, and less on tax incentives to attract investors.

- The ultimate impact of tax reform depends on many factors not apparent from the face of the final legislation: The relatively generous tax benefits available for capital investments under ACRS produced an investment climate in which high after-tax returns became the business norm. It is not possible to predict the ultimate impact of tax reform on capital formation until it can be seen whether the capital markets will adjust their expectations to accept lower after-tax return. If such an adjustment occurs, the loss of tax benefits may not be a significant impediment to the development of cogeneration and alternative energy projects.

Chapter 72

COGENERATION AND ENERGY MANAGEMENT IN THE NOT-FOR-PROFIT SECTOR: LEGAL AND FINANCIAL CONSIDERATIONS

R. D. Feldman

I. Introduction

A. Cogeneration and energy management as vehicles for enhancing not-for-profit financial capabilities.

B. Overview of regulatory framework affecting not-for-profit sector and energy supply projects.

C. Contractual considerations for not-for-profit entities when negotiating cogeneration and energy management agreements.

II. Regulatory Environment

A. The New York State hospital example - Regulatory concerns relevant to the not-for-profit sector.

1) The construction of capital improvements for hospitals requires prior approval of the State Commissioner of Health.

2) Upon the filing of a construction application, the Commissioner must determine whether there is a "public need" for the proposed construction.

3) Proposed construction or capital improvements must comply with general standards of construction, as enumerated in the New York State Code; certain waivers from these standards exist for projects proposing energy conservation measures.

4) Construction approval process includes determination of whether or not the project as described is financially feasible.

B. Regulatory concerns relevant to energy projects.

1) FERC NOPR's - resource matching and the advent of bidding.

2) Certification of project as a "qualifying facility" under PURPA, thus becoming eligible to sell power to local utility at "avoided cost rates."

3) Project ownership and the disposition of project energy may trigger additional regulatory scrutiny.

a) Analysis of applicable tarrifs.

b) Retail sales concerns.

c) Wheeling.

III. The Planning Process

A. Structuring Considerations

1) The "host" (i.e., the party contracting for energy services) must choose from an array of ownership structures.

2) Host must also determine its requirements concerning quality of service and desired availability factor for energy saving equipment.

3) Coordination of energy saving operations with the needs of the Host's business enterprise.

B. Risk Assessment - consideration of the probability and magnitude of project risks.

1) Risk of non-completion.

2) Risk of inadequate equipment performance.

3) Operating Risks.

4) Regulatory Risks.

C. Contractual Provisions - The Negotiation Process

1) Determination of ownership interest of developer and host and allocation of risks and benefits arising therefrom.

2) Developer's compensation.

3) Quantity and quality of energy to be delivered.

4) Assignability of the developer's obligations.

5) Procedure for disposition of the project equipment after the agreement terminates (either as a result of default or upon expiration of the agreement).

6) Insurance.

7) Indemnification.

8) Availability of consequential or special damages in the event of a project malfunction.

9) Coordination of project maintenance with host facility's operations.

10) Standard of care in operating the facility.

11) Dispute resolution procedures.

Chapter 73

PROJECT FINANCING ISSUES IN THE FUTURE COMPETITIVE BID WORLD

A. L. Hills

INTRODUCTION

On March 16, 1988, The Federal Energy Regulatory Commission (FERC) issued proposed rules to provide guidelines for the use of competitive bidding systems by state utility commissions to determine the price and sources of incremental generating capacity from either "qualifying facilities" (QFs) under PURPA and/or "independent power producers" (IPPs) under the Federal Power Act.

In announcing these proposed rules, FERC Chairman Martha O. Hesse said: "Providing more regulatory flexibility should enhance the competitive forces already emerging in the electric utility industry. Greater reliance on competitive forces is a means by which the Commission can meet its responsibility to ensure consumers adequate, reliable supplies of electricity at the lowest possible cost." (Emphasis added.)

This paper discusses the important repercussions of FERC's proposed bidding rules on how lenders and equity investors perceive the risks of project-supported financing arrangements and describes how project developers should in the future address such financing concerns in preparing competitive bidding proposals.

FINANCIAL IMPACTS OF BIDDING

1. Greater "Up-Front" Financial Risk for Developers

The immediate impact of formalized bidding systems (as opposed to administratively set avoided costs and individual utility contract negotiations) is to significantly increase the development risk capital that individual developers must expend on engineering and other costs related to bidding packages. This increased up-front financial exposure will force entrepreneurial developers to either affiliate or consolidate with more highly capitalized corporate developers and/or bring financial institutions and venture investment firms into co-development arrangements at the outset of bidding for specific project situations. This trend was accelerated by the October 1987 Stock Market Crash, which has to date, cut off the availability of public capital market debt or equity funds to the smaller independent power companies.

2. Earlier Development of Specific Financing Arrangements

In non-bidding circumstances, project developers undertake the negotiation of power sales agreements with the obvious goal of obtaining the highest possible avoided cost prices consistent with the ability to sell power on a continuous or nearly continuous basis. In such circumstances, only a general outline of the possible financing sources and their costs are necessary for a developer to determine the basic feasibility of a proposed project, at various possible power sales rates.

However, in a highly competitive bidding situation, where lowest possible power rates will be the dominant if not only criteria for selection of the winning bidders, it will be extremely important that developers work with their legal and financial advisors while preparing bid packages, to develop detailed ownership structures, financing sources and their costs. Only with such specific financial inputs is it possible for a developer to bid his lowest possible power prices that are consistent with successfully implementing a project if selected by the bidding process.

Without such initial involvement by the financial community, the risks are great that bidding for generating capacity will not result in meeting the FERC's stated goals of obtaining adequate quantities or reliable electricity supplies.

In addition, to the extent that various non-price qualitative factors are used in competitive bidding systems to narrow the field of responsible developers, it will become necessary for project sponsors and their advisors to negotiate with prospective lenders and equity investors to obtain formal financing commitment letters, subject to the successful selection of the proposed project with business and legal characteristics as proposed in the bid submission.

3. Necessary Changes in the Methods of Power Pricing for Utility Purchase Contracts

In addition to having a more precise understanding of the financing costs for bidding competitions, it will also be necessary for project developers, utility purchasers, regulatory commissions and financing sources to understand the interrelationships between the desired operating mode of a competitively bid power plant and necessary changes in the traditional avoided cost pricing mechanisms.

Specifically, the easily comparable "cents per kilowatt hour delivered" method of pricing for competitively bid power sources is incompatible in terms of project financing, with requiring bidders to accept large amounts of utility dispatchability. In simple terms, as utilities or regulatory commissions either require or give preference to IPPs or QFs that operate more similar to the utilities generating system, the power pricing terms for utility purchase contracts will also have to more closely mirror traditional utility costing methods. A highly dispatchable private power plant will not satisfy the credit criteria of project lenders or equity investors unless it can be clearly shown that all project fixed costs (including financing charges) can be recovered through an annual "capacity change" the payment of which is subject only to agreed upon levels of operating availability. Likewise, the variable costs of actually operating the project must be recoverable through an "energy charge" based upon actual power deliveries.

Obviously, the objective comparison of different project bidders utilizing different fuel sources and technologies and therefore different fixed and variable costs will be more difficult than for base-load power projects bid on a price per kilowatthour delivered basis.

4. Potential for Increased Risk Perceptions by Lenders and Investors

The most important general impact of the FERC's proposed bidding rules will be to increase the uncertainty facing project developers regarding the implementation of these rules and their economic impacts. This uncertainty affects both the potential economic viability of proposed projects as well as the perceived riskiness of lending to or investing in competitively bid projects.

Specifically, the tendency of all-out bidding to minimize the economic and cash flow margins of projects (as opposed to project economics based upon reducing costs below specified levels of administratively set avoided costs) may result in reduced levels of debt leverage for projects and/or increased interest rate spreads on loans. Similarly, equity investors may increase their required levels of return to compensate for increased amounts of equity investment or the greater risk of project financial difficulties as a result of narrowed profit margins.

As a consequence, if the actual implementation of the FERC proposals by individual states results in unrealistically predatory bidding practices, the end result will be either: (1) increased power costs to actually finance and build, economically viable, long-term power supplies as a consequence of increased financing costs, or (2) the failure of winning bidders to actually carry through and deliver the bid for power supplies at the expected costs and within the needed timeframes.

CONCLUSION

In general, from the perspective of project financing, the FERC proposals for competitive bidding and individual state procedures for implementing bidding systems, run the risks of either: (1) increasing the actual costs of successfully developed power supplies, (2) reducing effective competition by accelerating the consolidation trend in the independent power industry, and/or (3) frustrating the ability of successful bidders to obtain acceptable non-recourse financing for power project construction.

At worst, such bidding systems may fail in obtaining sufficient quantities of reliable power supplies and thereby hasten the return to the monopolistic utility power generation and pricing of the 1960's and 1970's.

However, the hoped for benefits of competitive bidding can be achieved, if the final FERC regulations and the subsequent state implementation efforts are adopted to properly address the financing concerns discussed in this paper.

BIOGRAPHY

ALAN L. HILLS, MANAGING DIRECTOR
COGENERATION CAPITAL ASSOCIATES, INC.
80 East Sir Francis Drake Boulevard
Larkspur, CA 94939
(415) 925-1564

Cogeneration Capital Associates is a private investment banking and venture capital firm located outside of San Francisco that specializes in the contractual negotiations and financing of cogeneration and small power projects. Over the past two years, Cogeneration Capital has raised more than $750 million of financing for such projects and independent power companies, as well as managing a merchant banking joint venture with First Interstate Bancorp.

Mr. Hills most recently arranged the financing for the $100 million Basic American Foods cogeneration project in California.

Prior to joining Cogeneration Capital, Mr. Hills was Vice President and Manager of the Alternative Energy Finance Group at Prudential Bache Securities in New York.

Mr. Hills has an M.B.A. from The Wharton School and was also a Certified Public Accountant. Mr. Hills is currently a member of the Board of Directors of the Northern California Cogeneration Association.

Chapter 74

ENERGY SERVICES— A WIN-WIN SITUATION

R. J. Heller

NEEDLESS WASTE

Billions of dollars are wasted every year in this country on energy related items. These billions are spent inefficiently not because technological or service-based solutions are unavailable to apply to the problems at hand but, because, in many cases, the institution with the problem simply lacks the resources (human or financial) to identify the problem, structure a workable solution, and implement it. This basic state of inefficiency, complicated by the lack of sufficient human and financial resources, is the main reason for the existence of the Energy Services industry.

THE WIN-WIN OF ENERGY SERVICES

The Energy Services industry is comprised of vendors whose resources and specialized expertise make it possible to implement cures for the aforementioned operational ills. These vendors comprise a new and exciting services industry. By skillfully addressing energy efficiency problems (or opportunities, as they appear to vendors) both parties to the transaction, buyer and the seller, win. The buying entity operates more efficiently and cost-effectively and the Energy Services vendor grows, as does the industry.

Remedying operational inefficiencies is very important to the buyer whether that buyer is a profit or non profit institution. It stands to reason that in an increasingly competitive world economy a for-profit institution cannot afford to be less productive than its counterparts here in the United States or in other parts of the world for that matter. A non profit institution should be similarly motivated. The effects of an increasingly competitive world economy have been themselves felt by these institutions via increasing downward pressure on their budgets. The goal of a non profit institution is the provision of a given level of service within budgetary constraints. Therefore, it makes sense for that entity to spend the money it receives as effectively as possible so that the highest possible level of service can be maintained.

The individual vendors in the Energy Services business, whether subsidiaries of large organizations or start-up ventures, all have before them an opportunity for substantial financial growth. The benefits of this growth are obvious in a capitalistic society.

The industry therefore offers a classic win-win opportunity. The buyer resolves operational inefficiencies by entering into an Energy Services contract thereby operating more cost-effectively and productively. The vendor, by leveraging his resources (human and financial) and specialized expertise, is able to sustain substantial growth.

FINANCIAL AND PERFORMANCE CONTRACTING SERVICES

Energy Services contracts provide the buyer with a financial services element, a performance contracting element, or both.

Historically, there have been instances in which the only impediment to the implementation of a productivity-enhancing energy retrofit was the absence of a financial services (or payment) arrangement tailored to the specific needs of the customer. There have been other instances in which the major impediment to implementation had to do with the degree of certainty associated with the positive cash flows (savings) "promised" by the retrofit. Finally, there have been cases in which the prospective buyer had a problem both managing the negative cash flows (payment stream) and feeling comfortable with the degree of certainty associated with the positive cash flows (savings) resulting from the energy efficiency retrofit. The Energy Services industry provides solutions for all three of these situations.

A variety of Lease and Installment Sales programs are available that help stretch and smooth out payment streams. These financial tools can be custom fit to the unique requirements of both private and public sector buyers.

With respect to assuring the positive cash flows associated with a project; Energy Services vendors, because of their experience and expertise, will contractually assure that some portion or all of the projected savings accrue to the buyer. This contractual guarantee typically states that, should the savings not occur as expected, the vendor will reimburse the buyer for the difference between projected and actual savings.

In a growing number of instances, the buyer receives a turn-key package consisting of a financial services vehicle tailored to his needs to address payment for the retrofit and a savings guarantee that is equal to or greater than the payment stream. This assures the buyer a break-even or positive cash flow throughout the term of the agreement.

MUTUAL UNDERSTANDING AND NEEDS ASSESSMENT

There have been those who have said, "it sounds too good to be true." There have also been those who have mistaken the Energy Services industry for a "free lunch" program. In answer to these concerns, yes, it is true; but, no, it's not a "free lunch" program.

The Energy Services industry, as the name would imply, is very much a services industry. It is incumbent upon the responsible Energy Services vendor to take the role of understanding buyer needs seriously.

Conversely, for the relationship to be successful (since it is a long-term services relationship) the buyer must also understand the needs of the seller.

The typical Energy Services contract is a long-term relationship. Entering into such an agreement should be a serious endeavor on the part of both the buyer and seller. Analogously, the process should be given the same degree of serious consideration as one would give to hiring or becoming a new employee or getting married. The long-term and the short-term needs of both the buyer and seller should be given serious consideration. An honest appraisal of one's own needs (as either buyer or seller) and the other party's ability to satisfy those needs while also satisfying their own must be made.

INDUSTRY GROWTH AND CHALLENGES FACED

The industry, while not in its infancy, is certainly not much past adolescence. Some of the growing pains experienced thus far have included:

- Confusion and frustration due to establishing mutually acceptable contractual terms and conditions.
- Confusion and frustration due to existing language barriers. (Does the Chief Financial Officer understand degree day adjustments? Does the Physical Plant Administrator understand discounted cash flow analysis?)
- Confusion and frustration due to inappropriate or less than optimal procurement processes. (Can a service offering be specified in detail just like a product is? Can one reasonably expect a number of different services vendors to respond to a solicitation document which requires absolute compliance to the letter of the request when in all likelihood each has a slightly different way of doing business?)
- Confusion and frustration due to high transaction costs on the part of both the buyer and the seller. (The industry is young. An Energy Services contract is <u>not</u> a common everyday transaction for most buyers. Can looking to eliminate every bit of uncertainty be justified if it raises transaction costs to a point where they become intolerable for both buyer and seller?)
- Confusion and frustration due to a long procurement cycle. (No one wins when the procurement cycle stretches out beyond reasonable limits. From the buyer's perspective the best time to start saving money is yesterday. From the seller's perspective, business is business and any cash flows generated today [if due to nothing other than the time value of money] are worth more than cash flows generated tomorrow.)

These are just some of the hurdles that have been faced and in most cases met. That's not to say that they don't still pop up on occasion or that when they do arise that they're ideally handled in all cases. As is true in many situations, however, knowing the problem is the first step to resolving it.

IMPORTANCE OF STRONG COMMUNICATIONS

As is also true in many cases, communications is the key to problem resolution. Earlier, the importance of a mutual understanding of both the buyer's and the seller's needs was stressed. This mutual understanding requires solid communications.

Parties inside the buying institution may need to learn new languages. The Physical Plant Administrator may have to learn to understand discounted cash flow analysis. The Chief Financial Officer may have to acquire at least a basic knowledge of the problem areas being addressed by the Energy Services vendor. Finally, the vendor must understand the language of both the Chief Financial Officer and the Physical Plant Administrator.

Much of the solution to resolving the confusion and frustration associated with less than optimal procurement processes can be addressed by keeping one simple fact in mind. Energy Services is a service business. Service, performance, and results are the things that count in a service business. Any solicitation or request for proposal should be written with this in mind.

Once improvements are made in communications, the mutual understanding of one another's needs, and with respect to procurement processes; the contract language problem tends to resolve itself. The overall result is then generally lower transaction costs and shorter procurement cycles. This really enhances an already attractive win-win situation.

THE FUTURE AND ENERGY SERVICES

What does the future hold for the Energy Services industry? Most certainly, over time, today's unusual will become tomorrow's routine. To the extent that the nature of the business transaction is better understood by both parties to the transaction the industry should enjoy a more rapid growth rate than it has in the past. This growth will be made possible because the very essence of the industry, its core, will become better understood by a wider audience.

Does this projected future stabilization imply stagnation? Hardly. As less time needs to be spent by both the buyer and seller on the very core of the offering, more time can be spent identifying and addressing specific customer needs. It's highly likely that contract terms and conditions will become much more standard and that the means of procuring Energy Services will become much more routine. It is also likely that the needs of specific customers or sets of customers (as in particular vertical markets - hospitals, schools, universities, etc.) will be better addressed by custom tailored product and service offerings.

PERFORMANCE

One thing will remain the same in the Energy Services industry. The business will remain a service business. Since it is and will always be a service business, performance and results (or the lack thereof) will distinguish the successful from the unsuccessful vendor and the satisfied from the unsatisfied buyer.

SECTION 16

PURCHASING STRATEGIES FOR NATURAL GAS

Chapter 75

DIRECT GAS PURCHASE IN LOCAL GOVERNMENT

J. A. Ventresca

INTRODUCTION

While direct gas purchase is common in the private sector, the public sector has been slow to respond. For example, when the Columbus program began in the fall of 1986, it was the only major Ohio city receiving natural gas through direct purchase. This is because direct gas purchase in the public sector requires overcoming significant institutional barriers that typically do not exist for the private sector.

In the public sector, several governmental agencies with conflicting missions must be coordinated to work together; public competitive bidding procedures must be adhered to; accounting procedures to pay for the gas must be developed to keep general fund tax revenues separate from self-sustaining revenues; and a consortium of many diverse commercial buildings must be created and managed. In addition, the public sector does not have the clear, direct, and measurable incentive of increasing profit. However, with the recently demonstrated successes, the public sector situation should improve rapidly.

Part of Mayor Dana Rinehart's policy to run the City government like an efficient business includes getting the best price for energy. Columbus has secured natural gas through direct purchase at a price 30% less than regular gas, saving thousands of dollars. To control government spending, public sector agencies must cut costs and boost efficiency.

This paper presents the Columbus experience with direct gas purchase in a step-by-step format. It is valuable for government administrators who are interested in direct purchase of natural gas, and also to vendors and consultants who are seeking business with the public sector.

DIRECT GAS PURCHASE CONCEPT

Historically, natural gas from the public utility, called "tariff" gas, has included the following in one bill;

1. The cost of the gas
2. The cost of the local pipeline transportation or "carriage"
3. The cost of the regional pipeline transportation or "carriage"
4. The cost of administration, meter reading, maintenance, regulation, etc.

With direct gas purchase, (also called Self-Help Gas, Transportation Gas, and/or Carriage Gas) one buys gas directly from a private sector supplier, and buys the gas transportation directly from the pipelines. The costs are separated out with separate contracts and billings. With the present gas supply bubble, reduced prices for the gas can be obtained. There are additional administrative costs for finding and selecting a supplier, estimating gas use, ordering the gas, paying multiple bills, etc. While there were significant administrative costs, the Columbus experience was a great success with savings far in excess of increased administrative costs.

METHODOLOGY

The steps taken by Columbus to purchase natural gas directly from the private sector are presented in the following order:

1. Simple Cost/Benefit Assessment.
2. Practical Research.
3. Coordination of Diverse Actors.
4. Maximize the Savings; Minimize the Work.
5. Stimulate Competition.

6. Interruptible Transportation Rates.

7. Securing Transportation.

8. Competitive Bid Purchase.

9. The Purchase Contract.

10. Consolidation of Accounts Into A Consortium.

11. Administration Costs.

1. Simple Cost/Benefit Assessment. In the very beginning of the program the most often asked question was, "is it worth it?" To quickly address this question the ballpark potential savings was determined. This was done by simply obtaining the total city-wide expense for natural gas from the budget document and assuming that the savings could be 10% to 30%. The potential annual savings was over $300,000. The potential cost was assumed to be no more than one additional full time staff person to handle additional administrative tasks. This favorable preliminary assessment helped to gain support from diverse governmental actors.

2. Practical Research. The next question was, "How do we go about doing this?" To answer that question, the State Public Utility Commission, local gas vendors, and local industries which were already using direct gas purchase were contacted. This research effort utilized only local sources and concentrated on specific details such as current prices and experience with transporting pipelines. The experience of local industries was particularly valuable.

3. Coordination of Diverse Actors. Very early in the project it became apparent that securing direct gas purchase would require action on the part of many diverse governmental actors. Not only the Administration, but also the City Attorney's Office, the Council, and the Auditor's Office, would all have significant roles. These agencies each have their own primary mission and are independent of each other. They are each headed by separate elected officials; they are not all of the same political party; and they had no knowledge of direct gas purchase. With this complex situation, coordination of these agencies began very early in the program with a meeting of a representative of each of the agencies. At this first meeting, lines of communication and contact persons in each of the agencies were established. The overall concept and potential savings and cost/benefit were explained. This early coordination and involvement was absolutely critical to the success of the program.

In addition to these separate governmental agencies, within the Administration itself the project required the assistance of: the Office of Management and Budget Department; the Facilities Management Division; the Water Division; the Waste Water Treatment Division; the Public Utilities Department; and the Purchasing Division. Communication links among these internal agencies were established and the program concept and potential savings were explained. This was done early in the program, and again, was essential for success.

4. Maximize the Savings; Minimize the Work. In planning out the tasks and timeline it became obvious that to secure gas for the coming winter, there was a tremendous time constraint. Also the nature of the project, being new and involving many diverse actors, pointed to a need to simplify the project as much as possible. The management goal was to obtain 80% of the results with the first 20% of the work. To do this, an incremental approach was developed. The program began with a Phase I or pilot for the 86/87 winter, and in the following year was expanded to a Phase II for the 87/88 winter.

In order to focus on the greatest results first, a review of the gas accounts revealed that the largest three accounts used about a third of the City's total gas; and the next largest twenty accounts used approximately a third of the total; while the smallest 150 accounts used only about a third of the total gas.

The three largest accounts are for water and waste water processing. Their total annual volume is about 100,000 MCF. Gas vendors were eager to supply this volume at good prices and the annual savings potential was about $150,000. Therefore, it was possible to greatly simplify the project for the pilot Phase I and still obtain significant savings by dealing with only the three largest accounts. Based upon the experience in the pilot phase, the program was expanded in Phase II to include a consolidated consortium (see step 10) and will continue to be expanded as long as it remains cost effective.

5. Stimulate Competition. In assessing the gas savings potential, it was noticed that significant savings could be obtained by utilizing #2 fuel oil. Almost all of the larger gas accounts in the City of Columbus were found to contain dual fuel burners. Many of the oil burners had not been used for five to ten years and in some cases the oil tanks no longer existed. However, in the three largest accounts oil backup was ready to go.

In our research of local industrial gas users, it was clear that they regularly switched to the lowest priced fuel, including #2 fuel oil, to maximize profits. However, governments were not as aggressive, and tended to avoid fuel switching to oil because of the additional maintenance.

To gain experience with the additional maintenance cost and operational changes required for burning oil, it was decided to turn off the gas and burn #2 oil as a test as soon as possible. The impact of this fuel switching allowed competition to enter the picture. It was clear that the City was aggressively pursuing a least-cost fuel use policy. As a result, cooperation from the gas utility improved.

In order to keep the City from wide spread switching to #2 fuel oil, the local gas company informed the City of a special oil replacement gas program which allowed the City to purchase natural gas at a reduced oil-replacement rate for a temporary time period while direct gas purchase contracts were negotiated.

6. Interruptible Transportation Rates. By assessing transportation rate options it was found that approximately eighty-five cents per MCF, or about 15%, could be saved on transportation costs if interruptible transportation rates were used instead of firm service rates. With interruptible rates, on cold days if the pipeline should not have sufficient capacity to meet all end user demands, the gas company can interrupt the gas flow. The City must then switch to #2 oil. This will actually help to hold down the price of gas to the residential customer because the City will use less gas on cold peak days when interrupted, which will make more gas available for homes. The City benefits from its capital investment in dual fuel capability, and so everybody wins.

However, there are potential problems with interruptible service. The transportation contracts specify that if the City does not interrupt the gas use when requested to do so, it is liable for penalties. Therefore it was very important to assess these penalties and maintain oil switch-over capability. We have found that maintaining oil switch over capability requires diligence and commitment. The oil systems are fired periodically throughout the year. Problems such as faulty oil pumps are discovered and corrected promptly. In addition, every spring approximately 1/4 of a tank of oil is burned at each site. Then, during the summer, when heating oil prices are typically their lowest, the tanks are filled. This must be done each year to assure that the oil in the tanks never becomes old and contaminated.

7. Securing Transportation. Securing transportation on the regional pipeline required approximately two months for the Phase I Pilot. Accurate monthly historical gas consumption data was necessary in order to complete pipeline transmission applications. Maximum and average annual totals and daily totals of gas used had to be calculated and incorporated as part of the transportation agreements.

For the expanded Phase II, it was decided not to seek transportation contracts between the City and regional transporting pipeline. Near the end of the Phase I pilot, take or pay crediting and imbalance penalties began to be implemented. This caused a significant increase in paperwork for checking the regional pipelines's transportation bills. Therefore, for Phase II the vendor was made responsible for both supplying the gas and transporting it on the regional pipeline(s) to the local distribution company (LDC). This was done to reduce the amount of administration paperwork for the City.

We were hesitant to give up the responsibility for the regional pipeline transportation. We were afraid that if the gas supplier was also made responsible for the regional pipeline transportation, this would block sufficient competition in the bidding process. However, there was still adequate competition, and we anticipate continuing in this manner with the gas vendor both supplying the gas and being responsible for the regional transportation up to the LDC.

8. Competitive Bid Purchase. Purchases over $5,000 require a formal competitive bidding process. Under this process a specification is written, competitive bids are secured, and the lowest bidder that meets the requirements of the bidding specifications is selected. This process works well when the item to be purchased can be clearly and accurately specified. Because there are many options for securing direct purchase gas, such as different geographic sources of the gas and different transporting pipelines, it was difficult to specify in advance exact purchasing requirements. Also the formal competitive bid process requires 60 to 90 days. Therefore, for the Phase I Pilot an alternate "informal competitive bid" procedure which only required three weeks was used.

This alternate procedure protected the City and the taxpayers by utilizing competitive bids, and it was efficient. In the manner of selecting professional consultants, a request for qualifications and competitive bid price was sent to a representative number of gas vendors. The best vendor was selected both on the basis of price and qualifications. This informal competitive bid procedure required the approval of the City Council to wave the formal bidding procedure.

With the bidding experience for the pilot, it was possible to write more specific formal competitive bid specifications for the Phase II bid. Some of the points of interest in the Phase II bid documents were:

- To simplify the evaluation of the bid and to assure that all vendors were bidding on an equal basis, the bid price of the gas was required to be specified into the same regional pipeline for all bids. The vendor is responsible for the cost of transporting the gas it to the specified regional pipeline. The gas itself may be from one source or a blend of sources. This procedure made the bid prices directly comparable regardless of the original geographic source of the gas. This eliminated the confusion that occurred with the pilot bid in determining the equivalent burner tip price for gases from different sources such as Texas Gas vs. Appalachian Gas vs. Gulf Gas etc.

- The price was fixed for a 12 month period. This was necessary because the amount of money required to purchase the gas for the whole year had to be known in advance so that it could be certified by the Auditor. This is a requirement of the City Charter.

- The total "equivalent burner-tip price" of the gas had to be clearly identified.

- In order to assess the vendor's qualifications, the average daily sales quantity supplied by the vendor for the previous five years was required. Also a listing of Municipal and Institutional customers, and specific customer references which had received gas from the vendor for at least one full year were required. With this information it was possible to determine if the vendor was well qualified and reliable.

9. The Purchase Contract. The gas vendors provided sample purchase contracts and several were reviewed by the City Attorney's Office. They then determined what items would be required in the contract. This approach worked well. The sample contracts were submitted to the City Attorney's Office months in advance of the actual bid award date. This was necessary to allow adequate time for a complete legal review.

A provision was included to allow for the cancellation of the gas purchase contract if the supplier failed to deliver at least 95% of the gas ordered for any month. We were aware of horror stories of unqualified vendors receiving bid awards because of very low bid prices and then failing to supply the quantity of gas ordered. This clause afforded the City some protection if the vendor failed to deliver, in that it allowed the City to cancel the agreement and be free to renegotiate a new agreement with the runner-up bidder.

We investigated entering into a back up agreement with the runner-up bidder. In theory this would provide the mechanism for very quickly switching to the runner-up bidder if the low bid provided poor service. This can be done and we will continue to investigate it, but we did not do it because the runner-up bidder would not enter into a back up agreement unless he was allowed to raise his bid price to a much higher back-up price.

The purchase contract makes the vendor responsible for any imbalance penalties. Imbalance penalties may be levied by the regional transporting pipeline if the amount of gas put into the pipeline by the gas vendor is greater or less than the amount ordered by the end user.

According to the purchase contract, the City is responsible for any banking penalties. Since the City determines the amount of the gas order, any banking penalties would be the result of the City's ordering to much gas and the City is responsible. Banks occur when the amount of gas actually used is less than the amount of gas ordered for a particular month. The excess amount of gas is then available for use in a future month and is said to be "in the bank." The bank maximum is set in the transportation contracts, and if it is exceeded banking penalties can result.

10. Consolidation of Accounts Into A Consortium. In Phase II the program was expanded to include several commercial government office buildings. All of these accounts also had No. 2 oil back-up capability, and therefore the transportation secured was the interruptible rate. As discussed in step #6 this provides a significant 15% savings over firm service regular tariff rates.

To supply direct purchase gas to the commercial office buildings, several commercial accounts had to be consolidated into one large consortium account. The consolidation of these accounts into a consortium was negotiated with the LDC. With the consortium, a single gas volume is ordered for all of the buildings. This amount of gas is then proportioned to each of the buildings within the consortium.

Our consortium was purposely kept small to minimize start up complications. There were only eight (8) buildings in the original consortium, with the plan to expand the consortium to 15 to 25 buildings in the near future. The consortium has not been expanded as rapidly as we planned because of the time consumed with administration. The time required to manually check the consortium bills for even these 8 buildings was prohibitive. However, the time was reduced to an acceptable level by developing a computerized spreadsheet to track the consortium bills. Three integrated spreadsheets were developed: 1. The bills-input spreadsheet; 2. The bills-verification spreadsheet; 3. The gas-order spreadsheet.

The bills-input spreadsheet is divided into sections for input of data from: 1) the gas purchase bill; 2) the regional transportation bill; 3) the local transportation bill; and 4) the regular tariff bill from the public utility gas company. The information from each of the 4 bills is transferred directly from the bills to the input spreadsheet. This simplifies data input and allows the data to be quickly checked against the bills for human errors. From the input spreadsheet the bills-verification spreadsheet calculates various unit costs, transportation costs, decatherms and MCFs delivered, etc. These calculated values are displayed next to the values from the bills and discrepancies are easily found. This has greatly reduced the time required to verify the bills. The gas-order spreadsheet calculates the amount of gas in the bank, the total amount of gas available for a given month, and the next month's order. Although some vendors will perform the calculation of the orders, we found the computerized system is still needed to verify and keep track of the bills.

The computer accounting system was also needed to calculate the exact charges for the portion of the total consortium order used at each building. This is especially important since some building's gas bills are paid from the "general fund" and others are paid from "self-sustaining funds". In government accounting, the general fund is all tax monies, while self-sustaining funds are monies from sources such as water and sewer fees. Therefore, the two types of funds must be kept separate.

In summary, we have found a computerized spreadsheet accounting system to verify the bills to the be absolutely essential for controlling the administrative cost associated with managing a consortium of accounts.

11. Administration Cost. Procedures had to be developed for reviewing, checking, and paying four bills, 1) the gas, 2) local transportation, 3) regional transportation, and 4) any regular tariff gas; instead of just one bill. Administration costs are also incurred in ordering the gas. The transportation contracts specify that if the amount of gas ordered is less than the amount used, the difference will be supplied as regular gas from the public utility at the regular higher tariff rate. On the other hand, if the amount ordered is more than the amount used, the excess is stored in a bank available for future use. However, the upper limit of the bank is set in the contract and if it is exceeded banking penalties may be incurred. Therefore, ordering the gas also requires keeping track of the bank and guessing what the weather will be in order to assure an adequate amount of gas and at the same time avoiding banking penalties.

The details of paying and checking all of the bills, ordering the gas, coordinating the activities of all the various actors, preparing and evaluating the purchase bid, and in general managing the entire program have consumed most of one full time staff person's activities for the first two years of the program. The Pilot Phase I was time consuming because everything was new. The administrative time for the pilot diminished as procedures were perfected. Then the administration time jumped back up again with implementing the consortium in Phase II. As discussed in Step 10, a computerized spreadsheet accounting system was an absolute necessity. With more experience with the computerized spreadsheet, it is felt that administration time will be reduced to half of one person's time. Even with the increased administration costs, the $500,000 total savings in Phase I and II demonstrates the program's success.

Chapter 76

ELEMENTS OF A PORTFOLIO APPROACH TO FUEL PROCUREMENT

W. G. Nelson, Jr.

The natural gas industry and its associated markets are undergoing vast, often rapid changes that are unprecedented in history. This evolutionary process is the direct or indirect result of legislation, regulation, supply and demand, and economic trends and conditions. The changes could be viewed as having a positive, negative, or neutral effect on the gas market and industry, depending upon one's perspective. Viewed from the standpoint of the industrial end-user of natural gas, these changes have brought about tremendous opportunities to take an active role in a market that had formerly been restricted, for the most part, to passive participation. Considerable savings on purchased fuel cost can accrue to those end-users buying natural gas directly in this market.

The Federal Energy Regulatory Commission (FERC) is the official governmental agency charged with the responsibility, amongst other things, of regulating the distribution of natural gas supplies from the areas of production (wellhead) to the areas of consumption (the end-users). Since mid-1983, they have been pursuing a more or less constant path aimed at providing greater access to the critical transportation facilities (predominately interstate natural gas pipelines) used to move gas from wellhead to end-user, for those persons wishing to do so. Although the path has been strewn with obstacles and they have suffered many setbacks, the ultimate results cannot be ignored. Third party transportation gas, as a percentage of all natural gas volumes transported in this country, has grown steadily since August of 1983 (the enactment of the first significant open access regulations - Order 234b), and judging from current regulations, the continued excess deliverability in the market, and the response of end-users, it will continue to grow for the foreseeable future.

The participants in the direct purchase market, especially those who became involved early on, have been traveling outward upon a long and sometimes difficult learning curve. They first had to master the rules, customs, procedures and even language of an industry that was, for the most part, foreign to them. Their experience was not made any easier by having the rules of the game changed almost every two years, thereby sending the entire market into a tailspin (witness the recent Order 500 debacle).

Those energy managers, buyers, engineers and purchasing personnel who survived the experience were rewarded handsomely for their efforts through the considerable cost savings that accrued to their companies; in some cases millions of dollars a year. Unfortunately factors are now at work in the market that threaten to reduce these savings. Increasing spot purchases of natural gas and impending contract demand (CD) conversion by local distribution companies (ldc's) lowers the overall system cost of gas, which, in turn, reduces their retail price of gas, the benchmark against which third party gas purchase cost reductions are measured. This, combined with the fact that wellhead prices are at or near bottom and are expected to turn up as the deliverability surplus or "bubble" diminishes, reduce the magnitude of the per-unit cost reduction available through direct gas purchases. Maintaining, let alone increasing cost savings under these circumstances will become increasingly difficult.

Does this mean an end to the era of direct purchase natural gas? Does it mean buying gas from the ldc at their dictated rate as was done in the past? Does it mean the loss of jobs for gas marketers, brokers, corporate energy managers and half the staff at the FERC? Hardly likely. What is more apt to occur is a shift in emphasis as the market itself shifts and adjusts to new conditions and regulations. Reliability of supply will become a more prominent concern as we move away from the highly one-sided buyer's market that has existed. As this happens, longer term deals will come back into vogue and the spot market, while it will always remain a part of the industry, should play a less dominant role than it has over the past several years. And, with this, a futures market lurks in the wings that holds in store for us its share of opportunities, challenges and surprises.

The state of flux that exists in the natural gas market demands of the purchaser a certain amount of creativity and flexibility in order to be able to compete effectively. Gone are the "good old days" of universal boiler-plate contracts, long-term fixed pricing, and high-percentage take or pay obligations. Further complicating the picture is the growing number of participants on both the marketing and purchasing side. In attempting to deal with these new trends and concerns, what may seem like a radical concept today could prove to be the industry standard in the next three months. An end-user must be prepared to negotiate diligently for those terms and conditions he feels necessary to protect himself against the vagaries of the market. At the same time, the end-user must adjust his priorities and keep an open mind to new strategies that may be dictated by changes taking place in the market.

One such strategy that will become attractive for the user with multiple facilities and moderate to large annual fuel consumption will be the portfolio approach to fuel procurement. When using the word portfolio in this context we are making a direct analogy to the portfolio approach to investing in securities. This is because the objectives, maximizing return while at the same time minimizing risk, or more simply optimizing financial position, and the mechanics of accomplishing these objectives, namely investing a fixed amount of capital in a mix of securities and altering the proportion of the holdings between them according to perceived or anticipated movements or changes in the larger financial market, are very similar.

In the stock market, an astute investor will take positions in different securities which he feels have different sensitivities to changes in the overall economy, and even different degrees of reaction to the movement of other securities in the market. He hopes that this will offset the potential for any great loss in one or two holdings by gains in counter sensitive industries or businesses. Thus, by mitigating undue risk, he can turn his attention to maximizing his position by shifting funds between holdings in response to anticipated changes in the market or economy.

Fuel procurement is very similar. There are any number of fuels that a purchaser can consider in his acquisition strategies, each with its own characteristics such as price, price stability, dependability of supply, heating value, substitutability, inventory cost, etc.

Perhaps less obvious, but also a legitimate way of diversifying risk, especially in the area of natural gas, is purchasing under different types of contracts. These contracts would differ primarily in the areas of price and length of term. Longer term fixed-price contracts assure dependability of supply, but not necessarily at the best price. Spot market sales allow one to buy at the best available price, but ensure neither price nor availability beyond that term.

A person or company may even wish to consider acquisition of their own oil, gas, or coal reserves. This can substantially reduce the delivered cost of fuel depending upon the proximity of the end-use location(s) to the reserves; perhaps by as much as 30-40% in today's market. Ownership of reserves, however, is anything but a straight forward and risk-free venture. There are dry holes, royalty owners, taxes, frost laws and all other manner and form of hurdles and surprises awaiting the would-be oil/gas/coal man.

It is now appropriate to turn our attention to some of the elements of fuel portfolio management. It should be stressed that a buyer's participation and, to a lesser or greater extent, effectiveness in the current market is situationally dependent. The number of facilities to be served, their locations and usages, the availability and price of alternate fuels (including traditional gas purchases through local utilities), and limitations or restrictions on the buyer's time and/or resources will all impact on the results of efforts to establish a portfolio approach to fuel procurement. What can be accomplished with seemingly little effort in some cases and situations will require considerable effort to accomplish, if indeed they can be accomplished at all, in others.

The portfolio approach is not, however, a viable option for all companies. The establishment of a portfolio program for fuel procurement is a business decision, just like deciding whether to exist as a partnership or file incorporation papers, to account for funds on a cash or accrual basis, or to expand by internal growth or acquisition. As such, it requires a comprehensive analysis of the individual situation before an intelligent decision can be reached. Attention will now be focused on some of the elements of such an analysis.

The first thing to look at is the distribution of facilities to be served. The total number of facilities is very important, because the more facilities there are, the more diversification of risk you can accomplish from a corporate standpoint. Also, the more facilities you have, the more fuel you'll consume, hence, the more leverage you'll have in negotiating purchase arrangements.

Geographical location is also very important, not only as a determinate of what fuels you can use, but also of their laid in cost. Fuels like oil and especially coal have a large portion of their cost associated with transportation. Hence, coal may be prohibitively expensive for a plant in Pocatello, Idaho, as would fuel oil be for a plant in Des Moines, Iowa.

The next step in the analysis is to develop an understanding of the fuel needs of the individual facility or facilities being considered. This entails more than just a review of past fuel bills to see how much money is being spent. First, if possible, a fuel consumption pattern must be determined and key factors that affect this pattern must be identified. These factors include seasonality of the product being manufactured or service being provided by the end-user, high heating/cooling demand as a function of geographic location, competitive market factors unique to the end-user's industry, or corporate policy on shutdowns for vacation or maintenance.

Next, and extremely important, is to determine what will be referred to as the criticality of the fuel supply. Criticality is used here as a relative measure of the negative impact an interruption or cessation of the fuel supply would have on the operation of the facility. The more negative the impact, the more critical is the fuel supply. This is an important point. The vast majority of direct-purchase gas sold in today's market is sold on a best-efforts basis and transported by the interstate pipeline systems on a fully interruptible basis. This means that supplies can and do get interrupted, to varying degrees, from time to time. Back-up to a direct-purchase gas supply is available, albeit at a greater cost in most cases. Fuel oil and other derivatives of crude, as anyone who can recall the Arab oil embargo of 1973-74 knows, is also subject to supply interruptions and rapidly escalating prices. Likewise, the United Mine Workers seem to be poised for a walkout every couple of years or so, and this wreaks havoc with coal supply planning. It is the relative likelihood of incurring a supply interruption that must be assessed for a particular location vis-a-vis other fuel choices or methods of purchase.

Criticality is many times a business decision or a result of the type of business being conducted at the facility. A central facility that is manufacturing an intermediary for distribution to several regional facilities for conversion into finished product, would in all likelihood be considered to have a critical fuel supply. Likewise, a hospital or a nursing home would also be considered to have a critical fuel supply. The plant chosen to manufacture a new product for introduction into the market could have its fuel supply rendered critical by decree, if there was enough riding on the success of the product or its timely introduction.

As pertains to gaseous fuels, the actual physical way in which the gas is consumed can have a marked effect on criticality of supply, and may also offer opportunities to reduce that criticality, albeit usually with an attendant investment of capital. Gas consumed in indirect heating, for example, such as in a steam boiler, is highly substitutable. It can be replaced by burning coal, fuel oils, electricity, liquified gases, wood and waste by-products. Nearly one-third of all the gas consumed is consumed by users having such a "dual fuel" capability. This capability drastically reduced the criticality of their supply.

Gas consumed in direct heating applications, such as a gas-fired drier, is a little more difficult to replace. Solid fuels, by the very nature of the system, are unacceptable. Fuel oils do not burn as cleanly as natural gas, many times giving off particulate matter or soot upon combustion. Thus, depending on finished product quality parameters and specifications, oil may or may not be an acceptable substitute. Only propane or butane could be considered as a comparable back-up.

At the opposite end of the spectrum on substitutability is natural gas used as a feedstock in the manufacturing process, such as the chlorination of methane to form carbon tetrachloride. In almost all cases of feedstock usage use of alternative fuels is extremely difficult and prohibitively expensive.

If direct-purchase natural gas is one of the fuel options under serious consideration (and since it is currently one of the more favorably priced fuel alternatives in most parts of the country it should be), one must consider all the transportation options and their associated costs in the evaluation. Fuel oil and coal are generally quoted on a delivered price per gallon or ton, respectively, with freight included. Direct-purchase natural gas is purchased at the wellhead, and then transportation arrangements are made, at additional cost, to get it to the end-use location. The costs and administrative requirements for this transportation service must be evaluated very carefully.

The first party to contact should be your utility or local distribution company (ldc). They can provide considerable information on direct purchasing and transportation of gas and save considerable time over gathering the information from various independent sources.

The first thing to ascertain from the utility is whether or not they have an approved tariff that provides for direct transportation service to end-users. Although the majority of ldc's have such tariffs, there are still some that do not. These tariffs contain rate sheets which state the charges for the transportation service, and also list any restrictions or qualifying standards that must be met in order to receive transportation service.

Different ldc's charge different rates for their service, and the end-user should make sure that all charges are fully explained for the service anticipated. This is essential to performing an accurate evaluation of the economic benefits of the direct-purchase option for gas supplies. Some ldc's charge a flat rate per dekatherm (dth) or MCF (thousand cubic feet) for all transportation irrespective of the volume transported. Others have step rates that reduce the rate for all gas being moved as certain minimum quantity levels are passed. Still others charge a block rate where all gas transported up to a certain amount is charged one rate, usually the highest or "bottom block" rate, all amounts in excess of bottom block and up to another pre-determined amount are charged another rate, usually lower, and this repeats for the number of blocks in the rate. This differs from the step rate in that once you exceed the bottom block amount, <u>all</u> incremental volume additions contribute to lowering the cost per dth of the gas being transported, whereas in the step rate, incremental additions do not affect the unit cost of service until that particular unit is added that puts the total volume into the next higher step or bracket.

The ldc is allowed, by regulation through its tariff, to retain a percentage (usually between 1.0-3.0%) of the gas the end-user has delivered to it to account for line losses in its system. This has the ultimate effect of raising the unit price of the gas. The price increase is usually in the range of a few cents and rarely much above a nickel. This could become significant, however, in facilities with high daily usage of gas.

Certain ldc's restrict transportation service to those customers whose usage or demand meets some minimum requirement level. Some require that the user have dual-fuel capability in order to qualify for transportation, or simply charge a higher rate to those who don't. higher rates are also sometimes charged for those customers who want a guarantee that they will be able to switch back to the ldc as a retail customer if their direct-purchase supply were to be interrupted for any extended length of time. Balancing requirements pose another possible restriction to the user and deserve special mention.

Balancing in direct-purchase gas situations is attempting to match the amount of gas asked for or nominated exactly to the requirements of the facilities; no excess and no shortfall. Perfect balance is a theoretical state that is almost impossible to attain. Balancing within certain limits, however, should be easily achievable.

The real problem with balancing comes into play when an end-user nominates more gas than the facility consumes. (In the case of undernominating, the utility simply makes up the shortfall to the end-user at the standard retail price in most cases). In this case, the excess or unused gas must be stored (banked) by the utility. This can be done either in storage or by retention of the gas in the utility's system. In either case, it represents an economic rent to the utility.

Most utilities give an end-user a certain period of time to eliminate this excess or "bank" of gas from the system. The method of accomplishing this is by reducing subsequent nominations below requirements, and the time period allowed is generally sixty days. (In all but the most horrendous imbalance situations thirty days should be more than sufficient to correct the problem). Any portion of the bank not eliminated in sixty days is either absorbed by the utility for system supply without compensation to the end-user or purchased from the end-user for system supply, at a cost generally below market.

Some ldc's, however, have balancing requirements that are rather punitive, demanding balance within certain limitations at the end of each billing period. Excess amounts above the limits are generally purchased at a cost below market, and shortfalls are sometimes sold to the end-user at a premium over the retail rate. this situation can be made even worse by limits placed on the number of times nominations can be changed during a given month or billing cycle. Caveat emptor!

The next step is to contact the interstate pipeline(s) that serve your ldc to get information on their costs and services. The person to contact would be the director or Manager of Transportation and Exchange (T&E). This individual should be able to answer all your questions about transportation, such as, whether or not the pipeline is currently offering interruptible transportation for end-users, how long it will take to receive a Transportation Agreement and start flowing gas, and whether or not their system is experiencing or expecting any capacity related curtailments and how limited capacity will be rationed if this is the case. They can also supply you with a copy of their tariff and rate sheets.

Particular attention should be paid to the rates since they are somewhat different in form from the ldc rates, and generally harder to calculate for a given deal. This is because the interstate rates are for the most part based upon distance traveled as well as volume moved.

Interstate rates generally fall into one of three categories. Mileage rates charge a fee directly in proportion to the distance traveled; usually in cents per dth per 100 miles. Zone rates split the pipeline system into geographic areas of zones, and there are specific rates for transportation initiating in any zone and moving to any other zone. Generally the more zones traveled through, the higher the cost. Finally, a few systems still have a flat rate. This means that it costs the same amount of money to move gas between any two points on the system irrespective of the distance of separation. This is sometimes referred to as a postage-stamp rate because of the analogy to the U.S. Mail system.

Pipelines have also begun the practice of discounting rates if they feel they are in competition for the load with another pipelines. Flexible rates are a feature of the Order 436 regulations designed to foster competition. The practice will in all likelihood become more widespread as more pipelines are granted 436 certificates.

Pipelines, like ldc's, are also allowed to retain a small percentage of the gas tendered to them for transportation to use as compressor fuel. The amount ranges from 1.5-5.0%, and is sometimes mileage-based.

It would now be appropriate to turn our attention to factors within the organization that will impact the direct-purchase decision. These can be broken down into physical factors and organizational factors. The physical factors will be addressed first, and they include size of facilities, number of facilities, location, and what the gas is used for at the locations.

In the natural gas arena, bigger is definitely better. It could almost be said that chances for success are directly proportional to the amount of business you can offer the pipeline carriers and the gas sellers. small volumes do not generate enough revenue to justify the FERC filings that the pipelines have to do. Large marketers are engaged in fierce competition for crown jewels such as the steel mills, fertilizer plants, chemical and utility business. Even the smaller marketers shy away from accounts below certain minimum sizes because they don't have the staffs to justify the time spent tracking down a customer whose account could generate less than $100 per month in the current over-supplied market. Grouping facilities by daily gas usage can help identify those facilities where you have the greatest chance of being able to secure a supply. A very generalized guide might be: over 1,000 dth/day - good chance; less than 250 dth/day - poor chance; 250-1,000 dth/day - situationally dependent.

Grouping facilities together can help increase load size. This is particularly helpful if the grouped facilities are served by the same pipeline or better yet, the same ldc.

Location also plays a big factor. Several Southeast and East Coast users are served by only one pipeline. This gives the pipeline a monopoly position and puts the end-users served by it at the mercy of its management. In these instances, dual fuel capability is almost essential. Midwest and Northeast locations, however, are generally served by more than one line and the end-user enjoys considerably more competitive leverage.

The effect of how the gas is used was discussed at length earlier when considering fuel usage requirements. To reiterate, the more substitutable the end-use, the more likely that efforts to establish a direct-purchase supply of natural gas will be successful.

Organizational factors affecting a direct-purchase program are staff size and line authority (centralized versus decentralized). Setting up any type of successful direct-purchase program on a corporate level for an organization with multiple facilities requires a considerable expenditure of time and effort. For any organization with more than a half dozen or so facilities, it is almost mandatory to devote an individual to the task on a full-time basis; at least until contracts are in place and gas flow is established. Centralizing the program, where possible, enhances the chances of success through economies of scale and the avoidance of duplication of effort.

The ultimate criterion for participation in all cases should be cost-benefit analysis. The benefits, in the form of direct cost savings on purchases, will be easy to calculate, or at least estimate, in most cases. The costs are more difficult to pin down, especially for the first few purchases when no historical yardstick for comparison exists.

The use of a net back pricing scenario can help to simplify the cost-benefit analysis. The first step is to select some level of cost savings at the facility or facilities involved that is felt to be adequate to cover the time and effort that will be required to achieve them. The delivered or burner-tip unit cost of the fuel currently in use is then determined. From this, the desired or target burner-tip price of the direct-purchase gas required to yield the estimated savings over some finite time period (usually a year) can be calculated. All costs charged by the ldc and interstate pipeline(s) involved in providing transportation are now subtracted from the target burner-tip price to get the required purchase price "netted back" to the fob purchase point of the gas (usually at an entry point into the first transporting pipeline). This price can be further refined or "netted back" to the well head by subtracting any gathering fees associated with moving the gas from the field or well to the pipeline. The well head or purchase price, thus calculated, can be compared with published data (such as that found in Oil & Gas Journal, Energy User News, Natural Gas Week, etc.) for typical prices found in the given area to determine whether or not the required savings appear realistic and possible to achieve.

Considerable time has been spent discussing direct-purchase natural gas because it currently seems to be the choice option in many areas, and it is a considerably more complex process than the purchase of fuel oil, propane, or even coal. However, mention must be made of some of the considerations involved in selecting one fuel over another.

Several factors differentiate one fuel from another. Some of these include price, heating value, regional availability, and storage characteristics and cost.

Heating value is important because it determines how much of a fuel must be purchased to meet requirements. Heating value is usually expressed in BTU's (heat producing capability) per unit of fuel. For comparison purposes we will use 1,000 cubic feet (1 MCF) of natural gas, which contains approximately 1,000,000 BTU's (1 MM BTU). The equivalent quantities of other fuels required to get the same heating value are:

Coal:	80 lbs.
No. 6 Fuel Oil:	6.67 gallons
Propane:	Almost 11 gallons

Therefore, a small to moderate natural gas user, consuming 1,000 MCF per day would need the equivalent of forty tons of coal, one tank truck (6,500 gallons) of fuel oil, or two tank trucks of propane.

This generates some interesting "price equivalents" or indifference prices for the various fuels (prices at which the user would be economically indifferent to choice, all other factors being equal). For natural gas at $3.00/MCF, the prices are:

Coal:	$72/ton
No. 6 Fuel Oil:	$0.45/gallon
Propane:	$0.27/gallon

Storage requirements and costs are perhaps one of the

most important considerations. With natural gas, of course, there are no storage costs; it's neatly waiting behind the value. Coal, on the other hand, requires approximately 42 cubic feet per ton. It must also be kept reasonably dry and suffer a minimum of handling to prevent excessive build up of fines.

No. 6 fuel oil must not only be stored but, it must be heated when temperatures drift below 60-70°F. When ready for use, it must then be pumped.

Propane must be stored in pressurized tanks, which are considerably more expensive than atmospheric vessels. Once ready for use, it is fed through a vaporizer, and from there can use the same exact lines as your natural gas supply.

It might appear that natural gas is the hands down choice because it doesn't have to be stored. What about an interruption, however? Without inventory, an interruption could mean no production or heat. The nice thing about inventory is once you've bought it, you own it and you can continue to use it until it runs out. The accountants, of course, will remind you that money tied up in inventory is a cost that serves to reduce profits.

Geographical factors were touched on before in relation to a facility's proximity to major fuel reserves and/or transportation facilities for moving fuel. This affects the costs of both traditional purchases and acquiring reserves in the ground.

It should be readily apparent that developing a portfolio approach to fuel procurement is not an easy task. The number of variables are far greater than those found in simple cost-based purchase decisions. Also, the decision maker is trying not only to minimize the cost of purchased fuels but, also, to maximize the reliability of multiple supplies; two objectives which could be partially self-exclusive. It's a far cry from direct-purchased gas where the worst down-side case is going back to the traditional purchase from the ldc and paying what one formerly paid.

Finally, as if that's not enough, the decisions require a good amount of evaluation of data to determine its validity, and most critically, interpretation of the data as to what it says about the future course of the complex and inter-related energy market.

It's not an easy task and the risks are considerable but, as things swing back to a seller's market and supplies get tight, it will be the only way to maximize your position vis-a-vis the market <u>and</u> your competition. As with most things, considerable rewards can accrue to a job well done.

SECTION 17

PURCHASING STRATEGIES FOR ELECTRICITY

Chapter 77

POWER MARKETING IN THE FUTURE

J. E. Davis

The electric utility industry has long been characterized as ultra conservative, low risk and financially stable. This was especially true in the 1960's and early 1970's when incremental costs were declining and unit costs of electricity were coming down. Working for an electric utility was even considered to be a mundane occupation and very seldom was there any press coverage. In fact, electric utilities were going to their regulators and of all things, requesting rate decreases. This was not big news.

After more than one and one-half decades of being battered with oil embargos, skyrocketing oil and gas prices and now a bust market for oil and gas prices, double digit inflation, increased environmental regulations, Three Mile Island, depressed load growth, utility bankruptcies and many other significant pressures, the electric utility industry still exists, but not as it did 10-15 years ago. The winds of change are blowing strongly. This change is being fueled, in part, by past deregulation of other traditionally regulated industries, regulatory changes such as the Public Utility Regulatory Policy Act of 1978, greater access to electricity suppliers through the growth of transmission interconnections over the past years, and increasing incremental costs and the widening of such costs among various electric suppliers. Because of these changes, many old ideas and phrases are taking on added significance in utility board rooms. Ideas such as competition; market driven, low cost producer; independent power producer; transmission access; market share; qualified facilities and deregulation are thoughts that will shape the way the electric utility industry will be directed, managed and operated in the future. Our industry is changing and those that are prepared to meet the challenge will survive, others won't.

A critical area of change in the electric utility industry is how power is being sold among utilities and to ultimate consumers. This change is the area I wish to focus attention. There are many other areas of change that can be predicted but none more important than how electric power will be bought and sold. I will discuss two specific examples. These examples may give us a glimpse of how marketing of electric power may be in the future.

Let's first focus our attention on the supply side of electric power. I am sure many of you have heard of the discussions taking place at the Federal Energy Regulatory Commission regarding the deregulation of certain types of power suppliers, collectively referred to as Independent Power Producers. These producers would be significantly deregulated and the prices would be more market based rather than cost based, as regulated prices are today. I am not going to dwell on Independent Power Producers but thought I would mention them since they represent a significant issue presently within the electric utility industry.

Specifically, I would like to discuss an experiment in the Western United States called the Western Systems Power Pool (WSPP). The experiment was approved by the Federal Energy Regulatory Commission for a two year period that ends May, 1989. Power pooling among electric utilities is not uncommon and is a practical method to reduce operating costs. However, the WSPP does have some unique charactertistics that set it apart from others and its operation may provide some yardsticks on how to measure the future interaction among electric utilities.

The WSPP covers most of the Western United States with the central computer located in Arizona Public Service Company's operation center located in Phoenix, Arizona. There are presently 24 members consisting of a mix of private, federal and municipal utilities. The WSPP is different than other power pools in that it has some noticeable absences such as spinning and operational requirements and specified installed generating capacity requirements. In fact, the WSPP might be more descriptively called the Western Systems Economic Pool, since it primarily directs its services toward economic considerations.

The services covered by the WSPP agreement include: (1) Economy Energy Sales, (2) Unit Commitment, (3) Short-Term Firm Power Sales and, (4) Transmission Service. Each of which have price quotes on a daily basis sent to the central computer by each member wishing to buy or sell such services. The central computer acts as an electronic marketing bulletin board. It also serves as a data bank to keep records of transactions performed under WSPP. The price quotes are released each day at 1:00 P.M. Mountain Standard time to all members for services the following day or a longer time period. After the price quote release, it is the responsibility of each member to negotiate its own separate arrangements. The major benefits of the WSPP are that it: (1) concentrates the market into a single bulletin board allowing more efficient market search; (2) for regulated utilities, it allows multiple transactions to be consummated under one agreement with pre-approved regulatory assurance; and (3) pricing for the various services can be based on market value rather than solely on costs. The Federal Energy Regulatory Commission was concerned about total deregulation of the experiment and would not approve the WSPP without price ceilings and, therefore, price ceilings do exist, but to my knowledge such ceilings have not impacted transactions under WSPP.

Of particular note is the transmission service provided under WSPP. The backbone of the WSPP is the transmission system and the pool, for the first time, allows market value pricing of transmission service. Open access transmission is a hot issue presently and the WSPP may provide to the Federal Energy Regulatory Commission and others some basic relationships between open access and market pricing.

Marketing power to the ultimate consumer has also been undergoing significant changes. It used to be that a customer had essentially one choice for service and that was firm service at system average rates. That served the electric industry for years and was not a bad deal for the consumer, especially when average rates were declining. Now the electric utility industry serves up all types of service and rate options; (1) Firm Service, (2) Instantaneous Interruptible Service, (3) Delayed Interruption Service, (4) Time-of-Use Service, (5) Economic Development and many others. I am sure that the list will continue to grow as competition becomes keener and customers evaluate their exact power needs more closely.

As an example of how retail power marketing has changed, I would like to give a brief description of a unique Retail Sales Agreement that Arizona Public Service (APS) has in place. Approximately 160-180MW of APS' load is copper mining. The load has historically been very good to APS in that it represented a sizeable steady revenue source for the Company. However, in 1984 due to extremely depressed copper prices, one large mine shut down. This shutdown essentially eliminated all jobs in a small town and a good revenue source to APS. In order to get the load back on line, APS entered into negotiations with the customer and the result was a unique agreement that combines 2 types of interruptible power service with firm power service, and the ultimate price of power is tied to the world copper price. Because of the fact that APS had another copper mine, and although still operating, the special service and rate package was also offered to it.

Many other special retail power arrangements have been used by APS to keep present load and gain new load.

The point of the two specific examples discussed briefly is that the electric power supply industry is undergoing many changes. This will mean new opportunities for electric utilities and customers of electric utilities and will certainly mean a much different future for those of us that are responsible for power marketing.

Chapter 78

APPLICATION OF STANDBY POWER GENERATOR UNITS

E. J. Matousek

Our previous speaker discussed the purpose and need for interruptible rates from the utility's viewpoint. I would like to discuss how a Standby Power Generator can integrate with interruptible rates of the utility while also benefiting the utility's other customers and, in particular, the customer owning the standby power generator.

Historical Background

But, before we get too deep into the details, let us look back into history and see what the driving forces are that have brought us to this point where utilities offer interruptible rates and accept cogeneration. I believe this is important as it shows that the conditions are not a passing fancy, but a condition that will last at least for the life of the equipment installed.

The electrical utility industry is a part of an even greater, more dynamic and influential field that has a major impact on our lives, i.e., world energy. Strangely enough, if I were to ask this audience of energy experts for a definition of "energy", I'm sure I would get a great variety of answers -- and why not; Physics is a study of it; Max Planck and Einstein proved its equal to mass times the square of the speed of light; it can't be created or destroyed; and on and on. For the purpose of this discussion, I would like to use Webster's simple explanation, "it is a substance that has the capacity to do work". In this capacity, look at Figure 1 to see what it has done to the way man lives.

Do you notice something in common about these curves? Right about 1850 all the curves take off. The reason is man has learned how to harness energy. Around 1850, James Watt invented the steam engine -- man, for the first time since the beginning of time, was no longer dependent upon himself or domesticated animals, or slaves to do work. The take off point is known as the "Industrial Revolution". Just look what is done -- Life Expectancy more than doubles in just one hundred years -- knowledge becomes widespread because of the number of books that can be printed -- explosive power, or the ability to put and control more energy in a smaller package, follows the same type curve. That could explain the need for the number of international conferences that were held. People could now find work in cities, they were no longer dependent on the land for survival. And last, because they learned to harness energy, they were able to travel at great speeds, no longer limited as they had from times beginning to the speed of a horse.

In just a few generations, going back to the beginning of time, man has been able to lift himself by his "energy" bootstraps to a new way of life.

This same definition of energy explains what happened to mankind since the industrial revolution. We have machines capable of doing work; the more work we do, the more we produce; the more we produce, the better we live. This is why a society's standard of living is directly related to the energy use per capita.

Let's look at the next set of curves (Figure 2). This first curve shows how energy consumption has increased in relation to population in just the last 30 years. The second curve shows how we in this country have used it since the start of the industrial revolution -- and why we live so well. The third curve shows we have become more efficient in our conversion of energy through the years and we still have a way to go. Our best steam turbines and engines are only about 40% efficient today.

The next curve shows how man became dependent on electro-mechanical energy, freeing himself and animals from doing the work. In a sense, our workers are no longer workers per se, they are directors of energy.

Electrical Utilities

Of all the forms of energy, electrical energy is the most versatile. It lights our homes and factories; it drives our motors at the turn of a switch or press of a button; it makes possible the use of our sophisticated controls, instruments and computers; and it allows us to communicate at great distances besides entertaining us with radio and TV.

The electrical utility industry for years has been the major converter and distributor of electrical energy. I say converter because, as you know, electricity does not occur in a natural usable form, it is converted from solar energy, water power, coal, gas, oil or nuclear fuels. Because electrical energy cannot be efficiently stored or stockpiled and because it is transferred at the speed of light, the utility had to have their systems regulated in order to efficiently distribute it. This resulted in a system of rates based upon "cost of service".

Figure 3 shows the cost per kwh divided into energy cost (mostly fuel cost) and demand cost (or capital cost) in relation to hours use per month. The customer cost is primarily the cost of metering and is minimal. Rates are designed to follow this curve, hence the customer, demand and energy portions of the utility billing structure.

This "cost to serve" concept worked very well during the period when utility costs were going down through "economy of size" and the development of more efficient conversion of fuels. However, when the EPA and the Clean Air Acts were mandated and OPEC capitalized on the confusion, the cost of new plant increased from lows of about $100/kw to the $2000 to $5000/kw ranges. Construction periods increased from 2 to 3 years to twenty, thus meeting growth trends which was not too difficult when plants could be built within two years became very difficult and risky when attempting to project twenty years ahead. At these very high prices of a new plant, management found themselves virtually "betting the company" whenever they decided to build new generation.

Electric Utility Problems

These problems did not appear overnight nor did they come as a surprise. In December of 1980, the Comptroller General made a report to Congress, excepts follow:

From '74 through '78 they pointed out:

184 of the nation's planned electric generating plants (were) cancelled --
- this represents about 26% of all electrical generating capacity available as of April '79

189 electric power plants placed in operation had delays --
- 149 (79%) had delays of 6 months to 23 months

Average delay 17 months

- Constructing an electric power plant is a complex undertaking --
  costs can exceed a billion dollars
  could take 8 to 14 years to build

As of March '79:

Of 330 electric generating plants projected on line before early '90's --
- 267 (81%) have experienced delays

Average delays, so far, 40 months

30.4 months for coal fueled plants
52.7 months for nuclear fueled plants
25.7 months for oil fueled plants

Delays were due to:
a. dramatic change in electricity consumption
b. difficulties in obtaining financing
c. difficulties overcoming regulatory problems
d. difficulties with growing uncertainty of nuclear power
e. construction problems

Continued power plant delays and cancellation will likely contribute to:
- increased oil consumption
  make us more dependent on uncertain foreign

- jeopardizing the utility industry's ability to provide
  uninterrupted electrical service

- increasing the future costs of electricity
  as consumers bear the added costs of power plant delays and cancellations

In March of 1981, Joseph C. Swidler, former chairman of the FPC, testified at a FERC conference on the financial status of electric utilities, excerpts follow:

"...Regulation is a blunt instrument

"...Financial deterioration of the utility companies

- a process that has been going on for almost 15 yrs.

- well known to investors who have watched their stake...decline in value...

(The electric utilities) " are cutting their commitments as fast as they can

- it has been 3 years since a firm order was placed for a nuclear plant

- last year only 3000MW of new capacity was ordered for the whole nation

- rate of return data demonstrate that the utilities will not recover the cost of new capital required to build generating plants

"...who is taking the responsiblity for the failure to build for the future when the result can only be power capacity shortages in the 1990's.

"...A special feature of the financial distress of utilities is:

- the accentuation of risk at the time earnings have declined

- the utilities building new plants are rightly worried

- they risk all or a major part of their equity on second guessing utility commissions as to whether (a new) plant should be built

- threats of disallowances of major investments (committed) have a chilling effect on investors

"...The tragedy is the impairment of utility credit

- costing the consumer dearly

- (and causing) defensive decision making

"...Several issues at stake

- whether private enterprise in the electric utility industry will survive

- whether the American system of regulation will survive...This constitutional arrangement is not working.

- whether this country will have enough efficient electric power capacity to fuel its economy in the 1990's...that capacity should be underway now

"...Considering nothing new is on the boards we face a national disaster in the 1990's."

Electric Utility's Solution

Under the circumstances, utility managements began to embrace the idea of conversation, interruptible rates and cogeneration were proposed (PURPA, 1978) and marginal costs rates were designed to mold the utility's load profiles. Let's view an average utility's daily load shape (Figure 4).

This figure shows the weekly peak load of an average utility for a year. The dotted line is the utility's estimate of the generating capacity needed to handle these peaks. As is obvious, they need a lot more capacity in the summer than in the winter. Now if one plotted the highest peak demand at the right hand side of a curve, the second highest next to it, and so on, to the lowest demand on the system, we would have a curve that looks like this next curve. It is called a load duration curve (Figure 4.1).

The load duration is an excellent means of determining where cogeneration and standby generation fit into the utility's picture. In a sense, cogeneration fills in for intermediate and baseload utility plants, i.e., the nuclear and coal plants shown on the curve. The cogenerators can compete by using a waste product for fuel, such as pecan shells, wood waste, or biomass; fuel costs are low or non-existent. In other situations where waste heat can be used in an on-site process the improved efficiency makes the cogenerator economically feasible. A reasonable ongoing profit is possible and the capital risk is prudent. The number of these type of situations, however, are limited due to the availability of the "waste" product fuels or a balance between electrical and steam use. We also have the really big players, like the GE's and the Becktels, the subsidiaries of utility holding companies, like Mission Energy (a part of SCE) who bid on large plants, the combined cycle units, or build hydro plants. They are the pros who did the design, manufacturing, and building for the utilities and in today's new environment must compete or close their operations. It makes things very competitive.

Standby Power Generation

With this background, I finally come to the application of Standby Power generation units. Looking back to the load duration curve, we see the very small part of the generation capacity that is allocated to peaking generation. It represents about 20% of the peak generator capacity needed by the utility, but it is used only about 3 to 4% of the time. In other words, you require a completely different piece of equipment. Instead of a high capital cost, low operating cost generator, you need a low capital cost generator. Since it is rarely run, the operating cost can be high. There are other important requirements, where nuclear and large coal plants take days for startup, the peaking plants must start up in seconds and the start must be reliable... over 99.5%. In addition, since they are used so seldom they must be automatic. From a utility's viewpoint, peaking plants are not very cost effective since they do not add much to their rate base, and with limited funds, investment in base load or distribution plant is more prudent. By offering interruptible rates, the utility can in effect gain peaking plant at no capital cost to them. On the other hand, industrial users concerned about reliability (and they should be if the projections and problems pointed out previously are true) can build standby generators on their site to carry their load. Payment for their investment in standby generation can be made through reduced power costs resulting from the lower interruptible rate. PURPA requires that interruptible rates be based upon cost of service. Since utility peaking plants cost about $650 to $700/kw, the interruptible rate's discount should compensate the customer for an investment of this magnitude. There is an additional bonus for the industry installing a standby generator. By having it on site, whenever the utility has an emergency due to acts of God, i.e., equipment failure or natural disasters, these on-site generators can be used to serve the customers load. FEMA, the Federal Emergency Management Agency, is very supportive of this concept as it is considered a "dispersed" generating source and is very useful in counteracting acts of terrorism.

The Standby generation, as described above, is not an on-going profit center for the builder as are cogenerators. It is, however, beneficial to the industry that gains the added reliability; to the utility who gains a peaker without a capital investment, and to other rate payers who gain lower rates from the utility's improved load factor. These benefits are possible through integrating the standby generator into the utility system as a peaker; thus doing double duty. In short, it is a more efficient means of designing this portion of the utility system.

Conclusion

Finally, to those industries requiring greater reliability of their power supply - and I refer to those industries becoming dependent on computers, robotics, "just in time production schedules", or continuous operating plants such as chemical or glass plants - the problems of brownouts and outages are approaching at a pace faster than anticipated. A number of recent studies (Applied Economic Research Co., NERC) point out that the middle Atlantic states face brownouts as early as this summer. The canceling of Shoreham, the shutdown of Peach Bottom and problems facing Seabrook plus the increased electrical demand in these states are factors responsible. Many utilities in these and surrounding states are soliciting bids for cogenerators and IPPs as a stop-gap measure, but for an industrial user who needs to be sure, who can't risk his plant's operation on plans and performance of others in these perilous times, the standby power generator is the answer. Coupled with interruptible rates, it is cost effective and can be installed within 6 to 8 months.

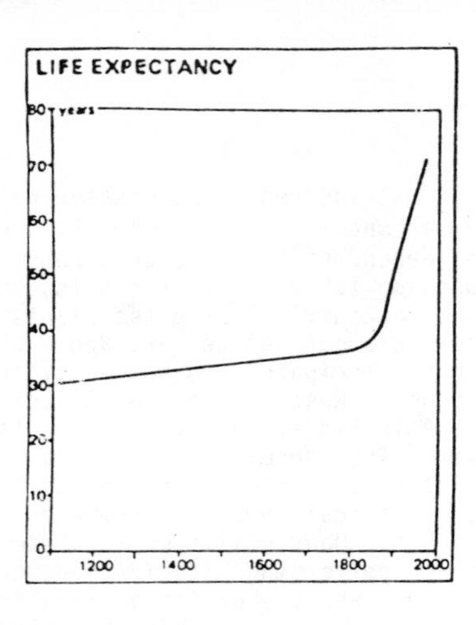
LIFE EXPECTANCY
years
1200
1400
1600
1800
2000

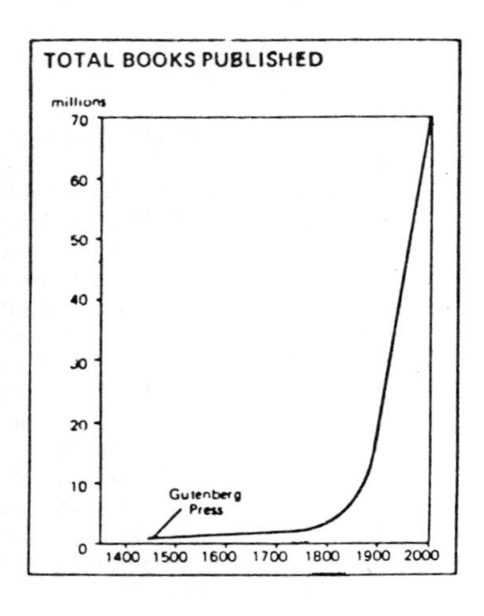
TOTAL BOOKS PUBLISHED
millions
Gutenberg Press
1400 1500 1600 1700 1800 1900 2000

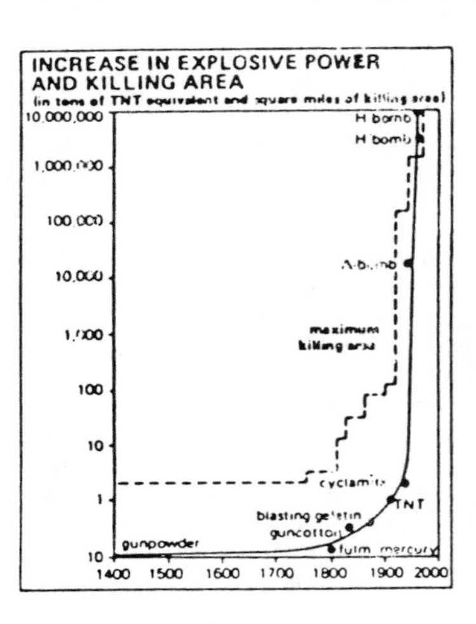
INCREASE IN EXPLOSIVE POWER AND KILLING AREA
H bomb
H bomb
A-bomb
maximum killing area
cyclamite
TNT
blasting gelatin
guncotton
gunpowder
fulm. mercury
1400 1500 1600 1700 1800 1900 2000

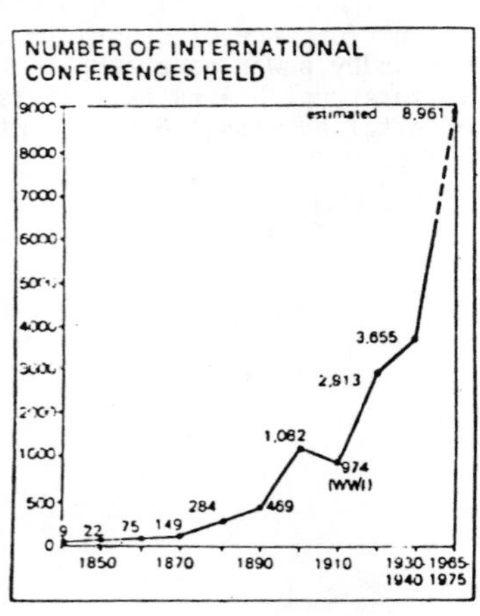
NUMBER OF INTERNATIONAL CONFERENCES HELD
estimated 8,961
3,655
2,913
1,082
974 (WWI)
469
284
149
75
22
9
1850 1870 1890 1910 1930-1940 1965-1975

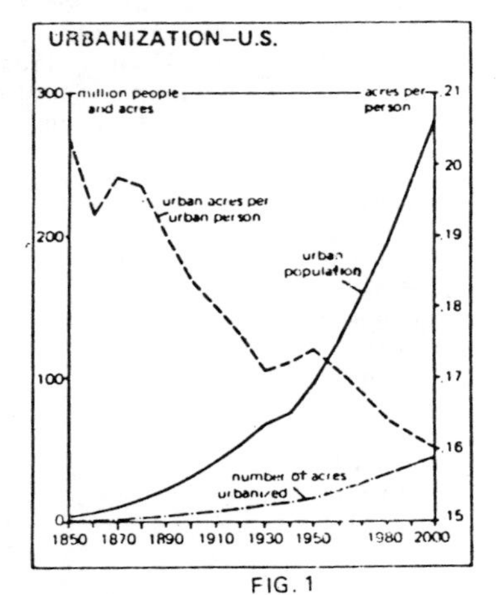
URBANIZATION—U.S.
million people and acres
acres per person
urban acres per urban person
urban population
number of acres urbanized
1850 1870 1890 1910 1930 1950 1980 2000

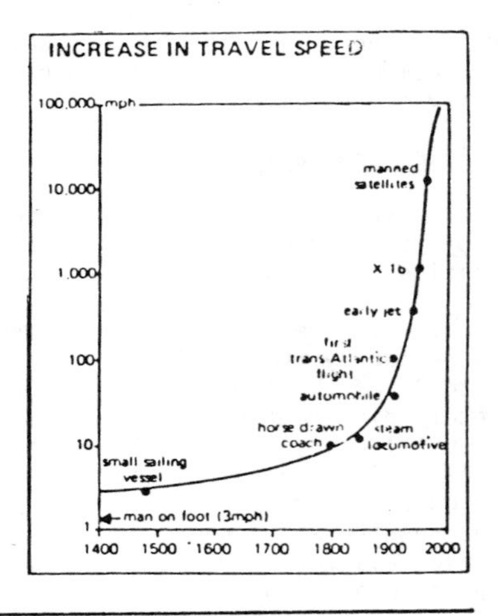
INCREASE IN TRAVEL SPEED
mph
manned satellites
X 15
early jet
first trans-Atlantic flight
automobile
horse drawn coach
steam locomotive
small sailing vessel
man on foot (3mph)
1400 1500 1600 1700 1800 1900 2000

FIG. 1

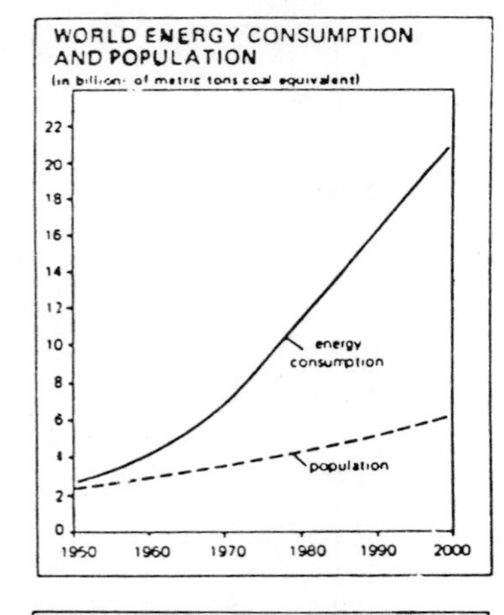
WORLD ENERGY CONSUMPTION AND POPULATION
(in billions of metric tons coal equivalent)
energy consumption
population
1950 1960 1970 1980 1990 2000

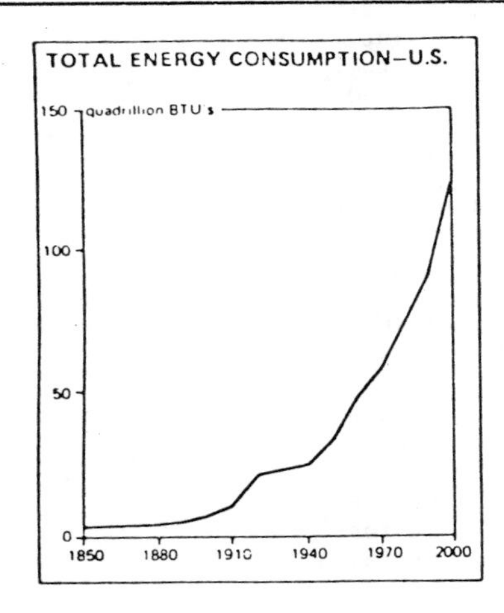
TOTAL ENERGY CONSUMPTION—U.S.
quadrillion BTU's
1850 1880 1910 1940 1970 2000

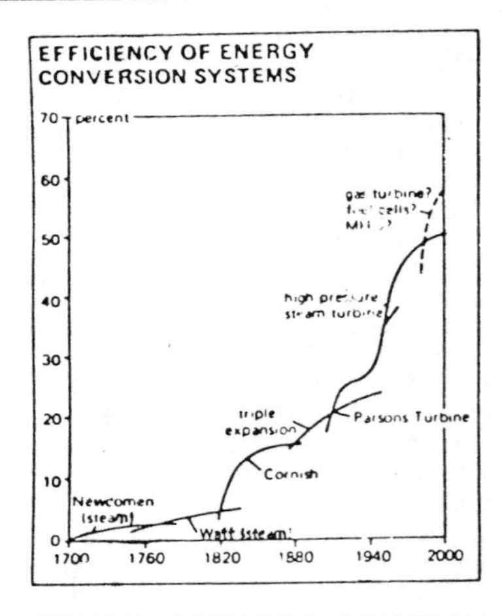
EFFICIENCY OF ENERGY CONVERSION SYSTEMS
percent
gas turbine?
high pressure steam turbine
triple expansion
Parsons Turbine
Cornish
Newcomen (steam)
Watt (steam)
1700 1760 1820 1880 1940 2000

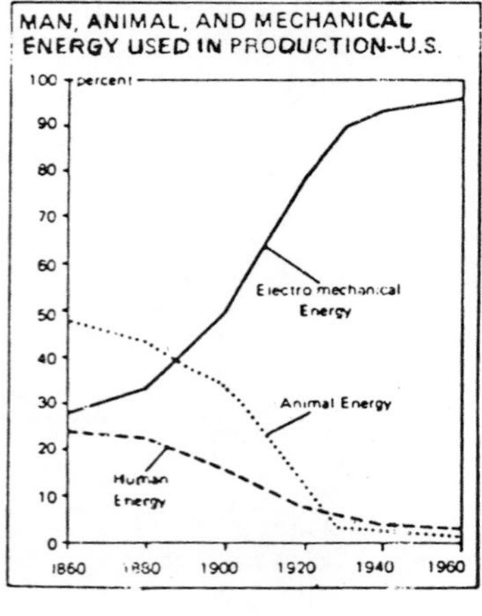
MAN, ANIMAL, AND MECHANICAL ENERGY USED IN PRODUCTION--U.S.
percent
Electro mechanical Energy
Animal Energy
Human Energy
1860 1880 1900 1920 1940 1960

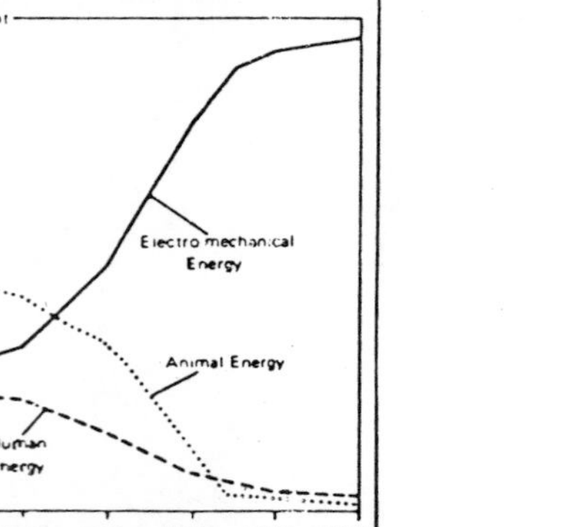

FIG. 2

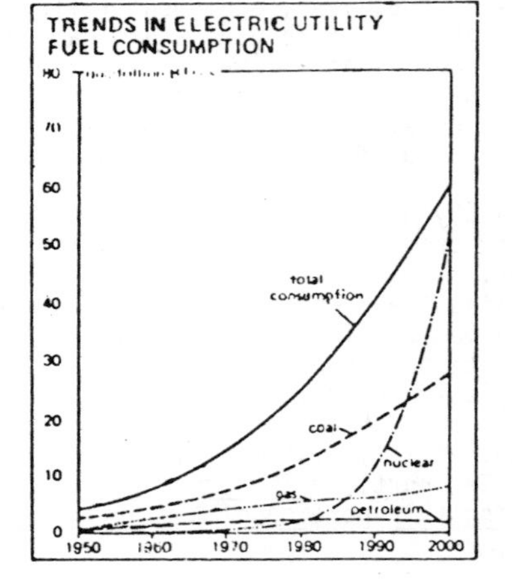
TRENDS IN ELECTRIC UTILITY FUEL CONSUMPTION
total consumption
coal
nuclear
gas
petroleum
1950 1960 1970 1980 1990 2000

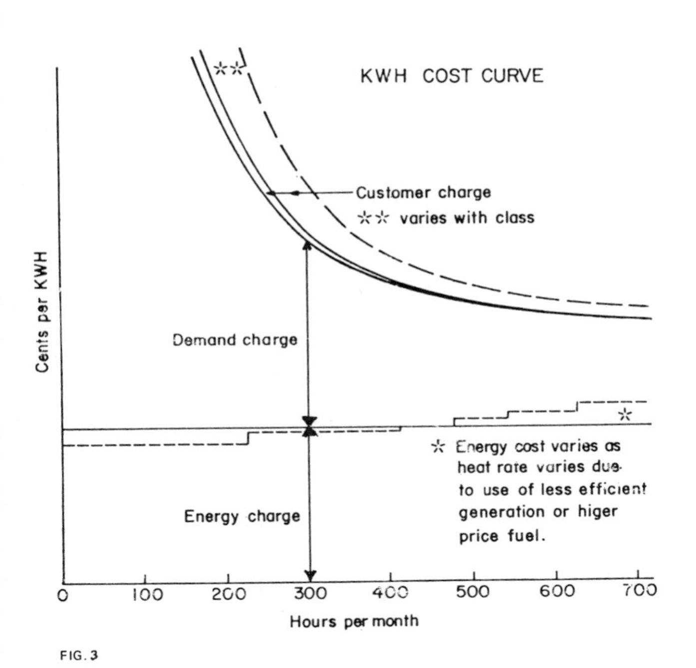

FIG. 3

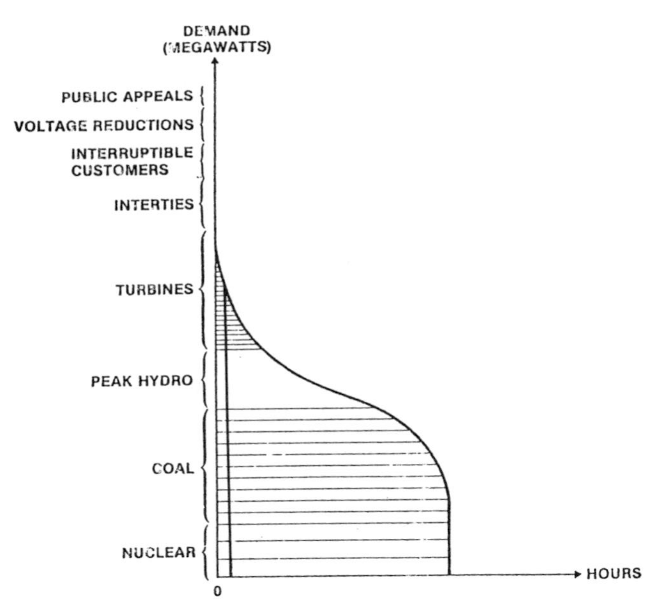

FIGURE 4-1. LOAD DURATION CURVE AND CAPACITY IN LOADING ORDER

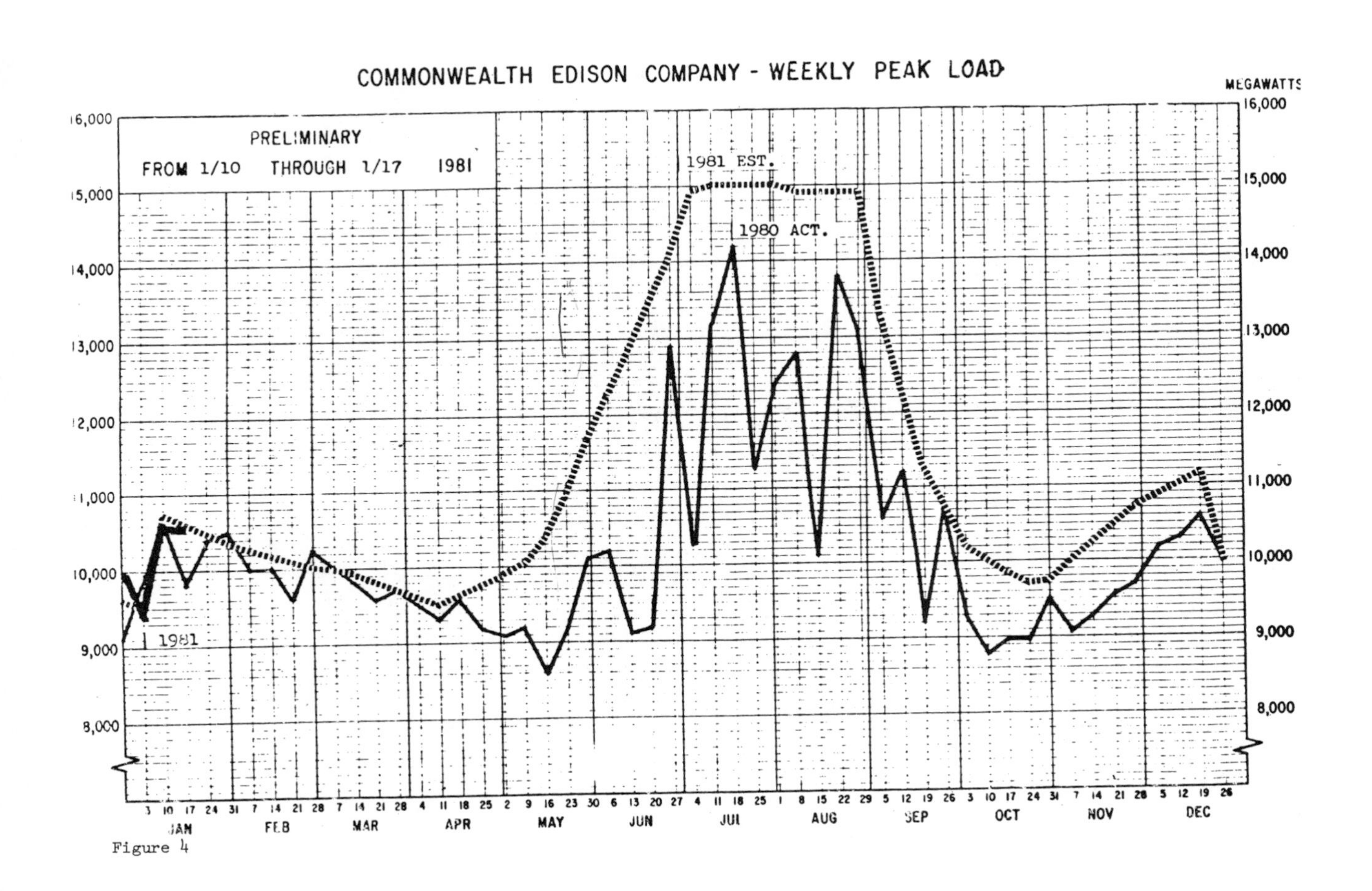

Figure 4

SECTION 18
PLANNING NEW ENERGY POLICIES

Chapter 79

ENERGY: THE ONCE AND FUTURE CRISIS

J. T. Mathews

In the short time available today I'd like to try to sketch the global energy picture for you, and outline its implications for U.S. policy.

Two premises underlie these comments. First, national energy policies--including a US policy--must be international policies. This is true both because energy markets themselves are now functioning international markets and because so many energy-related environmental problems are international if not global. Secondly, as energy issues reappear on the national policy stage, as they are about to do, it behooves us--this time--to attempt to formulate a lasting energy strategy, one that shapes a near term policy with the mid- and the long-term clearly in mind. There are important short term issues to be met, but it makes no sense to fashion a series of shortsighted policies each followed by a costly period of disarray and readjustment.

Global Energy Issues

The energy crisis is not over. It is alive and well today in the Third World and it is only in remission in the industrialized nations. The long term solution lies first of all in a demand side response: namely, a long term strategy of maximizing energy efficiency in both production and use and in both developed and developing countries. On the supply side, the key is the development and selective application of alternative fuels or of familiar fuels in new roles, among which natural gas is one of the most important.

There are at least four dimensions to the energy crisis. First, there is the immediate fuel crisis that afflicts 1.5 billion people today--namely, the fuelwood crisis. Secondly, is the energy-related debt crisis in the developing countries. Third, there is the short term problem looming for the U.S. and the other oil importing industrialized nations when, if corrective action isn't taken in the interim, the oil market will again tighten, and return OPEC to control of the market. And finally, there is the longer term challenge of the global greenhouse warming--posing a set of policy choices more difficult perhaps than any that have come before. Let me briefly outline each of these.

In 1980 almost a billion and a half people faced a fuelwood deficit, meaning either that they were using fuelwood faster than it could grow, or that they could not meet their minimum needs even by overcutting. Absent immediate action, by the year 2000 the number of people in this situation will double--reaching 3 billion people--half of the planet's expected population at that time.

The consequences are far reaching. Rural poverty deepens, health suffers, and deforestation lowers agricultural productivity and causes flooding, drought, and the siltation of expensive hydroelectric projects. The majority of fuelwood users are rural, but some are urban, and for them alternative fuels, including domestically produced natural gas, are a possibility for easing the pressure on crucial forests.

In the commerical sectors of developing country economies, there is an urgent need to find a resolution to the debt crisis that will allow these countries to resume economic growth. This, it should be noted, will do more than anything else to ease the U.S. trade imbalance. Many of these countries have very high projected energy demand growth, especially for electricity. With nuclear power looking less attractive for economic and safety reasons, the choices on the supply side are hydroelectric power or, for most countries, imported oil and coal. Both options require additional heavy borrowing. There are better alternatives.

The first is to exploit the nearly universally unrecognized potential for energy saving. Efficiency investments generally have a triple advantage over the construction of comparable joules or watts of new supply; they cost less, generate more jobs, and require less foreign exchange. In the seventies we learned that energy growth need not proceed hand in hand with economic growth. The eighties have produced a new understanding of comparable importance. It is that poverty causes more energy inefficiency than does great wealth. India, China, Egypt, Pakistan and many others all use more energy per dollar of GNP produced than does the United States. China uses more than twice as much. Thus there are opportunities for mining energy waste in the Third World as great as those we now recognize in the United States and other developed countries.

Second, if the developing countries are ever to grow out from under their foreign debt burden, future borrowing needs must decrease, not continue to rise, and that means that domestic resources must be used, and used efficiently. We know too little about the extent of global natural gas reserves, for example. We do know that gas is dispersed much more widely than coal or oil. The engineering challenge is to develop affordable means of exploiting a resource that is today unexploited or being flared in much of the world.

The U.S. Energy Picture

On the home front, an oil crisis lies in wait in the next decade. U.S. oil production is in the declining years of its life cycle. Projected annual production for the lower forty-eight states shows a steep decline, continuing the trend that has prevailed ever since 1970 despite a tripling of oil prices and of oil well completions. And this year Alaskan production will also begin to decline as the Prudhoe Bay field passes its halfway point.

With declining production, and demand rising in response to lower prices, rising imports are inevitable. The trend is already sharply evident. By the time present excess capacity is exhausted, the U.S. will likely rely on imports for at least 50 and perhaps as much as 70 or 75 percent of its needs, as compared to the 33 percent level of dependence that triggered the first oil crisis.

If growth continues to rise worldwide at 3 percent per year, today's large excess capacity would be quickly exhausted. If, instead, demand grows at 1 percent/year, there will be ample time to develop alternative energy sources and technologies. To accomplish this, efficiency initiatives would have to be taken in the face of market signals pointing strongly in the opposite direction. This makes the policy choices that much more difficult, but no less clearly in the national interest.

The world is facing a difficult transition period as the oil era begins to end. Energy prices and availability will be uncertain. The challenge before us is to embrace the future, not resist it. The active technological frontier is in high energy efficiency. One has only to look to at the most successful economies to see where the farsighted are investing their research dollars: high efficiency automobiles, photovoltaic research, permanent magnetic materials etc. In short, a central thrust of a strategy to restore U.S. economic competitiveness should be to recapture and expand the technological lead in high efficiency products, technologies and processes, for these are what will own future international markets.

Among the most promising of these technologies are several natural gas-fired power generation technologies which offer the combined benefits of high efficiency, low capital cost, modular size and short installation time. Though the potential of these technologies has received far too little policy level attention, these characteristics have not escaped the notice of utility planners plagued by high interest rates and highly uncertain demand. For them, gas-fired combined cycle plants are looking increasingly attractive.

If existing U.S. utility consumption of natural gas--primarily in inefficient peaking turbines--were replaced with new high efficiency technology, as much as 26 gigawatts of additional could be generated. New technologies nearly ready to enter the commercial market, such as inter-cooled steam injected gas turbines and regenerative gas turbines offer even greater promise.

The United States could also benefit considerably from the export of high efficiency gas turbines. Many developing countries with rapidly growing demand for power offer a promising market for low capital cost, high efficiency gas turbines that can be used in dispersed settings. The United States holds a competitive advantage in these new technologies because much of the relevant research has been done on jet aircraft engines, a technology that receives about $400 million a year from the Defense Department. We can market the spin-offs of this taxpayer-supported research to produce a significant competitive benefit. But unless we get moving, that advantage won't last long.

Global Environmental Concerns

1. The "Greenhouse Effect": Looking at the longer term future, we see that mankind is now embarked on a course that will profoundly alter the earth's atmosphere and thus, our climate. The "greenhouse effect" results from the fact that the planet's atmosphere is largely transparent to incoming solar radiation, but absorbs much of the longer wave length radiation that is re-emitted by the earth. Carbon dioxide, the product of all combustion, is the principal greenhouse gas, but it accounts for only about half of the total effect. A group of other gases each present in much smaller amounts, collectively account for the other half. These are methane, the chloroflurocarbons or CFCs as they are usually called, nitrous oxide and tropospheric ozone (ozone in the lower atmosphere). The concentrations of all of these are rising, some of them very rapidly.

Though there are many uncertainties, scientists believe that together these gases will force a rise in global average temperature of about 3 C by the 2030s. Both this temperature and the projected rate of greenhouse-induced warming would be unmatched in human experience and, indeed, in the past 100,000 years. Given the time it will require to set many of the emissions trends on new, lower trajectories, many actions to limit future climate change will have to be undertaken in the near future.

Except in the case of methane, the sources of the growth of greenhouse gas emissions are reasonably clear. With the exception of CFCs, they are principally the result of energy use. Thus, the primary options for limiting greenhouse emissions are to limit or eliminate CFC use and to control energy use. Similarly our energy options are of two types: improve end-use efficiencies and/or switch to carbon poor or carbon free fuels.

Some energy sources--solar, hydro, nuclear and others--produce no CO_2, but even among fossil sources, not all fuels are created equal. Natural gas has only about half the CO_2 content of coal and 70 percent that of oil. Allowing for the more efficient use of natural gas relative to coal, this benefit is still greater.

2. Stratospheric Ozone Depletion: Turning to other pressing atmospheric issues, the list is headed by stratospheric ozone depletion. As you know, an international agreement, the Montreal Convention, has been reached to cut CFC emissions in half. The agreement reflects a strong international consensus that ozone depletion is not a distant threat, but an immediate problem demanding redress. In fact, new data on the rate of depletion in the northern latitudes may lead to renegotiation of the agreement,

requiring a complete phaseout of CFC emissions, in the very near future.

In addition to CFCs, nitrous oxide contributes to ozone breakdown. Coal combustion produces substantial emissions, representing on the order of one-third of the total manmade emissions of nitrous oxide according to recent estimates by the World Meterological Organization. The concentration of nitrous oxide is projected to grow by twenty percent by 2030, causing a significant decrease in the ozone layer and an estimated increase in skin cancer rates of as much as 12 percent. Since skin cancer is currently far and away the fastest growing type of cancer, this is a serious concern.

Pulling all these various trends together, what are the implications for a sound energy policy? First, for a number of different reasons both the developed and the developing countries ought to be pursuing energy efficiency improvements very strongly. The developing countries need to buy energy as cheaply as possible and with as little foreign exchange as possible. Countries just now building large energy supply systems can capitalize on the insights gained in the industrialized nations over the past decade. For example, a detailed study of Brazil shows that an investment of $4 billion in high efficiency end use technologies (refrigerators, lighting, motors, etc.) would supplant construction of 21 gigawatts of conventional electrical capacity for a net saving of $19 billion between now and 2000. Beyond setting an example through the policies we adopt for ourselves, the U.S. ought to be pushing hard to get the international assistance agencies to do the necessary analysis and change direction on their energy lending, starting with U.S. AID. The same is true as regards international assistance to ease the fuelwood crisis.

The developed countries, and the United States in particular, also need to achieve high energy efficiency throughout their economies, both to avert another oil price crunch and to keep a lid on greenhouse gas emissions. We need to be working far more aggressively on alternative fuels and end use technologies. Between 1973 and 1985, U.S. GNP rose by 30 percent in real terms, while overall energy use actually declined slightly. But the truth is that we have only just scratched the surface of what is achievable and economically competitive. Despite the improvements of the past decade, the United States still stands near the very bottom of an international ranking of countries according to their efficiency of energy use. Japan still uses half as much energy per dollar of GNP produced.

Let me briefly illustrate some of the opportunities available to us.

- Electric motors consume two-thirds of the electricity used in the U.S. Energy savings of 30 percent or more of this can be achieved through the application of new materials and control systems (variable speed controllers, high performance magnetic materials).

- Another quarter of U.S. electricity use goes to lighting. Cuts of 50 percent or more are possible using existing technologies (high frequency ballasts, isotopic enhancement, advanced control circuits, etc.).

Together, improvements in these two areas could amount to a full third of current electricity use.

- We need energy efficiency codes for buildings just as we have them for wiring, pluming and safety. If the private sector process now underway does not produce them, the federal program repealed at the beginning of this Administration should be reinstated.

- Huge improvements can be achieved in the transportation sector, which accounts for two-thirds of our liquid fuel consumption. In the generally sluggish manufacturing sector, transportation--autos in particular--suffers from the worst case of technological inertia. Prototype vehicles now being tested using lightweight plastics, continuously variable transmissions and other innovations achieve from 70 to 124 mpg with roomy interiors, safety and high performance! But these vehicles are being produced by European and Japanese, not American, companies. Instead of pushing to raise the CAFE standards from 27.5 to 40 mpg, both Congress and the Administration have been convinced--in the face of concrete evidence to the contrary--that 27 mpg is as far as we can go, and indeed are rolling back the standards. American producers have already lost 30 percent of the domestic market to imports. Without competitive high mileage cars, they will lose another large slice when oil prices rise again. Hundreds of thousands of American jobs are at stake, as well as the health of the economy as a whole. If, on the other hand, we were to acheive a 40 mpg standard for new cars and 35 mpg for light trucks by 2000, the savings would amount to well over one million barrels of oil per day.

The aggregate effect of these and similar efficiency policies is enormous. A global energy futures analysis supported by the World Resources Institute looked at developed, developing, and newly industrializing countries and found that by applying the most efficient currently available technologies, in the context of optimistic pro- jections of economic growth, global energy use in 2020 need be only slightly higher than at present. This compares to the doubling or tripling of demand projected in conventional analyses.

With an aggressive research and development program whole new horizons appear.

- Photovoltaics, whose fuel source is unlimited and, which produce no CO_2, hold great promise. Japan is spending 50 percent more on matching support for photovoltaic development than the United States and last year exceeded U.S. production by a wide margin. Meanwhile this Administration has slashed federal R&D spending by a factor of four since 1981. The substantial technological lead enjoyed by U.S. photovoltaic firms in the 1970s is disappearing and may soon be gone.

Almost certainly, with a properly funded and sustained research effort, new efficiency and alternative supply options will appear that we can not now even guess at.

In sum, short, mid- and long-term trends all converge on the same policy. We are at the beginning of the end of the oil era. If demand continues to grow at 3 percent per year as it has recently, we could experience another oil shock as soon as the early to mid-1990s. Greenhouse warming is going to force us to wholly reshape our long term energy strategies. Third World economic growth--on which our own economic health ultimately depends--requires affordable supplies of energy with the least possible depletion of foreign exchange supplies. All of these trends place the highest premium on energy efficiency. This does not mean that there are not steps we can and should take on the supply side. We need a good research and development program on new oil production and exploration techniques and on enhanced recovery technologies, for example. We should continue to fill the Strategic Petroleum Reserve, and we need to be looking at new designs for inherently safe nuclear reactors. And there is much more. But the bottom line for energy policy --national and global--is that the long term answers lie on the demand side.

Chapter 80

THE PRESENT SITUATION AND FUTURE PROSPECTS FOR CHINA'S ENERGY PRODUCTION AND CONSUMPTION

G. Kai

China is a developing country with a population of over 1 billion. The growth of China's national economy will depend largely on the development of its energy intensive industries within the present century. The future prospects for China's modernization depends to a great extent on the supply and effective utilization of energy source, and the raw material industries with larger energy consumption shall be further developed. With only 7% of the world's arable land but 1/4 of its population, China needs a tremendous amount of energy for the modernization of its agricultural activities. Besides, the advancement of science and technology will also rely on the overall electrification so that the social productivity can be improved. People's material and cultural life in general is also an index symbolic of the consumption of energy. The realistic goal in this respect is that the energy consumption per capita will exceed 1 ton of standard coal in China by the year 2000. China has therefore given top priority to the energy industries which are regarded as having startegic significance in the development of China's national economy.

THE PRESENT PRODUCTION AND CONSUMPTION PATTERN

China is one of the few countries in the world relying largely on coal as the major energy source. In 1986, coal accounted for 76.0% of the aggregate consumption of commercial energy in China. At the present time, coal constitutes 72% of the power and fuel reguired by industries, 52% of the chemical materials, 92% of the fuel for civil use in the urban and rural areas (see Table 1). Of the total raw coal output of 894.04 Mt of 1986, 413.92 Mt were produced by the state-controlled coal mines; while 480.12 Mt were produced by the locally operated state-controlled coal mines scattered in the coal producing provinces, autonomous regions and cities, counties and by small mines owned by local townships, villages and individuals. Coal produced by minor coal mines are generally supplied to rural industries, sideline production and also used as household fuel. There are at present 151 coal preparation plants in China will an annual preparation capacity of 159 Mt, which means only 18% of the nation's total coal output can be washed before sales.

Petroleum industry is a new industry in China. There are a total of over 130 oil-fields and gas-fields being developed and a total of 16 oil-gas producing bases in China so far. In 1986, the annual total output of petroleum oil was 131.1 Mt and gas, 13,764Mm^3. There are a total of 33 refineries of large or medium size around the country with an annual oil processing capacity of 126 Mt. The oil and gas transmission pipelines which have been laid so far amount to 11,500 km. To date, 11 gas bearing structures have been discovered in Bohai Sea and Beibu Bay, and a large gas field discovered in the Yinggehai Basin is also ready for development. A marked increase in China's petroleum output can be expected in the 1990s.

The electrical power industry has been growing at a remarkable rate. In 1986, the nation's total installed capacity amounted to 93.8 GW and output 44.96 TW·h. The structure of electrical power consumption is as follows: 14.1% was consumed by rural population, 62.5% by industries, 1.0% by transportation, 8.2% by the urban population, while power consumed by power plants accounted for 14.2% of the total. In1986, a total of 137.5 Mtce of fuel were consumed by the nation's coal-fired power plants, of which 81.4% was coal, 17.5% was petroleum oil and 1.1% natural gas. Energy source (including hydropower) consumed by power generation accounted for 21.6% of the nation's total primary energy consumption.

TABLE 1. COAL CONSUMPTION AND ITS STRUCTURE PER MAIN APPLICATION

Item	1986	1985
Total consumption (10^4)		
Domestic consumption	86189	81416
Total consumption in production & construction:		
For power generation and heating	19592	17496
For coke making	6736	6352
For railway	2503	2624
For household use	19911	19187
Structure (%)		
Domestic consumption	100.0	100.0
Total consumption in production & construction:	76.9	76.4
For power generation and heating	22.7	21.5
For coke making	7.8	7.8
For railway	2.9	3.2
For household use	23.1	23.6

Till now, a nuclear industry system of considerable scale has been set up in China. Construction was begun in June, 1983 for the 300 MW Qinshan Pressurized Water Reactor (PWR) Nuclear Power Plant designed, developed and built by the Chinese and it is expected the project is scheduled for completion and will be connected to power grid in 1989. Construction for Dayawan Nuclear Power Plant began in the northeast part of Shenzhen city, Guangdong Province in April 1984, and it will be ready for power generation in 1992.

Among all forms of new and renewable energy in China, small hydropower stations and biogas are the most effectively developed to date. As at the end of 1986, the installed capacity of all small hydropower stations throughout the country totaled 10 GW, while 5.4 million biogas tanks are in effective use in rural areas.

For details about China's primary energy production, consumption and its structure,see Table 2 and Table 3.

TABLE 2. PRIMARY ENERGY OUTPUT AND STRUCTURE

Year	Total Energy Output (Mtce)	Percentage to Energy Output (%)			
		Coal	Crude oil	Natural gas	Hydropower
1985	855.38	72.8	20.9	2.0	4.3
1986	881.24	72.4	21.2	2.1	4.3

TABLE 3. PRIMARY ENERGY CONSUMPTION AND STRUCTURE

Year	Total Energy Output (Mtce)	Percentage to Energy Output (%)			
		Coal	Crude oil	Natural gas	Hydropower
1985	764.26	75.8	17.1	2.3	4.8
1986	808.82	76.03	17.06	2.26	4.65

ENERGY SUPPLY AND DEMAND, CHANGE IN ENERGY IMPORTS & EXPORTS

China has very rich and diverse energy resources. Statistics show that China's proven coal reserves amount to 749,600 Mt, prospective petroleum oil reserves are around 30,000 to 60,000 Mt, which are now undergoing geological serveying and prospecting. The exploitable hydroenergy amounts to 380 GW. In addition, considerable amount of uranium deposits have been proven in China. However, the average per capita energy reserve is low and what is more, energy reserves are not evenly distributed and they have only undergone lower degree of geological prospecting, all these present problems to the development, transportation and supply of energy.

Like many developing countries, energy shortage in China has become a serious obstacle to its economic development. The main problems are: lack of facilities for the production and transportation of primary energy, particularly coal, serious shortage of electrical power supply and shortage of fuels for the population in rural areas. It is estimated that over 20% of industrial production facilities in the industry are left idle because of the shortage of power supply. An estimation based on minimum standard of demand for energy consumption in the rural areas shows energy needed for daily consumption is still 20% short.

With the development of township coal mines in China during the past few years, coal, as the major energy in China, the supply of which has been relieved to some extent, and in some certain areas, cases of coal surplus do exist. However, viewed as a whold, the proven coal reserves in China are most concentrated in North China and Northwest China, but there is little coal reserves along the east coastal areas and East China where economy is quite developed. It is the reason why the nation's coal transportation setup is basically a eastward direction and southward direction over long distances. Besides, only a small portion of coal produced in China undergoes preparation process, while a major part of saleable coal is sold without washing for its higher ash and refuse content. Thus, a total of about 30 to 40 Mt of extra transport burden is added to railway transportation system. The proven petroleum deposits are also unevenly distributed with most of the oilfields located in north China and the region east of the Yantze River. So oil transportation is also a serious problem.

The present solutions to the contradiction between supply and demand of coal should be:

- to increase coal production and practies economical use of coal;
- to increase coal preparation capacity of the nation's coal industry so that more and more coal can be washed;
- to popularize energy conservation technology so as to improve energy utilization rate and to minimize environmental pollution and extra burden to transportation system.

China's energy exports have seen a major expansion over the past few years. Coal exports was limited to only about 2 Mt per annum during 1950s to 1970s. This figure was broken through in 1978 by an annual export of 3 Mt and coal exports have ever since been on a steady increase. In 1986, annual coal exports totaled 9.9 Mt. Petroleum exports have also experienced an even greater change. Before 1962, petroleum oil consumed in China was mainly imported from abroard. Thanks to the development of Daqing Oilfield, China became self-sufficient in petroleum oil supply since 1962. At the same time, China began to export a small amount of petroleum oil to foreign countries.

Petroleum exports exceeded 1 Mt in 1973, and have been on a steady increase in the following years. In1986, crude oil exports reached 26.5 Mt while exports of oil products were as much as 6.8 Mt. Exports in the above catagories totaled 33.3 Mt, which is 25.5% of the annual oil output of 1986. During the past few years, China's energy products exports accounted for approximately 1/4 of the nation's gross export value.

The future trends in energy exports in the future will be: final oil products in energy exports will increase while crude oil and primary products in energy will decrease; crude oil products exports will decrease while coal products exports will increase instead. The main market for China's energy exports will be Asian nations and Pacific Regions. According to estimations made by experts, it is almost unlikely for China to increase its petroleum oil exports unless a major breakthrough is achieved in petroleum oil exploration. It is predicted that China may become petroleum oil importing country by the turn of the century.

ENERGY TECHNOLOGY POLICIES

Energy is a very important material basis for social development. Energy industry should be given strategic priority in the national economic development so as to accomplish the goal of quadrupling the total output value of China's industry and agriculture by 2000.

According to preliminary prediction, the total output of primary energy in 2000 may be two times that of 1980, while the overall demand will be 2.27 times that of 1980, thus a great gap between energy supply and demand will come into being. The solution to this problem will be to implement the policy of paying equal attention to the development and economical utilization of energy on the basis of scientific and technological advancement, which are to be realized according to the actual conditions in China, i.e. along with the speeded exploration of energy sources, close attention shall be paid to rational and economical use of energy, adjustment of energy products structure and energy industrial structure and also sharp reduction in energy and raw material consumption.

Inorder to speed up energy development so as to improve the energy utilization rate and satisfy the need of the growing national economy, it's important to select appropriate direction for technological development and work out scientific policies concerning energy technology. To this end, the State Science Commission, State Planning Commission and State Economic Commission have organized departments concerned, officials and experts in related field to conduct discussions about the energy technology policies. The results of the discussions are:

1. Production of primary energy should be increased and the structure of primary energy improved;
2. Energy economic zones should be worked out so as to establish regional energy producing system with balanced distribution of energy production;
3. Coal exploration should be speeded up;
4. In the field of petroleum industry, priority should be given to oil reserve prospecting and exploration and to the better economic returns of oil-fields opened up;
5. Hydropower is renewable primary energy, therefore priority should be given to the exploitation of hydropower;
6. Nuclear power plants should be set up in economic developed areas where energy is in short supply;
7. Equal importance should be attached to the exploitation of natural gas and petroleum oil;
8. Development of electrical power industry should be speeded up;
9. New energy should be vigorously deve-

loped and utilized;

10. Energy utilization efficiency should be improved so as to save energy;
11. Petroleum oil and natural gas resources should be used rationally and oil refining process, oil products distribution should be improved;
12. Coal processing, combustion and conversion technologies and products distribution should be improved so as to realize the comprehensive utilization of coal;
13. Rural energy consumption structure should be set up as soon as possible so as to improve the situation of energy shortage in rural areas;
14. Energy consumption structure for urban population should be improved to satisfy as far as possible the reasonable energy demands from the urban population;
15. The quality and technical level of equipment for energy production should be further improved;
16. Close attention should be paid to environmental protection related to energy production and utilization.

PROSPECTS

Considering energy is the foundation of modern economy, the Chinese government attaches great importance to energy problem and has worked out relevant policies and measures to promote and quicken the pace of energy exploitation and effective utilization so as to ensure smooth progress of economic development and the realization of the four modernization.

During the past few years, some institutions and experts from both home and abroad carried out some predictions studies concerning China's energy demand at 2000. Table 4 shows the results of the studies.

TABLE 4. PREDICTION FOR CHINA'S ENERGY DEMAND & STRUCTURE AT 2000

Predictor	Time of the prediction	Total demand (Mtce)	Components (%)				
			Coal	Oil	Gas	Hydropower	Nuclear power
Energy Research Institute, State Economic Commission (minimum scheme)	1984	1469	66.87	23.36	3.17	6.60	
Nuclear Research Institute Tsinghua University (maximum scheme)	1985	1455	67.42	20.17	4.57	6.67	1.17
World Bank (minimum scheme)	1984	1428	78.71	14.01	0.91	6.37	
Jim Woodard (minimum scheme)	1980	1416	67.09	15.14	13.51	4.39	0.24

Notes:

1. Minimum scheme assumes that NGP at 2000 is quadrupled that of 1980 while maximum scheme assumes NGP at 2000 is 4.6 times that of 1980.
2. Prediction made by Nuclear Research Institute, Tsinghua University and World Bank are only for domestic demand; while predictions made by Energy Research Institute, State Economic Commission and Jim Woodard include both domestic and exports demand.

The prediction values in the Table are all minimum demand. Taking into consideration all-round prediction results, China's overall primary energy demands in 2000 will be around 1,400 to 1,700 Mt of standard coal, while the possible coal supply may be 1,300 to 1,500 Mt of standard coal. A great gap will therefore still exist between coal demand and supply.

Coal will account for over 2/3 of the overall

energy demand and supply in 2000, which means coal will still play the major role in China's energy sector. Hence a vigorous exploitation of coal, a full play given to China's coal resources and even better utility rate of coal are of great strategic importance to the energy development of China. Petroleum oil will still play an important role in China's energy structure in 2000. But in view of the present situation, it is unlikely to raise the oil output by a big margin since no major breakthrough has been achieved in petroleum exploration. So petroleum will only account for 20% of the overall energy production of China. Exploitation and utilization of natural gas will gain a major development and it is estimated that the percentage of natural gas in energy supply structure of 2000 will have a substantial expansion. Hydropower, although considered as a key part in the primary energy development, will not expand its percentage in the primary energy production in 2000 because of its presently lower percentage in primary energy production and longer time needed for the construction of hydropower stations. In the initial stage of development, China's nuclear power industry is now learning experiences from its first nuclear power station being built in Dayawan, Guangdong Province and Qinshan, in East China. Target has been set that the Chinese will try their best to complete nuclear power stations with a total capacity of 10 GW by 2000.

To conclude, energy problem is a major reastraint to China's modernization drive. The final realization of China's modernization will be hindered unless energy shortage is properly dealt with. China has attached great importance to energy problem and has tried hard to speed up the development of energy industry, and readjust the policies concerning energy development. All these have resulted in an initially quickened pace of construction in energy sector. We can therefore aptly say that China's energy industry is full of promises.

ACKNOWLEDGEMENT

The author wishes to thank his colleague, Senior Engineer Mr. Zhou Xiuwen and Senior Engineer Mr. Wang Qingyi at China Coal Scientific & Technical Information Institute for their valuable references and data for the preparation of this paper.

REFERENCES

1. Industrial Equipment & Materials No.4, Vol.7
2. P.R.China's Technology Policies (Blue Book issued by the State Science Commission, No.4), 1985
3. China Coal Industry Yearbook, 1986

SECTION 19

ENERGY EFFICIENT DESIGNS—CASE STUDIES

Chapter 81

ENERGY EFFICIENT DESIGN FOR THE NEW MADIGAN ARMY MEDICAL CENTER AT FORT LEWIS

A. G. Jhaveri

ABSTRACT

The task of designing an energy efficient facility such as a complex military hospital is difficult, to say the least, mainly because of the strict environmental, life safety, and regulatory requirements governing the diverse medical functional spaces like patient areas, ancillaries, logistics, and clinics.

With this challenge in mind, the design team for the new Madigan Army Medical Center (MAMC) concentrated its effort towards developing and recommending for implementation those life cycle cost efficient energy conservation measures (ECM) that would result in significantly lower energy costs, not only in each major functional area but also by individual end use categories, such as heating, cooling, hot water, lighting, and equipment. A quantitative measure of this total energy efficient design was provided by the estimated energy budget in British thermal units per square foot per year (Btu/sq. ft./yr.)

EXECUTIVE SUMMARY

As one of the largest regional military hospital construction projects in the United States, the design of MAMC was based on three major objectives, namely:

a. functional integrity (planning and code compliance),

b. cost effectiveness (based on life cycle costs), and

c. energy efficiency (energy conservation measures).

The last, but not the least, energy efficient design was based on a multidisciplinary team or systems approach, involving optimized architectural; mechanical; heating, ventilating and air-conditioning (HVAC); electrical power and building control's systems as well as equipment, elements, and processes.

The final MAMC design included, among other energy conservation features, the following innovative considerations:

(1) Stringent energy budget of 115,000 Btu/sq. ft./yr.

(2) Passive solar energy glazing and daylighting systems.

(3) Ground water for mechanical cooling/chillers.

(4) Maximum thermal insulation values (walls, floors, roof, etc.).

(5) Energy efficient lighting fixtures and electrical power systems.

(6) Sophisticated engineered smoke control system (ESCS) with independent fire alarm and energy management and control system (EMCS) components.

(7) Life cycle cost (LCC) justified building systems and equipment.

This paper provides a summarized description of the comprehensive design development and technical reviews accomplished during more than 3 years (1982 to 1985) of coordinated effort that resulted in the most energy conscious MAMC facility, which in turn, can provide an excellent opportunity to the designers of other similar complex health facility projects (public or private), in order to replicate the invaluable experience gained with the energy efficient design of MAMC.

INTRODUCTION

It is not uncommon with many large (500,000 sq. ft. or more) hospital facilities whose annual energy consumption exceed 250,000 Btu/sq. ft./yr., partly because of the lack of energy efficient design and partly due to uncoordinated operation and maintenance and control of major energy consuming building systems. At the same time, with state-of-the-art energy conservation/management technologies and innovative design approaches, it is realistic to achieve energy savings of as much as 50 percent in these large and complex health care facilities without compromising the critical life safety codes and functional requirements. This is precisely how the entire energy efficient design of MAMC was accomplished, including incorporation of cost effective ECM's. The entire MAMC complex, with approximately 1.2 million gross sq. ft. area, consists of four major building components, namely the nursing tower, logistics, ancillaries, and clinics. Currently, the new MAMC facility is designed for 414 beds. The total energy use was computed among such hospital systems as hot water, lighting and zone equipment, fans, chilled water, auxilliary equipment, chillers, and fossil fuel boilers. The process energy loads, such as medical equipment, were not included in the analysis.

METHODOLOGY

In order to accomplish the most challenging task of cost effective energy conscious design for MAMC, the following major tools and approaches were utilized.

a. Use of no more than 10 percent of north facing wall areas for glazing and no more than 15 percent of the total wall area for glass, windows, and doors.

b. Optimum thermal insulation (in terms of U-values) for walls, ceiling/roof, and floor assemblies with U-wall = 0.08; U-ceiling/roof = 0.03; and U-floor = 0.05.

c. Maximum advantage of passive solar and daylighting concepts, particularly in the patient rooms and skylighted atrium areas.

d. Variable Air Volume (VAV) with reheat system for clinic and noncritical administrative areas, that provides evenly distributed heat and variable air supply to match the cooling demand.

e. A combination of double bundle and single bundle centrifugal chillers and absorption chillers for noncritical and critical areas within the hospital, respectively.

f. Use of ground water from shallow aquifer wells (55 degrees F water) as condenser water for the chillers.

g. Hot water heating system, using steam (heat exchangers) from the central coal-fired plant with heaters and storage tanks to be supplemented with a variety of heat recovery systems.

h. Energy-efficient lighting fixtures for interior (e.g., fluorescent) and exterior (e.g., high-pressure sodium) use with photo-cell controls, time-clocks and two-level illumination control mechanisms.

i. Use of high-efficiency motors, (optional) variable-frequency motor drives, power factor controllers and 13.8 kV distribution power voltage for cost-effective HVAC (e.g., chillers) and smaller electric power (e.g., for transformers) operations.

j. All recommended Energy Conservation Measures (ECM's), based on life cycle cost (LCC) analyses of at least three (3) alternatives for major energy-consuming or energy-conscious design systems/concepts.

k. The total energy budget of the entire MAMC complex to remain below 115,000 Btu/sq.ft./yr., excluding process energy loads, with the use of Building Loads Analysis System Thermodynamics (BLAST) III computer program, simulating individual buildings and energy-consuming systems.

DISCUSSION

It is clear from the above that a comprehensive analysis of each major MAMC facility design component is essential, before the systems approach to overall energy conservation potential is estimated. This approach, therefore, not only targeted the major energy-consuming systems like HVAC and lighting systems, but also included architectural, controls and renewable energy resource systems' evaluation for design development. The following summarize those major ECM's that were found to be life cycle cost-effective and contributed to the total MAMC energy savings potential, as reflected by the Blast III computer simulation program energy budget estimate:

Architectural Energy Conservation: Among the specific ECM's considered for MAMC design incorporation include, site and landscaping, building massing and envelope, passive solar and daylighting. The 10-story nursing tower will be oriented on a north-south axis, thus minimizing exposure to prevailing winter winds from the south and maximizing psychological and therapeutic welfare of the patients with views of Mt. Rainier on the east and Puget Sound on the west as well as morning and evening solar heat during limited number of clear winter days. The other three functional structures-ancillaries, logistics, and clinics are low-rise building (maximum three floors) areas that are shielded from wind and solar by proper ground level treatment (e.g., berms) and landscaping. Figure 1 shows an overview of the MAMC facility as part of the artist's rendering for the project.

MAMC, as designed, has an optimum ratio of envelope-to-floor area due to massing building functions in compact multistory structures, resulting in spread of energy loss over many square feet of usable area for energy efficiency.

Exterior walls and glazing have been designed to meet or exceed optimum thermal insulation values (R- or U-values) to reduce heat loss, solar gain and cooling loads in the summer while increasing daylighting and solar gain during winter for energy-efficient HVAC and lighting systems operation. Significant design effort was expended in analyzing and recommending optimum window configuration that incorporated beneficial effects of sun's heat and light during winter, while minimizing glare and interior heat buildup during summer, using solar angle and shading technology.

HVAC Energy Conservation: Among the specific ECM's considered for MAMC design incorporation include, VAV with reheat, centrifugal and absorption chillers, well water for chiller condensers and hot water from the central plant. The VAV with reheat system design for clinics and noncritical administrative areas will result in roughly 70 percent reduction of air delivery before reheat coils are energized. Reheat fans will be equipped with inlet vanes or mechanical variable speed devices to control air flow. The return air fan, in turn, will be equipped with air monitoring devices for control to ensure that the air delivery of the supply air fan. With well insulated walls and restricted window areas, this system will evenly heat the areas without compromising the functional requirements with optimum energy conservation.

The outdoor design conditions for MAMC, as specified below, were based on the recorded climatic conditions from applicable weather stations in the vicinity:

	Summer		Winter	
	Degrees F Dry Bulb	Degrees F Wet Bulb	Degrees F Dry Bulb	Degrees F Wet Bulb
Medical	86	68	19	16
Nonmedical *	82	64	24	21

*Nonmedical areas are outpatient clinic areas, administrative areas, and public use areas.

The four centrifugal chillers, at approximately 600-ton capacity each, designed for MAMC, consist of large double-bundle type and smaller single-bundle type, while two smaller absorption type, with roughly 400-ton capacity each, that use steam. These later chillers were selected to serve the critical areas during power outages. Also, these absorption chillers will be used to provide peak cooling loads, thus reducing electric demand charges. One smaller centrifugal chiller will be operated for computer areas, using outside air for energy conservation and reduced operational costs.

The hot water heating system design was based on reduced costs and maximum energy efficiency. The clinic, ancillary and nursing tower each will have two hot water heaters with storage tanks. The kitchen supply will have a single hot water heating and storage tank. The use of many hot water systems will reduce the length of hot water piping and permit setting of the optimum water temperatures for each area. Basic water heating will be with central plant steam in heat exchangers. These hot water heating systems

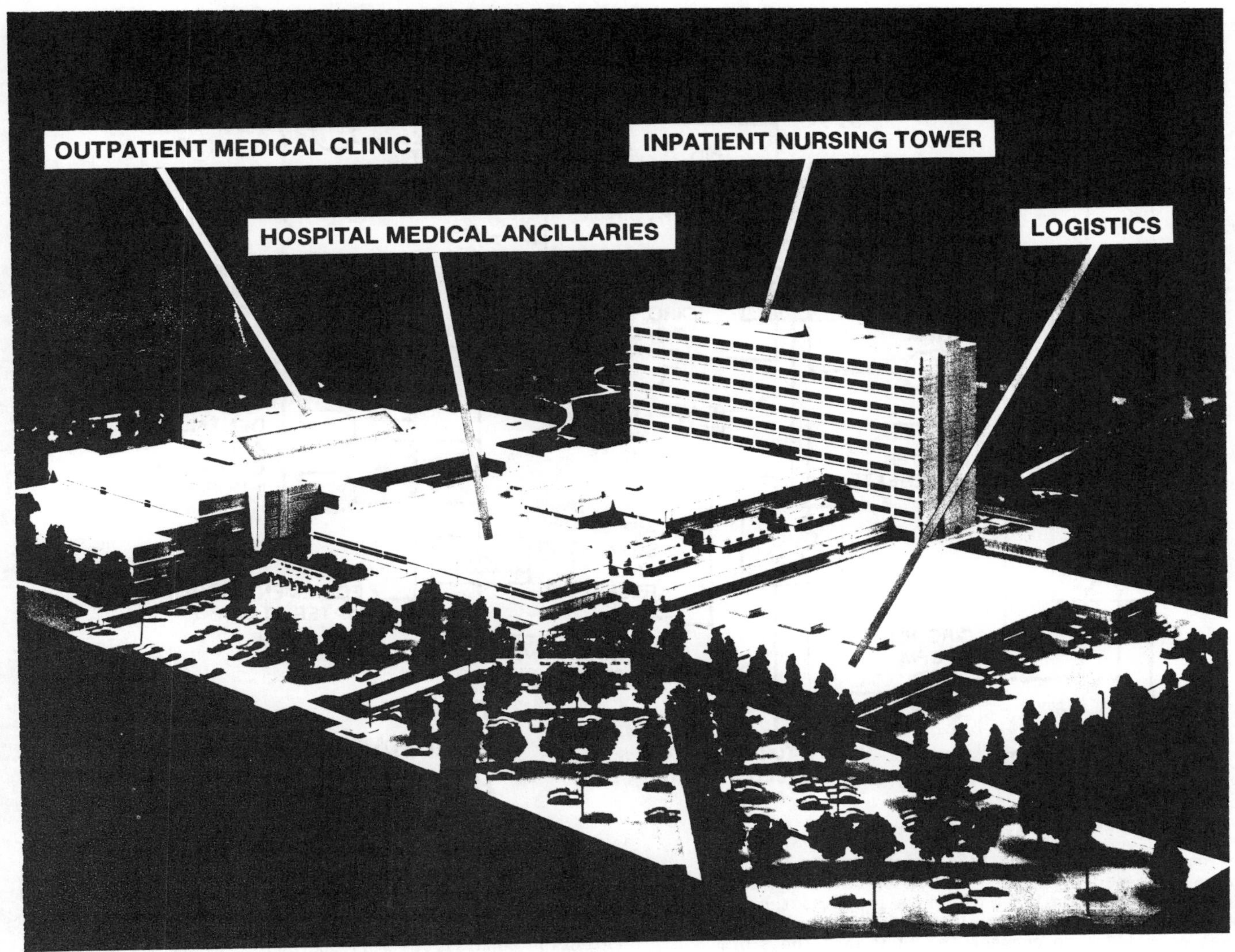

FIGURE 1 - OVERVIEW OF NEW MAMC FACILITY

will be supplemented with heat recovery from kitchen refrigeration, vacuum pumps and electric heat tape use. Figure 2 provides a schematic of the typical hot water system.

Lighting and Electric Power Energy Conservation: Since lighting is a second major energy user after HVAC, in a modern health care facility like MAMC, the following significant ECM's were recommended for design implementation:

- Outdoor lighting will be on photo cell control and will use energy-efficient high pressure sodium lamps.
- Two-level lighting will be provided for areas that have reduced activity during specified hours and where lighting levels exceed 50 footcandles.
- In general, fluorescent fixtures with required optimum brightness levels, energy-efficient ballasts and 3050 lumen lamps will be used; where fluorescent fixtures are inappropriate, high intensity discharge (HID) lamps will be used to the maximum extent possible. The use of incandescent fixtures will be minimized.
- Because of the potential electric power savings, a 13.8 kV distribution system will be used and highest available secondary voltage provided to the interiors for reducing transformers and transmission losses.
- Three-phase motors sized close to full load, will be used to provide optimum efficiency and power factor.
- For variable demand electric loads, variable-speed motors and/or controls for cyclic operation, will be used to reduce power consumption.
- When appropriate for energy savings, power factor correction capacitors will be provided for large motors and low power factor equipment.
- Lower lighting levels for specified interior functional areas, without compromising the minimum illumination requirements will be used for reduced power and air-conditioning loads.

Renewable Energy Resource and Innovative Energy Conservation: In addition to using shallow aquifer ground water for MAMC air-conditioning load, the

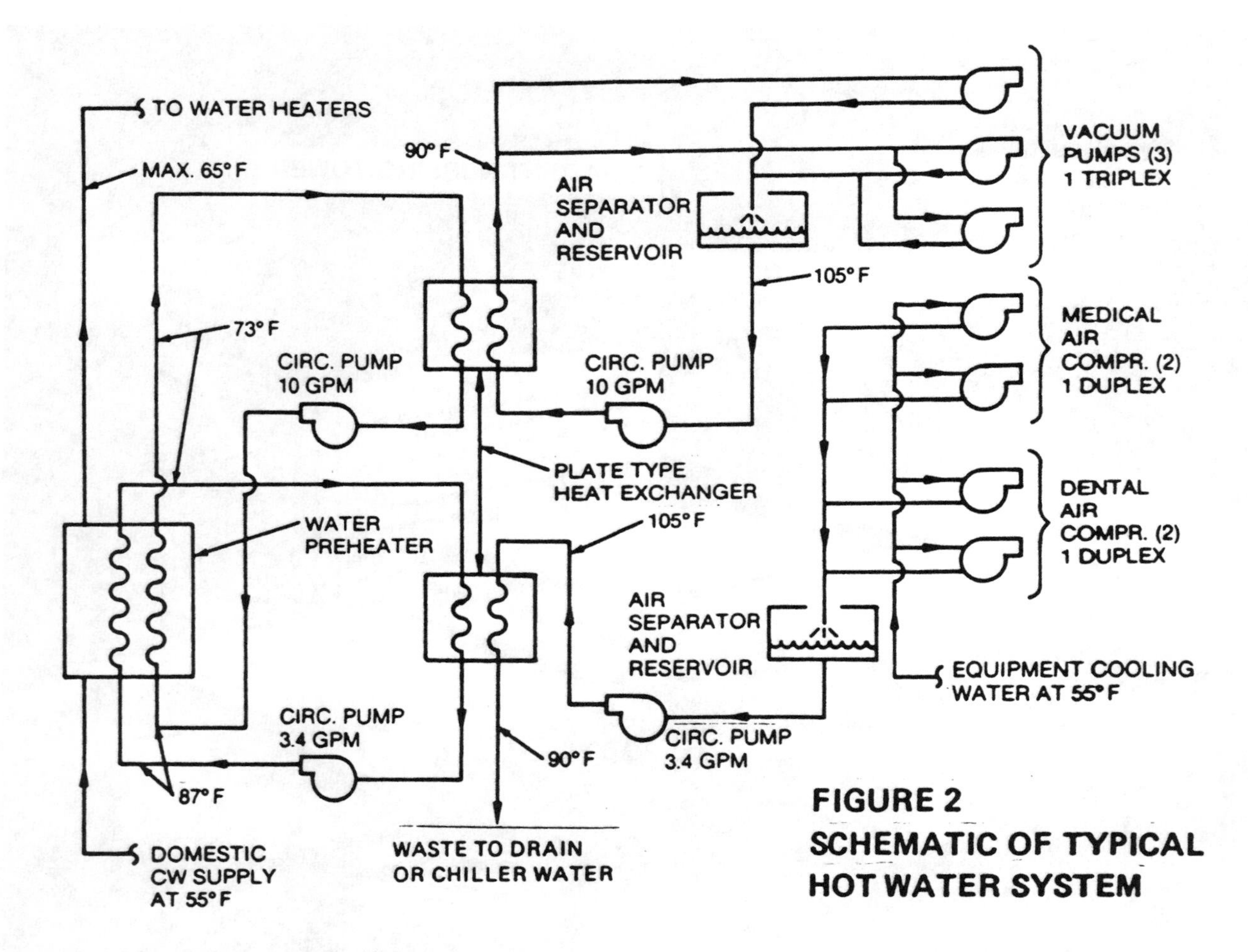

FIGURE 2
SCHEMATIC OF TYPICAL HOT WATER SYSTEM

design also recommended its application for cooling emergency generators, as a source of both fire protection water and domestic water in a postdisaster situation and as a source of irrigation water for landscaped areas as well as to augment the flow into the storm retention pond during dry summer period.

Condenser water for all seven chillers will be supplied from a constant pressure well water system, provided from three cooling wells producing a constant temperature of 55 degrees F ground water. This condensor water system is piped to all of the chillers.

The cooling system for MAMC requires a total of 4,500 gpm of water to function at its maximum load. A total of four wells are to be used in the close vicinity of MAMC, three of which to be drilled at approximately 120 feet depth and no closer than 250 feet from each other, so as to provide an average of 1,500 gpm per well. The fourth well will be drilled in a confined aquifer to prevent contaminants entering into the domestic water supply. During a critical postdisaster operating period, the following flows will be provided from the cooling water wells:

o	Two critical absorption chillers at approximately 400 tons	= 1,200 gpm
o	Domestic water supply	= 150 gpm
o	Vacuum pumps and air compressors	= 20 gpm
	Total required emergency flow	= 1,370 gpm

In addition to the above flow, four emergency generators will operate with a demand of 1,000 gpm. Generator cooling water will be obtained by routing the chiller discharge water to the cooling jackets. The retention pond system will operate to allow the warm chiller water, at 70 degrees F temperature, to infiltrate back into the shallow swayle of ground water system after natural evaporation.

Finally, the MAMC design provided for an innovative Engineered Smoke Control System (ESCS), which provides some features of the energy management and control system (EMCS) and many features of a smoke control and removal system. In addition, it will monitor the load shedding and filter status. The ESCS is based on the concept of distributive process for maximum reliability and speed. The system will be listed in accordance with Underwriters Laboratory (UL) standard 864 for proprietary fire alarm signalling systems. The ESCS will interface with the fire alarm system via contact closures at one designated location. Inputs to the ESCS will include four signals for each smoke zone to facilitate automatic smoke control. Three outputs to the fire alarm system from the ESCS will indicate the status of the smoke control process for a particular zone. When the fire alarm system detects smoke in a zone, it will stop the Air Handling Unit (AHU) and all exhaust fans serving that zone, and signal a "smoke alarm" condition to the ESCS. The ESCS will automatically put the smoke zone into the "exhaust" mode and all surrounding zones into the "pressurize" mode. The ESCS will interface directly with the pneumatic local loop temperature controls for fans, dampers, etc. Each Field Interface Device (FID) will contain full function Energy Management programs for the AHU's it serves. This will include optimized start/stop programs, occupied/unoccupied cycle programs, load reset and discharge temperature reset, chiller plant optimization, and optimum ventilation programs to ensure minimum energy consumption without affecting temperature and humidity comfort levels, minimum ventilation requirements, and occupancy schedules. Thus, a combined ESCS/EMCS control mechanism provides an innovative approach to safety and energy efficiency for MAMC. Figure 3 shows a simplified diagram of the facility systems interface, including ESCS, EMCS, fire alarm, and other building control functions.

The energy budgets for individual buildings and MAMC complex are shown below, indicating compliance with 115,000 Btu/sq.ft./yr. criterion:

(i) Nursing Tower	117,100 Btu/sq. ft./year
(ii) Logistics	74,436 Btu/sq. ft./year
(iii) Ancillaries	239,510 Btu/sq. ft./year
(iv) Clinics	55,793 Btu/sq. ft./year
Entire Complex (MAMC) (weighted total)	111,320 Btu/sq. ft./year

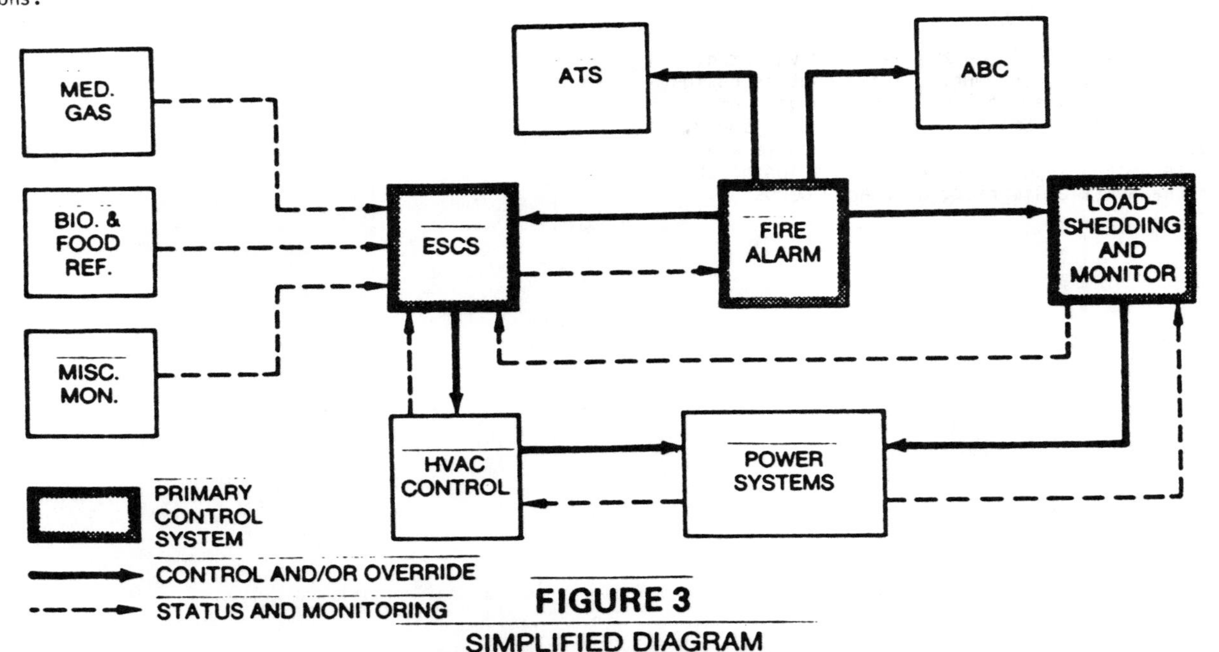

FIGURE 3
SIMPLIFIED DIAGRAM
FACILITY SYSTEMS INTERFACE

RESULTS AND CONCLUSIONS

Based on the selection of recommended ECM's in the MAMC design, the following table summarizes energy use by building and systems, as projected/estimated by the Blast III computer program:

System	Nursing Tower	Logistics	Ancillaries	Clinics	Entire Complex
Lighting and Zone Equipment (kWH)	2.6X10^{6}	5.5X10^{6}	1.7X10^{6}	1.5X10^{6}	9.9X10^{6}
Fan (kWH)	1.1X10^{6}	7.9X10^{4}	4.4X10^{5}	1.9X10^{5}	1.9X10^{6}
Hot Water (Btu)	8.3X10^{9}	2.4X10^{9}	1.1X10^{10}	2.5X10^{9}	1.8X10^{10}
Chilled Water (Btu)	1.5X10^{9}	1.9X10^{8}	6.0X10^{9}	1.4X10^{9}	5.5X10^{9}
Chiller (kWH)	3.9X10^{5}	2.0X10^{4}	6.6X10^{5}	2.2X10^{5}	5.6X10^{5}
Auxiliary Equipment (kWH)	8.8X10^{5}	2.1X10^{4}	9.7X10^{5}	5.9X10^{5}	1.0X10^{6}
Fossil Boiler Fuel (Btu)	2.3X10^{10}	3.9X10^{9}	7.4X10^{10}	1.1X10^{10}	9.3X10^{10}

Based on an average cost of between \$6 and \$7 per million Btu (MBtu) of either electrical or nonelectrical energy costs in real 1987 dollars, the annual MAMC energy expenses will be less then \$1 million for the estimated energy budget, barring any unexpected significant increases in fuel costs and electrical rates.

The MAMC project is under construction at this time, with tentative completion and occupancy scheduled for 1991. The phased construction began in 1986 with site work and is nearing the halfway point as of May 1988.

Therefore, it is fair to conclude that the energy efficient design of MAMC will result in significant cost and energy savings, without compromising the quality, functional integrity and life-safety requirements for this state-of-the-art complex military medical facility.

ACKNOWLEDGEMENTS

This paper relied heavily on the design development documents prepared by the entire joint venture design architect-engineer team of John Graham Company of Seattle and Sherlock, Smith and Adams of Montgomery, Alabama. The author wishes to thank all those who provided technical support in the preparation of this paper at the Seattle District, U.S. Army Corps of Engineers.

Chapter 82

AN ENERGY EFFICIENT DESIGN CONCEPT FOR THE MESQUITE INDEPENDENT SCHOOL DISTRICT

C. L. Maxwell

Mesquite is a suburb located on the east side of Dallas with a population of approximately 110,000. The school district covers over 60 square miles with a student population of 22,500. The 32 schools consist of 23 elementary, 5 middle and 4 high schools, with one additional elementary school scheduled to open in the Fall of 1988.

The school district facilities cover approximately 2.4 million square feet with an annual energy budget of $1.8 million (natural gas and electric).

Utility Rates:

Electric - Texas Utilities Company MS Rate $15/mo per meter + KWH charge. On peak monthly average (June 87 thru September 87) $.0655/KWH. Off peak monthly average (October 86 thru May 87) $.0481/KWH.

Natural Gas - Lone Star Gas Company - September 86 thru August 87 monthly average $3.71/MCF.

The outdoor design conditions of the Dallas, Texas area as taken from the ASHRAE Handbook 1985 Fundamentals Volume are: Summer: 100° D.B. and 78° W.B. and Winter: 18° D.B.

ENERGY MANAGEMENT PROGRAM

During the early 1980's Mesquite Independent School District (MISD) adopted an energy-efficient design concept for new schools built in the district. The features of the energy-conscious design concept include:

The use of Direct Digital Control.

Less than 5 percent area is glass.

The average U value is .05 $BTU/ft^2 h^o F$.

Thermal storage heating and cooling.

Variable speed drives on pumping and ventilation air handling units.

Closed circuit cooling towers.

Four pipe heating and cooling systems.

Hydropulse boilers for domestic hot water and space heating.

Heat recovery on walk-in freezers and refrigerators.

Occupancy sensors for lighting and HVAC control.

Electrical distribution system broken down into HVAC, lighting, and miscellaneous loads with each load submetered.

All the meters and submeters can be remotely read through the energy management system.

All exterior lighting is high-pressure sodium.

All interior lighting is 34w fluorescent with energy efficient ballasts and electronic ballasts are under consideration for future applications.

The use of a District Energy Management System.

Schools Occupied or Under Construction Using The MISD Design Concept

SCHOOL	SQUARE FEET	DATE OCCUPIED
Kimball E.S.	32,200	10/85
Poteet H.S.	114,642	6/86
A.C. New M.S.	93,780	8/87
Pirrung E.S.	35,400	9/87
Cannaday E.S.	45,000	9/88
Black E.S. (Addition)	12,000	9/88

Each of these schools utilize the following types of thermal storage.

Kimball E.S.	Static Ice Builder
Poteet H.S.	Chilled Water
A.C. New M.S.	Dynamic Ice Harvester
Pirrung E.S.	Static Ice Builder
Cannaday E.S.	Dynamic Ice Harvester
Black E.S.	Dynamic Ice Harvester

Cash Incentive Payments

The following cash incentive payments have been received, or will be received, from Texas Utilities Electric Company for Thermal Storage Projects from 1986 through 1988:

SCHOOL	KW REDUCTION	AMOUNT
Kimball E.S.	71	$ 17,750
Poteet H.S. Phases I & II	210	52,250
A.C. New M.S.	184	64,400
Pirrung E.S.	76	26,600
Poteet H.S. Phase III	140	49,000
Kimball E.S. Phase II	20	7,000
Black E.S.	37	12,950
Cannaday E.S.	90	31,850
TOTAL	828	$262,050

DR. RALPH H. POTEET HIGH SCHOOL

A project description of Poteet High School is that of a totally integrated energy conserving concept utilizing a weather impervious envelope, minimal internal loads, and highly efficient mechanical systems coupled with heating and cooling thermal storage to generate, store, and distribute thermal energy in the most efficient and cost effective manner.

By utilizing occupancy sensing equipment all elements of the energy consuming environmental comfort systems of the facility respond instantly to the utilization desires of its occupants never necessitating preplanning for equipment operation or seasonal operating mode selection. More importantly, all energy consuming equipment is automatically de-energized within a few minutes in the areas that occupants leave. The systems are so effective that a single individual can go to his or her office after hours without prior arrangements with total facility energy consumption only increasing proportionately.

A detailed listing of all the specific features are as follows:

1. Envelope

   Less than 5% thermal glazing, average U value for walls and roof is 0.05 with vestibules on all entrances with radiant heating.

2. Electrical Systems & Equipment

   A. Variable frequency drives on chiller, cooling tower fans, chilled water distribution pump and ventilation air unit.

   B. High pressure sodium fixtures used on all exterior lighting.

   C. Electrical distribution is divided according to function and submetered for HVAC loads, lighting loads, and process loads.

3. Mechanical Systems & Equipment

   A. Chilled Water Storage Tank- 280,000 gallon concrete vertical stratification without separation. Temperature measurement at six equally spaced elevations. Charged by chiller or hydronic vent cycle.

   B. Hot Water Storage Tank - Concrete lined steel 17,000 gallon with four headers arranged for dual temperature storage.

   C. Chiller - Variable frequency drive, 196 ton rated at 38°F. Sized for full site development peak day load.

   D. Boilers - Pulse fired high efficiency units sized for continuous operation at peak requirement. 440,000 BTU/HR output.

   E. Closed Circuit Cooling Tower - Operates in three modes. Rejects heat from chiller, generates cooling effect for facility directly or indirectly through charging of chilled water storage tanks.

   F. Distribution Pumping - Variable frequency drive heating and cooling pumps controlled to maintain proper differential across system load elements. Hot water pump operates below 70°F during occupancy periods or for facility warm up cycle. Chilled water pump operates during occupancy or cool down cycle. System differential set points varied according to supply water temperatures and outside air conditions.

   G. Ventilation - Variable frequency drive unit providing air to only the occupied rooms. Provides supplemental cooling anytime outside air is below 55°F. Supply temperature drops to 40°F before heating coil is controlled to maintain that minimum temperature. Gym and auditorium dampers are controlled for supplemental cooling. All ventilation is set for 5 CFM per occupant minimum.

H. Air Units/Fan & Coil Units - Units start automatically when area served is occupied.

I. DDC Control System - Controls equipment, provides system operation information, generates historical data files for meter readings and selected system points. The direct digital controllers are connected to a centralized energy management system through cabling from the local cable television company. The system automatically logs data on the status of HVAC and lighting systems for many schools. This allows a 9,600 baud data transmission rate. This enables examination of data to locate trouble spots in the operations.

The greatest innovation of this project is its ability to uniquely utilize existing technology to enable the school district to efficiently cope with the current rising energy costs. The project is a 114,642 square foot high school containing classrooms, a gymnasium and a fieldhouse.

The electrical systems feature high efficiency motors and variable frequency drives on the chiller, cooling tower fans, distribution pumps, and the ventilation air units. The majority of the interior lighting uses 34 watt lamps and high efficiency ballasts. Occupancy sensors control the lighting and air conditioning units for most areas.

Both the cooling load and hot water requirements are satisfied from storage systems. The chilled water storage tank uses stratification and low velocity headers to store and utilize 280,000 gallons, which represents 130 percent of the maximum daily cooling load. The hot water storage tank has dual storage at 190°F and 120°F, the upper-level is filled from the boilers, while the bottom-level is filled by waste heat recovery from kitchen refrigeration units. The chiller is sized to recharge the storage tank during a 16 hour full-load period from 8 p.m. to 12 noon. To increase efficiency, a Variable Frequency Drive for the chiller compressor takes advantage of lower night-time condensing temperatures rather than part-load conditions. A closed circuit cooling tower is used to reject condenser heat or to provide hydronic free cooling directly to the storage tank.

See Figures 1 and 2 for chilled and hot water system configuration.

A School Built in Phases

The school complex consists of Phases I and II that have been occupied since June of 1986 and Phase III which is under construction and scheduled for occupancy August 1988. The future plans for the site are two more additions (Phases IV and V) when required by the area student census.

Phases Occupied Since 6/86 (114,642 Sq. Ft.):

Phase I - two story academic wing, major mechanical equipment room, and hot and chilled water above ground thermal storage tanks.

Phase II - Gymnasium and Athletic Complex.

Phase Under Construction Scheduled for Occupancy 9/88 (73,600 Sq. Ft.):

Phase III - Kitchen, Cafeteria, Library and Vocational Wing.

Future Phases (Approximately 112,000 Sq.Ft.):

Phase IV - Auditorium, Band and Choral Facility and New Administration Area.

Phase V - Two Story Academic Wing

When all five phases are complete the complex will be approximately 300,000 square feet. The major mechanical equipment and the thermal storage tanks were designed for the complete facility (300,000 sq.ft.) and installed under the Phase I project.

COMPARISON OF ENERGY COSTS FOR POTEET HIGH SCHOOL AND THE SCHOOL DISTRICT'S THREE OTHER HIGH SCHOOLS

(Data for comparison from electric & natural gas bills - September 1986 through August 1987)

SCHOOL/SQ.FEET	BTU* SQ.FT/YR	$ SQ. FT/YR	SAVINGS** COST	BTU
Poteet (114,642)	40,029	.46	--	--
NMHS (262,915)	56,488	.55	16%	29%
WMHS (190,214)	58,002	.68	32%	31%
MHS (230,609)	78,146	.77	40%	49%
Three H.S. Avg.	64,212	.67	31%	38%

* Site BTU Consumption Data was used in calculations.
** Poteet High School savings per square foot compared to existing schools.

MESQUITE INDEPENDENT SCHOOL DISTRICT HIGH SCHOOL HVAC SYSTEMS

Poteet High School - Occupied 6/86 - 114,642 square feet. Four pipe hot water and chilled water thermal storage system.

North Mesquite High School - Occupied 9/69 - 262,915 square feet. Area cooling and ventilating air handling units with room electric reheat.

West Mesquite High School - Occupied 8/76 - 190,214 square feet. Two pipe hot/chilled water system with room diverting boxes (an early version of VAV boxes).

Mesquite High School - Occupied 9/63 - 230,609 square feet. Two pipe hot/chilled water system.

COMPARISON OF %MPE OF GENERAL CONTRACT SUM ON POTEET HIGH SCHOOL AND A NEIGHBORING SCHOOL DISTRICT HIGH SCHOOL THAT IS UNDER CONSTRUCTION AT THIS TIME

Poteet High School Phases I, II and III

General Contract Total**	$14,666,590
Mechanical, Plumbing and Electrical	3,587,250*

MPE = 24% of General Contract

*Includes cost of hot and chilled water storage tanks complete (also includes tunnel and piping from tanks to mechanical room).

**Under the Phase III contract a backup chiller and pumps were installed.

Neighboring School District High School

General Contract Total	$18,150,170
Mechanical, Plumbing and Electrical*	4,294,383

MPE = 24% of General Contract

*HVAC Two Pipe Hot/Chilled Water System.

NOTES: Bids for the Neighboring School District High School were taken about the same time that Poteet Phase III bids were taken. Data shown above is from the architect who designed both schools.

ECONOMIC IMPACT

The concepts utilized in the Dr. Ralph H. Poteet High School have unlimited transferability to all educational facilities, up to and including colleges and universities. In addition, they could be cost effectively applied to all other environmentally controlled facilities.

The economic impact is felt by all taxpayers in the school district, since this facility can operate on an energy budget that can be at least 30% less than for the existing schools in the same school district. In the future, when demand charges are added to the utility rate structure we anticipate it will operate at a savings of 50 to 60 per cent less.

AWARDS RECEIVED BY THE M.I.S.D. ENERGY MANAGEMENT PROGRAM

The Energy Management Program for the Mesquite Independent School District has received the following awards in 1987 and 1988:

1. Regional VIII Energy Award for New Commercial Institutional or Public Assembly Buildings, presented at the ASHRAE Chapter Regional Conference in Tulsa, Oklahoma - 1987.

2. First Place in State Competition in the National and State Awards Program for Energy Innovation. Top five award winners in the state were entered in national competition - 1987.

3. U. S. Department of Energy National Award in the National Awards Program for Energy Innovation, presented at a national ceremony in Washington, D.C. - 1987.

4. 1987 Texas Industrial Energy Efficiency Award in recognition of exceptional achievements in the application of energy conservation technology. Presented by the Energy Systems Laboratory of Texas A & M University at the 9th Annual Industrial Energy Technology Conference and Exhibition in Houston, Texas.

5. ASHRAE International Energy Award - Second Place in the New Commercial, Institutional, or Public Assembly Buildings category. Award presented at the ASHRAE Winter Meeting in January 1988 held in Dallas, Texas.

6. The Reuben Trane Energy Fitness Award for innovative comfort system engineering in the Joey Pirrung Elementary School presented to the MISD School Board in February 1988.

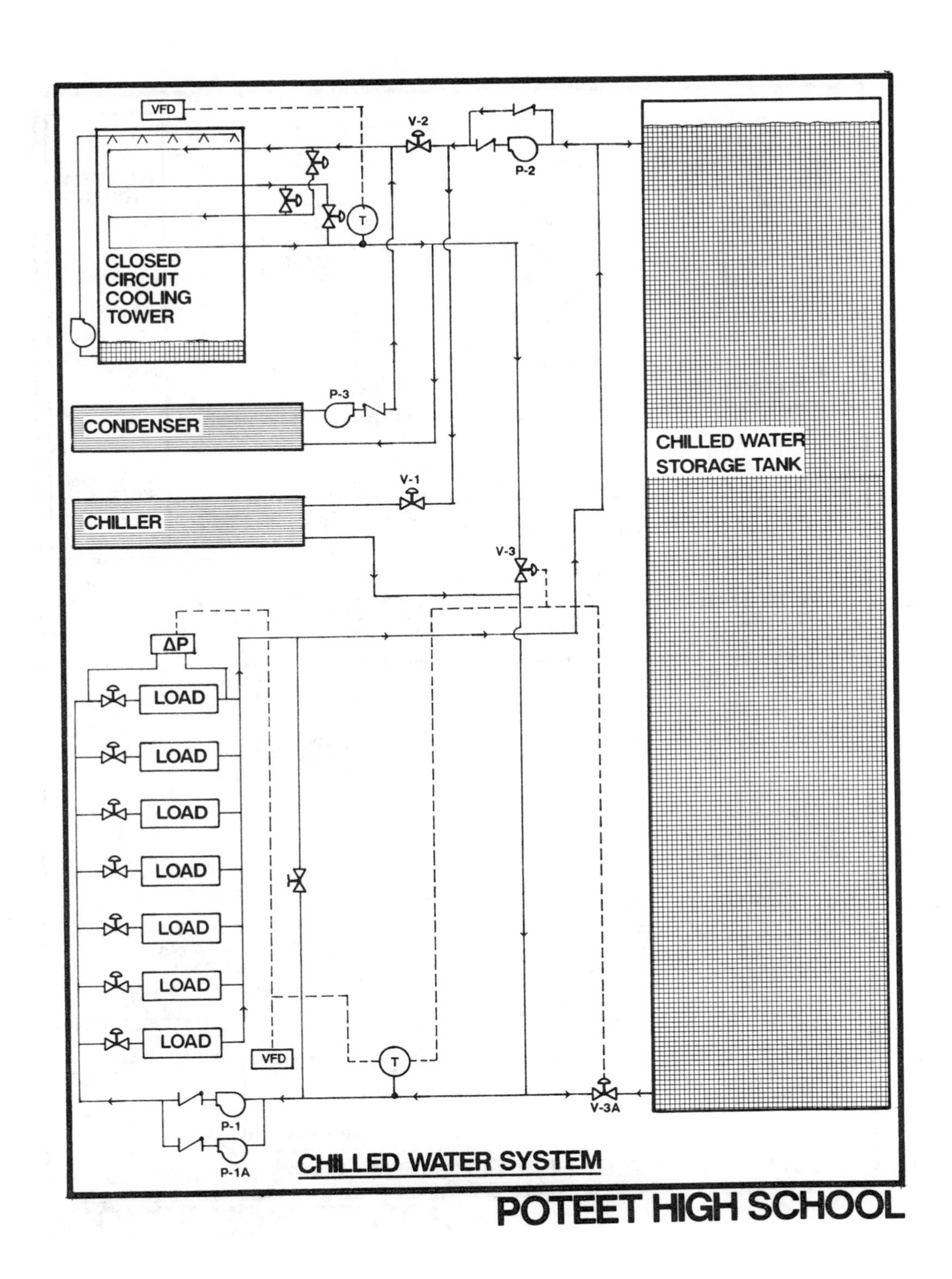

FIGURE 1. CHILLED WATER SYSTEM. THE CHILLED WATER STORAGE TANK USES STRATIFICATION AND LOW VELOCITY HEADERS TO STORE AND UTILIZE 280,000 GALLONS, WHICH REPRESENTS 130 PERCENT OF THE MAXIMUM DAILY COOLING LOAD.

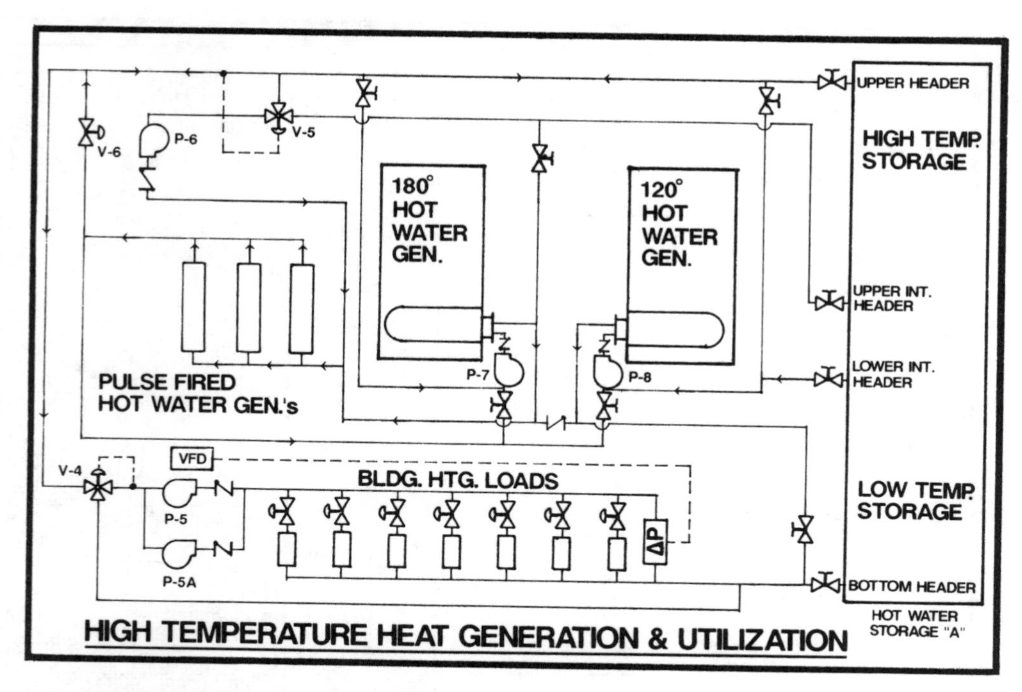

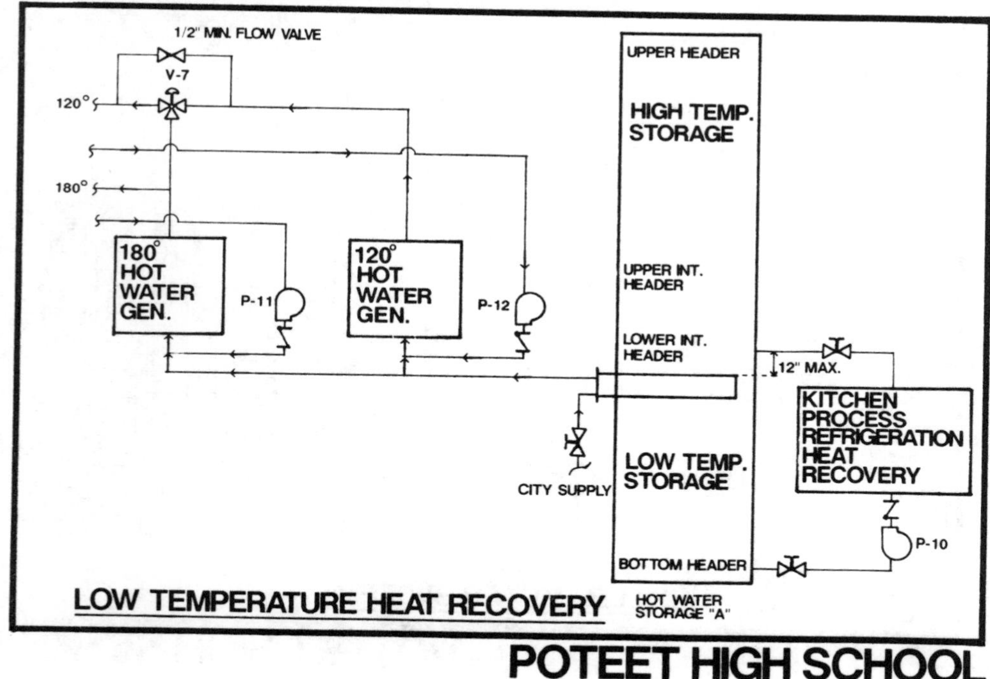

FIGURE 2. HOT WATER STORAGE SYSTEM. THE HOT WATER STORAGE TANK HAS DUAL STORAGE AT 190°F AND 120°F, THE UPPER LEVEL IS FILLED FROM THE BOILERS, WHILE THE BOTTOM-LEVEL IS FILLED BY WASTE HEAT RECOVERY FROM KITCHEN REFRIGERATION UNITS.

Chapter 83

SOLAR PONDS: A COST EFFECTIVE ENERGY SOURCE

T. J. McLean, A. H. P. Swift

On September 19, 1986 a solar pond in El Paso, Texas became the first such pond in the United States to generate electricity. This pond was also the first pond in the world to provide process heat to an industrial operation.

Most alternate energy sources that have been developed in the United States, while capable of delivering a substantial amount of energy, cannot do so at costs that are competitive with conventional fuels. In the southwest region of the United States where solar radiation is abundant and where salt is readily available, solar ponds are capable of providing energy at one-tenth the cost of other alternative energy sources. Low cost alternative energy sources are sorely needed in precisely these sparsely populated areas of West, Texas, New Mexico and Arizona. Such sources can provide power for pumping water for irrigation, desalination of brackish groundwater and as a peaking plant to reduce commercial electrical demand charges.

There are a number of experimental salt-gradient solar ponds in the U.S., ranging in size from $150m^2$ to $4000m^2$. The El Paso pond described in this paper is at $3355m^2$, the second largest U.S. pond. The size of all U.S. ponds pale in comparison to the ponds in Israel. The pond at Beit Ha'rava on the shore of the Dead Sea is 250,000m and is rated at 5 MW of electrical power.

The concept of a solar pond is to provide a sizeable body of water that will be heated by solar energy. As solar energy penetrates into a body of water, water is heated and the more bouyant hot water rises to the surface and heat is lost to the atmosphere. In order for a pond to be an effect energy store, energy loss to the atmosphere must be retarded. Very basically, and over simplistically, the tendency of the heated water to rise is offset by increasing the density of the water by adding salt. A stagnant layer of water is created in the upper portion of the pond to act as a highly transparent insulation layer to contain energy in the lower potion of the pond as shown in Figure 1.

Figure 1. Solar Pond Cross Section

Solar ponds function similar to a flat plate solar collector and are usually shallow bodies of water 2 to 5 meters deep. They are constructed and the salt concentration maintained so as to provide three separate and distinct layers as shown in Figure 2 [2].

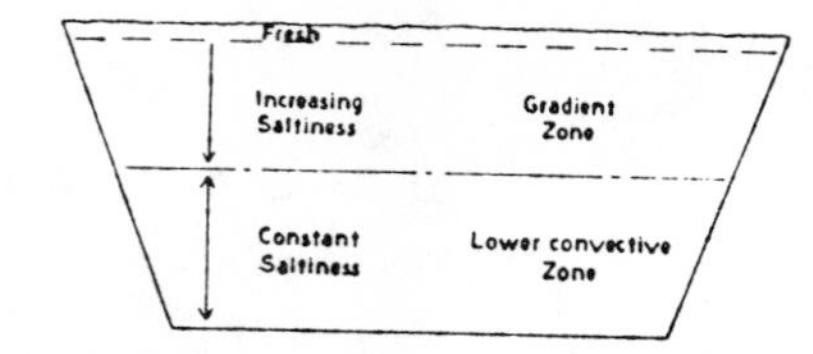

Figure 2. Salt Stabilized Solar Pond

A thin layer of low salt concentration water called the upper convective zone (UCZ) is maintained at the surface. Below this is a layer called the gradient zone (GZ) where the salt concentration increases with depth to near saturation. This gradient zone acts as the insulation for the bottom layer called the lower convective zone (LCZ) or thermal storage zone. The recoverable pond energy is stored in this lower convective zone. The hot brine is cycled through an external heat exchanger to extract the usable energy. Typical profiles of salt concentration and pond temperature as a function of pond depth are shown in Figure 3.

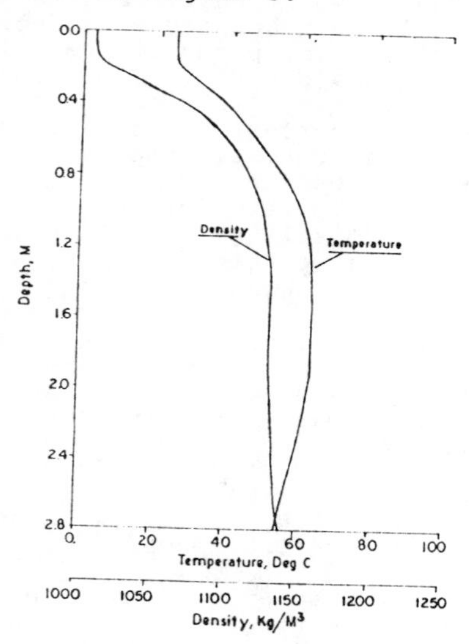

Figure 3. Salt Concentration and Temperature Profiles

Temperatures in the LCZ will typically range between 70° and 100° C. For a 5 meter deep pond, typical depths of the three layers would be: UCZ -.3m; GZ -1.2m and the LCZ - 3.5m. Changing the thickness of the LCZ changes the pond's thermal mass and provides some flexibility in the rate of heat extraction from the pond.

The size of pond required is a function of the energy that the pond will be expected to deliver. The energy delivery capability of a pond depends on the solar radiation received (called insolation), the angle at which the sun's rays are incident on the surface of the pond (a function of the latitude at which the pond is located) and the temperature difference between the pond and the environment [2]. A very rough approximation of the size pond required can be obtained from formula (1) where the variable are defined as:

SM = Area in square meters
TP = Average pond temperature (degrees C)
TA = Average ambient temperature (degrees C)
T = TP - TA
LT = Thermal load (watts)
I = Annual average insolation (watts/m^2)

$$SM = \frac{1.1T + \sqrt{1.21T^2 + LT(.3041 - .83T)}}{(.304I - .83T)}$$

The El Paso solar pond was started in 1983 as a cooperative effort involving the University of Texas at El Paso, The U.S. Bureau of Reclamation, the Texas Energy and Natural Resources Advisory Council (TENRAC) and the El Paso Electric Company. Using an abandoned pond that had originally been for fire protection, a salt-gradient solar pond was constructed. A flexible membrane polymer liner was installed over the existing liner and necessary piping as shown schematically in Figure 4 was installed.

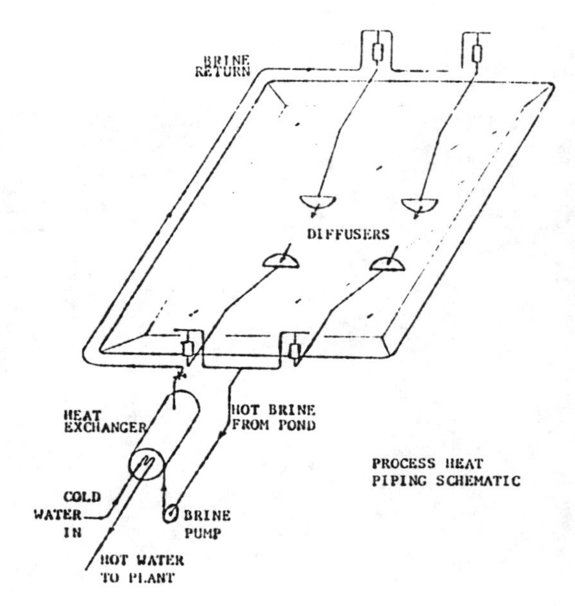

Figure 4 Process heat piping schematic.

By late 1984 the pond had been filled and salt had been dissolved to a uniform 20% concentration. During the summer of 1985 the pond was operated at 72°C (162°F) and heat was extracted to provide process heat to a canning plant at the site of the pond. Thus becoming the first solar pond in the world to provide process heat to an industrial plant.

During the summer of 1986 the pond was operated at 85°C (185°F) providing 332×10^6 BTU's to the food processing plant in the form of preheated boiler feedwater. In late July of 1986 installation of a 100 kW Organic Rankine Cycle engine was begun. Figure 5 shows the concept of electrical power generation from a solar pond while Figure 6 shows the piping schematic that permits operation with options of either process heat or electrical power generation at El Paso pond.

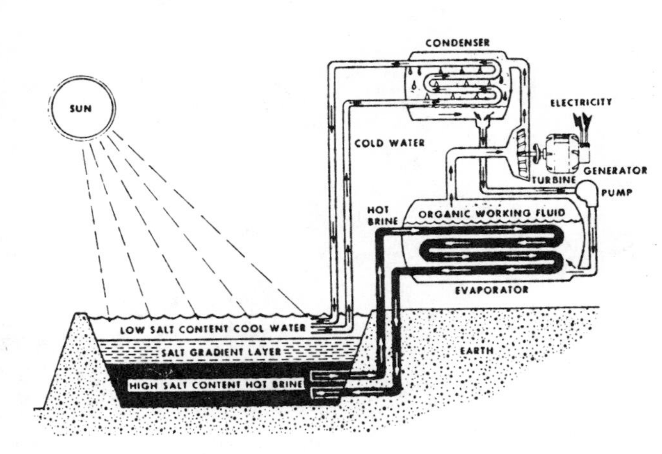

Figure 5 — Solar pond power generation concept.

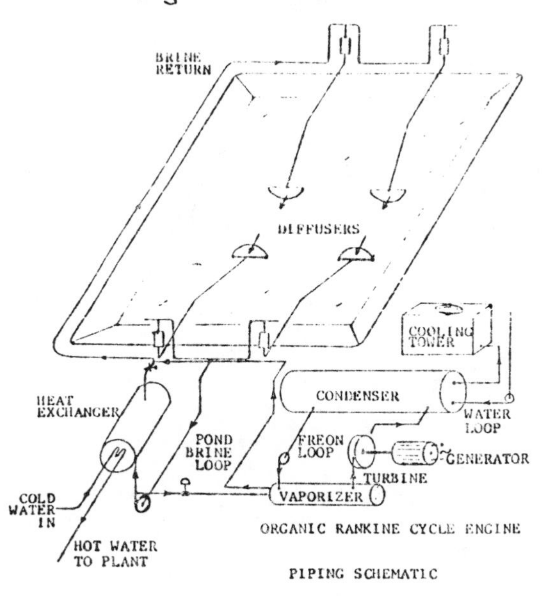

Figure 6: Organic Rankine cycle engine piping schematic.

On September 19, 1986, electricity was first generated by a pond in the United States. Since the pond power system is connected to the electrical grid and is within the property of the food processing plant, the plant receives full value for the electricity generated.

Another first in the U.S. was accomplished in June 1987 when a 24 stage, falling-film, low temperature desaltination unit began operating to produce high quality fresh water from brackish, non-portable water. This solar pond produces the heat and electricity required to efficiently produce potable water at the rate of 4,600 gallons per day. One of the major cost of other inland desaltination is disposal of waste brine, however, when coupled with a solar pond, the waste brine becomes an asset rather than a liability. The waste brine can be used as a source of salt for new ponds.

Installation of desalination unit.

Monitoring high quality fresh water produced by desalination unit.

Solar ponds have a definite roll to play in the area of alternative energy sources. This is particularly true in remote areas of the southwest. The comparatively low cost and the ability to deliver energy on demand are principal advantages of solar ponds over wind turbines and photovaltaic cells. In areas where the demand charge on the electric bill is very high, the electricity generated by a solar pond can be used to show the peak demands and substantially reduce the total charge for commercially supplied electricity. It is clear that solar ponds are technically and economically feasible in the southwest. However, there have not been enough large ponds constructed in the United States to develop optimal design and operating criteria.

References

1. McLean, T. J. and R. L. Reid, "Solar Ponds: A Viable Alternative Energy Source," 1983 Fall Industrial Engineering Conference Proceedings, November 13 - 16, 1983, Institute of Industrial Engineers, Norcross GA, 1983, pp. 136-141.

2. Reid, R. L., T. J. McLean and C-H Lai, "Feasibility Study of a Solar Pond/IPH/Electrical Supply for a Food Canning Plant," Solar Engineering - 1985, Bannerot, R. B., ed., ASME Press, New York, 1985, pp. 263-270.

3. Reid, R. L., A. H.P. Swift, W. J. Boegli, B. A. Castaneda, and V. R. Kane, "Design, Construction, and Initial Operation of a 3355m^2 Solar Pond in El Paso," Solar Engineering - 1986, Ferber, R. R., ed., ASME Press New York, 1986. pp. 304-315.

Biographical Sketch

Dr. Thomas J. McLean is a Professor of Industrial Engineering at the University of Texas at El Paso. He is past director of the Energy Management Division of IIE and is currently active in that division and the Operations Research Division. He is a member of the El Paso Energy Committee and has written numerous articles on energy and energy management.

Dr. Andrew H.P. Swift is an Associate Professor of Mechanical Engineering at the University of Texas at El Paso. He is also project director for the El Paso Solar Pond. He has 10 years experience in wind and solar research. He has authorized and co-authored numerous articles on alternative energy sources.

Chapter 84

AN INSTALLATION-WIDE ENERGY CONSERVATION DEMONSTRATION

L. Windingland

Introduction and Background

Energy consumption and the associated costs are increasing at many Army installations. These increases are often due to new facilities with greater demands for heating, ventilation, and air-conditioning (HVAC) and lighting. In an effort to curb rising consumption/costs the U.S. Army Construction Engineering Research Laboratory (USA-CERL) demonstrated several energy conservation measures on an installation-wide basis. Between Fiscal Year (FY) 83 and FY87, USA-CERL developed and implemented a research demonstration program to investigate energy conservation at Fort McClellan, Alabama.

Over the past 10 years, Fort McClellan has experienced growth in both facility square footage and population served. With this growth came an increase in energy use. In FY76, Fort McClellan had a reported thermal energy use of 99,526 Btu/sq ft and an electrical energy use of 30,623 Btu/sq ft. By FY79, energy use had risen to 99,899 Btu/sq ft and 39,091 Btu/sq ft for thermal and electrical use, respectively. The energy use for FY84 was 71,453 Btu/sq ft thermal and 34,904 Btu/sq ft electrical. It can be seen that a substantial energy conservation effort was already underway between FY79 and FY84. Data for FY86 showed a thermal usage of 61,491 Btu/sq ft and 31,495 Btu/sq ft for electrial. This is a 14.4 percent reduction since 1984. The projects listed below and described in detail throughout this paper contributed to this reduction. The activities described in this paper have potential for application at other installations.

Objective

The objective of this research was to demonstrate and evaluate the effectiveness of applying energy management techniques and energy conservation measures on an installation-wide basis to produce significant and predictable reductions in installation energy use and cost. Two reasons to choose Fort McClellan as a demonstration site are an Energy Engineering Analysis Program (EEAP) study (1979) conducted at Fort McClellan identified areas with significant savings potential. Also, the installation is small enough that energy reduction caused by this demonstration could be readily seen in the total utility bill and the energy cost per British thermal unit (Btu) was close to the national average.

Five major areas of energy conservation were identified and investigated: (1) pressure reduction in district steam heating systems; (2) summer shutdown of central plant heating equipment by utilizing on-site hot water generation; (3) reduction of outdoor air in HVAC systems; (4) replacement of oversized and inefficient motors in HVAC systems; and (5) combustion optimization of gas-fired heating equipment.

Data collected through metering and monitoring were evaluated before deciding whether or not to implement a project. This approach provided an accurate assessment of project viability based on current energy use in comparison to the retrofit designs and applications.

Energy Metering and Baselines

A variety of meters were used to define energy consumption patterns at Fort McClellan. To identify energy conservation opportunities in buildings served by district steam heating, coordinate meters were installed to measure the amount of steam, and hense, the energy, used by each building. The meters provide valuable information on building steam consumption. Of primary concern was correlating energy use before reducing the steam pressure with energy use after pressure reduction.

At Fort McClellan, condensed steam is collected from converters, air heating coils, and cooking and sterilization equipment in a receiving vessel at each building before being pumped back to the central plant. The pump is operated by a float valve within the receiving vessel. A positive displacement, non-resettable, totalizing flowmeter was installed between the receiving vessel and the pump at each building served by district steam. Through the use of flowmeters, condensate (and correspondingly pounds of steam) used by individual buildings was accounted for before being returned to the central plant.

If meter readings are taken on a regular basis, correlations of steam demand and such variables as occupancy and weather conditions can be developed. These correlations can be used to predict energy demand for future periods. Comparisons of predicted and actual steam consumption can also indicate when a building is using more steam than it has in the past, which may indicate problems such as failed traps, steam leaks, or poor energy management procedures.

Perhaps the most important information which can be gained through use of condensate metering is the overall efficiency of the district heating system. Records of fuel consumption at the plant can be compared to the heating energy consumed by all the buildings supplied by the system to determine the cost of supplying energy. This information is useful when making decisions on maintenance and replacement of heating equipment.

Electrical meters were installed on a small scale to evaluate potential energy conservation measures and to verify the actual savings of implemented projects. Short term, portable metering was used for project evaluation in areas such as electrical motor energy, air flow rates, and combustion efficiency measurements.

The Army does not currently have a firm policy regarding metering of energy use in buildings served by central plant heating systems. However, due to the number of such systems in existence throughout the Army, and the amount of information which can be gained, installing metering equipment in future construction as well as in existing buildings should be considered.

Steam Pressure Reduction in the District Heating System

Fort McClellan has four natural gas operated boiler plants that provide for the heating needs of various facilities. Three plants produce steam at 100 pounds per square inch gauge (psig). Plant No. 4 is a HTHW facility operating separately from the others. An analysis was performed to identify the major causes for heat loss in the steam district heating system. On the basis of this analysis, the annual efficiency of the system was determined to be 53 percent. The objective of this project was to reduce system conduction and live steam losses by implementing a reduction in steam pressure at the central boiler plants while maintaining the original heating capacity of the system.

One measure to improve fuel-use efficiency and also reduce operating costs is to reduce the boiler operating pressure where possible. At many Army facilities, the boiler operating pressure is maintained at a constant rate throughout the year regardless of demand. Ideally, the steam temperature/pressure is optimized so that reliability and customer satisfaction are maintained while heat transfer and live steam losses are minimized. The intent is to supply facilities and steam equipment with steam at the lowest pressure that will adequately handle the demand and equipment specifications. Typically, high pressure steam exiting the plant loses pressure in transit due to heat conduction, steam leaks, customer demand, and friction. When steam enters a building, it is throttled to the operating pressure of the equipment. For many applications, including domestic hot water production and building heat, this pressure is usually between 15 to 20 psig. Because this operating pressure is low, it is reasonable to consider supplying steam at a low pressure from the boiler plant. The following technical considerations were used to evaluate this change:

- Ensure that facility and steam equipment demands are met
- Ensure that steam velocities are well within specified standards
- Calculate the pressure drops at the reduced sendout pressure such that all demands are satisfied
- Survey all steam traps and evaluate their performance at the reduced pressure
- Maintain boiler operation within the manufacturers specifications

USA-CERL researchers determined that the pressure in the steam distribution system at Fort McClellan could be safely reduced without loss of steam availability. A reduction in steam pressure from 100 to 70 psig corresponds to a drop in temperature from 337 to 316 °F. This results in a 9 percent reduction in heat transfer losses. Similarly, a reduction in pressure from 100 to 50 psig yields a 14 percent reduction in heat transfer losses. Moreover, a considerable savings is expected where steam is being lost via leaks in valves, pipes, and faulty steam traps. Although these losses are difficult to quantify, the magnitude may be quite significant as many leaks were observed in mechanical rooms and along the main piping. When leaks occur, live steam losses are proportional to the square root of the pressure difference.

$$Q = AC\sqrt{2gh} \quad \text{[Eq 1]}$$

where Q = steam, loss (lbs/hr)
A = area of opening
C = discharge coefficient
g = gravity constant
h = pressure drop

It must be remembered that steam trap capacities are decreased by the system changes. A reduced inlet pressure results in a reduced capacity. A survey of the steam traps throughout the installation was performed to determine whether the traps had sufficient capacity at reduced operating pressure. Of the approximately 125 traps located in Fort McClellan's mechanical rooms, none of the traps were undersized for the reduced inlet pressure.

Each building on the district heating system is equipped with a pressure reducing station which has a known capacity at a given inlet pressure. For a reduced inlet pressure, the steam capacity will also decrease. Maintaining existing maximum steam capacities for each building when the pressure is reduced involves replacing pressure reducing valves (PRVs) with valves which have a larger flow coefficient (Cv). Consider the case where steam pressure is being reduced from 100 to 50 psig. Referring to the American Society of Heating, Refrigeration, and Air Conditioning Engineers (ASHRAE) saturated steam table, the enthalpy (heat content) of dry saturated steam is about 1,190 Btu/lb and 1,180 Btu/lb for 100 psig and 50 psig, respectively. Hence, the Maximum heat delivered to the equipment is 99 percent of the original maximum value although the temperature is approximately 40 °F lower.

To determine the new appropriate Cv, the Cv of the existing valve must be obtained (available from the valve manufacturer). Use Equation 2 to determine the existing steam capacity.

$$Q = 2.1(Cv)\sqrt{P1 - P2}\sqrt{P1 + P2} \quad \text{[Eq. 2]}$$

where Q = steam, flow (lbs/hr, capacity)
P1 = inlet pressure (pounds per square inch area [psia])
P2 = outlet pressure (psia)

Once the maximum steam capacity is determined for existing conditions, the new Cv is found by using Equation 2 with the new operating pressures. As a precautionary measure, the maximum capacities calculated in this manner were compared with coordinate measures at each building.

Predicted values for building steam demand and information gathered through condensate metering enabled USA-CERL researchers to develop an means of computing energy savings. The efficiency of the district heating system is approximately 53 percent. In 1982, the fuel use in boiler plants 1 through 3 was approximately 180,000 MBtu. Assuming a boiler efficiency of 75 percent, 135,000 MBtu enter the steam distribution system in the form of steam. Of this, 47 percent (63,450 MBtu) is lost to conduction between the pipes and the ground and as live steam from leaks and faulty equipment. Natural gas has a unit cost of $10.86 per MBtu. This results in total losses of $689,067 annually.

The savings due to steam pressure reduction were calculated by attributing 75 percent of the losses to conduction and the remainder to leaks and faulty valves. The project cost was $100,000; savings were $66,000, a simple payback of 1.5 years was achieved. It is recommended that other agencies with district heating systems operating at much higher than end-use pressures consider this type of retrofit. In any case, a detailed engineering analysis must be performed before implementation.

Boiler Plant No. 4 Summer Shutdown

This investigation showed the technical feasibility and implementation of shutting down the HTHW system in boiler plant No. 4 for up to 8 months a year. The shutdown coincides with a seasonal shift in heating equipment operation whereby local hot water or steam generators will supply the summer hot water needs of each building on the system. The reason for shutting down plant No. 4 is twofold. First, on-site heating can save energy when compared with large boiler part load performance and second, the same system of concrete trenches which contain the hot water piping also contain the chilled water pipes. The heat transfer to the chilled water lines was causing a significant increase in the load on the central chiller.

Currently, the system is operated continuously all year. During periods when space heating is not required in the buildings being served, domestic hot water and kitchen steam loads are met with the central system. The alternative operating scheme entailed using new heat generating equipment and laying up the central system equipment for these periods. This layup will impose conditions on the system which had not been experienced on a routine basis.

The investigation identified no significant technical constraint which would limit the ability to shutdown the HTHW system at Fort McClellan annually. Typically, district heating systems operate with an efficiency between 50 and 75 percent (Btu used by the customer/Btu entering the pipes less the return Btu). This energy represents the savings associated with use of the on-site boilers. Hence, for the months of interest, approximately 37,500 MBtu enter the system and assuming a district efficiency of 75 percent results in losses of 9,375 MBtu annually. A cost of $10.86 per MBtu results in savings of $101,812.50 annually.

In addition to significant energy and operating cost savings, a regular system shutdown will afford the opportunity for preventive and corrective maintenance. It is noted, however, that a potential increase in the number of system leaks may occur as a result of the thermal cycles on joints. These are expected to be relatively minor, and the additional maintenance required for repair will be more than offset by the savings of not operating the system during the extended shutdown period. Finally, the heat transfer from the HTHW piping to the chilled water piping will be alleviated.

Reduction in Outdoor Air Ventilation

The purpose of this project was to conserve energy by measuring and reducing the rate of forced outdoor air ventilation in large air handling systems. This information applies to all Army facilities with air handling systems that introduce outdoor air into the building environment for ventilation and/or free cooling.

Most air handlers are designed to introduce some outdoor air into the building for ventilation. While this air is required for health and comfort reasons, it can also cause significant energy expense during the heating or cooling season by contributing to the load. Since minimum ventilation requirements have decreased over the years, it is likely that many air handlers are introducing more outdoor air than necessary. Therefore, it is imperative from the standpoint of energy conservation that air be introduced at the minimum requirement established by the ASHRAE Standard 62-1981.

The researchers performed a survey on all air handling equipment at the installation to determine if outdoor air reduction was feasible. The survey identified the size and type of system (e.g., draw-through, multi-zone, variable air volume [VAV]) and suitable locations for temperature flow measurements. On the basis of this survey, 249 air handlers were selected for additional analysis.

For each unit, the number of occupants served by the system; temperature and flow rate of supply, return, and outdoor air; temperature rise across the fan; pressure rise across the fan; and the electrical characteristics of the fan including power, power factor, voltage, and current were determined. Measurements were cross-checked and the data were organized for additional conservation work. Of the 249 identified air handlers, 28 were introducing more outdoor air than required. Damper adjustments resulted in a reduction of 17,835 cubic feet per minute (cfm), or about 11 percent of the total outdoor air intake of the air handlers.

The annual savings for the heating season can be estimated by the following expression:

$$\text{Heating Savings} = k(R)(HDD)(C) \qquad \text{[Eq 3]}$$

where R = reduction in outdoor air ventilation rate (17,835 cfm)
HDD = heating degree days (2551)
C = delivered fuel cost per MBtu ($10.86)
k = a constant (2.592 E-05)

On this basis, the project will save an estimated $12,807 in annual heating costs. Note that this does not account for efficiency of the steam or HTHW distribution system or efficiency of local boilers if used.

For the cooling season, the calculations are less straightforward. However, using a method similar to Equation 3, the potential savings can be estimated by the following expression:

$$\text{Cooling Savings} = k(R)(CDD)(C)/(COP) \quad \text{[Eq 4]}$$

where
- R = reduction in outdoor air ventilation rate (17,835 cfm)
- CDD = cooling degree days to a base of 80 °F (600)
- C = electrical cost per kWh ($0.05)
- COP = coefficient of performance (1.5)
- k = a constant (7.60 E-03)

Using Equation 4, the potential savings for the cooling season are estimated to be $2,711. The total annual heating and cooling savings are $15,235.

Since the total project cost was $29,000, the simple payback period is less than 2 years. Outdoor air reduction to conserve energy at this installation was first suggested in a 1979 EEAP study. Based on their estimates of outdoor air flow rates, and adjustment in accordance with ASHRAE Standard 62-1973 (an earlier edition of the standard used in this study), the firm suggested a simple payback period of about 1 month. Since the current standard allows less outdoor air, thereby saving additional energy at virtually the same cost, one would expect a shorter payback period using the more recent standards. While the actual 2 year payback is reasonable, it does point out the value of accurate field measurements as opposed to blanket estimates.

Replacement of Inefficient Electric Motors

The electric motors that power fans in heating, ventilating, and air conditioning (HVAC) systems are a significant source of electrical power consumption. Proper matching of load and motor power is important, because an oversized motor consumes significantly more power than a motor which is properly sized for its application.

Of 201 motors surveyed at Fort McClellan, 68 were sufficiently oversized to warrant replacement. Some of the motors had operating efficiencies as low as 50 percent. Properly sized integral horsepower motors manufactured today operate at efficiencies of 75 to 95 percent, depending on the design and size.

Institute of Electrical and Electronic Engineering (IEEE) Standard 112-1984 specifies several methods for determining motor efficiency. These methods however, are intended for laboratory testing. They are much too involved, if not impossible, for field implementation. To facilitate the measurements, USA-CERL developed a method of approximating motor efficiency using an optical tachometer. A small strip of reflective tape is placed on the motor shaft, and the motor is placed in operation. The tachometer is aimed at the reflective tape, and the rotative speed is read on the instrument. The formulas developed below approximate the load and efficiency of the motor using this simple speed measurement and the motor nameplate data.

Because the majority of motors used in HVAC applications are of the NEMA type B design, equations were developed for that design only. Similar equations for other motor designs could be developed using the same principles.

Development of an equation for motor efficiency first requires a discussion of fractional load (FL) and its relationship to motor speed. This relationship, which is nearly linear except during startup, involves three quantities: the revolutions per minute synchronous (rpm_{sync}), rpm operating speed (rpm_{meas}), and rpm nameplate (rpm_{np}). Rpm_{sync} is a function of line frequency and motor design (number of poles); rpm_{meas} is the operating speed of the motor, which should be measured when the motor is under the maximum system load; rpm_{np} is the rpm stamped on the motor's nameplate and indicates the speed at which it operates when delivering its rated horsepower (hp_{np}). Fractional load (in percent) can be calculated using these parameters as follows:

$$FL(\%) = 100(rpm_{sync} - rpm_{meas})/(rpm_{sync} - rpm_{np}) \quad \text{[Eq 5]}$$

OR

$$FL = \frac{Bhp}{hp_{np}}$$

Kilowatt measurements indicated that 104 of the 201 motors evaluated were underloaded enough to yield a payback of 5 years or less. The rpm measurements confirmed that 78 of those 104 warranted replacement. Ten motors were not replaced because of restrictions on size, cost, enclosure type, or plant design. Replacement of 68 motors resulted in a 17 percent decrease in electrical power demand.

Combustion Optimization of Gas-fired Heating Equipment

Because of their relatively small size (50,000 to 250,000 Btu/hr), forced-air furnaces and hot water heaters of the type normally installed in barracks, mess halls, offices, and family housing units are often overlooked as candidates for installation-wide energy conservation measures. However, when the number of these heating units in operation at any installation is considered, their combined energy consumption is significant. At Fort McClellan, the combined heating output for all residential-size heating equipment is of the same magnitude as that produced by a central boiler plant.

Since the rapid increase in fuel prices experienced in the mid-1970's, numerous add-on devices and modifications for residential gas-fired heating equipment have been marketed as energy-saving measures. These devices include flue gas heat extractors, recirculators, and automatic vent dampers. While the concepts involved in the design of such devices are valid, very few operate effectively in the field. In a study of 20 such devices performed by the New York State Energy Research and Development Authority, it was determined that the most reliable and effective method of reducing fuel consumption in residential heating equipment is furnace derating. Furnace derating involves installing flue gas and vent dampers and adjusting the firing rate of the appliance. The effect is to reduce the output of the heating device. A reduced output is allowable because most furnaces are oversized for their particular applications. While typical efficiency improvements with furnace derating are approximately 8 percent, major modifications to existing equipment would have been too costly to perform in this demonstration. However, projects to adjust the firing rate, or optimize combustion were inexpensive and had a rapid payback.

Previously, the only way to optimize combustion in heating equipment of this type was by sight. Although it is possible to set the proper combustion rate by observing the color, intensity, and other characteristics of the flame, this method provides no information about the actual operating efficiency either before or after the adjustment. Orsat meters have been used to measure concentrations of oxygen and carbon monoxide in the flue gas, but this method is very time consuming and requires significant training and experience.

Recently however, several electronic micro-based devices have been marketed which allow rapid measurement of flue gas composition. These devices are portable and can be used in the field with a minimum of training. Costs range from $1,000 to $2,000 (1986 dollars), depending on the model and number of options present. Measurements of flue gas temperature, oxygen and carbon monoxide concentrations, and efficiency are taken either continuously or as a sample, and are displayed in the proper units on a digital display. Using this instrument, heating equipment can be adjusted, and the energy savings quantified, in less than 30 min per unit. This is a considerable improvement over methods which were used in the past and makes combustion optimization a very cost-effective, rapid payback project.

Combustion in any heating device occurs when fuel and air are ignited. The most efficient combustion occurs when the correct proportion of fuel and air are burned. The combustion products consist only of water vapor and carbon dioxide. If there is not enough oxygen present, the combustion process is incomplete. Some of the fuel does not burn and is lost through the exhaust gases. Some of the fuel that does burn forms carbon monoxide rather than carbon dioxide. Incomplete combustion is not only less efficient, but also presents a danger to the environment and possibly to the occupants of the heated space.

Because of the inefficiency of the combustion process, all heating devices require some excess air and carbon monoxide will always be present in the flue gas.

Most furnaces and water heaters contain a sleeve or disk that is used to adjust the amount of air introduced into the combustion process. It is relatively easy to fabricate an adjustment device for equipment that does not contain one.

An additional source of energy loss in residential heaters is air leakage after the combustion process. Because of the high temperature of the exhaust products, a stack effect is created and the exhaust rises into the flue and out to the atmosphere. This movement creates a partial vacuum in the heater, which causes additional air to be drawn into it. If the heater is located in the heated space of a building, the effect is to draw in heated air and eject it to the outdoors.

Although oxygen in the air is the driving force of the combustion process, flue gas carbon monoxide concentration is a more reliable predictor of combustion efficiency. This is because, as stated above, some excess air will always leak into the system and appear in the flue gas. Figure 1 shows a curve of typical carbon monoxide and oxygen measurements taken as a function of the position of the adjustable air gate in a furnace. As the air gate is closed off, the oxygen concentration tends to decrease and approach a bottom limit where the air leakage after the combustion process becomes the dominant source of oxygen in the flue gas. Measuring the oxygen concentration at this point provides little information on the efficiency of combustion.

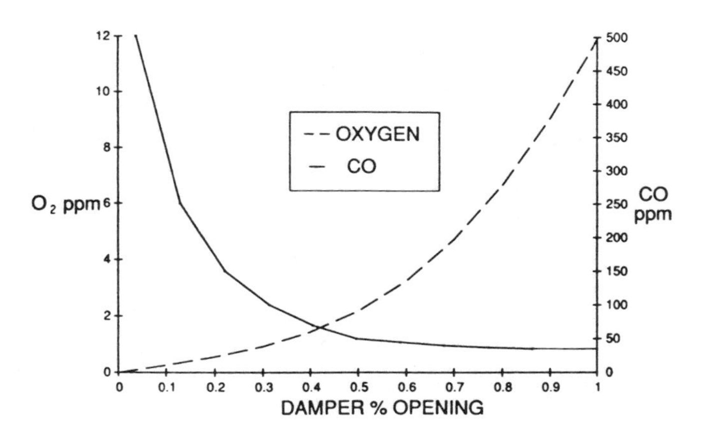

Figure 1. Oxygen and carbon monoxide vs damper position.

On the other hand, air leakage after the combustion process has very little effect on carbon monoxide concentrations in the flue gas. Figure 1 also shows that up to a certain point, the carbon monoxide concentration increases only slightly as the air damper is closed. This is due mostly to a decrease in dilution by the other combustion gases. However, as the air damper is closed further, the concentration of carbon monoxide begins to rise very rapidly because there is insufficient oxygen present in the combustion process. At this point, oxygen concentrations in the flue gas may still be quite high, but are the result of air leakage after the flame. Therefore, measuring carbon monoxide is more important than measuring oxygen to determine the combustion efficiency. Optimum combustion occurs just before the point at which carbon monoxide concentration begins to increase rapidly. Generally, a carbon monoxide concentration of 200 to 250 parts per million (ppm) indicates optimal combustion in the heating equipment normally encountered at military and civilian installations.

To demonstrate the effectiveness of combustion adjustment, 204 furnaces and hot water heaters at Fort McClellan were evaluated. These units had a combined heating capacity of 88,952,000 Btu/hr. The combustion efficiency of more than half of the units was improved, with efficiency increases of as much as 25 percent in some cases. The average improvement for all the units was 2.5 percent. Based on the average heating season and fuel costs for Fort McClellan, this project is estimated to have saved $16,000 with an investment of $17,000 in labor and equipment costs, hence a simple payback of 1.1 years was achieved.

Summary

The installation wide energy conservation demonstration at Fort McClellan contributed to the 14 percent reduction in baseline energy consumption since 1984. Throughout this demonstration, a consistent policy regarding metering and feasibility analysis was employed before implementing a project. Hence, a number of proposed projects with a modest return on investment were rejected. Generally the implemented projects have wide applicability to other Army installations.

Chapter 85

INNOVATION PRESERVES PROJECTIONS

D. G. Bergoust

INTRODUCTION

The winner of a 1986 National Department of Energy Award for Energy Innovation, St. Patrick Hospital, Missoula, Montana, has an energy efficient heating, ventilating, air conditioning (HVAC) and plumbing system for its new 240,000 sq. ft. hospital occupied in August of 1984. It is safe, quiet, cool, humidified and filtered. Heat is cascaded and recycled, leading to an energy intensive environment that is, first and foremost, comfortable.

With the exception of the food service area and some transfer fans, 85% of the hospital's space is served from one primary air handling supply system (see Figure 1).

The greatest advantage of a single air supply system is that the minimum outside ventilation air, which must be brought into the building to replace the air removed by the many exhaust air systems, can be specially treated and injected into the building at one point. Rather than becoming a liability in the cold northern climate, the outside air is used as the principal means for cooling the building during the spring, winter, and fall.

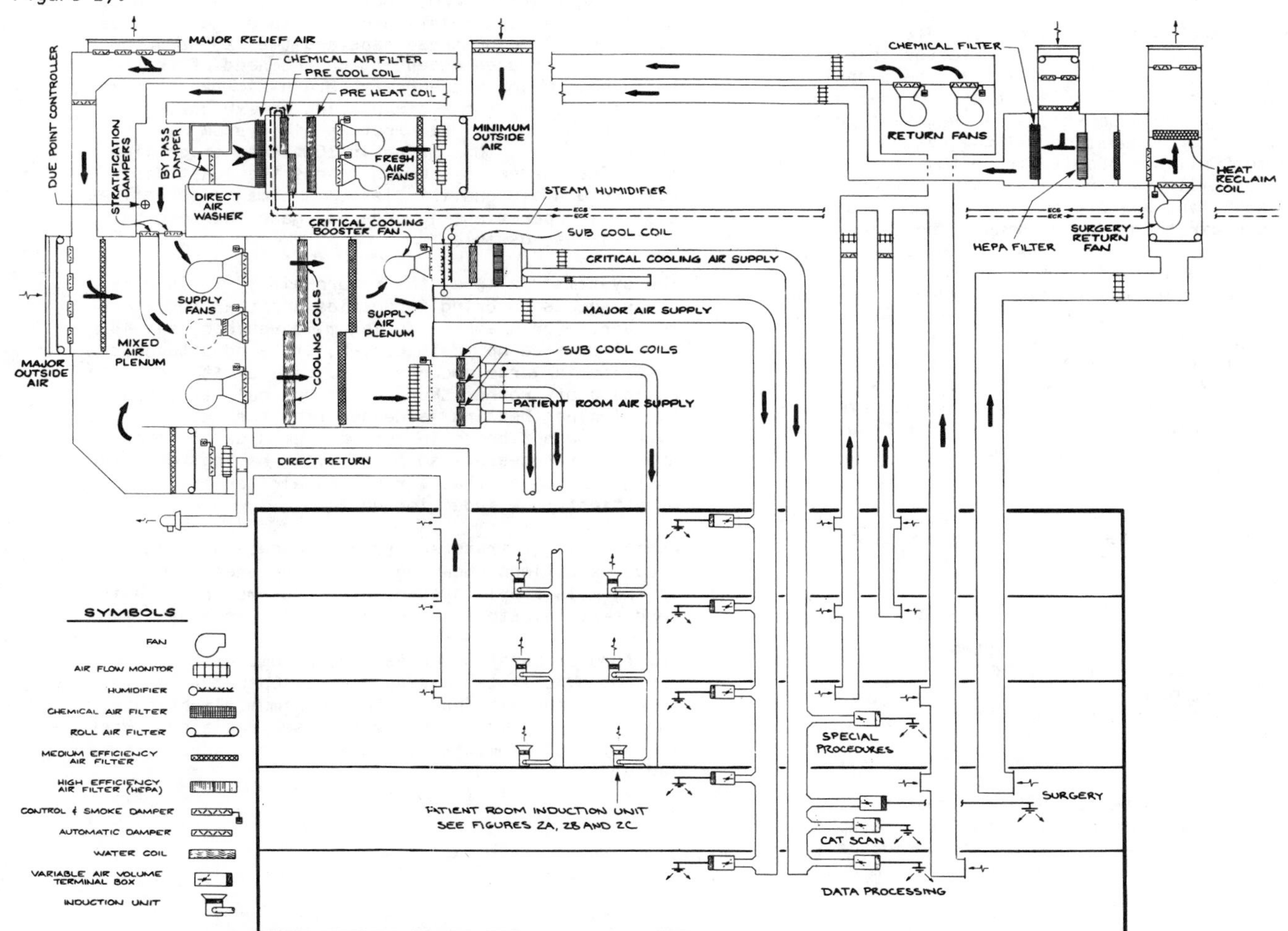

FIGURE 1. AIR SYSTEM SCHEMATIC

Conversely, with several air handling systems, each bringing in outdoor air to replace the exhaust air in that particular area, the ventilation air is usually more of a liability since it must be preheated more. A simple example is the relative amount of exhaust air that's removed from the kitchen. Rather than provide a separate air handling system that provides heated outside "makeup" air to the kitchen, the air is first introduced into the main hospital system and then filtered, "recooled", and transferred to the kitchen by transfer air units which "condition" the kitchen and offer the heat gained as "reclaimed".

Redundancy is built into the single air system by doubling up on fans (two fans in parallel), a booster fan in series, and cooling coils in series (for redundancy or subcooling), each served by separate components of the air conditioning plant.

If a supply fan is lost, a belt breaks or something, the booster fan (see Figure 1) will still deliver the amount of air needed for "critical cooling" even with only one main supply fan operating. "Critical cooling" areas are rooms such as surgery, catherization lab, data processing and the CT scanner. Furthermore, the remaining supply fan will not have to compete with the "sister" fan and instead of delivering half the air it will deliver about 70%.

A computer based multiform building automation system is provided for operating controls, alarm, and monitoring.

A smoke control system must be provided for such a central system whereupon, if a detector is activated in any zone, the computer will take the signal and go through a program to place the zone that's on fire (where there's smoke), under negative pressure by relieving or exhausting all the return air and cutting off all the supply air to that zone. Adjacent zones will have the return air shut off. All the exhaust fans operate and the return fans expel the return air from the building. The primary air system will operate with 100% outdoor air supply and the minimum outdoor air fresh air fan system is the means whereby sufficient heat is added at the preheat coil for the building during cold weather in order to permit the 100% outdoor ventilation air supply. If it is extremely cold, the variable air volume supply fans modulate down in air capacity so as not to freeze any water coils. Smoke detectors in each of the principal supply ducts stop all the fans in case smoke is being pumped into the building from the outdoors.

Several individual innovations can be seen in this integrated HVAC system. They are explored in greater detail below as the following are discussed:

- Heating Cooling Plant
- Supply and Return Fan System
- Heat Reclaim Chillers (HRC's)
- Air Terminal Units
- Patient Room Induction Units
- Fresh Air Fan System
- Critical Cooling System

These concepts can be applied in almost any climate with the possible exception of two stage evaporative cooling. Even that has application during many hours of operation in some of the more humid parts of the world.

HEATING COOLING PLANT

The air conditioning plant has a "conventional" 465 ton centrifugal chiller and "open" outdoor tower. There are two heat reclaim chillers (HRC's); a 100 ton reciprocating and a 150 ton screw type served by two "closed circuit" towers (with one tower inside for year around operation).

The heating plant, with a total capacity of 18 million BTU's per hour (BTUH), includes three hot water heating boilers and two steam boilers. One steam boiler serves as standby for the other (or any hot water boiler by incorporating a steam-to-water heat exchanger). The multiple hot water boiler approach allows each boiler to be sequenced on as demand requires, allowing them to operate at near peak capacities and efficiencies. Less overall capacity is installed because a smaller proportion of the total installed capacity is redundant for maintenance or repair shut downs.

The plant has space for another 465 ton chiller and tower and two more boilers for an additional 7 million BTUH heating capacity.

SUPPLY AND RETURN FAN SYSTEMS

The air system consists of two variable air volume supply fans (with provisions for a third) having a combined capacity of 139,000 CFM and is a variable volume type monitored with fixed, in-place air measuring stations whose output signals are used to maintain pressure relationships. Inlet guide vanes vary the volume on these fans and the 125 horsepower on each is reduced depending on the need of the building. Major and surgery return air fan systems provide a means for recirculating or exhausting air from the building. Each return air system may relieve air and thus bring in 100% outdoor cooling air. The surgery system which has the least desirable return air quality "leads" the major economizer system in operation, in order to make the hospital as fresh as possible.

The systems "relieve" the return air if it is more advantageous to bring in outside air than to return the air. For example, during mild weather, say 45 deg. F outside up to 75 deg. F, it is advantageous to bring in large amounts of outside air to cool the building off rather than run mechanical refrigeration (unless heating is needed from the HRC's). The system doesn't bring in any more outside air than absolutely necessary so as to avoid having to expend energy to reheat that air at the air terminal units or patient room induction units.

Additional air transfer systems cascade air and reduce additional heating or cooling energy needed in the main supply system. The systems are used in food service, storage, toilet and locker areas.

The food preparation areas have a completely independent HVAC system having several operating stages to correspond with unoccupied, minimum, light-duty, and heavy-duty modes in the food service area. Heat is reclaimed from kitchen exhaust air (except grease laden air).

The main supply fans and the sub-cooling coils mentioned above are so arranged that warm returning air and cooler outside air streams may be stratified and directed to the sides of the air plenum which need their respective temperature levels, thus reducing the amount of energy needed to heat or cool them and reach their required leaving temperatures. The three categories of air streams are as follows, with the "warmer" streams mentioned first: Patient floors (with windows), ancillary and support floors, and critical cooling spaces. The system creates a 6 deg. F difference from one side of the plenum to the other.

At the critical cooling side there is cold air. At the other extreme on the patient room side, warmer air is needed (most of the time) to offset the heat loss through the windows. The warmer air from the center of the building, extracted from the lights, is used to hug that side of the plenum, go through that supply fan and, by being 6 deg. F warmer, less energy is needed to heat the air at the patient room.

A refinement in the stratification control system can either direct the minimum fresh air along the cold side of the plenum or, by means of these stratification dampers, direct it to the center of the plenum.

HEAT RECLAIM CHILLER(S) (HRC's)

The heat reclaim chillers primary goal was to provide redundant (at least partial) cooling for patients and critical areas (surgery, etc.). This postponed the installation of a second 465 ton chiller. Also, at times these areas require colder air than the rest of the hospital due to sun shine, equipment, codes, etc., so the recool coils allow the outdoor air economizer systems and/or 465 ton chiller to give as warm air as possible to the majority of the hospital.

The available HRCs' condenser heat is always used in the building's HVAC and potable hot water systems for heating. During extreme cold weather or anytime additional heat is needed, the HRC's are augmented by the boilers. If the chillers' condenser heat is not needed for the HVAC system, it is used to preheat all the potable hot water from 49 deg. F to 105 deg. F.

Domestic water heat is considerable in a hospital. It happens 365 days a year. Building heat is needed mostly at night while domestic water is a daytime need, thereby leveling out the capacity of the HRC's. A 2000 gallon storage tank is used as a flywheel to preheat the incoming potable cold water (at 49 deg. F) to 105 deg. F using the heat off of a separate condenser on each heat reclaim chiller. Finally, if it is not needed by either system (HVAC or potable), the condenser heat is rejected at the cooling towers.

The HRC's extract heat energy from the internal loads of the HVAC system (and the exhaust air during winter months) and "cascades" or raises the temperature of the energy for use at the "minimum" outdoor air, zone control, perimeter heating coils, and the potable hot water system. The HRC's can make 3.5 million BTUH for the HVAC system as they are used to sub-cool air going to critical cooling areas, patient rooms on the sunny expósure, kitchen, elevator machine room, and control room.

The HRC's not only cool the air in those critical cooling ducts but they may also cool off all the exhaust air (only in the middle of winter) when the HRC's don't need to cool so much air for surgery or patient rooms but they do have to provide heat to the building. See "Heat Reclaim Coils" on Figure 1. Exhaust air is cooled only if the building needs the waste heat and the building subcool coils are not fully loaded.

The condenser heating water always goes through the indoor closed cooling tower so as to provide close head pressure control. However, the computer controls all the towers and keeps them off if the building is accepting the heat. If the building is not accepting the heat, provided we need air conditioning, the cooling towers will operate. The computer is programmed to continuously monitor and be sure that there is no artificial loading of the HRC's while rejecting the heat in the cooling towers. (The system does not cool off the exhaust air and run cooling towers at the same time.)

To keep the emergency generators from getting any larger, only the 150 ton HRC and the indoor tower (and associated pumps) were connected. Furthermore, it is possible to limit the input power requirement of the 150 ton screw machine, further limiting the maximum load on the generators.

AIR TERMINAL UNITS

The air system always delivers air cooler than the space temperature requires to each of the terminal units. The temperature of the air depends on the outdoor conditions and the mass of the building. Room thermostats will apply some terminal heat to warm the space if necessary. In the variable volume box, the volume of the air reduces before turning on the heating coil. The volume of cool air is reduced before turning on the heat. The computer will determine the temperature of the primary air going to each of the devices and also the temperature of the hot water to be supplied to the terminal heating coils.

PATIENT ROOM INDUCTION UNITS

The patient rooms are served from the supply system by means of hospital style induction units which only reheat primary air (see Figure 2).

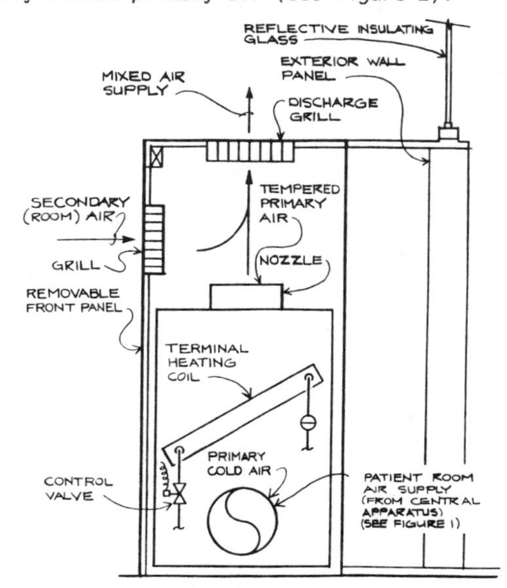

FIGURE 2. HOSPITAL PATIENT ROOM INDUCTION HVAC UNIT

On sunny days solar compensated sensors for each exposure of the patient tower direct the recool coils on the respective exposure (see Figure 1) to subcool the primary cold (supply) air to as cold as 38 deg. F. (see Figure 2C). The passive solar heat is captured in this manner via the HRC's.

Solar compensated sensors are used to determined the three primary cold air temperatures going to patient rooms (see Figure 1) as a function of how much sun is coming in the windows of the respective exposure as well as the outdoor temperature. Figures 2A, 2B, & 2C show three cases from cold outdoor air to hot with an infinite number of combinations in between.

The patient room air supply doesn't need a booster fan because the patient rooms are very close to the penthouse supply fans. Subcool coils offer redundancy by cooling that air down to 38 deg. F and sending it to the induction units. Using the colder than normal air results in tremendous savings in sheet metal duct and space for same.

The air is not discharged into the room at 38 deg. F. It is first mixed in a negative pressure chamber of the induction unit with some return air from the room (induced air) so that the mixed air supply temperature is raised from 38 deg. F to at least 55 deg. F.

When maximum heating is needed (see Figure 2A), the primary cold air is not recooled via the subcool coils,resulting in that air being 68 deg. F (the warm side of the plenum). If the individual room needs more heat, the room thermostat will blend some heating water into the terminal heating coil of the induction unit.

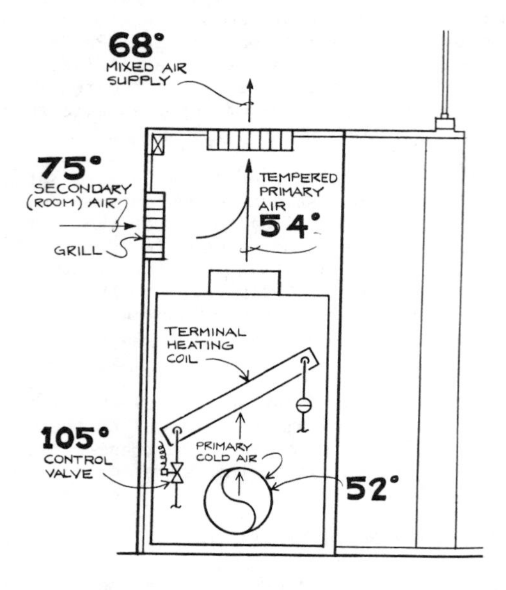

FIGURE 2B. +40 ABOVE "O" WITH SUNSHINE.

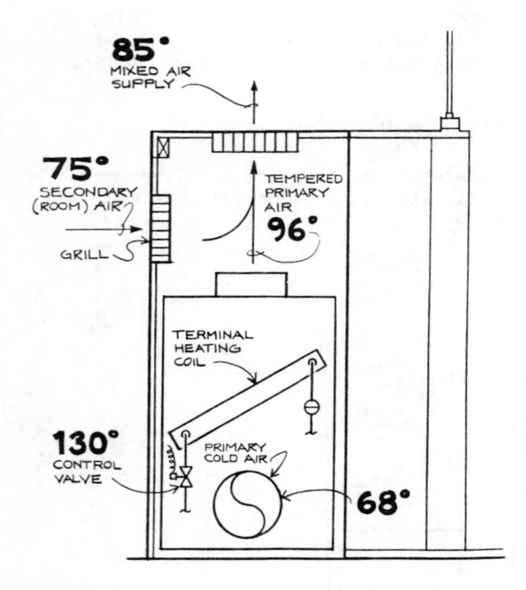

FIGURE 2A. -20 BELOW. NO SUNSHINE.

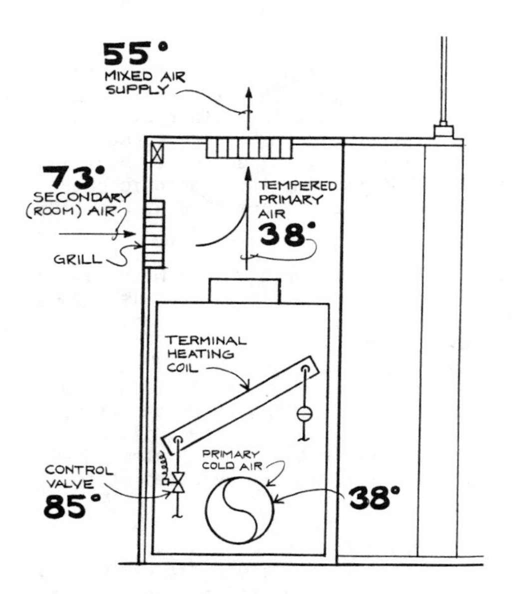

FIGURE 2C. +95 ABOVE "O" WITH SUNSHINE.

FRESH AIR FAN SYSTEM

A separate fresh air fan system (see Figure 1) allows the use of extra resistance items in the 60,000 cfm minimum outside air system without unnecessarily increasing the fan horsepower. The added resistance of a preheat coil, indirect evaporative pre-cool coil, chemical air filter, and direct air washer do not affect the power consumption of the supply fans.

The minimum outdoor ventilation air louver, as well as the major and minor outdoor air economy cooling louvers, are located on the side of the building away from the boiler plant, helicopter landing pad, kitchen exhaust, and emergency generators which are all sources of undesirable odors that must not be allowed to enter directly into the outside air intakes. The chemical filters are primarily used to capture the city's wintertime atmospheric air pollution.

The fresh air fan system capacity is matched to the exhaust fan system variable capacity (see Figure 3) and maintains a slight positive pressure in the building (with respect to atmospheric pressure) to avoid air infiltration to the building. Two stage evaporative water cooling is staged so as to save approximately 160 tons of refrigeration during peak summer conditions. This is done by first passing the fresh air through a precool coil served by cool water from an "open" air washer (indirect cooling) (see Figure 3) and then passing the air through a second open air washer (direct evaporative cooling) (see Figure 1). Reclaimed energy from the HRC's is used to preheat the incoming fresh air up to 32 deg. F (only as necessary) during bitter cold weather.

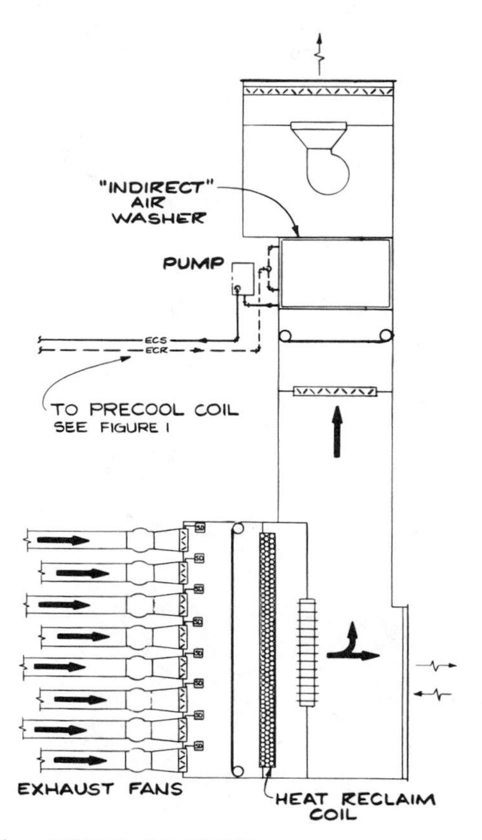

FIGURE 3. EXHAUST FAN SYSTEM

The exhaust system is variable depending on how many fume hoods are on in the lab, whether the toilet exhaust fan is on high speed or low speed and whether it is day or night. As the exhaust volume changes, the volume changes in the fresh air system as well which is always putting in the same amount of minimum outdoor air to equal the amount that is being exhausted from the building.

With a variable volume system it is important not to have doors standing open or have infiltration coming in cracks or through the vestibules. What air is exhausted is brought in. There are either air flow monitors that measure the amount of exhaust air (see Figure 3) or, in the case of kitchen "grease hood" exhaust fan "who energy is not reclaimed", an "on" signal tells the computer to add a predetermined air quantity to the fresh air fan capacity. As these fresh air fans vary in volume, their power consumption varies. The preheat coil gets most of its energy from the heat reclaim coil at the exhaust fans (see Figure 3). The HRC's take that exhaust energy and send it up to the preheat coil (see Figure 1).

In the summer the pre-cool coil cools the minimum outside air from 92 deg. F to 70 deg. F with no mechanical refrigeration. It employs evaporative cooling by using 63 deg. F water mode at the "indirect" air washer (see Figure 3) which is pumped to (and used by) the pre-cool coil to cool (without adding moisture to) the air. Then the system adds moisture in the direct air washer as the air is cooled from 70 deg. F to 53 deg. F and humidified (allowing the steam humidifier to be off in mild or hot weather).

The chemical filters are put in place in the wintertime. This "minimum" air must always be brought into the clean health environment. By removing the filters in the summertime, the fans don't have to overcome their air resistance, thereby saving energy since the system automatically compensates and measures the current air quantity whether the filters are in place or not.

The preheat coil uses an ethylene glycol mixture heated via a plate and frame heat exchanger connected to the heating hot water system.

CRITICAL COOLING SYSTEM

Fan redundancy and energy conservation are built into the "critical cooling" system serving the surgery and critical areas by means of a "booster" fan (see Figure 1) which overcomes the added resistance of the subcool coil, HEPA filter and steam humidifiers for these areas at the same time saving considerable power on the two supply fans. The booster is able to continue serving these spaces with full air volume in the event one of the main supply fans is disabled.

By investing 20 hp in a booster fan for just that component of the total air (approximately 20% of the total supply), we are able to save 50 hp apiece in each of the two supply fans and still keep the critical cooling system part of the primary system. If we lose the booster fan, there are doors that open up in the air plenums that will allow supply air through the critical cooling duct (by means of the supply fans) after reducing some air volumes to nonessential areas of the hospital and possibly even curtailing one or two of the surgeries.

Supply air humidification is provided, in the summer, by the direct evaporative cooling of the "minimum" outside air. For close control of humidity in the critical areas during cold weather (when the evaporative coolers are drained), a dual set of "dry" steam humidifiers is provided (in the main critical supply air plenum only).

To avoid unnecessary reheat in the remainder of the hospital the subcool coil is necessary so the HVAC system doesn't always make all the supply air a cold 50 deg. F. The hospital only needs 50 deg. F cold air supplied to surgery where it's often necessary to maintain a 68 deg. F room temperature.

SURGERY RETURN AIR

By putting the surgery rooms on a separate return fan the fears of returning air from same are lessened. Fears such as future code changes that may require exhausting same. If that ever happens, or, at the discretion of the building operator, only a software change is necessary here and not expensive duct revisions. Meanwhile the system saves on first cost and energy costs by saving the energy to heat or cool make up air if the surgery air were all exhausted during very hot or cold weather.

The surgery return air volume is reduced by 75% as it tracks the reduced supply air volume to surgery during the programmed "unoccupied" times.

Air returning from surgery areas is safely recirculated by filtering it with high efficiency particular air (HEPA) filters. Most codes permit return of surgery air; they do not require this type of filtering. This is an added safe guard.

A heat reclaim coil was added in the surgery relief air to lessen even further the energy losses whatever the reason for relieving the surgery air.

ECONOMICS

Besides insulating the building very well and providing reflective insulating glass, the architects, Mills, John & Rigdon of Seattle, WA, directed Bergoust Engineers & Company, Inc., of Missoula, MT, to design as efficient a mechanical/electrical system as possible. However, there were no additional moneys set aside for energy conservation equipment. The engineers were told to make the systems as efficient as possible within an ordinary budget for hospital work. A budget was established for the mechanical/electrical construction cost and that was not exceeded. All bids were within 2% of the budget and ran approximately $12,000,000 for general, $6,000,000 for mechanical and $3,000,000 for electrical in July of 1982. By keeping the size of more mundane work to a minimum the more exotic were made affordable. For example, first costs were saved with smaller duct to the patient room with the ability to supply 38 deg. F subcooled air to the induction units.

The building, primary fan system, and the central HVAC system are designed to accommodate future vertical and horizontal expansion of another 100,000 S.F.

The heat reclaim chillers operate 365 days a year and are needed for backup to the centrifugal chiller anyway. The added costs to achieve a heat reclaim capability were probably paid back in less than a year (an economic study was not commissioned). By adding up the power and gas consumptions, the building has been consuming, on the average, 179,000 BTU per square foot per year.

A prudent approach was taken on sizing the HRC's. If it gets down to 25 below zero, the primary heating pumps will take hot water from the boilers to blend in with the condenser heating water to give a little extra heat to avoid over sized HRC's.

SECTION 20

FULFILLING ENERGY ENGINEERING PROJECTIONS

Chapter 86

FIVE KEY ASPECTS TO ENERGY EFFICIENCY IN INSTITUTIONAL BUILDINGS

P. H. Rose

Because one of my major responsibilities in the Institutional Conservation Programs involves interacting with a variety of professional organizations such as the Association of Energy Engineers, I have a rather broad perspective on the range of interests in the energy efficiency field concerning institutional buildings.

I have also worked in this field for ten years now and that has given me an opportunity to observe the myriad approaches to managing energy in these institutions. In the first evaluation of the ICP matching grants program, conducted in 1981-83, and again in the second which has been completed this year, several salient points become obvious. To summarize the essential findings of these points in a positive way, I have reduced the material to five key aspects of energy efficiency which should be considered when approaching an institutional client.

Before getting to the list of five, one thing that should be emphasized is that there is no single right answer or approach that is best for all institutions. There are as many options for action as there are energy efficiency opportunities in most institutional buildings. Hard as we tried to find a magic "most effective" energy conservation measure, no one measure surfaced which filled that bill.

With that said, the following are my candidates for the five key aspects of energy efficiency.

One: Why do an energy program - what are the motivators?

For one thing, energy has the potential for returning revenue to the institution without cutting services. And it is really the only area of institutional management about which this can be said. Further, wasted energy makes no contribution to the mission of the institution. In fact, a large school or hospital, which may be the largest single energy consumer in a community, can contribute to utility problems in that community by pushing up the demand for power at peak times, thereby increasing the requirement for more capacity on the part of the supplier. This raises rates for all customers and wreaks havoc with long range planning on the part of the utility.

As an aside regarding utilities, frequently enough money can be saved by working with the utilities on rate structure and peak shaving that capital becomes available to invest in energy efficiency measures which require capital beyond that available in the usual facilities budget.

I won't bother to harp on all the patriotic rhetoric about imported oil and our need to reduce that reliance nor the environmental benefits from using less energy. But I'm sure there are institutional managers out there who are bothered by the waste of finite resources, no matter their origins.

Two: Good engineering as a basis for action - what kind of audit should there be?

The use of the term "energy audit" may ultimately be the biggest roadblock to the understanding and implementation of energy efficiency programs on the part of institutional managers. Engineers and institutions have been burned by the careless use of jargon in this field. And every school and hospital association's members have asked the question "What is an energy audit and how do I know when I have a good one?"

There can be legitimate differences in approaches to the energy audit. But these differences should be made clear to the building owner. Otherwise, how would the uninitiated know how to compare the free audit from their local utility to the professional services offered by the consulting engineer?

In any case, certainly "cookie cutter" engineering solutions to energy management in similar institutions should be viewed with a very jaundiced eye. Even if two institutions are housed in identical facilities, the people occupying and operating those buildings are not identical. Too often, in the quest for the engineering answers to energy efficiency questions, the functional and environmental needs of the people get lost. And, after all, the buildings are there to serve the people, not to serve themselves.

To sum up this point, when sound engineering and user considerations are the basis for energy management actions, energy efficiency programs have the best opportunity for success.

Three: Equipment - what are the technical capabilities of the institutional staff?

It serves no useful purpose to install a highly sophisticated energy management system in a small school which has a staff with limited technical capability. The system will need monitoring and fine tuning which will not be accomplished. Anticipated energy savings will not be achieved. And the frustrations of the staff will incite them to override the system. And, perhaps worst of all, depending on your perspective, the engineer who recommended the system and the vendor who supplied it will have done nothing to enhance their reputations.

Again, we get back to the people issue. When you have an understanding of the technical operating potential of the user, then apply the KISS principal, you improve the chances of having a success to your credit.

Four: Financing - what are your choices?

An investigation and thorough evaluation of all of the many financing options is a must for the happiest outcome in any major capital intensive energy efficiency improvement. All energy management programs should be win-win situations. In fact, it is the best indicator of the success of any energy program when the building occupant, the facilities manager, the institutional financial official and, in the case of third party financing, the outside party all feel they have achieved their objectives. If you have saved energy at the expense of environment or occupant comfort, malfunctioning auxiliary systems or less than necessary economic satisfaction, you only give energy programs a bad name. And once this happens it can be fatal to future initiatives, no matter how well planned or how well the lessons were learned from the previous errors.

Briefly, the financing choices include: to do it within the institutional budget, use savings from operation and maintenance changes, take advantage of utility rebates and other related incentives such as savings achieved from utility rate changes, use equipment leasing arrangements, and the performance (shared savings) contracts. In the evolution of this latter option several terminologies have been coined, and each has its own approach. Like anything else, these third party contracts can be as good or bad as the planning that goes into them, but they do offer an otherwise financially stressed institution an opportunity to improve its standing. Naturally, each of these contracts has its own split of the energy savings, and it has been a long time since 50/50 was considered viable from the financiers view. These contracts also can vary considerably on the sharing of risk.

In my experience in this field I have heard these third party contracts described in the extremes as the answer to an institution's prayers and as taking advantage of institutions which could have done more on their own and not shared so much of the savings. An institution's final assessment of their program will depend ultimately on how well they did their homework prior to signing a performance contract. And the need for the services of a consultant, either an engineer or a specialist in these types of contracts, may depend on the expertise available at, and the size and physical complexity of, the institution. As I said, ultimately the judgement of success and satisfaction of the institution will depend on the effort put into the front end of this type of contract. But 20 percent of something is better than 100 percent of nothing. And today, doing nothing about energy management costs you plenty.

Five: Commitment - how do you finish what you start?

For years we have heard that in communist countries the kiss of death for political leaders is to have a popular following based on a "cult of personality." In the case of institutional energy efficiency programs an enthusiastic personality with a dedicated following might be just what is needed to start a program with broad support and to maintain the momentum after the distinct projects have been started. It is not enough to impress building users and administrative types that their environments are more comfortable and first year savings projections have been achieved. In studies conducted over a three year life of installed energy efficiency measures it has been found that energy savings have, in many instances, declined from the first year to the third. Because it is reasonable to assume that properly installed and well maintained equipment would not be expected to deteriorate in this time period, it seems obvious that it was the people operating and calibrating/adjusting the equipment who failed. No matter how technically correct a piece of equipment may be for the energy problem to be solved, it will be of little use if there is no commitment on the part of all concerned to have the program succeed.

It should be well understood that, for the most part, decisionmakers in schools and hospitals are trained in education and healthcare. They have not been trained to be engineers, and most perceive energy issues as requiring more of their time to understand than they are willing to give. To get their commitment to any energy program requires their understanding of what is at stake for their institution in terms that have meaning in their realm of expertise. Before you grow impatient with their hesitancy to implement your recommendations, remember that from their perspective they are venturing into uncharted territory. And to get their commitment, they must believe in yours.

Well, that's my list of key aspects of energy efficiency in the institutional building sector. Each of the aspects relate to people and are not clearly delineated, but share common elements. Perhaps the most important point to remember is that technical solutions to energy efficiency challenges will work best when coupled with people-compatible solutions.

Chapter 87

WHERE ARE MY ENERGY SAVINGS?

R. K. Hoshide, W. F. Rock, Jr.

INTRODUCTION

This paper provides an overview of why institutions are not obtaining their projected energy and energy cost savings from energy conservation measures (ECMs). The appropriateness of some of these projects and various reasons why projections are not being realized are discussed. Actual case histories are included to illustrate difficulties others have experienced. Recommendations on how to avoid or correct these problems are presented. A better understanding on how to successfully participate in energy conservation is presented.

The Energy Technology Engineering Center (ETEC) has been providing technical assistance to the U.S. Department of Energy's Institutional Conservation Program during the past 9 years. ETEC's experience in the technical evaluation of ECMs under the "Schools and Hospitals" program results from having reviewed over 21,500 energy conservation projects valued at almost $400 million. Technical monitoring has also been conducted for various types of ECMs and problem ECMs.

BASIS FOR ENERGY SAVINGS

Fuel savings are usually presented in dollars saved per month or year. People understand savings in dollars but may have difficulty in relating to savings in British thermal units (Btu), kilowatt hours (kWh), therms, etc.

Some institutions may include other energy-related cost savings such as: operation and maintenance (O&M), disposal, replacement parts, insurance, depreciation, etc., in their savings. These costs can impact the overall life-cycle cost and the decision of the institution to go ahead or not to go ahead with the project.

There are economic and other considerations (Tables 1 and 2) that should be looked at prior to proceeding with an ECM.

Table 1. Typical Economic Considerations for ECM Selection

- Institution funds availability
- Availability of other funds (state, utility, and manufacturer's rebates, etc.)
- Simple payback of ECM
- Other life cycle cost criteria
  - Internal rate of return (IRR)
  - Savings to investment ratio (SIR)
  - Return on investment (ROI)
  - Return on assets (ROA)
  - Return on equity (ROE)
- Impact to other costs
  - Operational costs
  - Maintenance costs
  - Disposal costs
- Interaction with other equipment

Table 2. Typical Other Considerations for ECM Selection

- Comfort
- Health, safety, and service
- Reliability of equipment
- Flexibility of equipment
- Warranty of equipment
- Impact on personnel who keep equipment running
- Training requirements of facilities personnel
- Political (external) like radon, asbestos, air quality, etc.
- Equipment also used by school for education
- Energy incentive programs

REASONS FOR DISCREPANCIES

Based on ETEC's experience with the Schools and Hospitals program, many reasons have emerged as to why institutions are not getting their expected energy savings. Some are obvious, while others are not. An important purpose of this paper is to flag out those reasons. Based on ETEC's experience, here are what we consider to be the major reasons why these savings are not being obtained:

1. Changes in building load
2. Changes in utility rate design
3. Inadequate energy consumption monitoring
4. Poor technical assistance report
5. Inadequate operation and maintenance.

Changes in Building Load

The actual building load may have increased since the energy analysis was made. Increases in occupancy and hours of operation, building additions, addition of equipment such as computers, word processors, electric typewriters, etc., all increase energy consumption. The weather for the year may have been unusually hot in the summer and/or cold in the winter. These weather changes from the average would decrease the estimated energy savings.

Changes in Utility Rate Design

During the past several years, rules of the utility rate design have been changing considerably and have become more complex. Many utilities have or are planning to restructure their rate schedules. Terms such as time of use, on-peak and off-peak charges, maximum demand charge, annual demand charge, peak demand charge, summer and winter on-peak, mid-peak and off-peak demand, energy charges, volumetric charge, commodity charge, etc., are becoming the norm rather than the exception.

Because of utility rate structure changes, previous analyses may now be in error. The energy cost savings may have been based on average rates and lower demand charges. The time-of-day energy consumption may now be critical to energy costs. With these rate schedule design changes, savings in fuel costs can be drastically reduced. Many electric utilities use a declining rate schedule where the more electricity you use, the cheaper the rate gets. In conserving electricity, this may put you in a higher rate schedule because of the decreased electrical consumption. The savings in dollars would be less if the cost was based on the original average rate.

Inadequate Energy Consumption Monitoring

Because of changes in utility rate design, it is becoming increasingly important to monitor your energy consumption. One needs to obtain the energy load profile over the 24-hour period throughout the year where time of use and/or demand charges are used. Knowing how energy costs are being generated will assist you in deciding what to do. Although an ECM is more efficient than an existing unit, it may actually increase energy cost. An example of this is where an institution replaced strip heaters with heat pumps. The heat pumps were started at 6:00 a.m. to preheat the building. In doing this, an electrical spike raised the peak demand cost, which was ratcheted for the next 12 months. Steam lines were replaced and insulated, but a leak in the line was not detected for 3 months. A building had its incandescent lights replaced by fluorescent lights installed in an added drop ceiling. With monitoring, the electrical consumption showed an increase. Upon investigation, it was found that the incandescent lights were left on and never disconnected.

Poor Technical Assistance Report

The technical assistance (TA) report should provide the institution with a program for energy management. As such, it should provide an in-depth energy analysis that identifies all cost-effective ECMs. This TA report should describe the building energy use characteristics and O&M changes that result in energy savings. We can see that building load, utility rate structure, and energy consumption monitoring are all part of a good TA report.

Although energy use and saving calculations are straightforward and conform to accepted engineering practice, ECM analysis depends on assumptions of equipment performance and other factors influencing energy use. Because ECM analysis is not a precise science, different TA analysts will produce somewhat different results for the same ECM.

People tend to show higher energy savings and lower simple paybacks for recommended ECMs. This optimistic approach is sometimes used because it may make the selling, approval, and funding of the project much easier. Sometimes, the high savings estimates are just honest mistakes on the part of the TA analyst.

From our experience on reviewing TA reports, many reasons have surfaced as to why the estimated energy savings are not being obtained. Some of these reasons are:

1. Existing equipment consumption overestimated

2. ECM consumption underestimated

3. Existing system not compatible with ECM

4. Inadvisable ECMs installed

5. Interaction not considered

Existing Equipment Consumption Overestimated: By showing that existing equipment is less efficient than it actually is, the savings based on an energy-efficient replacement item will be larger. Many times, the existing equipment O&Ms have been neglected masking the actual performance. Operating time can also be exaggerated to increase energy consumption. One analyst factored in the burnout of incandescent lights in exit signs so that the actual existing usage was not exaggerated.

ECM Consumption Underestimated: There are various reasons why the consumption can be underestimated. Some of the typical examples we have run into are described below.

Lighting ECMs use lamp watts instead of the actual input fixture watts. For incandescents, the lamp wattage is about equal to the input fixture watts. For fluorescent and high-intensity discharge (both high and low pressure) lamps, additional watts are consumed by the ballast, etc. Typical values for high-intensity discharge lights are shown in Table 3. Because of the myriad of fluorescent lamp and ballast types on the market, no comparable table is shown. To be sure, the ballast wattage must be included in the input fixture watts when calculating energy consumption.

Table 3. Typical Lighting Fixture Wattage

Lamp Type	Wattage	Input Fixture Wattage
Mercury	100	118
	175	280
	250	285
	400	450
	1000	1075
Metal halide	175	201 to 210
	250	287 to 292
	400	455 to 470
	1000	1070 to 1090
High pressure sodium	35	43
	50	60
	70	82 to 88
	100	115 to 130
	150	170 to 188
	250	300 to 302
	400	465 to 484
	1000	1090 to 1210
Low pressure sodium	18	30
	35	62
	55	80 to 84
	90	125 to 131
	135	178 to 182
	180	220 to 228

Motor efficiency estimates were overstated. The overestimation may result from obtaining efficiency based on different test methods. The National Electrical Manufacturers Association (NEMA) recommends that IEEE Standard 112, Test Method B be used since it is by far the most accurate. Foreign manufacturers use different test methods and they do not include all of the losses. Experience has shown that the three techniques most widely used by foreign manufacturers* consistently overstate their efficiencies.

Other problems include using a constant value for efficiency regardless of horsepower and using only full-load efficiency and power factor values instead of the actual duty cycle.

Existing System Not Compatible With ECM: Many ECMs that appear worthwhile can end up with less energy savings when the entire system is analyzed. Some examples are given below.

Replacing an inefficient chiller with an energy-efficient unit resulted in decreased chilled water (CHW) requirements. This reduced flow shifted the CHW pump operating point to an off-design value where the pump efficiency was low. This also holds true for air-handling unit fans. Both motor-power factor and motor-load conditions also changed where efficiencies were lowered and energy costs increased.

A variable-speed condenser water pump was installed in parallel with two constant speed pumps. One of the constant speed pumps was shut down and the variable speed pump started when the cooling load decreased. No consideration was given to controlling the lower speed limit of the variable speed pump. Therefore, much of the variable speed pump operation was used to heat the condenser water.

Installing ECMs to existing equipment resulted in actual increase in energy consumption when the existing equipment was in poor working condition. Louver controls and other energy management system controls are ineffective if the louvers or other equipment to be controlled are inoperative. Roof insulation served no purpose once wet because of a leaky roof. Insulating windows did little good because doors were left open. Installing equipment like a new chiller unit is foolhardy if the condenser water loop is fouled by mineral deposits and scale. Equipment proposed must be able to withstand its environment and do its job over its useful life.

Inadvisable ECMs Installed: ECMs, especially those that disrupt the comfort and/or safety of the occupants, end up being disconnected or disabled from doing their job. An outlet damper control system that reduced outside air to uncomfortable and unhealthful conditions was disabled by other means. In one case, it was a 2x4 stuck between the louvers.

Another ECM walled up windows in classrooms to reduce heat loss and gain through the windows. By blocking off the windows, lighting levels dropped and necessitated increasing the lighting level, which in turn increased the air conditioning load. These two side effects resulted in increasing the overall energy consumption.

Interaction Not Considered: ECMs that interact with one another must be analyzed together so that energy savings are not duplicated. If each ECM was considered separately, the estimated energy savings would be larger. ECMs that reduce building or system loads should first be considered. ECMs that reduce distribution and generation loads should then be considered. One must calculate energy savings from the end of one ECM and continue from this point for the next ECM. Examples of interaction are given below.

Converting incandescent lights to fluorescent lights decreases the cooling load. By also replacing an inefficient air-conditioning unit with an energy-efficient unit, the energy savings must be based on the reduced-cooling load due to the lighting change.

Adding ceiling insulation reduces the heating requirements. By also replacing an inefficient furnace with an energy-efficient unit, the savings must be based on the reduced heating load due to the ceiling insulation. If more than one ECM is being planned for a given building, the building owner and the TA analyst must decide on which projects will be installed and the interaction among those measures must be analyzed. By not properly analyzing interaction, erroneous saving estimates have been obtained. We have even seen energy saving estimates of over 100%. Interaction among ECMs is very important as large differences in estimated and actual energy savings can result.

Inadequate Operation and Maintenance

Even with the proper engineering analysis and installation of a worthwhile ECM, without adequate operation and maintenance (O&M), the expected energy savings will not be realized. With time, the amount of savings will also decrease. A technical staff must be available, knowledgeable, and willing to conduct O&M procedures properly and on schedule. Some examples we have experienced are given below.

Even after installing ECMs that increased the overall efficiency of the heating, ventilating, and air conditioning (HVAC) system, the normal range of 68°F (heating) and 78°F (cooling) were changed to 70°F (heating) and 74°F (cooling). With these changes, heating cost increased 13% and the cooling cost increased 44%. This project that improved the HVAC system efficiency was negated by improper operation. An institution's technical staff checked out their new energy-efficient boiler by firing it up in the summer during their on-peak, time-of-use period. This caused an unnecessary utility charge that was ratcheted for 12 months. Under their normal boiler operation, a load leveling or demand limiting system would have been in operation.

There are ECMs, which originally were installed to save energy, that are now shut down and dormant because of either improper or lack of O&M being performed. The more complicated the system, the greater the chance of the system being shut down. We have seen projects like energy management systems and solar active space cooling that have fallen into this category.

The two most frequent reasons are the unavailability and lack of knowledge of the technical staff.

RECOMMENDATIONS

The whole area of energy conservation has gotten much more complex in recent years. We need to ask, which fuels should be used and what are the best ways to buy these fuels? Is the best electric rate schedule being used? Are the proposed ECMs up to date and applicable? Is the staff able to operate and maintain this equipment?

---

*Japanese Standard JEC-37, International Standard IEC 34-2, and British Standard BS 269

From our experience, we have listed five recommendations that will help your energy conservation program attain the desired goal of getting the expected energy and energy cost savings.

1. Prepare a comprehensive TA report
2. Perform operation and maintenance
3. Monitor energy consumption
4. Have good working relationship with utilities
5. Obtain commitment from everyone.

Prepare a Comprehensive TA Report

A good TA report will provide the institution with viable options on how to make their facility more energy efficient. The comfort, health, and safety of occupants must be maintained or improved along with saving energy. The equipment specified should be up to date, simple, easy to operate, control, and maintain, and be compatible with the existing system. Interaction must be considered for multiple ECM applications. Previous successful projects that have used proven equipment and concepts should be considered, while known failures should be avoided. Keep an open mind for potential candidate ECMs, no blinders or tunnel vision please. When you focus on one thing, you loose sight of the others. Keep your TA report updated as information like utility rates, fuel, equipment, labor costs, technology, etc., does change with time.

Perform Operation and Maintenance

The technical staff must be capable of operating and maintaining the ECM, no matter how much energy it will save or how low the simple payback. Good equipment will turn bad without proper O&M, and energy consumption will creep back up. If your staff does not understand how to operate the equipment, chances are the ECM will go dormant. Energy savings in the neighborhood of 25% can be obtained with proper O&M.

Monitor Energy Conservation

Monitoring energy conservation does not only mean reviewing your monthly utility bills, which are usually on master meters. It also means getting information and feedback to check that the ECM is operating as designed. A parallel condenser water (CW) pump system where only one of three pumps should have been operating showed three pumps rotating. Upon checking the pumps, it was found that one pump was rotating backwards and recirculating the condenser water. Its check valve had failed. In another building, one of the two CW pump motors was overloaded (120% of full load). The motor casing temperature was over 150°F. Upon checking the system, it was found that the other parallel CW pump inlet filter was clogged.

Monitoring also means checking items that will change energy consumption such as occupancy and utility rate schedule changes. Any large increase in energy consumption will alert your staff to check their equipment and make necessary corrections. If one does not realize what is happening, no action can be taken.

Monitors and controllers should be readily accessible. Controllers do not like varying temperatures or 100°F environments. Operators will not use or calibrate monitoring equipment located in hard-to-get-to places.

Have Good Working Relationship with Utilities

The utility rate design changes, which include new rates and rate structure schedules, necessitate that we keep informed of these changes through our utilities. To do this, a close working relationship should be established and maintained. Assistance from the utilities are available including any changes that impact the economics of ECMs. Fuel switching, rate schedule changes, recommended ECM changes, etc., may be necessary when utility rate design changes are made. Again, information and feedback from the utilities are necessary to maintain a successful energy conservation program.

Obtain Commitment from Everyone

For an energy conservation program to be successful, all people involved should be dedicated and committed to the program. Having or getting the right persons to do the job is very important. This is especially true for the operators and maintenance personnel.

By having people available before the ECM is installed, one can involve and teach them by example. If they get involved and do by example, they will understand the operation and maintenance requirements of the ECM. They will be able to put their arms around the project and become committed to the program. Complacency is defeated by commitment and dedication.

SUMMARY

There are institutions that look forward in seeing reduced utility bills with the incorporation of energy conservation measures. Some of the results have been disappointing and at times gruesome. Some reasons why savings are not being obtained are: (1) changes in building load, (2) changes in utility rates, (3) inadequate monitoring of energy consumption, (4) poor technical assistance report, and (5) inadequate operation and maintenance. Case histories have also been presented to stress these points.

Recommendations that will help your energy conservation program attain the expected energy and energy cost savings were discussed: (1) prepare a comprehensive TA report, (2) perform operation and maintenance, (3) monitor your energy consumption, (4) have a good working relationship with your utilities, and (5) obtain the dedication and commitment from all persons. By being familiar with the problems others have gotten into, and being aware of what to do, you should be able to get your projected energy and energy cost savings. You will know where your energy savings are.

Chapter 88

MAKING AN ENERGY CONSERVATION PROGRAM WORK

T. R. Todd

INTRODUCTION

This paper is based upon the author's experience, and that of his firm, in performing energy conservation studies, designs, construction management, testing and balancing and follow-up up services in over 2500 buildings and industrial facilities for a variety of public and private sector clients.

It has been observed in the course of this work that many, if not most, energy conservation programs fail to live up to their potential. In many cases they fail completely and in some cases they actually leave things worse than before they were begun. The reasons behind these failures vary. This paper attempts to identify some of these pitfalls and offers practical advice, based upon experience, of how to avoid them.

OVERVIEW OF THE ENERGY CONSERVATION PROGRAM

By its very name, the "Energy Conservation Program" is misleading. With the possible exception of a relatively few large governmental agencies (and those trying to obtain grants from these agencies), very few businesses or institutions have any real interest in energy conservation as such. What they are interested in is improving their performance in terms of productivity or, in the case of private enterprises, profitability. For many businesses and institutions utility costs are a major element of total operating cost. It might be said, therefore, that most enterprises are interested in "utility cost control," without regard as to whether energy is actually conserved. A dollar saved as a result of electric demand reduction is just as valuable as one saved as a result of KWH reduction.

It must also be recognized that, although utility costs are important, they generally are much smaller than those costs associated with the personnel employed by the enterprise. In the case of a typical office building utility costs may be on the order of $1 to $3/SF-YR. Personnel costs, on the other hand will likely be more in the $200 to $300/SF-YR range. It's clear that an Energy Conservation Measure (ECM) that reduces utility cost even 50%, but in the process results in a 5% decrease in productivity, would be a very poor investment indeed. Those responsible for controlling utility costs must be careful to consider <u>all</u> of the effects of their actions. A major goal of this paper is to point out that the best ECMs will almost invariably <u>improve</u> worker productivity (because they improve comfort conditions in the space).

The next important factor to be realized is that the enterprise is almost always faced with limited funds for investment in ECMs. It is the responsibility of the party making decisions, or his advisers, to lay out the program in a fashion which makes clear the "best" ECMs for any given level of expenditure. (Since there is often a strong synergism among ECMs, this can be a very difficult task.)

To accomplish these objectives (i.e., achieve a significant utility cost reduction without negatively impacting productivity and in the most cost effective manner possible), requires a unique combination of skills. Enterprises often will not have this type of expertise in-house and frequently employ the services of a consultant. (As an interesting footnote, it has been observed that the nature of the client's activity has a significant influence on his use of outside consultants. Larger enterprises tend to have more in-house technical capability and might, therefore, be expected to do more of the work without outside help. In fact, however, their facilities also tend to be more complex. As a result there is a tendency toward their being more accustomed to using outside consultants in significant roles.) To enhance the chance of success, the program is also generally tackled in phases with ample opportunity for review by the responsible parties at each step. This paper will, in turn, consider the major steps of the process. The comments made are equally valid whether the services are provided by in-house personnel or by outside consultants and whether phases are as shown or combined.

THE ENERGY CONSERVATION STUDY

<u>Purpose</u>: As with most things, a little money spent on the front end developing a plan can be a very wise investment. In recognition of this economic fact, it has become common for an enterprise interested in starting an Energy Conservation (i.e., Utility Cost Control) Program to first do an Energy Conservation Study. The purpose of this study is to (1) decide whether there appears to be sufficient room for improvement to justify the effort and expense of going ahead, (2) identify and develop the technical and economic feasibility of the various options and (3) put together the prioritized list of ECMs discussed earlier. Depending upon the size and complexity of the facility, the cost of this undertaking will range from $0.05 to $0.25/SF. Since this can be a considerable sum, the study will often be broken into two parts. In the first part the consultant will be asked to make a cursory inspection of the facility and determine whether it is advisable to do the full detailed study. To support his recommendation the consultant will generally be expected to give examples of the type of ECMs that he thinks might prove to be attractive and estimate their economics.

<u>Energy Audit</u>: This preliminary study is often referred to as an "Energy Audit". Without reservation, one can say that almost every enterprise should have an Energy Audit. A trained auditor will invariably discover enough low cost ECMs to more than pay for the cost of the audit. The same cannot be said for a

full detailed study. The purpose of the audit is to help the owner make this decision in a rational way and before he spends a lot of money.

Trouble occurs when an enterprise has an Energy Audit and thinks that it has had a full Study. It can suffer in two different ways. First, it may misunderstand the degree of risk in the identified ECMs and implement measures that are not well conceived, hence, wasting their energy conservation funds and, potentially, adversely impacting productivity. Second, the enterprise may feel that it has done all that it needs to do about "energy conservation" and, hence miss the attractive projects that would have come out of the more detailed Study.

Unfortunately services are not always named in accordance with what they actually are. An owner will contract for a full, detailed Study and, in fact, get a slightly expanded Audit. If one deals with reputable professionals and inspects their qualifications, it can generally be said that "you pay for what you get." If it seems too cheap, it probably is. It should be pointed out that owners often bring these problems upon themselves. They tend to view professional services as a commodity and ask for "bids" as they would for cleaning chemicals. The better engineers will not even respond, realizing that the scope of work is, of necessity, open ended and that they will be bidding apples against oranges. A qualification based selection procedure is much more appropriate. In this the owner first asks for qualifications with no mention of fee. He then ranks the firms who respond based upon their qualifications. He then sits down with the first firm and attempts to agree upon a well defined scope of services and a fee for doing the work. If he is not satisfied with this negotiation, he then moves on to the next firm on the list and tries again. This process has repeatedly been shown to result in the best value for the money. Many of the failures observed can be traced back to a bid type selection process which led to an unqualified firm being selected to perform a poorly defined scope of work.

<u>Study</u>: For purposes of evaluating the level of services purchased it is useful to go through the tasks which will typically be covered in a full, detailed study. Any study should begin with establishing the historical utility data for the facility. From this it is often possible to form important general conclusions as to the relative efficiency of the building as a whole and of its major energy consuming systems in particular. For example, the big summer peaks in electric use can be compared to those of the milder months to estimate the performance of the air conditioning equipment. (There are numerous computer programs which can help in this.) Even measures as simple as BTU/SF-YR can, if properly interpreted, give an indication of the savings potential of the facility.

Next, the engineer should make a thorough tour of the facility and talk with the occupants and the operating and maintenance (O&M) personnel. The purpose of this is to uncover existing comfort or operational problems and develop a general overview of the condition of the facility, its equipment and the capabilities of the O&M personnel. (The latter could significantly affect the type of modifications that are appropriate. One should not recommend a complex computer based control system if they are incapable of operating and maintaining it.)

The engineer should then make all pertinent measurements with respect to the heating, ventilating and air conditioning (HVAC) systems. The purpose of this is to ascertain the performance of the systems and diagnose any existing O&M or built-in design deficiencies. This must be done by personnel qualified to (1) take accurate measurements, i.e., having the right equipment and knowing how to properly use it and (2) form valid conclusions from what they observe. Among the items that will normally be included are: dry & wet bulb temperatures throughout the conditioned space and at all points in the air handling units (e.g., return air, outside air, mixed air, hot/cold decks, supply air, etc.); air velocities in the conditioned space; the temperature of chilled water, hot water/steam entering and leaving all coils; all major air flow CFMs (Note: Due to cost/benefit thinking this will not generally mean measuring the CFM at all supply air diffusers, but will cover main ducts, coils, return/outside air grilles, etc. A little ingenuity with temperatures can save a lot of money here.); pressures throughout the air and water systems; and electrical data on all major motors and loads. In addition, the controls must be put through their range of motion to determine if they work as intended. Beyond this the systems should be configured to simulate, as well as possible, their operation during other loading conditions. For instance, if the study is being done in the hottest part of the summer, the controls should be adjusted to investigate how the system would perform under lighter loads.

In addition, the field survey will be used to determine such information as the occupancy of the facility, dimensions and materials of its envelope, major internal loads, special condition requirement of areas such as computer centers, as well as collecting and verifying all As-Built drawings and submittal data on equipment. It is important to get good name plate information on all equipment so that cut sheets, e.g., pump curves, etc., which are not found in the submittal data, can be obtained from manufacturers.

The absence of a good, thorough field survey is probably the single greatest cause of disappointment in the performance of recommended HVAC ECMs. Put simply, without making appropriate measurements there is <u>no</u> way that <u>anyone</u> can diagnose an HVAC system and make sound recommendations for improving its performance. Do not be mislead. There is no computer program anywhere that can take the place of a proper field survey! To be done right, however, this requires time and costs money. Many times the field survey is the first item compromised in a low budget effort. The results are sad but predictable.

Particularly sad is the fact that the ECMs which pertain to the HVAC system are almost always the most significant and economically attractive. Their payback in terms of utility savings is generally better than that of utility savings associated with the building envelope for instance. In addition, HVAC ECMs will frequently have a positive effect on the the comfort conditions in the space and, hence, on the productivity of the people in the space. As pointed out above, this can be extremely important in considering the overall desirability of an ECM. The reason for this is quite simple. The very same problems which result in an HVAC system being energy inefficient often account for it providing poor comfort conditions. As an example of this, when an air conditioning system has been oversized (as they usually are), it will spend most of its time operating very low part loads. It will be energy inefficient and will usually do a poor job of maintaining comfort conditions (e.g., poor humidity control). By <u>properly</u> reducing its capacity both efficiency and comfort can be greatly enhanced. Another good example concerns controls. When controls do not work properly the unit will waste energy and do a poor job of conditioning the space. Repairing/modifying controls can be a tremendously attractive investment.

The point in all of this is that HVAC ECMs are repeatedly found to be the most attractive investments. They can be reliably developed only by qualified people working with a sufficient time budget to do proper field testing. This is no place to economize. (Note: The ECM that recommends raising the thermostat setting to 78 °F is not an HVAC ECM. It is an example of an "energy auditor's" attempt at making an HVAC recommendation. It's effect on productivity will almost always outweigh any utility cost it might avoid. Similar, but more subtle, is the recommendation that ventilation air be reduced to very low levels. This will generally decrease comfort and can even have some serious implications with respect to indoor air quality.)

The next step in the study is to analyze the information that has been collected. Since buildings are complex thermal systems, this suggests the use of computer programs which have been developed for that purpose. First, the engineer will try to make certain that he understands the facility in its current (or Base Case) condition. From this initial analysis and his field observations he will develop a list of all the ECMs which he feels could conceivably be attractive. The computer program, together with manual calculations when appropriate, can then be used to play "what if" and estimate the savings that would result from each option or group of options (i.e., carefully allowing for synergism).

The only missing element at this point is implementation cost. Next to poor field measurements there is probably no other part of the process which causes so much trouble. Most energy conservation projects consist of a number of small retrofit construction items. Estimating their cost is considerably different from estimating the cost of new construction or even larger retrofit projects. Many engineers have little experience in this area and develop cost estimates (upon which investment decisions are made) which ultimately prove to be inaccurate.

Finally, the results of the above must be collected into a plan of action and reported to the owner. This plan must provide sufficiently detailed technical and economic information to allow the owner to make an intelligent decision as to how to proceed. As discussed earlier it is important that this plan make clear the interrelations between ECMs and include a prioritized ranking of actions for each level of investment available.

THE DESIGN AND BIDDING PROCESS

Having developed a plan of action in the study phase, it is then time to implement it. How this can best be done depends upon the nature of the modifications that are to be made. In some cases the recommendations are simple enough that the study can be used directly to implement with in-house personnel or to contract out for their implementation. Be careful, however, in asking someone for advice who stands to profit from his recommendations. More frequently the recommendations are complex enough that they will require more formal bidding documents to insure that the installed modifications follow the intent of the study. (This is particularly true of the HVAC ECMs which have been shown to be so attractive.) Frequently this will mean that the owner will seek outside help from a design professional.

Selection of Designers and Contractors: Several words of advice in this respect are in order. First, the qualification based selection process described earlier is just as appropriate here as it was for the study. Next, hire a qualified professional and pay him a reasonable fee so that he can deliver a quality set of bid documents. It is money well spent. Be particularly careful to hire a design professional with experience doing the type of work represented in the project. Talk to their previous clients on similar jobs. How did they perform? One firm may be very good a new design but have little or no experience at energy conservation retrofits. Owners often remember the firm that designed their facility originally and, without thinking turn to them. This may not be wise. In fact it is sometimes very unwise. They may be poorly qualified for the work and may be less than enthusiastic about pointing out the things done in the original design that could have been done a better way. Next, hire a professional of the discipline which represents the largest segment of the work as the prime. This will keep the amount of subcontracting, and the associated double profits, to a minimum and help reduce the design cost. (It is always amazing to see a client hire an architect for a project which is almost exclusively mechanical or electrical.) Finally, when possible, have the same firm do the design that did the study. Who better understands what they intended by their recommendations. He also feels that his reputation is on the line. The last thing he wants is to have to say that a recommended action is not possible (or not possible within the estimated cost). For this reason he will work hard to overcome any difficulties that might arise in the course of the design. The owner benefits. This can also save a substantial amount in the design cost since it is almost impossible to do a study without at least doing a preliminary design. By changing to another firm, much of this can be lost. The owner paid for it once. Whenever possible it should be used again. This should not mean that the firm doing the study should always be employed for the design. Sometimes they are not qualified and it would be unwise to continue with them "out of their element." It should be mentioned that the attractive HVAC ECMs are not likely to have been developed in a study by a firm that was not also well qualified to design the required modifications. More and more it can be said that energy efficient design is nothing more than good design. It might be wise to make design capability a point in the selection criteria for choosing a firm to do the study.

In selecting a contractor to do the construction, the same basic concept applies. Recognize that the "low bid" is not necessarily the "best bid." First, be certain that the contractor is capable of doing the work. Insist on references for previous jobs of a similar nature and check them. Next, make certain that he fully understands what will be expected of him. Having a good set of bid documents is a start here but go the extra distance to make certain that he understands. Even though the contract may clearly state something and the contractor is legally obligated, misunderstandings always cost money. Finally, make certain that the contractor is really interested in the work. A big contractor, like the one who initially built the building, is not likely to be interested in a small energy conservation project, even if he says he is. He also is not likely to be well suited for the work. This will often become apparent once the work begins and has been the source of much disappointment and lost friendships. Unfortunately, like the engineer or architect who initially designed the facility, this is the person that the owner is likely to know best and think of first.

Quality: The key points in judging the quality of the services received is whether the ECMs perform as intended and go in within the budget. An important factor in staying within the budget is keeping the number and amount of change orders down. Contractors typically make a higher than normal profit on these items. The design and bidding documents should be well thought out and there should be a minimum number of necessary change orders. Less than 5% of the original construction cost is very good on projects of this nature. This amount should be covered as a contingency in the cost estimate. Clearly, the contractor's attitude will also be an important factor. If he is frustrated and wishes he never took the job, he is much more likely to ask for change orders for every little thing that arises and less likely to try to work past problems. (EJCDC standard contracts are probably a little better than AIA documents for these projects.)

Since these are modifications to an existing facility the designer often must plan for the continued operation during construction. Failing to do so can result in significant defects in the contract documents and major change orders. Another factor which can result in high change orders is the manner in which the design and contract documents provide for handling modifications that become necessary in the Testing, Adjusting and Balancing (TAB) phase. By its very nature, HVAC retrofits will have conditions arise that require that some decisions be made after construction has begun. As an example, consider the case of a chilled water pump. It is impossible for anyone to calculate the pressure drop of a piping system exactly. For this reason designers include a, hopefully small, safety factor. When the system is installed and tested one generally finds that the pressure drop was less than that which was expected. The normal response is to partially close a throttling valve. This is very poor practice. The added cost of disassembling the pump and having its impeller trimmed to the proper diameter will pay for itself within months. It is very important that the contract provide for these types of changes to the scope of work and insure that the cost to the owner is reasonable.

As a selection criteria, ask prospective designers and contractors to demonstrate their past performance with respect to change orders by giving specific examples that can be verified then do just that.

THE CONSTRUCTION PROCESS

The first and most important point in talking about the construction process for energy conservation projects is to recognize that these projects are often dramatically different from most other construction projects. An approach that might be appropriate for building a new building, for example, might not work very well. It is worth considering making several modifications to the process.

Designer Involvement: First, and probably most important, keep the designer involved in the project. The level of effort that is normally put forward in reviewing submittals, a few site visits and a final "coat and tie" inspection is totally inadequate for energy conservation retrofit projects. Being a grouping of small things and, hopefully containing some good HVAC retrofits, there are uncountable opportunities for things to go wrong. The owner needs a greater level of involvement by the designer. Traditionally this role has diminished over the years. As this has happened the fee for construction involvement by the designer has also diminished. Inspecting new buildings that have just been pronounced ready for acceptance, one almost always finds a long list of deficiencies (e.g., crossed zone thermostats, improper air and water balance, etc.). Unhappily many of these involve work that has been concealed to the extent that discovering the deficiency is difficult and correcting it, at a reasonable cost, almost impossible. Insist that the designer stay in touch with the project as it is constructed and be prepared to pay him a reasonable fee to do so. Like the field work in the study phase, this is no place to economize.

Testing, Adjusting and Balancing: Next, many energy conservation projects involve TAB. The sad fact is that the quality that has come to be accepted in this area is not very good. In energy projects it can, at times, be critical that high quality TAB be done. The construction process should encourage, not discourage, this. Most owners and designers have come to recognize that it is not a good idea to ask mechanical contractors to TAB their own work. In many cases though, the TAB contractor will be a subcontractor to the general contractor. This puts the TAB agency in an awkward position. He may have to say that work done by his boss (the man who processes his requests for payment) is defective. The TAB agency should always be employed directly by the owner. This will help to insure his absolute independence.

The next point also involves the TAB work. When this process is begun, the TAB subcontractor will usually start running into things that are not as expected in the design. It is important that the reaction to these issues be properly thought out. This does not necessarily mean that the designer has done a poor job. It can simply be a reflection of the nature of the work. For instance, in the case of the pump impeller sited above, it, in fact, represented good thinking. It is important that someone, who is capable, be available to say what the proper pump impeller diameter should be. The designer is a logical choice and this fits with the various other reasons for keeping him more closely in touch with the project. In some parts of the country it has become common for the designer to do the TAB work himself. On the one hand this might offer the benefit of using TAB more directly to insure that the intent of the design is preserved. On the other hand it might tend to diminish the independence of the TAB agency. It could be difficult for the designer (who is also doing the TAB) to report that the system, as designed, cannot be balanced as specified. The answer to this dilemma is not clear and probably depends upon the specifics of the design firm. In no case should a design firm be employed to do the TAB that is not qualified to do it.

THE POST CONSTRUCTION PHASE

When an energy conservation program is begun changes are made. There will usually be a substantial drop in utility cost. As time goes by one frequently finds that utility costs creep back up to their previous levels and things drift back toward the way they were done before the program was begun. This is a major problem in that it happens to a significant degree in probably 75% of the cases. Several critical elements can invariably be found in those cases where the initial gains are maintained, or even increased over time.

Training: Often a quality study is done, the recommendations properly designed and correctly installed and yet one finds a year or two later that much, or most, of the new equipment has been disconnected and operational modifications undone. The money spent has been wasted. This can often be traced back to the fact that the O&M personnel were never "sold" on the program and properly trained in the use of the new equipment. When problems occur, as they almost always do, the energy modifications will immediately be blamed. It is human nature. The equipment is either abandoned or partially or totally incapacitated. After construction has been completed and accepted the first thing that should be done is train the O&M personnel on any new equipment that has been installed. At the same time they are being trained it is smart to sell them on the benefits (mainly to them) of the project. Their enthusiasm is crucial to the long term success of the effort. If they feel a part of the program, they are much more likely to accept and, even, defend it against its critics. The importance of this cannot be overstressed! The contractor should be required in the construction contract documents to supply good quality O&M manuals on all equipment. Accurate As-Built drawings should be required. Care should be taken in specifying that the instructor on any training sessions be properly qualified to properly present the material and that adequate time be budgeted for these sessions. The sessions should thoroughly cover all of the material provided. It is also a good idea to have the designer involved in these training sessions to help explain the purpose of the modifications and make the sale as to why it is a good idea.

Since there tends to be a high turnover rate in O&M personnel, steps must be taken to insure that those who come on board later will also be trained in, and enthusiastic about, the program and its equipment. One way of helping with this is to videotape all training sessions. These tapes have been found to be good, not only for training new personnel, but also as a reference material for the existing staff. They are also useful in teaching refresher courses to the existing staff. This should also be recognized as necessary if the program is to be kept going as intended. People tend to forget things and need a little reminder from time to time.

Inspections: Independent periodic inspections are a powerful tool in keeping an energy conservation program on track. To be independent these should be done by someone who is not in the O&M chain of command and free of any possible political implications of his findings. To be effective the person performing the inspection should also be technically knowledgeable about the systems he is inspecting. Finding a person inside the organization with these attributes can be difficult. For this reason, an outside consultant will often be used. The person who designed the modifications is one logical choice. Be careful about using a service maintenance contractor who may not have participated in the original work. He may not understand the intent of the work and may not be "sold" on its importance.

In any event, the first step is to prepare a check list. This list should be given to the personnel who are responsible for the daily operation and maintenance of the facility and it should be made clear that this is what will be checked during the inspections. Give everyone copies of the checklist. (Ideally the O&M personnel will be asked to help in preparing this check list as this will make them feel more a part of the process and understand that its purpose is not to make them look bad but, instead, to get results.) Keep in mind that it is not only the performance of people that is being monitored, but also the performance of equipment. It is perfectly possible that the O&M staff will perform their duties and there will still be problems with the equipment. This is an important motive for performing the inspections and can help diffuse the natural tendency for the staff to start off in an adversarial relationship.

Next the visits should be scheduled with some fitting regularity. (Monthly or quarterly is reasonable for most facilities.) Everyone should be informed in advance of the dates of the inspections and surprise visits should not be made. Surprise visits promote the notion that the purpose of the program is to "catch" O&M personnel not doing their job. This is exactly what one wants to avoid. Remember that this must be their program. That is the only way it will succeed.

Predictably what will happen is that the O&M personnel will try to make the rounds before each visit and verify that everything is in order. This is good, not bad. In fact, it is exactly what one would want to have occur. Choose the items on the check list wisely and you have instituted a preventive maintenance inspection program. (Experience has shown that in many cases the O&M staff will make additional copies of the check list and fill them out themselves between visits. This should be encouraged but be careful not to push it. Let them come up with the idea. If they do not, "help" them think of it. By including the proper items on the list one can have an excellent plant log. Most importantly the O&M staff will feel that the log has a purpose and not that it's just another form that has to be filled out. As a result it will be kept with much greater care than is typically the case with plant logs that have been imposed upon the staff.)

Utility Monitoring: Ideally one would like to measure all of the effects of the program and track its progress. The impact on worker productivity attributable to the program, although of prime importance, is very difficult to measure and usually neglected. (For ECMs which improve comfort and enhance productivity this is a conservative assumption.) Sometimes there will be savings or costs that can be easily identified (e.g., an existing contract for chemicals that can be discontinued or a new maintenance contract on equipment being added). More commonly these effects will be considered at the time of implementation and only utility cost savings tracked thereafter.

One can monitor utility costs at various levels of detail. Most simply, the utility costs are merely gathered together in one place and logged. This log should be broken up by fuel and show both the energy and demand charges from which the bill is calculated as well as the actual dollar amounts for each fuel. (Spread sheets are ideal for this purpose and make it easy to quickly analyze the data in a number of ways.) The key is to have this done by the person who is responsible for the energy conservation program. It is surprising how many large organizations never collect utility data together in one place. Those that do often have it done in an accounts payable department far removed from anyone who can actually affect energy use. In case after case it has been shown that this simple act alone will produce a utility savings. In the case of enterprises with multiple branch locations the information should certainly be sent out to those persons in each branch with the responsibility for energy use.

The next level of tracking involves trying to allow for things that have changed since the program began which would be expected to affect energy use. For instance, if the building had been expanded to double its original area, one would surely want to allow for this when comparing utility bills before and after the program began. The addition of a major piece of energy consuming equipment would be another example. Certainly one would also want to allow for changes in utility rates when comparing dollars.

When a significant portion of the energy use is for space conditioning, as it is in almost all facilities one might, next, wish to allow for the effects of variations in the weather from year to year. Buildings are complex thermal systems. Their energy usage can be significantly affected by warm winters and cool summers, etc. There are various ways of attempting to allow for weather that range from simple "degree day" approaches to complex multiple linear regression analyses. In general, the accuracy of the method is almost directly proportional to its complexity.

The purpose of attempting corrections is to make the estimate of savings more accurate. This helps sell management on the fact that the program is a good investment by giving credibility to the claims of savings. It can also be useful if one wants to compare the performance of various similar locations. Systems in which branch locations are ranked by some formula have been found to be very effective in getting cooperation throughout a large organization. The energy conservation ranking can even be tied to various types of bonus plans and contests with good effect.

While one would like to be as accurate as possible, this tends to lead to more complex methods. These can sometimes be difficult to understand and explain to others. It can appear that the claims of savings are nothing more than a computerized slight of hand.

In addition to encouraging people to support the program, utility monitoring is useful in spotting things that have gone wrong. For instance, a failed economizer control on an air handling unit can cause a sudden increase in both heating and cooling costs. Often something like this will happen and go unnoticed for months. With a good monitoring system it will be spotted before too much waste can occur.

Finally, the utility monitoring system serves as a constant reminder to everyone that energy cost is important and that a program is underway to help control it. It establishes <u>accountability</u>.

The computer is a natural for this and there are myriad of computer programs available to monitor utility costs. Some of these are quite good and far cheaper than attempting a do it yourself version.

CONCLUSION

An Energy Conservation (i.e., Utility Cost Control) Program can work and can be a very attractive investment but it requires an uncommon attention to detail. If this is not done, it can fail miserably and hurt an organization in a multitude of ways. Before embarking on such a course think things out and go about it in an organized manner.

SECTION 21
RESTRUCTURING THE ELECTRIC POWER INDUSTRY

Chapter 89

COMPETITION, DIVERSIFICATION AND DISINTEGRATION IN ELECTRIC UTILITIES

M. A. Crew, K. J. Crocker

In telecommunication, electricity, gas, and even water there has been significant entry on the part of competitors and an increased trend toward diversification on the part of existing regulated utilities. Even before the break up of the Bell System on January 1, 1984 there was significant entry into the long distance business which had formerly been the exclusive province of AT&T. Since 1984 the seven Regional Holding Companies (RBOC's) have proceeded to diversify into other (usually closely related) businesses. In addition they have attempted to remove, relax and change the regulation of their traditional businesses. The same kind of trends have been evident, if to a lesser degree, in gas, water and electricity. In the case of electricity a further change has begun to emerge; the traditional vertical integration of generation, transmission and distribution is under attack. The process has been encouraged by in electricity by the Public Utility Regulatory Policies Act of 1978 (PURPA), which encouraged generation by independent producers and allowed traditional utilities to enter the independent generation business. Indeed, many utilities set up independent cogeneration subsidiaries and others even went so far as to disintegrate vertically, producing power through independent and unregulated subsidiaries.

The entry of non-utility generators, and the spinning off by utilities of their traditional generation business may be perceived as an attempt to avoid some of the effects of regulation. That such changes will result in a more efficient governance structure, however, is by no means proven. Some authors have not been at all sanguine about the prospects of deregulated power generation. For example, Herriott [1988] and Jurewitz [1988] note the difficulties associated with the loss of integration and coordination efficiencies as restructuring proceeds and argue that the problems of formalizing the cooperation requirement, either through a market exchange or long term contracts may be severe.

While the design of the contractual relationships to ensure efficient coordination may be difficult, the task is certainly not impossible, as illustrated by the existing power pools. The problem is that arms-length relationships which result from vertical disintegration leave open the prospect of severe opportunism which the original regulatory structure was designed to alleviate.

The purpose of this paper is to provide a first step at explaining these major changes. The paper is structured as follows. Section 1, using the New Institutional Economics,[1] analyzes the efficiency properties of regulated governance structures[2] for traditional natural monopolies. Section 2 examines the circumstances which led to it current disfavor. Section 3 analyzes the nature and efficiency of the governance structures emerging for electric utilities. Section 4 is by way of conclusion and implications for future research.

1. REGULATION: AN INSTITUTIONAL CHOICE

Regulation was originally regarded as a response to the (natural) monopoly that resulted from overwhelming economies of scale. To obtain the benefits of the economies monopoly was necessary. Regulation was perceived as a means of protecting the customer from the abuse of market power. Where overwhelming scale economies were not obvious deregulation sometimes[3] took place, for example, with airlines. As regulation become subject to more scrutiny the need for regulation was questioned even in traditional areas like electricity when it became clearer that the scale economies were primarily in transmission and distribution rather than in generation. The notion behind the widespread movement toward deregulation is that regulation is inherently flawed and can be replaced by a superior governance structure. Unfortunately, while regulation does have problems from an efficiency point of view, the alternatives may be worse. The hazards of rejecting regulation out-of-hand, at least in industries like electricity are given by Goldberg [1976]:

> "...some of the regulatory problems perceived by economists are largely illusory; they are the results of stacking regulatory outcomes against irrelevant standards generated by models

which suppress the contractual complexities inherent in the so-called natural monopoly sector." (p. 445)

An increasingly influential argument views regulation not as a response to a particular shaped cost curve but, rather, as a contractual response to the need for a long term relationship between the producers and consumers of electric power. As such, the guide for deregulators should not be the absence of declining costs. What counts is whether the particulars of the transaction require the actors be locked into a specific and mutually symbiotic relationship over the long term. In such an environment, the preferred mode of exchange is unlikely to be that of a classical spot market where agents engage in recurring and anonymous exchange guided by the market price.

I.1 Specialized Assets

The generation, transmission and distribution of electricity requires substantial investment in specialized assets which have little value in alternative uses or relationships. Williamson [1983] provides a useful descriptive taxonomy for classifying such relationship-specific assets which we apply to the particulars of the electricity industry.

Site Specificity: This refers to an irreversible location component of an investment. For example, a concrete dam on a river exhibits a high degree of site specificity: it serves to interrupt the flow of water on a particular river in a specific place, and cannot be easily moved to other rivers or other locations on the same river. Similarly, generation facilities are sited with an eye toward existing or planned transmission facilities or consumption patterns and the use of facilities to service other demands may require a large investment in alternative transmission and distribution network. Investments in the latter are even more site specific and may have no alternative to the intended application other than as scrap.

Physical Asset Specificity: refers to an investment by either the buyer or seller in an asset whose design and purpose is restricted to the particular relationship. Common examples of such specific investments would include specialized dies and machinery as well as unrecoverable expenditures or nontransferable research and development or production planning [see Klein, Crawford and Alchian 1978]. Much of the physical capital (generators, transformers, cables, poles and insulators) used in the electricity industry is relatively standardized and, therefore, transferrable to other uses. While moving the assets to an alternative use might entail some relocation costs, this is primarily a consequence of site specificity mentioned above, as opposed to any fundamental uniqueness in design or use of the assets.

Dedicated Assets: refer to generalized investment in production facilities for the purpose of dealing with a particular customer or supplier. While these assets would not be physical- or site-specific in nature, they would, nonetheless, not be purchased "... but for the prospect of selling a significant amount of product to a specific customer" [Williamson 1983, 526]. Inasmuch as the termination of the relationship could leave the actors with significant excess capacity, such assets would have lower value if used in alternative employment.

Dedicated assets are of particular importance in the electric power industry, particularly through the regulatory imposed requirements on adequate reserve margins in both generation and transmission. Since nonreliability of service may impose significant costs on customers, both directly (actual damage to flow process operations and machinery due to unscheduled outages) and indirectly (through individual attempts to mitigate direct damage through installation of alternative generating facilities and the like), reliability is of paramount concern to both utilities and regulators. But, in the uncertain world of winds, lightning and random equipment failure, reliability requires redundancy which, in turn, requires investment in capital which is designed to be used only under extreme circumstances.

I.2 Opportunism

Once an investment has been committed and the specialized asset is in place, a flow of potentially appropriate quasi-rents is generated equal to the difference between the value of the asset in the intended use and its value to the next-best user.[4] Consider, for concreteness, the case of an independent power producer who has constructed a 10 MW generation facility to sell power to a distribution utility. If the average total (amortized capital plus operating) costs are $12,000 per day, the producer would require a 5 cent per kWh rate to break even, assuming full utilization of the facility. But once the plant is in place, the producer may find himself exposed to opportunistic behavior as the purchaser attempts to gain access to the appropriable quasi-rents. Suppose that, absent sale to the intended user, the best the producer can do is wheel the power to another purchaser for the reduced (net of wheeling costs) rate of 3 cents per kWh. The appropriable quasi-rents are then 2 cents per kWh and the intended purchaser can potentially extract up to $4800 per day from the seller.

This problem cuts both ways. If, on the other hand, the distribution utility were to rely exclusively on the independent power producers for 10 MW of capacity, the utility would be subject to threat of opportunism as well. Suppose that the utility can replace the 10 MW in generation by purchased from another supplier, but only at the higher costs of 6 cents per kWh. This would imply that the utility would itself be vulnerable to opportunism, having appropriable quasi-rents of 1 cent per kWh for a total daily exposure of $2400. The total amount of quasi-rents under contention in this example is considerable, amounting to 3 cents per kWh or $7200 per day.

With such a large sum at stake, it is natural to expect that both parties would engage in privately remunerative but socially costly endeavors designed to alter the distribution of the gains from trade in their favor. Anticipating that such recurring opportunistic behavior could generate direct costs as well as dissipate the potential gains to exchange, the parties have the incentive to stipulate the terms of trade at the outset of the relationship in an attempt to restrict the scope for future strategic posturing. But such a relationship would be distinctly different from that of a classical spot market, and much closed to a long-term contractual relationship between two intimately familiar actors.

While long term contracts may serve to mitigate the problem of post-investment opportunism, they are by nature inflexible and may place stringent limits on the ability of the contracting parties to adjust to changing economic circumstances. Indeed, this inflexibility is both the raison d'etre for long term contacts as well as their principal limitation. Economic maladaption occasioned by an overly strictured contractual agreement may dissipate the gains from trade as effectively as opportunism. The challenge is to design a contractual relationship that attenuates opportunism, on the one hand, while permitting needed adaptation to a changing environment, on the other.

Goldberg [1976] argues that this is precisely the role adopted by regulation, which he views as a long term social contract between utilities and ratepayers. This contract is flexibly administered by regulators who have the responsibility of shielding consumers from monopoly excess and, at the same time, protecting utilities from expropriation of their quasi-rents by the public. The point is that regulation may be viewed as a contractual response to the particulars of a transaction requiring large, relationship specific investment and, consequently, subject to the continual threat of post investment opportunism. Eliminating the regulatory structure will not make the "problems" associated with regulation disappear; they are endemic to the relationship, not just to the regulatory governance structure.

2. WHAT WENT WRONG?

Rate-of-return regulation stood the test of time, performing well for over half a century. The strengths of the system were those described by Williamson and Goldberg-flexibility and the ability to make incremental adjustments to gradual changes in the economic environment. Indeed, regulation had evolved and adapted itself well to changes that occurred until the 1970's, when the price of oil increased fourfold, and with it other energy prices. The effect of the large changes in energy prices was to create a dramatic increase in the rents accruing to certain resource owners. Since energy markets are largely competitive, these rents could not be avoided by utilities and were passed on to them through the higher resource prices. Although there were some attempts to mitigate the dramatic accumulation of rents by infra marginal energy suppliers, most notably through direct oil price controls and extant wellhead price regulation of natural gas, these efforts generated severe allocative distortions and were short lived. The utilities were successful in passing through a portion of these costs through to ratepayers through fuel price adjustments, but these were never complete nor noncontroversial. Inevitably, the utilities were squeezed between the unavoidable increase in rents accruing to their suppliers, on the one hand, and the desire by politically sensitive regulators to minimize adverse price impacts to ratepayers on the other.

While owners of oil fields, gas fields and coal mines were the recipients of considerable rents, electric utilities with nuclear plants and a relatively large share of base load coal plants were, in addition, the recipients of large increases in quasi-rents. Such plants, because they used relatively little energy, were made dramatically more valuable by the fourfold increase in the price of oil. In contrast, consumers of electricity were considerably worse off since the massive rents created attracted a multitude of "rent seekers". For example, the rents to the nuclear power industry became the targets of large numbers of rent seekers. Consumer's through their regulators saw an opportunity to mitigate, at least partially, the effects of the oil price increase. The rents of the nuclear power industry were also the target of other regulatory-inspired groups. The nuclear industry became subject to extremely cumbersome oversight in its planning and regulation, which had the effect of dissipating the quasi rents among lawyers, regulators, inspectors, contractors, and even construction workers. The effect was that the construction costs of nuclear plants soared. Indeed the rents were over dissipated, with the result that planned nuclear plants and partially completed plants were abandoned.

The combined effect of large oil price increases and a high general level of inflation, required that regulators allow price increases of a magnitude never previously experienced resulting in significant adverse response by ratepayers and consumer groups. Thus regulators were bombarded by opposing forces which were much greater than anything previously faced, including a major revolt on the part of residential customers. Through their elected representatives, and through state appointed public interest interveners, the views of residential customers were strongly felt. The net effect was that regulators generally failed to compensate companies fully for the effects of inflation and fuel prices increases, effectively appropriating a portion of the utility's quasi-rents. By not imposing adequate price increases, the regulators did not do any immediate damage like turning off the lights, because they were eating into the quasi-rents, as long as the remaining quasi-rents were positive the electric utilities would still continue to produce.[5] The problem is that they had been held up, by the regulators' failure to provide adequate rate relief, just as effectively as in the classic case in the literature were one party deliberately expropriates the quasi-rents of the other party.

In addition to appropriation of the quasi-rents by regulators there is another aspect of regulation that favors residential customers. Traditionally residential customers have been the beneficiaries of cross subsidies from industrial customers. Nelson [1987] and Hayashi [1985,1987] examined electric utility prices relative to economically efficient or Ramsey prices. They found that residential customers were treated favorably relative to the economically efficient prices at the expense of the commercial and industrial customers. While this distortion may not have been very important when energy prices were low it became increasingly important with the major increase in energy prices, as large customers began to consider ways of conserving electricity and obtaining cheaper alternatives in the form of self generation and cogeneration.

With the increase in energy prices, cogeneration became relatively attractive. Cogeneration is an old technology that proved to be relatively uneconomical when faced with low fuel prices and central station power exhibiting significant scale economies. In addition to being relatively inefficient, cogeneration also faced barriers that were present as a result of the regulated monopoly governance structure. A cogenerator might be able to use all the heat he produced but not all the electricity. Sales to his local utility proved somewhat problematical, since he was facing a single buyer, or monopsonist, with whom he was hardly dealing on equal terms. Moreover the utility, at least traditionally, had an incentive provided by rate base regulation to offer a very low purchase price or even not make an offer. Cogeneration would displace its own generation and would therefore, at least potentially if not actually, reduce the amount of rate base upon which it could earn. In the days when fuel was cheap and therefore cogeneration was relatively inefficient it did not matter very much that regulation created an additional barrier to its implementation. However, the increase in fuel prices suddenly made cogeneration more valuable. Potential investors were attracted by the high returns from cogeneration and sought to change the regulatory governance structure so that the potential rents were attainable. The Public Utility Regulatory Policies Act, 1978 (PURPA) made it much easier for cogenerators to sell their power to the local utility. Indeed it set up a system that provided cogenerators with a bonanza, requiring utilities to pay the "avoided cost" to cogenerators. Avoided cost was set by regulatory commissions and often at a level that proved extremely attractive to cogenerators[6] and very expensive to electric utilities [Crew and Crocker, 1986].

As if these traumas were not enough for the regulatory governance structure, there was in addition the impact of increased environmental regulation. The increased costs of environmental regulation represented another instance when price increase were required of the regulator. In addition they presented another opportunity for appropriation of quasi-rents.

This hold up required a response on the part of electric utilities, but they were not equipped to respond immediately, as they were engineer-dominated organizations whose prime concern was delivery of reliable power. After years of being regulated they were slow to react to the kind of problems faced by unregulated companies, and they had very little understanding of competitive markets. Their first step was to continue to play the game that they knew - to file more and bigger rate cases. This met with limited success. Indeed, the rent seeking on nuclear power got so out of hand and the costs rose dramatically that regulatory commissions began to disallow large amounts of the costs incurred in the construction of nuclear plants.[7] Thus faced with limited success or even outright failure to attenuate hold up the electric utility industry had to consider a different direction. PURPA itself had provided one small concession to electric utilities. They could be 50% owners of cogeneration plants, known as "qualifying facilities", that were qualified to receive the avoided cost payment. Such activities could be placed in a subsidiary company and earn a return that was not regulated. Alternatives to depending essentially entirely on regulated returns became worthy of serious consideration. Electric utilities began to look at ways of diversifying and ways of getting more of their business out of the traditional regulatory process.

The response of electric utilities to hold up has only just begun. It has taken several forms, in addition to diversification. Electric utilities have sought ways of removing some of their most competitive business from the regulatory umbrella. They have sought, for example, to make special deals with large industrial customers and worked on more efficient pricing policies for large users such as real-time pricing. Such actions are an attempt to avoid losing their largest customers for whose business real competitive alternatives exist. Very few major power plants are being built and planned at least in the traditional way under the control of vertically integrated utilities. New plants may be built by independent generators or consortia which may include utilities. Such moves toward vertical disintegration will radically affect the structure of the industry as we will see in the next section.

3. PARTIAL REGULATION: A SOLUTION?

As partial deregulation of the electricity industry grows it will be necessary to devise alternate governance structure to replace regulation. The industry might involve a traditionally regulated portion consisting of distribution companies which also share ownership with other distribution companies of the transmission grid. Some of these companies will spin off their generating plants. Others will obtain power from their existing traditionally regulated plants, divested new plants, and plants owned by fully independent generators. To obtain the requisite investment in the unregulated plants will require a governance structure to replace regulation. In most instances spot market contracting will not be feasible primarily because long-lived specialized assets are required to produce electricity, and because there are complexities about transmitting energy through the transmission system.

Power plants are not like aircraft. Once

built they are essentially immovable and have little or no value in an alternative use. Unlike a manufacturing plant their output cannot simply be transported to where required. They have to be connected to a transmission grid. Thus spot market contracting is infeasible for most generating plants. A system of long term contracts would need to be devised, for example, as in the case of gas producers and pipelines. Such contracts are exceedingly complex, containing take-or-pay requirements, most-favored-nation provisions and price escalators which tie prices to producer price indices, various oil prices, and regulatory ceilings [Masten and Crocker 1985]. In spite of the extreme detail of these contracts they were inherently incomplete and were subject to costly renegotiation and even more costly adjudication by the courts when, with changes in circumstances, parties engaged in attempts to appropriate or protect quasi-rents.

The problems of long term contracting between gas producers and pipelines are likely to be even greater in the case of electricity, because of the complex technicalities of the transmission grid and the potential for externalities. The electric grid is much more complex than the gas grid. It is not just a simple matter of transporting the electricity produced from A to B. It may be that it is not physically possible to transport, or that transmitting the power between A and B may raise C's costs through grid externalities[8]. While it is beyond the scope of this paper to go into the technicalities of electric transmission networks, it is apparent that long term contracts are likely to be even more complicated to devise than in the case of the gas industry.

Moreover, the cooperation between utilities that currently exists may not be forthcoming in a deregulated environment. Now because of regulation, while profits, while limited are less uncertain, and the gains from opportunistic behavior between of utilities may not be very great. The engineer-dominated industry sees gains from cooperation in order that the technical quality of power delivery is high. By allowing the industry to earn on a rate base, within limits, it does not matter too much which company makes the additional investment in transmission required to keep the system reliable. Similarly, investments by utility A that improve utility B's reliability as well as its own are likely to be forthcoming fairly smoothly under the current system.

This may not be so under a vertically disintegrated system. The quasi-rents generated by the generation-transmission chain will no longer be internalized in a single, vertically integrated entity, but rather a source of contention among many independent and self seeking firms. Opportunism is likely to be rife. Because of considerable potential for gain distributors may hold producers up and conversely. The existence of long term contracts may help avoid some opportunism. However, attenuation of opportunism is limited because of the inability to write ex ante complete contingent contracts.

The courts will have to enter the picture to adjudicate disputes. Just as currently regulators administer contracts so the courts will have to arbitrate between parties. However, in view of its very nature court proceedings are likely to be more acrimonious than regulatory proceedings. In addition, courts are less well equipped to handle the problems than regulators, because they lack the specialized knowledge and experience of the regulators. In favor of the courts is that they are likely to be more detached and less politically motivated. Weighing the advantages and disadvantages of court adjudicated rather than regulatory administered contracts it is not clear that the advantages lie with the judicial solution.

4. CONCLUSIONS

The electric utility industry is facing an uncertainties of a magnitude not seen in recent years. This is true not just in the U.S. but also in the U.K. where major changes are taking place in the privatizing of the industry there.[9] It is by no means clear that the apparent success of deregulation of the airlines can be achieved in electricity. In telecommunications, which is closer to electricity but which does not have the complications of transmission access and grid operation it is not clear that the partial deregulation which has taken place has led to increased efficiency. In electricity, in view of the magnitude of the problems of major transaction specific investment, transmission access and grid operation we remain skeptical of the ability to devise partially deregulated governance structures that will perform efficiently.

However, partial deregulation is the direction in which events seem to be heading. Before things settle there will be major changes. Vertical disintegration may give electric utilities the incentive to engage in horizontal mergers. One way problems of opportunism and externalities in transmission and distribution can be resolved is to internalize them. There may, as a result, be many fewer distribution and transmission entities, in addition to being be less vertically integrated. Eventually the advantages of vertical integration may once again become apparent and a new age of regulation will be born. Meanwhile electric utilities have to manage the change and deliver the power. In the past they have proved capable to meeting this challenge.

NOTES

1. The New Institutional Economics originated primarily with the work of Williamson [1975], and Klein, Crawford and Alchian [1978], drawing on a number of famous antecedents such as Coase [1937].
2. Governance is the jargon for the framework in which exchange takes place.
3. Deregulation was not by any means a universal response to the absence of scale economies. Witness the continued regulation of taxicabs in major cities.
4. See Klein, Crawford and Alchian [1978] for an extensive discussion of this issue.

5. Norton [1988] found evidence to support the notion that utilities obtained addition revenue from their regulators to cover the increase in energy costs.
6. In part this may have been the result of lags in the regulatory process as well as a policy of some regulatory commissions to encourage cogeneration. To the extent that lags were involved the effect of the change from rapidly increasing to rapidly decreasing oil prices was not reflected.
7. Disallowances were widespread: Nine Mile Two, Seabrook, Diabalo Canyon, Peach Bottom, Zimmer, to mention but a few across the country.
8. John Jurewitz [1987] provides an explanation of such externalities
9. Privatizing Electricity: the Government's Proposals for the Privatization of the Electricity Supply Industry in England and Wales, HMSO, CM 322, 1988.

REFERENCES

Coase Ronald H., "The Nature of the Firm," Economica N.S., 4, 1937: 386-405.

Crew, Michael A., and Crocker, Keith J., "Vertically Integrated Governance Structures and Optimal Institutional Arrangements for Cogeneration", Journal of Institutional and Theoretical Economics, June 1986.

Goldberg, Victor P., "Regulation and Administer Contracts," Bell Journal of Economics, 7, Autumn 1976: 426-52.

Hayashi, Paul M., Sevier, Melanie, and Trapani, John M., "Pricing Efficiency Under Rate-of-Return Regulation: Some Empirical Evidence of the Electric Utility Industry," Southern Economic Journal, 51, January 1985: 776-92.

Hayashi, Paul M., Sevier, Melanie, and Trapani, John M. " in Michael A. Crew, (Ed.), Regulating Utilities in an Era of Deregulation, London: Macmillan Press, 1987.

Herriott, Scott, "Competition and Cooperation in Deregulated Bulk Power Markets" in Michael A. Crew, (Ed.), Deregulation and Diversification of Utilities, Boston: Kluwer Academic Publishers, forthcoming 1988.

Jurewitz, John L., "Deregulation of Electricity: A View from Utility Management," Contemporary Policy Issues, forthcoming 1988.

Klein, Benjamin, Crawford, R.A., and Alchian, A.A., "Vertical Integration, Appropriable Rents, and the Competitive Contracting Process," Journal of Law and Economics, 21, October 1978: 297-326.

Masten, Scott E., and Crocker, Keith J., "Efficient Adaption in Long-Term Contracts: Take-or-Pay Provisions for Natural Gas, American Economic Review, 75, December 1985: 1083-94.

Nelson, Jon P., Roberts, Mark J., and Troup, Emsley P., "An Analysis of Ramsey Pricing in Electric Utilities" in Michael A. Crew, (Ed.), Regulating Utilities in an Era of Deregulation, London: Macmillan Press, 1987.

Norton, Seth W., "Regulation, the OPEC Oil Supply Shock, and Wealth Effects for Electric Utilities," Economic Inquiry, 26, April 1988: 223-38.

Williamson, Oliver E., Markets and Hierarchies: Analysis and Antitrust Implications, New York, The Free Press, 1975

Williamson, Oliver E., "Credible Commitments: Using Hostages to Support Exchange", American Economic Review, 75, September 1983: 519-40.

Chapter 90

COMPETITION, REGULATION AND THE FUTURE OF ELECTRIC UTILITIES

J. C. Moorhouse

INTRODUCTION

The energy crisis and double digit inflation of the late 1970s has had a major and on-going impact on the electric utility industry. The effects have been protracted because the events of the late 1970s ushered in significant institutional and policy changes governing the industry. As a result, Investor-Owned Utilities (IOUs) are caught between growing competition from independent suppliers of electricity and increasingly outmoded and inflexible state utility regulation.

Reacting to the energy crisis, the federal government adopted new policies affecting the electric utility industry -- principally the passage of the Public Utility Regulation Practices Act of 1978 (PURPA) and the promulgation of more market oriented regulations by the Federal Energy Regulatory Commission (FERC). The goal of PURPA was to lessen the nation's dependence on high priced imported oil and nuclear generation. PURPA attempted to accomplish this by fostering the development of small scale plants using alternative fuels and by encouraging the conversion of waste energy into electricity through co-generation. PURPA opened markets for these independent producers by requiring utilities to purchase the power offered by "qualified facilities."

Long after deregulation of the domestic petroleum and gas industries contributed to ending the energy crisis, the legacy of PURPA, indeed the more significant and largely unanticipated result, has been to create a less stringently regulated independent generating sector linked to vertically integrated utilities subject to traditional regulation. Thus IOUs face market competition which they are ill-suited to meet given the regulatory constraints under which they operate.

The high and accelerating inflation of the late 1970s had an immediate adverse impact on electric utilities. First, inflation boosted construction and operating costs. Second, because anticipated inflation quickly became embedded in nominal interest rates (in the form of an inflation premium), the cost of obtaining capital to finance increasingly costly construction projects rose sharply. Moody's average yield on utility bonds peaked at 17.56 percent in September 1981 [20]. Third, the uncertainty caused by unanticipated changes in the rate of inflation itself led to the downgrading of utility bond ratings that, in turn, raised the risk premium on loans to electric utilities. Fourth, the inflationary increase in costs put a premium on obtaining timely approval from regulators of electric rate hikes and construction permits.

Significant as these effects were, the long-term results of double digit inflation are more important. Sharply rising costs and electric rates galvanized and then institutionalized consumer opposition to utilities' requests for rate hikes and approval of plans to add new generating capacity. In short inflation politicized utility regulation.

The 1980s have witnessed IOUs squeezed between growing competition on the one hand and increasingly politicized regulation on the other hand. As Charles G. Stalon, FERC Commissioner, observes, "...the current electric utility system is not sustainable" [19; p.5].

The above thesis is developed by discussing, in the next two sections, how the politicization of regulation has become a major source of uncertainty for IOUs and the implications of growing competition. The section following those summarizes the current financial plight of IOUs and how they are adapting to this adverse financial environment. Finally, the last section presents the positive case for deregulating electric generation.

THE POLITICIZATION OF UTILITY REGULATION

Prior to the economic dislocations of the late 1970s, there existed a reasonably well defined "compact" between utility regulators, acting on behalf of the public interest, and private electric utilities. Among other things, the compact provided that utilities would build large efficient central power stations and stand ready to meet all demands for electricity in a given service area, at reasonable (regulated) rates. In return, utilities were granted exclusive (monopoly) franchises and were guaranteed that they would recover the capital costs plus a fair rate-of-return on prudently made long-lived investment.

While it is well-known that this regulatory framework is inefficient, the point to be made here is that the compact provided a high degree of certainty [7, 14]. Even as utility managers coped with the problems of estimating future demand and costs when planning long-run investment, managers could rely on the basic ground-rules set forth by the regulatory compact. That certainty evaporated in the 1980s as Public Utility Commissions (PUCs) unilaterally modified the compact.

Inflation, higher (real) fuel prices, and new environmental and safety regulations all pushed the costs of generating and distributing electricity upward. These sharply rising cost trends, along with the failed promise of nuclear power, spawned a politically astute consumer movement that actively opposed utility rate requests and construction proposals at PUC hearings. Many PUCs permitted rate hearings to become political forums whereby the technical assessment of changing cost conditions degenerated into political negotiations over "fair" rate schedules. Consumer advocates effectively used procedural rules to force costly delays of hearings. The resultant rate shock and cost overruns only increased the political pressure brought to bear on PUCs.

Equity, however ill-defined, gradually supplanted efficiency at center stage. The regulatory agenda was shifted away from the traditional concern with the reliable efficient provision of electricity in the long run toward answering the question -- who should pay for the heavy capital investments of the industry. PUC deliberations took on some of the characteristics of a zero sum game -- where what we (ratepayers and taxpayers) gain, they (utility stockholders and bondholders) lose, and vice versa. Outcomes of the game (regulatory decisions) came to be heavily influenced by the short-run concatenation of political forces.

To the extent that the regulatory process was politicized, it gave rise to two unfortunate consequences. First, it diverted PUC attention toward a myopic balancing of the interests of various "groups" and away from assuring that a financially healthy industry could continue to meet future energy demands in an economically efficient and environmental sound way. Second, the regulatory process itself became a major source of uncertainty compounding the difficulties of IOU long-run planning. *Ex ante* it became very difficult to anticipate when and whether rate hikes and construction permits would be approved.

Worse, this regulatory induced uncertainty was exacerbated when PUCs began *ex post* reviews of construction projects. Such reviews, in many cases, led to abandonement or the disallowance of a portion of capital costs in the rate base. This second-guessing dramatically increases the uncertainty facing IOUs. Yet the *ex post* redistribution of sunk construction costs suits the politics of the situation perfectly. Indeed, it is a logical extension of pressure group politics because it appears to be a purely redistributive zero sum game.

Abandonement

The May 1988 abandonement of the Shoreham Nuclear Plant, completed in 1983 at a cost of $5.3 billion, represents only the most recent on a list of twenty major plant abandonements between 1980 and 1988. Others, including Seabrook owned by bankrupt Public Service of New Hampshire, seem likely to join the list. Some experts estimate the value of power plants abandoned during the decade at over $60 billion [2]. The waste of resources is staggering.

In the case of abandoned plants, the resources are in place, the costs are irretrievably sunk and the only question for the PUC becomes who shall pay for the unused plant. Martin Zimmerman persuasively argues that the willingness of PUCs to allocate the burden between ratepayers and stockholders introduces a systematic bias leading to the abandonement of socially useful power plants [25].

Let sunk costs of a partially completed plant equal S and the present discounted value (PDV) of the costs of finishing the plant and operating it over its lifetime equal M. In deciding whether to complete or abandon the plant, the socially optimal criterion would be to compare M to the PDV of the best alternative means of supplying electricity over the same time period. Let those costs, including conservation measures, equal A. Thus, if $M>A$, the plant should be abandoned and, alternatively, if $M<A$, the plant should be completed and operated. However, if PUC respond to interest group pressure that is not the criterion that is likely to be employed.

To see this, consider the payoff matrix recorded in TABLE 1 [from 25]. The choices are to continue or cancel the project and each element in the Table represents the costs to the utility or the ratepayers from the decision. The costs are defined above.

TABLE 1. PAYOFF MATRIX FOR UTILITY, RATEPAYERS AND SOCIETY

	UTILITY	RATEPAYERS	SOCIETY
Continue	$S+M-M$	$-(S+M)$	$-M$
Cancel	$-(1-\gamma)S$	$-(\gamma S+A)$	$-A$

Where a minus sign means an imposed cost and γ is that proportion of sunk costs that the PUC allows the utility to recover.

Assume for a given project $A>M$ and thus the plant would be socially optimal. Pressure by ratepayers (consumer groups) to cancel the project will still be exerted as long as: $S+M>\gamma S+A$, which seems likely. In fact, the lower γ the more likely that socially optimal plants will be abandoned. Only if $\gamma=1$, is the socially efficient criterion ($M \gtrless A$) utilized. The debate over who should pay for the power plant is really a debate over the value of γ. The more effective the pressure of consumer groups, the lower γ and the more resources that can be expected to be wasted. The world is not perfect. Decisionmakers make mistakes; resources can be wasted. But the phenomenon under discussion here is the wasting of potentially valuable resources precisely because the regulatory process has been politicized. The decisionmaking process that has evolved is not a sound institutional framework for reaching optimal decisions.

Prudency Reviews

As noted above, the compact guaranteed capital cost recovery plus a reasonable return on prudent utility investment. Presumably this meant that IOUs would seek prior approval of PUCs for proposed construction projects and would attempt to justify any cost overruns as construction proceeded. Evolving regulatory standards however, have turned an *ex ante* approval process into an *ex post* review. So-called prudency reviews are after-the-fact reassessments of a utility's construction program which may result in the PUC disallowing a portion of construction costs for inclusion in the rate base.

The rationale for prudency reviews is twofold: i) to check massive cost overruns and ii) to deny profits on construction of unneeded capacity. Such reviews have been interpreted as the regulatory substitute for the market test of investment. Responding to market signals, investors may undertake new investment. If they gauge the signals properly, investors earn profits. If mistakes are made, losses are generated. Open markets constitute a profit and loss system for organizing productive resources. But the above analogy must not be pushed too far. First, there is an asymmetry in the treatment of utilities by regulators. PUCs seldom permit utilities to earn supra-normal profits on unusually sound investments; even as they penalize utilities for mistaken investments. Second, utility managers are not responding to market signals but to an economic environment heavily influenced by regulators. Thus, regulators are responsible for much of the information set relied on by utilities in making long-run investment decisions. Third, if prior approval was sought and granted, PUCs cannot go back and rewrite history without generating additional uncertainty. Finally, fourth, the allocation of costs for completed projects between ratepayers and stockholders seems especially susceptible to political pressures.

Investment decisions are made *ex ante* given the information available at the time. During the 1970s, when the demand for electricity was growing rapidly (7-8 percent per annum), it is not surprising that utility managers looking ahead a decade (the average lead time for bringing new capacity on stream) over forecast the need for new plants. But that became known only after-the-fact as plants were completed during an era of reduced growth in demand. Prudency reviews, of course, change none of the historic costs -- the plants exist; the resources have been committed. *Ex post* reviews only serve to redistribute costs not to reduce them.

The real costs of prudency reviews, like those of abandonement, are largely hidden. Such second-guessing creates tremendous uncertainty about whether utilities can expect to recover the costs of future construction. As a senior official of Union Electric Company (Missouri) said, after $384 million of the company's Callaway project's $2.98 billion construction costs were disallowed, "We don't know where we are; we don't know how we're going to get this job done." In this respect, the San Onofre Plant case is instructive. In 1985, the staff of the California Public Utility Commission recommended that $1.77 billion of the $4.51 billion in construction costs of the San Onofre Plant be disallowed. The CPUC also announced that a complete review would take an additional four years (1990). By 1986, the review had already cost $30 million, and generated 10,000 exhibits and some 20,000 pages of transcripts. Meanwhile the CPUC adopted a "risk-sharing formula" which reduced the recommended disallowance to $1.13 billion. Part of which is explained by $293 million in cost overruns attributable to Nuclear Regulatory Commission licensing delays. This drawn-out maneuvering is part and parcel of the arbitrary uncertain economic climate generated by regulation [16; p.1]. In some cases, the situation is so bad, utilities simply negotiate an automatic disallowance rather than bear the cost and uncertainty of a prudency review. In other words, some utilities have been driven to plea-bargaining.

In summary, state utility regulation is seriously flawed when it becomes a major source of uncertainty. Politicized negotiations over the timing and extent of future rate hikes and unnecessary delays in obtaining approval for construction proposals make long-run planning by utilities difficult. But those difficulties are magnified when PUCs engage in after-the-fact reviews that increase the risk of capital recovery from investments already in place. Construction of power plants by regulated utilities has become a very risky business.

GROWING COMPETITION IN ELECTRIC GENERATION

The enactment of PURPA and the liberalized rules of FERC, combined with the inability and unwillingness of traditional utilities to build new generating capacity, has created a market niche for independent producers of electricity. Independent generating capacity is being built by three types of producers: co-generators, small scale plants using alternative fuels, and independent power producers.

By classifying co-generators and small scale plants as non-utilities, thus exempting them from inflexible state regulation, and requiring utilities to purchase the electricity offered by qualified facilities, PURPA provides the incentives for these independents to develop new generating capacity. Assured of markets for their power and freed of the high risks currently facing regulated utilities, co-generators and small plant operators are rapidly developing this market niche. By contrast, Independent Power

Producers (IPPs) enjoy few of the incentives provided by PURPA. Yet because of the regulatory constraints and concomitant risks faced by traditional utilities, market opportunities exist for IPPs to build capacity and sell power to IOUs. Unlike co-generators and scale plant operators, IPPs are building traditional generating stations. In addition, though utilities are not required to purchase electricity from IPPs, they do so because of the favorable terms negotiated under long-term contracts. Competitive pricing is emerging in the wholesale market for power.

TABLE 2.NEW SOURCES OF INDEPENDENT POWER (by year on line, in MW capacity)

State	1988	1989	1990-1992	TOTAL
Alaska	16	0	0	16
Arkansas	0	3	0	3
California	1564	1823	289	3676
Colorado	70	0	0	70
Connecticut	127	286	98	511
Florida	12	153	73	238
Hawaii	57	0	0	57
Illinois	93	0	20	113
Indiana	7	0	5	12
Iowa	194	0	0	194
Maine	39	55	85	179
Massachusetts	64	109	398	571
Michigan	11	136	1350	1497
Minnesota	0	37	0	37
Missouri	14	15	0	29
Montana	30	0	0	30
Nebraska	8	0	0	8
Nevada	48	0	0	48
New Hampshire	51	12	0	63
New Jersey	221	304	179	704
New York	62	185	178	425
North Carolina	0	5	73	78
Ohio	6	0	0	6
Oklahoma	0	106	0	106
Pennsylvania	612	292	570	1474
Rhode Island	0	0	20	20
South Carolina	0	10	0	10
Texas	249	417	0	666
Utah	0	0	45	45
Virginia	135	15	547	697
West Virginia	3	0	20	23
TOTAL	3693	3963	3950	11,606MW

Source: "New Sources of Independent Power" (a survey), Power Engineering, Volume Ninety-two, Number Five, May 1988, pp. 29-37. Independent projects for which neither capacity nor year on line are reported are excluded.

TABLE 2 presents estimates of new generating capacity to be brought on line between 1988 and 1992 by co-generators, small-scale plant and IPPs. The total amount of capacity to be built during this five year period is approximately 11.6 gigawatts. Though growth of independently owned and operated generating plant is rapid, it still represents a small proportion of total generating capacity in the U.S. Nonetheless this margin of independent generating capacity is exerting competitive pressure on IOUs. By-passing the system by large industrial customers, self-generation, and the necessity of competing with IPPs to supply independent distribution companies, are all sources of emerging competition in the electric power industry.

Some utilities are joining the independents by selling off generating capacity and creating unregulated subsidiaries or by establishing co-generation subsidiaries. An example of the former is Tucson Electric Power sale of its Alamito Plant in 1985 [18]. Alamito is an independent wholesale power company that sells electricity to TEP and San Diego Gas and Electric under long-term contract. An example of the latter is Southern California Edison's creation of its co-generation subsidiary Mission Energy. Mission Energy is involved in joint ventures operating 915 MW of generating capacity with an additional 500 MW under construction in California and 600 MW in other states [22]. Thus traditional utilities are finding ways to participate in this largely unregulated market.

Benefits of Independent Power Generation

The significance of these developments should not be underestimated. Independents are wholesaling power at negotiated rates. Because these rates reflect the expectations of both buyer and seller about future demand and supply conditions, they convey a great deal of market information. (By contrast, regulated rates are robbed of any information content.) To survive and prosper, an independent must provide reliable service at favorable rates in order to maintain present customers and to win new customers. In the short run, the more skillfully an independent can configure the time specific set of customer demands to its load capacity, the more net revenue earned. Its ability to sell under contract, at negotiated rates, and in spot markets makes obtaining an optimal configuration more likely. How very different this pricing process is from that of the regulated utility, where rate schedules are determined after costs are estimated -- a backward looking procedure. For the independent, dynamically flexible rates, responding to both changing demand and supply factors, becomes a device for optimizing costs.

In the long-run, market prices provide guidance for further investment. For example, growing demand in a regional market would increase the load on cycling and peak capacity. Because power would be produced at higher marginal costs, electric rates and net revenues would rise. In turn, higher profits would signal the desirability of building new lower cost base load capacity. Falling demand would generate equally appropriate price signals influencing future investment.

There are additional benefits from having independents tied into a regional power pool. Fuel diversity is enhanced and that provides a margin of protection against widespread disruption or rate shock should the supply of a particular fuel become excessively tight and its price volatile. Moreover, there is some evidence that plants with less than 600 MW capacity are more reliable than larger scale plants [4]. The sudden loss of 1200 MW of generating capacity is more likely if the

1200 MW is produced in a single central station than if it is provided by four 300 MW interconnected plants. Dynamic reliability is an important factor, along with static efficiency (scale economies), in the operation of a power pool. The final advantage of having small plants connected to a grid is that the law of large numbers reduces reserve requirements. This follows because networking facilitates less costly scheduling of maintenance, increased load diversity, and lower marginal costs in meeting daily demand fluctuations [13].

Problems of Independent Power Generation

Ten years after its passage, PURPA and attendant FERC rulings have created a two-sector electric generation industry. One sector consists of largely unregulated independents and the other highly regulated utilities. The co-existence of regulated and unregulated generators is problematic.

First, PURPA guidelines require utilities to purchase power from qualified facilities at "the incremental cost to the utility of alternative electric energy". However, the determination of appropriate rate schedules consistent with the avoided cost standard is left to state PUCs. In some cases, state PUCs have interpreted the standard to mean avoided fuel and operating costs with less consideration of capacity costs. This in turn provides an incentive for independents to construct fuel intensive (capital saving) relatively small, short-lived facilities with short payback periods. In other cases, avoided cost formuli put a premium on burning fuels that are currently cheap, natural gas for example, but may not be economical over the long-haul [9].

Second, efficient co-generation converts a thermal residual into electricity as a by-product of its primary production of steam or hot water. Current federal regulation requires a qualified co-generator to sell a minimum of five percent of its thermal load as steam or hot water. As a result, fossil fuel plants are being designed to produce electricity for resale to utilities while meeting the minimum requirement. These "PURPA machines" are not constructed to take advantage of the potential efficiency of co-generation. Indeed, while they are qualified facilities, some PURPA machines are not as efficient as modern combined cycle plants which, under present rules, do not qualify. In short, federal guidelines as interpreted by PUCs do not necessarily encourage the use of the best available technology [24].

Third, some PUCs require utilities to entertain bids from independents for the construction and operation of new generating capacity. If an IOU cannot build a similar facility at the same or lower cost, presumably it must contract with the independent. Under present state utility regulation, few utilities are going to be able to compete with unregulated independents in building new capacity. There is nothing wrong with a system of competitive bidding, but in this case, some potential participants, regulated utilities, are severely handicapped. To the extent that a utility cannot or does not participate in building new capacity by establishing an independent subsidiary, that utility will suffer the prospects of a shrinking, depreciating rate base and reduced earnings over time.

Independents can rely on flexible rates to optimize operations and can earn rates-of-return commensurate with the market risks associated with building capacity. By comparison, regulated utilities face a high degree of policy induced risk, must live with relatively inflexible administered rates, and earn regulated rates-of-return too low to encourage the building of new power plants. The irony is that competition between these two sectors may not lead to developing the most socially efficient and environmentally sound inventory of power plants. While the industry is being opened up by market forces, remaining regulations prevent utilities from building capacity and induce independents to build less than the most efficient plants possible.

This undesirable and unstable situation is likely to end in one of two ways. One, policymakers will deregulate electric generation and allow meaningful competition to replace regulation as the governing principle in the electric power industry. Two, little or nothing will be done and the industry will evolve out from under outmoded state regulation. The transition period will be wasteful and drawn out. Over time independents and unregulated utility subsidiaries will own and operate an increasing proportion of generating capacity. Utilities will operate and gradually retire an antiquated, depreciating inventory of fossil fuel and nuclear plants. *Ad hoc*, piecemeal regulatory reforms, popular with short-sighted risk averse policymakers and politicians, would only slow the process and waste additional economic and environmental resources.

A specific deregulation proposal is outlined in the last section of the paper. But first the current financial plight of IOUs, along with their methods of adapting to it, are reviewed.

THE FINANCIAL SQUEEZE ON INVESTOR-OWNED UTILITIES

The consequences of inflexible regulation in the face of changing technology and economic conditions are readily apparent and fully reflected in financial markets. By any index, the financial health of IOUs has deteriorated, and for some utilities, seriously, since the mid-1970s. Utility stocks have consistently sold below book value since 1975. Price earnings ratios have trended downward for twenty years. What was once thought of as a conservative investment now appears relatively risky. (See Table 3). This higher perceived risk increases the cost of capital to electric utilities.

In debt markets, new issues of electric bonds now command higher average returns than for all new corporate issues. (See Table 4). Where once utility bonds sold at a premium, compared to the corporate average, they now go at a discount. Both major bond rating services downgraded the quality of electric utility bonds between 1976 and 1986. Of 39 electric utilities rated, Forbes rated 14 (36 percent) as having low or very low stability in their future expected earnings [8; p.120]. Again, this evidence reflects the higher risk associated with providing capital to the industry. (See Table 5 for Moody's rating of electric utility bonds).

TABLE 3. EQUITY MARKET FINANCIAL RATIOS (For 24 Electric Utilities in Moody's Index)

(1) Year	(2) Price/Earnings	(3) Price/Book Value
12/87	7.5	--
86	5.8	--
85	6.9	0.77
80	6.1	0.53
75	6.6	0.60
70	11.5	1.17
65	19.8	2.22

Source: Moody's Public Utility Manual, vol. 2, 1986.
Moody's Public Utility News Report, Dec. 29, 1987.

TABLE 4. DEBT MARKET

(1) Year	(2) Utility Rates	(3) Corp. Rates	(4) Diff. Col. 2 - Col. 3
1985	11.83%	11.71%	+0.12
1980	13.46	12.60	+0.86
1975	9.97	9.42	+0.55
1970	8.89	8.85	-0.06
1965	4.61	5.59	-0.98
1960	4.72	4.82	-0.10

Source: Moody's Public Utility Manual, vol. 2, 1986.

TABLE 5. RATINGS OF ELECTRIC UTILITY BONDS

Rating	1976 Proportion of Bonds (maturing 1/1981-12/1990) n = 567	1986 Proportion of Bonds (maturing 1/1991-12/2000) n = 801
Aaa	4.7%	4.0%
Aa	33.2	25.9
A	34.0	37.3
Baa	25.4	27.2
Ba	2.3	3.5
B	0.3	2.0
Caa	0.0	0.1
TOTAL	100%	100%

Source: Moody's Public Utility Manual, vol. 2, 1986.

The financial plight of IOUs has reduced capital expenditures sharply. The industry has no current plans for construction of future base load capacity. Current investment expenditures are being made to complete generation, transmission and distribution systems planned a decade ago. Real expenditures have fallen annually since 1980 and have declined as a proportion of total (non-residential) private investment since 1975 (currently the proportion is 5.5 percent compared to the historic average of 10 percent). Even these investment figures are exaggerated because much of the capital expenditures by electric utilities in the last decade and a half were for anti-pollution equipment not productive capacity. This can be seen by the downward trend in the growth of new generating capacity since 1975. Between 1970 and 1975, the average annual rate of growth of new capacity was 8.7 percent; between 1980 and 1985, that rate had fallen to 2.2 percent. (See Table 6).

TABLE 6. ELECTRIC UTILITY INVESTMENT

(1) Year	(2) Total E.U. Investment (Billions)	(3) Prop. of Total Non-Residential Investment	(4) Real Total E.U. Investment
1987	$26.50	5.5%	$9.49
1986	33.60	7.3	12.53
1985	36.11	7.9	13.60
1980	28.12	8.7	13.78
1975	17.00	10.4	12.04
1970	10.65	10.1	10.65
1965	--	--	--
1960	--	--	--

(1) Year	(5) Installed I.O.U. Capacity (in gigawatts)	(6) Aver.Annual Growth Rate of Capacity
1987	--	--
1986	--	--
1985	530.4	
	>	2.2%
1980	477.1	
	>	3.7
1975	399.0	
	>	8.7
1970	262.7	
	>	8.1
1965	177.6	
	>	6.7
1960	128.5	

Source: Cols. 2, 3, & 4 - Federal Reserve Bulletin, various years.
Col. 5 - Moody's Public Utility Manual, vol. 2, 1986.
Col. 4 - represents total Electric Utility investment deflated by GNP deflator, 1970=100.

Some commentators are optimistic about the near future for IOUs because of recent up-ticks in several financial indices of the industry. But this may be a serious misreading of the evidence. Currently the electric power industry has excess generating capacity. (Although, it should be noted, reserve margins differ widely by geographical region.) And, as noted previously, electric utilities have no current plans for future construction of base load capacity. The industry already has swallowed tremendous cost overruns and some estimated $38 billion in costs disallowed by prudency reviews. Under such circumstances, even modest growth in the demand for electricity, at (say) 3.0 percent per annum, promises to improve net earnings over the next five years. But this may prove a temporary respite for the industry. The real crunch comes in the mid-1990s when reserve margins are normal or even low (by historic industry standards) and no new capacity is being readied to bring on stream. Thus the modestly improving financial health of the industry in the immediate future must not obscure the very real problems that lie ahead. This is especially important for an industry with such long lead times in planning additions to productive capacity.

Projections of future installation of future capacity have also been revised downward. (See Table 7).

TABLE 7. PROJECTED ADDITIONS TO NAMEPLATE CAPACITY

(1) Year	(2) Dec. 1981 Projection	(3) Dec. 1986 Projection	(4) Diff. Col.3-Col.2
1987	13.6	16.0	+2.4
1988	11.4	4.9	-6.5
1989	12.3	6.2	-6.1
1990	6.2	5.8	-0.4
1991	8.9	5.2	-3.7
TOTALS	52.4	38.1	-14.3

Source: Inventory of Power Plants in U.S., Energy Information Administration, 1986 and 1981. (August 1987 and August 1982.)

Methods of Adaption

The financial squeeze induced by inflexible regulation has forced electric utilities to adapt in ways that may not serve the consumer and utility owners, that are inefficient, and that may leave the industry ill-equipped to meet the future demand of this increasingly desirable form of energy.

Capital Minimization: Given the uncertain future, many electric utilities are reducing their exposure to risk (arising from increased uncertainty over whether or not they can recoup capital costs) by: drastically reducing capital expenditures, selling off existing capacity and capacity under construction, and participating in joint ventures whereby two or more utilities finance new construction. Not all of these adjustments are economically unsound, of course. Presumably, plants and transmission lines are acquired by firms that are capable of using the capacity more effectively or operating the plants at lower costs. Thus the market for power plants serves to increase the efficiency with which electricity is generated and distributed. By the same token, joint ownership is an obvious device for diversifying risk. But that the industry is engaging in "capital minimization" is not a healthy sign. Sharp reductions in capital investment may be interpreted as a sensible short-run adjustment to excess capacity. But given the long lead times necessary to bring new capacity on stream, this strategy entails future risks to the nation's economy.

Demand Management: To forestall shortfalls and to meet summer and winter peak demands, utilities have adopted a variety of practices including: the (limited) adoption of seasonal and time-of-day pricing; importing electricity; purchasing power from other utilities, and attempting to influence demand through a variety of conservation programs. Again, these measures are not necessarily inefficient. It depends on the circumstances. Wholesale markets for electricity undoubtedly rationalize capacity and increase efficiency. The same is true for importing electricity from Canada and Mexico when the negotiated price is less than or equal to the marginal cost of the utility producing the same amount of time specific power. On the other hand, buying old appliances, performing energy audits, and providing low or zero interest loans to residential customers to finance the installation of insulation, solar hot-water systems, and heat pumps is patently inefficient. Conservation is better achieved by consumers responding to higher electricity prices (which fully reflect the economic and environmental costs of production and distribution) by reducing their own demands for electricity and deciding what conservation measures to adopt, given their personal preferences and circumstances, rather than adopting a particular measure because a local utility happens to subsidize it. Where higher electricity prices work a demonstrable hardship, a case can be made for direct subsidies to affected families from the state. But it is inefficient and burdensome to management for electric utilities to administer subsidy programs.

Diversification: Another way utilities are responding to low and uncertain returns is to diversify [3, 23]. Typically acquired firms join a family of companies under the umbrella of a utility dominated holding company or become wholly owned subsidiaries of a utility. The pattern of investment in non-traditional areas is quite varied but seems explicable in terms of: 1) elements of vertical integration; 2) reliance on the service expertise utilities have developed, or 3) by-products of electricity generation. The first involves closely related ventures in fuel exploration and transportation. The second includes power plant construction and operation, energy management services, telecommunication, oil and gas pipelines, (presumably based on expertise of managing a regulated utility), real estate development (expertise in land acquisition -- an underutilized department in an era of no growth) and timber management. The third area includes, for example, the establishment of fish hatcheries.

Given the unfavorable regulatory environment discussed above, utility diversification can be explained as an attempt to move the center of the business away from the regulated sector. If returns are in fact suppressed by regulation, this would explain diversification into both related and unrelated businesses. Moreover, the possibility of realizing cost savings through vertical integration or by producing and marketing two or more related products (economies of scope) provides an additional incentive to acquire closely related enterprises. Finally, risk averse utility managers may opt for conglomerate acquisitions to reduce the volatility of their compensation.

Electric utility diversification raises many regulatory issues. For example, economies of scope are realizable only through central management, but consolidated management of regulated and unregulated enterprises may present regulators with an impossible situation. How could regulators meaningfully determine whether subsidiaries were paying an appropriate share of common costs? Would not managers have an incentive to charge common costs against the utility? Similarly, vertical integration raises questions about genuine arm's-length dealing with subsidiaries that supply resources to the utility. Furthermore, how are ratepayers to be protected from being forced to subsidize financially troubled subsidiaries? More generally, how will the economic fortunes of diverse enterprises within a single corporate shell be kept separate? These are real problems and raise issues about which regulators have comparatively little experience.

It is not clear just how successful diversification is proving. The trend is too new to have generated much evidence one way or the other; no utility is so diversified that that is having a significant impact on its financial statement, and, finally, accounting conventions and tax provisions seem likely to obscure the underlying economic realities of diversification for some time. The consequences of utility diversification remain difficult to assess [5].

By the same token, to the extent that farflung diversification has been spurred by an unfavorable regulatory environment, there is cause for concern. If the trend accelerates, as seems likely under current conditions, the electric power industry could look very different in the year 2000. Already serious policy questions have been raised. Clearly, diversification should not be allowed to divert capital that might otherwise be used to finance the modernization and expansion of the industry. Furthermore, diversification cannot be allowed to distract utility executives from meeting the challenges that lie ahead for the power industry.

DEREGULATION OF ELECTRICITY GENERATION

Increasing experience with the competitive forces already unleashed in the electric power industry; growing recognition that electricity markets are regional, even national, rather than local, and growing dissatisfaction with outmoded and inflexible regulation will lead to some form of deregulation. Though many proposals for deregulation exist, given what is known, the most feasible reform appears to be that of deregulating generation capacity.

The typical proposal has the following elements [1, 13, 17, 21]:

i) separation of the generation, transmission, and distribution functions into independent corporate entities. With some exceptions, this would involve divestiture and the end of the vertically integrated electric utility.
ii) ending all state price and rate-of-return regulation of independent generating companies.
iii) ending all territorial franchises thus permitting entry into relevant markets.
iv) the federal regulation of transmission companies as common carriers with open access to all; continued state regulation of distribution companies as natural monopolies.
v) ending all direct and indirect subsidies to government-owned generating capacity.
vi) continued compliance with all applicable environmental protection regulation.

Restructuring the Industry

The above proposal contemplates a fairly radical restructuring of the industry. The cornerstone of the proposal is to substitute competition for vertical integration and regulation as the governing relationship between generation and distribution. Independent generating companies tied into regional power pools, would compete to sell electricity to independent distribution companies and large industrial users of electricity. Contract terms, including time specific prices and contract length, would be negotiated. Contract price schedules would not necessarily be fixed but could include escalator clauses or even provide that spot prices be charged.

Electric utilities already have considerable experience in negotiating long-term contracts. Many utilities have contracted for a portion or all of their fuel requirements for periods up to fifty years. Utility managers have learned how to design contracts with provisions covering price escalation, quantity and quality of fuels, contract duration, and methods for adjusting one or more of the above provisions over time [11,12]. Standalone distribution companies have a long history of purchasing electricity by contract. Furthermore, evidence from the natural gas industry suggests that independent suppliers can work well with transmission and distribution companies in serving customers reliably and efficiently [6].

In addition, generating companies might opt to sell some electricity on spot markets.

Because distribution companies could shop around, generators would have an incentive to maintain service standards and offer reasonable rates. A generator's ability to negotiate contract prices and to sell in spot markets, would afford it a great deal of latitude in configuring customer (time specific) demands to its own (time specific) load capacity. The more skillfully it designs its pricing and service schedules, the more efficient its operations and the higher its net revenues.

Self-generation by heavy industry; access to the power pool by co-generators and small scale plants; the option of importing electricity; the threat of entry by new companies proposing to build generating plants, and competition with alternative energy sources all would serve to place an external competitive check on the pricing and service of the independent generating company. Moreover, competition would spur the use of the best available technology including that of combined cycle generating plants and the practical application of developing technologies such as fluidized and gasified bed combustion and more efficient transmission techniques.

Competition at the generating stage would be complemented by coordination at the transmission stage. Indeed, assuring the access of independent generating companies and co-generators to the transmission grid is crucial to the success of this deregulation proposal. The transmission company would provide the physical linkage among the generating companies and the distribution companies in the regional power pool. As such, the transmission company would be a regulated common carrier, providing access to the grid, and charging a regulated markup on electricity transmitted and sold by the company. Its main function would remain that of providing efficient central dispatch.

In this scenario, distribution companies would remain regulated. Retail electric rates for residential and commercial customers would not fluctuate with spot prices in wholesale markets, although, retail prices might vary with the time of day, depending on metering costs.

Consequences of Deregulation

While studies indicate that for most areas of the country, workably competitive wholesale markets for electricity are sustainable under deregulation, the studies also suggest that retail electric rates would climb from 15 to 30 percent [10]. The rate hikes are accounted for by two principal factors: i) the possibility of higher tax payments to the federal government after the financial restructuring of the industry. ii) the ending of consumer subsidies. These rate increases would be mitigated by the increased efficiency in the generation of electricity. Moreover the rate increases, though potentially significant, represent a short-run transitory phenomenon. In the long-run, studies suggest retail rates would be lower than under the present regulatory framework. The next few years offer an unusually propitious time to deregulate because current excess capacity could ease upward pressure on electric rates.

Some critics of deregulation argue that the substitution of market determined rates for regulated rates will increase rate volatility and risk. Assume that is true. Taking a cue from recent financial market deregulation and innovation, it is easy to envision the development of forward markets in electricity that would permit market participants to hedge against risks associated with rate fluctuations. Indeed, buy and sell options could serve as insurance protecting generating companies and distribution companies from losses arising from contract terms that are no longer appropriate given changed economic circumstances. Meanwhile defective provisions could be renegotiated.

The major advantages of deregulation include: i) more efficient generation of electricity; ii) more rational long-run investment planning; iii) better opportunities for utilities to compete successfully with independents; iv) a higher probability of entry into the industry; v) a decreased interest on the part of utilities to diversify; vi) elimination of regulatory lag and a reduction in the costly delays associated with construction of new facilities, and vii) stimulation of research and development of new energy producing and saving technologies.

The chief problems associated with deregulation are threefold. One, the divestiture implied by this deregulation proposal raises a number of questions about the financial restructuring of the industry. How are the assets of a vertically integrated utility to be separated so as to equitably protect the interests of current bondholders and stockholders? Are equity and debt markets ready to finance huge new construction projects in a newly competitive industry? Two, if regulators permit distribution companies to pass through the average cost of buying electricity from independent generating firms, that blunts the incentive for distribution companies to search for the lowest wholesale prices for electricity. But what option exists, short of having regulators reaching back into the generation stage and monitoring wholesale prices? A policy must be formulated that permits distribution companies to share in the gains arising from the purchase of electricity from the lowest cost supplier. Three, formidable interest groups will mount a political campaign to prevent deregulation.

In the electric power industry, old institutional arrangements are giving way to new economic realities. Emergent market forces, though not fully formed, already are providing incentives to be more efficient. However, regulation of electric generation continues to impede the process of transition, to create inequities, and to waste resources. Traditional utility regulation, largely unchanged since the 1920s, is too inflexible to cope with changing economic and technological conditions; has itself become a major source of uncertainty, and cannot serve as the basis for dealing with emerging regional markets. The introduction of meaningful competition, by deregulating electric generation, now seems technologically feasible and economically desirable.

REFERENCES

1. Alternative Models of Electric Power Deregulation. Economics Division, Washington, D.C.: Edison Electric Institute, May 1982.

2. "Are Utilities Obsolete?" Business Week (21 May 1984): 116.

3. Beck-Dudley, Caryn L. and J. Robert Malko, "Diversified Regulatory Approaches to Electric Utility Diversification," Proceedings of the American Business Law Association, 1987.

4. Berlin, Edward; Charles J Cicchetti; and William J. Gillen. Perspectives on Power: A Study of the Regulation and Pricing of Electricity. Cambridge, Mass.: Ballinger Publishing Co., 1974.

5. Brewer, H.L. and Morteza Rahmatian. "Risk Adjusted Performance Measures of Diversified Public Utility Firms," Journal of Energy and Development, Vol. XII, #2, Spring 1987, 185.

6. Canes, Michael E. and Donald A. Norman. "Long-Term Contracts and Market Forces in the Natural Gas Market," Journal of Energy and Development, Vol. X, #1, Autumn 1984, 73.

7. Crew, M.A. and P.R. Kleindorfer. The Economics of Public Utility Regulation. Cambridge, Mass.: The MIT Press, 1986.

8. "40th Annual Report on American Business," Forbes, January 11, 1988, 120.

9. Henderson, Yolanda K, et.al. "Planning for New England's Electricity Requirements," New England Economic Review, Jan./Feb. 1988, 3.

10. Hobbs, Benjamin F. and Richard E. Schuler, "An Assessment of the Deregulation of Electric Power Generation Using Network Models of Imperfect Spatial Markets," Papers of the Regional Science Association, Vol. 57, 1985, 75.

11. Joskow, Paul L. "Price Adjustments in Long-Term Contracts: The Case of Coal," Journal of Law and Economics, Vol. XXXI, #1, April 1988, 47.

12. _______. "Vertical Integration and Long-Term Contracts: The Case of Coal-Burning Electric Generating Plants," Journal of Law, Economics, and Organization, Vol. 1, #1, Fall 1985, 33.

13. Joskow, Paul L. and Richard Schmalensee. Markets for Power. Cambridge, Mass.: The MIT Press, 1983.

14. Moorhouse, John C., editor, Electric Power: Deregulation and the Public Interest. San Francisco: Pacific Research Institute, 1986.

15. "New Sources of Independent Power," Power Engineering (92 May 1988): 29.

16. "Prudency Reviews," Wall Street Journal, October 2, 1986, 1.

17. Schmidt, Ronald H., "Deregulating Electric Utilities: Issues and Implications," Economic Review of the Federal Reserve Bank of Dallas, September 1987, 13.

18. Smith, Vernon L., "Currents of Competition in Electricity Markets," Regulation, #2, 1987, 23.

19. Stalon, Charles G. "On Deregulation," Electric Potential, Vol. 3, #3, May-June 1987, 3.

20. Statistical Yearbook of the Electric Utility Industry/1982. Washington, D.C.: Edison Electric Institute, 1983.

21. Stelzer, Irwin M., "Electric Utilities -- Next Stop for Deregulators?," Regulation, July/August, 1982, 29.

22. "Update," Power Engineering (92, May 1988): 7.

23. York, Stanley and J. Robert Malko, "Utility Diversification: A Regulatory Perspective," Public Utility Fortnightly, January 6, 1983, 3.

24. Zweifel, Peter and Konstantin Beck, "Utilities and Co-generation: Some Regulatory Problems," The Energy Journal, Vol. 8, #4, October 1987, 1.

25. Zimmerman, Martin B., "Regulatory Treatment of Abandoned Property: Incentive Effects and Policy Issues," Journal of Law and Economics, Vol. XXXI, #1, April 1988, 127.

Chapter 91

WHEELING ISSUES

R. F. Shapiro

I. FEDERAL ENERGY REGULATORY COMMISSION AUTHORITY OVER RATES AND TERMS FOR, AND ACCESS TO, TRANSMISSION SERVICES

The FERC possesses exclusive jurisdiction over the rates for the sale for resale and transmission of electric energy in interstate commerce. FPC v. Florida Power & Light Company, 404 U.S. 453 (1972); FPC v. Southern California Edison Company, 376 U.S. 205 (1964); New England Power Co. v. New Hampshire, 455 U.S. 331 (1982).

Recently, in Florida Power & Light Company, Docket No. EL87-19-000, 40 FERC ¶ 61,045 (July 20, 1987), the FERC held that it had exclusive jurisdiction to determine the reasonableness of terms and conditions of a utility's transmission (wheeling) service on behalf of qualifying facilities in Florida. Florida Power & Light Company ("FP&L") sought a declaratory order from FERC that would inform the Florida Public Service Commission that the Florida rules, which purported to give Florida the jurisdiction over wheeling terms and conditions, were preempted under the Federal Power Act. The FERC agreed, on the same basis that it had previously determined that the Florida Commission was preempted from establishing the rates for wheeling service. Florida Power & Light Co. and Florida Public Service Commission, 29 FERC ¶ 61,140 (1984). As in the prior order, the FERC declined to address the issue whether the state has the authority to direct a utility to provide transmission service in the first instance.

To date, there is no FERC or court decision that clearly establishes the line between appropriate and preempted activity by states in the area of transmission services. However, FERC has traditionally taken the view that, outside of Texas, Alaska and Hawaii, there is no such thing as intrastate wheeling for purposes of setting the appropriate rates for service. See e.g., Consolidated Edison Co. of New York, 15 FERC ¶ 61,174 (1981).

While FERC's authority over the setting of reasonable rates and terms of transmission service is well established, the scope of its authority to direct utilities to provide transmission service is far less certain. Prior to the passage of PURPA, the FERC possessed no express authority to order a utility to wheel power for another utility. Those that wanted to obtain wheeling service from a recalcitrant utility resorted to the argument that it was unlawful for a utility to wheel for one party and not for others, or to restrict the scope of a particular transmission service. This argument did not get the complainant very far at FERC or in the Courts. New York State Electric & Gas Co. v. FERC, 638 F.2d 388 (2nd Cir. 1980); Florida Power & Light Co. v. FERC, 660 F.2d 668 (5th Cir. 1981); Richmond Power & Light v. FERC, 574 F.2d 610 (D.C. Cir. 1978).

However, in dicta, the D.C. Circuit in Associated Gas Distributors v. FERC, 824 F.2d 981 (1987), a case relating to FERC's Order No. 436 requirement of open access transportation for gas pipelines under the Natural Gas Act, rejected the contention that prior Court of Appeals decisions barred FERC from imposing an open-access condition on electric utilities in all circumstances. The Court's discussion suggested that FERC might be able to use open access conditions for transmission service as a remedy for anticompetitive or discriminatory conduct. 824 F.2d at 999. Several parties have petitioned the United States Supreme Court to review the D.C. Circuit's holding on open-access transportation by natural gas pipelines in the AGD case. S. Ct. Nos. 87-976 et al.

PURPA added sections to the Federal Power Act (Sections 211 and 212) to give FERC extremely limited authority to order a utility to transmit power for someone else. No qualifying facility has yet applied under these sections for an order directing a utility to provide transmission service. In fact, the language of these sections does not even make clear that a qualifying facility (other than a geothermal facility) is eligible to seek wheeling service. Assuming the eligibility to apply, a qualifying facility would face enormous obstacles in attempting to satisfy the requirements under the statute. FERC would have to hold a hearing, and could only issue an order directing a utility to wheel if it was shown that there would be a significant benefits to society, the order would preserve the existing competitive relationships (i.e., would not cause the transmitting utility to lose load), the order would not interfere with the state's authority over retail service, the order would not authorize direct transmission service to an ultimate consumer, and the order would not result in an undue burden or economic hardship to the utility or impair the reliability of any utility affected by the order. In short, it is nearly impossible to get a wheeling order out of FERC.

The Federal Power Act sections that apply, directly or indirectly, to wheeling include Sections 203 and 204 (pre-approval for securities issuances, acquisitions and dispositions -- FERC may have authority to condition approval on wheeling); sections 205 and 206 (ratemaking sections -- FERC may have authority to order wheeling to remedy anticompetitive or discriminatory conduct); and sections 211 and 212 (the limited wheeling authority added to the FPA by PURPA).

II. STATES THAT REQUIRE WHEELING FOR QUALIFYING FACILITIES

Several states have either statutory provisions or administrative regulations requiring electric utilities to wheel electricity generated by qualifying facilities. Maine has a statutory provision which requires wheeling to an electric utility as well as wheeling to an affiliated non-utility entity (self-service wheeling). 35 M.R.S.A., tit. 35(a), § 3182. According to one member of the Maine PUC Staff, it is unclear whether the PUC has authority to require interstate electric utilities to wheel. Both statutory provisions are untested. Connecticut also has a statutory provision granting authority to the state's Department of Public Utility Control to order self-service wheeling and wheeling to an electric utility, Conn. Gen. Stat. § § 16-243a and b, provided it can be shown, inter alia, that wheeling would not result in an adverse impact on the wheeling utility's customers.

Florida has an administrative regulation, Rule 25-17.88, which provides for wheeling of qualifying facility power to electric utilities. As previously noted, the Florida wheeling regulation has been subject to some dispute at FERC. See Florida Power & Light Co., 40 FERC ¶ 61,045 (July 20, 1987). Florida's regulations also provide for self-service wheeling, but only if the applicant can show that the wheeling will not have an adverse economic effect on the utility's general body of ratepayers. Rule 25.17.882. This provision has been narrowly construed so far to apply only to an affiliate that has the same owner as the owner of the qualifying facility. See Petition of Metropolitan Dade County. Docket No. 860786-EI, Order No. 17510 (May 5, 1987).

Other states have statutes and regulations requiring utilities to wheel qualifying facility power to another utility. Indiana requires utilities to enter into long-term contracts to "purchase or wheel electricity or useful thermal energy" from projects less than 80 MW in capacity. Burns. Ind. Code Ann § 8-1-2.4-4(a). The Indiana Commission defines wheeling as "transmission . . . to a purchasing electric utility." 170 I.A.C. 4-4.1-1. The Massachusetts Department of Public Utilities ("DPU"), which has a similar requirement, defines wheeling as transmission "to another utility." See § 8:03(5)(a) of the DPU's rules, Order No. 84-276-B (August 25, 1986). The Texas Commission requires the wheeling of firm qualifying facility power to another utility. See 16 T.A.C. § 23.66(d)(4).

III. FERC'S REGULATIONS IMPLEMENTING PURPA AS THEY AFFECT WHEELING

FERC's rules implementing PURPA Section 210 provide that a utility can avoid the obligation to purchase cogenerated power by wheeling the power to another utility. That second utility would then have the obligation to purchase the cogenerated power. 18 C.F.R. 292.303(d). The FERC regulations do not require any utility to wheel power for cogenerators. It would be a voluntary arrangement between the parties. However, if the local utility wants to transmit the power to a second utility, and the second utility does not offer as attractive a rate for the power, the cogenerator has the option of rejecting the transmission service. See Florida Power & Light Co., 29 FERC ¶ 61,140 (1984).

IV. PRICING OF TRANSMISSION SERVICES

FERC has traditionally priced transmission services on a fully-distributed, "rolled-in" basis. This means that all of the fixed costs of the utility's entire transmission system would be included in the determination of the appropriate wheeling rate. The entity obtaining the wheeling service would then be allocated an appropriate share of the total fixed costs of the system, based on the entity's requirement compared with the total demand on the system. See Sierra Pacific Power Co. v. FERC, 793 F.2d 1086 (1986).

There have been several attempts by parties to modify the "rolled-in" concept (1) by eliminating certain assets, like generation feed lines and subtransmission lines, from the total; (2) by establishing a joint transmission rate over more than one utility system to lower the total rate; or (3) by designating certain services as "non-firm" and therefore entitled to a lower rate. No such attempts have been successful thus far. See e.g., Florida Power & Light Company, 21 FERC ¶ 61,070 (1982), reh. denied 22 FERC ¶ 61,012

(1982), aff'd, Ft. Pierce Utilities Authority v. FERC, 730 F.2d 778 (1984).

Recently, in Baltimore Gas & Electric Company ("BG&E"), 40 FERC ¶ 61,170 (1987), the FERC finally gave a green light to a transmission pricing by auction scheme. BG&E was allowed to auction off a portion of its entitlement to use the Pennsylvania-New Jersey-Maryland ("PJM") pool transmission system. The potential bidders are the other joint owners of the PJM transmission system. FERC had previously rejected BG&E's auction proposal because of, inter alia, a lack of demonstrated efficiency gains and the use of a sealed bid arrangement. Under the new proposal that FERC approved, BG&E explained how its proposal would enhance efficiency (it will improve regional efficiencies among the PJM members) and agreed to an open bid method in lieu of sealed bids. FERC also granted BG&E's request for waiver of the filing of cost-of-service data. It remains to be seen whether this auction approach represents the wave of the future in transmission pricing.

V. THE FUTURE OF WHEELING SERVICES

1. Wheeling in Exchange for Market Pricing

In a recent decision, Pacific Gas & Electric Company, 42 FERC ¶ 61,406 (1988), FERC approved an agreement between Pacific Gas & Electric Company ("PG&E") and the Turlock Irrigation District that largely relieves PG&E of an obligation to serve Turlock in exchange for giving Turlock certain limited rights to use PG&E's transmission system to acquire alternative supplies. The agreement also permits PG&E to charge flexible prices to provide so-called Coordination Services to Turlock. Coordination Services are non-firm services. This case indicates that FERC may be willing to consider favorably proposals that allow utilization flexible or market prices in exchange for an agreement to provide wheeling services.

2. Wheeling for QFs Under PURPA

The FERC is under an obligation to issue rules to promote cogeneration and small power production (see Section 210(a) of PURPA). Congress passed Section 210 of PURPA because it believed that the increased efficiency of cogeneration and the use of renewable resources with small power production facilities were of value to the nation and must be encouraged. This value cannot be fully realized without transmission.

As previously noted, FERC's existing regulations give utilities the option of purchasing electricity from qualifying facilities or transmitting it to a neighboring utility. Under the regulations, qualifying facilities have no other market besides the local utility unless the local utility agrees to deliver the power to others through use of its transmission system. If potential qualifying facilities are limited to a single market for their product (i.e., the local utility), in many cases such facilities will simply not be constructed, even if they can economically produce power below the incremental (avoided) cost of a utility* interconnected to the local utility.

It can be argued that, when the regulations were first promulgated, FERC was not obligated to require wheeling for qualifying facilities because a competitive market for power did not exist at that time and neither FERC nor any other expert body could foresee the changes that the electric industry would undergo over the succeeding eight years. However, FERC and the industry itself have now recognized the opportunities for competition in the industry, and the need for alternative supply sources to be included to reduce the utility's avoided cost. FERC can and should use its authority under PURPA to require wheeling for QFs in order to satisfy its twin obligations under PURPA: (1) the encourage cogeneration and small power production; and (2) to establish the appropriate avoided cost rate.

As the Commission's existing regulations recognize, PURPA does not limit the purchase requirement to the local utility with which the cogenerator is interconnected. Thus, if the Commission determines that cogeneration needs to be encouraged and/or that an accurate avoided cost can only be determined by having the distant utility, which needs the power more than the local utility, purchase the cogenerated power, the Commission possesses the authority under PURPA Section 210 to issue rules that would require the local utility to take the necessary steps to transmit the power to the more distance utility. It is currently unclear whether FERC will take this step in the future.

3. Recent FERC Rulemakings Dealing With Alternative Generation of Electricity

On March 16, 1988, FERC issued three proposed rulemakings to clarify and revise the manner in which qualifying facilities sell power to utilities and to encourage independent power producers (IPPs) that do not satisfy the requirements of qualifying facilities. One of those rulemakings encourages the use of competitive bidding by a utility to acquire needed electricity from all sources of supply. Docket No. RM88-5-000. The utility is allowed to make a bid to supply itself, along with other potential suppliers.

*The "avoided cost" is the basis for utility payment for electricity provided by qualifying facilities.

The competitive bidding rulemaking contains a proposal called "Wheeling-In/Wheeling-Out." Under the "wheeling-in" concept, a utility bidding to supply power to another utility must agree to provide transmission service to the winning bidders located within the bidding utility's service territory or that "are capable of reaching one of its interconnection points." (mimeo at 87). Under the "wheeling-out" proposal, the utility bidding to supply its own power needs must agree to wheel the power of the unsuccessful bidders to utilities that border its service area." (mimeo at 89). Comments on these proposed rulemakings are due to be filled in the summer of 1988, and final rules are expected by the end of the year.

It remains to be seen whether FERC will adopt its wheeling-in/wheeling-out proposal when the final rules are issues.

Robert F. Shapiro
Chadbourne & Parke
June 10, 1988

Chapter 92

PROTECTING THE RIGHT TO BE SERVED BY REGULATED UTILITIES SUBJECT TO COMPETITION: A CRITICAL ASSESSMENT

D. L. Weisman

1. INTRODUCTION

Competition is a common theme among many regulated utilities today. In the electric power, natural gas and telecommunications industries, customers often have competitive alternatives. This transition to competition has been more evolutionary than revolutionary. As a result, much of what has transpired in these industries may have gone unnoticed by the public at large. But in much the same manner that afternoon turns to evening and evening to night, at some point in time it becomes necessary to turn on the lights in order to see just exactly where it is we are going. And so it is in similar fashion with regard to competition for regulated utilities. Competition lies at the very foundation of our capitalist economic system in the United States. It is not an orthodoxy easily challenged without risk of condemnation. We are taught that competition fosters innovation, consumer choice and economic efficiency. And when all things are equal, this is generally true. But when competition is fueled primarily by handicapping market incumbents with artificial restrictions, there is no definitive relationship between competition and economic efficiency. The concern here, of course, is not with the ideal of competition itself but with the way it has been nurtured in these industries. Regulated utilities today are required to operate under common carrier obligations, while their competitors may enter the market at will under terms and conditions of their choice. It is this asymmetry of regulation that questions the authenticity of the competition we have experienced as well as the consumer benefits such competition supposedly represents.

Competitive entry into markets formerly served exclusively by franchised providers will necessitate a renegotiation of the regulatory contract under which utilities provision service. Carrier-of-last-resort obligations, geographically averaged rate schedules and rate designs serving "social engineering" objectives will all be tempered at the hands of competition. The relationship between regulation and competition may have been envisioned as one of "master" and "canine", but it is far less clear today just who is walking whom. Once competition is allowed in a market, regulation cannot be used as the avenue through which to limit its encroachment. This would be like trying to transport water in a wire cage. And yet we continue to press on with more rules and regulation in hopes that we can accomplish what fewer rules and less stringent regulation could not. Time will prove us wrong and history will not be kind.

Certainly one of the most critical observations that should be made at the outset of this analysis concerns the nature of the cost structure underlying the utilities' production process. Specifically, the costs incurred by the utilities are those primarily associated with provisioning the option of use as opposed to actual use. In other words, the ratio of variable cost to total cost is generally very small for these industries. This observation bears significant implications for the sustainability of any particular pricing structure under competition. The importance of these cost structure characteristics is far less critical under a regulated monopoly market structure than it is under competitive provisioning because the mapping between option of use and actual use is very close to one-to-one. With competitive entry, however, demand uncertainty is increased and this mapping may diverge significantly from one-to-one.

The primary question confronting the utilities and their regulators today concerns how the regulatory contract should be renegotiated in order to account for competitive entry. This paper is concerned primarily with the terms under which a renegotiated regulatory contract should address the utilities' carrier-of-last-resort obligation in a competitive market. In particular, to what extent does the loss of the insurance policy (franchised right to serve) for the utility warrant a loss of the insurance policy (right to be served) for the customer? Throughout the analysis, the objective is to secure a greater understanding of the process through which regulation both leads and lags competitive entry. It is only through such an understanding that it will be possible to ascertain the degree to which competition is merely an artifact of asymmetric regulation and the degree to which it is the natural result of market evolution and technological change.

The format of the remainder of this paper is as follows. Section 2 explores the early foundations of the regulatory contract from an economic and public policy perspective. Section 3 advances an analysis of competition in regulated utilities with emphasis on the electric power industry. The carrier-of-last-resort obligation is taken-up in Section 4 in terms of how this obligation has been administered historically as well as how it should be administered in the future under competitive market conditions. Section 5 presents a case study of the carrier-of-last-resort issue for a specific case of competitive entry in telecommunications. Section 6 provides a summary and conclusion.

2. THE REGULATORY CONTRACT: ITS ORIGIN AND ITS DEMISE

The first and most obvious question to consider with regard to the regulatory contract is why utilities should be regulated at all? There are perhaps two primary reasons for regulating utilities. The first is that the industry is perceived to be a natural monopoly. The second is that the public interest deems it necessary that the terms of service provisioning include some "social engineering" objective that may not be satisfied naturally in a competitive marketplace. [7, pp. 167-169] The inevitable conflict between these two objectives of regulation has been a source of cyclical volatility in these markets that has created tremendous and often wasteful excess capacity. [2, pp. 93-94; 12, p 3.2; 24, p. 14]

The economic foundations of a natural monopoly center on the technical concept of subadditivity. [6; 25, pp. 2, 4-7] Specifically, an industry cost structure is said to be subadditive when market demand can be satisfied at least cost by a single as opposed to multiple providers. That in fact, the provision of service by multiple providers would be economically inefficient because it results in wasteful duplication. What is frequently misunderstood with respect to the application of this concept is that competitive market forces respond to market prices not firm-specific costs. If the pricing structures mandated by regulators are not allowed to reflect the underlying cost efficiencies of a natural monopoly, competitive entry will occur, which from society's point of view will represent an inefficient and wasteful allocation of resources. This may seem like an obvious point, but it has been overlooked time and time again with resultant losses in consumer welfare.

The second primary justification for regulation is the perceived need to satisfy certain so-called social engineering objectives. These objectives primarily relate to rate structures, rate levels and protections for service provisioning. Geographically averaged rate structures for electric power and telephone service are cases in point. Rate structures that flow cross-subsidies either between customer classes, such as between business and residence customers, or between distinct services are quite common as well. Regulators have also taken measures to ensure that no one is denied service by charging the franchised provider with a carrier-of-last-resort obligation. Generally, this obligates the carrier to provide service anywhere in its certificated geographic territory on a timely and non-descriminatory basis.

The regulatory contract bestowed an exclusive franchise upon the utility to serve and thereby restricted competitive entry. In exchange, the utility accepted the carrier-of-last-resort obligation which restricted market exit. The entire foundation of the regulatory contract was irrevocably changed when regulators sanctioned competition in these markets. [7, 14, 16] The need to renegotiate the terms of the regulatory contract today arises largely in response to the regulators unilateral modification of this agreement.

Before proceeding further with our historical overview of the regulatory contract, it is necessary that we stipulate on the definition of a contract. Webster's Encyclopedic Dictionary of the English Language defines a contract in part as follows:

> An agreement or mutual promise upon lawful consideration or cause which binds the parties to a performance. [33, p. 183]

The foundation of any contract rests upon the willingness of the parties to limit their future options, foresaking uncertain prospects for either gain or loss, in exchange for a performance that is guaranteed with virtual certainty. The critical point being that the parties future rights and obligations are balanced by the contractural agreement in a manner that seemingly improves their relative positions based upon the information that is known or knowable at the time the contract was entered into.

In the context of public utility regulation, this balance centers on the utilities' obligation to serve and the customers' right to be served. [11] The foundation of the regulatory contract is premised upon the regulators' ability to limit both entry and exit in the marketplace. In order to ensure that there would be at least one provider of the service, the regulator, through the exclusive franchise, had to ensure that there would be only one. It was this precommitment for service on the part of the customer base, underwritten by the regulator, that ensured the requisite capital was brought forward in order to provide service. In this role, the regulator was serving as an agent for the customer base (i.e., the principals) in negotiating the most favorable terms of service with the utility.

This agency relationship is of paramount importance because it suggests that the regulator, as a governmental representative, had the power to award the property rights for service provisioning on behalf of the customer base. Presumably, in the absence of this agency relationship, the utilities (monopoly provision no longer guaranteed) would have been forced to enter into individual contracts with each customer desirous of service. This process would have resulted in very high transaction costs that would ultimately be reflected in service costs higher than would otherwise be necessary. [3; 5] In sponsoring this precommitment on behalf of the customer base, the regulator effectively reduced the financial risk associated with ubiquitous capital deployment through restrictions on competitive entry and thereby afforded the franchised carrier a reasonable opportunity for financial returns commensurate with the risks assumed. That is, regulation allows for a fair opportunity, though not a guarantee of cost recovery. [30] The critical issue then centers on whether sanctioned competitive entry in the market precludes the common carrier from a reasonable opportunity to recover costs. In particular, should the regulator sanction competitive entry while continuing to enforce common carrier obligations on the franchised provider, has not the regulator unilaterally increased the financial risks associated with the franchised carriers provisioning of service and thereby precluded a reasonable opportunity for cost recovery.

The capital markets will ultimately respond to the regulators' unilateral modification of the regulatory contract (i.e. sanctioned competitive entry) with an effective increase in the cost of debt and equity. In addition, the utilities face an increased financial liability today because their depreciation schedules have generally failed to keep pace with the rate of technological advance and were never designed to recover capital costs in a competitive market. This deficiency constitutes a true financial liability because the rate adjustments necessary to remedy this deficiency may not be sustainable under competitive market conditions. Since this financial liability arose in part from a breach of the regulatory contract, there is a legitimate question as to whether the utilities are entitled to some form of compensation for damages incurred. This issue is taken up futher in Section 4.

3. COMPETITION IN REGULATED UTILITIES

Competition has become more the rule than the exception for regulated utilities today. The electric power, natural gas and telecommunications industries today are increasingly subject to competition. Consequently, the terms and conditions of the regulatory contract will change dramatically in the next few years, as regulators, voluntarily or otherwise, surrender their powers to the natural play of market forces.

In the electric power industry, co-generation and self-generation are displacing en masse the generating capacity previously supplied by the regulated utilities. The Public Utility Regulatory Policies Act of 1978 (PURPA) mandated, among other things, that utilities purchase electric energy and capacity made available from qualifying cogeneration and small power production facilities at just and reasonable rates while being required to provide stand-by and back-up service to these facilities at non-discriminatory rates. [4; 9; 13; 23, p. 6 26; 38]

In the natural gas industry, large users are connecting directly to the interstate gas pipelines and thus bypassing the local distribution companies. In addition, the so-called "open access" transportation regulations require natural gas utilities to serve a dual role as both sellers and transporters of natural gas which has been purchased by natural gas consumers directly from third parties. This practice is commonly referred to as mandatory wheeling and affords end-users the opportunity to shop for least-price gas on the spot market without being tied into a long-term contract with the local distribution company. The competition in natural gas arises in large measure through regulation that mandates distribution facilities be made available to end-users for gas shopping in geographically diverse markets. [18; 19] What has not been adequately addressed through these regulations designed to spur competition, however, is the degree to which the local distribution companies are required to stockpile gas reserves in order to satisfy their obligation to stand-by as the carrier-of-last-resort for end-users purchasing gas supplies on the spot market from third parties.

In telecommunications today, it is difficult to find a sector of the industry that is not experiencing significant competitive inroads. The divestiture of the Bell System was designed, in theory, to separate competitive from so-called bottleneck or natural monopoly markets, though it is far from clear that this is what was accomplished. [8] The competitive long distance business of the former Bell System went to AT&T, whereas the bottleneck local distribution part of the business went to the Bell Operating Companies. What has become apparent, however, is that the so-called bottleneck facilities are rapidly becoming competitive. [12, p. 3.2] Large and medium users are bypassing the local distribution companies either directly to the interexchange carriers or through the establishment of private networks providing self-contained end-to-end telecommunications. In addition, metropolitan area networks comprised primarily of fiber optic rings around city cores are springing up in the larger metropolitan areas. These metropolitan networks connect primarily business locations with state-of-the-art telecommunication services. Shared tenant service providers are moving into office buildings and industrial campus environments and reselling local service in direct competition with the local exchange carriers (See Section 5). The local exchange carriers, however, continue to be subject to restrictive common carrier obligations that require them to stand-by as carriers-of-last-resort with statewide average rate schedules.

The electric power industry today stands at the crossroads of open competition in much the same way that the natural gas and telecommunications industries stood years before. [28] Under current rules and regulations governing cogeneration, cogenerators are only permitted to sell electric power back to regulated electric utilities, regardless of jurisdiction. Should regulators invoke mandatory wheeling on an unrestricted basis, as has been done in the natural gas industry, the competitiveness of the electric industry would increase significantly. [19; 28, p. 72] Unrestricted mandatory wheeling would allow end-user customers to utilize regulated utilities' transmission capacity to shop for least-price power across diverse geographic markets. Excess capacity in these markets could be made available to end-user customers located in markets with excess demand.

In the natural gas industry, so-called "open access" transportation regulations require natural gas utilities to serve as both suppliers of gas and transporters of gas that have been purchased by natural gas customers directly from third-party sources. [19] In the telecommunications industry, customers who deploy their own private networks are permitted to resell excess capacity on these networks on a for-profit basis without being subject to common carrier obligations. [29] Regulators in telecommunications have not as yet moved to require local exchange or interexchange carriers to purchase excess capacity on privately owned networks (i.e. the telecommunications analogue of cogeneration). A multitude of compatibility problems related to the technical interface between networks could arise in the absence of a common set of standards.

The recurring theme that can be drawn across all three of these industries concerns the degree to which asymmetric regulatory policies coexist with sanctioned competition. The regulated utilities are required to stand-by as carriers-of-last-resort while their customers are granted the flexibility to pursue competitive alternatives. In addition, these utilities are frequently required to price their services on an average rate basis without allowance for geographic or volume deaveraging. In other words, these rates are not generally based on underlying economic costs. The problem with this practice is that it propagates a subsidy flow between customer classes that may not be sustainable under competition. [7; 13; 15] It is apparent that these regulatory restrictions constitute a de facto double penalty for the utilities. Their economically inefficient rate structures drive bypass of their services while their carrier-of-last-resort responsibility obligates them to stand-by for bypassers with facilities in place with little or no opportunity to generate revenues. The economic harm arising from the coexistence of asymmetric regulation and competition can be seen quite clearly with respect to the cogeneration issue. Under PURPA (1978) rules and regulations, cogenerators may sell their excess power back to the regulated electric utility at its avoidable cost. This includes both capital and running costs. Recall, however, that the electric utility is required to stand-by with sufficient operating capacity margins for both cogenerators and self-generators. In addition, cogenerators may at some point be able to sell power directly to end-use customers. To the extent that these uncertain demand requirements, stemming from the carrier-of-last-resort obligation, serve to raise the regulated utility's avoidable cost higher than it would otherwise be, cogenerators may be able to collectively influence, albeit indirectly, the rates paid to them for power by the utilities. These perverse economic incentives may have resulted in a level of cogenerated power higher than could otherwise be justified on the basis of economic efficiency or conservation alone. [16, pp. 62-64; 26; 28, p. 72] It is in this sense that the "competition" observed is very much the product of a self-fulfilling prophecy.

In light of these developments, it is not surprising that while capacity plans for the utilities' generating capacity have been scaled back dramatically since the 1970's, cogeneration has been expanding at an accelerating rate. This development reflects in large measure the utilities desire to transform capital costs arising from building traditional generating capacity into variable costs arising from the purchase of cogenerated power. The significant ratebase disallowances in the late 1970's and early 1980's have incented the regulated utilities to avoid significant additional capital cost outlays. While this increased reliance on cogenerated power has tended to appease the "shell-shocked" capital markets, the issue remains as to whether cogeneration will constitute a stable and economically efficient source of electric power in the future. [9, pp. 14-15; 16, p. 64]

4.0. THE CARRIER-OF-LAST-RESORT OBLIGATION FOR REGULATED UTILITIES: A TIME FOR REASSESSMENT

One of the key provisions of the regulatory contract is the requirement that the franchised utility serve as the carrier-of-last-resort. This means that the utility bears the obligation to stand-by with the option of providing service on demand to any customer on a timely basis anywhere in its franchised service territory. As previously discussed, this obligation to stand-by causes the regulated utility to incur costs in plant and equipment that are largely independent of actual usage levels. That is, the characteristics of these production processes cause the majority of costs to be incurred in providing the option of use. In a marketplace wherein the regulatory contract is enforced, the mapping between prospective use of the system and actual use of the system approximates one-to-one. The issue then of whether costs should be recovered on a usage-sensitive basis or a flat-rate basis, while bearing perhaps significant economic welfare implications, is probably not critical to the question of sustainability. Under competition, however, demand uncertainty increases and this mapping may diverge significantly from one-to-one. Consequently, the regulated utility may not be able to fully recover its costs under usage-sensitive pricing.

The "free rider" problem created by this pricing structure will encourage inefficient market entry. Competitors can exploit the utilities carrier-of-last-resort obligation by utilizing the franchised carriers' distribution network for redundancy and stand-by instead of building this capacity into their own distribution network. This redundancy provides the customer with a greater degree of reliability in the event of system failure or excess demand than would otherwise be possible. The problem, of course, is not that the regulated utilities' distribution system is being used for stand-by, but that it is not being compensated for that stand-by on the proper terms -- namely for the insurance that it provides competitors and in turn their customers.

This issue of franchised utilities serving as the carrier-of-last-resort under non-discriminatory terms is one of the most hotly debated issues in the electric power industry today. [21; 28; 41] This issue has also been brought to the fore in the natural gas industry, [18] and quite recently in the telecommunications industry. [27] The fact that regulated utilities are now raising this issue of the carrier-of-last-resort obligation should not come as a surprise. Recall that the obligation to serve as the carrier-of-last-resort was a provision of the regulatory contract, symmetrical performance of which obligated the regulator to restrict market entry. Now that regulators have sanctioned competition, the utilities' performance obligation under the abrograted regulatory contract is unclear.

There are in reality two separate dimensions to this carrier-of-last-resort issue. The first issue is of an historical nature in that it concerns the damages incurred by the regulated utility as a result of the breached regulatory contract. The second issue concerns the renegotiation of the regulatory contract as it relates to the carrier-of-last-resort issue under competition. These two questions are now addressed in turn.

4.1 The Carrier-of-Last Resort Issue In Historical Perspective: Issues of Compensation

The issue of possible damages due the regulated utility as a result of breach of the regulatory contract is certainly a very complex legal and economic question. The complexity arises in large measure because there are actual costs associated with addressing this issue in the form of a remedy as well as future costs associated with ignoring the issue altogether. First, should the abrogation of the regulatory contract result in no loss of economic value to the plant and equipment deployed by the regulated utility in order to satisfy its obligation to serve, there would appear to be no valid argument for damages. This would be the case if competitive entry caused no measurable dimunition in the discounted stream of earnings on the plant and equipment deployed by the utility under its obligation to serve. The Market Street Railway Case held that the regulator cannot insure the value of the rate base against the operation of economic forces. At the same time, utilities are seemingly shielded from "governmental destruction of existing economic values", in violation of the 14th Amendment due process clause. [31, p. 780] In other words, utilities are protected against so-called capricious acts by regulators that have the effect of reducing the existing economic value of their rate base. Whether a regulatory body sanctioning competition in a market previously served by a franchised provider, while holding that provider to common carrier obligations constitutes a capricious act must await a ruling on a case not yet filed. The critical issue in such a case would center on the degree to which the regulator artificially propagated the competitive market forces by which the utilities' rate base was devalued.

Suppose now that a case of this type were filed and affirmed as a capricious act on the part of the regulator. A possible remedy could require that the customer (competitor) exercising his competitive options compensate the franchised provider for the unamortized value of the plant and equipment deployed under the auspices of the regulatory contract. A compensation mechanism of this type has been suggested as a possible remedy in the context of competitive entry in telecommunications markets. [22; pp. 13-14] Alternatively, a more all-encompassing compensation mechanism of this same general principle could take the form of an accelerated depreciation schedule coupled with the concommitant rate adjustments sufficient to cover the incremental depreciation expense.

Professor Goldberg has analyzed this issue generically in the context of "property rules" for protecting an entitlement. [10, pp. 145-56; 1] Specifically, if customers were assigned an entitlement for the right to be served by the regulated utility, the utility could presumably terminate service upon payment of an agreed upon damage claim to its customers. Conversely, suppose the utility was assigned an entitlement for the right to serve these customers. Customers who then exercise their competitive options and terminate service with the utility would presumably pay damages. Both of these scenarios constitute "buy-outs" or termination agreements for ending a certain contractual obligation. These buy-outs may actually be welfare enhancing if they allow for an accurate assessment of the relative efficiencies of service provisioning. In the presence of price/cost disparities far competing services, however, the "theory of second best" suggests that the efficiency gains may well be indeterminate. In the context of the discussion in Section 2, it is appropriate to view the regulatory contract as an assignment of symmetrical entitlements. The utility is entitled to a protected right to serve, while customers are entitled to a protected right to be served. Under this interpretation, either party desirous of exit from the other party's entitlement would presumably be obligated to pay compensation.

4.2 The Carrier-of-Last-Resort Futuristic Perspective: Issues of Competitive Provisioning

The question as to the terms and conditions under which the utility, or any provider for that matter, should serve as the carrier-of-last-resort may be somewhat more straightforward in a futuristic sense than it is in a historical sense. The firm providing the carrier-of-last-resort service should be entitled to compensation for the costs it incurs in standing by with plant and equipment necessary to provide the option of use. Whether this stand-by service is optional or non-optional may depend upon whether the previously franchised utility is the designated provider.

The provider serving as carrier-of-last-resort would optimally require each stand-by user to report peak-period demand requirements. This would be necessary for two reasons. First, it would allow the carrier to deploy plant and equipment in the most efficient manner possible. In turn, there is a self-policing dimension to this peak-period demand reporting requirement in that the customer would bear the burden if demand forecasts were either too high or too low. Second, if regulators proceed with designation of submarkets for competitive/non-competitive provisioning, they would likely require that the non-competitive submarkets be protected from the "overspray" of competitive market forces. Specifically, regulators would not allow misreporting of demand requirements for stand-by customers to adversely affect general customers, either as a result of insufficient peak period capacity or because general customers would be indirectly paying for plant and equipment deployed on a stand-by basis to serve both cogenerators and self-generators. The pricing structure for stand-by service is critical for an efficient transition to a competitive marketplace. As discussed previously, a major share of the regulated utilities' costs are incurred in providing for the option of use. This implies that the costs associated with providing stand-by service are largely independent of actual usage levels. These costs are, however, completely and inextricably dependent on forecasted usage levels. Consequently, from an economic efficiency perspective, stand-by charges should be levied on a flat-rate rather than a usage sensitive basis. [27] Presumably, long term contracts with buy-out options or termination agreements could serve to protect utilities from deploying capital with little or no opportunity to generate revenue. In addition, there are serious social equity issues at play here. Failure to implement a flat-rate vis-a-vis usage sensitive pricing structure could, under rate-of-return regulation, actually result in a subsidy flow from regular users to stand-by users of the system.

The use of varying grades of service will also tend to foster flexibility in meeting customer needs. Stand-by charges would then ideally reflect the degree of reliability necessary to serve the customer base. For instance, transponder capacity on satellites can be purchased on a protected, unprotected or preemptible basis. The rates vary widely across these service classes. [20] A service ranking of this type is also important from a regulatory perspective because regulators will want rankings defined in a manner that will not adversely affect general customers. This service priority classification may also affect the manner in which power supplies are pooled in order to meet customer demand. Cogenerators are both suppliers and demanders of electric power. Self-generators, while producing the majority of their own power, continue to rely on the regulated utility for stand-by service. This, as discussed previously, may tend to impose complex load forecasting problems for the regulated utilities because of the difficulties associated with estimating intermittent demand requirements. The use of service priority classifications could serve to allocate the risk of power shortages to those customer classes responsible for the greatest variance in demand. For example, suppose it is required that all stand-by power for cogenerators and self-generators be met out of sell-back supplies of cogenerators. This practice could serve to balance the service requirements for the utilities. Pooling power supplies in this manner may benefit the utilities in terms of minimizing capital deployment and optimally spreading risk over those entities causally responsible. The efficiency gains from power pooling will serve to minimize capital deployment in much the same manner that highways traversing the country will be fewer when they are allowed to interconnect. This occurs because load balancing resulting from interconnection will tend to minimize aggregate capital deployment outlays since peak periods on individual facilities will not necessarily lead to excess demand on a system-wide basis under conditions when there is excess capacity elsewhere on the power grid. This stand-by service option is in essence insurance or protection for service provisioning. Since pooling of risks is the principle that gives rise to insurance markets, it is reasonable to believe that the market for stand-by service should be structured in much the same manner. This implies, of course, that those customers responsible for the largest variance in demand over time would pay the highest per unit charges. [17, pp. 231-263]

5. A CASE STUDY OF THE CARRIER-OF-LAST RESORT ISSUE: THE SHARED TENANT SERVICES CASE IN MISSOURI

On May 11, 1984, Southwestern Bell Telephone Company filed a petition with the Missouri Public Service Commission for an investigation of provision of local exchange telephone service by entities other than certificated local service providers. [36] In this case, Southwestern Bell sought clarification as to the meaning of the local exchange franchise in terms of rights and responsibilities. This case was filed in response to market entry by so-called shared tenant services (STS) providers. STS providers are typically landlords or real estate developers who provide telephone service to the tenants of a building or office park. STS is generally provided through a customer-owned PBX. (A PBX or private branch exchange is a switching device, similar to the local company's central office, which is capable of switching voice and data communications between stations in a single building, tenants in a building, or in other continguous areas such as an office park or campus.) The owner of the PBX will generally purchase PBX trunks from the local telephone company and resell local service to the various subscribing tenants in the building.

On September 23, 1985, the Missouri Public Service Commission ruled that STS was not a regulated telephone company within the statutory definition of that term because they did not offer telephone service to the general public. [36] The Commission's ruling effectively sanctioned competitive entry, albeit selectively, in the provision of local exchange telephone service. The Commission simultaneously ruled that Southwestern Bell was no longer required to serve as the carrier-of-last-resort for tenants who voluntarily decide to occupy an STS building. The Commission apparently perceived that requiring Southwestern Bell to serve as the carrier-of-last-resort in a competitive market was a residual regulatory obligation that unfairly impeded the Company's ability to compete.

In what was in essence the second round of the first STS case, [32] the STS industry sought reversal of the Missouri Public Service Commission's Policy on the carrier-of-last-resort issue. In particular, the STS industry argued that its growth was "chilled" by the STS users' inability to require the local telephone company to stand ready to provide back-up capacity. [39] Southwestern Bell contended in opposition that the Commission had correctly decided the carrier-of-last-resort issue in the first STS case. Specifically, the Commission had recognized that imposing an obligation to stand-by on Southwestern Bell in the face of competition was itself anti-competitive. Southwestern Bell argued that it should not be required to provide stand-by facilities to serve tenants in an STS building in the event the STS provider should liquidate operations or otherwise cease to serve the tenants. It remained Southwestern Bell's policy that it should not be required to serve as the carrier-of-last-resort without due compensation. [40; 37, p. 51] The Company further contended that it is inequitable for general ratepayers to subsidize the STS providers' financial ventures by indirectly paying for plant and equipment deployed in a stand-by mode with little or no prospect for revenue generation. [40]

On May 11, 1988, the Missouri Public Service Commission reversed its initial position and ordered Southwestern Bell to serve as the carrier-of-last-resort for STS locations on a non-discriminatory basis. The Commissions wording in the report and order is particularly instructive:

> The Commission has determined that an LEC should stand ready to provide service to any customer within its service area who requests service. This is the foundation of universal service and is a keystone of this nation's and states economic and social progress [35, p. 12]

The Commission's final ruling in the STS case reflects the central problem addressed in this paper. Namely, the Commission has sanctioned competition in the market while continuing to enforce common carrier obligations on the regulated utility. The Commission cites the value inherent in customers being afforded a protected right to be served, while ignoring the terms of the regulatory contract under which such protections initially found their genesis. Whether this is a sustainable market structure remains to be seen.

6. CONCLUSION

Competition is a market reality today in electric power, natural gas and telecommunications -- a reality that will undoubtedly accelerate over time. The regulatory contract that continues to constrain regulated utilities to common carrier obligations in the face of sanctioned competitive entry must be renegotiated. By its very definition, a contract limits future options of both parties. And yet herein lies the problem. The regulatory contract holds the utility to a precommitment to serve the customer base without enforcing a corresponding precommitment on the part of the customer base to be served. This asymmetry is the fundamental cause of the demise of the regulatory contract.

The question of the carrier-of-last-resort issue must be bifurcated into historical and futuristic perspectives. From an historical perspective, it has been suggested that while the regulatory authority is not charged with protecting utilities from the operation of economic forces, it cannot act capriciously, in violation of the 14th Amendment due process clause, in pursuing actions that serve to devalue the ratebase. The question then as to compensation for damages when regulators artificially create the economic forces that reduce the value of the utilities' rate base or property is a question that should be given careful and relentless scrutiny.

The forward-looking problem of the manner in which the carrier-of-last-resort issue should be managed in a competitive marketplace maybe somewhat more straightforward. The carrier-of-last-resort should be compensated on a flat-rate basis under contract for costs incurred in providing the option of use. This pricing structure is economically efficient because the utilities' costs are largely independent of actual usage levels. Failure to implement a flat-rate vis-a-vis usage-sensitive pricing structure could, under rate-of-return regulation, result in a subsidy flow from regular users to stand-by users of the system.

The issues underlying markets in competitive transition are not easy ones -- either for utilities or their regulators. The challenge before us is to search for solutions that offer the most equitable balance of bringing the full benefits of competition and innovation to the marketplace without sacrificing long-run stability of service provisioning. The great unknown, of course, is the degree of competition this represents. The renegotiation of the regulatory contract must attempt to restore a symmetry of entitlements between utilities and their customers. This symmetry will ensure that we have allowed for a fair test of the competitive experiment and, in turn, discovered the market structure that best serves the public interest.

REFERENCES

[1] Calabresi, Guido and Douglas A. Melamed, Property Rules, Liability Rules and Inalienability: One View of the Cathedral. Harvard Law Review, Vol. 85, Number 6, April 1972.

[2] Chaffee, David C. Rewiring of America, Academic Press Inc., Orlando, Florida: 1988.

[3] Coase, Ronald. The Problem of Social Cost, 3, Journal of Law and Economics. 1 (1960)

[4] Culler, Floyd L. Technical Opportunities and Issues for Electric Power. Public Utilities Fortnightly, May 28, 1987.

[5] Demsetz, Harold. Some Aspects of Property Rights. Journal of Law and Economics. Vol. 9, October 1966.

[6] Demestz, Harold. Why Regulate Utilities? Journal of Law and Economics. Vol. 11, (April 1968).

[7] Egan, Bruce L. and Dennis L. Weisman. "The U.S. Telecommunications Industry in Transition", 10, Telecommunications Policy, 164, (1986).

[8] Faulhaber, Gerald. Telecommunications in Turmoil; Ballinger Press, Cambridge: 1987.

[9] Geist, Jerry. The Electric Industry Charting a Course to the Future. Public Utilities Fortnightly. May 23, 1987.

[10] Goldberg, Victor. Protecting The Right To Be Served By Public Utilities. Research In Law and Economics, Vol. 1, (1979).

[11] Goldberg, Victor. Regulation and Administered Contracts. Vol. 7, Bell Journal of Economics and Management Science. 426 (1976).

[12] Huber, Peter. The Geodesic Network: 1987 Report on Competition In The Telephone Industry, published by The Antitrust Division, Department of Justice.

[13] Johnson, L. Competition and Cross-Subsidization In The Telephone Industry (Rand Corporation Publication No. R-2976-RC/NSF) (1982).

[14] Kahn, Alfred E. "Who Should Pay for Power Plant Duds?" Wall Street Journal, August 18, 1985, at 24, Col. 3.

[15] Kahn A. and W. Shew. Current Issues in Telecommunications Regulation: Pricing, Vol. 4, Yale Journal on Regulation 191 (1987).

[16] Kalt, Joseph P. and et. al. Reestablishing The Regulatory Bargain in the Electric Utility Industry. John F. Kennedy School of Government. Energy and Environmental Policy Center. Discussion Paper Number E-87-02, Harvard University, March 1987.

[17] Knight, Frank. Risk, Uncertainty and Profit. Houghton-Mifflin: Chicago (1921).

[18] Lambert, Jeremiah D. Bypass In The Natural Gas Industry: The Fruit of Regulatory Change. Public Utilities Fortnightly, April 3, 1986.

[19] Lookadoo, Phillip G. What Happened In The Gas Industry Can Happen In The Electric Industry, Public Utilities Fortnightly, May 28, 1987.

[20] Nelson, Satellite Appraisals, Telecommunications, June 1986.

[21] Pace, Joe. Increased Competitiveness and The Obligation To Serve: An Oxymoron, Public Utilities Fortnightly, January 7, 1988.

[22] Panzar, John C. "The Continuing Role for the Franchise Monopoly In Rural Telephony," Prepared for The Rural Telephone Coalition, March 21, 1987.

[23] Paul, Bill. Cogeneration, After Slow Start, Quickly Coming of Age, Wall Street Journal, March 2, 1987, at 6, Col. 1.

[24] Porter, Michael E. Competition In The Long Distance Telecommunications Market: An Industry Structure Analysis. June 1987.

[25] Sharkey, William. Theory of Natural Monopoly. Cambridge University Press: Cambridge 1982.

[26] Stewart Robert D., Jr. The Law of Cogeneration In Oklahoma. Public Utilities Fortnightly. November 27, 1986.

[27] Weisman, Dennis L. Default Capacity Tariffs: Smoothing the Transitional Regulatory Asymmetries In The Telecommunications Market. 5, Yale Journal on Regulation, Vol. 149, Winter 1988.

[28] The Electric Utility Executive's Forum. Public Utilities Fortnightly, May 28, 1987.

[29] First Report and Order, Docket No. 83-426, released April 1, 1985, 50 Fed. Reg. 13338 (April 4, 1985).

[30] Federal Power Commission vs. Hope Natural Gas Co., 320 vs. 591 (1944).

[31] Market Street Railway Co. vs. Railroad Commission of California, 324 vs 548, 567 (1945).

[32] Missouri Public Service Commission Case No. TO-86-53.

[33] The New Websters Encyclopedia Dictionary of the English Language Consolidated Book Publishers, Chicago MCMLII.

[34] Public Utility Regulatory Policies Act of 1978 (PURPA), Pub. L. No. 95-617, 92 Stat. 3117 (codified at 16 U.S.C. 260-1-2708 (1982)).

[35] Report and Order Of The Missouri Public Service Commission In Case No. TO-86-53, May 11, 1988.

[36] Report and Order Of The Missouri Public Service Commission In Case No. TC-84-233, September 23, 1985.

[37] Report and Order of the state of Vermont Public Service Board in Docket No. 4946, (February 21, 1986).

[38] Report of Arizona Commission Staff on Topics and Issues in Cogeneration (1987).

[39] Testimony of Charles B. Blumenkamp in Mo. Public Service Commission Case No. TO-86-53, June 1987.

[40] Testimony of Dennis L. Weisman in Mo. Public Service Commission Case No. TO-86-83, June 1987.

[41] Trends and Topics: Stand-by Service Options and Rates Public Utilities Fortnightly, September 3, 1987

ABSTRACT

Competitive entry into markets for regulated utility services, such as gas, electric and telephones, has brought to the fore a number of complex economic and legal issues. The "regulatory contract" under which regulated utilities enjoyed an exclusive right to serve is currently in a rather dramatic state of transition. As a result, the highly averaged rate structures and carrier-of-last-resort obligations operative on the utilities impede them in their ability to effectively compete in the marketplace. Regulators are thus confronted with a difficult dilemma: How to bring society the full benefits of competition without depriving general ratepayers of the protections they enjoy under franchised provision of the utilities services? Regulators must thus confront the economic reality that choice has its price. When sanctioned competition was allowed in these markets, the dye was cast for a transition to end residual regulatory obligations on incumbent firms. Unless regulators recognize this trade-off, they will do little more than inflict the greatest economic harm on that segment of society they are striving so desperately to protect.

The ideas and opinions expressed herein are those of the author and do not necessarily reflect those of Southwestern Bell Telephone Company.

The author would like to thank Peter Grandstaff, Dale Lehman and Henry Peluffo for kindly providing many insightful and thought-provoking comments on an earlier draft of this paper.

Chapter 93

THE STRATEGIC DIRECTION OF THE ELECTRIC UTILITY INDUSTRY

L. E. De Simone

For over fifteen years, rising costs have been a principal characteristic of the electric utility industry -- a characteristic that has emphatically and profoundly changed the economics and the structure of the electric utility industry. As costs have risen, new energy sources have become more competitive; and regulations have been established lowering the barriers of entry into the electric energy production business. All of this has been occurring in a market that has continued to grow (in terms of energy and demand) at a high rate. So, there are sound economic reasons for major change in the operating environment of the industry.

Currently, this environment can be examined in terms of three major forces:

- competition
- regulation
- technology

Competitive forces encompass not only the traditional battle of electricity versus natural gas in retail markets and electric company versus electric company in wholesale markets, but some encroachment by traditional electricity producers and new entrants into wholesale and retail electric markets. Attempts by neighboring systems to poach large industrial customers and municipal customers are beginning to surface throughout the country, requiring regulators and courts to reconsider longstanding charter, franchise, and obligation-to-serve concepts.

Regulation is forcing change in the industry largely because of significant uncertainty surrounding recent utility investments. The emergence of perfect hindsight prudence reviews and sizeable imprudence disallowances has undermined the historical ratepayer/shareholder compact. As a result, few utilities in the future will undertake the risk of large capital expenditures unless the regulations regarding cost recovery are clarified.

Technology is also a major driving force for change because of its impact on electricity options-particularly on the customer side of the meter. Technological advances not only facilitate the penetration of cogeneration in the industrial sector, but also could lead to substantial penetration of mini-cogeneration or fuel cells in the commercial sector, and ultimately personal generators or even solar photovoltaic breakthroughs in the residential sector. Technology, as in the case of bypass in the telecommunications industry, presents electric utilities with tremendous uncertainty in predicting future loads and revenues to support investment.

In assessing the impact that these driving forces-competition, regulation, and technology-have had on other industries, such as airlines, financial services, natural gas, telecommunications, steel, automotive, and others, it is clear that the electric utility industry is facing the potential for substantial changes.

If we evaluate these trends, look at the implications for our industry, and assess the concerns currently identified by most utility managements, we can conclude that there are four broad critical and emerging issues of immediate concern:

- certification/cost recovery
- competitive positioning
- delivery capability
- corporate/industry restructuring

Certification/cost recovery includes utility concerns in dealing with the political and regulatory environment. Competitive positioning covers utility efforts in dealing with the realities of a more competitive marketplace. Delivery capability focuses on utility efforts to improve the delivery of the product and service. Finally, corporate restructuring addresses a utility company efforts to realign itself in anticipation of structural changes in the entire industry.

The first critical issue is cost recovery/certification. This issue is a major concern of utility management because of the precedent being set on perfect hindsight prudence reviews and because of the significant disallowances associated with major capital expenditures. It is my view that these prudence reviews will be extended to fuel, purchased power, and operation and maintenance expenditures as the current construction cycle winds down and commission staffs look around for things to do. Alternatives to traditional front-end loaded cost recovery schemes, such as multi-year phase-in or more exotic value-based pricing mechanisms, in which new baseload facilities are afforded cost recovery comparable to a cogeneration facility, have surfaced over the last few years. The approaches adopted for large new generating units today appear to be consistent with a path toward incremental deregulation of future generating capacity. Finally, several utilities face considerable exposure through ownership of existing plants. Nuclear shutdown initiatives, like those in Oregon, Maine, and Sacramento,

California, as well as acid rain legislation could lead to substantial stranded investment in our industry. This stranded investment could also result from some of the more recent least-cost initiatives, which require an exhaustive review of existing generation, as well as new generation, in the supply and demand portfolio.

The second critical issue is competitive positioning. Competitive positioning has become a necessity as utilities recognize that the industry is undergoing a basic shift from being construction driven to being market driven. If, in fact, our markets are becoming more competitive, electric utility efforts to survive or excel will depend on three fundamentals:

- o knowledge of customers
- o knowledge of competition
- o knowledge of own company

As utilities continue to compile detailed information on customers, such as revenue, sales, cost, profitability, end-uses, et cetera, in response both to internal needs and to external least-cost filing requirements, we must recognize that our tendency to overpublish as an industry provides substantial information to potential competitors. In fact, it is not unusual to hear of one utility relying on the documents of another utility in assembling comparative profiles, targeting specific marketing programs toward specific customers on neighboring systems, and assessing the feasibility of merger/acquisition strategies.

A second, but equally important, concern regarding competitive positioning is the natural inclination of state regulatory commissions to support conservation and load management strategies, but not marketing programs. It is not uncommon to observe commissions supporting programs that chase a kilowatt or a kilowatt-hour away from the system based on a naive perception that any and all such programs are, by definition, good. In reality, significant benefits accrue to all customers through improved load factor and more efficient utilization of utility plant that can be achieved through selective load enhancement strategies. Consequently, conservation, load management, and marketing can, and should, coexist as an integral part of least-cost planning. Using a combination of such programs, one can envision a utility extracting kilowatts and kilowatt-hours from less efficient, low load factor, unprofitable end-uses, like residential air conditioning, and diverting these same kilowatts and kilowatt-hours to more efficient, higher load factor, more profitable uses, such as those in the industrial sector. This transfer can be accomodated without any significant change in utility plant. Increased kilowatt-hour consumption, particularly in off-peak periods, could provide all customers with lower rates by allowing the utility to spread fixed charges over a broader kilowatt-hour sales base.

The third critical issue for the utility industry is delivery capability. This issue has become a significant concern for both the short-run and long-run. Most utilities recognize that we are no longer in a cost-plus situation in which all expenses can be passed onto customers through higher rates. In fact, as competition becomes more intense, utilities will strive to be mean and lean with an expectation that the future will belong to the lowest cost producers. Efforts to improve productivity and efficiency, to contain costs, and to maintain reliability will force utilities to squeeze more out of their existing systems. Efforts to lower breakeven points and to be better leveraged on sales will result in significant downsizing activity. Over the longer term, utilities are unlikely to assume the financial risks associated with large conventional construction projects. We already see considerable focus on bridge technologies, such as increased interconnections, power pooling, power purchases, cogeneration repowering and economic life extension. These options will buy some time and provide a transition to more advanced technologies, such as coal gasification combined cycle, fluidized bed, fuel cells, and more exotic options like energy storage and solar photovoltaics.

The significant point is that, as we embark upon the next construction cycle, we will be looking at shorter lead-time options than the conventional plants going into service today. In addition, we will have a handful of viable, credible independent power producers to turn to who have a proven track record of constructing and operating generating facilities at low cost and high reliability. These parties will include, but not be limited to Mission Energy, Dominion Energy, Cogentrix, Applied Energy Services and others who survive the shake out of the next five years. Capacity additions will be more contractual in nature than in the past. Consequently, we must ensure that planning guidelines provide for competitive bidding without compromising the process with undue disclosure of information through regulatory review.

The fourth critical electric utility issue is corporate/industry restructuring. Several companies, under pressure to increase share-holder value, are looking beyond traditional schemes to increase earnings, dividends and cash flow. Those companies are undertaking a serious effort to restructure their businesses culturally, legally, and financially. No one restructuring plan will prevail. Different strategies fit different companies. Restructuring alternatives include formation of a holding company, separate generation, transmission, and distribution business units diversification into non-regulated business ventures, divestiture of utility operations, mergers, acquisitions, and financial restructuring (ranging from sale lease backs to Chapter 11 filings).

These restructuring strategies are being pursued just as the entire industry is undergoing a transformation. While deregulation will proceed gradually and transmission access/wheeling will be the sticky issue, we can anticipate some realignment of the fragmented pieces of the current industry into different consolidations. Most of these consolidations will be based on economic fundamentals that will ensure lower cost electricity supplies than under existing structures.

In Conclusion, it is apparent that the electric utility industry is undergoing a major transformation as a result of three major driving forces - competition, regulation and technology. The industry is shifting from what historically has been a construction-driven industry to what has become a market-driven industry. Rate increases are no longer viable or sufficient action for financial stability. Competitive positioning strategies based on a careful assessment of customers, competitors and a company's own capabilities, coupled with efforts to increase productivity and contain costs, point to a future which will belong to the lean and mean low-cost producers. Federal and State legislative and regulatory initiatives pave the way for new entrants and increased competition. Restructuring of several companies within the industry, as well as restructuring of the electric utility industry itself, is inevitable.

SECTION 22
DEMAND-SIDE MANAGEMENT OPTIONS

Chapter 94

AN APPROACH TO LEAST-COST UTILITY PLANNING

L. A. Buttorff

BACKGROUND

Utilities and regulatory agencies across the country are increasingly concerned about the cost and availability of new generation to meet growth in customer demands. If new generation or purchased power could be deferred, the resultant savings could be used to implement the demand-side management (DSM) strategies that would allow the deferral. In addition to the capital savings potential, the demand-side strategies, direct or indirect, may also change the load shape so that production cost savings could be realized. Thus, investigation of demand-side planning strategies and determination of the costs for implementation versus the potential benefits are becoming an integral aspect of a generation expansion plan for a utility.

Traditionally within the utility industry, emphasis has been on increasing the supply of generating capacity to meet the demands of customers at the lowest cost (least-cost supply planning). Least Cost Utility Planning or Integrated Resource Planning (LCUP) requires a broader perspective as demand-side and supply-side options are evaluated equally.

This paper focuses on an approach to determine the least cost of providing service considering both demand and supply options on an equal basis under the constraints of reliability and risk to the consumer and stockholder. The interrelated nature of the process is illustrated Figure 1.

FIGURE 1

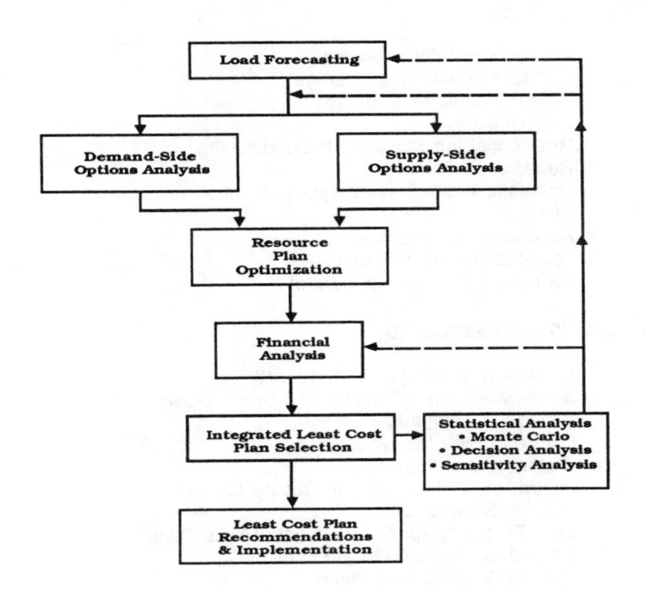

The four key components in this process are load forecasting, analysis of demand-side options, analysis of supply-side options, and financial and risk analyses. This LCUP approach integrates these components into a structured process which forms an integrated utility resource plan.

METHODOLOGY

The first step in establishing a LCUP is to establish a detailed research plan which incorporates the methodologies that should be considered in meeting the utility's strategic objectives. Following the development of a work plan, the forecasting and demand-side analyses commences, followed by the supply-side analysis. The final step integrates the demand and supply sides and properly evaluates the financial impact on the utility.

LOAD FORECASTING

A common problem in most of the utility industry is that traditional load forecasting approaches have not kept up with rapidly changing energy markets. Conservation impacts still appear strong and are likely to influence future energy consumption patterns at least over the near term.

The solution to forecasting uncertainty lies in a greater emphasis on market analysis, including customer surveys, detailed end-use forecasts, and greater utilization of load research data. Unfortunately, this approach is time-consuming and market research requirements must compete with other equally pressing needs for allocation of scarce resources.

An end-use model provides a more flexible methodology than an econometric approach. In this approach, a forecast is built based upon saturation rates of electric intensive appliances, such as air conditioners, water heaters or space heaters. Saturation rates are forecast based on price and economic factors subject to data availability. This approach allows more flexibility in analyzing the impacts of changing load patterns, changing rate structure, and load management. The drawback to an end-use model is that the required data, such as load research data and detailed multi-year customer surveys, are often unavailable. In addition, multiple assumptions are required to account for average use per appliance, changing appliance efficiencies, and average age of the appliance stock. In addition, the methodology can be refined and enhanced over the longer term as improved data allow the implementation of new and more advanced techniques.

DEMAND-SIDE ANALYSIS

Once the load forecasts are completed the demand-side management (DSM) tasks can commence. In this approach DSM data is collected, a load modification strategy is identified, applicable load shape modification programs are determined, models to screen DSM activities are set up, and a ranking procedure is developed. The ranking allows the comparison of the most cost beneficial DSM programs on an equal basis with supply-side options. This is important so that DSM programs are not implemented when a supply-side program may be more cost beneficial. This approach is illustrated in Figure 2.

FIGURE 2

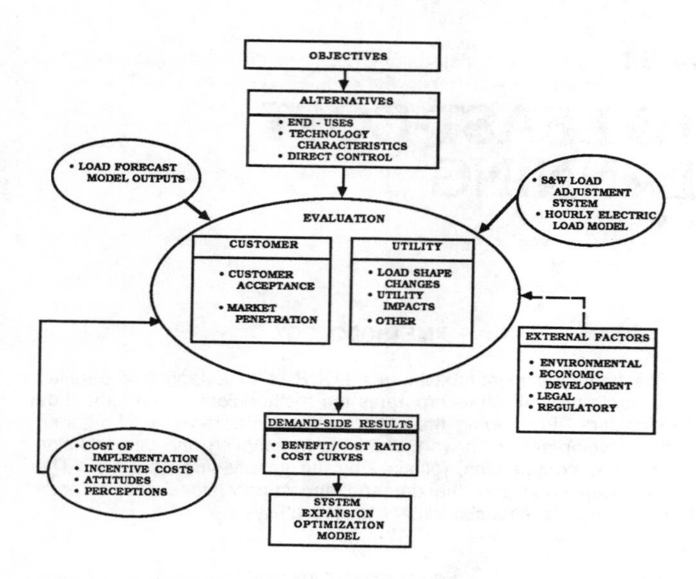

For the utility there are several impacts that should be quantified. In the short-run DSM programs change the daily operation of generation units and maintenance scheduling, then these changes impact the level of fuel usage, the number of plant start-ups and shutdowns, the units providing spinning reserve and the level of losses on the system. In the long-run the generation, transmission and distribution facilities will change if the DSM program is effective.

Other utility impacts include maintenance and administration expenses, and DSM equipment costs. Benefits that may occur include reduced financial risks, increased operating flexibility, improved cash flow, increased revenues and more options for customers.

The first step in the demand-side analysis is to select the objectives of the DSM program. These objectives should incorporate the utility's overall financial objectives, as well as load shape, and operational and planning objectives. The overall objectives might incorporate financial goals that should be met, while the load shape objectives include the basic strategies identified by EPRI which include: peak clipping, valley filling, load shifting, strategic conservation, strategic load growth and flexible load shape. Once the objectives are identified, specific DSM programs are listed that could be used to accomplish the targeted load shape changes. Some of the sponsored programs by EPRI are listed on Table 1.

Although the tendency of DSM programs, strategies or options to increase the load factor of the utility system is usually a desirable action, care must be taken that system reliability is not compromised. If system reliability must remain constant, the percent margin must also go up as load factor increases. Within the normal load factor increases caused by DSM the reliability of the system only changes slightly. These changes do need to be considered and evaluated, however. Another consideration is that DSM controlling hardware is not normally included properly in the reliability calculation, and therefore, the system Loss of Load Hours (LOLH) are not properly accounted for in the evaluation. These reliability concerns must be considered and formulated into a DSM objective.

The second step is to determine the alternatives involved in selecting the end-use to be controlled, the control devices available and the market implementation methods. In the identification of DSM programs, different strategies also have to be considered; such as the demand modification and control method, pricing policy, hardware ownership and incentives paid to the customer. Both direct and indirect load control alternatives should be evaluated. The direct control programs include appliances such as water heating, air-conditioning, space heating, motors and industrial load. Indirect controls include rate alternatives, conservation, and voltage reduction programs.

EPRI DSM PROGRAMS
TABLE 1

Peak Clipping

High Efficiency Air Source Central Heat Pump
Groundwater Source Heat Pump
Ground-Coupled Heat Pump
Multizone Heat Pump
Room Heat Pump
Dual-Fuel Heating Systems
Active Solar Space Heating
Task Heating
Zoned Resistance Heating
Heat Pump Water Heater
Heat Recovery Water Heater
High-EER Air Conditioner
Receiver Switches
Domestic Water Heater Cycling Control
Air Conditioner Cycling Control
Variable Service Level Devices
Timers
Appliance Interlocks
Programmable Controllers
Temperature-Activated Switches
Load Management Thermostats
Swimming Pool Pump Control

Valley Filling

Dual-Fuel Heating Systems
Add-On Heat Pump
Central Ceramic Heat Storage
Room Ceramic Heat Storage
Slab Heating
Residential Ice Storage Air Conditioning

Load Shifting

Active Solar Space Heating
Central Ceramic Heat Storage
Room Ceramic Heat Storage
Slab Heating
Residential Ice Storage Air Conditioning
Receiver Switches
Domestic Water Heater Cycling Control
Timers
Appliance Interlocks
Load Management Thermostats
Swimming Pool Pump Control

Strategic Conservation

Insulation (Ceiling, Wall, Floor)
Storm/Multipane Windows and Storm Doors
Window Treatments
Duct and Pipe Insulation
Water Heater Blanket
Infiltration and Indoor Air Quality Control
Passive Solar Design
High Efficiency Air Source Central Heat Pump
Groundwater Source Heat Pump
Ground-Coupled Heat Pump
Multizone Heat Pump
Room Heat Pump
Active Solar Space Heating
Task Heating

EPRI DSM PROGRAMS
TABLE 1 (continued)

Strategic Conservation

Heat Pump Water Heater
Heat Recovery Water Heater
Solar Domestic Water Heating
High-EER Air Conditioner

The third step is the screening analysis where the most appropriate DSM programs are identified. Screening matrices are set up to compare DSM programs with the load shape objectives, the end-uses evaluated and the marketing methods. Then, the DSM programs are summarized in terms of system costs and customer responses regarding satisfaction.

The list of EPRI demand-side programs are used as a starting point for screening the demand-side alternatives to a manageable number so that the alternatives can be evaluated on an equal basis with the supply-side alternatives. First, programs are eliminated that are not applicable. Then, various alternatives are screened using a spreadsheet program which calculates a cost/benefit ratio for each DSM alternative. Those programs whose cost/benefit ratio falls well below the 1.0 level perhaps could be eliminated. After the 150 programs are screened, the most viable programs are left for detailed evaluation.

The screening analysis evaluates the costs and benefits to the utility, participants and nonparticipants. A cost/benefit analysis is particularly useful in that it offers a framework for organizing data to evaluate certain objectives, and for understanding a decision situation. All too often in DSM program evaluations, certain consequences of the DSM control are ignored because they are difficult to quantify (i.e. public relations, inconveniences, reliability of service, environmental impacts).

Customer impacts must also be evaluated and quantified in evaluating a DSM program. The net benefit of electric service to customers depends on the value and cost of electricity, the inconvenience due to the DSM program, the customer investment required, O&M expenses, taxes, incentive investment payments, and unplanned outages. Uncertainty analysis can also be incorporated in the screening analysis to develop probability distributions rather than a single cost/benefit ratio.

After the screening process is completed, a detailed evaluation of the most desirable DSM programs is completed. The development of the end-use load shapes used for the DSM evaluation is the first step in this process. If end-uses are directly metered, this data is used as a starting point. Otherwise load research data must be disaggregated into its components by means of systematic comparisons. As a starting point, the class load shape data from the load research information, the load forecast end-use estimates and the end-use load shape data that is already available from industry sources are used.

After the end-use hourly load shapes are developed, the amount of controllable load is determined. For each program evaluated, market implementation methods are incorporated into the analysis. The residential appliance saturation survey serves as the basis for analysis. The survey provides the information needed to determine the residential end-use saturation statistics necessary; such as, appliance stock, type of housing structures, housing characteristics, etc. In the commercial and industrial sectors, borrowed data can be used for items such as the type of business, end-use energy applications, lighting by fuel type, and other customer characteristics.

Also, the customer acceptance of the various marketing approaches and cycling strategies will need to be determined via a survey or based on estimated data. Before DSM programs can be implemented, a customer acceptance survey should be completed to determine how the DSM programs should be implemented.

How customers' attitudes affect customers' participation in DSM programs should be considered in this analysis, because customers' attitudes can make or break a DSM program. Answers that need to be obtained are:

- Did or would the customers like or dislike the program?
- Do the customers know enough about the program?
- Intentions and goals of customer.
- What are the barriers to energy conservation?
- How and when does the consumer make purchasing decisions and how does this affect energy conservation and sales?

The second step in the evaluation of customer acceptance will be the tabulation of customer withdrawal rates. If a survey is not completed, then borrowed utility data will have to be used. Several past studies have shown that customers do not vary significantly by region and territory.

Other areas of concern are determining the hardware necessary for the demand-side control and deciding what incentives are necessary to motivate the customers to participate. Such factors as differences in financing costs and whether the utility or customer owns the device can make a difference in the analyses. Also, the incentives, such as special rates, credits or rebates, low interest loans and tax credits, can affect the results of the cost benefit analysis.

Once the saturation and customer participation rates are estimated, forecasted hourly load shape is modified on an hour-by-hour basis. If analysis of the load shape changes has shown that load can be shifted so as to effect a reduction in peak, it becomes necessary to address the cost of implementation. There are three general components of cost associated with the implementation of a demand-side management program. These include the fixed charges on the capital investment, operation and maintenance costs, and the administrative and general expenses. These costs incorporate control equipment costs, customer incentives and inconvenience costs, additional transmission and distribution costs, customer service expenses, and possibly reduced energy sales. The experience of several utilities indicates that significant financial incentives are required to attract customers to a voluntary DSM program. The incentive can be in the form of a monthly bill credit, a lump sum annual payment, or a lower energy rate.

The resultant modified load curves and costs developed in this task are used in the integration of the demand and supply-side alternatives.

The supply-side analysis incorporates generation expansion alternatives as well as detailed production costing attributes. In this analysis it is important to examine the array of alternatives that could be considered and select those that should be utilized in determining the expansion plan. Many supply-side alternatives are available for any utility to consider over the study period. These alternatives include conventional technologies - oil, gas, coal, nuclear and hydroelectric; nonconventional and renewable resources - solar, wind, biomass, fluidized-bed combustion and others such as generating unit life extensions, deferred retirements and purchases of capacity from outside of the utility.

The array of alternatives are then considered and those that are selected should be utilized in determining the expansion plan. Conventional technologies include pulverized coal-fired steam units, combustion turbines and combined cycle oil or gas-fired units and nuclear power plants. Nonconventional and renewable resource technologies are also included. The potential for cogeneration opportunities in the utility's service territory is estimated. Timing and size for each of these facilities are noted and utilized in the study as appropriate. Technologies such as fluidized-bed combustion, coal gasification combined cycle, solar, wind, photovoltaics, MHD, biomass and refuse-derived fuel (municipal solid waste) are reviewed to determine their commercial viability and projected costs.

Many options exist in meeting the future supply requirements of a system. Inherent in the capacity alternatives are the units 15 to 20 years or older that have been relegated to a lesser role in delivering energy to the system because of obsolescence, high operating costs, low reliability and general overall operating condition. These units require some capital investment to upgrade their operating condition and extend their life. However, that investment usually is considerably less than the cost of new capacity.

Generating unit life extensions are gaining in popularity since the site is already available and the capital investment is usually significantly lower than that required for new capacity. However, life extension modifications are still subject to an economic justification process to determine if the dollars expended are justified in terms of additional life and generation obtained from the unit as compared to other alternatives.

A thorough examination of each of the potential planning alternatives and alternative mechanisms of meeting load growth such as life extensions results in estimated capital, fuel and O&M costs for each. These costs may result in certain technologies being dropped from further consideration via a screening analysis.

The most likely options for planning alternatives determined from the screening analysis are then used in the integrated demand- and supply-side analysis.

INTEGRATION AND EVALUATION

For the integration process two methodologies are available. One method is to modify the expected (no load management) hourly load shape to account for load changes resulting from any of the demand-side options considered to have potential overall savings. The modification also subtracts the incremental losses when load is removed in an hour and add the losses back on an incremental basis when load is returned to service. Then, the modified load curve is used for the evaluation of the supply-side alternatives.

The second method involves using the nondispatchable technology capability where the demand- and supply-side options are evaluated simultaneously. Thus, the cost savings (demand and energy) and reliability differences between different alternatives are evaluated simultaneously. The benefits derived from each scenario are calculated by a comparison of the discounted cash flows and present values. The shifts in load patterns are evaluated for costs as well as benefits. The components of costs are the charges for the capital investment, the operation and maintenance expenses for the equipment and the administrative and general expenses.

Since the primary economic benefits from the installation of direct control load management on a power system are expected to result from the deferral or elimination of new generating capacity, it is important to use an optimization program for this process. The cost savings associated with this deferral or elimination should be a major factor in offsetting the costs of installing and operating a direct control system. In order to realistically reflect the cost factors involved, the evaluation of DSM controls must account for these deferrals consistent with the system's normal generating capacity planning practices.

FINANCIAL ANALYSIS

The next step in the LCUP is to assess the financial and rate impacts for the plan that has been developed. Financial statements that are used in this analysis include the balance sheet, income statement, cash flow, and sources and uses of funds. As another step in the analysis, the changes in customer rates are studied by class.

A plan that is optimal from a long-term economic viewpoint may not always be financially feasible for any utility thus, it is very important to test these plans using a financial model. The financial evaluation is used to evaluate the level and timing of financing requirements, rate adjustments, interest coverages, level of earnings and rate of return, to mention a few. From the financial model, a number of customized financial reports, financial ratios and rates can be developed depending on the utility's requirements. An illustration of these models is shown in Figure 3.

FIGURE 3

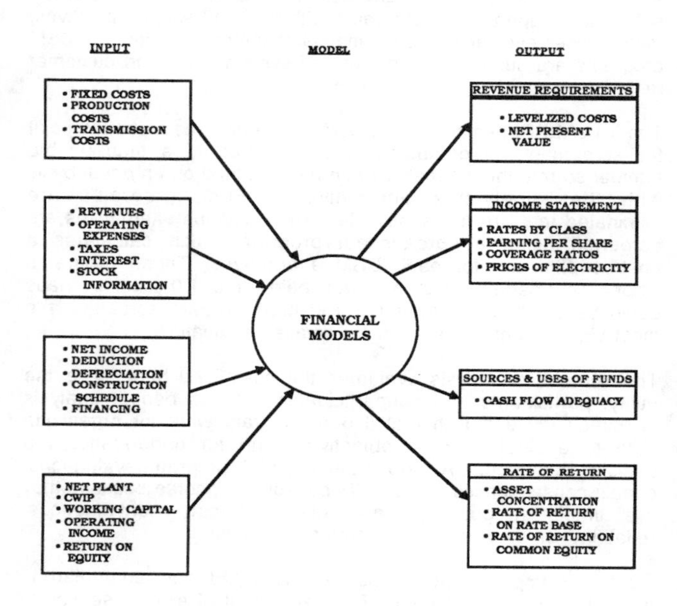

From the financial model a number of financial reports, financial ratios, and cost-of-service rate levels can be developed, depending on the utility's requirements. For example, the rates charged to the customer may be held constant while the models solve for changes in other factors. The results derived explicitly recognize the changes in the value of service. This is the single most overlooked aspect of composite evaluations. Also a qualitative evaluation must be done which assigns weighting factors for various degrees of inconvenience that the demand-side scheme may impose on the customer.

RISK ANALYSIS

The components discussed previously enable the utility to develop revenue and financial requirements for the LCUP. Because of the uncertainty about the future, the variability of the plans should also be tested. In developing the least cost plan a sensitivity analyses on all major independent variables including the load forecast, capital and fuel costs, cost of money and others is then completed. These sensitivities are combined into scenarios to calculate the associated revenue requirements, and evaluate each scenario using a risk analysis. Risk analysis requires the assignment of probabilities to each scenario developed during the sensitivity analysis. In this type of analysis, uncertain variables are varied over ranges specified from low to high to measure their impact on the net benefit differences of various demand- and supply-side options.

This analysis can incorporate Monte Carlo or decision tree concepts, so that when a particular decision is tested, a probability answer can be derived.

Either of these uncertainty analyses can be performed to assess alternative scenario impacts. A flowchart of the integration of these alternatives is presented on Figure 4.

FIGURE 4

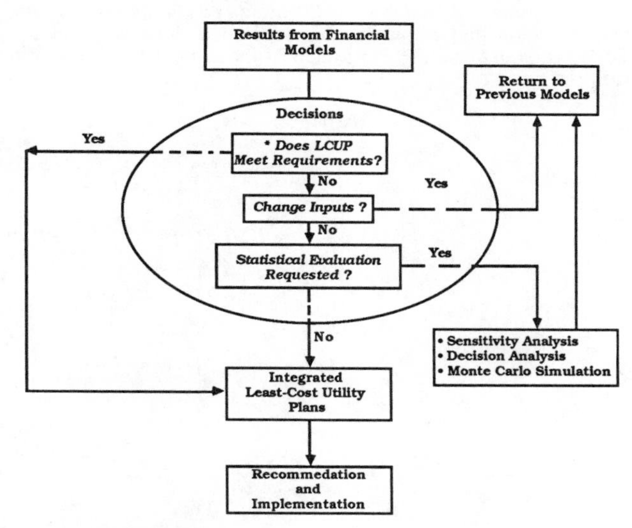

In summary, this methodology to evaluate demand and supply options in a Least Cost Utility Plan enables that utility to select a plan that meets the needs of the customers, as well as the utility's strategic objective.

Chapter 95

PERFORMANCE MONITORING OF A EUTECTIC SALT COOL STORAGE SYSTEM

R. H. Sterrett, E. Steudtner

ABSTRACT

Southern California Edison (SCE) is promoting the use of cool storage systems within their service territory to reduce the summer on-peak electrical demands. As part of their Thermal Energy Storage (TES) program, they have hired Science Applications International Corporation (SAIC) to monitor a cool storage site in Irvine, California. The effort is part of Electric Power Research Institute's (EPRI) Commercial Cool Storage Monitoring Program which monitors and evaluates the performance of cool storage systems throughout the country.

This paper describes the monitoring and performance evaluation of one of the three sites monitoried by SCE in 1987. The site is an office building which uses a eutectic salt storage system to cool the building during the on-peak hours.

The cool storage system was instrumented and the performance of the major system components are being evaluated. Data is analyzed with COOLCALC, a program developed as part of the EPRI project. Performance parameters are calculated using 15-minute average data, and include the chiller COP, storage utilization and efficiency, building cooling load, and cooling system on-peak electrical demand. The program uses the data to simulate a conventional chilled water cooling system and to determine the on-peak demand shift provided by the storage system.

The preliminary monitoring results were evaluated and several operating flaws were identified. They included occasional on-peak operation of the chillers, high specific energy use during the charging cycle and low storage utilization.

A meeting was held with the building operator and storage system vendor to discuss the problems and possible changes. It was determined that only one of the two storage tank compartments was being used, and a change to the EMS software resolved this problem. A recommendation was made to delay the charging cycle and to finish charging the tank while directly cooling the building during the morning hours. This would increase the chiller load, improving its specific energy usage.

Storage system operation showed improvement over the next several months. Average specific energy usage during the charging cycle decreased by 17%, from 1.8 to 1.5 kW per ton. On-peak electric demand shift increased to 365 kW in August, resulting in a $3,000 savings in the electrical bill.

Performance results indicate that when properly operated, cool storage systems do effectively reduce on-peak demand. Through evaluation of the system performance, specific operating problems were identified and corrected. The improved system operation benefits both the utility and the building owner/operator.

Introduction

Cool storage is a technology for transferring on-peak electrical demand associated with air conditioning to off-peak periods. Off-peak electricity is used to make chilled water or ice which is stored and used later to meet all or part of the on-peak air conditioning requirement. In the Southern California Edison (SCE) service territory where air conditioning constitutes a significant part of the peak demand, cool storage can be an attractive load management technique. With appropriate rate structures, cool storage can also benefit the SCE customer by lowering the customer's electric bill.

Despite the potential advantages of cool storage, its widespread use is hampered by a lack of verifiable data on system performance, cost, and reliability. To eliminate this obstacle, SCE and other electric utilities are installing and monitoring demonstration cool storage systems as part of the Electric Power Research Institute's (EPRI) "Commercial Cool Storage Field Monitoring Project". SCE is monitoring three commercially available cool storage technologies: a single tank multicompartment chilled water system, an ice harvester, and a eutectic salt storage system. During 1986, SCE contracted with Science Applications International Corporation (SAIC) to instrument and evaluate the performance of these cool storage systems. This paper describes the 1987 monitoring results of the eutectic salt storage system.

The site is a six-story concrete and glass office building located in Irvine, California. The building has 190,000 square feet of floor space, and is occupied by approximately 500 people. The building is normally cooled from 7 A.M. through 6 P.M. on weekdays.

The cooling system is designed to meet an annual peak cooling load of 580 tons. The cooling system is oversized and consists of two 350-ton centrifugal chillers, a cooling tower and a 2800 ton-hr eutectic salt storage tank. The storage system is designed to meet the entire summer on-peak (12 P.M.-6 P.M) cooling load, making it a full storage system. The storage medium is eutectic salt with a freezing point of 47°F and is contained in plastic containers. Eutectic salts, like ice, undergo a phase change which results in a higher energy storage capability for a given volume than chilled water. The salt freezes during charging and is melted during discharging.

Approach

The approach of the EPRI Cool Storage Monitoring Program is to instrument the cool storage system, evaluate the performance of the system, and simulate a conventional chilled water air conditioning system using the measured cooling loads and ambient conditions.

The performance of the cool storage system is determined by monitoring the system inputs and outputs. The system input is the electrical consumption of the cool storage system, and the output is the cooling supplied to the building. Figure 1 shows the instrumentation installed at the site.

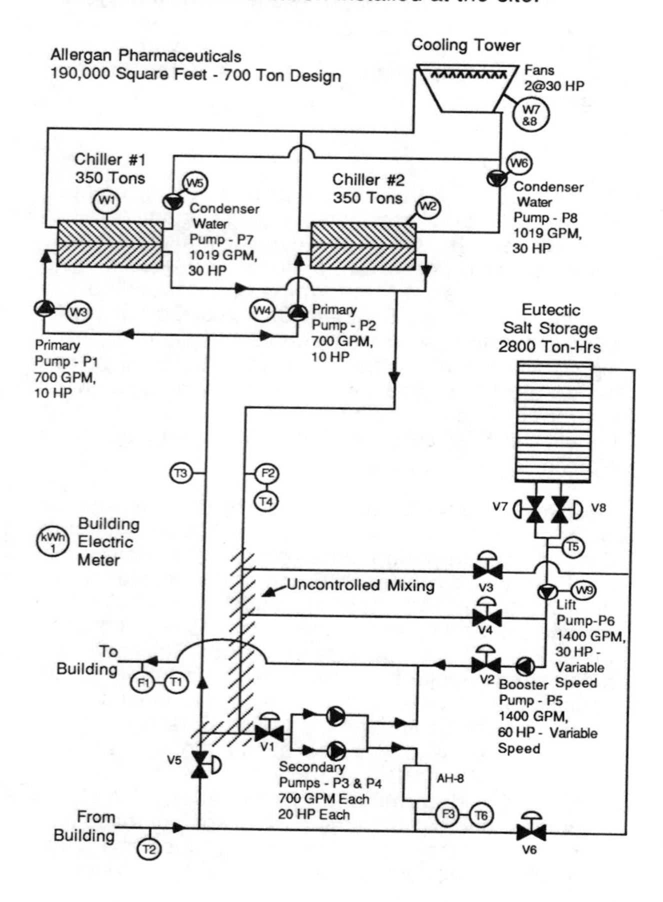

FIGURE 1. SYSTEM DESCRIPTION - SCHEMATIC

The thermal output from the chiller is determined by measuring the evaporator water inlet and outlet temperatures and flow rate. The building cooling requirement is determined by measuring the chilled water supply and return temperatures of the building along with the flow rate. The electrical consumption of the chillers, primary chilled water pumps, condenser water pumps, cooling tower fans, and storage lift pump is being measured. The total cool storage electrical consumption is the sum of the chiller consumption plus the parasitic load plus the cool tower fan and condenser water pump loads.

The parasitic load are those that would not be present if the building had a conventional air conditioning system. In this case, the parasitic load is the primary chilled water and lift pumps that circulate the water from the chiller to the storage tank.

In order to fully evaluate the reduction in electric air conditioning demand during the on-peak hours, the cool storage system demand must be compared to the demand of a conventional system. To do this, the building cooling load is measured and the simulated conventional system is modeled to supply the measured building cooling requirement. The conventional system simulation is performed by the EPRI software program called COOLCALC.

Preliminary Results

A meeting was held in July to discuss the preliminary performance results with the building operators and storage system vendor. The results indicated that the chillers were operated during the on-peak period several times a month, the storage capacity was not being utilized, and the chiller performance during the end of the cycle was quite low. The operator felt that most of the on-peak operation was due to a situation where the tank was not fully charged the previous night. The low chiller performance was due to low load of the chiller at the end of the charging cycle. During the discussions it was determined that only half of the available storage was being used due to a control valve that was not opened.

There were a number of recommendations made regarding the operation of the system. They included:

1. Fix EMS software and control valve so both halves of storage tank can be used.
2. Delay start of charging cycle to midnight, and use the charging/direct cooling mode at the end of the charging cycle. This would increase the load on the chillers, and avoid the low chiller performance.

The performance results during August showed a marked improvement, mainly due to the changes made to the EMS software and the storage system control valve. Figure 2 characterizes the chiller operation before and after the meeting with the building operator. Before the meeting, the high specific energy usage is a result of the low chiller loading during the end of the charging cycle (less than 20%). By resolving the problem with the storage control valve, the chiller loading was maintained above 50 percent during most of the charging cycle. This helped to reduce the average specific energy use from 1.67 kWh/ton-hr to 1.24 kWh/ton-hr

Performance Results

The performance monitoring results have been used to evaluate the system performance for specific days and compile monthly averages of component performance parameters. The daily evaluations provide information on how the system is operated. The cooling load profile for a hot day is shown in Figure 3.

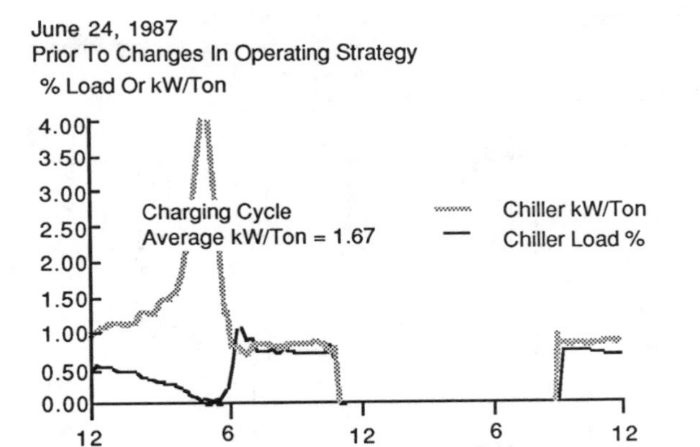

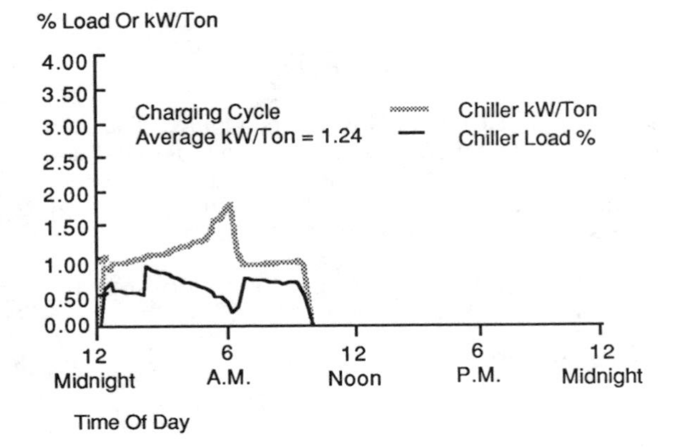

FIGURE 2. CHILLER PERFORMANCE BEFORE AND AFTER MEETING WITH BUILDING OPERATOR

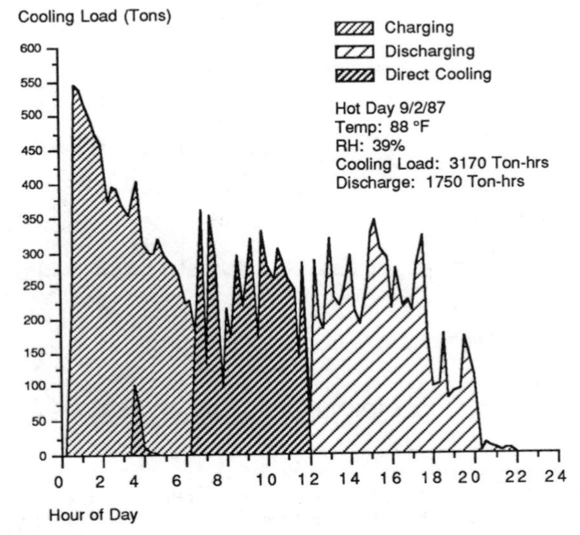

FIGURE 3. COOLING LOAD PROFILES FOR A HOT DAY

The storage system is charged from 12 A.M. through 6 A.M., and the building is cooled directly with the chillers from 6 A.M. through 12 P.M. The storage system is used to cool the building from 12 P.M. to 8 P.M. which includes SCE's on-peak period (12 P.M.-6 P.M.) The electrical demands of the cool storage system and the simulated conventional system are shown in Figure 4.

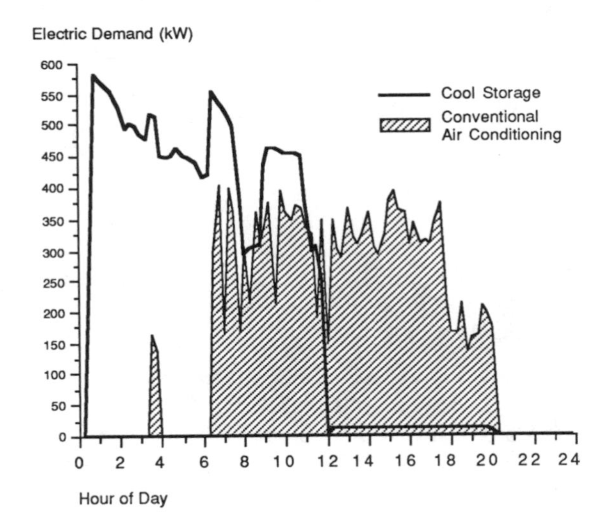

FIGURE 4. COMPARISON OF COOLING SYSTEM ELECTRICAL DEMANDS FOR A HOT DAY

During the on-peak period the storage systems demand is limited to the storage lift pump (30 kW) reducing the on-peak demand by over 365 kW.

The monthly performance plots verify the improvements in the system operation after the meeting with the building operator. During July there were problems with the instrumentation, and the performance parameters could not be accurately calculated. The average specific energy use is shown on Figure 5.

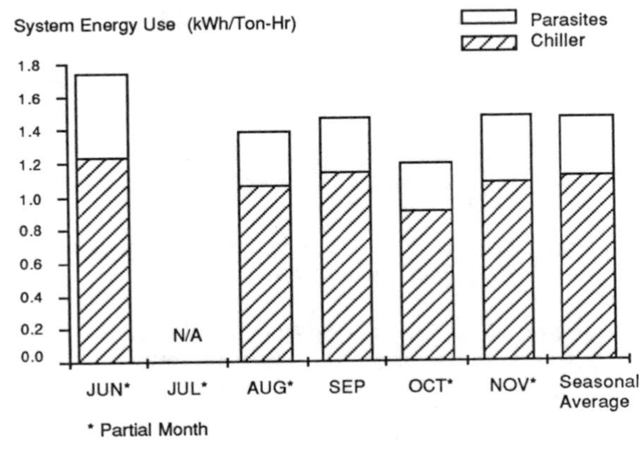

FIGURE 5. COOL STORAGE SYSTEM SPECIFIC ENERGY USE

The changes in the operation resulted in a 0.37 kWh/ton-hr (25%) reduction in the specific energy use between June and August. The specific energy use of the conventional system ranged between 1.25 to 1.43 kWh/ton-hr for this period.

The storage sizing factor is a measure of how much of the nominal storage capacity is being utilized. The average monthly sizing factor and the peak daily sizing factor are shown in Figure 6.

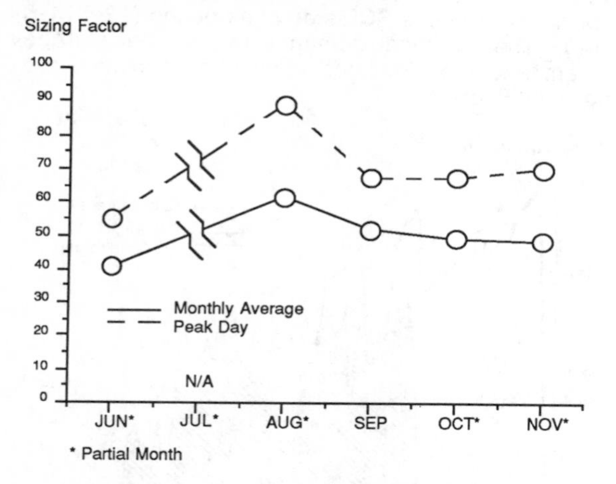

FIGURE 6. COOL STORAGE SYSTEM SIZING FACTOR

After the problem with the storage tank's control valve was resolved, the storage system usage increased. This helped to improve the operation of the system and reduce the on-peak demand. The storage utilization dropped off during the winter due to the later on-peak period (5 P.M.-9 P.M.) and the lower cooling loads.

The reduction in the building's on-peak demand is plotted in Figure 7.

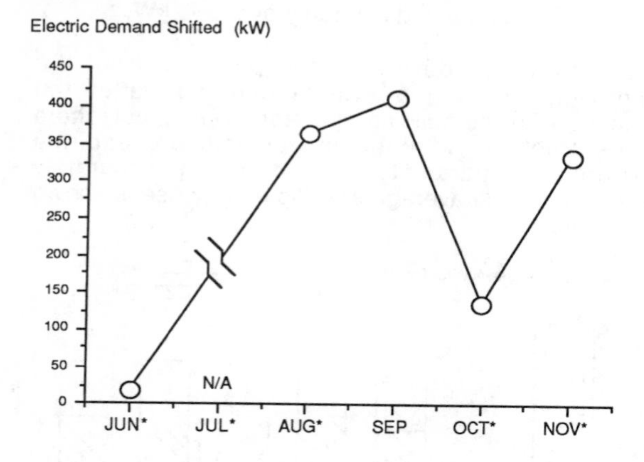

FIGURE 7. COMPARISON WITH CONVENTIONAL SYSTEM PEAK ELECTRIC DEMAND

The demand reduction ranges from 19 kW to 411 kW. During the summer months, when the system was operating as a full storage system, the demand reduction ranged between 300 kW and 400 kW.

The monthly electrical consumption data for the cool storage system and the simulated conventional system is shown in Figure 8.

The electrical consumption on this plot is for partial months so month-to-month comparisons should not be made. The cool storage system was able to effectively limit its on-peak usage. Only 4% of the electrical energy used by the cool storage was on-peak, compared to 33% of the simulated conventional system's usage.

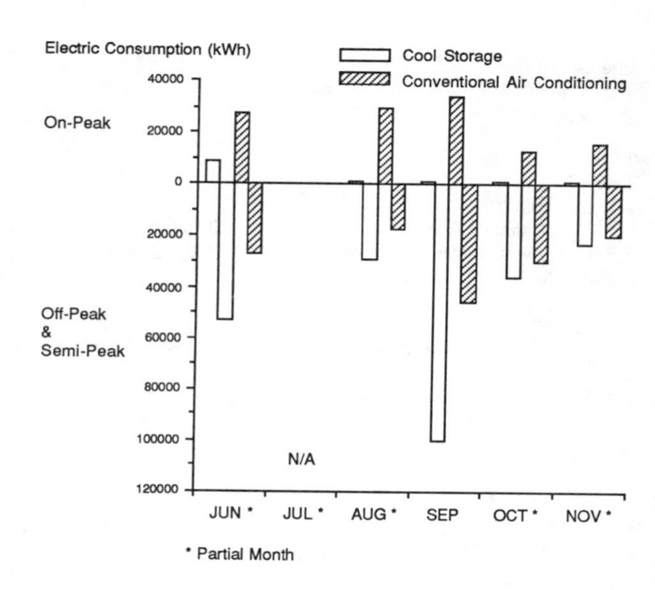

FIGURE 8. COMPARISON TO CONVENTIONAL SYSTEM ELECTRICAL ENERGY USE

The site is on SCE's TOU-8 rate schedule. This is a time-of-use rate schedule which has both demand and energy charges. The demand charges are relatively low, but there is a substantial difference in energy charges between the on- and off-peak periods.

The billing impact resulting from the use of the cool storage system is shown in Figure 9.

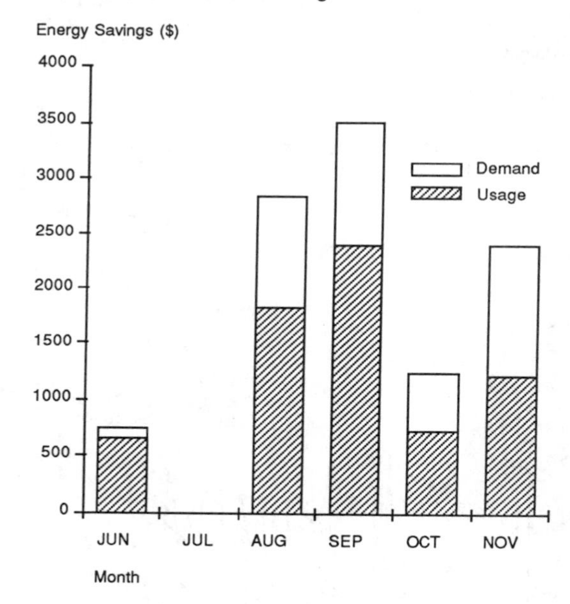

FIGURE 9. COOL STORAGE SYSTEM BILLING IMPACTS

The usage has been scaled up to represent the consumption for complete months. The cool storage system was able to reduce the electrical bill by $11,000 for the five months evaluated. Although the savings are less than the original projections, the estimated payback period is still less than five years.

Conclusions

The cool storage system at the site was able to effectively reduce the on-peak electrical demand and usage of the building. Initially there were some operating problems with the cool storage system, but these were resolved during a meeting with the building operator, storage system vendor, SCE and SAIC.

The effectiveness of the storage system to reduce the on-peak demand is dependent on proper operation. Prior to the performance results meeting the system was operated on-peak several times per month. This reduced both demand and energy savings. The importance of proper operation was stressed heavily and the performance of the system showed marked improvement after the meeting.

The performance results at the site indicate the need of providing the operator with a measure of the system performance. Once the operator was made aware of the importance of proper operation and given a measure of the performance he paid closer attention to the proper operation of the cool storage.

Chapter 96

A METHOD TO DESIGN COOL STORAGE UNITS

A. Solmar, J. Persson, D. Loyd

ABSTRACT

In cooling systems for industrial and commercial buildings a cool storage with phase change material should preferably be utilized. During the design process we determine the impact of flow, geometry and materials on cooling capacity during changing conditions. With rapidly or irregularly changing conditions normal design methods are insufficient. By using the proposed design method such situations can easily be handled.

A simulation model is developed parallel with laboratory experiments. Three types of encapsulations of interest for cool storage are options in the model. The engineer can choose between spheres, cylinders or rectangular pads. In the laboratory tests the capsules are keept in a container which has a latent heat capacity of about 30 kWh. Any phase-change material can be calculated if the heat capacity versus temperature for the material is known.

A general purpose simulation program is used. Once the actual design proposal is defined the simulation is operated by commands. The results are directly shown in diagrams. If needed a parameter or an initial value can easily be changed and the simulation repeated.

An effective utilization of the cool storage is important for both technical and economical reasons. The simulation model can be used for analysing applications of cool storage units. Measured values of flow and temperature from industrial or commercial applications can be used in the model to generate results of performance. The model can also use defined input data and consequently be an excellent tool in the development of cool storage techniques.

INTRODUCTION

Energy management in resent years has vitalized the development of new products and encouraged engineers to improve analyses and design methods. The cool storage is a component with great potential for energy managment and on which interest is focused, (2) and (4). Load management in the industrial and commercial area is possible with a component which can utilize off peak rates. Industrial processes needing peek cooling capacity can be supplied with a cool storage which show better pay back than low utilized cooling machinery. A common technique to achive cooling capacity with a cool storage is to pass a fluid through a material at low temperature. The storage material must have high specific heat capacity or as an alternative a physical phenomena like a phase-change can be utilized. For a phase-change material the transition temperature should occur at an appropriate level which usually is governed by the cooling and heating sequences.

When designing a cool storage the engineer uses some kind of method, (6). Using heat and mass transfer analyses a desktop method can result in a realistic design. As an opposite method a cool storage can be developed completely guided by experiments in a laboratory. A mixture of theoretical analysis and experiments can be very successful.

The method presented in this paper is a numerical simulation method with which the designer has the facilitie to interact with experiments, see Figure 1. The method also gives the engineer a possibility to simulate a design proposal or a marketed cool storage unit, together with actual measured or defined conditions from the system where the storage shall be fitted. The application will show a well matched cool storage in regard to the conditions used in the simulation, (5). The design method is advantageous especially when the conditions are changing rapidly or irregulary.

DESIGN METHOD

The design method is a based on numerical simulations and if needed experiments, see Figure 1. The designer has to follow a cool storage specification for writing conceptual initial parameter values to use in the simulation. The parameters concerns the cool storage insulation, the storage material with encapsulation and the cooling fluid. The complete parameter specification give a conceptual design of the storage unit. The cooling fluid is specified more in detail by flow rate and temperature. These values can be supplied either from experiments, application measurements or assumed by the designer. Simulation results can be stored as time variations of about 120 variables. An analysis of the results must be simple and usually the designer prefers only some informative result diagrams. If the cool storage performance has to be improved one or more parameters are changed and new simulations are made. At this level an experienced designer may have results to describe the thermal performance of a cool storage manufactured from the parameter values.

Sometimes theory and practice differ, a realized storage unit may perform not according to the simulation results. If this is the situation an experiment with parameter indentification can give very useful information. In order to classify modifications of an experimental unit more simulations are made and a final design can be accepted.

Parameter Specification

The initial parameter set up of values has to be defined by the user. The way to find realistic values can be from specifications of cooling power or from specifications of stored heat during a working cycle. Depending on the combination of storage material and cooling fluid either power or the stored energy will determine the size of the cool storage. With an assumed storage material and encapsulation the mass of storage material can be roughly estimated as well as the number of capsules. With knowledge of the packing or mounting of the capsules the size of the storage container is calculated. From this conceptual design the initial parameter specification is made. In the following comments are given on some of the parameters.

Container: The function of the container is to support the storage material and to prohibit heat flux between the storage material and the environment. The simulation model needs an overall heat transfer coefficient between the heating fluid and the environment. The coefficient is assumed to be constant. With a well insulated cool storage the insulation layer is the governing resistance factor which justify the assumption. Further the free inlet area is specified together with the specific outside container area, which is the circumfirential area devided with the storage length in the fluid flow direction.

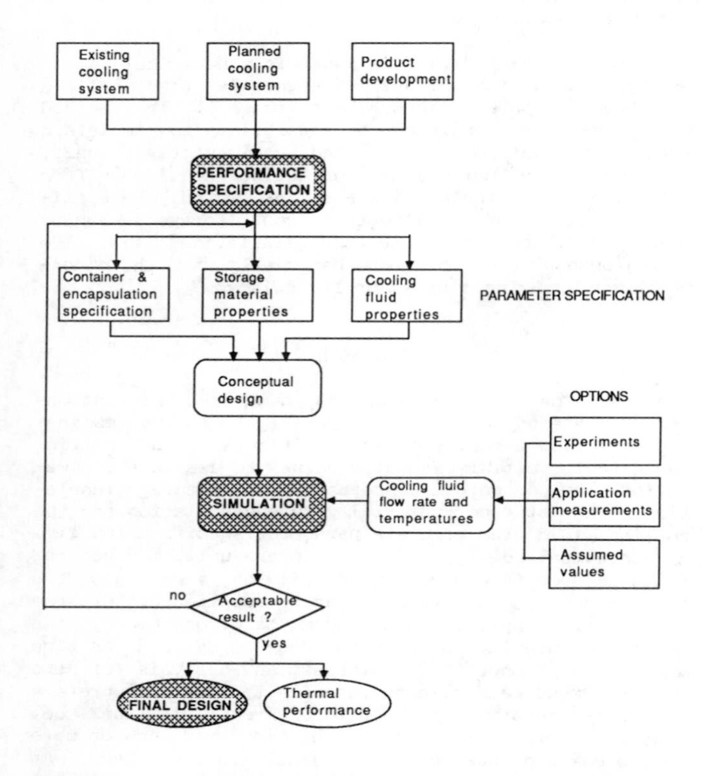

FIGURE 1. THE STRUCTURE OF A METHOD TO DESIGN COOL STORAGE UNITS.

Encapsulation: The simulation model has facilities to use encapsulation of the storage material as spheres, cylinders or rectangular pads. The spheres do not need any sofisticated fixture and the spheres can be packed in a bed which is an advantage. The cylinders and the rectangular pads can be manufactured easily by conventional techniques which make them attractive when considering encapsulation geometry. The simulation model needs values of the size of the choosen geometry.

Storage material: Storing heat and providing cooling capacity can be performed both with sensible and latent heat materials. Phase-change materials, PCM's, are latent heat materials and they are attractive when a cool storage with high storage capacity is required. Using the heat of transition a cool storage with small dimensions can be designed.

The storage material is specified by a table defining the temperature variation of the specific heat. The latent heat of fusion is included in the specific heat. A sensible material will have a nearly constant specific heat and a latent material will have an increasing specific heat at every phase transition. The temperature interval in which the transition occurs can be more or less wide. A differential scanning calorimetry measurement is one of the best methods to prepare the specific heat parameter input. The thermal conductivity as a function of temperature and the density of the storage material must also be specified.

Cooling fluid: The heat transfer between the cooling fluid and the storage material is forced convection and in some cases with large flow cross section characterized by rather low Reynold numbers. The convective heat transfer coefficient is continously adjusted in the model according to the temperature and the velocity of the cooling fluid. The adjustment uses the Reynold and Prandtl numbers and the temperature variations of the kinematic viscosity and Prandtl number of the cooling fluid are specified. The variations are specified in similar tables as the specific heat.

The cooling fluid is also described by flow rate and storage inlet temperature. There are two possibilties to give this information, either measurements from an existing cooling system or assumed values. The first possibility is also used when parameters are identified from experiments.

SIMULATION

When all initial parameters are defined the first simulation is prepared. The results of a simulation are shown directly in diagrams. Since many variables are stored there is no meaning to show all of them. Usually in the first simulations the inlet and simulated outlet temperatures are studied in a diagram. Later more specific result diagrams are convenient to use. It is possible to studie the performance of the complete cool storage unit as well as sections of the storage.

As mentioned before a cool storage specification often includes a storage capacity in values of stored energy or cooling power. Both variables can be presented in diagrams during the simulation. If the results indicate that the storage does not meet the specifications the user has to decide to change one or more parameters and to run a new simulation. If the cooling power has to be increased the encapsulation area can be increased or another storage material can be used etc. The results of the simulation are automatically stored and it is suitable to store or print the initial parameter and variable setup.

Parameter Identification

When a parameter i.e. the convective heat transfer coefficient is difficult to determine, a parameter identification is suitable. The determination is a combination of experiments and simulations. The cool storage is cycled with different inlet temperatures and cooling fluid flow rates. The temperatures at inlet and outlet and the cooling fluid flow are recorded and used as input for simulation. By minimizing the error between simulation and measurement an identified parameter value can be achived. In the simulation model the error between the real and simulated outlet temperature is calculated and stored as a result.

SIMULATION MODEL

The reality described by the simulation model is a dynamic process where heat is transfered between a fluid and a storage material. The process is rather complex since the heat flux is a function of both time and space, (3). A modified lumped heat capacity method is used where the storage is divided into five segments. The length of a segment is one fifth of the storage length in the fluid flow direction and consequently 20% of the capsules, containing the storage material, is located in each segment. Further each capsule is divided into five concentric elements with equal volume. The model includes now 25 types of volumes which are connected to each other.

A general interactive simulation program, SIMNON, is used. SIMNON (SIMulation language for NON-linear systems) is an interactive simulation program for dynamical systems described by differential and difference equations SIMNON, (1), is used for research and education in e.g. automatic control, biology, chemical engineering, economics and electrical engineering. SIMNON also exists in a version for personal computers, which has been used for the calculations in this paper. The key concept of SIMNON is a system, which corresponds to a mathematical model of the reality the user wants to simulate. In SIMNON a system is a sequence of statements in a language with Algol flavour. There are continous systems (differential equations), discrete systems (difference equations) and hybrid systems from continous and discrete subsystems. Once a system is defined, it can be operated by simple commands. The simulation procedure is first to draw diagram axles, select variables for plotting and start the simulation over an appropriate time interval.

Assumptions

The major assumptions can be summarized as

1. The specific heat capacity and density of the cooling fluid are constant.
2. The thermal resistance of the encapsulation material is included in the convective heat transfer coefficient.
3. The heat transfer in the phase-change material is only conduction.
4. The storage is divided into five segments.
5. Each capsule with storage material is divided into five concentric elements with equal volume. All capsules in a segment have equivalent energy distribution.
6. The influence of the radiation heat transfer inside the storage is neglected.

Equations

The heat transfer between the cooling fluid and the storage material can be described as

$$\frac{\partial P_s}{\partial x} = h_{fs}\left[T_f - T_s\right]a_s \qquad (1)$$

where T_f is the temperature of the fluid, T_s the temperature of the capsules in the segment, h_{fs} the convective heat transfer coefficient between fluid and capsule, a_s the area of the capsules per unit length, x the coordinate in the fluid flow direction and P_s the thermal power. The heat transfer between the fluid and the environment is

$$\frac{\partial P_e}{\partial x} = U_{fe}\left[T_f - T_e\right]a_e \qquad (2)$$

where T_e is the temperature of the environment, a_e the area per unit length and U_{fe} the overall heat transfer coefficient between fluid and environment.

The coefficient h_{fs} depends on e.g. fluid velocity, temperature, viscosity, Prandtl number and geometry. The following standard expression is used for the Nusselt number

$$Nu = N\,(w\,D/\upsilon_f)^n\,Pr^{1/3} \qquad (3)$$

where the coefficients N and n depend on the geometry. D is a characteristic length of the capsule and υ_f the kinematic viscosity. The mean value of the convective heat transfer coefficient is

$$h_{fs} = Nu\,k_f/D \qquad (4)$$

where k_f is the thermal conductivity of the fluid.

The sum of the equations (1) and (2) must balance the temperature change of the cooling fluid which gives the following differential equation

$$C_{pf}\,\dot{m}_f\,\frac{\partial T}{\partial x} = -\left[\left(h_{fs}\,a_s + U_{fe}\,a_e\right)T_f - \left(h_{fs}\,a_s\,T_s + U_{fe}\,a_e\,T_e\right)\right] \qquad (5)$$

The temperature of the fluid at time 0 is $T_f(0) = T_0$.

Equation (5) can now be solved and gives the temperature of the cooling fluid

$$T_f = (T_0 - T_\infty)f(x) + T_\infty \qquad (6)$$

where

$$T_\infty = \frac{h_{fs}\,a_s\,T_s + U_{fe}\,a_e\,T_e}{h_{fs}\,a_s + U_{fe}\,a_e} \qquad (7)$$

$$f(x) = e^{-gx} \qquad (8)$$

$$g = \left(h_{fs}\,a_s + U_{fe}a_e\right)/C_{pf}\,\dot{m}_f \qquad (9)$$

When the fluid passes a segment with the length L the thermal power P_f is

$$P_f = \dot{m}_f\,C_{pf}[T_0 - T_L] \qquad (10)$$

The part of the thermal power P_f which is related to the storage material of a segment is

$$P_s(L) = \int_0^L\left(\frac{\partial P_s}{\partial x}\right)dx \qquad (11)$$

and with equation (1) and (6)

$$P_s(L) = h_{fs}\,a_s\left[[T_\infty - T_s]L + \left(T_0 - T_\infty\right)\cdot[1 - e^{-gL}]/g\right] \qquad (12)$$

The part of the thermal power P_f which is related to the environment is consequently

$$P_e = P_f - P_s \qquad (13)$$

Equations (1) to (13) describe the situation of the cooling fluid. The following equations describe the heat transfer inside the storage material. The heat transfer from centre outwards is

$$P(r) = -\,k_s\,A(r)\,\frac{\partial T}{\partial r} \qquad (14)$$

The co-ordinate r originates from the centre, or centre-line, of the capsule; see Figure 2. The co-ordinate r for the layers is calculated as

$$r_i = R\ (i/5)^{1/\beta}$$

$$i = 0,1,\ \ldots,\ 5,\ \begin{cases}\beta=1 \text{ for rectangular pads}\\ \beta=2 \text{ for cylinders}\\ \beta=3 \text{ for spheres}\end{cases} \quad (15)$$

where R is the inner co-ordinate of the encapsulation. The thermal power between two layers is for $1 \le i \le 4$

$$P(r_i) = k_{s,i}\ A(r_i)\ \frac{\partial T}{\partial r}\Big|_{r=r_i} \quad (16)$$

The derivative is approximated as

$$\left(\frac{\partial T}{\partial r}\Big|_{r\,=\,r_i}\right) = 2\left[T_{i+1/i} - T_{i/i-1}\right]/\left[\,r_{i+1} - r_{i-1}\right] \quad (17)$$

where $T_{i+1/i}$ is the temperature of the element between r_i and r_{i+1}. The thermal power is

$$P(r_i) = 2\ k_{s,i}\ A(r_i)\left[T_{i+1/i} - T_{i/i-1}\right]/\left[\,c_{i+1} - r_{i-1}\right] \quad (18)$$

where

$$A(r_i) = A\ (i/5)^{[(\beta-1)/\beta]} \quad (19)$$

where A is the inner area of the encapsulation

$$k_{s,i} = \frac{k_{s,i+1/i} + k_{s,i/i-1}}{2} \quad (20)$$

The motive for using a mean value of the thermal conductivity is the large difference between the solid phase and the liquid phase for some phase-change materials.

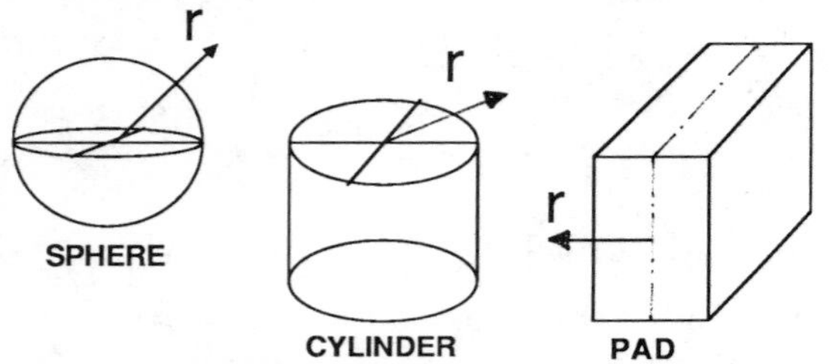

FIGURE 2. THE CO-ORDINATE "r" FOR THREE DIFFERENT TYPES OF ENCAPSULATION.

The equations above give the physical base of the model when the heat storage is heated. Similar equations describe the cooling of the storage. The model divides the storage into five segments with five layers of storage material and procedures are needed to connect the resulting 25 volumes. The simulation program, SIMNON, included facilities to connect the volumes and to define physical property tables.

EXPERIMENTS

Storage unit

A storage unit with about 600 kg phase-change material was used. Three types of encapsulations were tested: spheres, cylinders and rectangular pads. The spheres with a diameter of 7 cm were packed in a bed and the cylinders with a diameter of 4 cm and a length of 1 m were placed in crossflow with mountings at the ends. The rectangular pads with the dimensions 0.04x0.4x0.6 m^3 were placed horisontally with distance elements of corrugated sheet aluminium.

Equipment

An air flow equipment for the full scale storage unit was used. The operating temperature is from 5 to 90°C. The heating and cooling fluid is air and the flow can vary from 1500 to 15 000 m^3/h. Figure 3 shows the design of the equipment and the main components.

The control system can use inlet temperaturs and flow rates from an existing cooling system as set values. If a system is suggested to be supplemented with a storage a test can be made to evaluate the impact of a suggested storage unit. It is possible to define any combination of temperatures and flows within the specifications of the equipment.

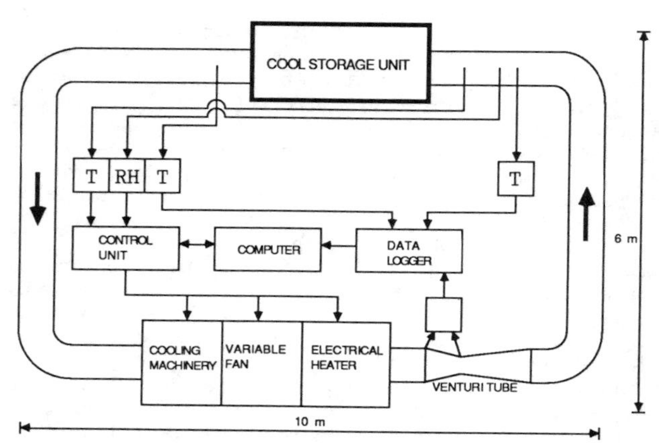

FIGURE 3. EXPERIMENTAL EQUIPMENT.

RESULTS

The numerical model is validated for some phase-change materials with air as cooling fluid. For other materials than tested a parameter identification should be included in the design method. The validations are made with a full scale storage unit. In figure 4 the results are shown for experiments and simulations with a test unit packed with spheres. The storage material is calcium chloride hexahydrate with additives which has a melting point of 29°C. The temperature may too seem high for a cool storage but in some industrial applications the melting point is even too low.

The showed experiment is a stepwise change of both storage inlet temperature and cooling fluid flow. The temperature is varied between 12 and 70°C and the flow is varied between 3200 and 6800 m^3/h. The unit included 2300 spheres with a diameter of 7 cm which makes a mass of 660 kg phase-change material. The experiments preceeded the simulations which gave the possibility to use experimental values of the inlet temperature and the air flow in the simulations.

The results show very good agreement between experimental and simulated storage outlet temperature. A slight discrepancy is observed during solidification periods but a more accurate parameter specification should improve the situation. In Figure 4 the last two diagrams show the stored thermal power and the stored energy according to the simulation. The variation of the thermal power is an important characteristic of this latent cool storage.

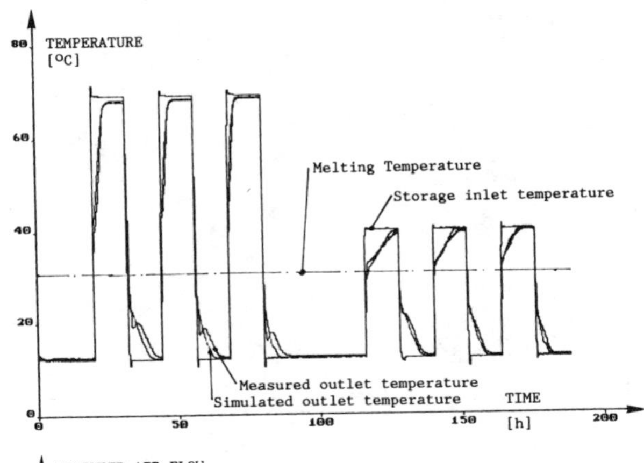

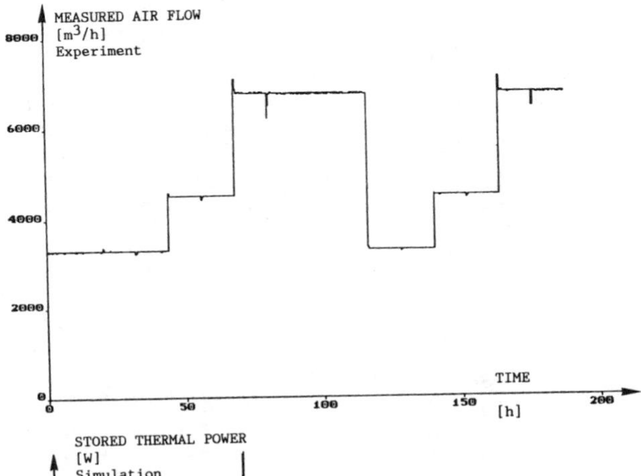

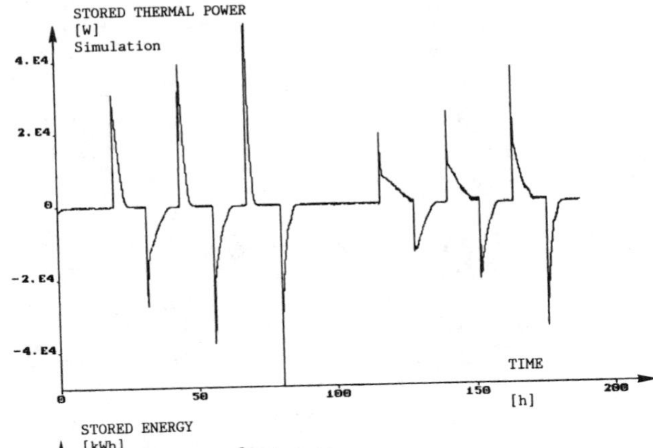

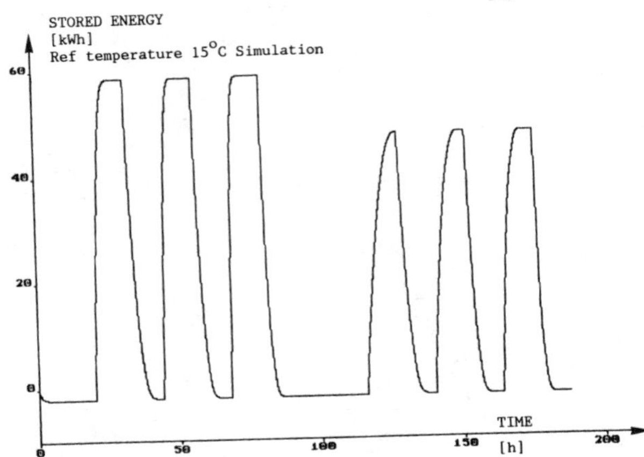

FIGURE 4. AN EXAMPLE OF EXPERIMENT AND SIMULATION WITH A FULL SCALE COOL STORAGE UNIT.

CONCLUSION

It is possible to study the impact of flow, geometry and materials on the performance of a cool storage unit. Situations with rapidly and irregulary changing conditions can be calculated with a general simulation program for nonlinear systems. The numerical simulations can be supplied with measurements from experiments or applications providing a fast, low cost and accurate design and evaluation method. Both sensible and phase-change materials can be handled with the design method as well as different encapsulation geometries.

ACKNOWLEDGMENTS

This work was supported in part by the Swedish National Board for Technical Development, STU. The authors are very grateful to Dr. B. Bengtsson for his help during the work and to Ms A. Thuresson for layout and word processing.

NOMENCLATURE

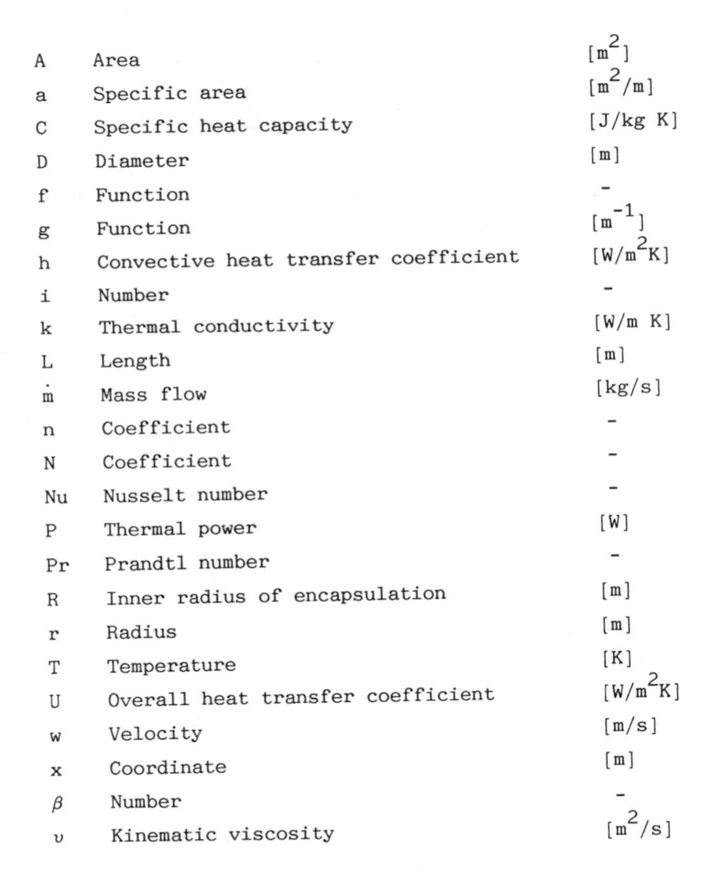

A	Area	[m^2]
a	Specific area	[m^2/m]
C	Specific heat capacity	[J/kg K]
D	Diameter	[m]
f	Function	-
g	Function	[m^{-1}]
h	Convective heat transfer coefficient	[W/m^2K]
i	Number	-
k	Thermal conductivity	[W/m K]
L	Length	[m]
$\dot{m}$	Mass flow	[kg/s]
n	Coefficient	-
N	Coefficient	-
Nu	Nusselt number	-
P	Thermal power	[W]
Pr	Prandtl number	-
R	Inner radius of encapsulation	[m]
r	Radius	[m]
T	Temperature	[K]
U	Overall heat transfer coefficient	[W/m^2K]
w	Velocity	[m/s]
x	Coordinate	[m]
β	Number	-
υ	Kinematic viscosity	[m^2/s]

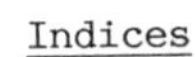

Indices

e	Environment
f	Fluid
s	Storage material
i	Number
p	Constant pressure

REFERENCES

(1) Elmqvist, H., Åström, K.J. and Schönthal, T. SIMNON, Users Guide for MS-DOS Computers. Department of Automatic Control, Lund Institute of Technology, Lund, Sweden, 1986.

(2) Johansson, T. and Solmar, A. Climatization of Industrial Premises by use of Short Term Storages. IEA-Workshop on Latent Heat Stores - Technology and Applications, Stuttgart,Federal Republic of Germany, March 7th - 9th, 1984.

(3) Loyd, D. and Solmar, A. (1986). Heat Transfer at a Latent Heat Storage, Proceedings of the 3rd International Conference on Computational Methods and Experimental Measurements (Eds. Keramidas, G.A. and Brebbia, C.A.), Springer Verlag, Berlin.

(4) Solmar, A. Energy Storage in a Bakery - Full Scale Experience. 2nd BHRA Fluid Engineering Conference on Energy Storage, Stratford-upon-Avon, England, May 24th - 25th, 1983.

(5) Solmar, A. and Loyd, D. Limitation of Cooling Water Temperature by using a Latent Heat Storage. Proceedings of International Conference Envirosoft 88, Porto Carras, Greece, 27-29 September, 1988.

(6) Söderström, M. Evaluation Method of Industrial Thermal Storages. Linköping Studies of Science and Technology, Dissertions. No 146. Linköping University, Sweden, 1986.

Chapter 97

PERFORMANCE OF COOL STORAGE AT A WAREHOUSE IN SWEDEN, TECHNICAL AND ECONOMIC EVALUATION

J. Persson, A. Lewald, A. Solmar, B. Karlsson

ASTRACT

In this paper an installation of a eutectic salt cool storage at a combined office and warehouse building in Sweden is presented.

A performance test of the cool storage is carried out. Results from these measurements are used together with a method to optimze dynamical energy systems to suggest alternate sizing of the storage. Measurements carried out under operating conditions are presented.

The results show that charged and discharged energy meets with specifications given by the manufacturer. Cooling power level at the discharge period achives desired cooling peak power reduction.

The installation of the cool storage lowers the investment cost compared to an ordinary system.

In case of an retrofit installation, that is, existing refrigeration machines, a peak power demand charge or cool storage inducement program of 430 US $ / kW is required to achive a three year pay back. The demand charge is within the range of applied inducement programs in the US. In Sweden however, a retrofit installation of a cool storage must be coordinated with increased cooling demand to show acceptable economics.

INTRODUCTION

Diurnal and industrial thermal energy storage can serve as a most useful part of an energy system. Energy audits carried out in Sweden show that large amounts of energy can be saved on a national basis [1]. However, in many cases the reduction of peak power demand for an energy system may be the main reason for installation.

Facing the the fact that use of electricity is increasing many utilities have started to adopt load management measures. In many cases thermal energy storage can be used for load management.

In Germany a substantial load shift from day to night is done by water and ceramic heat storages. In the US most utilities are summerpeaking why the market for cool storage has developed rapidly. Several utilities have adopted programs to encourage cool storage installations. Today more than 25 utilities with active cool storage inducement programs support installations with 100 $ - 500 $ per kW of load shift [2]. A number of these utilities also support industrial load management in general. Although cool storage has its major market potential in a hot climate it can be successfully adopted in Sweden.

In this paper an installation of a cool storage at a combined office and warehouse building in Sweden is presented.

To evaluate the installation on site measurements has been carried out. The results shows how the installation performs under operation conditions. To find the most profitable size of cool storage a combination of numerical simulation of the storage and a method to optimize dynamical energy systems are used. In short the following subjects are treated;

- A simulation model of the cool storage
- A method to optimize dynamical energy systems
- The cool storage installation
- Results from on site measurements
- Optimization of the cool storage size
- Economical outcome of the installation

SIMULATION MODEL OF COOL STORAGE

General

Numerical simulation aims to calculate the the temperatures of the dynamic process of charging and discharging the storage. Knowledge of inlet and outlet temperature ables us to calculate power levels and amount of energy stored and discharged. A lumped heat capacity model is used. See e.g. Holman, [3].

The storage material, in this case eutectic salt or phase change material (PCM), may have three geometries of encapsulation, sphere, cylinder or a plate. Here the spherical encapsulation is used to study the model. The lumps of a one sphere are defined as 5 equal volumes. The most inner volume is a sphere and the other 4 are layers in radial direction. To reduce the complexity of the model the storage is divided into 5 segments. It means that a storage with 10 000 spheres have 5 segments with 2000 spheres. Se figure 1. The heat transfer in the PCM is assumed to be by heat conduction. At the surface heat is transfered by forced convection.

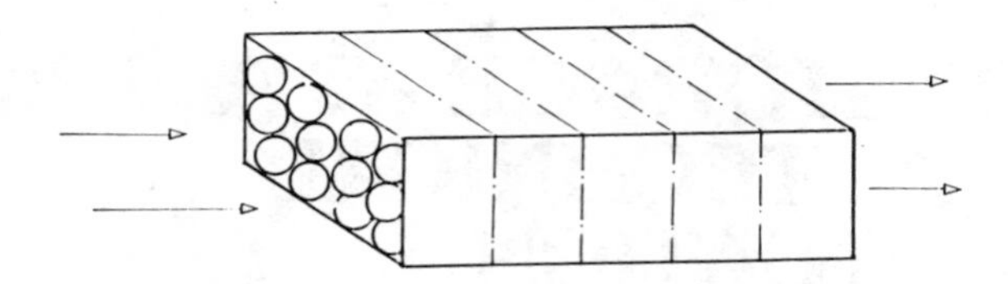

FIGURE 1. THE EUTECTIC SALT COOL STORAGE IS DIVIDED INTO FIVE SEGMENTS.

The heat capacity of the heating fluid is constant. Density of the heating fluid may vary if the fluid is a gas. Due to influence on the convective heat transfer the viscosity of the heating fluid may vary in the model. The convective heat transfer coefficient is assumed to be influenced by velocity, geometrical dimensions of encapsulation and viscosity of heating fluid. Heat capacity of the encapsulation material is neglected and the thermal conductivity is infinite.

Heat capacity and heat of transition of the PCM are defined as a heat capacity versus temperature function. Therefore materials with one or more transitions over temperature intervals may be described.

The assumptions above combined with governing heat transfer equations gives information how the temperature of the heating fluid is changed when it passes the storage.

The simulation model can be used to optimize an installation, provided that the application can be described as a source-sink pair [4]. A more detailed description of the model is given in [5].

Simulation program

The program used for simulating the PCM storage is Simnon [6]. Simnon is a special programming language for simulating dynamical systems. Systems may be described as ordinary differential equations, as difference equations or as combinations of such equations. The program is a general purpose program, that is, there is no energy system component library. The language has an interactive implementation which makes it easy to work with. Simnon allow optimization, introduction of experimental data and parameter fitting.

METHOD FOR ENERGY SYSTEM OPTIMIZATION

General

Existing methods has been studied by Backlund [7]. Mostly they are not optimizing, others can not use TOU-rates. Therefore a model has been elaborated which uses linear and integer programming methods for optimizing. Different measures adopted to the energy system can be studied in order to find the minimum cost for a specified time period. That is, the sum of investment costs and future costs are minimized.

The model can be used for:

- Studying the feasability for energy storage
- Different energy price increase
- Different TOU-rates
- Comparing different economical viewpoints
- Considering the dynamics of the energy system
- Considering changes in the technical system characteristics

The model

The model is based upon well-known optimizing techniques, linear and integer programming [8]. The basic idea is to set up an energy balance for characteristic time periods. Here, seasonal changes and TOU-rates constituate the different time periods. An energy balance, i.e.

Energy in = Energy out + Energy stored

is set up for each period. The model is divided into different parts. Every period is divided into two sides; the energy supply side and the energy use side. Each of these can then be divided arbitrary times. In the upper left part of Fig 2 the different energy sources and the possible heating and distribution systems can be seen, thus, the supply side of the period. In the upper right part the use side is shown. The figure also shows the borderline to another period. The two lines that cut the border indicate the possibility to store energy from one period to another. The node named free stands for possible energy comming from, for example: solar gains, persons in the building and appliances.

When using the model a definition of the number of energy sources, distribution systems and their connection is made. Also their cost and characteristic, in different time periods, must be defined. On the demand side, possible reduction of the energy use, is defined by the cost and gain, in U-value, if it is cost for insulaton. The function for the cost versus the gain, in U-value, is often a step function followed by a nonlinearity, see Fig 3. In the model a linearisation is made, see Fig 4. First there is a step function then a piecevice linearisation.

The model chooses one of a number of, in the problem prescribed occations, when a system or a quality change is possible, e g should I buy a new water heater this year or five years later (expected remaining life time).

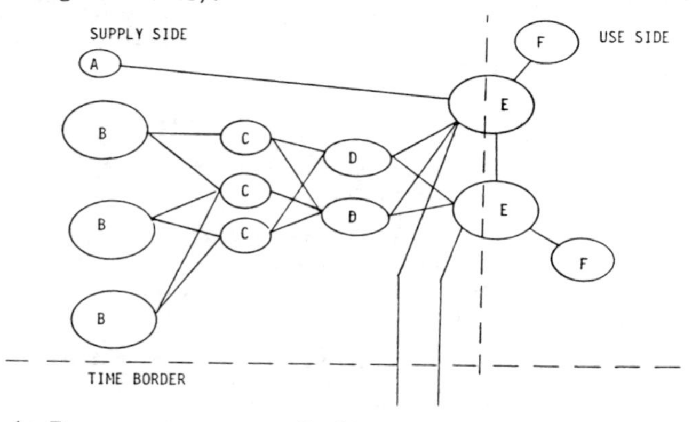

A: Free energy
B: Energy sources
C: Heating devices
D: Distribution systems
E: Energy loads
F: Conservation measures

FIGURE 2. SCHEMATIC DRAWING OF THE MODEL.

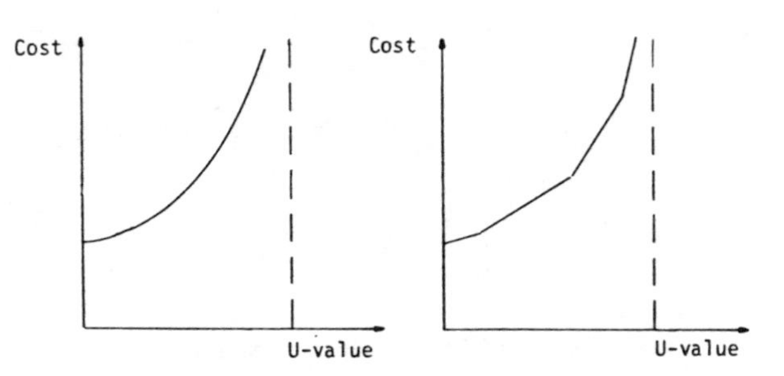

FIGURE 3. INSULATION COST FUNCTION

FIGURE 4. LINEARISED INSULATION COST FUNCTION.

There is no limitation of the number of nodes in the supply side and options in the use side.

The model is general and can be applied to a variety of energy systems. The energy supply side could as well be a district heating system with different fuels and network, and the use side could be different housing areas with their possible energy need reduction.

COMBINED OFFICE AND WAREHOUSE BUILDING

The building is situated in Stockholm which is in the southern part of Sweden. The climate varies substantially from winter to summer with lows in the -25 °C range and highs in the +25 °C range. The warehouse part is used for storing delicatessen, i.e. salami, ham and crawfish.

The volume of the building amounts to 25000 m^3, meaning about 5000 m^2 of space. The warehouse part of the building needs to be kept at a range of low temperatures, -25 °C - +15 °C (-13 °F - 59 °F). Offices are supplied with brine for individual temperature adjustment. Heating of the building is done by condensor heat and electric heater. The cooling system consists of refrigeration machines and a cool storage connected in parallel to the cooling load, see figure 5.

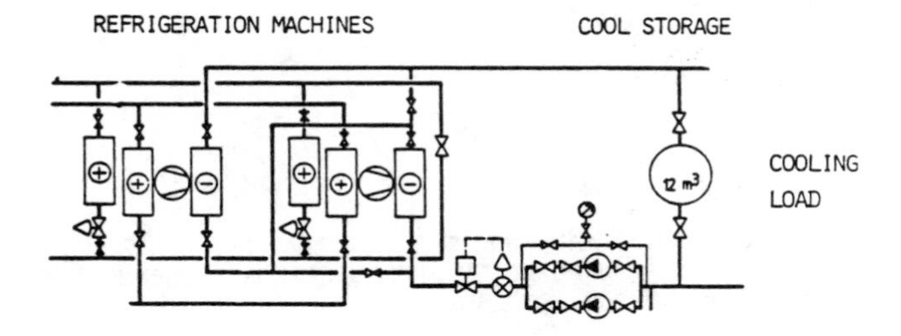

FIGURE 5. REFRIGERATION MACHINES AND COOL STORAGE.

The cool storage is a cylindrical container with 12 m^3 of eutectic salt nodules having a melting point of -10 °C (14 °F). This volume equals about 600 kWh of latent heat.

Some of the reasons why a cool storage was installed is mentioned below.

- Reduction of installed power to the building
- The number of starts and stops of the refrigerant compressors are minimized with longer lifetime of the compressors as a result
- Smaller and cheaper refrigerant compressors can be choosen as the cool storage covers part of the peak power cooling demand
- The total cost of cooling system including storage is of the same magnitude as an ordinary system

In sizing the cool storage the daily cooling load curves for the building were used. In figure 6 the cooling load curve for 25 °C is shown.

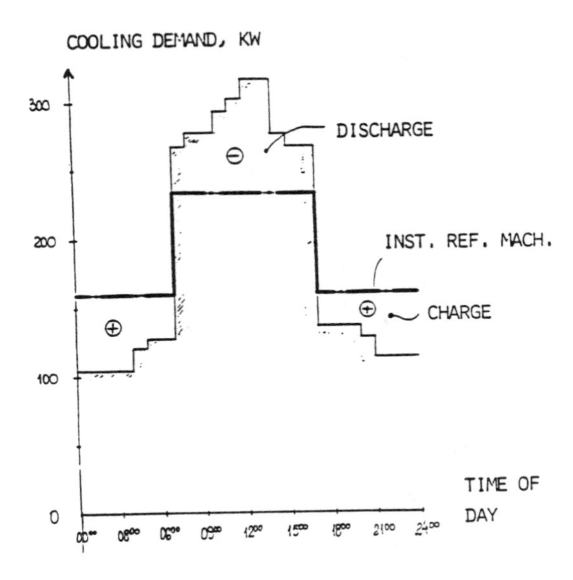

FIGURE 6. COOLING LOAD CURVE FOR BUILDING AT 25 °C OUT DOOR TEMPERATURE, NO HEATING IS REQUIRED.

RESULTS FROM ON SITE MEASUREMENTS

Measurements on the cool storage has been carried out in two periods. A performance test was done during a period when the system were about to be taken in operation. Continous monitoring of the building will continue for a one year period.

Performance and Simulation of the Cool Storage

A charge and discharge cycle of the cool storage has been recorded. Data have been collected in 15 minute intervals. Figures 7,8 and 9 show measured and simulated values of temperatures, energies and powers for the cool storage. The volume flow of the fluid is at charge about 28 m^3/h and at discharge about 5 m^3/h.

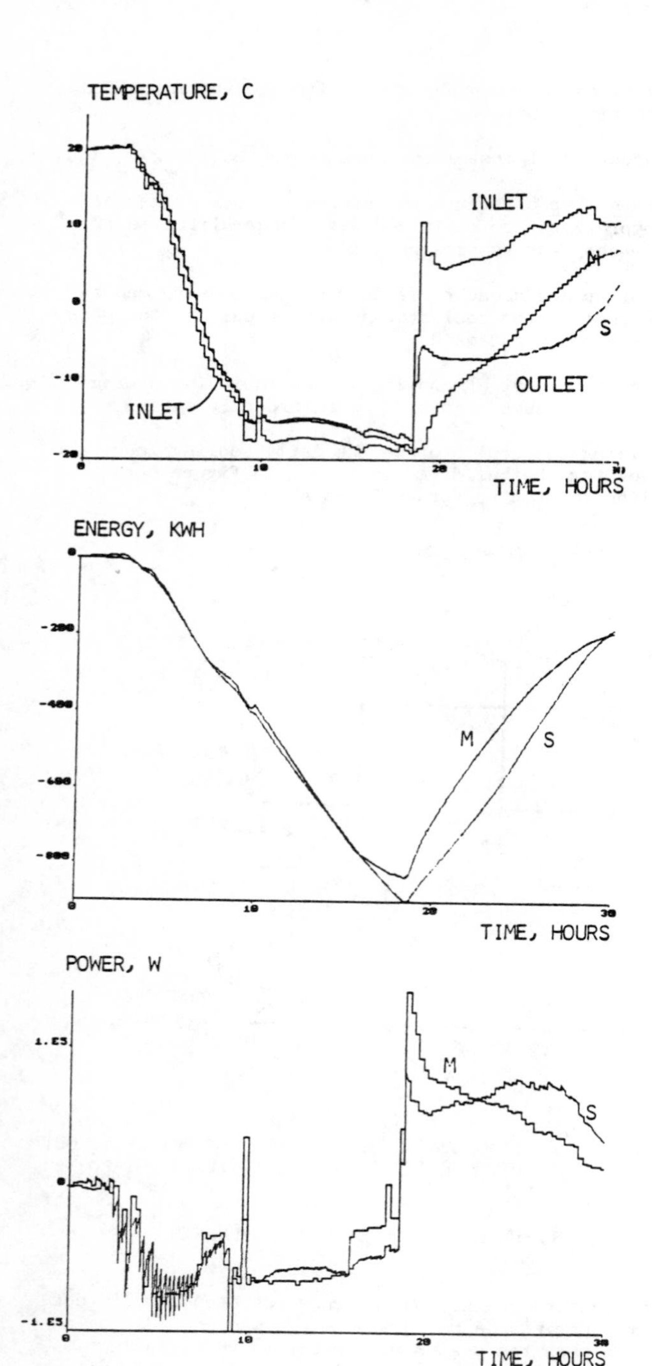

FIGURES 7-9. TEMPERATURES, ENERGIES AND POWER LEVELS, MEASURED (M) AND SIMULATED (S) VALUES OF THE COOL STORAGE.

The results show that charged and discharged energy meets with specifications given by the manufacturer. Cooling power level at the discharge period achives desired peak power reduction.

The simulation show good accordance at charging the storage. However at discharge the discrepancy is to large to be accepted. Ideas for possible explenations are as follows.

- The simulation model does not account for the fact that the flow direction is reversed at discharge.

- Natural convection inside the PCM is not represented.

- The volume flow at discharge is very low, i.e. the flow distribution is uncertain.

Measurements under Operating Conditions

Measurements under Operating Conditions has been carried out starting June 1:st 1988. During this period the outdoor temperature has reached levels which was set as dimensioning for the cooling system. Still the peak cooling power demand has been recorded substantionally lower than predicted when dimensioning the cooling system. This can be explained to some extent by the fact that the complete building is not in full operation. Another reason is that the peak cooling load is expected during the latter part of the summer as the body of the building has warmed up. However this cannot explain the large discrepancy between predicted peak cooling demand and recorded. Further analyses of recorded data is required and the addition to cooling demand as the building is fully used is yet to be seen.

Below is reported results from measurements as of June 17:th, the outdoor temperature reached about 28 °C this day. In order to get the cool storage in operation half of the refrigeration machine capacity were shut down. Figure 10 show cooling demand for the building (B) and how this is supplied by refrigeration machines (R) and cool storage (C).

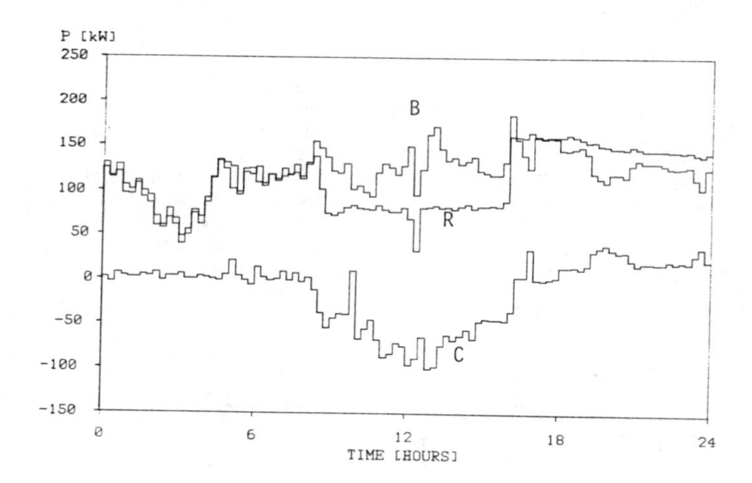

FIGURE 10. COOLING DEMAND FOR THE BUILDING (B) AND HOW THIS IS SUPPLIED BY REFRIGERATION MACHINES (R) AND COOL STORAGE (C).

The discharge of the cool storage is achived by reversing to flow of brine. Se figure 5 showing the cooling system. In practice this is achived by; Increased flow through the cooling load, simoultaneous reduction in the the flow through the refrigeration machines controlled by increased temperature of returning brine from the cooling load. Combined this causes the cool storage to discharge as in figure 10. Figure 11 below show how the brine flows are altered during the discharge process.

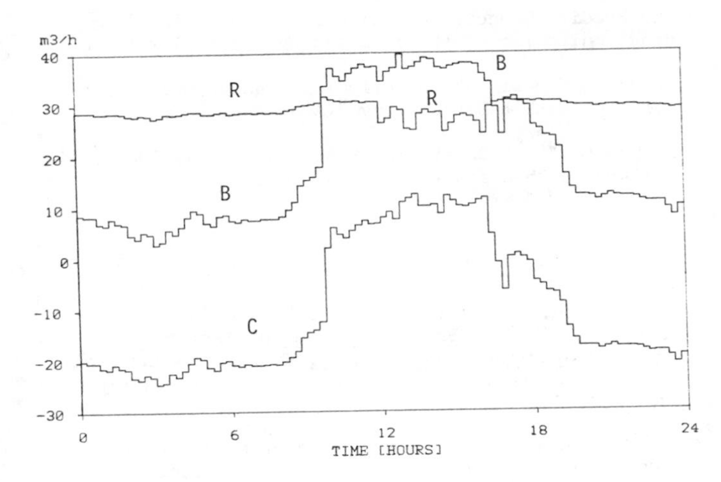

FIGURE 11. BRINE FLOWS THROUGH THE COOLING LOAD (B), REFRIGERATION MACHINES (R) AND COOL STORAGE (C).

OPTIMIZATION OF COOL STORAGE INSTALLATION

Having validated the model described above it can be used in combination with the optimization method.

The daily cooling load curve is used together with the simulation model to establish a cost function for the latent heat storage. The cost function found by simulation is expressed in SEK or US $ per kWh of latent heat storage capacity.

Important to study from the simulation results are cooling power levels from the cool storage. In applications with a need for high cooling power during short periods the simulations may show that it is necessary to oversize the storage to meet the power demand.

Below an optimization of the cool storage installation is carried out. Here a cost function derived from figures given by the cool storage manufacturer is used. Other cost functions for the refrigeration machinery, installation costs and utility rate structure are given by companies and the utility in question.

The energy system in the building is shown in figure 12.

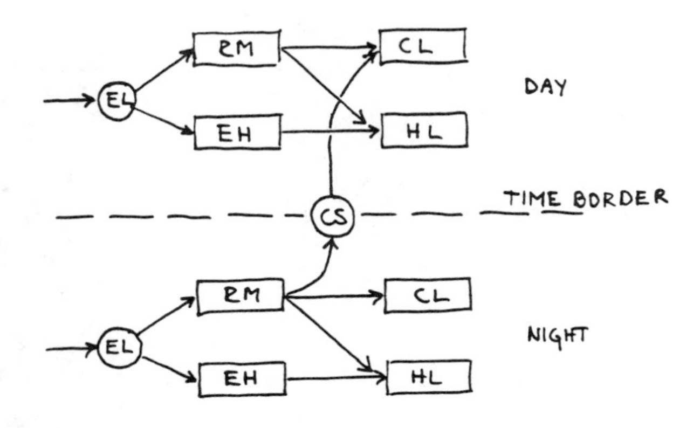

FIGURE 12. ENERGY SYSTEM OF THE BUILDING.

Electricity (EL) is supplied to refrigeration machines (RM) and electric heater (EH). At daytime the refrigeration machines can supply cooling and heating load (CL and HL). Heating load can also be supplied by the electric heater. At nighttime the refrigeration machines also supply a possible cool storage (CS). The cool storage supplies the daily cooling load as shown in the figure by crossing the time border.

Figure 12 is used to establish the energy and power balances used in the subsequent optimization.

Cooling and heating loads are given by daily load curves. Here, daily load curves for three outdoor temperatures, -20 °C, 0 °C and 25 °C, are used. The TOU rate used in this case has its peak period during November thru March. The daily load curves for the three outdoor temperatures are distributed over the full 12 month period. This means that figure 12 is repeated four times to account for all time periods in our optimization problem.

The TOU rate applied is as follows (1 US $≈6 SEK):

- November thru March, weekdays between 7 AM and 9 PM, 0.52 SEK/kWh other time, 0.25 SEK/kWH.
- April thru Oktober, 0.25 SEK/kWh.
- Demand charge, all year 65 SEK/kW, peak period, 225 SEK/kW.

This gives cost functions for electric power. Included is also subscription charge and installation cost for different sizes of switch gear. The cost functions used in the optimization are:

Cool storage	600	SEK / kWh
Refrigeration machines (elec. power)	10000	SEK / kW
Electric heater	100	SEK / kW
Demand charge, peak	337	SEK / kW
Demand charge, other time	65	SEK / kW

Optimization is carried out for a one year period, that is, the result reflects investment cost, demand charge and energy cost during one year.

ECONOMICS OF THE INSTALLATION

Calculation according to above gives a cool storage of 807 kWh and refrigeration machines of 100 kW electric power. Total cost amounts to 1.6 MSEK. Ruling out the cool storage and doing a new optimization results in refrigeration machines of 162 kW electric power and a total cost of 1.8 MSEK.

This shows that by installing a cool storage the cost for a one year period including installation costs, is <u>less</u> than when not using cool storage. Here the single factor of utmost importance is the cost of the refrigeration machines. In the warehouse a storage of 600 kWh was installed. This means a cost higher than the optimum. However, it is <u>lower than if no cool storage were to be used</u>.

Now suppose that you are studying an old building where refrigeration machines which cover the full cooling load are already installed. What demand charge (or cool storage inducement program) is required to achieve a three year pay back period on a cool storage installation ?.

That is, the installation of a cool storage needs to be covered by demand charges and TOU rate. The latter is of minor importance. Optimization is carried out with a stepwise increasing demand charge. The result is shown in figure 13.

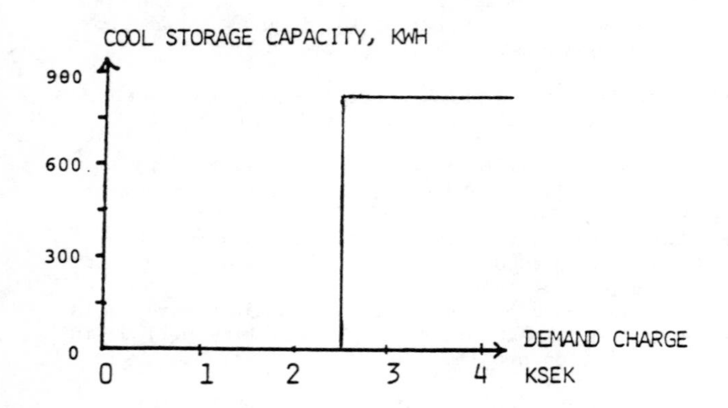

FIGURE 13. STORAGE SIZE FOUND BY OPTIMIZATION AS A FUNCTION OF APPLIED DEMAND CHARGE.

In case of an retrofit installation, that is, existing refrigeration machines, a peak power demand charge or cool storage inducement program of 2600 SEK/kW is required to achive a three year pay back. The optimum size of storage in this case will be 807 kWh. The demand charge is within the range of applied inducement programs in the US. In Sweden however, a retrofit installation of a cool storage must be coordinated with increased cooling demand to show acceptable economics.

CONCLUSIONS

The installation of a cool storage in the warehouse actually lowers the installation costs compared to an ordinary system, e.g. negative pay back period.

A performance test of the eutectic salt cool storage show that charged and discharged energy meets with specifications given by the manufacturer. Cooling power level at the discharge period achives desired peak power reduction.

The cooling peak power demand of the building is recorded lower than predicted. By decreasing the refrigeration machines cooling power it is shown that the cool storage adds cooling power as predicted to the system.

The combination of numerical simulation and dynamical optimization may prove to be a powerful tool to promote the efficient use of PCM thermal energy storage in energy systems.

ACKNOWLEDGEMENTS

The authors whiches to thank;

- The Swedish Board for Technical Development for funding a research project on Industrial use of PCM Heat Storages including the initial measurements in this report

- The Swedish Council for Building Research for funding monitoring of the cool storage installation

- Bröderna Hartwig AB for allowing unautorized personnel mess with their cooling system

- Bo Lundfeldt of Cristopia AB Sweden for invaluable assistance when starting the project

REFERENCES

1. J. Persson, M. Söderström, T. Johansson and B. Karlsson, (1987), Energy Storage in Swedish Industry, Potential ans Case Studies. First International Seminar on Energy Storage, Sophia Antipolis, France, 1987.

2. L. MacCannon, (1988), ITSAC Newsletter, January issue, International Thermal Storage Advisory Council, San Diego, USA.

3. J. Holman, (1981) Heat Transfer, 5th ed., McGraw Hill, Tokyo.

4. J. Persson and B. Karlsson, (1987), Industrial Case Study of a Latent Heat Storage, Innovation for Energy Efficiency Conference, Newcastle, England.

5. A. Solmar, J. Persson and D. Loyd, A Method to Design Cool Storage Units, 11th World Energy Engineering Conferance, Atlanta, USA, 1988.

6. H. Elmqvist, K. Åström, and T. Schönthal, (1986). SIMNON, Users Guide for MS-DOS Computers, Department of automatic control, Lund Institute of Technology, Lund.

7. L. Backlund, (1988), Optimization of Dynamic Energy Systems with Time Dependent Components and Boundary Conditions, Linköping Studies in Science and Technology, Dissertations, No. 181, Linköping Institute of Technology, Department of Mechanical Engineering, Linköping.

8. Foulds, L.R., (1981), Optimization Techniques, Springer-Verlag.

SECTION 23
SUPERCONDUCTORS

Chapter 98

SUPERCONDUCTIVITY—YESTERDAY AND TODAY

C. H. Rosner

INTRODUCTION

Rarely are scientific discoveries embraced with such universal and exuberant enthusiasm by as diverse a combination of physicists, chemists, materials scientists, engineers and governmental R&D agencies, as well as by the technical and popular press. Such was, and still is the case with the discovery of higher temperature superconductors.

As the name implies, "Superconductors" are a class of certain materials that are able to conduct or transmit electric currents in an extraordinarily efficient fashion. In fact, the phenomenon of "Superconductivity" (SC) represents the true absence of electrical resistance, such that zero electric power losses are possible when these materials are cooled below their critical transition temperature, referred to as Tc.

Until 1986, almost all known superconducting materials were either pure metals, alloys or various intermetallic compounds that had Tc's below 23°K and thus needed to be cooled in liquid helium to near absolute zero (-273°C = 460°F) before many practical applications became feasible.

The scientific discovery and breakthroughs that have been achieved since 1986, demonstrate that certain ceramic type materials could also become "superconductors" and most importantly, that this could occur at a temperature up to 100 degrees higher than was possible with the earlier metallic based materials.

The enormous worldwide attention subsequently paid to superconductivity may be somewhat surprising if one realizes that the discovery of the SC phenomenon occurred already in 1911, and is due to a Dutch physicist, Heike Kamerlingh Onnes. At that time he tested mercury to establish its electrical properties down to near absolute zero and found the sudden disappearance of all resistance at 4.16^{o}K which is the Tc for mercury.

During the intervening 75 year period, research and development carried out around the world led to the discovery of many elements and over a thousand superconducting metallic alloy and compound material combinations. However, only a handful of these could be fabricated into useful shapes and had sufficiently practical properties such that commercially attractive applications could be pursued.

Nevertheless, since the early 1960's several billion dollars have been expended, principally by Government agencies, universities and industry, both in the U.S. and abroad, in further research, development and commercialization efforts, involving the lower Tc or conventional metallic superconducting materials and devices.

Many successes were achieved in providing high field magnets to such diverse uses as High-Energy Physics Research, Fusion Power Research, as well as for physics, chemistry and biological research areas.

In this presentation a brief overview is given of the desirable characteristics of superconductors and a discussion of some of the more significant accomplishments in demonstrating practical applications of this remarkable phenomenon. Also addressed are the possible additional benefits that would be obtained from successful development of the newer, higher-temperature superconductors now under intense scrutiny in many countries around the world. The prospect that these types of materials can be cooled in liquid nitrogen instead of helium or the possibility of materials being discovered that might even operate at room temperature, has caused an unprecedented upsurge in research activities in virtually every university and technologically advanced country. Annual expenditures are now estimated to exceed 500 million dollars and encompass governmental organizations as well as the electrical, electronics, ceramics and chemical industries in particular. It is clear that the pay-off from these expenditures will extend over 5-10 years and most likely involve new material discoveries. On the near-term horizon, existing lower-temperature technology should see expanded uses, which could then gradually be replaced by incorporation of the newer materials.

SUPERCONDUCTING MATERIALS

The on-going intense focus on the critical transition temperature Tc of the higher temperature superconductors and wide-spread speculation in the popular press as to the impending revolution in store for electronic devices and electrical equipment, often overlook the fact that besides Tc, other SC material properties really determine whether or not a material with higher Tc ultimately becomes useful.

Aside from obvious economic and material cost considerations, as well as manufacturability, there are typically several additional parameters that must complement a high Tc of a superconductor to establish its more general usefulness. These are:

Magnetic Field Range
characterized by Critical Magnetic Field -- Hc

Electric Current Range
Critical Current -------------------------- Ic

and more generally

Mechanical Properties must be suitable to withstand fabrication and device operational requirements, especially the existence of high strength over a large temperature range and high strain tolerance.

Next to Tc the most significant parameter would be Ic, or more importantly Jc, i.e. the current density often denoted in Amperes per unit surface area or A/cm^2. This particular parameter addresses both the economic and practical usefulness of a superconductor. For most applications a value of much less than 10,000 A/cm^2 (10^4 A/cm^2) is marginally interesting, since many copper or semi-conductor based devices operate quite satisfactorily at or below this level, requiring only modest cooling costs which are adequate to cover a vast range of room temperature or lower temperature based applications.

If higher Tc materials are to have the potential for replacement of these conventional non-SC materials with superconductors, a level approaching 100,000 A/cm^2 (10^5 A/cm^2) or higher should be attainable before the hoped-for revolutionary introduction of superconductors into broad areas of commerce at multi-billion dollar levels can be realized. Only at that point would we achieve more efficient and cost-effective utilization of the newer high Tc materials.

In the meantime, it is encouraging that several of the conventional metallic-type superconductors such as niobium-titanium (Nb-Ti) and niobium-tin (Nb_3Sn) do indeed exhibit the ability to support such high current densities. Hence, they became the only really practical superconducting materials during the past several decades. Already, some of the newer high Tc materials have been demonstrated to support similar high current density levels, albeit primarily in thin-film form.

Generally, almost all materials that exhibit higher Tc also have a high Hc. The overall relationship between Tc, Ic and Hc characteristics of most SC materials can be seen in Figure 1. This is also referred to as the "Critical Surface" of a SC material.

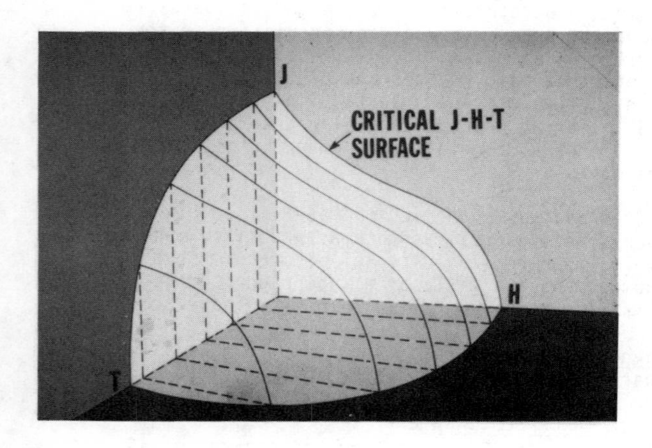

FIGURE 1
"CRITICAL SURFACE" OF SC MATERIALS

Variations in processing and manufacturing procedures influence these two more intrinsic characteristics of typical superconductor materials only to a very minor extent.

It is in the area of current carrying capability Jc where material processing and fabrication methods have a dramatic impact on a material's ability to exhibit high levels of current carrying capacity. The reason for this is related to the corresponding microstructure developed in the material, since high current levels require the presence of so-called "flux pinning" sites, the existence of which control a material's current carrying capacity in the presence of high magnetic fields.

Once high Tc's of several materials are verified, the selection process as to which material would be the object of further R&D, to develop enhanced current carrying capability, then evolves quite naturally from its workability or perceived manufacturing difficulties if large quantities of superconductors are needed.

These combined factors, but especially processing procedures that enhance Jc, have led to the development of niobium-titanium (Nb-Ti), which became the universally preferred superconductor. Even though other materials than Nb-Ti have either higher Tc or higher Hc, and even higher Jc capability, Nb-Ti is an alloy that has been shown to be readily workable in combination with copper and can be extruded and drawn into long lengths of wire. Other SC materials with Tc's that are twice as high as the 10^oK Tc of Nb-Ti are typically brittle intermetallic compounds, such as niobium-tin(Nb_3Sn) or niobium-germanium (Nb_3Ge). These characteristics are also shown in Figure 1. Corresponding with Figure 1, Figure 2 details the Tc and Hc relationship for selected SC materials.

S/C MATERIALS PROPERTIES

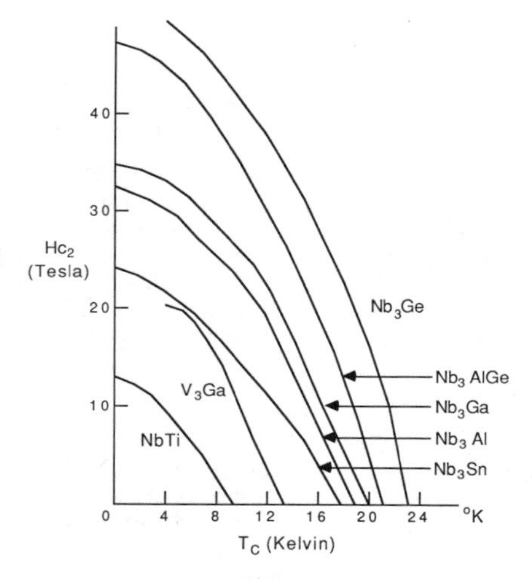

FIGURE 2
Tc AND Hc RELATIONSHIP FOR SELECTED SC MATERIALS

It should be noted that practical implementation of superconductive wires into devices has required not only the addition of copper in intimate contact with the superconductive materials, but also a subdivision of the SC materials into many fine filaments, each of which in turn is surrounded by copper. This has been necessary in order to provide for thermal and electromagnetic stability under device or power equipment operating conditions. An example of such a multifilamentary Nb-Ti superconductor is shown in Figure 3.

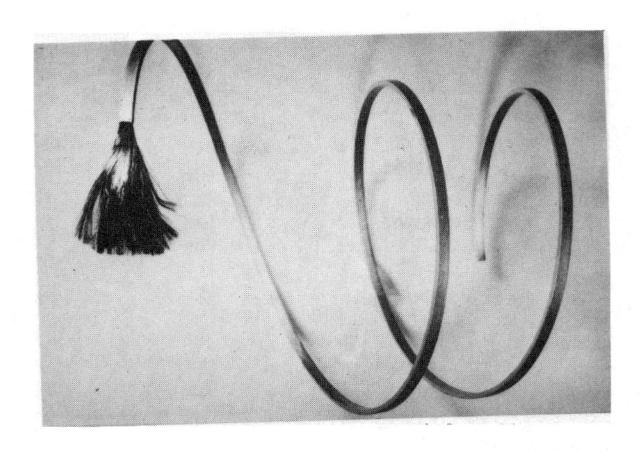

FIGURE 3
COPPER CLAD MULTIFILAMENTARY NB-TI SUPERCONDUCTOR

Similarly, for electronic elements or devices, thin films provide the typical configuration of useful superconductors for sensing or switching applications, as well as a whole range of electronic devices listed below.

SUPERCONDUCTING MAGNETS

The most widely useful application of superconductors is in the form of electro-magnets that can generate very high magnetic fields. The strength of such magnetic fields corresponds directly to the applied electric currents (I) and the number of turns (N) in what is also called a "solenoid". Almost any desired magnetic field (H) at the center of a round solenoid is thus obtainable simply by coiling a length of wire around a tube or coil-form in accordance with the following relationship

$$H \propto NI$$

where the magnetic field units are measured in Tesla or Gauss (1 Tesla = 10,000 Gauss).

For reference, the earth's magnetic field is approximately one-half gauss, whereas a "stick-on" refrigerator magnet has a field strength of a few hundred gauss.

The usefulness of magnetic fields is pervasive in our everyday lives, originating with the compass invented by ancient Chinese thousands of years ago, to modern day products such as motors, generators, TV's, etc., and more recently, medical diagnostic imaging equipment. The compass and most conventional low-field magnets utilize either ferromagnetic materials such as iron or so-called permanent magnets such as the "stick-on" magnet types mentioned earlier.

Until the advent of practical superconductors, all electro-magnets were made with copper wire. Since copper and similar metals exhibit resistance to the flow of electric currents, the generation of high magnetic fields is typically limited by the heating accompanying the flow of current through such normal resistive conductors, leading to large power losses and extensive cooling requirements.

It is precisely the absence of resistance in superconductors that make them so attractive, since nearly unlimited currents can be passed causing virtually no heating or power losses, except in lead-ins or resistive connections.

In practical terms, the currents and corresponding magnetic fields are limited to Ic and Hc, referred to earlier, beyond which the superconductor reverts to a normal resistive state or "quenches" from the superconductive state, even though Tc may not have been reached.

Thus, we experience that superconductors can "quench" by exceeding either their Tc, Ic or Hc limit, or corresponding to points on the critical surface shown in Figure 3.

Magnet design approaches thus include safety margins that establish operating points well within the critical surfaces, often only at one-half to three quarters of the continued current-field Ic-Tc limit, so as to avoid unexpected quench transitions.

APPLICATIONS

While the scientific aspects of superconductivity continue to intrigue, puzzle and challenge research scientists and experimentalists alike, the broader interest by thousands of other individuals around the world focuses on the application potential of superconductive materials and devices. This is the outgrowth of a general realization that technological progress in the 20th Century has, to a large extent, successfully exploited the combined fields of electricity, magnetism and electronics. In these, the ever increasing uses of electric power, electronic devices, computers, etc. are generally based upon a few materials such as copper, aluminum and semi-conductors, primarily silicon.

It is now expected that the discovery of higher-temperature superconductors may change all of this and dramatically expand the benefit of these technologies in environments of significantly reduced power losses and operate at higher levels of efficiencies. Since the early 1960's when the lower temperature superconductors were discovered to have useful critical field and critical current characteristics, similar excitement and expectations for large scale applications were raised. At that time, many corporations mounted extensive R&D programs to search for higher temperature materials, while many electromagnetic and electronic applications were pursued.

Gradually, it became apparent that all known superconductors were essentially limited to lossless DC current operations only and required liquid helium environments; i.e. 4°K above Absolute Zero, much of the interest waned. Except for a smaller group of organizations, much of this work was abandoned by the early 1970's.

In the intervening period, materials and device developments reached a stage where many applications were demonstrated. A listing of most of these appears in Table 1.

	INTERMAGNETICS' ACTIVITIES
COMMERCIAL & INDUSTRIAL USES	
Magnetic Resonance Imaging	
Nuclear Magnetic Resonance (NMR)	X
Magnetic Resonance Imaging (MRI)	X
Magnetic Resonance Spectroscopy (MRS)	X
Gyrotrons	X
Magnetic Separation	X
R&D Magnets	X
PHYSICS MACHINES & DEVICES	
High Energy Colliders	
Beam Line & Detector Magnets	
Smaller Machines	
Superconducting RF Cavities	
ENERGY RELATED APPLICATIONS	
Magnetic Fusion	X
Superconducting Generators/Motors	X
Magnetohydrodynamic Power Generation	
Superconducting Power Transmission	X
Superconducting Magnetic Energy Storage (SMES)	X
Current Limiters	
Superconducting Transformers	
TRANSPORTATION	
High Speed Trains	X
Ship Propulsion	
SUPERCONDUCTING COMPUTERS, INSTRUMENTATION	X

TABLE 1
PRINCIPAL SUPERCONDUCTOR APPLICATIONS

Yet, the increasing availability of high magnetic fields was welcomed by the physics community in particular. The area of high energy physics research was literally transformed by the availability of high-field magnets. Initially, large bubble chambers were built, followed by accelerator magnets and culminating in the construction and successful operation of Fermilab, a 1000 GeV energy saver located outside of Chicago. A view of the area is shown in Figure 4.

Aerial view of the Fermi National Accelerator Laboratory (Fermilab) at Batavia, Illinois. The four-mile main ring, originally a 400 GeV fixed-target proton synchrotron, has been upgraded twice. First, a ring of superconducting magnets was added, doubling the proton energy. Second, an antiproton source was built, so that both protons and antiprotons could be injected into the main ring, and accelerated around it in opposite directions. This new machine, the Tevatron, now functions as a collider, and, with a total energy of 2 TeV, is the world's most powerful accelerator. (Courtesy of Fermi National Accelerator Laboratory).

FIGURE 4
FERMILAB NATIONAL ACCELERATOR LABORATORY

Larger scale demonstrations were also carried out by constructing magnets for Fusion power experiments and building an experimental SC power transmission line at Brookhaven, N.Y., among others.

The most successful and first significant commercial application occurred in using SC magnets for medical diagnostic purposes where the phenomenon of nuclear magnetic resonance (NMR) is used to provide cross-sectional images of human organs.

A cross-section of such a MRI magnet is shown in Figure 5.

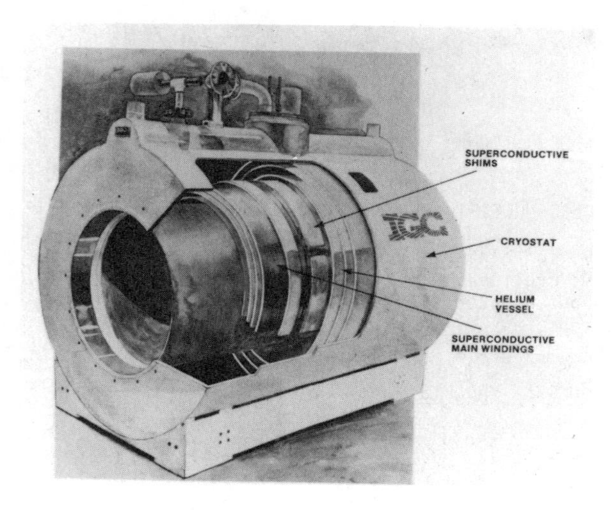

FIGURE 5
CROSS SECTION OF MRI MAGNET

Another rather intriguing application was pursued by Japan, where a high speed magnetically levitated train utilizing SC magnets was demonstrated and operated at speeds exceeding 500 km per hour. See Figure 6.

FIGURE 6
MAGNETICALLY LEVITATED TRAIN - COURTESY JNR

Experimental verification was also achieved in many other areas demonstrating that superconductive motors, generators and energy storage systems were possible. Renewed interest in all of these applications can now be expected.

HIGH TEMPERATURE SUPERCONDUCTORS

The several newly discovered ceramic-type materials have now exhibited Tc's up to $125^{0}K$. Already thin film structures have operated at liquid nitrogen temperatures and single electronic sensors and switching devices - Josephson Junctions - have been fabricated.

Much uncertainty still surrounds the exact conduction mechanism of the ceramic superconductors. The presence of oxides has been an important requirement, especially in the form of copper-oxide. This contrasts with the well understood BCS theory covering metallic superconductors, for which Bardeen, Cooper and Schrieffer received the Nobel prize in 1972. It should be remembered, however, that complete theoretical understanding has not deterred experimental progress on practical devices. The most promising near-term applications of higher Tc superconductors should be emerging in electronic devices. Here small scale and a variety of chemical thin film deposition techniques can circumvent the brittle and often unstable nature of SC ceramic oxides.

These materials require only to be cooled to a liquid nitrogen temperature range; i.e. almost 75^{0} higher than the lower liquid helium based superconductors. However, their primarily ceramic nature requires much more sophisticated fabrication methods, compared with the earlier known metallic superconductors. In addition, their inherently brittle nature has so far made it very difficult to fabricate bulk or wire configurations that exhibit useful critical current properties.

Large scale efforts have been mounted in the United States, Japan and Europe to exploit the potentially enormous commercial potential that is inherent if present day helium based SC technology could be replaced by devices and equipment operating at liquid nitrogen. Even greater opportunities would become apparent if superconductors would be found or synthesized that could operate at or above room-temperature. Already, a number of such claims have surfaced, but none could be verified or repeated to-date.

A most desirable near-term objective of today's R&D would be the ability to combine existing semi-conductor technology with the superconducting ceramic material within a single device or chip structure. Especially desirable would be the ability to include superconducting interconnect leads within large scale logic or computer semi-conductor arrays.

CONCLUSION

For many decades, superconductivity held a fascination only for relatively few individuals. Suddenly, the emergence in 1987 of higher temperature superconductors into the public consciousness via the newsmedia, raised enormous near-term expectations.

A more sobering assessment of the potential impact of higher-temperature superconductivity on technology and commercial opportunities is now emerging. Initial projections of a 1-2 year time frame for wide-spread utilization of ceramic-type superconductors has given way to a 3-10 year forecast that will more likely be required to prove and exploit the very attractive properties of the newer materials and resultant devices.

The universal acclaim that has greeted the discovery of high-temperature superconductors, already recognized by the rapid award of the 1987 Nobel prize to Bednorz and Mueller, has seemingly galvanized Japan more so than any other country. There, a carefully organized coalition of universities, industry and government groups has been formed to systematically explore and exploit the scientific and commercial opportunities presented by a multi-disciplinary SC technology. The pervasive impact that success in these endeavors could have is perhaps best illustrated by the forecasts made by Japanese scientists, shown in Figure 7.

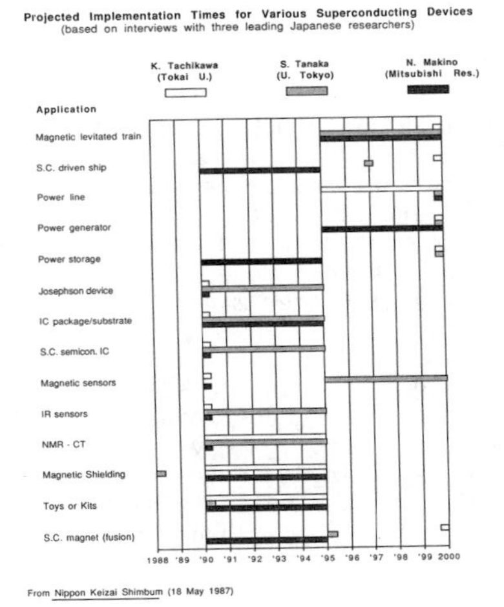

FIGURE 7
PROJECTED EMERGENCE OF SC DEVICES
AS COMMERCIAL PRODUCTS

On the chart, the three recognized authorities have given their assessment as to when various technological opportunities would be commercialized and each achieve multi-billion dollar levels.

In Europe and the United States a proliferation of scientific papers, newsletters and market studies are similarly attempting to lift the veil of missing knowledge, that would transform a clear scientific breakthrough in superconductivity into broadly-based commercial success.

Efforts to achieve a wider consensus of priorities that should be established for U.S. based SC activities, are still underway. A major assessment by the Federal Government is also being conducted. For a while it seemed that projects such as the Superconducting SuperCollider (SSC), - See Figure 8 - estimated to cost 4-6 billion dollars over a 8-10 year time frame, should be delayed to await cost savings expected through the use of higher temperature superconductors. Closer scrutiny of potential economic benefits from such a delay have proven to be minor, if not illusory.

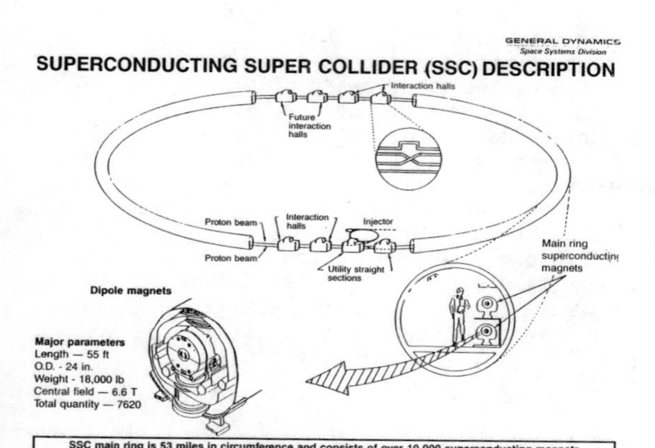

FIGURE 8
SUPERCONDUCTING SUPER COLLIDER (SSC) DESCRIPTION

Similarly, it is more likely that any transition from lower temperature, commercially attractive superconductor technology to that operating at higher temperature levels will happen more gradually than suddenly, extending over many years, if not a decade or two.

Most encouraging is the thought that we are still at the threshold of understanding and absorbing the fact that superconductive technology may hold many surprises in store for spawning improvements in technology and betterment of mankind's standard of living in the world, in as yet unimaginable new products.

Finally, it may be said that just as the second half of the 20th Century saw an explosion in electronic and computer based technology, so may the first half of the 21st Century see enormous strides in commercialization and exploitation of superconductive based technology.

Chapter 99

GENERAL DYNAMICS' EXPERIENCE WITH LARGE SUPERCONDUCTING MAGNETS AND FORESEEABLE FUTURE APPLICATIONS*

R. W. Baldi, R. A. Johnson, E. R. Kimmy, J. F. Parmer

ABSTRACT

General Dynamics is the largest producer of large superconducting magnets for U. S. Government applications. The Energy Programs group of the Space Systems Division has won over 65% of the superconducting magnet application contracts awarded by the U. S. Department of Energy over the past 12 years. These applications include thermonuclear fusion, magnetohydrodynamics, isotope separation, and accelerator magnets. Overall, General Dynamics has built 29 large superconducting magnets with a total contract value of over $83 million.

At present, we have contracts involving space-based and ground-based magnetohydrodynamics, superconducting magnetic energy storage, and thermodynamic aspects of accelerator magnets.

Potential near-term production of large-scale applications include superconducting magnets for the Superconducting Super Collider (SSC) and the Superconducting Magnetic Energy Storage (SMES). Potential far-term production programs may include thermonuclear fusion, magnetohydrodynamics, and possibly electromagnetic launch.

INTRODUCTION

In 1975 we sought to expand our division's product line to include superconducting magnets. The combination of skills in structures, cryogenics, and manufacturing appeared to be the right blend for designing and building superconducting magnets. We envisioned a facility in which engineering, manufacturing, quality, and purchasing would all be co-located; and we were pleased to move into that facility in 1983. Our magnet facility is located on San Diego Bay in California. It consists of a 5,000-square-foot office building and over 60,000 square feet for manufacturing. It even includes a barge dock for shipping magnets by sea. We also have a superconducting test laboratory located adjacent to our office building.

A collage highlighting the superconducting magnets we have produced is presented in Figure 1. As shown, we built several superconducting magnets for experimental thermonuclear fusion facilities, isotope separation for uranium enrichment, and accelerator magnets for the Superconducting Super Collider (SSC). Over the past 12 years, we have won 65% of the Department of Energy's business for large-

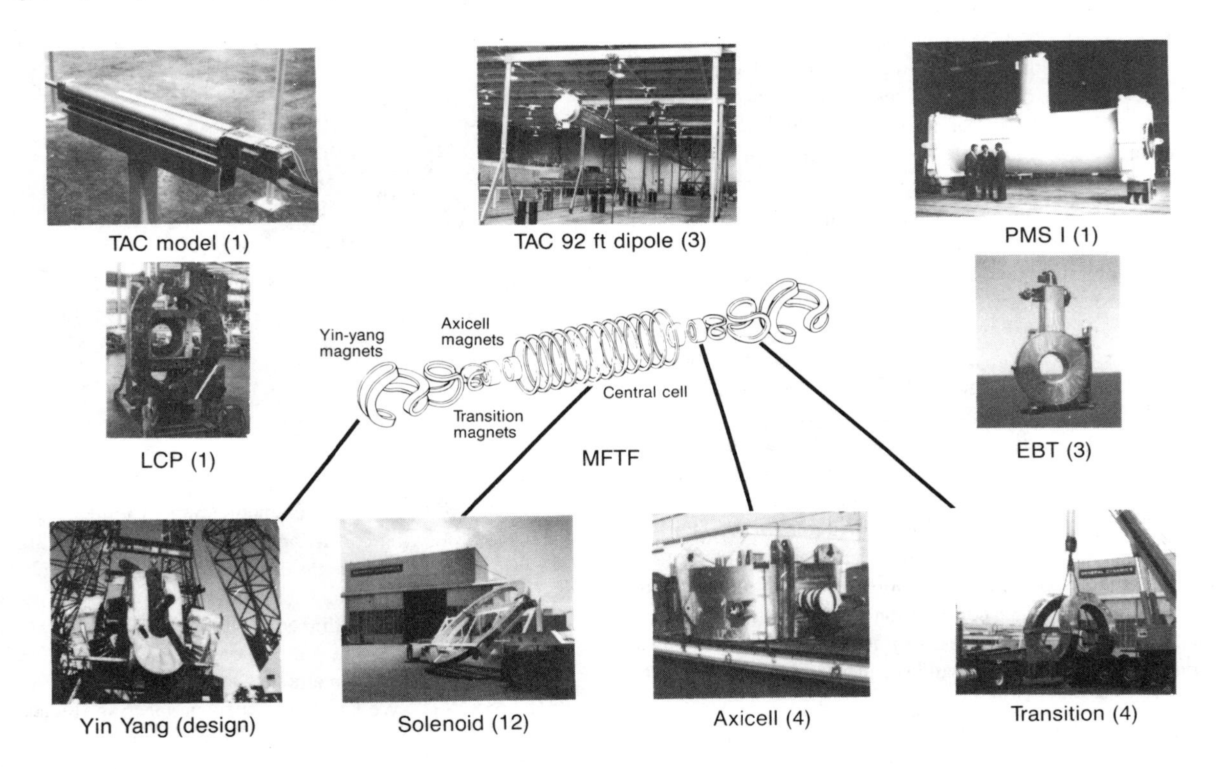

Figure 1. We have successfully fabricated 29 large superconducting magnets, and all of them met performance specifications.

*The work described herein was sponsored in part by the U.S. Department of Energy and the Defense Nuclear Agency.

scale applications. That amounts to about $83 million, and we are proud that all magnets met performance specifications. This paper provides greater details regarding these magnet programs. All of the superconducting magnets described herein were made with liquid helium-cooled, niobium-titanium superconductor.

ISOTOPE SEPARATION

The Prototype Magnet System (PMS) was a magnet system for isotope separation. This magnet was a one-meter-bore solenoid magnet built for TRW. Its application was for uranium enrichment. Its magnetic field strength was 2 tesla (20,000 gauss) and it weighed 44 tons. TRW installed and commissioned this magnet in 1981, and it has operated continuously since delivery.

THERMONUCLEAR FUSION

Another area in which we're very active is thermonuclear fusion. To support experimental facilities in this area, we built superconducting magnets for the Large Coil Program (LCP), Elmo Bumpy Torus (EBT), and the Mirror Fusion Test Facility (MFTF).

The Large Coil Program was performed for Oak Ridge National Laboratory (ORNL). This was an international experiment in which General Dynamics, General Electric, Westinghouse, and sources in Germany, Japan, and Switzerland all produced magnets for that program. Each superconducting magnet produced a magnetic field of 8 tesla (80,000 gauss) and weighed about 45 tons. This was an experiment in large coil design and fabrication. It was never intended for experimentation with fusion plasma. The machine was completed by ORNL last year and fully acceptance tested. Each coil was charged to 9 tesla. All told, this was a very successful program, and ORNL will publish the final report this spring.

Elmo Bumpy Torus (EBT) was another fusion program performed for ORNL. To support this program, we built three superconducting magnets. Each of them produced 7.4 tesla and weighed 1-1/2 tons. The intended facility consisted of 36 magnets to levitate a plasma in the torus created by that magnetic field, but this facility was never built.

The largest fusion program we have worked on was the Mirror Fusion Test Facility (MFTF). An artist's rendition of this facility is shown in the center of Figure 1. This facility was actually built, and the assembly of magnets measured 59 meters from end to end. We designed all 26 of the primary superconducting magnets. That includes the baseball-seam-shaped yin-yang magnets on the ends, the baseball-seam-shaped transition coils, the axicell coils, the solenoid coils and also two 12-tesla niobium-tin coils in the axicell region. We built 20 of these superconducting magnets. They include the transition, axicell, and solenoid coils. Lawrence Livermore National Laboratory built the yin-yang magnets and 12-tesla niobium-tin coils. They also built 16 trim coils, which are not shown in Figure 1. All told, MFTF consisted of 42 superconducting magnets. Total stored energy is 1,800 megajoules, which is the largest amount of stored energy of any superconducting magnet facility. Final construction of MFTF was completed in 1985, and the magnets were successfully acceptance tested. Unfortunately, fusion research funding dwindled, and this project is at present mothballed.

An attractive notion regarding high-temperature superconductor is the possibility of very high field applications. This property of high-temperature superconductor will surely create new challenges to engineer structures that adequately contain the electromagnetic Lorentz forces. As an example of this challenge, even with the limitations of present day superconductors, the actual operating stsresses in the MFTF Yin-Yang magnets are over 80,000 psi. Accommodating higher magnetic field allowables will require equally dramatic breakthroughs in strength allowables for magnet structural materials.

ADVANCED FUSION REACTOR CONCEPT

In 1982 we assisted Lawrence Livermore National Laboratory and TRW with a group of other universities, national laboratories, and industry to design a reactor of the future, based on the mirror fusion principle. This study was called the Mirror Advanced Reactor System (MARS). It is illustrated in Figure 2. As shown in this figure, the fuel of fusion is sea water. Our oceans contain sufficient fuel (heavy hydrogen) to last for probably 30,000 years at present energy demands. From each gallon of sea water, the equivalent of 300 gallons of gasoline can be derived with minimal nuclear waste. Magnetic confinement thermal nuclear fusion power plants appear to be cost competitive with nuclear fission. Devices such as this have projected costs of $3,000 to $4,000 per kilowatt.

An important factor regarding magnetic confinement for fusion is that use of superconductivity is an absolute must. Without superconducting magnets (but rather conventional magnets), a device such as the MARS facility would consume over ten times more power than it would ever generate. Therefore, superconductivity is a must even if type-II liquid helium-cooled superconductors are used. High-temperature superconductor will make such a facility up to 25% more economical. Perhaps more importantly, liquid helium refrigeration can be eliminated, improving reliability and system availability.

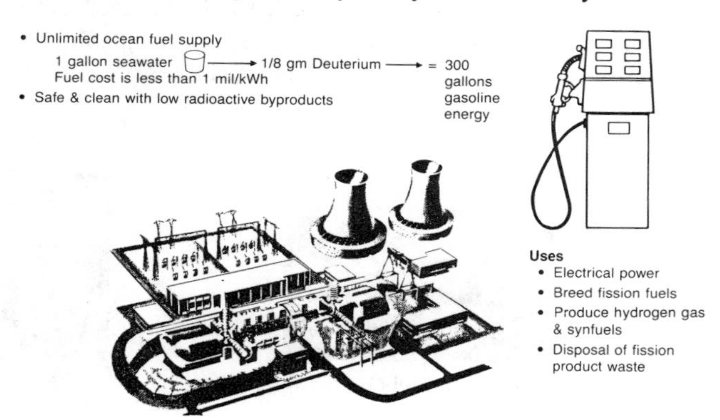

• Unlimited ocean fuel supply
1 gallon seawater → 1/8 gm Deuterium → = 300 gallons gasoline energy
Fuel cost is less than 1 mil/kWh
• Safe & clean with low radioactive byproducts
Uses
• Electrical power
• Breed fission fuels
• Produce hydrogen gas & synfuels
• Disposal of fission product waste

Figure 2. Superconducting magnets are mandatory for economic operation of presently envisioned magnetic confinement thermonuclear fusion reactors.

ACCELERATORS

The Department of Energy is currently planning to build an enormous accelerator called the Superconducting Super Collider. This accelerator is a 53-mile ring consisting of 10,000 superconducting magnets. The principal magnets are the dipoles — 7,620 of them, 6.6 tesla, two feet in diameter, 55 feet long, and weighing about 16,000 pounds each. These magnets curve the flight of the proton beams as they are counter-rotated in this device. The protons, traveling at near the speed of light, eventually collide in the interaction halls. We have actively worked on SSC since 1983. We have performed engineering studies, cost estimates, and magnet fabrication. According to a study by the Central Design Group for the SSC, high-temperature superconductor would save about 3% on the construction of this device ($57 million) and about 10% in the operating cost per year ($25 million).[1] Cost reductions are principally achieved by replacing liquid helium refrigeration with liquid nitrogen. The SSC Central Design Group study included the need to increase the bore tube of the dipole magnet to allow for degraded vacuum without liquid helium cooling.

Figure 3 shows one of three world-record, 92-foot-long superferric dipole magnets which our group built for Texas Accelerator Center. This has to be the world's most photographed shipments of a magnet. The truck shown was about 110 feet long from front bumper to tail flag. That magnet was safely shipped from San Diego, California, to the Texas center in Houston, Texas. As a matter of interest, all of the magnets we built for Lawrence Livermore National Laboratory, Oak Ridge National Laboratory, TRW, and Texas Accelerator Center were built in San Diego and shipped to their respective customers.

Figure 3. The 92-foot-long SSC dipole magnets are of world-record length, and they were safely shipped from San Diego to Houston.

HYBRID PULSE POWER TRANSFORMER

A new, revolutionary device called the Hybrid Pulse Power Transformer (HPPT) is illustrated in Figure 4. Basically, this device can store energy which can be slowly charged and rapidly discharged. It has a superconducting primary with many turns and a conventional secondary with a single turn. The primary is quenched by driving the superconductor either over-temperature, over-current, or over-field, at which time all of its stored energy transfers quickly to the secondary. HPPT has potential SDI applications such as rail guns and for commercial applications such as welding. This device was invented in our group in 1986. Proof-of-principle testing has been performed, and we have applied for a patent. New high-temperature superconductor will really aid this device, because quenching a liquid helium superconducting magnet boils off liquid helium. Having the ability to have a high-temperature superconductor in liquid nitrogen or even water-cooling would be much less expensive to operate.

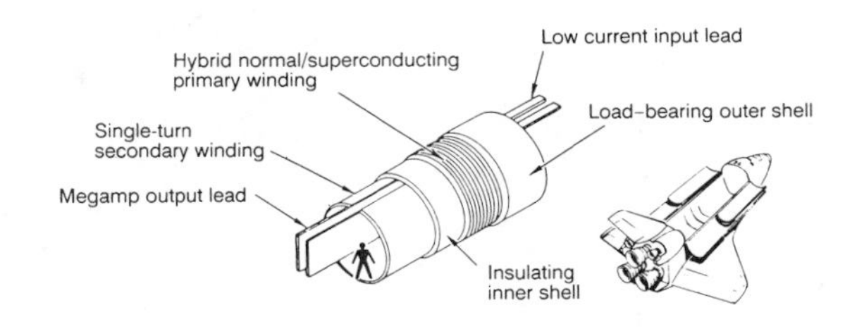

Application

- SDI ground & space-based weapons – hybrid normal/ superconducting transformer

Advantages

- Eliminates high-power, high-current switch
- High transfer efficiency
- Low voltage/low insulation weight
- Lightweight transformer

Figure 4. A Hybrid Pulse Power Transformer is a revolutionary device involving superconductivity.

MAGNETOHYDRODYNAMICS

During the late 1970s and early 1980s, we performed many studies on magnetohydrodynamics (MHD). This technology was quite popular during the gasoline shortage. Oil shortages caused the utilities to study switching from oil to coal. Since coal generates less BTUs in power plants, the utilities were studying MHD topping cycles. In this particular case, hot flue gas seeded with potassium or cesium would be flowed through the bore of an MHD magnet. Basically, MHD is a generator principle in which the conductive medium flowing through the magnetic field create a current. This technology has never been commercialized but has led to some interesting military applications. At present, we are subcontractor to Avco for Pittsburgh Energy Technology Center (PETC) to design a space-based MHD unit for SDI applications, as shown in Figure 5. The requirements for this device include one mega-ampere of current for 100 seconds to power directed energy weapons. This particular magnet is niobium-titanium liquid-helium cooled.

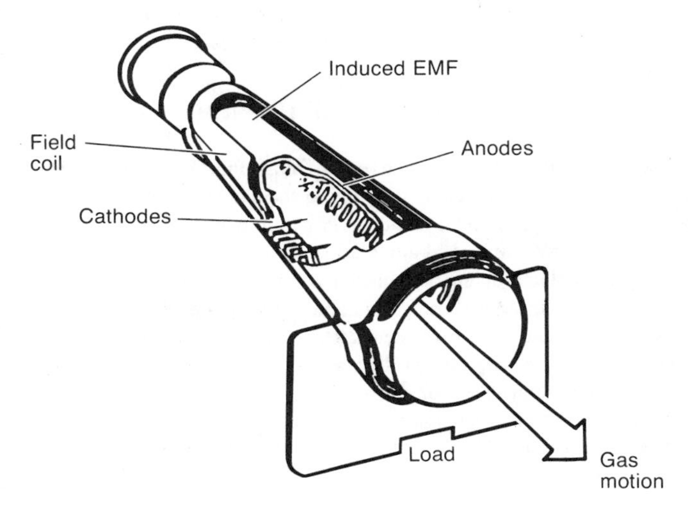

- Lightweight magnets for space-based SDI
- Magnets for ground-based SDI or utility retrofit

Figure 5. Superconducting magnet applications are being studied for magnetohydrodynamic applications.

SUPERCONDUCTING MAGNETIC ENERGY STORAGE

A new area that is gaining popularity is superconducting magnetic energy storage. We are subcontractor to Bechtel Corporation, in conceptually designing a 20-megawatt-hour, 140-meter-diameter superconducting magnet to power a free electron laser. The superconducting magnet is illustrated in Figure 6. We are currently performing conceptual design and critical component testing. Construction go-ahead to build this unit is expected in two years. Eventually, units for utility applications up to a 1,000 meters in diameter are envisioned. These units will store 5,250 megawatt hours of electrical energy - enough to run a medium-size city for about two hours. Utilities are very interested in these for diurnal loading in which they will store energy at night and discharge it in the day during peak load hours. Eventually, it is expected that as much as 50 units at about $1 billion each will be needed over the next 30 years. Present units are being designed with liquid-helium cooled, type-II niobium-titanium superconductors. The benefits of high-temperature superconductors have been studied for SMES units, and the savings in capital cost appears to be quite small. Perhaps the improved availability due to elimination of helium refrigeration systems will prove to be the major attraction.

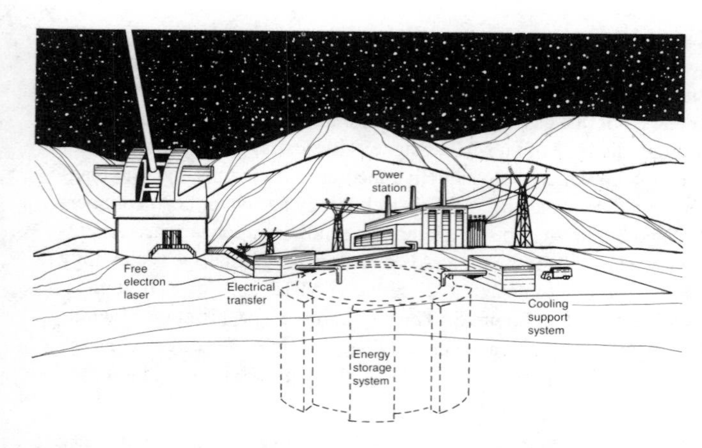

Figure 6 Conceptual design and critical component testing are now being performed for superconducting magnetic energy storage.

ELECTROMAGNETIC LAUNCH

Perhaps electromagnetic launch could augment chemical rocket launch for economical access to space in the future. At $2,500 per pound, it costs $1million to launch a 55-gallon barrel of water to the space station. We've looked at a number of ways of improving that, and certainly electromagnetic launch, as shown in Figure 7, is an attractive option. In the early stages, rail gun-type systems could launch to space items like ice, frozen steaks, bolts, nuts, or structures, under very high-g force conditions. High-temperature superconductor, in our opinion, is a must for such applications. Many of the concepts are based on quench guns and coil guns, which involve quenching superconducting magnets. The liquid helium boiloff would be immense and very costly. But applications involving high-temperature superconductor solve that problem.

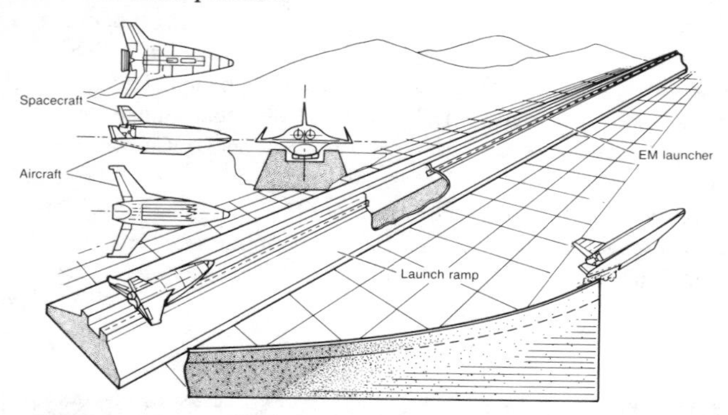

Figure 7. Superconducting magnets may offer an economical solution for electromagnetic launch.

FUTURE LARGE-SCALE APPLICATIONS

Potential near-term production of large-scale applications includes magnets for the Superconducting Super Collider (SSC) and Superconducting Magnetic Energy Storage (SMES). Potential far-term production programs may include magnets for thermonuclear fusion, magnetohydrodynamics, electromagnetic launch, as well as many other concepts yet to be discovered.

REFERENCES

1. M. S. McAshan, Peter Vander Arend, "A Liquid Nitrogen Temperature SSC," SSC Central Design Group Report SSC-17, April 1987.